Dark Matter in Astroparticle and Particle Physics

Dark 2007

Proceedings of the
6th International
Heidelberg Conference

Dark Matter in Astroparticle and Particle Physics

Dark 2007

University of Sydney, Australia
24 – 28 September 2007

Editors

Hans Volker Klapdor-Kleingrothaus
Heidelberg, Germany

Geraint F. Lewis
University of Sydney, Australia

World Scientific

NEW JERSEY • LONDON • SINGAPORE • BEIJING • SHANGHAI • HONG KONG • TAIPEI • CHENNAI

Published by

World Scientific Publishing Co. Pte. Ltd.

5 Toh Tuck Link, Singapore 596224

USA office: 27 Warren Street, Suite 401-402, Hackensack, NJ 07601

UK office: 57 Shelton Street, Covent Garden, London WC2H 9HE

British Library Cataloguing-in-Publication Data
A catalogue record for this book is available from the British Library.

DARK MATTER IN ASTROPARTICLE AND PARTICLE PHYSICS
Proceedings of the 6th International Heidelberg Conference

ISBN-13 978-981-281-434-0
ISBN-10 981-281-434-5

Printed in Singapore by Mainland Press Pte Ltd

PREFACE

The Sixth HEIDELBERG International Conference on Dark Matter in Astrophysics and Particle Physics, DARK 2007, took place at the University of Sydney, Australia, September 23–28, 2007. It was, after Cape Town 2002 and Texas, the third conference of this series held outside Germany. The earlier meetings that started in 1996, were held in Heidelberg.

Dark Matter is still one of the most exciting and central fields of astrophysics, particle physics and cosmology. The conference covered, as usual for this series, a wide range of topics, theoretical and experimental.

We are confident that the Proceedings of this conference will provide an useful overview of this exciting field of research — its current status and the future prospects — and of its fundamental connections to various frontier disciplines of particle physics and cosmology. We hope that this book can also serve as a kind of handbook for students.

The Organizers express their thanks to all colleagues from many countries, who contributed so actively to the success of the meeting.

Thanks go to the University of Sydney for the generous financial support. We thank all people who contributed in one way or another to the organization of the conference, and in creating a pleasant and inspiring atmosphere during the meeting. We are indebted in particular to the graduate students Matt Francis, Madhura Killedar, Brendon Brewer and Berian James for their help in the organisation. Particular thanks go to the Scientific Secretary, Dr. Irina Krivosheina, who was responsible for creating most of the exciting program of the conference, and to whom we are also indebted for preparing this Proceedings volume.

Last but not least, one of the Co-chairmen would like to give his personal thanks to Professor Geraint Lewis for making this successful event possible, while the other Co-chairman would like to thank Professor Hans V. Klapdor-Kleingrothaus for choosing the University of Sydney as the conference site for this third DARK conference outside Germany.

H. V. Klapdor-Kleingrothaus
Chairman of HEIDELBERG
DARK Matter Conferences
Heidelberg, GERMANY

I. V. Krivosheina
Scientific Secretary of DARK 2007
Nishnij Novgorod, RUSSIA

G. Lewis
Co-chairman of DARK 2007
Sydney University
AUSTRALIA

December 2007

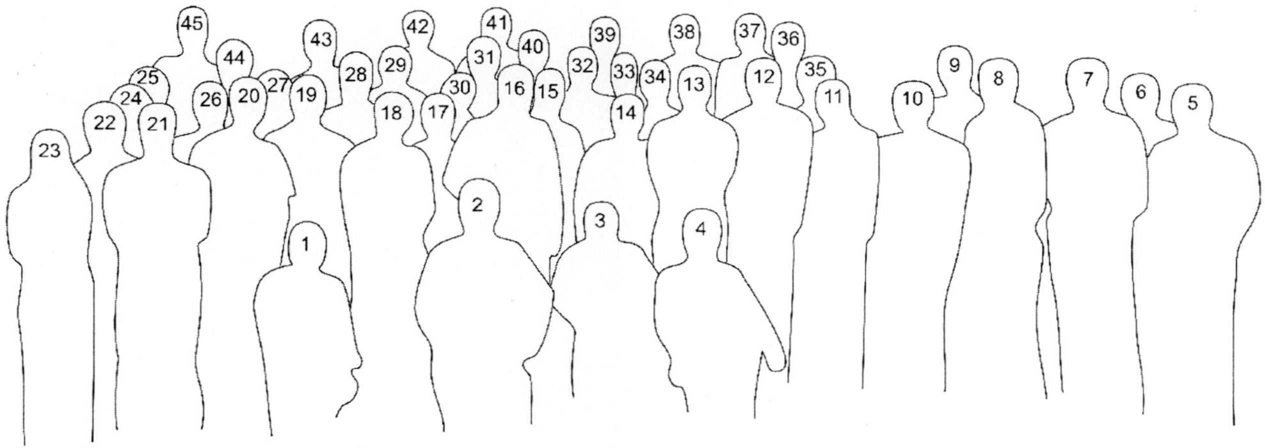

Fig. 1. 1. Alexander Zakharov 2. Matthew Francis 3. Margarida Nesbitt Rebelo 4. Dimitri Schritt 5. Neil Spooner 6. Burkhard Fuchs 7. Raoul D.Viollier 8. Roland Allen 9. Tim Schrabback-Krahe 10. Jihn E. Kim 11. Dharamvir Ahluwalia 12. Hans V. Klapdor-Kleingrothaus 13. Geraint Lewis 14. Irina V. Krivosheina 15. Benoit Famaey 16. Eduardo Guendelman 17. Marusa Bradac 18. Chris Power 19. Ricardo Cerulli 20. Tommy Ohlsson 21. Dai G. Yamazaki 22. Kazuhiko Kojima 23. Madhura Killedar 24. Jonghee Yoo 25. Motohiko Kusakabe 26. Brendan Brewer 27. Henrique Araujo 28. Patrick Scott 29. Colin R.L. Montgomery 30. George Georgevits 31. David Wiltshire 32. Habib Khosroshahi 33. Tony Noble 34. Ishwaree Neupane 35. Masayuki Nakahata 36. Edward Drobyshevski 37. Patrick Tisserand 38. Ernst Otten 39. Wim de Boer 40. Manfred P. Leubner 41. – 42. Brian Winer 43. Steffen Knollmann 44. – 45. Berian James

CONTENTS

BERKELEY

BERKELEY

HET!

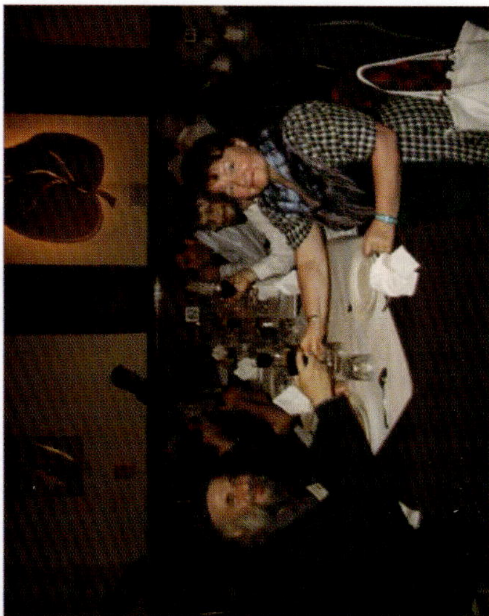

BERKELEY

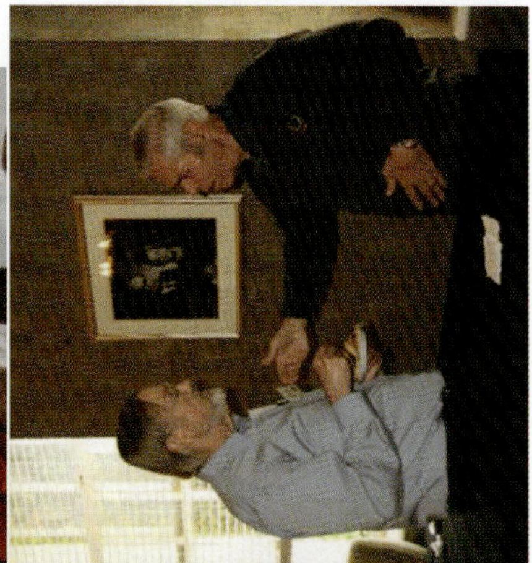

PART I

Direct and Indirect Search for Cold Dark Matter

RESULTS AND DEVELOPMENTS WITH LARGE MASS HIGHLY RADIOPURE NaI(Tl) AT LNGS

R. BERNABEI, P. BELLI, F. MONTECCHIA, F. NOZZOLI

Dip. di Fisica, Università di Roma "Tor Vergata" and INFN, Sezione di Roma Tor Vergata, I-00133 Rome, Italy

F. CAPPELLA, A. INCICCHITTI, D. PROSPERI

Dip. di Fisica, Università di Roma "La Sapienza" and INFN, Sezione di Roma, I-00185 Rome, Italy

R. CERULLI

INFN - Laboratori Nazionali del Gran Sasso, I-67010 Assergi (L'Aquila), Italy
E-mail: riccardo.cerulli@lngs.infn.it

C. J. DAI, H. L. HE, H. H. KUANG, J. M. MA, X. D. SHENG, Z. P. YE

IHEP, Chinese Academy, P.O. Box 918/3, Beijing 100039, China

The low background DAMA/NaI experiment ($\simeq$ 100 kg highly radiopure NaI(Tl)) at the Gran Sasso National Laboratory of the I.N.F.N. had the unique feature to effectively investigate the presence of Dark Matter particle in the galactic halo by exploiting the model independent annual modulation signature. This experiment has collected data over seven annual cycles for a total exposure of more than 10^5 kg $\times$ day and has pointed out at 6.3 σ C.L. a modulation effect satisfying all the many peculiarities of the signature. Neither systematics nor side reactions able to account for the observed effect were found. Several (but still few with respect to the many possibilities) corollary model dependent quests for the candidate particle have also been carried out. At present the second generation DAMA/LIBRA set-up ($\simeq$ 250 kg highly radiopure NaI(Tl)) is in operation. A R&D towards possible ultimate 1 ton NaI(Tl) experiment is also in progress.

Keywords: Scintillators; NaI(Tl); Dark Matter.

1. The DAMA/NaI experiment

The low background DAMA/NaI experiment was located in the underground laboratory of Gran Sasso; it has been part of the DAMA project, which is also composed by several other low background set-ups, in particular: i) DAMA/LXe [1]; ii) DAMA/R&D [2]; iii) the new second generation radiopure NaI(Tl) DAMA/LIBRA set-up (see later); iv) DAMA/Ge for sample measurements, which is located in the low background LNGS Ge facility [3]. A detailed description of the DAMA/NaI set-up regarding its performances, radiopurity, hardware procedures and the determination of the experimental quantities has been given e.g. in refs. [4,5].

DAMA/NaI was proposed in 1990 [6], designed and realized having the main aim to investigate the presence of Dark Matter (DM) particles in the galactic halo by exploiting the model independent annual modulation signature [7]. In fact, as a consequence of the Earth annual revolution around the Sun, a larger flux of DM particles cross the laboratory around roughly June 2nd (when the Earth velocity is summed to the Sun velocity with the respect to the Galaxy) and by a smaller one around roughly December 2nd (when the two velocities are subtracted). This annual modulation signature offers many peculiarities since a seasonal effect induced by DM particles must simultaneously satisfy all the following requirements: (i) the rate must contain a component modulated according to a cosine function; (ii) with one year period; (iii) a phase roughly around 2nd June; (iv) the modulation must only be found in a well-defined low energy range, where DM particles can induce signals; (v) it must apply just to those events in which only one detector in a multi-detectors set-up actually "fires" (*single-hit* events), since the probability that DM particles would have multiple interactions is negligible; (vi) the modulation amplitude in the region of maximal sensitivity has to be $\leq 7\%$ for usually adopted halo distributions, but it can be significantly larger in case of some possible scenarios such as e.g. those of refs. [8,9]. To mimic such a signature either spurious effects or side reactions should be able not only to account for the observed modulation amplitude but also to contemporaneously satisfy all the requirements; none has been found nor suggested by anyone over about a decade ([4] and refs therein).

Other approaches for DM particle investigation have been exploited by DAMA/NaI as well and, in addition, several other rare processes have also been studied [10].

2. The DAMA/NaI model-independent result

The DAMA/NaI experiment took data over seven annual cycles collecting an exposure of 107731 kg $\times$ day. A clear annual modulation has been observed in the measured rate of the *single-hit* events in the lowest energy region, satisfying the many peculiarities of a DM particle induced effect. In Fig. 1 the time behaviour of the residual rate of the *single-hit* events in the (2-6) keV energy interval is reported. Fitting the data with a cosine-like

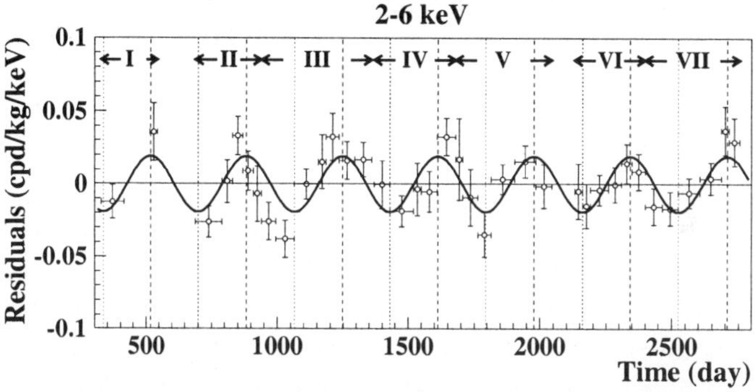

Fig. 1. Experimental residual rate for *single-hit* events in the cumulative (2–6) keV energy interval as a function of the time over 7 annual cycles (total exposure 107731 kg $\times$ day); end of data taking July 2002. The experimental points present the errors as vertical bars and the associated time bin width as horizontal bars. The superimposed curve represents the cosinusoidal function behaviour expected for a DM particle signal with a period equal to 1 year and phase exactly at 2nd June; the modulation amplitude has been obtained by best fit. See ref. [4].

function the presence of annual modulation is favoured at 6.3 σ C.L. with an amplitude equal to (0.0200 ± 0.0032) cpd/kg/keV, a phase $t_0 = (140 \pm 22)$ days and a period $T = (1.00 \pm 0.01)$ year. The period and phase agree with those expected in the case of an effect induced by DM particles in the galactic halo ($T = 1$ year and t_0 roughly at $\simeq 152.5^{th}$ day of the year). The χ^2 test on the (2–6) keV residual rate disfavours the hypothesis of unmodulated behaviour giving a probability of $7 \cdot 10^{-4}$ ($\chi^2/d.o.f. = 71/37$). The same data have also been investigated by a Fourier analysis, where a clear peak corresponding to a period of $\simeq 1$ year is present. For details on other energy intervals see refs. [4] and references therein.

A careful quantitative investigation of all the known possible sources of systematic and side reactions has been regularly carried out and published at time of each data release. No systematic effect or side reaction able to account for the observed modulation amplitude and to satisfy all the requirements of the signature has been found; for details see ref. [4].

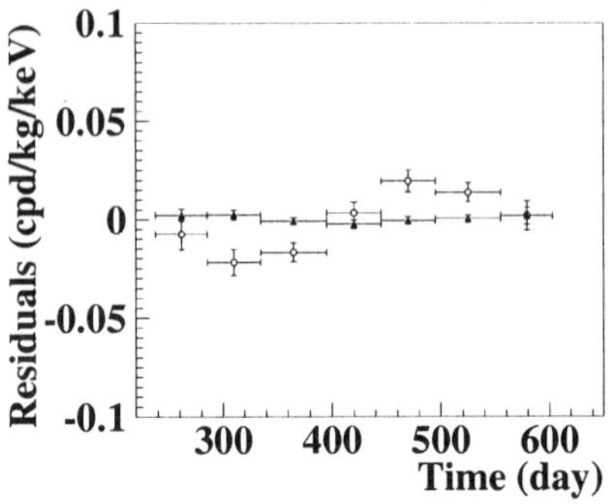

Fig. 2. Experimental residual rates over seven annual cycles for *single-hit* events (open circles) – class of events to which DM particle events belong – and over the last two annual cycles for *multiple-hits* events (filled triangles) – class of events to which DM particle events do not belong – in the (2–6) keV cumulative energy interval. They have been obtained by considering for each class of events the data as collected in a single annual cycle and using in both cases the same identical hardware and the same identical software procedures. The initial time is taken on August 7th.

A further investigation has been performed in the last 2 annual cycles (DAMA/NaI-6 and 7) when each detector was equipped with its own Transient Digitizer and a dedicated renewed electronics has been installed; this has allowed to record the pulse profiles of the *multiple-hits* events (i.e. events in which more than one detector fire in coincidence). This class of events have been studied and analysed by using the same identical hardware and the same identical software procedures as for the case of the *single-hit* events (see Fig. 2). The *multiple-hits* events class – on the contrary of the *single-hit* one – does not include events induced by DM particles since the probability that a DM particle interacts in more than one detector is negligible. The fitted modulation amplitudes are: $A = (0.0195 \pm 0.0031)$ cpd/kg/keV and $A = -(3.9 \pm 7.9) \cdot 10^{-4}$ cpd/kg/keV for *single-hit* and *multiple-hits* residual rates, respectively [4]. Thus, evidence of annual modulation with proper features is present in the *single-hit* residuals (events class to which the DM particle-induced signals belong), while it is absent in the *multiple-hits* residual rate (event class to which only background events belong). Since the same identical hardware and the same identical software procedures have been used to analyse the two classes of events, the obtained result offers an additional strong support for the presence of DM particles in the galactic halo, further excluding any side effect either from hardware or from software procedures or from background.

In conclusion, the presence of DM particles in the galactic halo is supported in a model independent way by DAMA/NaI at 6.3 σ C.L.. Except the presently running DAMA/LIBRA, no other experiment whose result can be directly compared with this one is available so far in the field of Dark Matter investigation.

3. Corollary model-dependent quests

On the basis of the obtained 6.3 σ C.L. model-independent result, corollary investigations can also be pursued on the nature of the DM candidate particle. This latter investigation is instead model-dependent and – considering the large uncertainties which exist on the astrophysical, nuclear and particle physics assumptions and on the parameters needed in the calculations – has no general meaning (as it is also the case of exclusion plots and of the DM particle parameters evaluated in indirect detection experiments). Thus, it should be handled in the most general way, as pointed out with time passing [4,9,11,12].

For simplicity, the results of the corollary analyses are presented in terms of allowed volumes/regions obtained as superposition of the configurations at given C.L. for the considered model frameworks. Details on the followed procedure can be found e.g. in ref. [4,9,11,12]. It is worth to note that although we have taken into account some of the existing uncertainties of the models and parameters needed in the calculations, the inclusion of other existing uncertainties would further extend the allowed volumes/regions and increase the sets of obtained best fit values. In addition we remind that the results briefly summarized here and the several other ones available in literature are not exhaustive of the many scenarios possible at present level of knowledge.

3.1. *Some corollary quests for WIMP class candidates*

For the case of WIMP class candidates, it has been considered so far low (of order of few GeV) and high mass (up to many hundreds of GeV) candidates interacting with ordinary matter via: i) mixed spin-independent (SI) & spin-dependent (SD) coupling; ii) dominant SI coupling; iii) dominant SD coupling; iv) preferred SI inelastic scattering. A detailed discussion on the volumes/regions allowed by the DAMA/NaI data for these candidates in some given model frameworks can be found in ref. [4]. This analysis has been extended in ref. [11] by including the possible contribution arising from a non thermalized DM particle component in the dark halo; in particular, the Sagittarius Dwarf Elliptical Galaxy (SagDEG) has been considered. In Fig. 3, examples of slices of the allowed region for purely SI candidate

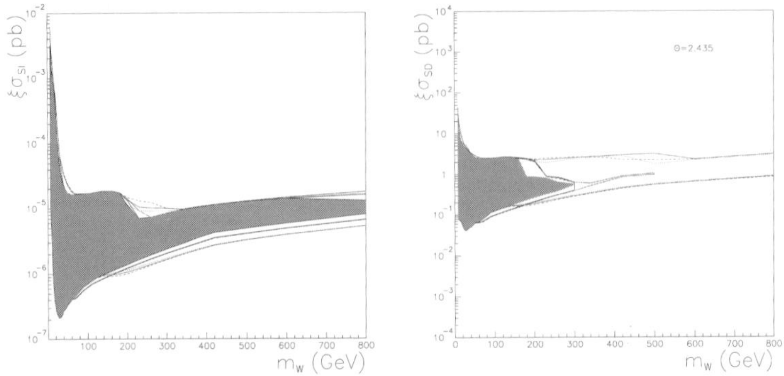

Fig. 3. Left: Region allowed in the $(\xi\sigma_{SI}, m_W)$ plane in the considered scenarios for pure SI coupling. Right: a slice of the 3-dimensional allowed volume in the $(\xi\sigma_{SD}, m_W)$ plane for $\theta = 2.435$ in the considered scenarios and for pure SD coupling. The filled region has been determined for no SagDEG contribution as in the previous subsection [4], while the areas enclosed by lines are obtained by introducing in the analysis the SagDEG stream. Different lines refer to the considered possibilities for the SagDEG stream velocity and velocity dispersion [11]; see ref. [11] for more results on the various couplings and for details.

and of the 3-dimensional allowed volume for a candidate with purely SD coupling are shown for the considered scenarios [11]. The filled region has been determined for no SagDEG contribution while the areas enclosed by lines are obtained by introducing in the analysis the SagDEG stream with DM density not larger than 0.1 GeV cm^{-3}. Different line types refer to the various considered possibilities for the SagDEG stream velocity and velocity dispersion. For details refer to [11]. Other non-thermalized substructures, such as the Canis Major satellite galaxy or the streams in the halo models with caustics, can be considered in near future.

Recently, it has also been investigated the role of the electromagnetic contribution produced in the interaction of the WIMP with target nuclei [12]. Since the 40's, in fact, it has been shown that a recoiling atomic nucleus can induce ionization and excitation of bound atomic electrons. As a consequence, a certain quantity of electromagnetic radiation (made of escaping electron and of X-rays and/or Auger electrons) arises from the rearrangement of the atomic shells. This radiation is fully contained in a detector of suitable size. This effect, named hereafter Migdal effect [13], was not yet considered in any investigations on WIMP-nucleus elastic scattering. Note that, because of its electromagnetic (e.m.) nature, this part of the signal is lost in all those approaches based on discrimination procedures of the electromagnetic component of the measured counting rate. Although the Migdal effect appears quite small, the e.m. unquenched nature of this contribution – with respect to the behaviour of the energy distribution of the nuclear recoils – can have appreciable impact in the DM direct searches when interpreted in terms of WIMP candidates, with particular regard for the low mass WIMP. Note that in order to point out just the impact of the Migdal effect, the same scenarios as in ref. [4] have been considered without any inclusion of the possible SagDEG contribution. Some examples of the results obtained by including the Migdal effect analyses for the WIMP quest are reported in Fig. 4 . As it can be seen, the accounting for the electromagnetic aspects of the interactions provides in the considered scenarios

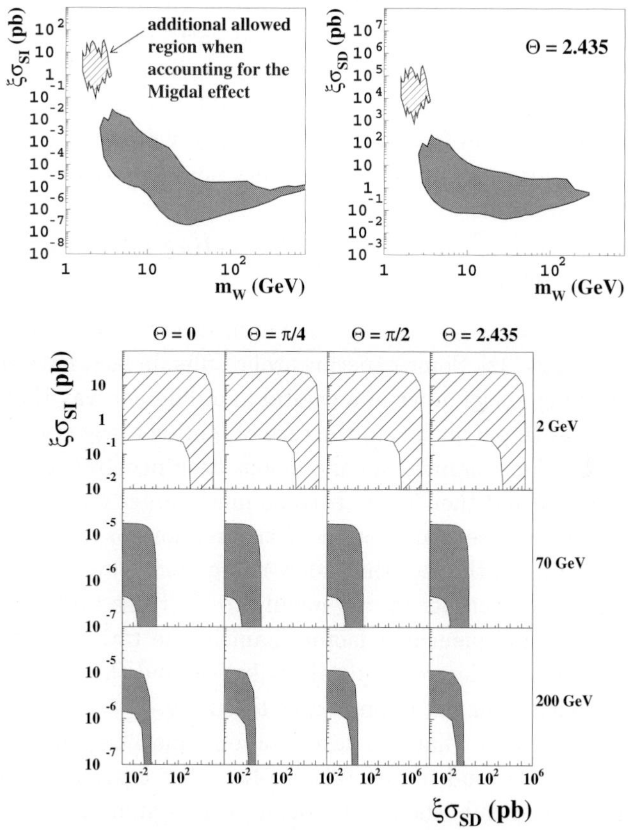

Fig. 4. Examples of region/slices allowed in the considered model frameworks for: i) pure SI coupling (upper-left); ii) pure SD coupling for the value $\theta = 2.435$ (note that θ can range from 0 to π) (upper-right); iii) mixed SI&SD coupling for some mass and θ values (bottom). The hatched region appears when accounting for the Migdal effect. Note that the inclusion of other contributions and/or of other uncertainties on parameters and models, e.g. the SagDEG contribution or more favourable form factors, would further extend the region and increases the sets of the best fit values. See ref. [12] for details.

additional volumes/regions at GeV mass scale, not topologically connected with the remaining allowed parts. This depends on the behaviour of the expected energy distributions at low masses (where the Migdal contribution is appreciable) with respect to that at higher masses, where recoils dominate.

It is worth to note that we remind that GeV mass DM particles have been widely proposed in literature in order to offer a mechanism able to account for the Baryon Asymmetry in the Universe and to explain other cosmological topics [14–16]. Among the GeV mass WIMP candidates we remind: i) the H dibaryon, already predicted within the Standard Model of particle Physics [16]; ii) the *Darkon*, a real scalar field in an extended Standard Model [17]; iii) the light photino early proposed in models of low-energy supersymmetry [18]; iv) the very light neutralino in Next-to-MSSM model [19]; v) the scalar GeV mass DM candidates of ref. [20]; vi) the mirror Deuterium in frameworks where mirror matter interactions [21] with ordinary matter are dominated by very heavy particles.

3.2. *Possible implications of the channeling effect in NaI(Tl) crystals*

Recently, we have also investigated the implication of the channeling effect in NaI(Tl) crystals [22]. Some experimental results, in fact, have shown that ions and recoiling nuclei move in a crystal in a different way than in amorphous materials. In particular, in the case of motion along crystallographic axes and planes, ions manifest an anomalous deep penetration into the lattice of the crystal and their range become much larger than the maximum they would have in case of motion in other directions or in amorphous materials. This is due to the fact that a low-energy ion, entering in the lattice into a channel, transfers its energy mainly to electrons rather than to the nuclei and, thus, its quenching factor (namely the ratio between the detected energy in keV electron equivalent [keVee] and the kinetic energy of the recoiling nucleus in keV) approaches the unity.

It is worth to note that the results about quenching factor obtained for NaI(Tl) crystals (and in general also for other crystal detector) using neutron source and published in literature can contain channeled events, but considering the low-statistics of these measurements, the small effect looked for and the energy resolution of the detectors they cannot easily be identified. Moreover, the channeling effect becomes less important at increasing energy and its contribution results more suppressed. Therefore, there is no hope to single out the channeling effect in the already-collected neutron data [22].

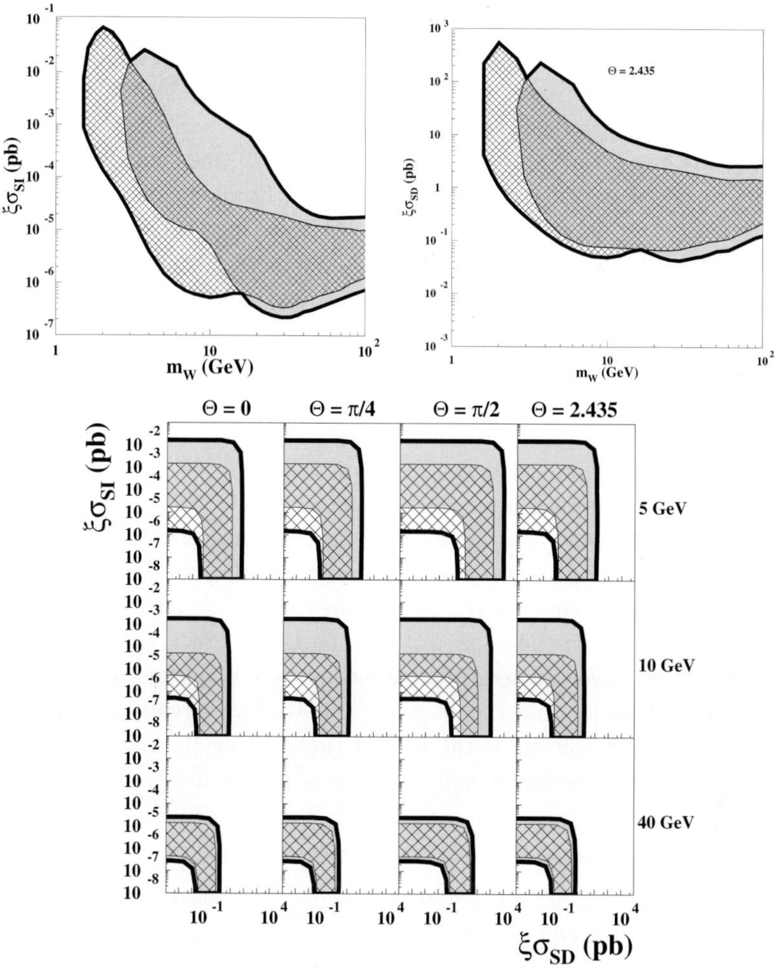

Fig. 5. Examples of region/slices allowed in the considered model frameworks for: i) pure SI coupling (upper-left); ii) pure SD coupling for the value $\theta = 2.435$ (note that θ can range from 0 to π) (upper-right); iii) mixed SI&SD coupling for some mass and θ values (bottom). The dotted region is obtained in absence of channeling effect, while the dashed one is obtained when accounting for it. The dark line marks the overal external contour. It is worth to note that the inclusion of other contributions and/or of other uncertainties on parameters and models, such as e.g. the possible SagDEG contribution and the Migdal effect or more favourable form factors, different scaling laws, etc., would further extend the region and increases the sets of the best fit values. See ref. [22] for details.

The inclusion of this existing effect gives an appreciable impact in corollary analyses in terms of WIMP (or WIMP-like) candidates since, as mentioned, the quenching factor is a key quantity to derive the energy of the recoiling nucleus after an elastic scattering [22]. In particular lower cross sections are explorable in given models for WIMP and WIMP-like candidates by crystal scintillators, such as NaI(Tl). Similar situation holds for purely ionization detectors, as Ge (HD-Moscow - like experiments), while a loss of sensitivity is expected when pulse shape discrimination is used in crystal scintillators (such e.g. in KIMS) since the channeled events – having $q \simeq 1$ – are probably lost. Moreover, no enhancement can be present either in liquid noble gas experiments (DAMA/LXe, WARP, XENON, ...) or in bolometer experiments; on the contrary some loss of sensitivity can be expected when applying discrimination procedures, based on $q_{ion} << 1$, since some events (those with $q_{ion} \simeq 1$) are lost.

In Fig. 5 are reported some examples of regions/volumes allowed by the DAMA/NaI data when the channeling effect has been taken into account. It has been considered cases of a DM particle with pure SI, pure SD and SI&SD coupling in some given model frameworks. Also for this analysis, in order to point out just the impact of the channeling effect, the same scenarios as in ref. [4] have been considered without any inclusion of the possible SagDEG contribution and the existing Migdal effect already discussed.

The results further show the role of the existing uncertainties and of correct/complete description and inclusion of all the involved processes.

3.3. *Results for light bosonic candidates*

The DAMA/NaI model-independent evidence has also been investigated by considering as a DM candidate a light ($\simeq$ keV mass) bosonic particle, either with pseudoscalar or with scalar coupling [9]. For these candidates, the direct detection process is based on the total conversion in NaI(Tl) crystal of the mass of the absorbed bosonic particle into electromagnetic radiation. Thus, in these processes the target nuclei recoil is negligible and is not involved in the detection process; therefore, also signals from these light bosonic DM candidates are lost in experiments applying rejection procedures of the electromagnetic contribution to the counting rate.

In all the main processes involved in the detection of a DM light bosonic particle – both in the pseudoscalar and in the scalar interactions – the total (including the secondary processes: X-rays and Auger electrons) energy release, E_{rel}, in the detector matches the total energy of the a particle, $E_a \simeq m_a$, since the a velocity is of the order of 10^{-3}c. For details refer to [9].

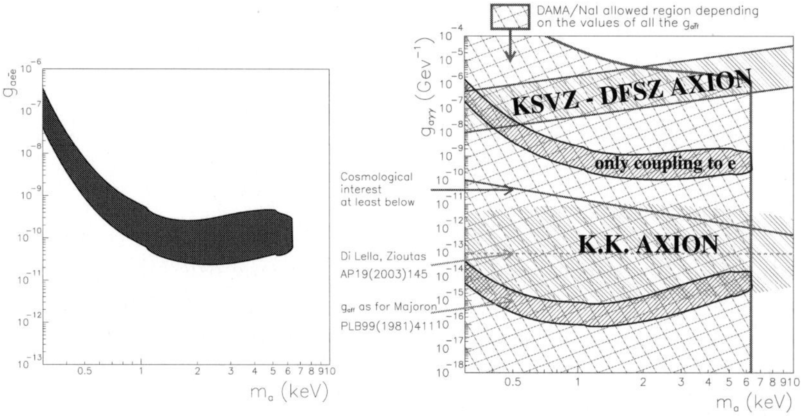

Fig. 6. Case of a pseudoscalar boson candidate (here named a). Left: region allowed at 3σ C.L. in the plane $g_{a\bar{e}e}$ vs m_a by the DAMA/NaI annual modulation data in the considered model framework. Right: allowed region in the plane $g_{a\gamma\gamma}$ vs m_a (crossed hatched region). All the configurations in this region can be allowed depending on the values of all the $g_{a\bar{f}f}$ (couplings to fermions). In the plot examples of the expectations for some models are also shown. The solid line corresponds to a particle with lifetime equal to the age of the Universe; at least all the $g_{a\gamma\gamma}$'s below this line are of cosmological interest. See ref. [9].

In Fig. 6 just the results obtained for the case of a pseudoscalar boson (here named a) are shown. In particular, in Fig. 6-left the allowed region in the plane $g_{a\bar{e}e}$ (coupling constant of the particle to the electrons,) and m_a (the particle mass) is reported. The allowed configuration can be also of cosmological interest as shown in Fig. 6-right.

Furthermore, in the case of a scalar boson (here named h), only a coupling to hadronic matter can be considered in order to account for the DAMA observations [9] and the result is an allowed multi-dimensional volume in the space defined by m_h and by all the $g_{h\bar{q}q}$, coupling constants to quarks.

For related results and discussion see ref. [9].

3.4. ... *and more*

Various possibilities for some of the many possible astrophysical, nuclear and particle Physics scenarios can be considered and are also available in literature, such as e.g. refs. [8,23–32]. Recently, the DAMA Collaboration have also investigated a DM candidate interacting only with electrons [33]. An investigation of the DAMA/NaI effect in terms of sterile neutrino as DM candidate is also under consideration.

4. Some comparisons in the field

No experiment is available so far – with the exception of the presently running DAMA/LIBRA – whose results can be directly compared in a model independent way with that of DAMA/NaI. Thus, claims for contradictions have intrinsically no scientific meaning. Some discussions can be found e.g. in ref. [4,11].

In particular, as regards some claimed model-dependent comparisons presented so far we just mention – among the many existing arguments – that the other experiments available so far: i) are insensitive to the annual modulation signature; ii) use different exposed materials; iii) release just a marginal exposure (orders of magnitude lower than the one by DAMA/NaI) after several/many years underground; iv) exploit strong data selection and strong and often unsafe multiple rejection techniques of their huge counting rate, becoming at the same time insensitive to several DM candidates; v) consider a single model fixing all the astrophysical, nuclear and particle Physics assumptions as well as all the theoretical and experimental parameters at a single questionable choice; vi) generally quote in an incorrect/partial/not updated way the implications of the DAMA/NaI result.

Hereafter few considerations will be summarized. For the WIMP case they do not account for the existing uncertainties on the real coupling with ordinary matter, on the spin-dependent and spin-independent form factors and related parameters for each nucleus, on the spin factor used for each nucleus, on the real scaling laws for nuclear cross sections among different target materials; on the experimental and theoretical parameters, on the effect of different halo models and related parameters on the different target materials, etc. Moreover, large differences are expected in the counting rate among nuclei fully sensitive to the SD interaction (as ^{23}Na and ^{127}I) with the respect to nuclei largely insensitive to such a coupling (as e.g ^{nat}Ge, ^{nat}Si, ^{nat}Ar, ^{nat}Ca, ^{nat}W, ^{nat}O) and also when nuclei in principle all sensitive to this coupling but having different unpaired nucleon (e.g. neutron in case of the odd spin nuclei, such as ^{129}Xe, ^{131}Xe, ^{125}Te, ^{73}Ge, ^{29}Si, ^{183}W and proton in the ^{23}Na and ^{127}I). Obviously, when the detection of the DM particles involves electromagnetic signals (see, for example, the case of the light bosons discussed above, but also electromagnetic contribution in WIMP detection arising e.g. from known effect induced by recoiling nuclei, etc.), all the other experiments do lose the signal in their data selection and multiple "rejection" procedures of the electromagnetic contribution to the counting rate.

In addition, the other experiments present many critical points e.g. regarding their energy threshold, their energy scale determination in the energy range of interest for the investigation of DM candidate particles and their multiple selection procedures, on which their claimed "sensitivities" for a "single" set of assumptions and parameters' values (ignoring both experimental and theoretical uncertainties) are based. In addition, critical items in the used "rejection" procedures are the related stabilities and efficiencies, the systematics in the evaluations of the rejection factors (ranging from 10^{-4} to 10^{-8}), stabilities and monitoring of the spill-out factors, ...

For completeness, it is also worth to further note that no results obtained with different target material can intrinsically be directly compared even for the same kind of coupling, although apparently all the presentations generally refer to cross section on the nucleon. The situation is much worse than the one in the field of double beta decay experiments when different isotopes are used.

As regards the indirect searches, a comparison would always require the calculation and the consideration of all the possible DM particle configurations in the given particle model, since it does not exist a biunivocal correspondence between the observables in the two kinds of experiments. However, the present positive hints provided by indirect searches are not in conflict with the DAMA/NaI result.

Finally, it is worth to note that – among the many corollary aspects still open – there is f.i. the possibility that the particle dark halo can have more than one component; some example have already been considered in literature.

5. The new DAMA/LIBRA and beyond

In 1996 DAMA proposed to realize a ton set-up and a new R&D project for highly radiopure NaI(Tl) detectors was funded at that time and carried out for several years in order to realize – as an intermediate step – the second generation DAMA/LIBRA experiment (successor of DAMA/NaI) with an exposed mass of about 250 kg. Thus, as a consequence of the results of this second generation R&D, the new experimental set-up DAMA/LIBRA (Large sodium Iodide Bulk for RAre processes), $\simeq$ 250 kg highly radiopure NaI(Tl) crystal scintillators (matrix of twenty-five $\simeq$ 9.70 kg NaI(Tl) crystals), was funded and realized. In fact, after the completion of the DAMA/NaI data taking on July 2002, the dismounting of DAMA/NaI occurred and the installation of DAMA/LIBRA started. The experimental site as well as many components of the installation itself have also been implemented [34]. DAMA/LIBRA started to take data in March 2003 and the first data release will, most probably, occur at end 2008. At present it is also in progress a new R&D for a possible ultimate radiopure 1 ton NaI(Tl) set-up.

6. Conclusion

DAMA/NaI experiment pointed out at 6.3 σ C.L. the presence of DM particles in the galactic halo by investigating the model independent annual modulation signature over seven annual cycles. No systematic effect or side reaction able to account for the observed effect has been found. Several corollary quests for the investigation on the nature of the DM candidates have also been pursued and shortly reminded here, others are available in literature considering various DM candidates and scenarios. At present the second generation DAMA/LIBRA (a $\sim$ 250 kg highly radiopure NaI(Tl) set-up) is in operation [34]. The first data release will, most probably, occur at end 2008. Finally, a third generation R&D for a possible ultimate radiopure 1 ton NaI(Tl) detector, proposed in 1996, is also in progress.

References

1. P. Belli et al., *Astropart. Phys.* **5**, 217 (1996); *Il Nuovo Cim.* **19**, 537 (1996); *Phys. Lett. B* **387**, 222 (1996); *Phys. Lett. B* **389**, 783 (1996)(err.); *Phys. Lett. B* **465**, 315 (1999); *Phys. Rev. D* **61**, 117301 (2000); R. Bernabei et al., *Phys. Lett. B* **436**, 379 (1998); *New Journal of Physics* **2**, 15.1 (2000) ; *Phys. Lett. B* **493**, 12 (2000); *Nucl. Instrum & Meth. A* **482**, 728 (2002); *Eur. Phys. J. direct C* **11**, 1 (2001); *Phys. Lett. B* **527**, 182 (2002); *Phys. Lett. B* **546**, 23 (2002); *Eur. Phys. J. A* **27 s01**, 35 (2006).
2. R. Bernabei et al., *Astropart. Phys.* **7**, 73 (1997); *Il Nuovo Cim. A* **110**, 189 (1997); P. Belli et al., *Astropart. Phys.* **10**, 115 (1999); *Nucl. Phys. B* **563**, 97 (1999); R. Bernabei et al., *Nucl. Phys. A* **705**, 29 (2002); P. Belli et al., *Nucl. Instrum & Meth. A* **498**, 352 (2003); R. Cerulli et al., *Nucl. Instrum & Meth. A* **525**, 535 (2004); R. Bernabei et al., *Nucl. Instrum & Meth. A* **555**, 270 (2005); *Ukr. J. Phys.* **51**, 1037 (2006); *Nucl. Phys. A* **789**, 15 (2007). *ROM2F/2007/13* to appear on *Phys. Lett. B. ROM2F/2007/17* to appear on *Phys. Rev. C.*
3. P. Belli et al., *Nucl. Instrum & Meth. A* **572**, 734 (2007), P. Belli et al, in the volume *"Current problems in Nuclear Physics and Atomic energy"*, ed. INR-Kiev, 479 (2006).
4. R. Bernabei el al., *La Rivista del Nuovo Cimento 26* **1**, 1 (2003), R. Bernabei et al., *Int. J. Mod. Phys. D* **13**, 2127 (2004) and references therein.
5. R. Bernabei et al., *Il Nuovo Cim. A* **112**, 545 (1999).
6. P. Belli, R. Bernabei, C. Bacci, A. Incicchitti, R. Marcovaldi, D. Prosperi, *DAMA proposal to INFN Scientific Committee II*, April 24th 1990.

7. K.A. Drukier et al., *Phys. Rev. D* **33**, 3495 (1986); K. Freese et al., *Phys. Rev. D* **37**, 3388 (1988).
8. D. Smith and N. Weiner, *Phys. Rev. D* **64**, 043502 (2001); D. Smith and N. Weiner, *Phys. Rev. D* **72**, 063509 (2005).
9. R. Bernabei et al., *Int. J. Mod. Phys. A* **21**, 1445 (2006).
10. R. Bernabei et al., *Phys. Lett. B* **389**, 757 (1996); *Phys. Lett. B* **408**, 439 (1997); *Il Nuovo Cim. A* **112**, 1541 (1999); P. Belli et al., *Phys. Rev. C* **60**, 065501 (1999); *Phys. Lett. B* **460**, 236 (1999); R. Bernabei et al., *Phys. Rev. Lett.***83**, 4918 (1999); F. Cappella et al., *Eur. Phys. J. direct C* **14**, 1 (2002); R. Bernabei et al., *Phys. Lett. B* **515**, 6 (2001); *Eur. Phys. J. A* **23**, 7 (2005); *Eur. Phys. J. A* **24**, 51 (2005).
11. R. Bernabei et al., *Eur. Phys. J. C* **47**, 263 (2006).
12. R. Bernabei et al., *Int. J. Mod. Phys. A* **22**, 3155 (2007).
13. A. B. Migdal, *J. Phys. USSR* **4**,) 449 (1941; G. Baur, F. Rösel and D. Trautmann, *J. Phys. B: At. Mol. Phys.* **16**, L419 (1983).
14. R. Kitano and I. Low, *Phys. Rev. D* **71**, 023510 (2005).
15. N. Cosme, L.L. Honorez and M.H.G. Tytgat, *Phys. Rev. D* **72**, 043505 (2005).
16. G. Farrar and G. Zaharijas, *Phys. Rev. Lett.* **96**, 041302 (2006); G. Zaharijas and G. Farrar, *Phys. Rev. D* **72**, 083502 (2005); G. Zaharijas *astro-ph/0510088.*
17. X.G. He et al., *hep-ph/0701156.*
18. D.J.H. Chung, G.R. Farrar, E.W. Kolb, *Phys. Rev. D* **56** 6096 (1997); G.R. Farrar, E.W. Kolb, *Phys. Rev. D* **53**, 2990 (1996).
19. J.F. Gunion, D. Hooper, B. McElrath, *Phys. Rev. D* **73**, 015011 (2006).
20. C. Bœhm and P. Fayet, *Nucl. Phys. B* **863**, 219 (2004).
21. Z. Berezhiani, *Int. J. Mod. Phys. A* **19**, 3775 (2004).
22. R. Bernabei et al., *arXiv:0710.0288*, to appear
23. A. Bottino et al., *Phys. Rev. D* **67**, 063519 (2003); A. Bottino et al., *Phys. Rev. D* **68**, 043506 (2003).
24. A. Bottino et al., *Phys. Rev. D* **69**, 037302 (2004).
25. A. Bottino et al., *Phys. Lett. B* **402**, 113 (1997); *Phys. Lett. B* **423**, (1998); *Phys. Rev. D* **59**, 109 (1999) 095004; *Phys. Rev. D* **59**, 095003 (1999); *Astrop. Phys.* **10**, 203 (1999); *Astrop. Phys.* **13**, 215 (2000); *Phys. Rev. D* **62**, 056006 (2000); *Phys. Rev. D* **63**, 125003 (2001); *Nucl. Phys. B* **608**, 461 (2001).
26. K. Belotsky, D. Fargion, M. Khlopov and R.V. Konoplich, *hep-ph/0411093.*
27. R. Foot, *hep-ph/0308254.*
28. S. Mitra, *Phys. Rev. D* **71**, 121302 (2005).
29. E.M. Drobyshevski et al., *arXiv:0704.0982.*
30. E.M. Drobyshevski, *arXiv:0706.3095.*
31. C. Arina and N. Fornengo, *arXiv:0709.4477.*
32. A. Bottino et al., *arXiv:0710.0553.*
33. R. Bernabei et al., in preparation.
34. R. Bernabei et al., in the volume *Frontier Objects in Astrophysics and Particle Physics* Vol. **90**, 581 (2004).

THE DARK CONNECTION BETWEEN THE EGRET EXCESS OF DIFFUSE GALACTIC GAMMA RAYS, THE CANIS MAJOR DWARF, THE MONOCEROS RING, THE INTEGRAL 511 keV ANNIHILATION LINE, THE GAS FLARING AND THE GALACTIC ROTATION CURVE

W. DE BOER

University of Karlsruhe
Postfach 6980, D-76131 Karlsruhe, Germany
E-mail: wim.de.boer@cern.ch

The EGRET excess of diffuse Galactic gamma rays shows all the key features of dark matter annihilation (DMA) for a WIMP mass in the range 50-100 GeV, especially the distribution of the excess is compatible with a standard halo profile with some additional ringlike substructures at 4 and 13 kpc from the Galactic centre. These substructures coincide with the gravitational potential well expected from the ring of dust at 4 kpc and the tidal stream of dark matter from the Canis Major satellite galaxy at 13 kpc, as deduced from N-body simulations fitting to the Monoceros ring of stars. Strong independent support for this substructure is given by the gas flaring in our Galaxy. The gamma rays from DMA are originating predominantly from the hadronization of mono-energetic quarks, which should produce also a small, but known fraction of protons and antiprotons. Bergström et al. an antiproton flux far above the observed antiproton flux and they conclude that the DMA interpretation of the EGRET excess can therefore be excluded. However, they used an isotropic propagation model, i.e. the same diffusive propagation in the disk and the halo. It is shown that an anisotropic propagation model is consistent with the EGRET gamma ray excess, the antiproton flux and the ratios of secondary/primary and unstable/stable cosmic ray particles. Such an anisotropic propagation is supported by the large bulge/disk ratio of the positron annihilation line, as observed by the INTEGRAL satellite. In this case no need for new sources specific to the bulge are needed, so the claimed evidence for strong DMA in the bulge from these observations is strongly propagation model dependent. In the framework of Supersymmetry cross section predictions for direct dark matter searches are presented taking into account the EGRET data, the WMAP data and other electroweak constraints.

1. Introduction

Cold Dark Matter (CDM) is well established by the high rotation speeds in galaxies and clusters of galaxies. Recent cosmological measurements yield a dark matter (DM) density of $22 \pm 2\%$ of the energy of the Universe.[1] If this DM is created thermally during the Big Bang the present relic density is inversely proportional to $\langle \sigma v \rangle$, the annihilation cross section σ of DM particles, usually called WIMPS

(Weakly Interacting Massive Particles), times their relative velocity. The average is taken over these velocities. This inverse proportionality is obvious, if one considers that a higher annihilation rate, given by $\langle\sigma v\rangle n_\chi$, would have reduced the relic density before freeze-out, i.e. the time, when the expansion rate of the Universe, given by the Hubble constant, became equal to or larger than the annihilation rate. The relation can be written as:[2]

$$\Omega_\chi h^2 = \frac{m_\chi n_\chi}{\rho_c} h^2 \approx \left(\frac{3 \cdot 10^{-27} cm^3 s^{-1}}{\langle\sigma v\rangle} \right). \tag{1}$$

For the present value of $\Omega h^2 = 0.105 \pm 0.008$, as measured by WMAP,[1] the thermally averaged total cross section at the freeze-out temperature of $m_\chi/22$ must have been around $3 \cdot 10^{-26}$ cm^{3}s^{-1}. Note that $\langle\sigma v\rangle$ as calculated from Eq. 1 is independent of the WIMP mass (except for logarithmic corrections) as can be shown by detailed calculations.[3] If the s-wave annihilation is dominant, as expected in many supersymmetric models, then the annihilation cross section is energy independent, i.e. the cross section given above is also valid for the cold temperatures of the present universe.[4] Such a large cross section will lead to a production rate of mono-energetic quarks[a] in our Galaxy, which is 40 orders of magnitude above the rate produced at any accelerator. The fragmentation of these mono-energetic quarks will lead to a large flux of neutrinos, photons, protons, antiprotons, electrons and positrons in the Galaxy. From these, the protons and electrons disappear in the sea of many matter particles, but the photons and antimatter particles may be detectable above the background, generated by cosmic ray interactions with the gas in the Galaxy. Such searches for indirect Dark Matter detection have been actively pursued. Recent reviews and references to earlier work can be found in Refs.[5–7]

Gamma rays have the advantage that they point back to the source and do not suffer energy losses, so they are the ideal candidates to trace the dark matter density and have a spectral shape characteristic for mono-energetic quarks. The charged components interact with Galactic matter and are deflected by the Galactic magnetic field, so they do not point back to the source. Therefore the charged particle fluxes have large uncertainties from the propagation models, which determine how many of the produced particles arrive at the detector. For gamma rays the propagation is straightforward: only the ones pointing towards the detector will be observed.

An excess of diffuse gamma rays compatible with dark matter annihilation

---

[a]The quarks are mono-energetic, since the kinetic energy of the cold dark matter particles is expected to be negligible with the mass of the particles, so the energy of the quarks equals the mass of the WIMP.

(DMA) has indeed been observed by the EGRET telescope on board of NASA's CGRO (Compton Gamma Ray Observatory).[8] Below 1 GeV the CR interactions describe the data perfectly well, but above 1 GeV the data are up to a factor two above the expected background. The excess shows all the features of DMA annihilation for a WIMP mass between 50 and 70 GeV.[9] Masses up to 100 GeV are possible if one assumes the cosmic ray energy spectrum to vary in the Galaxy. The excess was observed in all sky directions, which would imply that DM is not dark anymore, but shining in gamma rays.[9] Of course, such an important observation needs to be scrutinized heavily. Among the most important criticism was a paper by Bergström et al.[10] claiming that the antiproton flux from DMA, using the DM distribution from the EGRET excess, would be an order of magnitude higher than the observed antiproton flux. However, they use an isotropic propagation model with the *same* propagation in the disk and the halo. For the expected anisotropic propagation models everything can be made consistent by having a faster propagation perpendicular to the disk than in the disk, in which case the antiprotons from DMA in the halo do not return to the disk. This demonstrates the large uncertainties from propagation models for indirect dark matter searches.

It is interesting to note that anisotropic propagation models can easily explain a large bulge/disk (B/D) ratio of the positron annihilation, as observed from the intensity of the 511 keV annihilation line of thermalized positrons. These positrons are e.g. produced in the β-decays of radioactive nuclei from supernova explosions. A large B/D ratio is simply due to the different geometries of the disk and the bulge: in the bulge the positrons and electrons can annihilate and radiate before escaping to the halo, in the disk they enter the halo much faster without having time to radiate and/or annihilate. In contrast, in isotropic propagation models one can only explain the large bulge/disk ratio by a new source for positrons for the bulge alone. Anisotropic propagation models have enough freedom to explain the results without new sources. So the claimed evidence for DMA from these observations[13] are strongly propagation model dependent.

The paper is organized as follows: in section 2 we summarize the DMA interpretation of the EGRET excess. In section 3 we discuss the problems with the isotropic propagation models. An anisotropic propagation model, which simultaneously describes the antiproton flux, the EGRET excess and the observations concerning primary and secondary cosmic rays and cosmic clocks is discussed in section 4, while in section 5 the consequences for the INTEGRAL excess of the 511 keV line in this anisotropic propagation model are discussed. In section 6 the constraints from direct dark matter searches are discussed. Section 7 summarizes the results.

2. The DMA Interpretation of the EGRET Excess of Diffuse Galactic Gamma Rays

It is well known that the EGRET satellite data on diffuse gamma rays shows an excess above 1 GeV in comparison with the expectations from CR interactions.[8] Below 1 GeV the CR interactions describe the data perfectly well. The excess shows all the features of DMA annihilation for a WIMP mass between 50 and 70 GeV.[9] Especially, the two basic constraints expected from any indirect DMA signal are fulfilled:

- the excess should have the same spectral shape in all sky directions.
- the excess should be observable in a large fraction of the sky with an intensity distribution corresponding to the gravitational potential of our Galaxy. The latter means that one should be able to relate the distribution of the excess to the rotation curve.

The analysis of the EGRET data is simplified by the fact that the spectral shapes of the DMA contribution and the background from CR interactions with the gas of the disk are well known from accelerator experiments: (i) the DMA signal should have the gamma ray spectrum from the fragmentation of mono-energetic quarks, which has been studied in great detail at LEP.[14] (ii) the background in the energy range of interest is dominated by CR protons hitting the hydrogen of the disk. Therefore the dominant background spectral shape is known from fixed-target experiments. Given that these shapes are known from the two best studied reactions in accelerator experiments allows to fit these *known* shapes to the observed gamma ray spectrum in a given sky direction and obtain from the fitted normalization constants the contribution of both background and annihilation signal. So in this case one does not need propagation models to estimate the background, since the data itself calibrates the amount of background. A typical spectrum is shown in Fig. 1, which clearly shows the rather distinct shapes of DMA and background, so the two normalization constants are not strongly correlated. The shape of the DMA contribution shifts to the right for heavier WIMP masses. The preferred WIMP mass is between 50 and 70 GeV with a maximum allowed value of 100 GeV.[9] Repeating these fits over 180 independent sky directions showed that indeed: (i) the *shape* of EGRET excess is consistent with a 60 GeV WIMP mass in each sky direction (ii) the *intensities* of the EGRET excess in various sky directions are as expected from the gravitational potential, which could be proven by reconstructing the rotation curve from the EGRET data after adding the known distribution of visible matter to the DM halo.[9] A parametrization of the best fitted DM halo is shown in Fig. 1 on the right hand side, which clearly shows the ringlike structures at 4 and 13 kpc, as determined from the enhanced intensity of the EGRET excess

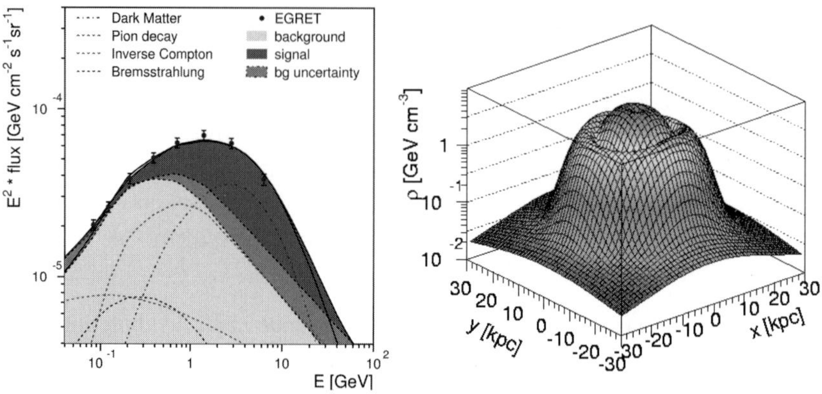

Fig. 1. Left: Fit of the shapes of background and DMA signal to the EGRET data in the direction of the Galactic centre. The dark shaded (red) area indicates the signal contribution from DMA for a 60 GeV WIMP mass using the shape from high energy electron-positron annihilation experiments, while the area below it represents the background with the various indicated contributions. The (blue) area between the dotted lines just below the shaded area is the estimated uncertainty in the background, which is dominated by solar modulation. The normalization of both the background and the DMA contribution have been left free in the fit of the known shapes. Right: Parametrization of the dark matter density profile as determined from the distribution of the EGRET excess in the sky. Both pictures are taken from Ref.[9]

in these regions. The ring at 4 kpc (inner ring) coincides with the ring of dust in this region. The dust is presumably kept there because of a gravitational potential well, which is provided by the ring of DM. The ring at 13 kpc (outer ring) is thought to originate from the tidal disruption of the Canis Major dwarf galaxy, which circles the Galaxy in an almost circular orbit coplanar with the disk.[15–19] Three independent observations confirm this picture of the ring originating from the tidal disruption of a dwarf galaxy:

(i) a ring of DM is expected in this region from the observed ring of stars, called Monoceros ring, which was discovered first with SDSS data.[20,21] Follow-up observations[22] found that this structure surrounds the Galactic disk as a giant ring (observed over 100 degrees in latitude) at Galactocentric distances from 8 kpc to 20 kpc. Tracing this structure with 2MASS M giant stars,[23] suggested that this structure might be the result from the tidal disruption of a merging dwarf galaxy. N-body simulations show indeed that the overdensity in Canis Major is indeed a

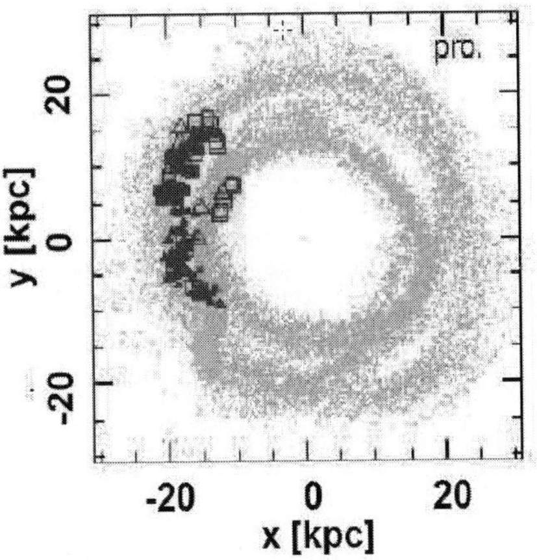

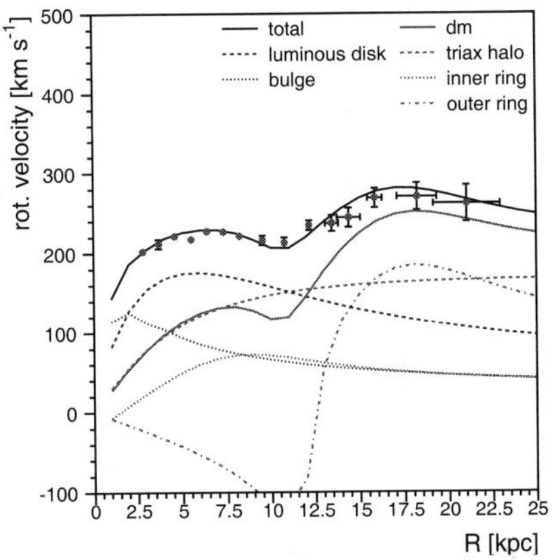

Fig. 2. Left: results of an N-body simulation of the tidal disruption of the Canis Major dwarf Galaxy, whose orbit was fitted to the observed stars (symbols). The simulation predicts a ringlike structure of dark matter with a radius of 13 kpc. From Ref.[16] Right:The rotation curve with the contributions of the bulge, the disk, the triaxial dark matter halo and the two ringlike structures. The outer ring causes the peculiar change of slope in the rotation curve at about 11 kpc. From Ref.[9]

perfect progenitor for the Monoceros stream and they predict a DM ring at 13 kpc with a low ellipticity and almost coplanar with the disk, as shown in the left panel of Fig. 2. The orientation of the major axis at an angle of 20 degrees with respect to the axis sun-Galactic centre and the ratio of minor to major axis around 0.8 agrees with the EGRET ring parameters given in Ref.[9] This correlation with the EGRET excess lends both support to the DMA interpretation of the EGRET excess *and* the interpretation that the Monoceros stream originates from the tidal disruption of the Canis Major (CM) satellite galaxy, thus rejecting the interpretation that the overdensity of stars forming the Canis Major dwarf is a warp of the Galactic disk (see discussions e.g. in Refs.[24,25]). A further rejection from the hypothesis that the CM overdensity is just due to the warp comes from the gas flaring discussed below, which shows that the Monoceros stream is connected to an enormous amount of dark matter.

(ii) Such a massive ring structure influences the rotation curve in a peculiar way: it decreases the rotation curve at radii inside the ring and increases it outside. This is apparent from the change in direction of the gravitational force from the ring on a tracer, since this force decreases the force from the galactic centre for

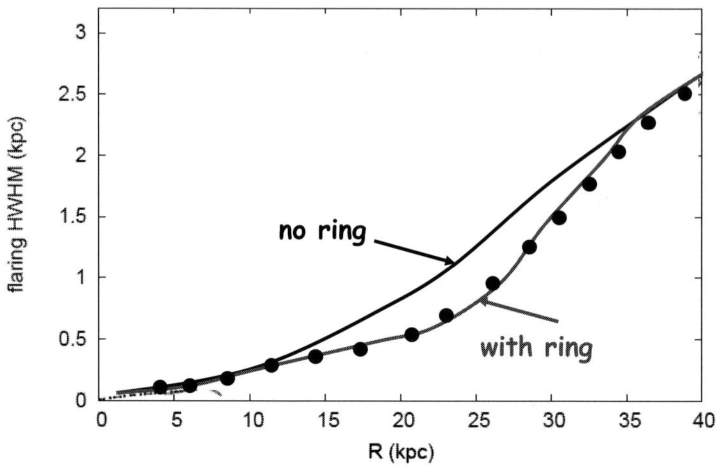

Fig. 3. The half-width-half-maximum (HWHM) of the gas layer of the Galactic disk as function of the distance from the Galactic center. Clearly, the fit including a ring of dark matter above 10 kpc describes the data much better. Adapted from data in Ref.[26]

a tracer inside the ring, but increases it outside the ring. This is indeed observed as shown in the right hand panel of Fig. 2, where the negative contribution of the outer ring is clearly visible.

(iii) A direct proof of the large amount of DM mass in the outer ring comes from a recent analysis of the gas flaring in our Galaxy.[26] Using the new data of the LAB survey of the 21 cm line in our Galaxy led to a precise measurement of the gas layer thickness up to radii of 40 kpc. The increase of the half width of the layer after a decrease to half its maximum value (HWHM) as function of distance is governed by the decrease in gravitational potential of the disk. The outer ring increases the gravitational potential above 10 kpc, which is expected to reduce the gas flaring. Only after taking the ring like structure into account the reduced gas flaring in this region could be understood. The effect is shown in Fig. 3. A fit averaged over all longitudes requires a DM ring with a mass of 2.10^{10} solar masses, in rough agreement with the EGRET excess.

Clearly, these three independent astronomical observations need a ringlike DM structure above 10 kpc, thus providing independent evidence for the DMA interpretation of the EGRET excess.

These independent observations *cannot* be explained by alternative explanations of the EGRET excess, of which the strongest one is provided by the "optimized" model.[27] In this case the cosmic ray spectrum of protons and electrons is not taken to be the locally observed one, but modified to increase the gamma ray spectrum at high energies. This requires a strong break in the injection spectrum of electrons and protons in order not to change the gamma ray spectrum below 1 GeV, but only above 1 GeV. However, the gamma ray spectra at intermediate and high latitudes are created by the *local* cosmic rays. Still they show the excess. Furthermore, spectral shape differences are not expected, since diffusion is fast compared with the energy loss time, so diffusion equalizes the spectrum everywhere in agreement with the observation that the gamma ray spectra in all directions can be described with the *same* CR spectrum in all directions, i.e. the same spectral shape everywhere in the Galaxy.

Another explanation is provided by tuning the efficiency of the EGRET spectrometer to simulate DMA.[28] However, this requires the efficiency already to be modified around 1 GeV and reaching a change in efficiency of 80% at 10 GeV in clear disagreement with the calibration error in a photon beam before launch[29] and the residual uncertainties below 20% during the flight after correcting for time dependent effects.[30] Although their is some uncertainty in the efficiency of the veto counter at higher energies because of the backsplash from the calorimeter, this effect should not start at 1 GeV. Even the authors agree that the considered effects are too small individually and it is not clear that if one adds the errors all linearly that one gets effects up to 80%. And if these effects add up, it would be a remarkable coincidence that the excess corresponds exactly to the very specific sharply falling spectrum from the fragmentation of mono-energetic quarks! An even more remarkable coincidence is that the distribution of the excess in the sky follows the gravitational potential, as proven by the gas flaring and the rotation curve. Therefore the excess is not as isotropic as suggested by these authors, if one observes in finer sky bins, which reveals ringlike structures.

3. The Antiproton Flux from DMA in an Isotropic Propagation Model

As mentioned in the introduction, a serious objection about the DMA interpretation concerns the antiproton flux. However, this depends strongly on the propagation model. Here we summarize the concepts used for the "conventional" propagation model, discuss its priors and its alternatives.

The main features of our Galaxy are a barred central bulge with a diameter of a few kpc and a large spiral disk of about 15 kpc. Most of the gas is distributed in the disk, which extends to radii of 15-20 kpc, while the supernovae remnants (SNR) peak at a distance of a few kpc from the centre. They are thought to be the source of the cosmic rays (CRs) with energies up to 10^{15} eV. These CRs form a plasma of ionized particles, in which the electric fields can be neglected by virtue of the high conductivity and the magnetic fields form Alfvén waves, i.e. a traveling oscillation of ions and the magnetic field. The ion mass density provides the inertia and the magnetic field line tension provides the restoring force. The wave propagates in the direction of the magnetic field with the Alfvén speed, although waves exist at oblique incidence and smoothly change into magnetosonic waves when the propagation is perpendicular to the magnetic field. If the wavelength of the Alfvén waves equals a multiple of the gyration radius of a CR, resonant scattering occurs, which leads to a change in pitch angle of the CR (pitch angle scattering) without changing its energy.[31] Such a process leads to a random walk of CRs, which can be described by a diffusion equation (see review in Ref.[32]). If the B-field has no preferred direction, i.e. if the turbulent small scale component is much stronger than the regular large scale component, the waves propagate randomly in all directions and the diffusion of CRs is isotropic. From the isotropy of CRs one usually assumes the propagation to be isotropic.

Most primary nuclei show a power law spectrum falling with energy like $E^{-2.7}$. This can be easily tuned by selecting the injection spectrum of the primary particles accordingly. However, since the inelastic cross sections for secondary particle production are usually not strongly energy dependent at higher energies, this would lead to rather flat spectra for the secondary/primary ratios in contrast to the observed B/C ratio, which shows a maximum at about 1 GeV/nucleon and decreases as $E^{-0.6}$ towards higher energy. This can be accommodated by assuming energetic particles diffuse faster out of the Galaxy, i.e. the diffusion constant is proportional to $E^{0.6}$. This reduces the high energy part of the B/C spectra. The decrease at low energies can be accommodated by diffusive reacceleration, which shifts the spectrum to higher energies.[32] Alternatively, one can have a strong increase of the diffusion coefficient at low energies because of the damping of the Alfvén waves,[33] thus reducing the B/C ratio at low energies as well.

Both stable and unstable nuclei are produced in supernovae (SNe) explosions with a ratio given by their known production cross sections. From the decay time and the remaining amount of the unstable nuclei one can reconstruct the time of CRs needed for their journey from the source to our local cavity, so they act as “cosmic clocks”. Such measurements yield an average residence time of CRs in the Galaxy of the order of 10^7 yrs. Since they travel with relativistic speeds the long residence time requires that they cannot move rectilinear from the source to us or to outer space, but the CRs must be scattering many times without loosing too much energy, i.e. the diffusion must be effective. During their journey CRs may interact with the gas in the Galaxy and produce secondary particles. This changes the ratio of secondary/primary particles, like the B/C ratio. From the residence time and the amount of secondaries one can estimate the grammage, i.e.

the amount of matter traversed by a CR during its lifetime t_{CR}, which is given by $\rho c t_{CR}$, where $c t_{CR}$ is the path length for a particle traveling with the speed of light c. It was found to be of the order of $10 g/cm^2$, which corresponds to a density of about 0.2 $atoms/cm^3$.[34,35] This is significant lower than the averaged density of the disk of 1 $atom/cm^3$, which suggests that CRs travel a significant time in low density regions, like the halo.

The most advanced program providing a numerical solution to the diffusion equation is the publicly available GALPROP code.[36–38] The basic parameters are the injection spectrum parameters, the diffusion coefficient, the convection speed, the Alfvén speed and the size of the halo. The latter determines the CR residence time inside the Galaxy, since as soon as they pass the border, they are assumed to escape to outer space. By tuning these parameters to the secondary/primary spectra and the unstable/stable spectra one obtains a self-consistent propagation model of our Galaxy. The amount of secondary CR particles and gamma rays are described by this model by the cross sections of the interactions of the primary and secondary CR with the gas of the disk using a network with more than 2000 cross sections. This is one of the great triumphs of GALPROP.

Remaining problems are connected with the large scale structure of the propagation in the Galaxy, as sampled by gamma rays. The problems are twofold. First of all the gamma rays in the GeV range, which are mainly produced by inelastic collisions of CRs with the gas of the disk, show a too small radial gradient: large amounts of gamma rays are produced at large longitudes, i.e. towards the Galactic anticentre. This is unexpected, since the main sources of CRs are assumed to be the SNe explosions, which are preferentially located at radii of a few kpc. In addition the gas decreases at large radii and the gamma ray flux is proportional to the cosmic ray density times the gas density. This problem can be remedied by assuming much more gas at large radii than determined from the molecular hydrogen (H_2) tracer, which is the λ = 2.6 mm line for the J=1 to J=0 transition of the carbon monoxide (CO) molecule. This is a good tracer since CO gets excited by collisions with H_2 molecules. Explaining additional gas at large radii requires a strong radial dependence of the ratio $X_{CO} = N(H_2)/N(CO)$, i.e. X_{CO} increases by an order of magnitude from the inner Galaxy to the outer Galaxy.[39] An alternative explanation is provided by the Galactic wind models, in which case the transport from the disk to the halo is provided by the mass outflow from the disk to the halo from the SNe explosions, which has a strong radial dependence because of the strong radial dependence of the observed SNR.[40] This leads to strong anisotropic propagation with the convection being strongest in the regions with the highest cosmic ray pressure from the SNR, i.e. at distances between 4 and 10 kpc. At smaller radii the gravitational potential of the bulge limits the outflow, while at larger radii the CR pressure decreases.[40] This reduces the CR intensity in the disk and the corresponding gamma ray production at the position of the sources.

The second GALPROP problem is its failure to describe the EGRET excess of gamma rays, which can be remedied by dark matter annihilation, as discussed in section 2. If one attributes the EGRET excess to DMA one runs into the problem of a too large flux of antiprotons, as discussed in detail by Bergström et al.[10] We have implemented the DMA as a source term into the publicly available GALPROP code[b] and find a similar result, as shown in the left panel of Fig. 4 This is not surprising, since GALPROP uses the same priors as the program used by Bergström et al.:[10] (i) the propagation is dominated by diffuse scattering, which is assumed to be isotropic, i.e. the same in the halo and the disk (ii) the gas in the disk is smoothly distributed (iii) the influence of observed regular magnetic fields can be neglected. These priors fulfill the basic picture of the origin and propagation of cosmic rays discussed above. The main reason for the large flux of antiprotons from DMA is *not* that DMA produces so many antiprotons, but the fact that the residence time of charged particles is required to be of the order of 10^7 yrs, as determined from the cosmic clocks. The last requirement can be fulfilled in a model with isotropic diffusion only with a large halo, so CRs do not escape, but they perform random walks in a large volume. In this case there is no difference between primary particles produced by SNe or primary particles produced by DMA, so *all* CRs are stored inside the Galaxy. In this case DMA increases the averaged density of antiprotons by orders of magnitude, so the flux of antiprotons becomes of the same order of magnitude as the EGRET excess. Note that the production ratio of antiprotons/gammas from DMA is only at the percent level, as is well known from accelerator experiments for the fragmentation of mono-energetic quarks. If one assumes that the propagation is *not* isotropic, the picture completely changes: the DMA antiprotons may be transported quickly to the halo by a combination of convection or fast diffusion along the regular magnetic fields depicted in Fig. 5. Such an anisotropic propagation model will be discussed in the next section.

4. The Antiproton Flux from DMA in an Anisotropic Propagation Model

The propagation picture with isotropic propagation, as discussed above, is based on hydromagnetic wave theories, in which the turbulent small-scale components of the magnetic field dominate over or are at the same order of magnitude as the regular large scale components. However, the turbulence is expected to be different in the halo and the disk, since the disk is not fully ionized in contrast to the

[b]The GALPROP code can be obtained from http://galprop.stanford.edu/.

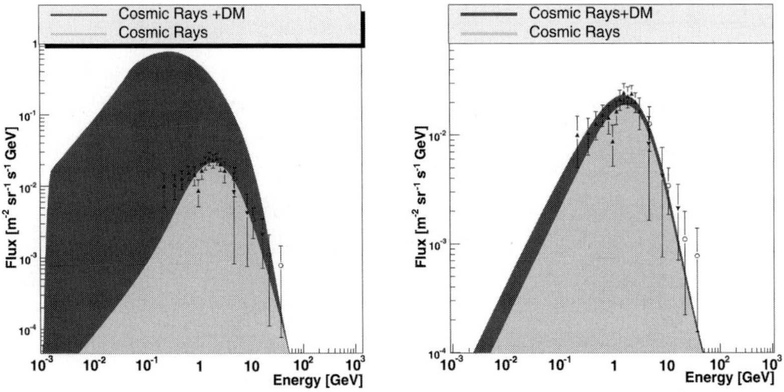

Fig. 4. Comparison of the antiproton production including DMA in the conventional model with isotropic propagation (left) and a model with anisotropic propagation (right).

halo. This implies that the Alfvèn waves are efficiently dissipated in the disk by ion-neutral damping. Furthermore the most important contribution to the random field in the disk is by turbulent mass motions, induced by supernova explosions and other stellar mass loss activity, which leads to a different wave spectrum in the disk as compared to the halo, thus leading to different diffusion coefficients in the halo and disk.[40] In addition one expects the diffusion along the regular magnetic fields to be an order of magnitude faster than the diffusion perpendicular to the magnetic fields, even if the turbulent component is of the same order of magnitude (see e.g.[40] and references therein). These analytical estimates of fast diffusion along the regular magnetic field lines as compared to the transverse diffusion were confirmed by following the trajectories of CRs in realistic magnetic fields with both a regular and turbulent component. Given the toroidal field in the disk[41,42] found that the diffusion along the azimuthal magnetic field is one to two orders of magnitude faster than the transverse diffusion. This means that CRs preferentially diffuse along the toroidal fields just above or below the disk or into the halo via the poloidal fields sketched in Fig. 5.[43–45] In such a picture our local cavity is situated between two thick pancake-like structures of higher CR density only 300 pc apart in the z-coordinate. This leads to a reduced grammage compared with the CR density, which is maximal in the disk, a longer residence time, a reduced radial gradient in the gamma ray flux and an isotropization of the CR flux, all features difficult to explain simultaneously in an isotropic propagation models with a source distribution located towards the centre.

Another possible effect concerning charged particles diffusing fast along reg-

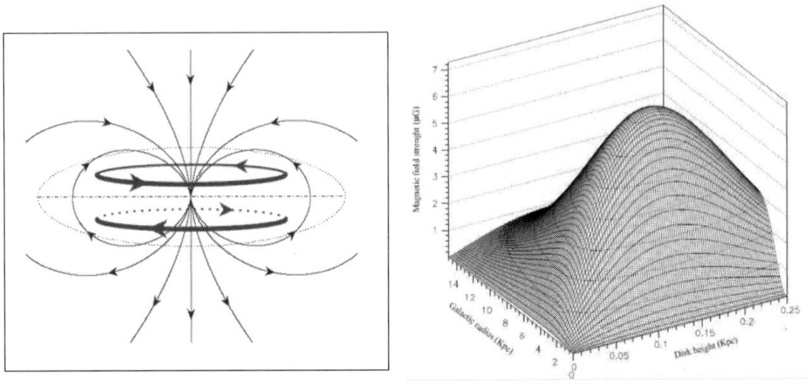

Fig. 5. Left: a schematic picture of the magnetic fields in the Milky Way, consisting of a dipole (poloidal) field component (strongest in the center) and a circular (toroidal) field component, which is strongest just above and below the disk at a distance of 150 pc (from Ref.[45]). Right: a parametrization of the toroidal component (from Ref.[42]).

ular magnetic field components is related to the molecular clouds: the gas density in the disk varies from 10^{-3} cm^{-3} in the warm ionized medium to 10^2-10^3 cm^{-3} in clumps of cold gas with a size of a few pc. In the centre of these clumps the density may be as high as 10^7 cm^{-3} in dense molecular clouds (MC), where star formation occurs. On average the gas density is 1 cm^{-3} in the disk. Inside these MC magnetic fields far above the random components have been observed (see Ref.[46] for an excellent review). What is more important, these fields seem to be correlated with the observed static magnetic fields outside the MCs.[47] This can only be understood, if the MCs remember the large scale magnetic fields in the interstellar medium, i.e. if during the contraction flux freezing occurs. In this case the magnetic field lines from the ISM will become highly concentrated near the MCs and CRs in the ISM, which preferentially diffuse fast along these field lines, will be reflected by the higher density of the field lines near the MCs. So, as worked out by Chandran.[48] the MC act like magnetic mirrors for CRs, just like the concentration of magnetic field lines near the poles from the earth trap the CRs in the famous Van Allen radiation belts. The large distances (10-100 pc scale) between the MC allows to trap particles up to the TeV scale, thus increasing the grammage and the residence time, which are now increased by the trapping time between the MC in the disk, *not* by how often they pass from the halo to the disk, as is the case in the isotropic propagation model. So the halo size is not a sensitive parameter anymore and particles, once in the halo, will be preferentially transported away from the disk by a combination of convection or fast diffusion along the regular field lines in the halo. It should be noted that only a small fraction of the CRs enter the MCs, if these act as magnetic mirrors. This could explain why positrons mainly annihilate in the gas *between* the MCs, not with electrons from molecular hydrogen *inside* the MCs, as deduced from the annihilation line shape.[49,50] Of course, one could argue that although the MCs make up the largest mass fraction, they have the lowest filling factor. But if they have strong magnetic fields, the CRs are expected to be tunneled towards the MCs by the high concentration of magnetic field lines.

How can one implement such a propagation model? Existing programs are not suitable. E.g. the CR tracing programs calculating the trajectories in a magnetic field do not calculate all secondary particles and the GALPROP program with isotropic diffusion does not have a regular magnetic field in the propagation of charged particles. However, the basic modifications needed are: the propagation follows preferentially the regular magnetic field lines, which are toroidal in the disk and poloidal in the halo. Such a propagation would require tuning the 3D version of GALPROP. However, this takes an excessive amount of CPU time. Therefore transporting the CRs to the halo by a combination of fast diffusion and convection in the z-direction in the 2D version is much more effective and leads to the same effect: CRs in the halo will hardly come back to the disk. But it should be kept in mind that the parameters are effective parameters in a simplified axisymmetric 2D version of the 3D reality, which has also preferred diffusion directions in the disk.

We have modified the publicly available source code of GALPROP by (i) allowing for a diffusion tensor instead of a diffusion constant, thus modifying the diffusion equation and the Crank-Nickelson coefficients accordingly; (ii) allowing an inhomogeneous grid in order to have step sizes below 100 pc in the disk region and large step sizes in the halo; (iii) implement the dark matter annihilation as a source term of stable primary particles, especially antiprotons, positrons and gamma rays. The dark matter distribution was taken to be the one obtained from the EGRET excess, as discussed above. The grammage and escape time were adjusted for charged particles to account for the fact that secondary particles are now produced largely locally, since particles produced far away from the solar system are likely not to reach our local cavity. If in addition the trapping between molecular clouds is effective, only a small fraction of CRs will penetrate the MCs and most of them will be reflected. But the grammage in between the MCs will be enhanced by the multiple passages from the mirroring, so one would expect it to increase the grammage and residence time by the same factor. This turns out to be working, so this was simply introduced in GALPROP as a constant g, called grammage parameter, multiplying the HI and HII gas densities. A grammage paramater

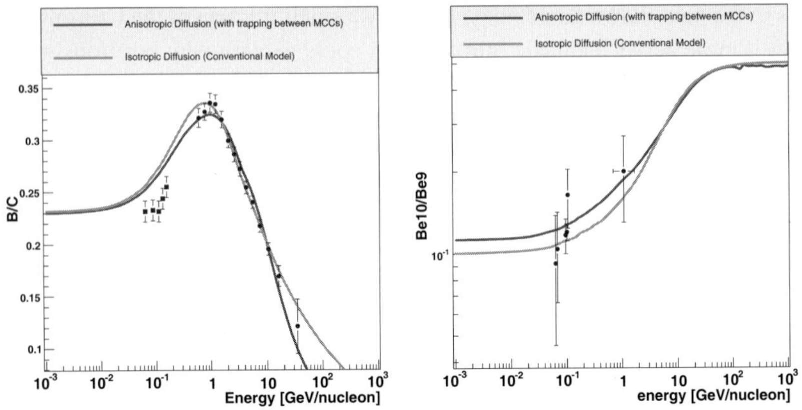

Fig. 6. Comparison of the secondary/primary (B/C) ratio and unstable/stable ($^{10}Be/^{9}Be$) ratio in the conventional model with isotropic propagation and model with anisotropic propagation.

of about 12 is needed to describe the B/C ratio combined with the $^{10}Be/^{9}Be$ ratio. Note that this grammage parameter determines the local production of charged particle, so this grammage parameter is not necessarily the same as the grammage needed for the gamma ray production, since the latter is determined by the large scale gas densities. Given the large fluctuations in and strong radial dependence of the gas densities the difference can be large. Also the CR density is expected to have a radial dependence because of the convection having a radial dependence, as discussed before. The CR density can also have strong variations from the fact that Alfvén waves can be damped in high density gas regions, thus leading to strong variations in the diffusion coefficient and corresponding CR density variations.

The transport from the disk to the halo is quite uncertain. It should be noted that the average scale height of SNIa is expected to be about 300 pc (thick disk) and the ejecta connect to the halo in chimney like structures (see e.g. Ref.[51,52] and references therein), which can drive magnetic field lines towards high altitudes ($\approx$ 1 kpc), thus facilitating the transport to the halo by the fast parallel diffusion. This was simulated as an enhanced convection term starting at $v_0 = 250km/s$ at 100 pc above the disk and then increasing with the distance z above the disk as $dv/dz = 37km/s$. The remaining GALPROP parameters can be found in Ref.[53] As shown in Figs. 4 and 6 the antiproton flux, the B/C ratio and the $^{10}Be/^{9}Be$ ratio are all well described by this set of parameters. Clearly, the excess of diffuse gamma rays can be well described together with the B/C ratio and the antiprotons.

In summary, we do not believe that the DMA interpretation of the EGRET is "excluded by a large margin" because of the overproduction of antiprotons as claimed by Bergström et al.[10] Such a statement is only valid *within their propagation model based on isotropic propagation.* Anisotropic models with different propagation in the halo and the disk can perfectly describe all observations including DMA.

5. Positron Annihilation in Our Galaxy

Additional support for the propagation picture with anisotropic propagation, as discussed in section 4, comes from the positron annihilation signal observed by the INTEGRAL satellite.[11,12] Since positron annihilation is only efficient at non-relativistic energies, the positrons must have energies in the MeV range. Sources of such positrons are largely coming from the decay of radioactive nuclei expelled by dying stars, especially SNIa, since in this case the core makes up a large fraction of the mass. This makes it easier for the positrons to escape from the relatively thin layer of the ejecta. Light curves, which are sustained first by the gamma rays in the shock waves and later by the electrons and positrons, suggest that only a few percent of the positrons escape from the ejecta and can annihilate outside after thermalization. Positrons annihilating inside the ejecta will produce also gamma rays, but these will not be visible as a single 511 keV line because of further interactions in the shock wave.

The main observation is that the 511 keV line from the annihilation between thermalized positrons and electrons (either free or bound in nuclei) is largely confined to the bulge with a bulge/disk (B/D) ratio of a few,[11,12] although additional data suggests a lower ratio.[54] Taking the dominant source to be SNIa, one would expect a B/D ratio to be well below one, because of the higher mass in the disk and the higher rate of SNIa explosions expected in the thick disk as compared to the bulge.[55] An additional problem presents the observation of the 1.8 MeV line from the ^{26}Al radioactive isotope, which has clearly been observed, both in the bulge *and* the disk by the Comptel detector on NASA's CGRO observatory.[56] These nuclei are thought to be produced by nucleosynthesis in massive stars in the thin disk and yield in their decay on average 0.85 positrons. Because of their non-relativistic speed and high charge ^{26}Al nuclei loose their kinetic energy rapidly and decay with a half life time of about 10^6 years close to the position where they were created. The observed flux of positron annihilation in the disk seems to be saturated already by the positrons from ^{26}Al and ^{44}Ti decays.[11] So what happened to the positrons from SNIa explosions in the disk?

In an anisotropic propagation model a large B/D ratio for the positrons is expected, since the bulge is a much more extended object than the disk, so the particles have much more time to thermalize and annihilate in the bulge than in the disk before reaching the halo. Once in the halo they move away from the disk, where there is hardly any gas for them to annihilate. As mentioned before the fountain like structures of SNR (see e.g. Refs.[51,52] and references therein) can drive magnetic field lines towards high altitudes ($\approx$ 1 kpc), thus facilitating the transport to the halo by the fast parallel diffusion for the relativistic positrons.

A fraction of the positrons escaping the disk may move by the poloidal field to the bulge, thus enhancing the B/D ratio.[55] So the problem of the large B/D ratio for positron annihilation is intimately related to the propagation of the positrons. In a conventional propagation model without magnetic fields and homogeneous gas distributions the positrons annihilate near their source[49,50] and one must resort to new positron sources specific for the bulge, like DMA of very light WIMPS.[13,57] The WIMP masses have to be below of few MeV, since else they would be visible by synchrotron radiation above the limits set by Comptel and EGRET.[58,59] In anisotropic propagation models the low strength of the annihilation signal in the disk is simply a consequence of the fact that the disk is so thin, so positrons can escape easily to the halo, where they find no partners to annihilate. So it is the same solution required by the ratio of antiprotons and EGRET excess of gamma rays: fast diffusion perpendicular to the disk either by the regular magnetic fields or convection. In this case no unnatural light WIMPs are needed. Such light WIMPs would need additionally new gauge bosons to have a large enough annihilation cross section compatible with Eq. 1.

6. Constraints from Direct Dark Matter Search Experiments

Direct detection experiments search for signals of dark matter particles elastically scattering off the nuclei in their detector. The event rate is given by $\Gamma = <\sigma v> n$, where n is the local number density of WIMPS, v the velocity between detector and WIMP and σ the scattering cross section. The brackets indicate that the average over the velocity distribution and corresponding cross sections has to be taken. No positive results have been reported and experimental limits on the cross section have been derived under the assumption that the local density of WIMPS corresponds to 0.3 GeV/cm^3, as estimated from the rotation curve (see e.g. a recent review by Spooner[60]). The best limit obtained by the XENON10 experiment is about 5.10^{-8} pb at 90% C.L. for a 50 GeV WIMP.[61] This is below the cross section limit expected from the EGRET data, if one assumes the minimal supersymmetric model to be valid. The region allowed by all constraints (WMAP, LEP limits on charginos and Higgses, $b \to s\gamma$ branching ratio, EGRET mass limits) is shown in Fig. 7 together with the limits from some typical direct dark matter search experiments.[62] No signal was found sofar, so the regions above these lines are excluded. Naively one could conclude that the DMA interpretation of

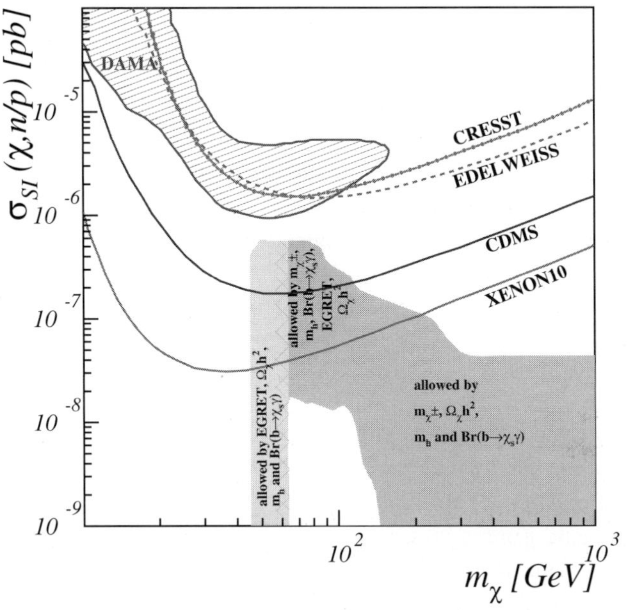

Fig. 7. The spin-independent cross section for neutralino-nucleon scattering compared with experimental constraints: the region above the XENON10 limit is excluded. Details can be found in Ref.[62]

the EGRET excess combined with all electroweak and relic density constraints (green region in Fig. 7) is excluded by the XENON10 data. However, there are severe caveats. First of all, the event rate in these experiments is proportional to the cross section *and* the local WIMP density. Both have large uncertainties: the cross section is model dependent and has only been calculated in the minimal supersymmetric models. Although the EGRET data is perfectly consistent with supersymmetry, it does not prove supersymmetry and in e.g. extra dimension models the cross sections could be very different. Secondly the local relic density has large uncertainties, because the densities obtained from the rotation curve do not say anything about the clustering of dark matter, but yield only an averaged density. From N-body simulations one knows that dark matter is clumpy, as can be expected already from the fact that galaxies are formed by dark matter starting to pull together from the initial density fluctuations in the early universe. The number of clumps is very large because of the steep decrease of the mass spectrum with increasing mass ($\propto M^{-2}$) and the lightest clumps are very light (about $10^{-6} M_\odot$, see Ref.[63]). But even with this large number of clumps the probability of finding a clump in the solar neighbourhood is small, so the direct searches are more likely to observe first interactions from the diffuse component, which originates from the tidal stripping of the clumps. How many are disrupted is an ongoing debate (see e.g. Refs.[64,65]), but it certainly depends sensitively on the dark matter profile of the clumps. These profiles cannot be calculated yet reliably from N-body simulations. It is conceivable that the density of the diffuse component is an order of magnitude less than the clumpy component, so any statements concerning exclusions from direct dark matter detection experiments should take this into account.

7. Conclusion

With an anisotropic propagation model the amount of antiprotons expected from DMA annihilation can be reduced by one to two orders of magnitude. Therefore the claim by Bergström et al.[10] that the DMA interpretation of the EGRET excess of diffuse Galactic gamma rays is excluded "by a large margin" is strongly propagation model dependent. It only applies for their simplified propagation model, which assumes the same diffusion in the halo and in the disk. An anisotropic propagation model with different propagation in the halo and the disk can reconcile the EGRET excess with the antiproton flux and the ratios of secondary/primary and unstable/stable nuclei. In addition the difference in geometry between bulge and disk leads to much more radiation and annihilation of positrons and electrons in the bulge as compared to the disk, thus alleviating the need to introduce new sources of positrons and electrons for the bulge to explain the INTEGRAL excess of the 511 keV line in the bulge.[13] Direct dark matter search experiments reach the cross section range expected from the EGRET excess. However, the experimental limits are inversely proportional to these cross sections times the local relic density. Both are uncertain: for the cross section one assumes the WIMPS are the lightest supersymmetric partners of the minimal supersymmetric model with gravity inspired breaking of supersymmetry, while for the local relic density one assumes the WIMPS are smoothly distributed instead of the expected distribution in clumps. The clumpy nature can drastically reduce the local density if we are not located inside a clump.

In summary we consider DMA is a viable explanation of the EGRET excess of diffuse Galactic gamma rays, especially since it is observed with the same shape of the fragmentation of mono-energetic quarks in all sky directions and the intensity distribution of the excess traces the DM profile, as shown independently by the rotation curve, the gas flaring and the N-body simulation of the disruption of the Canis-Major satellite galaxy.

Acknowledgements

I wish to thank the organizers for the kind invitation to this splendid conference. Furthermore I thank P. Blasi, D. Breitschwerdt and N. Prantzos for helpful discussions and my close collaborators I. Gebauer, D. Kazakov, M. Weber and V. Zhukov for their contributions.

This work was supported by the BMBF (Bundesministerium für Bildung und Forschung) via the DLR (Deutsches Zentrum für Luft- und Raumfahrt).

References

1. D.N. Spergel et al., ApJS, **170** (2007)377.
2. G. Jungman, M. Kamionkowski, and K. Griest, Phys. Rep., **267**(1996)195.
3. E. Kolb, and M.S. Turner, M.S. 1990, The Early Universe (Addison Wesley)
4. W. de Boer, C. Sander, V. Zhukov, A.V. Gladyshev and D.I. Kazakov, Phys. Lett. B **636** (2006) 13, arXiv:hep-ph/0511154.
5. L. Bergström, L., Rep. Prog. Phys., **63** (2000) 793; arXiv:hep-ph/0002126.
6. G. Bertone, D. Hooper, and J. Silk, Phys. Rept. **405**(2005) 279; arXiv:hep-ph/0404175.
7. T.J. Sumner, 2002, www.livingreviews.org/Articles/Volume5/2002-4sumner; Living Reviews in Relativity published by the Max Planck Institute for Gravitational Physics, Albert Einstein Institute, Germany.
8. S.D. Hunter, et al., ApJ, **481**(1997) 205.
9. W. de Boer, C. Sander, V. Zhukov, A.V. Gladyshev, and D.I. Kazakov, A&A 444 (2005) 51 , arXiv:astro-ph/0508617.
10. L. Bergström, L. et al., JCAP **0605** (2006) 006, arXiv:astro-ph/0602632.
11. J. Knödlseder, *et al.*, Astron. Astrophys. **441**(2005) 513, arXiv:astro-ph/0506026.
12. G. Weidenspointner *et al.*, A&A, **450** (2006)1013, arXiv:astro-ph/0601673.
13. C. Boehm, D. Hooper, J. Silk, M. Casse and J. Paul, Phys. Rev. Lett. **92** (2004) 101301, arXiv:astro-ph/0309686.
14. W.M. Yao, *et al.* [Particle Data Group], J. Phys. G **33**2006 1.
15. N.F. Martin, R.A. Ibata, M. Bellazzini, M.J. Irwin, G.F. Lewis and W. Dehnen, Mon. Not. Roy. Astron. Soc. **348**(2004) 12, arXiv:astro-ph/0311010.
16. J. Penarrubia et al., Astrophys. J. **626** (2005) 128, arXiv:astro-ph/0410448.
17. D. Martinez-Delgado, D.J. Butler, H.W. Rix, Y.I. Franco and J. Penarrubia, Astrophys. J. **633** (2005)205, arXiv:astro-ph/0410611.
18. M. Bellazzini, R. Ibata, N. Martin, G.F. Lewis, B. Conn and M.J. Irwin, Mon. Not. Roy. Astron. Soc. **366**(2006) 865, arXiv:astro-ph/0504494.
19. Penarrubia, J., Martinez-Delgado, D. and Rix, H.W. 2007, arXiv:astro-ph/0703601.
20. H.J. Newberg et al. ApJ **569** (2002) 245.
21. B. Yanny et al. ApJ, **588** (2003) 824 [Erratum-ibid. **605** (2003) 575]; arXiv:astro-ph/0301029.
22. R.A. Ibata, M.J. Irwin, G.G. Lewis, A.M.N. Ferguson and N. Tanvir, MNRAS, **340** (2003) L21; arXiv:astro-ph/0301067. [arXiv:0705.3655 [astro-ph]].
23. H.J. Rocha-Pinto, S.R. Majewski, M.F. Skrutskie, and J.D. Crane, ApJ **594** (2003) L115 and arXiv:astro-ph/0307258.

24. M. Lopez-Corredoira, Y. Momany, S. Zaggia and A. Cabrera-Lavers, arXiv:0707.4440 [astro-ph].
25. B.C. Conn et al., Mon. Not. Roy. Astron. Soc. **376** (2007) 939, arXiv:astro-ph/0701664.
26. P.M.W. Kalberla, L. Dedes, J. Kerp and U. Haud, A&A, **469** (2007) 511, arXiv:0704.3925 [astro-ph].
27. A.W. Strong, I.V. Moskalenko and O. Reimer, ApJ, **613** (2004) 962; arXiv:astro-ph/0406254.
28. F.W. Stecker, S.D. Hunter and D.A. Kniffen, arXiv:0705.4311 [astro-ph].
29. D.J. Thompson et al., IEEE Trans. Nucl. Sci., **34** (1987) 36.
30. J.A. Esposito et al., ApJ **123** (1999) 203.
31. R. Kulsrud and W.P. Pearce, ApJ **156** (1969) 445.
32. A. W. Strong, I. V. Moskalenko and V. S. Ptuskin, Ann. Rev. Nucl. Part. Sci. **57** (2007) 285, arXiv:astro-ph/0701517.
33. V.S. Ptuskin, I.V. Moskalenko, F.C. Jones, A.W. Strong and V.N. Zirakashvili, Astrophys. J. **642**(2006) 902, arXiv:astro-ph/0510335.
34. R. Schlickeiser, R., 2002, "Cosmic ray astrophysics," Springer Verlag, Berlin.
35. V.S. Berezinsky, S.V. Bulanov, V.A. Dogiel, V.L. Ginzburg and V.S. Ptuskin, 1990, "Astrophysics of Cosmic Rays" Amsterdam, North-Holland Publishing.
36. A.W. Strong and I.V. Moskalenko, ApJ, **509** (1998) 212; arXiv:astro-ph/9807150.
37. I.V. Moskalenko, A.W. Strong and O. Reimer, A&A, **338** (1998) L75; arXiv:astro-ph/9808084
38. A.W. Strong, I.V. Moskalenko and O. Reimer, ApJ, **537** (2000) 763 [Erratum-ibid. **541** (2000) 1109]; arXiv:astro-ph/9811296; Details on the latest GALPROP versions can be found at http://www.mpe.mpg.de/ aws/propagate.html.
39. A.W. Strong, I.V. Moskalenko, O. Reimer, S. Digel and R. Diehl, 2004, A& A **422** (2004) L47, arXiv:astro-ph/0405275.
40. D. Breitschwerdt, V.A. Dogiel and H.J. Völk, A&A, **385** (2002) 216, arXiv:astro-ph/0201345.
41. D. De Marco, P. Blasi and T. Stanev, JCAP **0706** (2007) 027, arXiv:0705.1972 [astro-ph].
42. A. Codino and F. Plouin, arXiv:astro-ph/0701498.
43. J.L. Han, R.N. Manchester, E.M. Berkhuijsen and R. Beck, A&A, **322** (1997) 98
44. J.L. Han, Ap&SS, **278** (2001) 181, arXiv:astro-ph/0010537.
45. J.L. Han, Chin. J. Astron. Astrophys. **6** (2006) 211, astro-ph/0603512v2.
46. C. Heiles and R. Crutcher, in "Cosmic Magnetic Fields", 2005, Eds. Richard Wielebinski and Rainer Beck. Lecture notes in Physics Volume 664, publ. by Springer, Berlin, Germany, p. 137, arXiv:astro-ph/0501550.
47. J.L Han and J.S. Zhang, A&A **464** (2007) 609, arXiv:astro-ph/0611213.
48. B.D.G. Chandran, Space Sci. Rev. **99** (2000) 271, arXiv:astro-ph/0010105.
49. P. Jean, *et al.*, Astron. Astrophys. **445** (2005) 579, arXiv:astro-ph/0509298.
50. N. Guessoum, P. Jean and W. Gillard, A&A **436** (2005)171, arXiv:astro-ph/0504186.
51. M.A. de Avillez and D. Breitschwerdt, Astron. Astrophys. **425** (2005) 899, [arXiv:astro-ph/0407034].
52. M.A. de Avillez and D. Breitschwerdt, Proc. of the XXXIXth Rencontres de

Moriond, La Thuile, Aosta Valley, Italy, March 21 - 28, 2004, Eds. T. Montmerle, A. Chalabaev, J. Tran Thanh Van, arXiv:astro-ph/0502078.
53. I. Gebauer, arXiv:0710.4966 [astro-ph].
54. G. Weidenspointner *et al.* 2007, arXiv:astro-ph/0702621.
55. N. Prantzos, A&A **449**(2005) 869, arXiv:astro-ph/0511190.
56. Diehl, R. *et al.*, Nature **439** (2006) 45, [arXiv:astro-ph/0601015].
57. Y. Ascasibar, P. Jean, C. Boehm and J. Knödlseder, Mon. Not. Roy. Astron. Soc. **368**(2005) 1695, arXiv:astro-ph/0507142.
58. J.F. Beacom, N.F. Bell, and G. Bertone, Phys. Rev. Lett. **94**(2005) 171301,899 arXiv:astro-ph/0409403.
59. C. Boehm and P. Uwer, arXiv:hep-ph/0606058.
60. N. J. Spooner, arXiv:0705.3345 [astro-ph].
61. J. Angle *et al.* [XENON Collaboration], arXiv:0706.0039 [astro-ph].
62. J. Deger-Glaeser, "Kombination von direkter und indirekter Such nach dunkler Materie, Diplomarbeit, Univ. of Karlsruhe, 2007.
63. J. Diemand, B. Moore and J. Stadel, Nature **433**(2005) 389 arXiv:astro-ph/0501589.
64. B. Moore, J. Diemand, J. Stadel and T. Quinn, arXiv:astro-ph/0502213.
65. V. Berezinsky, V. Dokuchaev and Y. Eroshenko, Phys. Rev. D **73**, 063504 (2006), arXiv:astro-ph/0511494.

DENSITY WAVE THEORY APPLIED TO THE SPIRAL ARMS IN THE MILKY WAY PREDICTS A MASSIVE OUTER GALACTIC DISK

B. FUCHS

Astronomisches Rechen-Institut am Zentrum für Astronomie der Universität Heidelberg,
Mönchhofstraße 12 - 14, 69120 Heidelberg, Germany
E-mail: fuchs@ari.uni-heidelberg.de

Keywords: Galaxy: kinematics and dynamics.

Observations of the spiral structure of the Milky Way do not show a well ordered grand-design spiral but a semi regular multi-armed pattern of relatively short spiral arms[7] . The dynamical evolution of such spiral arms is described by the swing-amplification theory[2–4] of shearing density waves. When such density waves grow they take part in the shear due to the differential rotation of the Galactic disk. They reach maximal amplification at trailing orientation, decay then, and are replenished by freshly growing waves.

The density wave theory makes a specific prediction of the circumferential wave length $\lambda_{\rm max}$ at which density waves preferentially grow. This preferred wave length depends on the rotation curve of the Galaxy and the surface density of the Galactic disk[2,6] . The expected number of spiral arms m is then given by how often $\lambda_{\rm max}$ fits onto the circular annulus,

$$m = \frac{2\,\pi\,R}{\lambda_{\rm max}}\,. \tag{1}$$

This conjecture has been tested successfully on numerical simulations of the dynamical evolution of galactic disks[1] .

If the observed distribution of mass in the Galactic disk and the observed rotation curve are used to evaluate eq. (1), four spiral arms are predicted in the vicinity of the solar annulus and in the inner disk of the Milky Way, precisely as observed[7] (cf. also the contribution by Bodaghee et al. in this volume).

However, beyond the solar annulus there is a serious discrepancy between the theoretically predicted and observed morphology of the Milky Way's spiral arms. These can be traced in the HI distribution in the Galactic disk out to galactocentric radii of 25 kpc[5] as four arms. But, if conservative estimates of the disk mass and the observed outer rotation curve of the Milky Way are inserted into eq. (1), many more than the actual four spiral arms are predicted. A quantitative analysis (Fuchs, in preparation) shows that an excess mass of $3.8 \cdot 10^{10}\, M_\odot$ is required to ensure a four-armed spiral between 8.5 and 20 kpc.

References

1. Fuchs B., in: H.V. Klapdor-Kleingrothaus and R.D. Viollier (eds.), *Dark matter in astro- and particle physics* (Springer, Berlin 2002) p. 28.
2. Fuchs B., *A&A* **368**, 107 (2001).
3. Goldreich P. & Lynden-Bell D., *MNRAS* **130**, 125 (1965).
4. Julian W. H. & Toomre A., *ApJ* **146**, 810 (1966).
5. Levine E.S., Blitz L., Heiles C., *Sci*, **312**, 1773 (2006).
6. Toomre A., in: S.M. Fall and D. Lynden-Bell (eds.), *The structure and evolution of normal galaxies* (Cambridge Univ. Press, Cambridge 1981) p. 111.
7. Valée J.P., *ApJ*, **566**, 261 (2002).

NEW CONSTRAINTS ON SPIN-DEPENDENT WIMP-NEUTRON INTERACTIONS FROM HDMS WITH NATURAL Ge AND ^{73}Ge

H.V. KLAPDOR-KLEINGROTHAUS*,
I.V. KRIVOSHEINA** and V.A. BEDNYAKOV***

**Heidelberg, Germany, prof.klapdor-kleingrothaus@hotmail.de*
http://www.klapdor-k.de
***Radiophysical-Research Institute, Nishnij-Novgorod, Russia*
****Joint Institute for Nuclear Research, Dubna, Russia*

In the mixed spin-scalar coupling approach the data from *both* HDMS (Heidelberg dark matter search) experiments performed in the years 1996, 1997, 2001-2003, with *natural* Ge and with *enriched* ^{73}Ge were *simultaneously* analyzed. This analysis together with a new procedure for background identification and subtraction from the measured ^{73}Ge spectrum allowed us to obtain a significant (about one order of magnitude) improvement for the limits on the WIMP-neutron spin-dependent coupling. As a result the HDMS experiment is now giving the most sensitive limits on the WIMP-neutron spin coupling for WIMP masses larger than 60–65 GeV/c^2.

Detection of Weakly Interacting Massive Particles (WIMP), which still are leading candidates for the non-baryonic dark matter (DM)[1] in our Galaxy, is a challenge for modern particle physics and astrophysics (see, for example[2–10]).

It was for the first time demonstrated by the DAMA collaboration[11–13] and confirmed latter in[9,14] that for analyzing data from DM detectors with spin-non-zero targets one should use the so-called mixed spin-scalar coupling approach. In particular, this procedure allows one to extract information about both the spin-dependent (SD) and spin-independent (SI) WIMP-nucleon cross sections *simultaneously*. Therefore one can expect that such a procedure allows an improvement of the exclusion curves relative to corresponding curves obtained in the one-coupling dominance approach.

In this letter on the basis of the above-mentioned approach we present results of our analysis of both background spectra from the two HDMS setups (prototype — operating only a natural Ge detector, and final —

operating an enriched ^{73}Ge detector), measured in the period 02.2001 to 07.2003. The HDMS (Heidelberg Dark Matter Search) experiment with an enriched ^{73}Ge inner detector (enrichment 86%) is the first (and the only one) setup with a high-spin ($J = 9/2$) Ge target isotope for direct DM search (for details, see[15–18]). A new analysis in the one-coupling dominance approach has been recently published for HDMS in[18] . The idea of our *combined* analysis relies on the unique possibility that two targets with different isotope (from natural Ge and enriched ^{73}Ge) were used as inner detector in the *same* HDMS setup under the same outer background conditions at the Gran Sasso Underground Laboratory (LNGS). Furthermore, we use a new procedure for determination of background in the measured spectrum of enriched ^{73}Ge. This background identification and subtraction allowed a further — about one order of magnitude — improvement of the WIMP-neutron spin-dependent exclusion curves.

The extraction of the WIMP-nucleon SD or SI cross section follows, in general, the conservative assumption that the whole experimental background spectrum consists of WIMP-induced events (see, for example[16–19]). This means that for any given WIMP mass m_χ and for all available recoil energies $E_{\rm R}$ above the threshold energy ϵ both the WIMP-nucleon cross sections for spin-independent and for spin-dependent interaction, $\sigma_{\rm SI}$ and $\sigma_{\rm SD}$, are excluded if

$$\frac{dR}{dE_{\rm R}} = \kappa_{\rm SI}(E_{\rm R}, m_\chi)\,\sigma_{\rm SI} + \kappa_{\rm SD}(E_{\rm R}, m_\chi)\,\sigma_{\rm SD} \;>\; \frac{dR^{\rm data}}{dE_{\rm R}}(\epsilon, E_{\rm R}) \equiv R(\epsilon, E_{\rm R}). \tag{1}$$

The differential event rate (see, for example,[3]) for nonzero velocity of the Earth ($v_E \neq 0$) and the finite Galaxy escape velocity ($v_{\rm esc} \neq \infty$) is

$$\frac{dR(v_E, v_{\rm esc})}{dE_{\rm R}} = \tag{2}$$

$$F^2(E_{\rm R})\frac{R_0}{E_0 r}\frac{k_0}{k_1}\left\{\frac{\sqrt{\pi}}{4}\frac{v_0}{v_E}\left[{\rm erf}\left(\frac{v_{\rm min}+v_E}{v_0}\right) - {\rm erf}\left(\frac{v_{\rm min}-v_E}{v_0}\right)\right] - e^{-v_{\rm esc}^2/v_0^2}\right\},$$

$$\frac{k_1}{k_0} = {\rm erf}\left(\frac{v_{\rm esc}}{v_0}\right) - \frac{2}{\sqrt{\pi}}\frac{v_{\rm esc}}{v_0}e^{-v_{\rm esc}^2/v_0^2},$$

$$\frac{R_0}{E_0 r} = \frac{N_{\rm Av}}{A}\frac{\sigma^A(0)}{v_0\sqrt{\pi}}\frac{\rho_\chi}{m_\chi}\frac{(m_\chi + M_A)^2}{m_\chi^2 M_A}, \qquad r = \frac{4 m_\chi M_A}{(m_\chi + M_A)^2}.$$

Here $v_{\rm min} = \sqrt{2E_{\rm min}/m_\chi}$, and $E_{\rm min} = E_{\rm R}/r$ is the smallest WIMP energy which can produce the recoil energy $E_{\rm R}$, $N_{\rm Av}/A$ is the number density of target nuclei. For the WIMP mass density the value $\rho_\chi = 0.3$ GeV/cm^3 is used. We also used in (2) the following velocity parameters

$$v_0 = 220 \text{ km/s}, \qquad v_{\rm esc} = 600 \text{ km/s}, \qquad v_E = 232 \text{ km/s}. \tag{3}$$

Therefore, as in[18] for the calculation of the expected WIMP spectra we used a truncated Maxwellian velocity distribution in an isothermal WIMP-halo model. However, it should be mentioned that other models exist, and that varying the halo model can affect the results significantly (see, for example[12,20–29]).

The WIMP-nucleus cross section $\sigma^A(0)$ multiplied by the nuclear form factor $F^2(E_{\rm R})$ can be presented as a sum of SI and SD contributions:

$$F^2(E_{\rm R}) \cdot \sigma^A(0) \equiv F^2_{\rm SI}(E_{\rm R}) \cdot \sigma^A_{\rm SI}(0) + F^2_{\rm SD}(E_{\rm R}) \cdot \sigma^A_{\rm SD}(0); \tag{4}$$

$$\sigma^A_{\rm SI}(0) = \frac{\mu_A^2}{\mu_p^2}\sigma^p_{\rm SI}\, A^2, \qquad \mu_A = \frac{m_\chi\, M_A}{m_\chi + M_A};$$

$$\sigma^A_{\rm SD}(0) = \frac{\mu_A^2}{\mu_p^2}\sigma_{\rm SD}\left[\frac{4}{3}\frac{J+1}{J}\left(\langle S_p\rangle\cos\theta + \langle S_n\rangle\sin\theta\right)^2\right]. \tag{5}$$

Here the effective spin WIMP-nucleon cross section $\sigma_{\rm SD}$ and the coupling mixing angle θ were introduced[12,13] in such a way, that the SD WIMP-proton and SD WIMP-neutron interaction cross sections have the form $\sigma^p_{\rm SD} = \sigma_{\rm SD}\cos^2\theta$, and $\sigma^n_{\rm SD} = \sigma_{\rm SD}\sin^2\theta$. The parameter $\langle S_{p(n)}\rangle$ is the spin of the proton (neutron) averaged over all nucleons in the nucleus A with the target nucleus mass M_A. For the neutron-odd group isotope ^{73}Ge we neglect the sub-dominant contribution from WIMP-proton spin coupling $\langle S_p\rangle$ (due to $\langle S_p^{73}\rangle \ll \langle S_n^{73}\rangle$) and use $\cos\theta = 0$.[9] Therefore in our consideration $\sigma_{\rm SD} \equiv \sigma^n_{\rm SD}$. We use $\langle S_n^{73}\rangle = 0.496$ from[30] to be consistent with the best SD exclusion curve from.[18] We also assume $\sigma_{\rm SI} = \sigma^p_{\rm SI} = \sigma^n_{\rm SI}$. The coefficients $\kappa_{\rm SI}(E_{\rm R}, m_\chi)$ and $\kappa_{\rm SD}(E_{\rm R}, m_\chi)$ from (1) can be easily obtained from the formulas given above. To reproduce the SD WIMP-neutron exclusion curves from[18] (obtained in the one-coupling dominance approach) we used the following form factor given in:[3]

$$F^2_{\rm SD}(E_{\rm R}) = j_0^2(qr_A) \qquad \text{for} \qquad qr_A < 2.55, \quad qr_A > 4.5, \tag{6}$$

$$F^2_{\rm SD}(E_{\rm R}) \approx 0.047 \qquad \text{for} \qquad 2.55 \le qr_A \le 4.5,$$

calculated in the so-called thin-shell approximation and corrected so that the first zero of the Bessel function is partially filled with the value of the function at the second maximum. Here $q = \sqrt{2M_A E_{\rm R}}$ is the momentum transfer and r_A is the nuclear radius.[3] As well as in[18] for the SI WIMP-nucleus coupling we used the Helm approximation of the Bessel form factor:

$$F^2_{\rm SI}(E_{\rm R}) = \left(\frac{3j_1(qr_A)}{qr_A}\right)^2 e^{-(qs)^2}$$

where $s \approx 1$ fm is the nuclear skin thickness. It is known that for a semi-conductor germanium detector one has to take into account the ionization efficiency. For the HDMS Ge setup in the (visual) energy interval $4 < E_{\rm R} < 50$ keV we use a simple relation between the visible energy in our Ge detector and the real recoil energy, $E_{\rm vis} \equiv E_{\rm R} = 0.14E^{1.19}_{\rm real\ recoil} \approx QE_{\rm real\ recoil}$[16–19] with $Q = 0.33$ being the quenching factor for Ge.

Finally, we note that for any WIMP mass, m_χ, and any target mass, M_A, due to kinematics one has no event at all with

$$E_{\rm R} > {E_{\rm R}}^{\rm max} = Q\frac{2v^2_{\rm max}M_A m^2_\chi}{(M_A+m_\chi)^2} = Q\frac{2v^2_{\rm max}}{M_A}\mu^2_A, \qquad v_{\rm max} \approx v_{\rm esc}. \qquad (7)$$

For example, ${E_{\rm R}}^{\rm max} = 4$ (50) keV, for $m_\chi = 12$ (77) GeV/c^2. It is clear from (7) that for fixed $v_{\rm max}$, M_A and the detector energy threshold ϵ there are undetectable WIMPs (with rather light m_χ) if their maximal recoil energy is smaller than the threshold: ${E_{\rm R}}^{\rm max} < \epsilon$. Therefore one has two restrictions for the theoretical event rate (see eq. (1)) as function of the WIMP mass m_χ:

$$R(E_{\rm R} > {E_{\rm R}}^{\rm max}(m_\chi)) \equiv 0 \qquad \text{and} \qquad R({E_{\rm R}}^{\rm max}(m_\chi) < \epsilon) \equiv 0. \qquad (8)$$

These restrictions have been taken into account in our analysis. The first one of them allows us to determine the background (for the WIMP-nucleus signal) in the ^{73}Ge measured spectrum, which leads to drastic improvement of the deduced exclusion curves.

To find both $\sigma_{\rm SI}$ and $\sigma_{\rm SD}$ for given m_χ we numerically minimize the relevant χ^2-functional using the experimental spectra published previously in[16] for natural Ge and in[18] for enriched ^{73}Ge. For enriched ^{73}Ge we used the most clean (runs 721–1000) spectrum from.[18] As previously in[15–18] a sliding variable energy window approach is used to check the excess events above the experimental spectrum. The minimum among the cross section values obtained via the multiple fits is taken as the cross section limit for the corresponding WIMP mass. Therefore simultaneous determination of the SD and SI WIMP-nucleon limits for a given WIMP mass m_χ (exclusion curves for $\sigma_{\rm SD}(m_\chi)$ and $\sigma_{\rm SI}(m_\chi)$) is obtained by means of minimization of the discrepancy between the expected rates calculated with (1) and (2) and the above-mentioned experimental spectra.

First, we carried out the one-coupling dominance fit of the enriched ^{73}Ge spectrum. The result is the thin (black) exclusion curve labeled with "SD only fit" in Fig. 1. The curve is consistent with the previous analysis of,[18] which is labeled with "HDMS-2005/SD" in Fig. 1. The small discrepancy between the "HDMS-2005/SD" and "SD only fit" curves for low WIMP masses is mainly due to our restrictions (8).

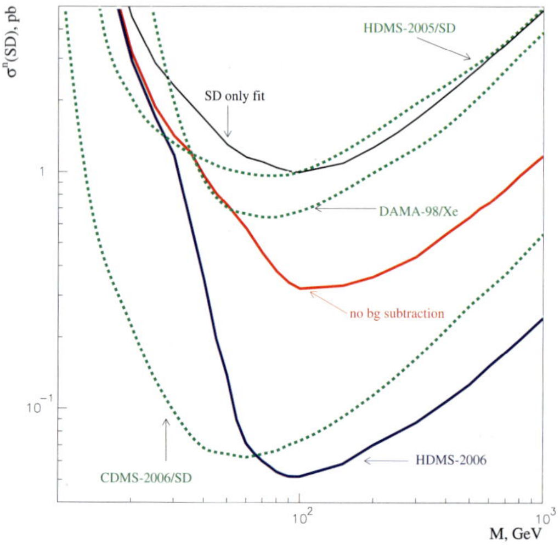

Fig. 1. Spin-dependent cross section upper limits $\sigma_{\rm SD}$ as function of WIMP mass from data analysis of both HDMS spectra. In our approximation $\sigma_{\rm SD} \equiv \sigma^n_{\rm SD}$. The label "no bg subtraction" shows our best exclusion curve (in red) obtained from simultaneous minimization of both natural Ge and the best enriched ^{73}Ge spectra within the mixed spin-scalar coupling approach *without* any background subtraction from the ^{73}Ge spectra. The curve (in blue) labeled with "HDMS-2006" is from the same minimization of both natural Ge and ^{73}Ge spectra *after subtraction* of a background of 0.11 events/kg/day in the last spectrum (see text). The thin (black) exclusion curve "SD only fit" from the traditional one-coupling dominance fit of the enriched ^{73}Ge spectrum only is given for comparison with the previous analysis of[18] labeled with "HDMS-2005/SD". The best exclusion curve for WIMP-neutron spin coupling from the CDMS collaboration[31] is labeled with "CDMS-2006/SD" (dashed green). The dashed line with label "DAMA-98/Xe" corresponds to the DAMA results from.[32]

Next, we performed our analysis within the mixed spin-scalar coupling approach.[9,12,13] To fit in this approach simultaneously both measured spectra we have taken the same energy threshold of 4 keV, which corresponds to the real threshold of the HDMS final setup with enriched ^{73}Ge.[17] The curve labeled in Fig. 1 with "no bg subtraction" shows our exclusion curves obtained from the above-mentioned fit of both natural Ge and the best enriched ^{73}Ge spectra. One can see a 3–4 fold improvement in comparison with the one-coupling dominance result presented in Fig. 1 by curves "SD only fit" and "HDMS-2005/SD".

Furthermore, with the measured spectrum of enriched ^{73}Ge and the first relation from (8) one can give a conservative estimate of the background which can not be a result of any WIMP-nucleus interaction (see aslo[33]). In accordance with (7) and (8) for any m_χ, fixed $v_{\rm max}$ and M_A there is maximal recoil energy ${E_{\rm R}}^{\rm max}(m_\chi)$ for which WIMP-nucleus interactions are unable to produce a signal if $E_{\rm R} > {E_{\rm R}}^{\rm max}(m_\chi)$. Therefore, for fixed m_χ the measured recoil spectrum in the region $E_{\rm R} > {E_{\rm R}}^{\rm max}(m_\chi)$ gives directly this extra background index, which we assume to be a constant function of energy. We have estimated these constants for each allowed m_χ (until ${E_{\rm R}}^{\rm max}(m_\chi) < 50$ keV) and assumed the minimal of these constants (which is equal to 0.11 ± 0.02 events/kg/day/keV) to be the mean background for all measured $E_{\rm R}$ and all m_χ.

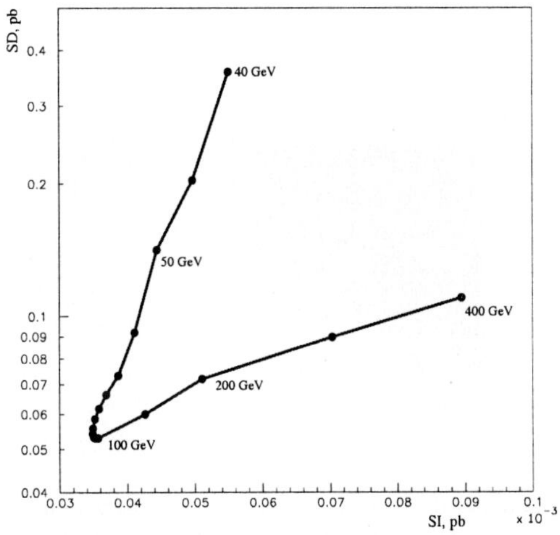

Fig. 2. Spin-dependent cross section upper limit $\sigma_{\rm SD}$ as function of spin-independent cross section upper limit $\sigma_{\rm SI}$ obtained from fit of the data from both (natural and enriched) HDMS setups. Points running over the curves mark the relevant WIMP mass values. For example, a point marked with filled circle and labeled with "200 GeV", shows the simultaneous upper limits for SI and SD WIMP-nucleon interaction cross sections $\sigma_{\rm SI}$ and $\sigma^n_{\rm SD}$ for $m_\chi = 200$ GeV/c^2.

Finally, with common energy threshold of 4 keV, the minimization of both natural Ge and the best ^{73}Ge spectra *with* subtraction of the above-mentioned extra background of 0.11 events/kg/day yields the SD exclusion curve, labeled with "HDMS-2006" in Fig. 1. One can see that this SD exclusion curve improves (at least for $m_\chi > 60$ GeV) currently the best exclusion curve (labeled with "CDMS-2006/SD") for the WIMP-neutron spin coupling from the CDMS collaboration.[31] The CDMS result is still the best one for $m_\chi < 60$ GeV due to their smaller energy threshold.

Despite of the traditional form of presentation of the SD exclusion curves in Fig. 1 as function of WIMP mass one should keep in mind that the obtained $\sigma_{\rm SI}$ and $\sigma^n_{\rm SD}$ constraints for fixed WIMP mass are correlated. This correlation is presented explicitly in Fig. 2, where the dependence of the WIMP mass is given by means of the points running over the curves and marked with the relevant WIMP mass values.

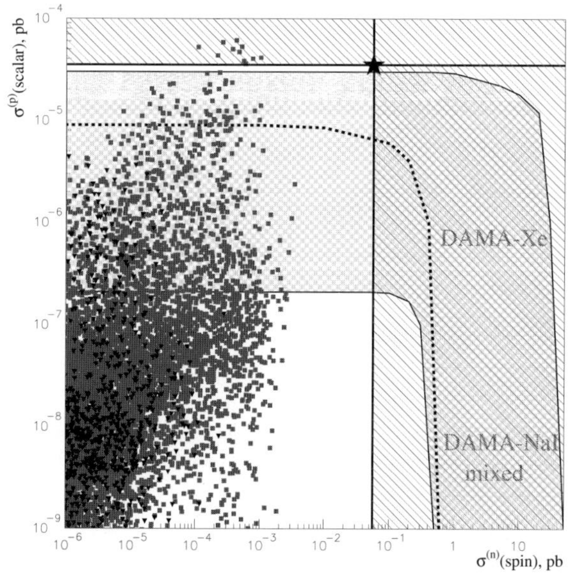

Fig. 3. The DAMA-NaI allowed region (inside yellow band) for SD WIMP-neutron coupling versus SI WIMP-nucleon coupling is from[12] and corresponds to 40 GeV/$c^2 < m_\chi <$ 110 GeV/c^2. The scatter plots from[9] give correlations between $\sigma^p_{\rm SI}$ and $\sigma^n_{\rm SD}$ in the effMSSM for $m_\chi <$ 200 GeV. The squares (red) correspond to sub-dominant relic neutralino contribution $0.002 < \Omega_\chi h_0^2 < 0.1$ and triangles (black) correspond to WMAP relic neutralino density $0.094 < \Omega_\chi h_0^2 < 0.129$. The dashed line from[13] shows the DAMA-LiXe (1998) exclusion curve for $m_\chi = 50$ GeV/c^2. The star gives our simultaneous upper limits for $\sigma_{\rm SI}$ and $\sigma^n_{\rm SD}$ for $m_\chi = 80$ GeV/c^2. Therefore values of $\sigma_{\rm SI}$ above the horizontal line and of $\sigma^n_{\rm SD}$ located right from the vertical line are excluded by our analysis.

To perform a comparison with the DAMA allowed region of $\sigma_{\rm SI}$ and $\sigma^n_{\rm SD}$ (Fig. 3) we have chosen for illustration only one WIMP mass m_χ = 80 GeV/c^2 (the star in the figure). Our upper limits for all other WIMP masses, 40 GeV/c^2 < m_χ < 110 GeV/c^2, are *very close* to this point, and thus also within the DAMA allowed region in Fig. 3. For this WIMP mass, values of $\sigma_{\rm SI}$ located above the horizontal line and of $\sigma^n_{\rm SD}$ located to the right side of the vertical line are excluded. Therefore, for any fixed WIMP mass in the domain 40 GeV/c^2 < m_χ < 110 GeV/c^2 SUSY-like calculations should give simultaneously $\sigma_{\rm SI}$ below the relevant (to the fixed WIMP mass) horizontal line and $\sigma^n_{\rm SD}$ to the left side of the relevant vertical line, respectively. Our limits improve the DAMA-Xe limit significantly (about one order of magnitude) and exclude a DAMA allowed region of large spin-dependent WIMP-neutron cross sections.

In Fig. 3 we compare our new HDMS-2006 limits on the SD and SI WIMP-nucleon interactions with the relevant DAMA constraints extracted from measurement of the annual signal modulation with a NaI target[12] as well as with our calculation in the low-energy effective MSSM.[9]

In conclusion, the mixed spin-scalar coupling approach was applied to analyze the data from both HDMS experiments with natural Ge and with the neutron-odd group high-spin isotope ^{73}Ge. Due to fit of data with two parameters, spin-dependent $\sigma^n_{\rm SD}$ and spin-independent $\sigma_{\rm SI}$ cross sections, a clear improvement of the limits for the spin-dependent WIMP-nucleon coupling is achieved in this approach relative to the traditional one-coupling dominance scheme.

Furthermore, a new procedure for background identification and subtraction in the measured ^{73}Ge spectrum allowed us to obtain a further more than one order of magnitude improvement for the WIMP-neutron spin-dependent coupling. Our best limits for WIMP-neutron spin-dependent coupling improve the relevant limits from the CDMS collaboration[31] for WIMP masses larger than 60–65 GeV/c^2. They also exclude a rather large part of the DAMA-allowed range of spin-dependent WIMP-neutron cross sections.

V.A. Bednyakov thanks the Max Planck Institut für Kernphysik for the hospitality and RFBR (Grant 06–02–04003) for support.

References

1. M. W. Goodman and E. Witten, *Phys. Rev.* **D31**, 3059 (1985).
2. G. Jungman, M. Kamionkowski, and K. Griest, *Phys. Rept.* **267**, 195 (1996).
3. J. D. Lewin and P .F. Smith, *Astropart. Phys.* **6**, 87 (1996).
4. V. A. Bednyakov, H. V. Klapdor-Kleingrothaus, S. Kovalenko, *Phys. Rev. D* **50**, 7128 (1994).
5. V. A. Bednyakov, H. V. Klapdor-Kleingrothaus, S. Kovalenko, *Phys. Rev. D* **55**, 503 (1997).

6. V. A. Bednyakov, H. V. Klapdor-Kleingrothaus, *Phys. Rev.* **D59**, 023514 (1999).
7. V. A. Bednyakov and H. V. Klapdor-Kleingrothaus, *Phys. Rev.* **D63**, 095005 (2001).
8. V. A. Bednyakov, S. G. Kovalenko, H. V. Klapdor-Kleingrothaus, and Y. Ramachers, *Z. Phys.* A **357**, 339 (1997).
9. V. A. Bednyakov and H. V. Klapdor-Kleingrothaus, *Phys. Rev. D* **70**, 096006 (2004).
10. P. Gondolo, hep-ph/0501134, and Proc. *DARK 2004*, Texas, 2-9 Oct. 2004, edited by *H. V. Klapdor-Kleingrothaus and R. Arnowitt* (*Springer, Heidelberg*, 2005) p. 610.
11. R. Bernabei *et al.*, *Phys. Lett.* B **480**, 23 (2000).
12. R. Bernabei *et al.*, *Riv. Nuovo Cim.* **26**, 1 (2003).
13. R. Bernabei *et al.*, *Phys. Lett.* B **509**, 197 (2001).
14. V. A. Bednyakov and H. V. Klapdor-Kleingrothaus, hep-ph/0504031, and Proc. DARK2004, Texas, 2-9 Oct. 2004, edited by *H. V. Klapdor-Kleingrothaus, and R. Arnowitt* (*Springer, Heidelberg*, 2005) p. 583.
15. H. V. Klapdor-Kleingrothaus *et al.*, hep-ph/0103077, in Proc. of *IDM2000*, York, England, 18-22 Sept., 2000 (*World Scientific, 2001*) p. 415.
16. C. Tomei, *PhD Thesis*, Univ. of L'Aquila (Italy) 2004; L. Baudis, A. Dietz, B. Majorovits, F. Schwamm, H. Strecker and H. V. Klapdor-Kleingrothaus, *Phys. Rev. D* **63**, 022001 (2001); H. V. Klapdor-Kleingrothaus *et al*, Proc. *Dark2000*, Heidelberg, 10-14 July 2000, edited by H. V. Klapdor-Kleingrothaus (*Springer Heidelberg, New York, 2001*), p. 553.
17. H.V . Klapdor-Kleingrothaus *et al*, *Astropart. Phys.* **18**, 525 (2003).
18. H. V. Klapdor-Kleingrothaus, I. V. Krivosheina, C. Tomei, *Phys. Lett. B* **609**, 226 (2005).
19. Heidelberg-Moscow Coll., *Phys. Rev. D* **59**, 022001 (1999).
20. A. Kinkhabwala and M. Kamionkowski, *Phys. Rev. Lett.* **82**, 4172 (1999).
21. F. Donato, N. Fornengo, and S. Scopel, *Astropart. Phys.* **9**, 247 (1998).
22. N. W. Evans, C. M. Carollo, P. T. de Zeeuw, *Mon. Not. Roy. Astron. Soc.* **318**, 1131 (2000).
23. C. J. Copi and L. M. Krauss, *Phys. Rev.* D **63**, 043507 (2001).
24. P. Ullio and M. Kamionkowski, *JHEP* **03**, 049 (2001).
25. C. J. Copi and L. M. Krauss, *Phys. Rev. D* **67**, 103507 (2003).
26. A. Kurylov and M. Kamionkowski, *Phys. Rev. D* **69**, 063503 (2004).
27. D. Tucker-Smith and N. Weiner, *Phys. Rev. D* **72**, 063509 (2005).
28. C. Savage, P. Gondolo and K. Freese, *Phys. Rev. D* **70**, 123513 (2004).
29. P. Gondolo and G. Gelmini, *Phys. Rev. D* **71**, 123520 (2005).
30. M. T. Ressell *et al.*, *Phys. Rev.* D **48**, 5519 (1993).
31. D.S. Akerib *et al.* [CDMS Collaboration], Phys. Rev. D **73**, 011102 (2006).
32. R. Bernabei *et al.*, Phys. Lett. B **436**, 379 (1998).
33. C. E. Aalseth *et al.*, Phys. Atom. Nucl. **63**, 1268 (2000); [Yad. Fiz. **63**, 1341 (2000)].

NEUTRINOS FROM WIMP ANNIHILATIONS

TOMMY OHLSSON

Department of Theoretical Physics, School of Engineering Sciences,
Royal Institute of Technology (KTH) – AlbaNova University Center,
Roslagstullsbacken 21, SE-106 91 Stockholm, Sweden
E-mail: tommy@theophys.kth.se

Royal Swedish Academy of Sciences (KVA),
P.O. Box 50005, SE-104 05 Stockholm, Sweden

In this talk, we make an improved analysis on the production of neutrinos coming from WIMP annihilations inside the Sun as well as the Earth. We treat both neutrino interaction and oscillation effects in a consistent three-flavor framework. Our numerical simulations are performed in an event-based setting, which is useful for both theoretical studies and creating neutrino telescope Monte Carlo simulations. We find that the yield of muon-type neutrinos is enhanced or suppressed depending on the dominant WIMP annihilation channel. In addition, we find that oscillations can significantly affect the neutrino yields from WIMP annihilations in the Sun. Effectively, the neutrino flavors are mixed. Finally, for the Earth, the oscillations have no large impact at energies for the new neutrino telescopes such as IceCube, ANTARES, and NESTOR.

Keywords: WIMP annihilations; neutrinos; neutrino interactions; neutrino oscillations.

1. Introduction

Cosmological and astronomical observations show that the energy and matter in the Universe consist of 4% ordinary baryonic matter, 23% dark matter, and 73% dark energy.[1] One of the most plausible dark matter candidates are so-called weakly interacting massive particles (WIMPs). In particular, neutralinos are promising such WIMP candidates.

WIMPs in the Milky Way halo can scatter in the Sun/Earth and be gravitationally bound to the Sun/Earth.[2–6] Eventually, they will scatter again and sink to the core. In the core, WIMPs will accumulate and can annihilate and produce neutrinos.

2. Simulations

We simulate 13 different annihilation channels using PYTHIA 6.400.[7] The annihilation channels are

$$\chi\chi \to u\bar{u}, d\bar{d}, s\bar{s}, c\bar{c}, b\bar{b}, t\bar{t}, \tau^+\tau^-, W^+W^-, Z^0Z^0, g\bar{g}, \nu_e\bar{\nu}_e, \nu_\mu\bar{\nu}_\mu, \nu_\tau\bar{\nu}_\tau$$

and the simulations are performed with a full three-flavor Monte Carlo code including both neutrino interactions and neutrino oscillations. The code is event-based, and thus, useful for both theoretical analysis and building a neutrino telescope Monte Carlo. We perform simulations for the following 19 WIMP masses: 10, 25, 50, 80.3, 91.2, 100, 150, 176, 200, 250, 350, 500, 750, 1000, 1500, 2000, 3000, 5000, and 10000 GeV. For each mass and annihilation channel, we perform $2.5 \cdot 10^6$ annihilations. In addition, we perform the simulations for seven different sets of neutrino oscillation parameters. We collect all neutrinos $(\nu_e, \nu_\mu, \nu_\tau)$ and antineutrinos $(\bar{\nu}_e, \bar{\nu}_\mu, \bar{\nu}_\tau)$, keeping track of the flavors separately.

3. Neutrino Interactions and Neutrino Oscillations

3.1. *Neutrino interactions*

On the way out of the Sun, neutrinos can participate in both charged- and neutral-current interactions. Neutral currents degrade the energy of the neutrinos, whereas charged currents give charged leptons: electrons and muons are stopped before they can give neutrinos, but taus will decay and give neutrinos (regeneration). Simulations are performed using CTEQ6 DIS parton distribution functions.[8]

3.2. *Neutrino oscillations*

A completely general three-flavor neutrino oscillation framework (with matter effects included) as well as the realistic Standard Solar Model[9] (for propagation in the Sun) and the Preliminary Reference Earth Model[10] (for propagation in the Earth) are used, *i.e.*, we calculate the total evolution operator for neutrino propagation using accurate matter density profiles for both the Sun and the Earth. In matter, we decompose the total evolution operator S in k factors as

$$S = S_k S_{k-1} \dots S_2 S_1,$$

where S_i is the evolution operator in layer i with constant matter density.[11] The total evolution operator contains the information of evolution for the neutrinos and includes of course both amplitudes and phases. The probability of neutrino flavor transition after propagation through matter is then given by

$$P_{\alpha\beta} = |S_{\beta\alpha}|^2.$$

At the surface of the Sun, at the distance of 1 AU, and at an actual detector on Earth, we obtain the neutrino yields in a general format (with amplitudes and phases). The neutrino oscillation parameters used are: $\theta_{12} = 33.2°$, $\theta_{13} = 0, 5°, 10°$, $\theta_{23} = 45.0°$, $\delta = 0$, $\Delta m^2_{21} = 8.1 \cdot 10^{-5}\,\mathrm{eV}^2$, and $\Delta m^2_{31} = \pm 2.2 \cdot 10^{-3}\,\mathrm{eV}^2$ (normal and inverted mass hierarchies).

4. Neutrino Production

Although we have preformed simulations and computed spectra of neutrino production for a number of annihilation channels, WIMP masses, as well as for seven sets of neutrino oscillation parameters, we present only results for the annihilation of 250 GeV WIMPs into $\tau^+\tau^-$, using the following set of values for the neutrino oscillation parameters (std. osc.) $\theta_{12} = 33.2°$, $\theta_{13} = 0$, $\theta_{23} = 45.0°$, $\delta = 0$, $\Delta m^2_{21} = 8.1 \cdot 10^{-5}\,\mathrm{eV}^2$, and $\Delta m^2_{31} = 2.2 \cdot 10^{-3}\,\mathrm{eV}^2$ as well as no oscillations, in this talk. Nevertheless, the other spectra are publicly available, see Ref.[12]

In Figs. 1 and 2, we show the neutrino yields at production in the center of the Sun and at the surface of the Sun (using matter neutrino oscillations in the Sun for the propagation from the center of the Sun to the surface of the Sun), respectively. From these two figures, two main effects can be observed:

(1) Oscillations effectively mix muon and tau neutrinos, whereas the electron neutrinos remain essentially unmixed.
(2) There is a pile-up of events at low energies.

The first effect is due to the electron neutrino being equivalent to one of the matter eigenstates when propagating in matter with large matter density, whereas the second effect comes from charged- and neutral-current interactions.

Next, we propagate the neutrinos from the Sun to the Earth, using vacuum neutrino oscillations. In Fig. 3, we show the neutrino yields at the distance of 1 AU, *i.e.*, the distance between the Sun and the Earth. Again, two main effects can be observed:

(1) Electron neutrinos are effectively mixed with muon and tau neutrinos

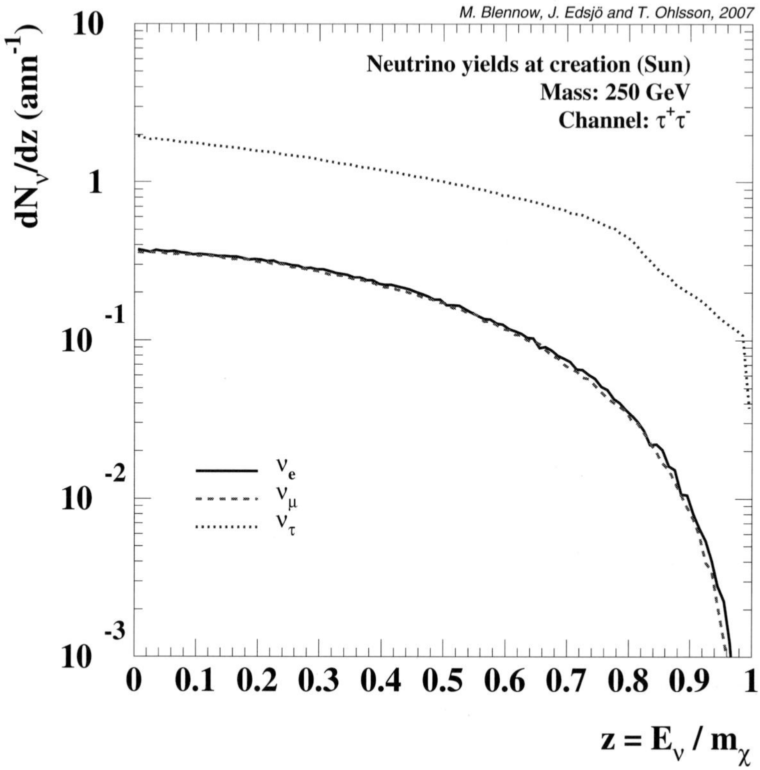

Fig. 1. Neutrino yields at creation in the center of the Sun. The figure has been adopted from Ref.[13]

solely due to vacuum oscillations governed by the small mass squared difference Δm^2_{21}.

(2) The yields of muon and tau neutrinos essentially coincide due to fast oscillations governed by the large mass squared difference Δm^2_{31} and the leptonic mixing angle θ_{23}.

Then, we propagate the neutrinos through the Earth to an actual detector, using matter neutrino oscillations in the Earth. The simulations are performed with a time stamp to include effects of the Earth's distance to the Sun (due to the eccentricity of the Earth's orbit around the Sun) as well as rotation (which affects the distance traversed through the Earth). In Fig. 4, we present the neutrino yields at a fictive detector at latitude

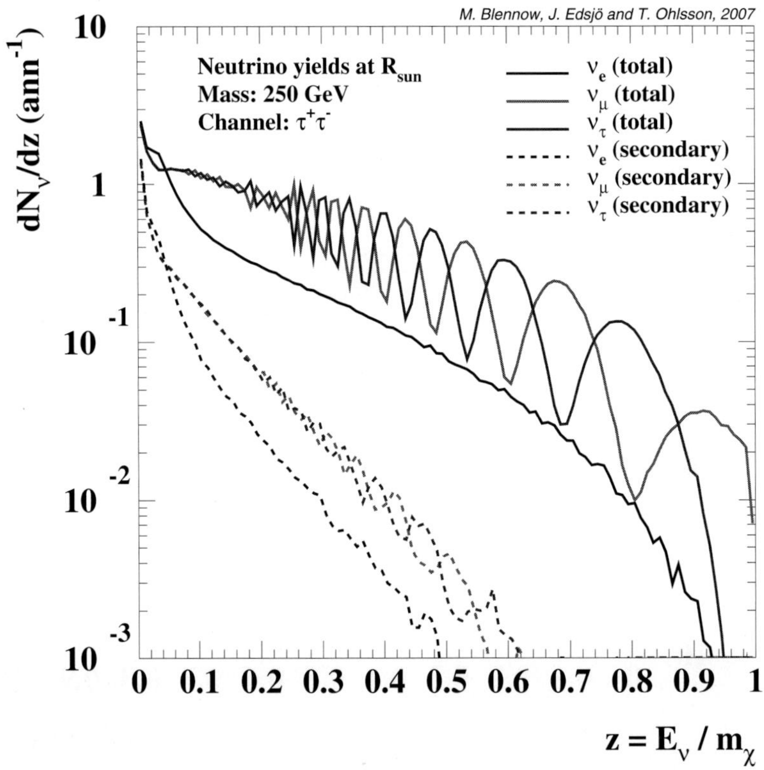

Fig. 2. Neutrino yields at the surface of the Sun. The secondary neutrinos are coming from $\tau^{\pm}$ decays after $\nu_\tau/\bar{\nu}_\tau$ charged-current interactions. The figure has been adopted from Ref.[13]

$-90°$, which could *e.g.* be the IceCube neutrino telescope, and we have averaged the result of the simulations over half a year. In this case, the main effect is that the oscillations that were present before are now almost completely washed out. In fact, matter effects from the passage through the Earth are negligible and the main reason for this "averaging" comes from the eccentricity of the Earth's orbit.

Finally, we consider production of neutrinos from WIMP annihilations in the center of the Earth. In Fig. 5, we present the neutrino yields at a possible detector. The effects of oscillations are seen as a decrease of the tau neutrino yield and an increase of the muon neutrino yield at low energies (below about 50 GeV). For higher energies, that are of most interest for

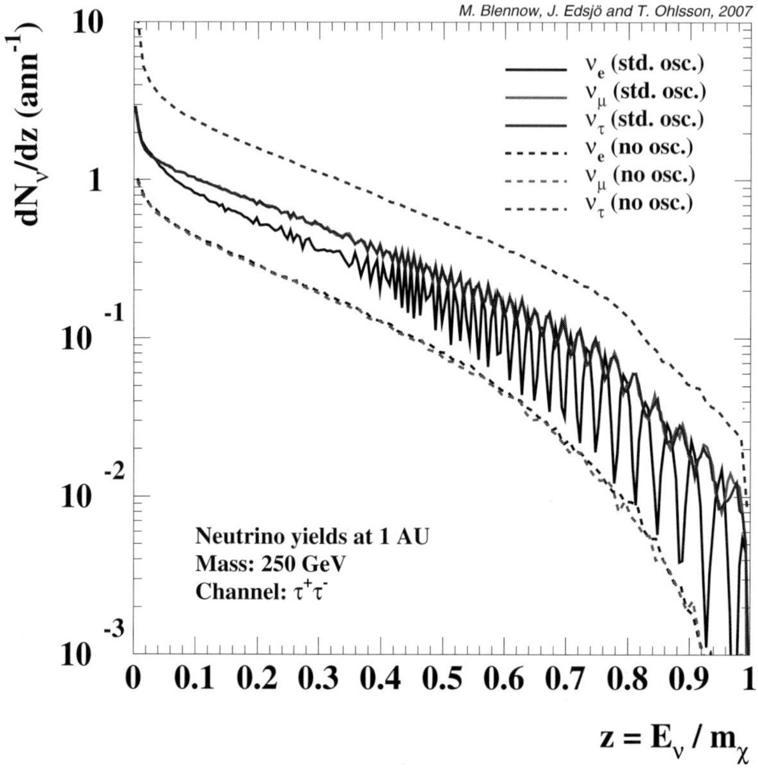

Fig. 3. Neutrino yields at the distance of 1 AU for both with and without neutrino oscillations. The figure has been adopted from Ref.[13]

the currently planned neutrino telescopes such as IceCube, ANTARES, and NESTOR, neutrino oscillations do not have a large effect on the neutrino yields from annihilations of WIMPs inside the Earth.

5. Conclusions

For WIMP annihilations in the center of the Sun, using typical WIMP masses [*i.e.*, $(10 - 10^5)$ GeV], neutrino oscillations effectively average muon and tau neutrinos on the way out of the Sun, average electron and muon/tau neutrinos on the way through the Earth, and wash out the remaining oscillation pattern in the spectra due to the eccentricity of the Earth's orbit around the Sun.

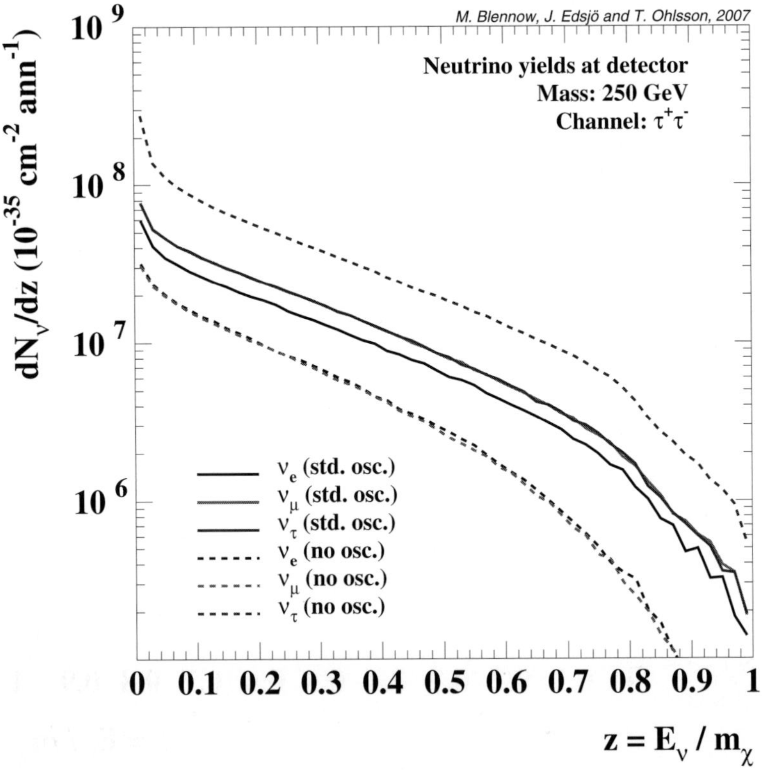

Fig. 4. Time-averaged neutrino yields at a detector at latitude $-90°$ for both with and without neutrino oscillations. The figure has been adopted from Ref.[13]

For WIMP annihilations in the center of the Earth, there are essentially no effects of neutrino oscillations (except below 50 GeV where muon and tau neutrinos are mixed).

Acknowledgments

I would like to thank Mattias Blennow and Joakim Edsjö for useful collaboration that led to the publication[13] upon which this talk is based. In addition, I would like to thank the organizers of Dark 2007 for giving me the opportunity to give this talk.

This work was supported by the Royal Swedish Academy of Sciences (KVA) and the Swedish Research Council (Vetenskapsrådet), Contract No. 621-2005-3588.

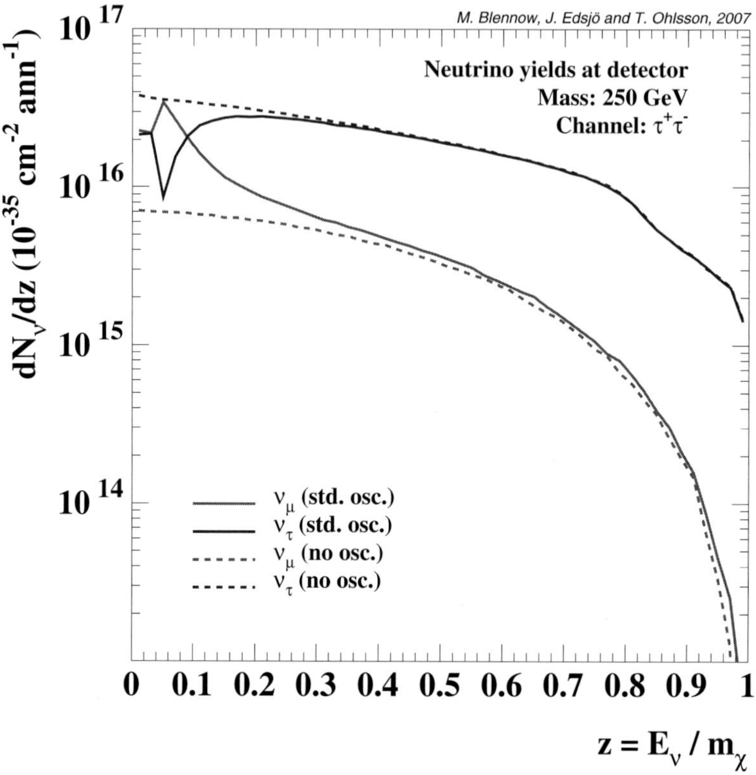

Fig. 5. Neutrino yields (from WIMP annihilations in the center of the Earth) at a detector at the surface of the Earth for both with and without neutrino oscillations. The electron neutrino yields are not plotted, since they are identical to the muon neutrino yields without neutrino oscillations. The figure has been adopted from Ref.[13]

References

1. D. N. Spergel *et al.* [WMAP Collaboration], Astrophys. J. Suppl. **170**, 377 (2007) [arXiv:astro-ph/0603449].
2. W. H. Press and D. N. Spergel, Astrophys. J. **296**, 679 (1985).
3. K. Freese, Phys. Lett. B **167**, 295 (1986).
4. L. M. Krauss, M. Srednicki, and F. Wilczek, Phys. Rev. D **33**, 2079 (1986).
5. T. K. Gaisser, G. Steigman, and S. Tilav, Phys. Rev. D **34**, 2206 (1986).
6. J. Lundberg and J. Edsjö, Phys. Rev. D **69**, 123505 (2004) [arXiv:astro-ph/0401113].
7. T. Sjöstrand, S. Mrenna, and P. Skands, JHEP **0605**, 026 (2006) [arXiv:hep-ph/0603175].
8. J. Pumplin, D. R. Stump, J. Huston, H. L. Lai, P. Nadolsky, and W. K. Tung, JHEP **0207**, 012 (2002) [arXiv:hep-ph/0201195].

9. J. N. Bahcall, A. M. Serenelli, and S. Basu, Astrophys. J. **621**, L85 (2005) [arXiv:astro-ph/0412440].
10. A. M. Dziewonski and D. L. Anderson, Phys. Earth Planet. Interiors **25**, 297 (1981).
11. T. Ohlsson and H. Snellman, Phys. Lett. B **474**, 153 (2000) [arXiv:hep-ph/9912295].
12. Our results are publicly available at `www.physto.se/~edsjo/wimpsim`
13. M. Blennow, J. Edsjö, and T. Ohlsson, arXiv:0709.3898 [hep-ph].

UNDERGROUND COMMISSIONING AND PROSPECTS FOR ZEPLIN-III, A TWO-PHASE XENON DETECTOR FOR DARK MATTER WIMPs

HENRIQUE ARAÚJO (for the ZEPLIN-III Collaboration)*

Imperial College London & STFC Rutherford Appleton Laboratory, UK
E-mail: h.araujo@imperial.ac.uk

**Edinburgh University (UK), Imperial College London (UK), ITEP-Moscow (Russia), LIP-Coimbra (Portugal), STFC Rutherford Appleton Laboratory (UK)*

ZEPLIN-III has been deployed at the Boulby Underground Laboratory and is now undergoing final commissioning tests ahead of the first fully-shielded run in search of dark matter WIMPs. We aim to improve on the discovery potential of other two-phase xenon detectors by means of lower background construction and better electron-nuclear recoil discrimination, partly aided by operation at a higher electric field. In this article we give an update on the status of the experiment and its future prospects in terms of sensitivity to WIMP-nucleon scalar interactions. We also highlight the potential of this technology for present and next-generation searches for rare events with small energy deposits.

Keywords: ZEPLIN-III; Dark Matter WIMPs; Two-Phase Xenon Detectors.

1. Introduction

ZEPLIN-III is a two-phase xenon experiment just deployed 1100 m underground at the Boulby mine (North Yorkshire, UK) which aims to detect galactic dark matter WIMPs.[1,2] It operates on the principle that electron and nuclear recoils, produced in liquid xenon by different particle species, generate different relative amounts of scintillation light and ionisation charge. WIMPs are expected to scatter elastically off Xe atoms (much like neutrons), and the recoiling atom will produce a different signature to gamma rays, which create electron recoils. In this hybrid detector, the vacuum ultraviolet (VUV) photons generated by the scintillation and, indirectly, the ionisation channels are measured by the same array of 31 photomultiplier tubes immersed in the liquid xenon.

The ZEPLIN-II detector, its predecessor, has just measured a WIMP-nucleon cross-section exclusion limit of 6.6×10^{-7} pb to 90% c.l.[3,4] With ZEPLIN-III we hope to improve on this result significantly to explore deeper into the currently-favoured SUSY parameter space which WIMPs are expected to inhabit. We expect to achieve this improvement through a combination of lower background construction using radio-pure materials, improved light collection for scintillation, high electric field uniformity, wall-less 3D fiducial volume definition and higher electron-nuclear recoil discrimination than achieved in other xenon two-phase systems to date.

Fig. 1. Left: The ZEPLIN-III photomultiplier array, which sits at low temperature inside the liquid xenon target vessel. All associated electronic components are located outside of the detector to reduced backgrounds as well as xenon contaminants. Right: The detector located in the hydrocarbon neutron shield at Boulby. The lead gamma shield is presently being assembled.

At the core of the detector, inside the copper-made cryostat and xenon vessels, a 40-mm thick disk of liquid xenon is viewed from below by the array of 31 photomultipliers immersed in the liquid. This maximises light collection for the scintillation signal (S1), which is also used to set the energy scale, by eliminating unnecessary optical interfaces and by exploiting total internal reflection at the liquid surface. The fully-assembled array can be seen in Fig. 1 (left).

A strong electric field ($\sim$5 kV/cm) in the liquid is obtained by applying a high voltage ($\sim$20–25 kV) between a top electrode mirror in the gas phase and a wire grid above the array. This field extracts the ionisation electrons from the interaction site, drifting them upwards and extracting them across the liquid surface into a thin (3–5 mm) layer of xenon vapour, where it generates electroluminescence photons with high efficiency. These are detected by the phototube array as a larger burst of 175 nm photons (S2); the time delay between S1 and S2 is proportional to the depth of the interaction in the liquid, which can thus be recovered with sub-mm precision. A reverse-field region is created by a second grid located just above the phototubes in order to suppress ionisation signals from the bottom 5 mm of liquid, and so reduce low-energy gamma- and x-ray backgrounds from the array. This grid, tied to the photocathode potential, also prevents distortion of the internal field at the input stage of the photomultipliers.

2. Low-Energy Sensitivity of Two-Phase Xenon Detectors

Energy thresholds already achieved by two-phase xenon detectors are quite impressive. Moreover, we expect this to be the case also for larger systems, since neither the scintillation nor the ionisation channels need be too affected by up-scaling. The high scintillation yield of liquid xenon is one main reason. This was found to be at least 60 photons/keVee at zero field in both ZEPLINs. The zero-field photoelectron yield of ZEPLIN-III is ≈5 phe/keV;[1] at the operating field we observe some 2 phe/keVee (hereafter, 'keVee' and 'keVnr' stand for 'electron' and 'nuclear recoil' equivalent energy, respectively). This suggests a very realistic energy threshold of 2 keVee (or 4 photoelectrons), which translates to approximately 5 keVnr (assuming 20% relative scintillation efficiency to ^{57}Co gamma rays). This energy threshold is also being reported by the XENON10 experiment.[6] Although the *attenuation* length for xenon scintillation photons is known to be affected by Rayleigh scattering, *absorption* lengths of order 1 m have been consistently observed, including in ZEPLIN-II.

The sensitivity of the ionisation channel is even more impressive: with ZEPLIN-II we were able to demonstrate efficient detection of *single* ionisation electrons extracted from the liquid.[5] These electrons were attributed to photoionisation of a yet-undetermined species in the liquid bulk by the 175 nm photons. Besides providing a striking example of the amplification achieved in the ionisation channel, this result also suggests that the mean free path for the absorption of scintillation photons is potentially rather large (∼km) — assuming photoionisation to be the main photon absorption mechanism. This sensitivity to single charge carriers even exceeds that required for the detection of ∼keV recoils (which is instead limited by the energy threshold set by the scintillation channel) and may indeed find applications beyond WIMP detection, with coherent neutrino-nuclear scattering being suggested. Significantly, the ionisation response will not degrade with increasing target masses, provided that high voltages can be applied.

3. Electron-Nuclear Recoil Discrimination in ZEPLIN-III

The ZEPLIN-II and XENON10 experiments operating at moderately low drift fields of order 1 kV/cm have shown electron-nuclear recoil discrimination at the ~99% level. This may turn out to be sufficient for very large detectors, where self-shielding of an inner fiducial region by the surrounding liquid can decrease internal electron-recoil backgrounds to a suitable level – provided that xenon-intrinsic backgrounds, such as ^{85}Kr, can be eliminated. However, existing systems in the ~10 kg mass scale are still affected significantly at this level of discrimination.

We hope to demonstrate soon that ZEPLIN-III can achieve a better level of discrimination: i) by using a stronger electric field, it is possible that statistical fluctuations in the extraction of ionisation from the primary track can be reduced, even if the relative *mean* charge extracted from electron and nuclear recoils remains constant; ii) by using an 'open-plan' design together with 3D position reconstruction of the interaction vertex, a fiducial region can be apportioned with boundaries located well away from lateral detector walls; radioactivity from surface contamination can thus be rejected — such as plating with ^{222}Rn progeny either deposited before assembly or carried in the xenon; iii) no extraction grids are used near the liquid/gas interface which might otherwise achieve less-than perfect transparency to the drifting charge cloud, create a position-dependent S2 gain or light obscuration; iv) reduction of optical cross-talk by use of a copper matrix into which the photomultipliers are embedded, capped by a copper plate overhanging the phototubes (visible in Fig. 1); this prevents cross-talk both directly and through the active volume, for interactions occurring in the liquid xenon surrounding the phototubes.

One class of events has been identified which could challenge a high level of discrimination: gamma-ray double scatters where one vertex occurs in the reverse field region below the cathode and the second in the active volume. These will necessarily show decreased S2/S1 ratios and therefore leak into the nuclear-recoil acceptance region. These events have seriously effected the XENON10 experiment.[6] We believe that by exploiting the light pattern of the S1 signal together with a consistency test between the reconstructed (x,y) coordinates for S1 and S2 (i.e. taking into account that charge drifts vertically) we should be able to reject these and other unwanted event topologies down to an acceptable level.[2]

4. Present Status of the Experiment

The recently-funded 'exploitation' stage of the ZEPLIN-III programme consists of two phases. In phase I the system is deployed underground as tested on the surface, with the addition of passive neutron and gamma-ray shielding. The aim is to achieve a proof-of-principle result of the discrimination potential at high electric field with a relatively short exposure of ~240 kg*days. In phase II, the photomultiplier array will be retro-fitted with compatible phototubes currently being manufactured — the existing ones were procured several years ago and dominate our background by a large factor. In addition, an active scintillator veto, already well advanced in its construction, will replace part of the passive neutron shield to achieve extra mitigation of the neutron background. A year-long run (3000 kg*day exposure) will then be undertaken until the end of 2009 to achieve a world-competitive sensitivity to the WIMP-nucleon scalar cross-section.

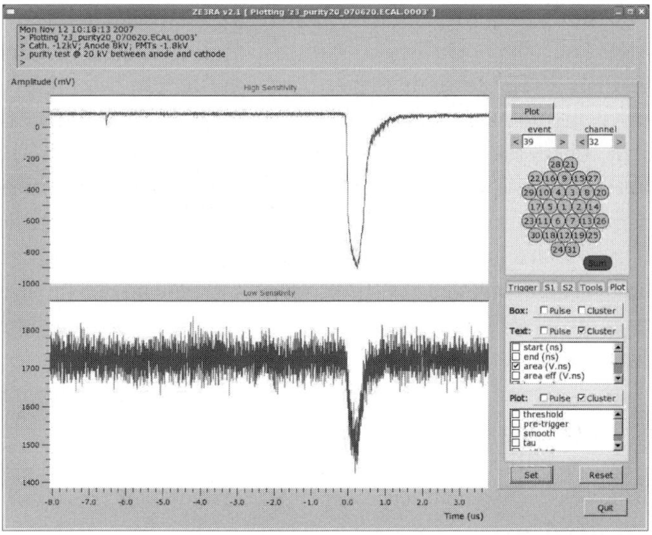

Fig. 2. ZEPLIN-III event displayed with our 'ZE3RA' data reduction package showing a 662 keV gamma-ray interaction at high field (upper and lower panels represent high- and low-sensitivity scales in the dual range DAQ); note the very large difference in the relative sizes (areas) of S1 and S2 pulses.

Transport of the instrument and ancillary systems from Imperial College to Boulby was followed by a recommissioning campaign underground. Fig. 1 (right) shows the detector in the 30-cm thick neutron shield. The very delicate process of relocating the detector, gas handling and purification system and control and acquisition electronics was confirmed by successful two-phase tests. A two-phase signal (summed waveforms) of a ^{137}Cs gamma ray obtained at ~5 kV/cm electric field in the liquid is shown in Fig. 2.

Testing is now underway to optimise and automate the system, to implement the stringent safety requirements for underground operation and to calibrate the detector. In parallel, the gamma shield, consisting of a 20 cm thick lead castle, is being installed. A calibrated, fully-shielded run concludes phase I early in 2008.

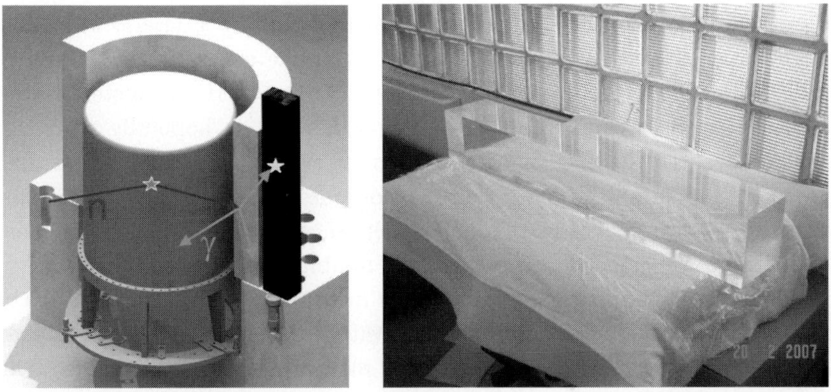

Fig. 3. Left: Veto design showing detector inside a Gd-loaded plastic shield surrounded by scintillator modules (one is shown, in dark). Right: a scintillator slab tested at ITEP.

Preparation for phase II is well underway, with installation expected by mid-2008. The active veto design consists of plastic scintillator modules around and above the target coupled to photomultipliers encased in the lower shield section. Inside this assembly, Gd-loaded hydrocarbon captures neutrons with high efficiency, producing energetic gamma-rays which can be detected by the surrounding scintillator (Fig. 3). The scintillator modules, produced by the Kharkov Institute (Ukraine), are due to arrive at Boulby shortly for assembly and testing. Light collection measurements and simulations suggest that a ~100 keVee neutron detection threshold is realistic, allowing the tagging of 75% for neutrons entering the shield.

New photomultipliers, tailored specifically to our needs and pin-by-pin compatible with the existing array, are being manufactured by ET Enterprises Ltd (formerly ETL). A ~30-fold reduction in neutron and gamma rates is anticipated, maintaining excellent photon detection performance.

5. Predicted Backgrounds and WIMP-Nucleon Sensitivity

The phase I phototube array is expected to produce 25 single-scatter events per year above 10 keVnr in a 8 kg fiducial volume. Previously[2] we have calculated the corresponding WIMP cross-section sensitivity by raising the nuclear recoil energy threshold to 14 keVnr in order to reject 99.999% of electron recoils. In this instance one expects an average of one event in 240 kg∗days (one month at 100% duty cycle), the phase-I planned exposure. The corresponding sensitivity is 1×10^{-7} pb at 60 GeV WIMP mass.

To enable a more straightforward comparison with other experiments, the acceptance region could be lowered to 5–50 keVnr at the expense of admitting more nuclear and electron recoils (assuming a deteriorating discrimination at lower energies). To offset this, one can sacrifice the upper half of the nuclear recoil acceptance region in the S2/S1 parameter, in line with the analysis carried out in ZEPLIN-II,[3] and so hopefully reject all electron events. The new expected rate is 1.8 events in 5–50 keVnr for that exposure, including the lower (50%) recoil detection efficiency. Actual observation of two events in this new acceptance region produces the same 90% c.l. limit as mentioned above at the minimum of the sensitivity curve.

Ongoing component-level radiopurity measurements and simulations of the upgrade phototubes place a conservative *upper limit* of 1.2 evt/year above 10 keVnr. Ceramic feedthroughs may add a similar number. The 400 kg of OFHC copper in the detector contribute only at the level of 0.1 evt/year, similar to muon-induced neutrons. Our latest estimates total ∼3 events in 5–50 keVnr in the 3000 kg∗day phase II exposure, leaving only a mean of one event remaining unvetoed with 50% recoil acceptance. Observing one event in phase-II would produce a 90% c.l. limit of $\sim6\times10^{-9}$ pb.

ZEPLIN-III should be able to compete with larger systems expected to deliver results in the same timescale. In addition, its innovative design will provide important guidelines for future xenon experiments.

References

1. D. Yu. Akimov *et al.*, *Astroparticle Phys.* **27** 46–60 (2007).
2. H. M. Araújo *et al.*, *Astroparticle Phys.* **26** 140–153 (2006).
3. G. J. Alner *et al.*, *Astroparticle Phys.* **28** 298–302 (2007).
4. G. J. Alner *et al.*, *Phys. Lett B.* **653** 161–166 (2007).
5. B. Edwards *et al.*, *arXiv:0708.0768v1* [physics.ins-det] (2007).
6. J. Angle *et al.*, *arXiv:0706.0039v1* [astro-ph] (2007).

CDMS EXPERIMENT: CURRENT STATUS AND FUTURE

JONGHEE YOO (for the CDMS collaboration)
Fermi National Accelerator Laboratory, PO Box 500
Batavia, IL 60510, USA
E-mail: yoo@fnal.gov

We present the current status of the Cryogenic Dark Matter Search (CDMS). The five tower detector array, total 30 detectors, are running stable since October 2006. We have accumulated more than 900 kg-days of low background data. We also summarize the prospect of SuperCDMS project.

Keywords: Dark matter, WIMP, CDMS, SuperCDMS

1. Introduction

The observations of rotational curve of galaxies, cosmic microwave background, large scale structure, galaxy clusters, gravitational lensing, bullet clusters provide that nonluminous, nonbaryonic mass components may constitute most of the matter in the Universe.[1] Cosmological relic abundance arguments suggests that it is reasonable to probe weak interaction scale and around 100 GeV of mass. Weakly interacting massive particles (WIMPs) are currently the most popular dark matter candidate. Interestingly, supersymmetric scenarios provide stable massive neutral particles, the neutralinos, in the weak interaction scale. Future collider experiments such as LHC and ILC may reveal the characteristics of the supersymmetirc particles. However it should be pointed out that the dark matter is not necessarily a supersymmetric particle, and need not to have an electro-weak interaction[2].[3] Therefore direct detection of the dark matter is a crucial step to understand its nature.

The Cryogenic Dark Matter Search (CDMS) experiment is designed to detect WIMP signal through nuclear recoil by elastic scattering. The detector is capable of reading out both phonon-energy and ionization-energy of an interaction in Ge or Si crystals. The CDMS-I experiment was carried out in Stanford underground laboratory and the performance of detectors had been demonstrated. The CDMS-II experiment moved to the Soudan underground laboratory which substantially reduced cosmic-ray induced neutron background. There are three phases to the CDMS-II experiment. The first phase was a single tower (4 Ge and 2 Si detectors) operation and the second phase was a two tower (6 Ge and 6 Si detectors) operation. The third and current phase of the CDMS-II experiment, five tower detector array and total 30 detectors, are running stable since October 2006. We accumulated more than 900 kg-days of data. The five tower operation will be kept until the summer of 2008.

2. CDMS Experiment

CDMS ZIP detectors (depth(Z)-sensitive Ionization and Phonon detectors) is able to discriminate WIMP-nucleus recoil energy by measuring both ionization and phonon signal from the crystal.[6] The ZIP detector is made of pure Ge (250 g) or Si (100 g) crystal in a cylindrical shape of 1 cm thickness and 7.6 cm in diameter. A tower consists of a vertically stacked 6 ZIP detector-array. The detectors are cooled by a dilution refrigerator down to 50 mK. The low temperature configuration prevents background fluctuations from thermal excitations in the crystal. The ionization signal is the interaction that excite the electron-hole pairs of the semiconductor crystal. The electron and hole pairs are separated by an electric field through the crystal. The ionization signals are then read out by inner and outer electrodes. The inner electrode covers 85% of ionization side of the detector. The outer electrodes provides discrimination power of the edge events. The outer electrode is used to discriminate those edge-events. The phonon signals that are produced by vibration of crystal lattice are read out by total 4144 per detector of Quasiparticle-assisted Electrothermal-feedback Transition-edge sensors (QETs). Each QET consists of a 1 μm wide strip of tungsten connected to 8 superconducting aluminum collection fins which cover the phonon sensor side of the crystal. The tungsten strips from the Transition-Edge-Sensors (TESs) are voltage biased, with the current through them monitored by a high-bandwidth SQUID array.

When an interaction occurs in the crystal, huge amount of phonons are produced. Almost half of the phonon energy goes into thermal phonons which may not read out by the QETs. Most of the athermal phonons that reach the surface of phonon sensor area can scatter into the aluminum fins. The phonons, energetic enough (above 340 μeV) to break Cooper pairs in a superconducting state of aluminum fins, produce quasiparticles. The quasiparticles enter into TESs. The interaction between the quasiparticles and conduction electrons in the TESs increase the system temperature and hence increase the tungsten resistance. The increase of resistance decreases the current supplied by the voltage bias. The reduction of Joule heating from the voltage bias lowers the tungsten temperature. This strong electro-thermal-feed-back guarantees that the power deposited into the TES is exactly compensated for by a reduction in Joule heating. The energy deposited can then be measured by reading out change of currents.

Detector calibrations are carried out during the normal data-taking period. A ^{133}Ba gamma-ray source is used to calibrate energy scale and to characterize detector response to electron recoils. A ^{252}Cf neutron source is used to characterize the detector response of nuclear recoils produced by neutrons. The nuclear recoil characteristics is important to determine WIMP signal region. Results of Monte Carlo simulations of ^{133}Ba and ^{252}Cf calibration show excellent agreement with data.

3. Backgrounds

The CDMS detectors demonstrated excellent performance of the particle identification. The ionization signals producing a nuclear recoil, which WIMPs may cause, are suppressed compared to electron recoils while phonon signals are not. The ratio between ionization and phonon (recoil) energy, termed as ionization yield, provides strong discrimination power of background events. However, the events in the detector surface suffer ionization yield suppression, therefore some of those events could be misidentified as nuclear recoils. The phonon pulse time information, such as start and rise of phonon pulse, helps to reject 97% of these surface events while keeping 70% of true nuclear recoils. Figure 1 shows ionization yield versus timing parameter for calibration data of ^{133}Ba and ^{252}Cf sources. The ionization yield shows clear separation between bulk-electron recoil events from nuclear recoil events. The surface-electron recoil events can be excluded by setting a timing parameter cut. The squared area in the figure is defined as a WIMP signal region before looking at WIMP search data sample.

The WIMP is supposed to interact weakly ($<10^{-42}$cm^2) and the mean free path in Ge is order of 10^{10} m. Therefore WIMP will scatter a single time while neutrons can multiple scatter in the detector array. Choosing only single scatter events, the neutron events can be reduced. Since WIMPs are assumed to transfer small momenta to the nucleus they will scatter off coherently with whole nucleus. Therefore the interaction rate of WIMPs in nuclei is expected to be proportional to mass-square ($\propto$A^2) of the target nuclei. Accordingly the WIMP interaction rate in ^{73}Ge is $\sim$6 times larger than ^{28}Si while neutron interaction rates are similar for both crystals. Once we observe handful amount of WIMP candidate events in several detectors, by comparing event rates between Ge and Si detectors will provide statistical discrimination power between neutrons and WIMPs.

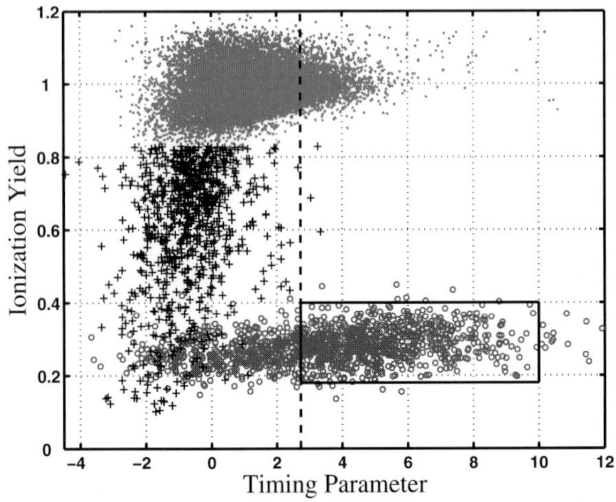

Fig. 1. Ionization yield versus timing parameter for calibration data in a Ge detector with the recoil energy range 10–100 keV. The ionization yield shows clear separation between bulk-electron recoil events (dots and yield near 1.0; from ^{133}Ba calibration source) from nuclear recoil events (circles and yield near 0.3; from ^{252}Cf calibration source). The surface-electron recoil events (crosses) from ^{133}Ba show wide distribution along the ionization yield and non-negligible amount of events are leaked into nuclear recoil area. The timing parameter has discrimination power of those electron leak events. The vertical dashed line shows the minimum allowed timing parameter for WIMP candidate event. The squared area is an allowed region of nuclear recoil.

The CDMS-I experiment had been carried out at the Stanford Underground Facility (SUF), California, USA at a depth of 10.6 m (17 meters water equivalent). The expected cosmic-ray muon background rate in this shallow site is 50 muons $\mathrm{sec}^{-1}\mathrm{m}^{-2}$. Neutrons, the most serious background source at SUF, gave a rate of $\sim$1 event $\mathrm{kg}^{-1}\,\mathrm{day}^{-1}$. A single tower, total 6 detectors (4 Ge + 2 Si), was operated a total of 65.8 detector live days and 28.3 kg-day of net exposure. 20 nuclear recoil events are observed from the WIMP search data sample and those agreed with the expected neutron background rate.[11]

The first phase of the CDMS-II experiment, a single tower operation with the identical detector configuration at SUF, was carried out in the Soudan Underground Laboratory, Minnesota, USA, at a depth of 780 m from the surface (2090 meters water equivalent). The cosmic muon rate was reduced down to 0.25 muons min^{-1} m^{-2}. Neutron background rate is estimated to be 1 event kg^{-1} $year^{-1}$. The rate of unvetoed neutron induced recoils is 3×10^{-4} kg^{-1} day^{-1} and is estimated by Monte Carlo simulations. A total of 52.6 detector live days and 19.4 kg-day of net exposure were achieved. The dominant source of background in Soudan Underground Laboratory is surface electrons from the radioactive contaminants. The number of electron-recoil events expected to be misidentified as nuclear recoils were estimated 0.7 ± 0.3 in Ge detectors. The analysis results found one nuclear-recoil candidate event at 64 keV, consistent with the expected background event[12].[13]

The second phase of the CDMS-II experiment, a two tower operation, was carried out with six more detectors. The operation periods was from March 25 to August 8, 2004 total 74.5 detector live days and 34 kg-day (13 kg-day) net exposure for Ge (Si) detectors.

The third phase of the CDMS-II experiment, a five tower operation, is now up and running since October 2006. We have accumulated almost 800 kg-day of detector exposure data. We expect the results of the first half of data analysis in the early 2008.

4. Data Analysis

Since the five tower data blind analysis is currently in progress, we describe general schematics of CDMS data analysis based on two tower analysis.

To prepare good WIMP search data samples, we exclude data sets with known problems such as events triggered by noise burst, non-operational channels and fails off-line diagnostics. For recoil energies above ~10 keV, events due to background photons are rejected with > 99.99% efficiency. Electromagnetic events very near the detector surface can mimic nuclear recoils because of reduced charge collection. These surface events, however, are rejected with > 96% efficiency by using additional phonon pulse shape information.

The band-width of the electron-recoil and nuclear-recoil in the ionization yield parameter space (ionization energy / recoil energy) are taken to be $\pm 2\sigma$. The veto coincident events are removed. Then we choose single scatter events, which has a signal only in one detector out of 12 detectors.

Events with low ionization yield in the ^{133}Ba calibration data, which presumably come from surface electron recoils, were used to develop rejection cut criteria. Two phonon pulse timing quantities, the time delay of the phonon signal relative to the fast ionization signal and the phonon pulse rise time, in the quadrant with the largest phonon signal or local-quadrant phonon signal are used to discriminate surface events. The sum of delay and rise time forms a timing parameter, where energy corrections to the delay and rise time are applied to achieve energy-independence (see Fig. 1).

The surface event backgrounds at WIMP search signal region (nuclear recoil band) are estimated and the expected background rate is based on the passing fraction between 10–100 keV in the WIMP-search data. The number of surface events expected to pass the timing cuts are 0.4±0.2(stat)±0.2(syst) between 10–100 keV in Ge detectors and 1.2±0.6(stat)±0.2(syst) between 7–100 keV in Si detectors. From simulations, the expected background from unvetoed nuclear recoils due to impinging cosmogenic neutrons is 0.06 events in Ge and 0.05 events in Si.

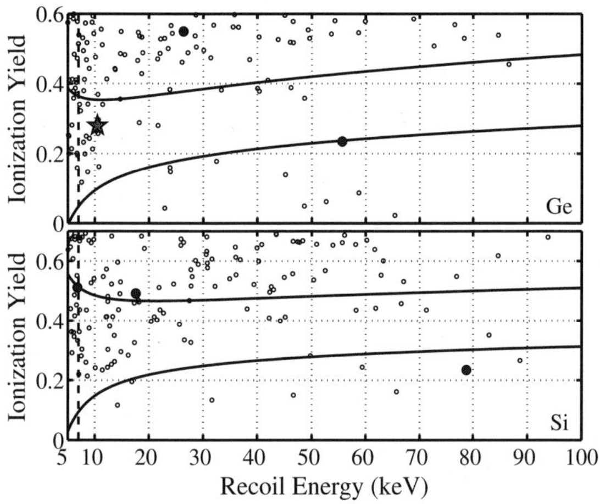

Fig. 2. Ionization yield versus recoil energy events for Ge (upper figure) and Si (lower figure) detectors before (small circles, solid dots and star mark) and after (solid dots and star mark) surface electron recoil rejection cut (timing parameter cut). The 7 keV energy threshold is shown in vertical dashed line in each figure. The ionization yield between curved lines in each figure is WIMP signal region. One candidate event is found in WIMP search signal band in Ge detector with recoil energy of 10.5 keV (shown in star mark in upper figure).

Figure 2 shows the unmasked WIMP search data in ionization yield versus recoil energy for both Ge (upper figure) and Si (lower figure). We observe one WIMP candidate event which turned out to have occurred in a detector while it suffered inefficient ionization collection. Therefore the single event indeed not a WIMP signal. Although we count the event as a WIMP candidate, it is still consistent with the rate of expected background. The results exclude WIMP-nucleon cross section above $1.6\times10^{-43}\,\text{cm}^2$ from Ge detectors and $3\times10^{-42}\,\mathrm{cm}^2$ from Si detectors, for WIMP mass $60\,\mathrm{GeV/c}^2$ with 90%C.L..[14]

The five tower (19 Ge and 11 Si, total 5.85 kg) detector array is currently running stable. The accumulated detector exposure is more than 900 kg-day and it is planned to operate until the summer of year 2008. The five tower operation will provide additional factor ten in sensitivity of WIMP-nucleon cross section in near future.

5. SuperCDMS

The CDMS experiment is currently the most proved equipment to carry out Dark Matter search with background free configuration. The background rejection power provides the sensitivity of the experiment directly proportional to the mass of the detector and exposure time (MT) rather than square root of them ($\sqrt{MT}$). Therefore increasing mass of the detector is the most efficient way to prove smaller WIMP-nucleon cross sections.

In the summer of 2008, we are planning to replace the first two towers of the normal detector to two towers of the super detector (1 inch thick) arrays. The experiment is to focus more on the detector performance test for the 25 kg operation at the existing Soudan facility. The two SuperTowers with

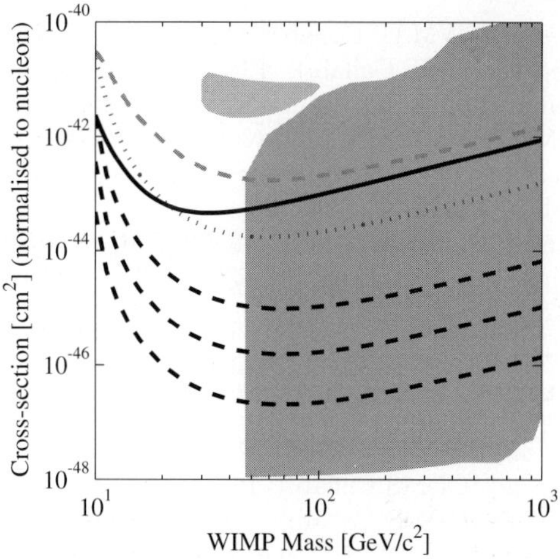

Fig. 3. WIMP search sensitivity (projected). The gray-dashed line at the top is the CDMS-II two tower limit. The black-solid line is the XENON10 limit. The gray-dotted line at the second is the projected limit for third phase (five tower) of CDMS-II. The consecutive dotted, dashed and solid lines are projected sensitivity for SuperCDMS phase A (25 kg), B(120 kg) and C(1 ton) respectively.

$\sim$8 kg of Ge operated during 2008-09 will have the WIMP-nucleon cross section sensitivity down to $\sim 7 \times 10^{-45}$ cm^2 at 60 GeV of WIMP mass.

The SuperCDMS[15] project proposes to install super towers in the SNO-Lab, total mass 25 kg, the seven SuperTowers of six 2.5-cm-thick ZIP detectors during 2011-13. With advanced surface rejection analysis technique and radio active background handling, the SuperCDMS 25 kg SNOLab phase is expected to reach a sensitivity of cross-section of 1.3×10^{-45} cm^2. Figure 3 shows projected WIMP search potential for each phase of detector operation. The SuperCDMS experiment can scan most of the region of WIMP-nucleon cross-section and mass parameter spaces.

Acknowledgments

This work is supported by the National Science Foundation under Grant No. PHY-9722414, by the Department of Energy under contracts DE-AC03-76SF00098, DE-FG03-90ER40569, DE-FG03-91ER40618, DE-FG02-94ER40823, and by Fermilab, managed by the Fermi Research Alliance, LLC, under Contract No. DE-AC02-07CH11359 with the Department of Energy. The ZIP detectors were fabricated in the Stanford Nanofabrication Facility operated under NSF.

References

1. D. N. Spergel *et al.*, (WMAP Collab.), Astrophys. J. Suppl. **148**, 175 (2003); M. Tegmark *et al.*, (SDSS Collab.), Phys. Rev. D **69**, 103501 (2004).
2. G. Steigman and M.S. Turner, Nucl. Phys. **B253**, 375 (1985).
3. G. Jungman, M. Kamionkowski, and K. Griest, Phys. Rep. **267**, 195 (1996);
4. P. Salucci and A. Borriello, Lect. Notes Phys. **616**, 66 (2003); T. Broadhurst *et al.*, Astrophys. J. **621**, 53 (2005).
5. J. D. Lewin and P. F. Smith, Astropart. Phys. **6**, 87, 1996.
6. K.D. Irwin *et al.*, Rev. Sci. Instr. **66**, 5322 (1995); T. Saab *et al.*, *AIP Proc.* **605**, 497, 2002.
7. Y. G. Kim *et al.*, JHEP **12**, 034 2002.
8. J. Ellis *et al.*, Phys. Rev. D **71**, 095007 (2005).
9. E. A. Baltz and P. Gondolo, JHEP **10**, 052, 2004.
10. R. Bernabei *et al.*, Phys. Lett, **B389**, 757, 1996.
11. D. S. Akerib *et al.*, (CDMS Collab.) Phys. Rev. D **68**, 082002 (2003).
12. D. S. Akerib *et al.*, (CDMS Collab.) Phys. Rev. Lett. **93**, 211301 (2004).
13. D. S. Akerib *et al.*, (CDMS Collab.), submitted to PRD. arXiv:astro-ph/0507190.
14. D. S. Akerib *et al.*, (CDMS Collab.), submitted to PRL. arXiv:astro-ph/0509259.
15. R. W. Schnee *et al.*, (SuperCDMS Collab.), astro-ph/0502435; P. L. Brink *et al.*, (SuperCDMS Collab.), astro-ph/0503583.

STATUS of XMASS EXPERIMENT

MASAYUKI NAKAHATA (for the XMASS collaboration)
Kamioka Observatory, ICRR, University of Tokyo,
Higashi-Mozumi, Kamioka, Gifu 506-1205, Japan
e-mail: nakahata@suketto.icrr.u-tokyo.ac.jp

XMASS is a multi-purpose experiment using liquid xenon which aims to detect cold dark matter and low energy solar neutrinos, as well as search for neutrinoless double beta decay. So far a R&D study has been performed using a prototype detector. An 800 kg size liquid xenon detector is going to be constructed in a few years. Results of the R&D study and prospects of the 800 kg detector are described.

1. Introduction

The name of the project, "XMASS", stands for either "Xenon detector for weakly interacting MASSive particles", "Xenon MASSive detector for solar neutrinos", or "Xenon neutrino MASS detector". As indicated by those names, XMASS is a multi-purpose experiment using liquid xenon for dark matter, solar neutrinos and neutrinoless double beta decay studies.[1,2]

Various astronomical observations suggest that there is dark matter (non-luminous matter with mass) in the universe. One of the most likely candidates for dark matter is a weakly interacting massive particle (WIMP). It is quite important to reveal the characteristics of dark matter. If its nature is elementary particles, it will open a new field beyond the standard model. Actually, in the supersymmetric extension of the standard model, the neutralino (mixture of Higssino, Wino and Bino) is a possible WIMP candidate. The density of dark matter is estimated to be about 0.3GeV/cm^3, and if it is made of WIMPs, then they are traveling with a speed of about 300 km/sec. The XMASS experiment searches for WIMPs by a direct detection method. The WIMP interacts with the xenon nucleus, which recoils and produces scintillation light in the liquid xenon. The recoil energy is less than about 100 keV, and the visible scintillation signal of the WIMP is less than about 30 keV because of a quenching factor of 0.2.

Solar neutrinos have been measured by the Homestake, Kamiokande, SAGE, GALLEX, Super-Kamiokande(SK) and SNO experiments, which have shown a deficit of fluxes(called the "solar neutrino problem"). The precise ^{8}B solar neutrino measurements by SK and SNO demonstrated that the solar neutrino problem is due to neutrino oscillations and the global analysis of all solar neutrino experiments shows that the neutrino oscillation parameters are $\Delta m^2 = (5-9) \times 10^{-5} eV^2$ and $\sin^2\theta = 0.27-0.33$ (so called Large Mixing Angle (LMA) solution). The LMA solution was confirmed by the KamLAND experiment using reactor neutrinos. ^{8}B neutrinos contribute only 0.01% to the total solar neutrino flux. The majority of solar neutrinos are below 1 MeV, but so far they are measured only by radio-chemical methods as a partial contribution to their counting rates. The solar neutrino plan for the XMASS project is to measure sub-MeV neutrinos by neutrino-electron scattering. With 10 tons of liquid xenon, we expect about 6.5 pp solar neutrino events per day. The high statistics measurement of pp neutrinos, combined with the precisely predicted flux (from the luminosity constraint), should enable us to determine the θ_{12} oscillation parameter quite precisely.

The evidence for neutrino oscillations, from atmospheric, solar, reactor, and accelerator neutrinos, shows that neutrinos have masses. However, neutrino oscillation experiments measure only neutrino mass square differences, and the absolute masses of neutrinos are not known yet. So far, the direct neutrino mass experiments give an upper limit of 2 eV, which is already much smaller than the masses of quarks and charged leptons. The see-saw mechanism explains the smallness of the neutrino masses by introducing right-handed heavy neutrinos. In the see-saw mechanism, the ordinary neutrinos are Majorana type. So, it is fundamentally very important to search for double beta decay, not only for measuring the absolute mass of neutrinos, but also for identifying the type of the neutrino mass (either Majorana or Dirac type). An advantage of xenon is that ^{136}Xe is a double beta decay nucleus, and its natural abundance is relatively high (8.87%). Furthermore, enrichment is easily possible by a centrifugal method. The large scintillation light yield of liquid xenon is able to give good energy resolution to discriminate $2\nu\beta\beta$ background from the real signal of $0\nu\beta\beta$.

Figure 1 shows the strategy of the XMASS project. So far we have studied the feasibility of the project using a small prototype detector, which has 3 kg of fiducial volume. As will be explained in the next section, the study by the prototype detector proved our basic ideas for the background reduction and demonstrated the feasibility of the future ton-scale detectors.

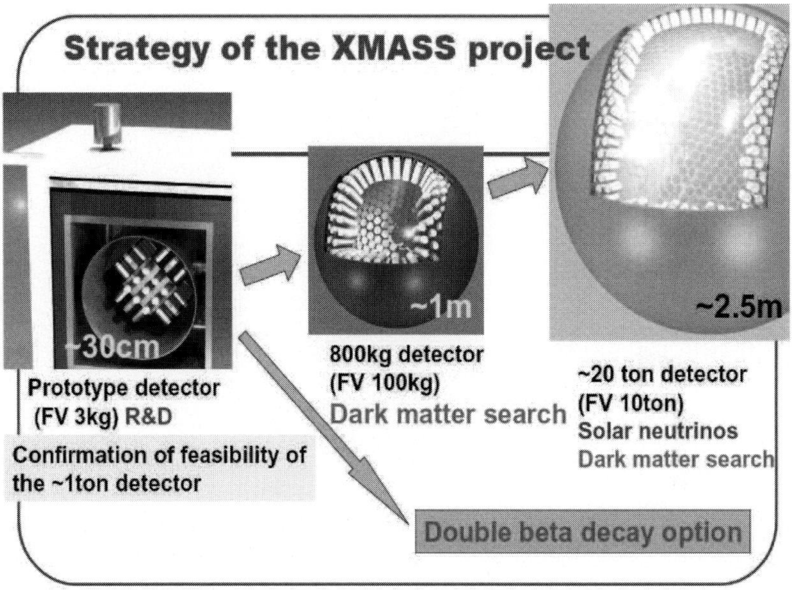

Fig. 1. Strategy of the XMASS project.

We are now at the stage of constructing the 800 kg detector. It is a detector mainly for the dark matter search as described in section 4. Based on the results and experiences which will be obtained by the 800 kg detector, we plan to make a 10 ton class detector for the study of low energy solar neutrinos and the further investigation of dark matter. In parallel to this stream, we will investigate a method to perform a high sensitivity neutrino mass measurement (to a level of 0.01 eV) by double beta decay.

2. Key Idea of the Project

This section describes the key idea of the XMASS project. In order to perform the physics described in the previous section, a low background technique is necessary. The key idea of the background reduction in XMASS is "self-shielding". The background gamma rays can be absorbed by the liquid xenon itself. As an example, Fig.2 shows a simulation of ^{238}U-chain gamma rays injected on the surface of a liquid xenon volume. The outer surface of the liquid xenon volume absorbs the external gamma rays within 10-20 cm of thickness, and realizes a low background environment in the central volume. WIMPs and neutrinos, however, interact throughout the

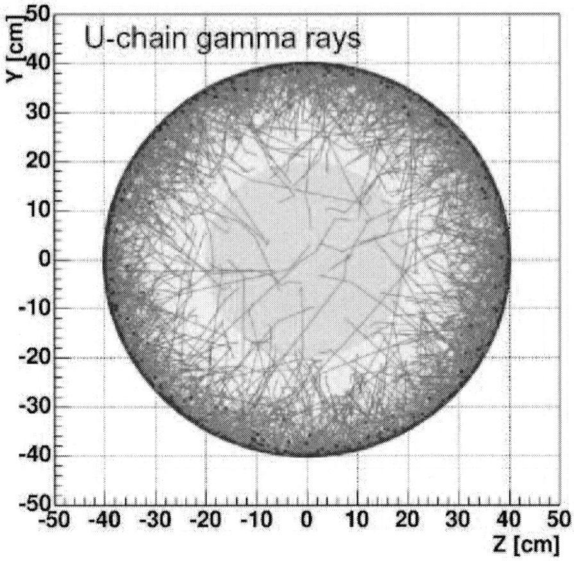

Fig. 2. A simulation of ^{238}U-chain gamma rays injected on the surface of the liquid xenon volume. The tracks show trajectory of gamma rays. The diameter of the liquid xenon volume is 80 cm.

detector. Therefore, WIMPs and neutrinos can be observed in a low background environment by extracting the events in the central fiducial volume. The event vertex can be reconstructed by the photon intensity pattern observed by many photomultipliers(PMTs) mounted on the inner surface of the detector. Liquid xenon has the following advantages:

- With the high atomic number of xenon ($Z = 54$), and the high density of liquid xenon ($\sim 3\,\mathrm{g/cm^3}$), external gamma-rays can be absorbed in a short distance from the detector wall.
- A large light yield of 42,000 photons/MeV, which is as good as NaI(Tl) scintillator, enables good event reconstruction as well as detection of small energy signals like dark matter recoil.
- 175 nm scintillation light of liquid xenon can be read out by typical PMTs with a bi-alkaline photocathode and a quartz window.
- Purification is easier than with other materials (e.g. distillation is possible).
- Isotope separation is possible. It is possible to enrich ^{136}Xe for double beta decay or deplete ^{136}Xe for solar neutrino measurements.

3. R&D Study by the Prototype Detector

Figure 3 shows the 3 kg fiducial volume prototype detector. It is a 30 cm

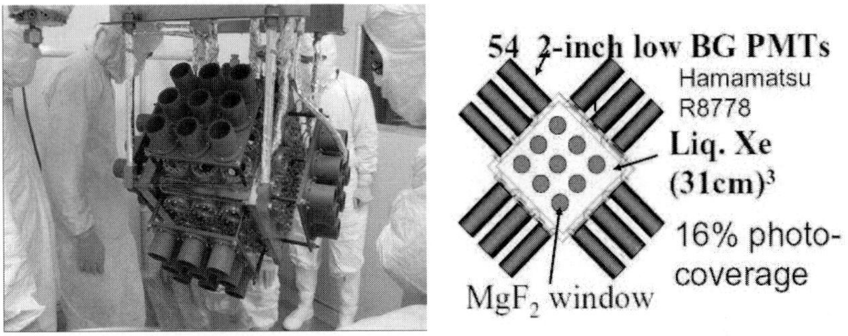

Fig. 3. The prototype detector with 3 kg of fiducial volume. The left figure shows a photograph of the detector and right figure shows schematic view of the detector.

cubic detector viewed by 54 2-inch low background PMTs. The PMTs are Hamamatsu R8778's, which are developed by the XMASS group and Hamamatsu photonics. The detector itself is made from Oxygen Free High purity Cupper (OFHC), and MgF_2 windows are placed in front of the PMTs. The total mass of the liquid xenon is about 100 kg, and the fiducial volume is defined as the central 10cm cubic volume, which is 3 kg of xenon. The detector is placed in a low background setup which consists of 5 cm-thick pure copper, 15 cm-thick lead, 10 cm-thick boric acid, and 15 cm-thick polyethylene shielding for reducing gamma ray, neutron and radon backgrounds.

The vertex position is reconstructed using the pulse height pattern of hit PMTs. In order to do that, Monte Carlo events are generated at various positions in the detector and hit pattern maps are made. The vertex reconstruction program essentially searches for the map which reproduces the observed pulse height pattern of PMTs. The pulse height and arrival time of photon of each PMT are measured by charge sensitive ADCs and TDCs. The total sum signal of all PMTs is monitored by Flash-ADCs(FADC). The TDC and FADC data is being used for the study of possible discrimination of gamma-ray backgrounds and nuclear recoil signals using the pulse shape.

Figure 4 shows the reconstructed vertex position distribution of gamma-ray source(^{137}Cs) data. Three small holes with a diameter of 1 cm are opened in the outer copper and lead shields, which then work as collimators for radioactive sources outside the shield (of course, they are closed when background data are taken). The figure shows the X-Y projection of the reconstructed vertex. The upper figures show the observed data and the lower figures show the corresponding Monte Carlo simulation. The data and Monte Carlo agree with each other very well, which demonstrates that the vertex is reconstructed as expected. Figure 5 shows the reconstructed

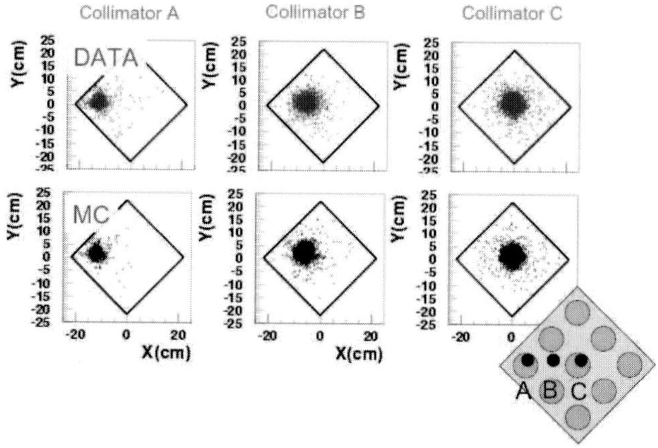

Fig. 4. Upper figures show the reconstructed vertex position on the X-Y plane for a gamma radioactive source data taken at three collimator positions on the detector. Lower figures show the corresponding Monte Carlo simulations. The definition of the coordinates and injected position of the gamma-ray source are shown by the drawing at the right bottom corner.

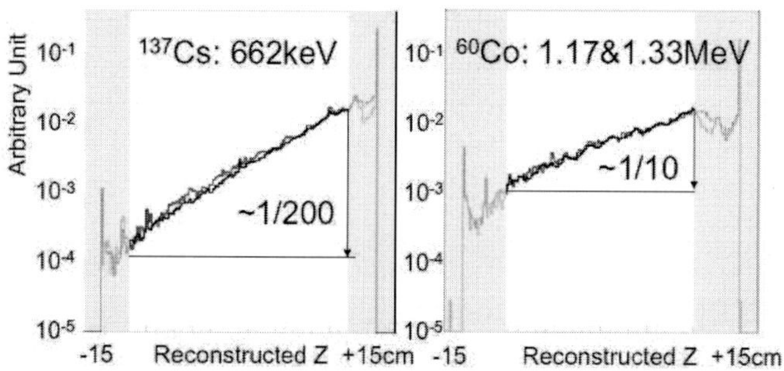

Fig. 5. Reconstructed vertex position distributions along the direction of the injected gamma ray. Left: ^{137}Cs source, right: ^{60}Co source. The gray line shows the observed data and the black line shows the corresponding Monte Carlo simulation. The shaded regions of the histograms (right and left ends) are not relevant because the ADC of the PMTs near the vertex position saturates for the events in these regions.

vertex position distribution along the direction of the injected gamma ray. The gray line shows the observed data and the black line shows the corresponding Monte Carlo simulation. The number of events decreases as a function of the distance from the injected position of the gamma-ray, which is expected from self-shielding. ^{137}Cs 662 keV gamma rays are attenuated by a factor of about 200 over the distance of 24 cm, and there is a factor of 10 attenuation for 1.17 and 1.33 MeV gamma rays from the ^{60}Co source. The distribution of reconstructed vertex positions agrees well with that of the Monte Carlo simulations.

"Background" data of the prototype detector was taken in December 2003 and August 2004.[3,4] Here the "background" data means the data taken without putting the radioactive sources near the detector. Figure 6 shows

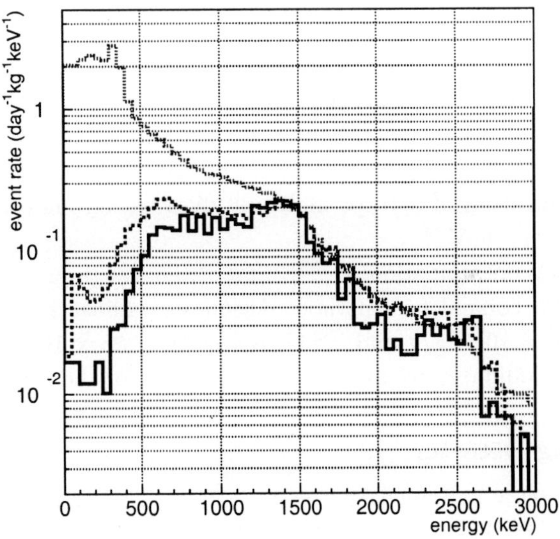

Fig. 6. Background spectrum measured by the prototype detector. The dashed, dotted and solid histograms show the rate of the events which were observed in the full volume, 20 cm, and 10 cm cube at the center of the detector, respectively.

the latest measurement of the background distribution. A large reduction of backgrounds in a smaller detector volume is observed at several hundred keV, which again demonstrates the self-shielding. The background of the inner-most 10 cm cube is about 10^{-2} day^{-1}kg^{-1}keV^{-1} for the low energy range (<100 keV). The observed background spectrum is consistent with the Monte Carlo simulation, which takes into account gamma ray emission from the uranium, thorium decay chains, and ^{40}K gamma rays from the PMTs. The radioactivity of the PMTs was independently measured using a germanium detector. The measured activities are $(1.8\pm0.2)\times10^{-2}$Bq/PMT for the uranium-chain, $(6.9 \pm 1.3) \times 10^{-3}$Bq/PMT for the thorium-chain, and $(1.4 \pm 0.2) \times 10^{-1}$Bq/PMT for the ^{40}K.

As for the internal source of backgrounds, ^{85}Kr in xenon is one potential source. In order to remove ^{85}Kr from xenon, a distillation system was constructed which is able to process 0.6 kg Xe/hour and reduce the krypton to 1/1000 of the original concentration. The distillation technique efficiently removes krypton from xenon because the boiling temperature of xenon and krypton are quite different(165 K for xenon and 120 K for krypton at 1 atmosphere). The system has 99% efficiency for collecting xenon (i.e. only 1% of original xenon is rejected). In March 2004, the system processed 100 kg of xenon. The krypton concentration in the produced xenon was measured by a high sensitivity measurement technique developed by the XMASS group and Taiyo-Toyo-Sanso Co. The measured concentration was 3.3 ± 1.1 ppt mol/mol of krypton. The xenon before the distillation had 3 ppb of krypton. So, the removal of krypton by a factor of 1000 was demonstrated. The krypton concentration of 3.3 ppt mol/mol gives background at a level of $7 \times 10^{-5}\,\mathrm{day^{-1}kg^{-1}keV^{-1}}$.

The internal contamination of ^{238}U and ^{232}Th daughters are measured using the Bi-Po coincidence method. Assuming radioactive equilibrium, ^{238}U is $(9 \pm 6) \times 10^{-14}$g/g and ^{232}Th is less than 23×10^{-14}g/g. These radioactive contaminates are near the requirements for the 800 kg detector.

4. The 800 kg Detector

The 800 kg detector is funded and we are going to construct the detector within 2 or 3 years. The detector will mainly search for dark matter. Figure 7 shows a schematic view of the 800 kg detector. The detector uses about 800 kg of liquid xenon in the volume inside the PMT photocathod plane, which is about 80 cm diameter. The fiducial volume of the detector is a 40 cm diameter sphere, and its liquid xenon weight is 100 kg. A new version of the R8778 PMT is being developed that has a hexagonal photocathod shape. With 812 hexagonal PMTs mounted as shown in Fig.7(right), a photocoverage of 70% can be achieved, and the scintillation light yield of the detector will be about 5 photoelectrons/keV. We will reduce the radioactive contamination of the new R8778 PMT by a factor of 10 from the current R8778.

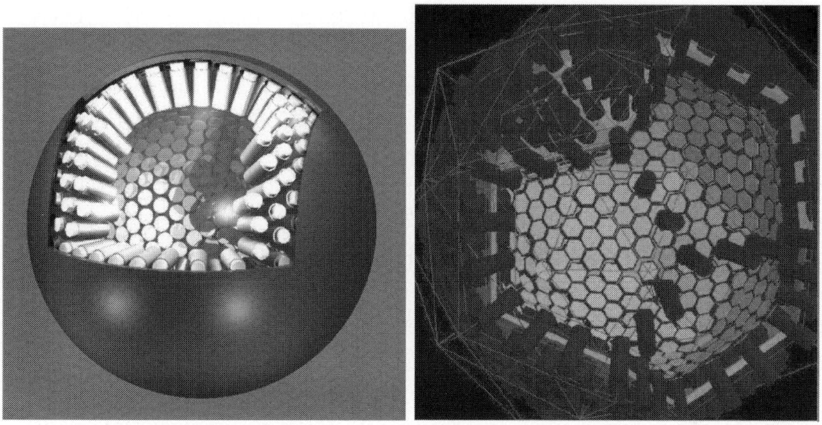

Fig. 7. Conceptual design(left) and the detailed geometry design(right) of the 800 kg detector.

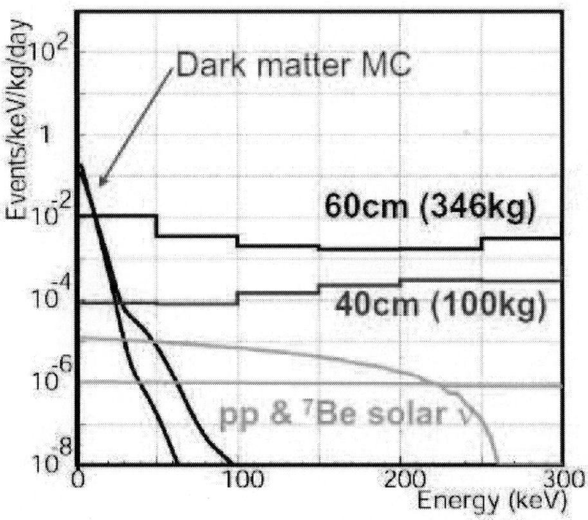

Fig. 8. Simulated background level for the 800 kg detector. The histogram labeled as 40 cm (100 kg) shows the background level in the fiducial volume of the detector. For a reference, background rate in 60 cm diameter is also shown. The hard spectrums show expected energy spectrums from dark matter with 50 GeV and 100 GeV WIMP masses. The soft spectrums show expected energy spectrum of solar neutrinos.

Figure 8 shows the background spectrum simulated by GEANT4, assuming the reduced radioactivity of PMTs (i.e. assuming 1.8×10^{-3}Bq/PMT for uranium-chain, 6.9×10^{-4}Bq/PMT for thorium-chain, 1.4×10^{-2}Bq/PMT for ^{40}K, and 5.5×10^{-3}Bq/PMT for ^{60}Co). Background levels as low as 10^{-4} /kg/day/keV can be achieved in the 40 cm diameter fiducial volume. Requirements on internal background are $^{238}U < 1 \times 10^{-14}$ g/g, $^{232}Th < 2 \times 10^{-14}$ g/g and $^{85}Kr <$ 1ppt. These requirements are based on a simulation in which their contribution to the background is negligible compared with that from PMTs. As described in the previous section, the internal background level achieved by the prototype detector has not yet reached the required level for the 800kg detector, but they are close, within a factor of 10. We think they can be achieved in the real detector by the filters for removing radon and by designing a distillation system with slightly longer tower length.

Figure 9 shows the expected sensitivity of the 800 kg detector for spin independent interactions (SI), and spin dependent interactions (SD), assuming 5 years of data with the 100 kg fiducial volume, The detector is sensitive down to $\sim 10^{-45}$ cm^2 for SI and $\sim 10^{-39}$ cm^2 for SD. It is almost two orders of magnitude better than existing experiments.

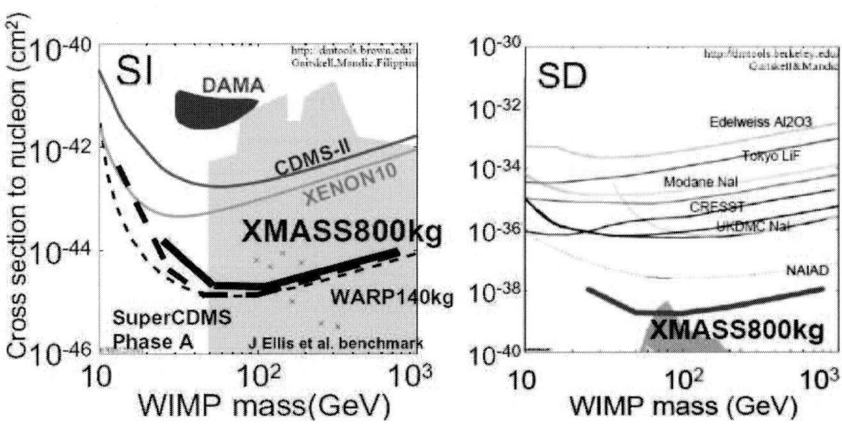

Fig. 9. Sensitivity of the 800 kg detector for the dark matter search for spin independent interactions(left) and spin dependent interactions(right). Five years of data with 100 kg fiducial volume is assumed. The statistical significance level of the sensitivity is 3σ discovery level. The thick solid curve shows the sensitivity of the 800 kg XMASS detector and thin solid curves show existing results from other experiments. Dotted and dashed curves show future prospects of other experiments. The light dark regions show the expectation from SUSY models.

In order to reduce the environmental radioactive background, we will put the 800kg detector in a water shield. Simulation of gamma rays and fast neutrons show that about a 200 cm thick water layer is enough to shield the environmental background to a level below that of the PMT induced background. So, a water tank ~10 m in diameter and ~10 m high is enough for this purpose. The whole setup of the 800 kg detector will be placed in the Kamioka mine. New underground areas for accommodating the 800 kg detector and some other small experiments are being excavated now, as shown in Fig.10. The new area will be ready for installing the detectors by summer 2008.

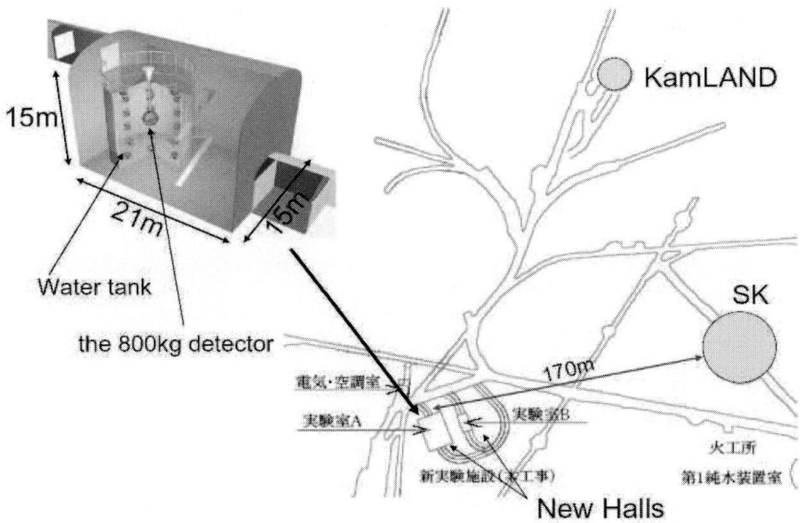

Fig. 10. The new underground area being excavated now (as of September 2007) in the Kamioka mine. The 800 kg XMASS detector equipped with the water shield will be located in the bigger hall.

5. Conclusion

XMASS is a multiple purpose experiment aiming at detection of dark matter, low energy solar neutrinos, and double beta decay. So far the feasibility of the experiment has been studied using a prototype detector. A real detector with 800 kg liquid xenon (~ 100 kg fiducial volume) is going to be constructed in a few years and will search for dark matter. The sensitivity of the detector is about a factor of 100 better than the current experiments in the world as shown in Fig. 9.

References

1. Y. Suzuki et al., "Low energy solar neutrino detection by using liquid xenon", Aug. 2000, hep-ph/0008296.
2. S. Moriyama, "XMASS EXPERIMENT I", proceedings of the International Workshop on Technique and Application of Xenon Detectors, Kashiwa, Japan, 3-4 Dec. 2001.
3. Y. Takeuchi, "Recent status of the XMASS project", presentation in the International Conference on High Energy Physics (ICHEP), Beijing, China, 16-22 Aug. 2004.
4. S. Moriyama, "XMASS experiment", proceedings of the International Workshop on the Identification of Dark Matter, Edinburgh, Scotland, 6-10 Sep. 2004.

DREAMS AND REALITY OF USING NAKED Ge DETECTORS IN LIQUID NITROGEN, STATUS (LONG-TERM STABILITY) OF GENIUS-TF

IRINA V. KRIVOSHEINA

Radiophysical-Research Institute, Nishnii-Novgorod, Russia, irinaKV@web.de

HANS V. KLAPDOR-KLEINGROTHAUS*

Heidelberg, Germany, E-mail: prof.klapdor-kleingrothaus@hotmail.de,
Home-Page: http://www.klapdor-k.de

GENIUS-TF-II is a setup of *six* naked high purity Ge detectors (15 kg) in liquid nitrogen in Gran Sasso. It has been installed in October, 2004 - after the first four naked Ge detectors had been installed on May 5, 2003 (GENIUS-TF-I). The GENIUS-Test-Facility (GENIUS-TF) is **the first** and up to now **only** setup ever testing the novel technique aiming at extreme background reduction in search for rare decays in particular underground. The goal of GENIUS-TF was to test some key operational parameters of the full GENIUS project in 1997.[1-6] Simultaneous physical goal was to search for the annual modulation of the Dark Matter signal.[12,25]

After operation of GENIUS-TF over three years with finally six naked Ge detectors (15 kg) in liquid nitrogen in Gran Sasso we realize serious problems for realization of a full-size GENIUS-like experiment: 1. Background from ^{222}Rn diffusing into the setup, on a level far beyond the expectation. 2. Limited long-term stability of naked detectors in liquid nitrogen as result of increasing leakage current. None of the six detectors is running after three years with the nominal leakage current. Three of the six detectors do not work any more at all.

The results of our three years of investigation of the long-term stability casts serious doubt on the possibility to perform a full GENIUS project - or its copies GERDA or CAMEO/GEM.

Keywords: High purity Ge detectors; HEIDELBERG-MOSCOW experiment.

*Spokesman of HEIDELBERG-MOSCOW, GENIUS-TF and HDMS Collaborations.

1. Introduction

Some years ago, the status of cold dark matter search, of investigation of neutrinoless double beta decay and of low-energy solar neutrinos all required new techniques of *drastic* reduction of background in the experiments. For this purpose we proposed the GENIUS (GErmanium in liquid NItrogen Underground Setup) project in 1997.[1–6] The idea of GENIUS (and GENIUS-TF) is to operate 'naked' Ge detectors in liquid nitrogen (as applied routinely already for more than 20 years by the CANBERRA Company for short term technical functions tests,[8] and later 'rediscovered'[24]), and thus, by removing all materials from the immediate vicinity of the Ge crystals, to reduce the background considerably with respect to conventionally operated detectors. The liquid nitrogen acts both as a cooling medium and as a shield against external radiactivity.

After the success of the HEIDELBERG-MOSCOW experiment in $0\nu\beta\beta$ decay,[17–21] GENIUS is no more needed for $\beta\beta$ decay experiments with ^{76}Ge, but may still be required for dark matter and solar neutrino experiments using Ge as target.[16] Therefore, we continued the research on the GENIUS Test Facility. *Monte Carlo simulations* for the GENIUS project, (and for GENIUS-TF) and investigation of the *new physics potential* of the project have been performed in great detail, and have been published elsewhere.[1,2,6,7,12,15] We were **the first** to show (in our HEIDELBERG low-level facility already **in 1997**) that such device can be used for *spectroscopy*[1,3,5] at least over short measuring times.

The small scale version of GENIUS, the GENIUS-Test-Facility has as main goal to test the long-term stability of the detectors under liquid nitrogen conditions, and also other operational parameters. A detailed description of the GENIUS-TF project is given in.[12,14]

The GENIUS-Test-Facility has been approved by the Gran Sasso Scientific Committee in March 2001.

Additionally to investigation of some key operational parameters of GENIUS, the GENIUS-TF was planned to extend our work on WIMP-nucleon cross sections with the HEIDELBERG-MOSCOW and HDMS experiments,[32–34] and aims at testing of the claimed evidence for WIMP dark matter from the DAMA experiment.[35] The relatively large mass of Ge in the full scale GENIUS-TF compared to existing experiments would permit to search directly for a WIMP signature in form of the predicted seasonal modulation of the event rate.[25]

GENIUS-TF which has been operated over three years in Gran Sasso (see Fig.1) is the only up to now existing test facility underground for a project operating naked Ge detectors in liquid nitrogen such as GENIUS[2] and its copies (Cameo, Gerda).

The first four naked detectors (in total 10 kg) were installed on May 5, 2003 (GENIUS-TF-I). This has been reported in CERN Courier[13] and in.[14] In[14] first energy calibration spectra are shown which demonstrate the (initial) good energy resolution obtained.

Fig. 1. Upper part: Left: Location of GENIUS-TF-I was the building on the right (car in front), opposite to the HEIDELBERG-MOSCOW experiment building (left side). Location of GENIUS-TF-II is the building of the HEIDELBERG-MOSCOW experiment (left). Right: from the right to the left - Irina Krivosheina, S. Karpov, H. Strecker (December 2006). Lower part: Left: The first four contacted naked Ge detectors in the vacuum box during transportation from the previous GENIUS-TF-I localization. Right: Hans V. Klapdor-Kleingrothaus (second from the right) together with his colleagues - Sergey Karpov (right), Herbert Strecker (on front from the left) and Theo Apfel.

2. The GENIUS-TF-II and III Setups

In October 2004 we have installed a new setup GENIUS-TF-II (see Figs. 1,2,3), containing now **six naked** Ge detectors (in total 15 kg), and, as technical improvement **a second copper vessel**, for further shielding of the Radon. That ^{222}Ra diffusing into the setup has been a problem for GENIUS-TF-I, has been described by analysis of the measured spectra in detail in.[11] The inner shielding by bricks of (5-10) cm super-clean polycristalline Germanium (∼300 kg) was used also in this setup forming the inner highly efficient shield of the Ge detectors (see Fig. 3).

The thin wall (1 mm) inner copper box containing the liquid nitrogen is made of high-purity electrolytic copper and is thermally shielded by 20 cm

Fig. 2. Left - View from the top on the GENIUS-TF-II setup during installation in October 2004. Right - The six contacted naked Ge detectors.

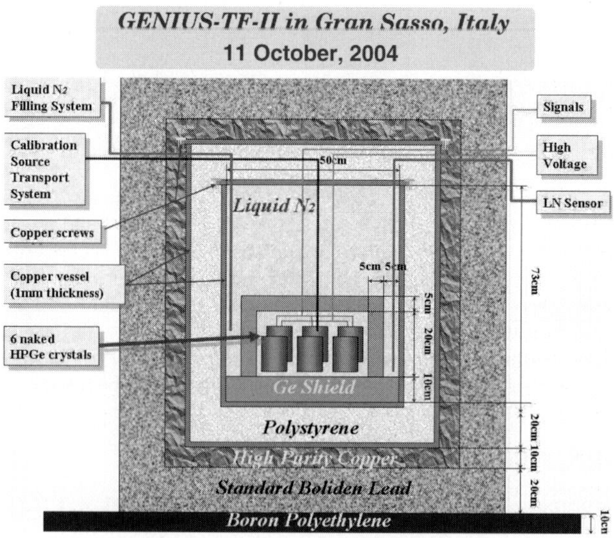

Fig. 3. Schematical view of the TF-II setup.

of special low-level styropor, the outer copper box (also made of electrolytic copper) is followed by a shield of 10 cm of electrolytic copper (15 tons) and 20 cm of low-level (Boliden) lead (>35 tons).

The high-purity liquid nitrogen used, is produced by the BOREXINO nitrogen plant, which has been extended for increase of the production capacity to be able to provide enough nitrogen also for GENIUS-TF (see[11,14]).

GENIUS-TF-III started operation in beginning of March 2005 after the inner shield of polycrystalline Ge bricks had to be removed according to contract with Kurchatov institute.

3. Development of Operational Parameters

3.1. *Background from* ^{222}Rn

The unexpected (according to our Monte Carlo simulations[12]) high background from ^{222}Rn in GENIUS-TF-I (see[11]) has been reduced in GENIUS-TF-II by about a factor 2. This ^{222}Rn background is still compatible with the goal of GENIUS-TF to search for dark matter,[25] but will be a *serious problem* for any full GENIUS-like experiments looking for neutrinoless double beta decay, because the ^{222}Rn leads to the 'famous' background lines from ^{214}Bi near the Q-value for double beta decay of ^{76}Ge.[18,23]

Table 1. The high voltages applied to the detectors after installation in GENIUS-TF I, II and III as function of time, and the nominal voltages.

Det-s	D1	D2	D3	D4	D5	D6
	GENIUS-TF-I, from 10.12.2003, till 25.09.2004					
10.12.03	2404	2603	2879	2301	n. inst.	n. inst.
06.04.04	2600	2220	2879	2301	-	-
04.05.04	2600	2220	3200	2500		
	GENIUS-TF-II, from 18.11.2004, till 28.02.2005					
08.10.04	250	1296	261	954	1253	502
18.11.04	364	2200	347	2298	3501	1015
20.01.05	"	"	"	"	"	1000
	GENIUS-TF-III, from 15.03.2005					
15.03.05	80	1802	20	2153	3501	980
03.05.05	0	1700	0	1500	3501	980
09.03.06	-	1700	-	-	3500	911
19.05.06	-	1600	-	-	2100	850
Nom.	**3000**	**2600**	**3200**	**2500**	**3500**	**2000**

3.2. *Long-Term Stability*

The most dramatic result is obtained for the long-term stability of the detector operation in liquid nitrogen. It is shown in Table 1. As a result of increasing leakage currents, finally from initially six detectors only three were still working in 2006 and *not one of them* with the nominal high voltage (see Table 1). The reasons for the deterioration of the surface purity with time which lead to the increasing leakage currents need further investigation.

There might, however, be also other reasons. For some detector experts a surface–purity problem may not come very surprising.[26–30] From their experience, partly in detector producing companies, they expect this for a naked detector, having no shield of its surface against an open surrounding (the liquid nitrogen) other than a coating by some passivant of some sort, as used[29] by most manufacturers including the company having produced the GENIUS-TF detectors (ORTEC). In fact it would not be surprising that the high voltage of several thousand volts lying at the detector surface would attract ions from the surrounding liquid nitrogen. It is further well known that recently crushed germanium pieces have a highly active surface and are a very active getter They have been used as such in deionized water to further improve the purity of the water, and will be an active getter for impurities also in liquid nitrogen.[29,30] Perhaps this suspicion might be checked by mass spectroscopic investigation of detector surface material. The energy resolution on long terms also decreases systematically[31] (see[9,10]).

4. Perspectives

An important task is to improve further the presently used method of pulse shape analysis. In particular the efficiency of the method at low energies has to be carefully determined. The inner polycrystalline Ge shield which had to be removed according to contract with Kurchatov institute, at beginning of March 2005 should be replaced by a shield of monocrystalline Si, of which a sufficient amount is available.

For the purpose of testing the modulation signal measured by DAMA the amount of detectors has to be increased to about 40 kg to reach a comfortable time scale of measurement of a few years.

5. Conclusions

GENIUS-TF is the *only setup with naked Ge detectors worldwide running underground and over a longer time period.* It has lead to important insight into the conditions of technical operation of naked Ge detectors in liquid nitrogen. The relatively large background from ^{222}Ra diffusion is a problem unsolved up to now. The main problem realized, is, however, the increase of leakage current after long running of the detectors, caused by increasing surface impurity of the crystals. This led to serious restrictions of the high voltages applicable and finally to destruction of the detectors. *The information GENIUS-TF delivered after almost three years of operation on the possibility of long-term operation of such experiments, may cast some doubts on the possibility of such experiments on larger scale in general.*[9,22]

Acknowledgement

The authors would like to thank their colleagues from MPI Heidelberg: Herrn H. Strecker, T. Apfel, M. Reissfelder, M. Saueressig for their help during installation of GENIUS-TF-II. The authors would like to thank the technical staff of the Max-Planck Institut für Kernphysik and of the Gran Sasso Underground Laboratory. We are also thankful to Dr. Sergej Karpov for his help at the last time of existence of the GENIUS-TF experiment at Gran-Sasso and to Prof. Vadim Bednyakov for permanent support. We acknowledge the invaluable support from BMBF and DFG, and LNGS of this project.

References

1. H.V. Klapdor-Kleingrothaus, *Int. J. Mod. Phys.* **A13** (1998) 3953.
2. H.V. Klapdor-Kleingrothaus, J.Hellmig, M.Hirsch, *GENIUS-Proposal*, 20 Nov. 1997.
3. J. Hellmig and H.V. Klapdor-Kleingrothaus, *Z. Phys.* **A359** (1997) 351-359, and nucl-ex/9801004.
4. H.V. Klapdor-Kleingrothaus, M. Hirsch, *Z. Phys.* **A359** (1997) 361-372.
5. H.V. Klapdor-Kleingrothaus, J. Hellmig, M. Hirsch, *J. Phys.* **G24** (1998) 483- 516.
6. H.V. Klapdor-Kleingrothaus, *CERN Courier,* **Nov. 1997**, 16- 18.
7. H.V. Klapdor-Kleingrothaus et al. **MPI-Report MPI-H-V26-1999**, and *Preprint: hep-ph/***9910205**, and in Proceedings of the 2nd Int. Conf. on Particle Physics Beyond the Standard Model BEYOND'99, Castle Ringberg, Germany, 6-12 June 1999, edited by H.V. Klapdor-Kleingrothaus and I.V. Krivosheina, *IOP Bristol* (2000) 915 - 1014.
8. J. Verplancke, *CANBERRA Company*, priv. commun. 5.03.2004.
9. H.V. Klapdor-Kleingrothaus, I.V. Krivosheina, *Nucl. Instr. Meth.* **A 566** (2006) 472-476.
10. I.V. Krivosheina, H.V. Klapdor-Kleingrothaus, *Phys. Scr.* **T127** (2006) 52.
11. H.V. Klapdor-Kleingrothaus et al., *Nucl. Instr. Meth.* **A 530** (2004) 410-418.
12. H.V. Klapdor-Kleingrothaus et al., *Nucl. Instrum. Meth.* **A 481** (2002) 149.
13. H.V. Klapdor-Kleingrothaus, *CERN Courier* **43 Nr.6** (2003) 9.
14. H.V. Klapdor-Kleingrothaus et al., *Nucl. Instr. Meth.* **A 511** (2003) 341; H.V. Klapdor-Kleingrothaus, CERN Courier 43 Nr.6 (2003) 9; H.V. Klapdor-Kleingrothaus et al., Proc. of the 3nd Intern. Conf. on Particle Physics

Beyond the Standard Model, BEYOND02, Castle Ringberg, Germany, 2002, IOP 2003, ed. H.V. Klapdor-Kleingrothaus, 499.
15. H.V. Klapdor-Kleingrothaus, "60 Years of Double Beta Decay - From Nuclear Physics to Beyond the Standard Model", WS (2001) 1281 p.
16. H.V. Klapdor-Kleingrothaus, in Proc. of 2th Int. Workshop on Low Energy Solar Neutrino Detection, Tokyo, Japan, 4-5 Dec. 2000 (World Scientific, Singapore 2001) p.116.
17. H.V. Klapdor-Kleingrothaus and I.V. Krivosheina *"The Evidence for the Observation of* $0\nu\beta\beta$ *Decay: The Identification of* $0\nu\beta\beta$ *Events From the Full Spectra", Mod. Phys. Lett.* **A21** (2006) pp. 1547-1566.
18. H.V. Klapdor-Kleingrothaus, I.V. Krivosheina, A. Dietz et al., *Phys. Lett.* **B586** (2004) 198-212.
19. H.V. Klapdor-Kleingrothaus, I.V. Krivosheina, A. Dietz et al., *Nucl. Instrum. Meth.* **A522** (2004) 371-406.
20. H.V. Klapdor-Kleingrothaus et al. hep-ph/0201231 and *Mod. Phys. Lett.* **A 16** (2001) 2409-2420.
21. H.V. Klapdor-Kleingrothaus, in Proc. "2nd Scandanavian Neutrino Workshop (SNOW 2006)", Stockholm, Sweden, 2-6 May 2006, *eds* T. Ohlsson et al., Stockholm, Roy. Swed. Acad. Sci., *Phys. Scripta* **T127** (2006) pp. 40-42.
22. I.V. Krivosheina and H.V. Klapdor-Kleingrothaus in Proc. "2nd Scandanavian Neutrino Workshop (SNOW 2006)", Stockholm, Sweden, 2-6 May 2006, *eds* T. Ohlsson et al., *Phys. Scripta* **T127** (2006) pp. 52-53.
23. K.Ya. Gromov et al., *J.Part. Nucl. Lett.* **3** (2006) 30-41.
24. G. Heusser, *Ann. Rev. Nucl. Part. Sci.* **45** (1995) 543-590.
25. C.Tomei, A.Dietz, I.Krivosheina, H.V. Klapdor-Kleingrothaus, *Nucl. Instrum. Meth.* **A508** (2003) 343-352.
26. G.F. Knoll, "*Radiation Detection and Measurement*", second ed. 1989, John Wiley & Sons.
27. E. Flyckt (*Philips Photonics Comp.*, Brave, France) private comm. May 2006.
28. A.B. McDonald (SNO collaboration), Queen University, Canada, private communication, May 2006.
29. T. Raudorf, ORTEC Company, private communications 7.08.2006.
30. W.L. Hansen et al., IEEE Trans. on Nucl. Sci. NS-27, No.1, (1980); W.H. Brattain, P.J. Boddy, J. Electrochemical Soc. 109 (1962) 574; D. Frankl, Electrical Propoties of Semicanductor Surfaces, Pergamont Press (1967).
31. H.V. Klapdor-Kleingrothaus, I.V. Krivosheina et al., *Gran Sasso Ann. Report* 2005, p.125-133.
32. HEIDELBERG-MOSCOW collab., *Phys. Rev.* **D 59** (1998) 022001.
33. H.V. Klapdor-Kleingrothaus, I.V. Krivosheina and C. Tomei, *Phys. Lett.* **B 609** (2005) 226-231.
34. V.A. Bednyakov, H.V. Klapdor-Kleingrothaus and I.V. Krivosheina, *"New Constraints on Spin-Dependent WIMP-Neutron Interactions from HDMS With Natural Ge and* ^{73}Ge*", Phys. Atom. Nucl.* **71, Nr. 1** (2008), (see also these Proccedings).
35. R. Bernabei et al., *Phys. Lett. B* **424** (1998) 195; **450** (1999) 448; **480** (2000) 23; *Riv. Nuovo Cim.* **26** (2003) 1-73.

DARK MATTER SEARCH WITH NOBLE LIQUIDS: THE ARDM EXPERIMENT

S. NAVAS

Dpto. Física Teórica y del Cosmos & C.A.F.P.E., University of Granada, Granada E-18071, Spain
E-mail: navas@ugr.es
www.ugr.es/~navas

Weakly Interactive Massive Particles (WIMPs) are one of the most attractive candidates of Dark Matter in the Universe. The goal of the ArDM collaboration is to develop and operate a liquid argon detector to search for direct evidence of WIMPs. The technical design of the detector aims at independently measuring the scintillation light and the ionization charge generated by the recoil of target argon nuclei after the WIMP collision. A Large Electron Multiplier system will be used to collect the charge while an array of photomultipliers will detect the light. A 1-ton scale prototype is currently under development to demonstrate the feasibility of such a direct detection experiment with large target masses. In this paper, the principle of the experiment and the conceptual design of the detector are described.

Keywords: dark matter; WIMP; liquid argon; cryogenic detector; ArDM.

1. Introduction

The understanding of the nature of Dark Matter (DM) is one of the most exciting problems in Particle Physics. Astronomical observations give strong evidence for the existence of non-luminous and non-baryonic matter, presumably composed of a new type of elementary particle. Common models of DM distribution describe it as consisting of a sea of heavy weakly interacting massive particles (WIMPs), forming a cold thermal relic gas. The detection of these WIMPs is based on the capability of measuring the recoils of target nuclei with kinetic energy in the range of 10–100 keV. Due to the weak coupling and the low imparted energy, WIMP signals are rare and elusive events. The total interaction rate depends on underlying models whose parameters are numerically poorly known, even in the context of well defined extensions of the Standard Model like e.g. SUSY, and is therefore not yet predictable. Noble liquid detectors like argon or xenon provide advantageous features for direct WIMP detection like high density and low ionization potential, resulting in a high ionization (charge) and scintillation (light) yields. A main subject of any such detector is the method of the readout for the ionization and scintillation.

Within the ICARUS[1] R&D program, it was first shown that noble liquid TPCs can act as particle detectors at very large scales. More recently, the XENON[2] and WARP[3] collaborations have reached the first competitive limits on WIMP search using liquid xenon and liquid argon detectors, respectively, but they both rely exclusively on photomultipliers for their readout. The possibility to directly detect the ionization charge is less well developed although it might provide alternative and potentially large benefits.

The Argon Dark Matter project (ArDM),[4] initiated in 2004, aims at developing a 1–ton liquid argon (LAr) detector with *independent ionization and scintillation readout* to demonstrate the feasibility of a ton-scale liquefied noble gas Dark Matter detector. The detection principle of ArDM is shown in Fig. 1. A WIMP–argon elastic collision in the detector active volume would produce a nucleus recoil with kinetic energy of tens of keV. This recoil argon generates excitation and ionization of the medium along the path, producing free electrons and scintillation photons. The idea is to use Large Electron Multiplier systems (LEMs) and photomultipliers to collect the charge and the light, respectively. The experimental technique is based on the comparison of the relative charge and light yields induced by different types of interactions, which allows particles causing events in the detector to be identified. Additionally, the time dependence of the scintillation light can be used to further discriminate between WIMP and background signals. In addition, background radiation must be reduced to a very low level by locating the experiment underground and by the use of low radioactive detector materials.

2. Detection Principle and Conceptual Design

Despite of the much lower cost, the choice of liquid argon for the initial ton scale detector instead of xenon is motivated by the fact that the event rate in argon is less sensitive to the threshold on the recoil energy than for xenon because of form factors (Fig. 2). For a threshold of $\sim$ 30 keVr, the rates on Xe and Ar per mass are similar. With such a threshold a WIMP-nucleon cross-section of 10^{-44} cm^2 yields about one event per ton per day. In any case, the relevance of obtaining data on argon and xenon is given by the fact that, due to kinematics, the recoil spectra in Xe and Ar are different, providing an important crosscheck in case of positive signal.

The conceptual ArDM detector design is illustrated in Fig. 3. The detector is contained in a cylindrical vessel where the liquid and vapour of the argon are in equilibrium. The three technical key points are the charge readout on the top, the drift field on the central region and the light readout on the bottom.

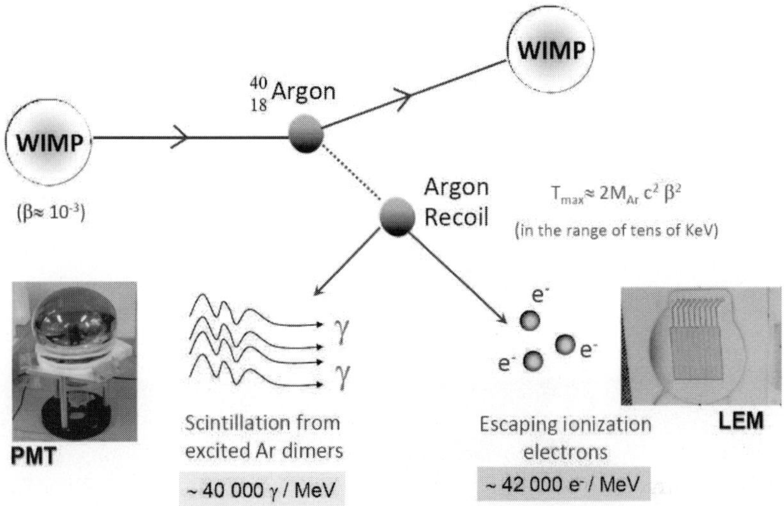

Fig. 1. The WIMP elastic interaction in liquid argon produces argon nuclei recoils: free electrons and scintillation photons are generated and detected with LEMs and photomultipliers.

If a particle interaction occurs in the active volume free ionization electrons and scintillation light ($\lambda \sim$128 nm) are produced (see Fig. 4). The former are pulled away from the interaction point and drifted upwards thanks to a strong electric field provided by a Greinacher/Crockroft-Walton high-voltage circuit, designed to reach up to 4 kV/cm. The maximal drift length is ~120 cm. When the electrons reach the level of the liquid argon surface, they are extracted into the gaseous argon volume on the top of the detector where the readout of the ionization charge takes place. Two Large Electron Multiplier systems (LEM, based on the GEM[5] concept) provide the needed electron amplification by means of a high electric field generated in small cylindrically shaped holes (500 μm diameter, 800 μm spacing). This defines the location of the event in space together with the prompt light signal.

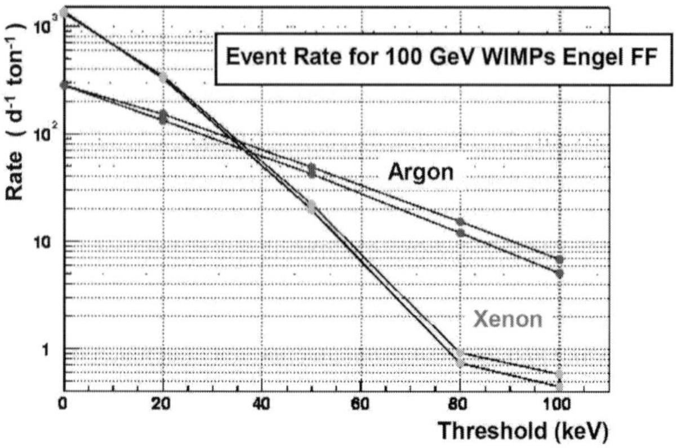

Fig. 2. Event rate per day per ton on argon and xenon targets for winter and summer periods as function of the detection threshold for a WIMP-nucleus cross-section of 10^{-42} $cm^2 = 10^{-6}$ pb and a WIMP mass of 100 GeV.

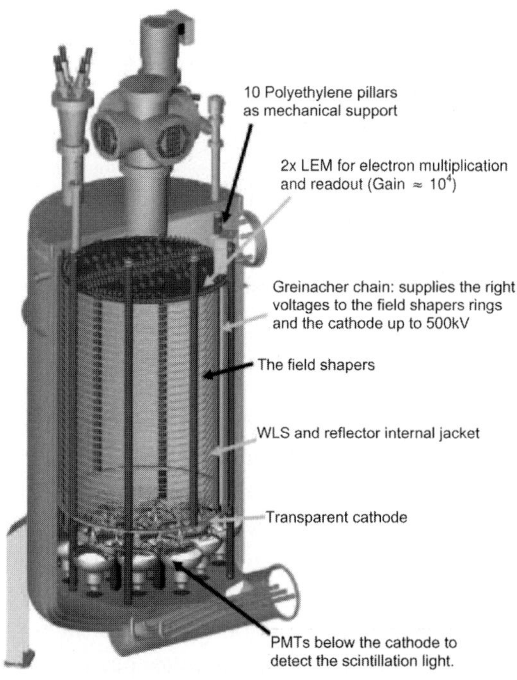

Fig. 3. Drawing of the ArDM detector.

On the other hand, scintillation light is readout with an array of 14 photomultiplier tubes (PMT) located at the bottom of the detector, just behind the semi-transparent cathode. The vessel wall is also covered with foils of Mylar coated with a thin aluminium layer to reflect the light, increasing the photon detection efficiency. The primary VUV scintillation light of argon needs to be shifted to visible light in order to match the PMT sensitivity window. This is achieved by coating both the photocathode and the foils with a wavelength shifter (see Fig. 5).

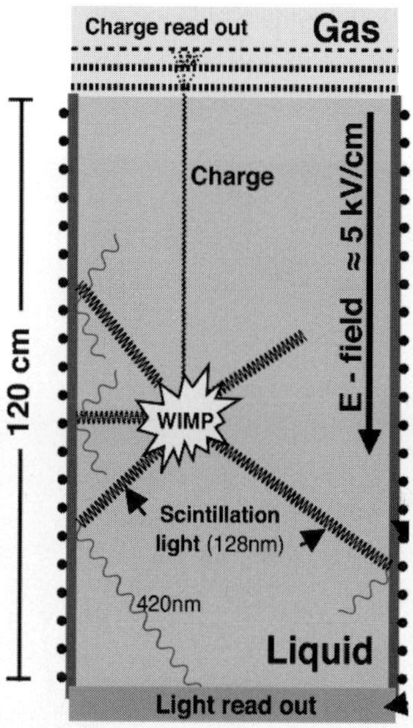

Fig. 4. ArDM detection principle. Free electrons are drift upwards, extracted to the argon gas phase and detected with LEMs. Scintillation photons suffer reflections in the wavelength shifter walls and are detected thanks to an array of photomultipliers.

The amount of charge Q and light L depends on the energy release E and the applied electric field ϵ according to the laws:[6]

$$Q(\epsilon, E) = Q_0(E) \cdot \frac{C}{\epsilon} \cdot \ln\left(1 + \frac{\epsilon}{C}\right) \quad ; \quad L(\epsilon, E) = L_0(E) - Q(\epsilon) \qquad (1)$$

where C is a constant strongly dependent on the ionization power of the particle[7] (large for high stopping power, e.g. like for nuclear recoils). Hence, by comparing the charge and scintillation yields, one should be able to distinguish the nuclear recoils from e/γ signatures with high discrimination power.

In addition to the elements described above, a series of slow control devices (levelmeters, temperature and pressure sensors, etc.) are foreseen in order to on-line monitor some basic internal detector parameters. The levelmeters are crucial in the region near the LEM since the gas argon gap between the liquid argon and the LEM drives the charge amplification. High precision capacity levelmeters have been custom made. They are made of parallel Invar plates in a Macor structure. Precisions of $\sim$0.03 mm on the LAr level have been reached in cryogenic laboratory tests. An array of PT10K resistors all along the pillars has been installed for temperature monitoring.

Fig. 5. Detail of the ArDM internal skeleton. The polyethylene pillars, the field-shape rings and a section of the wavelength shifter (blue panel on the left) are visible.

3. Expected Background

One of the most important backgrounds comes from the ^{39}Ar isotope, present in natural argon liquefied from the atmosphere.[8] It is a beta emitter with lifetime 269 years and Q value 565 keV. The decay rate is approximately 1 kHz per ton of liquid argon.[3]

In principle, the combination of charge and light readout should provide a powerful enough background rejection. Since nuclear recoils due to WIMP or neutron interactions have a much higher characteristic light over charge ratio than electron and gamma interactions, these types of events are in principle well distinguishable if this ratio is measured precisely (see Fig. 6). In addition, the comparison on fast ($\tau_{singlet} \sim 6$ ns) and slow ($\tau_{triplet} \sim 1.7$ μs) light components can also be used to further discriminate between nuclear recoils and background events. Due to ionization densities, the population of argon excimer states is different for nuclear recoil and electron/gamma events, leading to characteristic contributions to the fast and slow scintillation light components.[9]

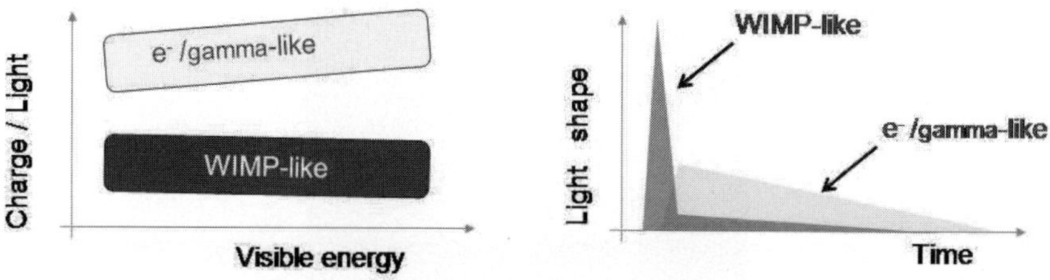

Fig. 6. WIMP signal and electron/gamma background can be disentangled thanks to different patterns on charge over light ratio (left) and scintillation light (right) shapes.

The alternative possibility of using ^{39}Ar-depleted argon extracted from underground natural gas rather than from atmosphere is currently also under study. This would provide a reduction of this background although its cost is to be estimated. On the other hand, the ^{39}Ar decays, evenly distributed in the target, provide a precise calibration and monitoring of the detector response as function of time and position.

Another important background is neutron-induced nuclear recoils which are hardly distinguishable from WIMP events. The neutrons from radioactivity arise mainly from contamination of materials or rock by the two radioactive elements uranium and thorium. In addition, their decay chain-daughters contain α particles (energies of 3.5–11 MeV) which undergo (α,n) reactions producing neutrons in the MeV range. Besides (α,n) reactions, neutrons can arise from spontaneous fission. Neutron background can be heavily reduced by requiring single interactions (neutrons are likely to suffer more than one scattering given the big size of the detector) and cuts in the total visible energy of the event.

Muon-induced neutrons arise from cosmic muon interactions with surrounding materials. Neutrons are produced by spallation or photonuclear processes, or by secondary interactions of muon-induced hadronic showers. The shielding of the detector optimized for natural radioactivity is less efficient in this case and faster neutrons can penetrate inside the detector fiducial volume. Even though highly energetic muons are able to penetrate deep underground, we expect this background to be small.

Fig. 7. Photo of the ArDM cryostat (big stainless-steal cylinder) and the internal structure (suspended on the top) at the CERN workshop.

4. Outlook

The first goal of the ArDM collaboration is to build and test a 1-ton liquid argon prototype to demonstrate the feasibility of the current design. The key points of the operation are (1) the LEM-based charge readout in gaseous argon, (2) the generation of a very high and stable electric field and (3) the efficient readout of the argon scintillation light. A full test programme will be carried out at CERN, where the prototype is currently under construction (Fig. 7).

The operation also involves the design, acquisition and run of a cryogenic system and a complex liquid argon purification system. The first milestone to be achieved during the next incoming months is a proof of principle and stability studies, as well as optimization of the detector performance regarding γ-ray and β-electron background rejection versus nuclear recoils. In a second phase, the design of a shielding against neutrons from detector components, surrounding facilities and muon-induced neutrons will be addressed.

After the successful testing of the prototype, an underground operation is planned in order to minimize background radiation induced by cosmic rays. With a recoil energy threshold of 30 keV, a WIMP-nucleon cross section of 10^{-42} cm^2 would yield approximately 100 events/day/ton, accessing the WIMP-nucleon cross section region of 10^{-6} pb. By improving the background rejection power and further limiting the background sources, a sensitivity of 10^{-8} pb would become reachable. Due to scalable technologies used, an enlargement of the volume is a realistic prospect.

Acknowledgments

The help of all ArDM colleagues from Granada University, ETH Zürich, University of Zürich, CIEMAT, University of Sheffield and Soltan Institute Warszawa is greatly acknowledged.

References

1. F. Arneodo *et al.*, *Nucl. Instrum. Meth. A* **527**, 329-410 (2004).
2. J. Angle *et al.*, *e-Print: arXiv:0706.0039 [astro-ph]* (2007).
3. P. Benetti *et al.*, *Nucl. Instrum. Meth. A* **574**, 83-88 (2007).
4. `http://neutrino.ethz.ch/ArDM/`
5. P. Sauli, *Nucl. Instrum. Meth. A* **386**, 531 (1997).
6. J. Thomas and D.A. Imel, *Phys. Rev. A* **36-2**, 614 (1987).
7. T. Doke *et al.*, *Nucl. Instrum. Meth. A* **134**, 353 (1976).
8. The concentration in ^{40}Ar is 8.1×10^{-16} from H.H. Loosli, *Earth and Planetary Science Letters*, **63** (1983).
9. C. Regenfus, *Proc. 6th Int. Workshop on the Identification of Dark Matter (IDM 2006).*

THE DRIFT DIRECTIONAL DARK MATTER DETECTOR AND FIRST STUDIES OF THE HEAD-TAIL EFFECT

N.J.C. SPOONER and P. MAJEWSKI (for the DRIFT II Collaboration)

Department of Physics and Astronomy, University of Sheffield, Sheffield, UK
E-mail: P.Majewski@sheffield.ac.uk, N.Spooner@sheffield.ac.uk

Measurement of the direction of the elastic nuclear recoil track and ionization charge distribution along it, gives unique possibility for unambiguous detection of the dark matter WIMP particle. Within current radiation detection technologies only Time Projection Chambers filled with low pressure gas are capable of such measurement. Due to the character of the electronic and nuclear stopping powers of low energy nuclear recoils in the gas, an asymmetric ionization charge distribution along their tracks may be expected. Preliminary study of this effect, called Head-Tail, has been carried out here using the SRIM simulation program for Carbon and Sulfur in 40 Torr carbon disulfide, as relevant to the DRIFT detector. Investigations were focused on ion tracks projected onto the axis of the initial direction of motion in the energy range between 10 and 400 keV. Results indicate the likely existence of an asymmetry influenced by two competing effects: the nature of the stopping power and range straggling. The former tends to result in the Tail being greater than the Head and the latter the reverse. It has been found that for projected tracks the mean position of the ionization charge flows from 'head' to 'tail' with the magnitude depending on the ion type and its energy.

Keywords: Dark Matter; Nuclear Recoils; Head-Tail; DRIFT; SRIM; WIMP.

1. Introduction

It has been demonstrated by the Dark Matter searching experiment DRIFT[1] using Time Projection Chambers (TPC) filled with low pressure CS_2 that the direction of few mm long nuclear recoil tracks can be measured. However, due to the limited spatial resolution of the readout and ionization charge diffusion in the large detector volume, its distribution along the low energy nuclear recoil track is difficult to reconstruct.

In this work preliminary results of simulations of the energy loss and ionization charge distribution along tracks of Carbon and Sulfur ions in 40 Torr CS_2 (ρ=1.67$\cdot$10^{-4} g/cm^3) are presented. The so-called Head-Tail effect, describing the location of the majority of the ioinisation charge either at the beginning (Tail) or at the end (Head) of the track as a function of ion energy and diffusion has been studied. SRIM[2] and TRIM programs were used to calculate stopping powers for ions in gases, to investigate with Monte Carlo generators the energy loss and ionization charge distribution along the ion track on an event-by-event basis.

2. SRIM

The Stopping and Range of Ions in Matter (SRIM) simulation program calculates as a function of energy: the electronic (S_e) and nuclear (S_n) energy loss, range and longitudinal and lateral straggling of ions moving in matter. They have been calculated here for ion energies up to 500 keV. To understand the behaviour along the tracks they are plotted in Fig. 1 and Fig. 2 against ion range, the ion energies marked here on the plot at 50 keV intervals.

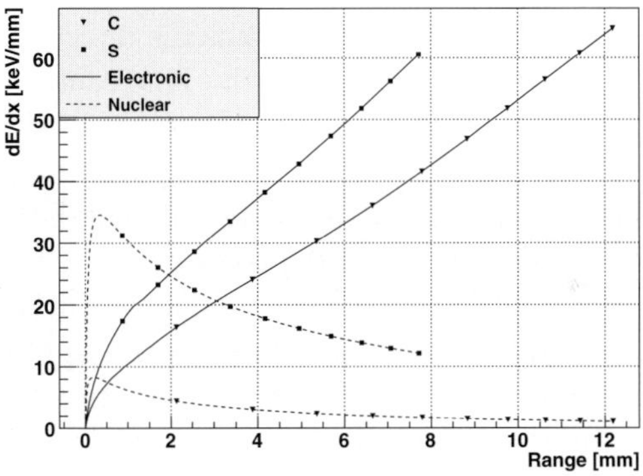

Fig. 1. Calculated with SRIM the ionization energy loss dE/dx as a function of ion range in the medium. Energy losses in electronic (solid line) and nuclear (dashed line) channels are shown for Sulfur(■) and Carbon(▼) ions in 40 Torr CS_2. The ion energy is marked up to 500 keV with an interval of 50 keV.

As can be seen at higher energies the electronic energy loss is dominant and decreases with energy where at lower values the nuclear energy loss becomes greater than the electronic. For example for Sulfur ions the nuclear energy loss starts to dominate already at an energy of 100 keV. This means that energy loss along the ion track is not continuously decreasing. This has an impact on the density of the ionization charge along the track and makes the head-tail effect energy dependent.

In the nuclear interactions secondary nuclear recoils are created. Although 3-4 times higher energy is needed to create an electron-ion pair by nuclear recoils than by electrons, they are a source of ionization.

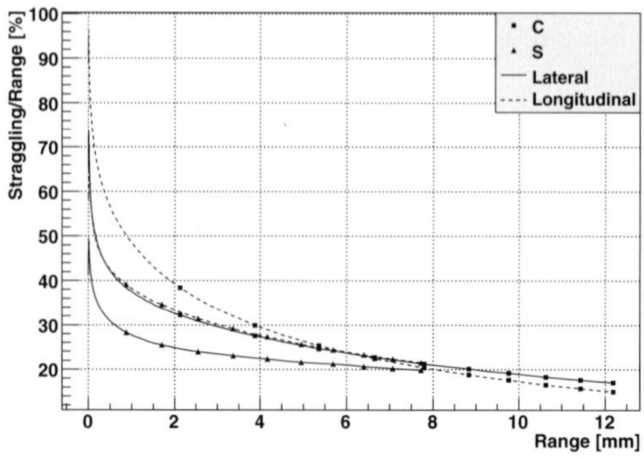

Fig. 2. Lateral (solid line) and longitudinal (dashed line) straggling/range as a function of range for Carbon (■) and Sulfur (▲) ions with energy is marked up to 500 keV with an interval of 50 keV.

Due to continuous collisions with gas atoms the direction of the moving ions become deviated from the original path. This causes fluctuations in the ion's range, described by the lateral and longitudinal straggling and depending on ion energy (see Fig 2). One can see that ions slowing down experience increasing straggling relative to the drift length they have ahead, which rises sharply at the very end of the track.

Tables calculated by SRIM are used in the Monte Carlo generator program called TRIM. In addition to on-line histogramming of various quantities, TRIM results are also recorded event by event in two different formats. The first is included in the file called *COLLISON.TXT* generated either in quick simulation, where for every ion collision its current energy, position in 3D and type of secondary recoils with their initial energy is recorded, or in full nuclear recoil cascade simulation, where information about the energies and positions of all created nuclear recoils is added. TRIM creates also an output file called *EXYZ* which, though with superior spatial resolution to the other outputs, delivers information only about ion energy loss along the track. The latter format has been chosen here for the Head-Tail effect analysis.

3. TRIM Simulation and Results Analysis

For each ion with a given energy 10^4 events were recorded and further analysed. The energies used were for Sulfur: from 10 to 100 with a step of 10 keV then up to 250 with a step of 25 keV and up to 400 with a step of 50 keV and Carbon from 10 to 100 with a step of 10 keV then up to 200 keV with a step of 25 keV and up to 300 with a step of 50 keV. Each event in the TRIM output *EXYZ* file contained on average 100 collision points with information about: current ion energy, the 3D coordinate of the collision point and energy loss, for which the sum equals the initial energy of the ion. For Carbon and Sulfur ions the number of ionization charges was calculated using W-values taken from.[3] For each ion, a W-value was used at every collision according to its current energy. Note that the actual value of W at energy below 20 keV is not well known and values used here are based on an extrapolation of the function obtained from the polynomial fit to the existing data.

Based on this approximation the Head-Tail effect has been studied for three different configurations of ion interaction points: along the track, projected onto the line between the beginning and the end of the track and projected onto the axis of the initial ion direction. Because of the range straggling of ions with the same energy, the position of each interaction projected point has been normalised to the projected range of the ion.

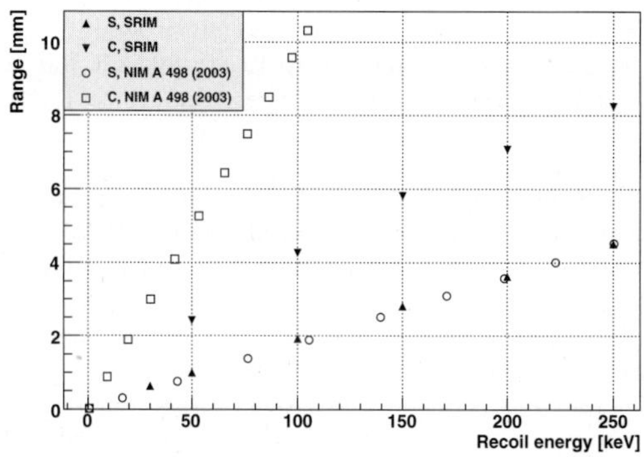

Fig. 3. Sulfur and Carbon ion range in CS_2 as a function of the energy, calculated with SRIM (▲ and ▼) and taken from[3] (○ and ❑).

3.1. *Ion Ranges*

Using coordinates of the first and the last point of the ion track its projected range has been calculated. Fig. 3 shows results for Sulfur and Carbon ions in CS_2 in comparison with data from .[3] There is very good agreement for Sulfur but a large discrepancy for Carbon ions for which results from this paper are more than twice smaller than in.[3]

In the conversion of the energy loss into ionization charge, the number of electrons attached with no loss to CS_2 molecules, creating CS_2^- and called Negative Ions Pairs (NIPs), was summed for each Carbon and Sulfur ion energy. Fig. 4 shows, as a function of NIPs, the range of nuclear recoils together with the range of alpha particles and electrons calculated with TRIM and ESTAR,[4] respectively.

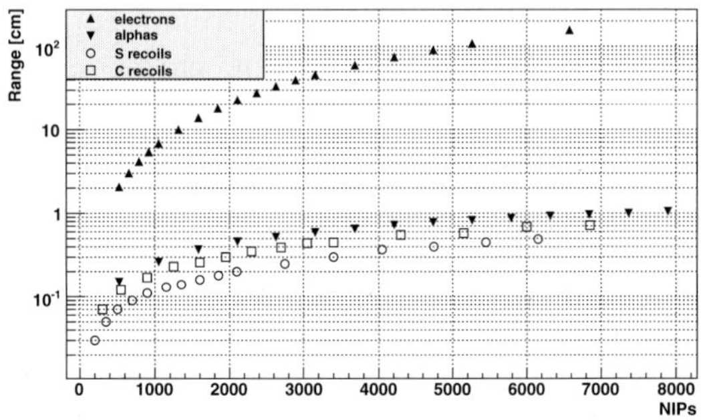

Fig. 4. Sulfur (O), Carbon (❑), alpha (▼) and electron (▲) range in CS_2 as a function of ionization charge (NIPs) created. NIPs for Sulfur and Carbon were calculated using the W-values from [3] while for alpha particles and electrons a constant value of W=19 eV was used. To calculate the range of electrons the ESTAR program was used for energies from 10 to 125 keV with steps 2.5, 5 and 10 up to 20, 60 and 100 keV, respectively. Results for the alpha particle ranges are for energies from 10 to 150 with a step of 10 keV.

Table 1. Ranges of 103 keV ^{206}Pb recoils in pure and binary gases at STP measured and simulated with TRIM. Values from measurement are taken from.[5] 1σ of the simulated range distributions is 10% of the tabulated mean values for all gases except Xe for which 1σ is 15%. Measurement uncertainty is $\pm$ 2μm.

Gas	Ar	Xe	CH_4	C_2H_4	Air	N_2
Measurement [μm]	79	44	84	58	83	80
SRIM [μm]	73	36	95	61	80	74

The NIP values were calculated using W-value=19 eV. As one can see the range of electrons with the same number of NIPs as nuclear recoils is one order of magnitude greater than the range of recoils. This demonstrates that a TPC operating with 40 Torr CS_2 has great separation power between electron and nuclear recoils. However, it shows also that the range of nuclear recoils and alpha particles are close to each other, with a difference not greater than 1-2 mm for tracks a few mm long, often also elongated due to the charge thermal diffusion.

To check the consistency of the SRIM results with existing measurements, the range of 103 keV ^{206}Pb recoils in several pure and binary gases at STP was calculated and compared with results from.[5] The results are shown in Table 1 indicating in most cases less than a 10% difference. Similar differences between measurements and simulation were observed for 5-6 MeV alpha particles from Rn and Po decays in 40 Torr CS_2.[6]

3.2. *Head-Tail*

Along with the event-by-event information on ion energy loss, TRIM also delivers histograms accumulated on-line and showing distributions of average values of several quantities as a function of target depth. With default initial TRIM settings the target depth corresponds to projection onto the axis of the initial direction of ion motion. One of the histograms shows distributions of the average ionization energy loss by ion and secondary recoils. An example of such a distribution, for 1000 Sulfur ions with energy of 100 keV, is shown in Fig. 5. One can see ionization energy loss by ion,

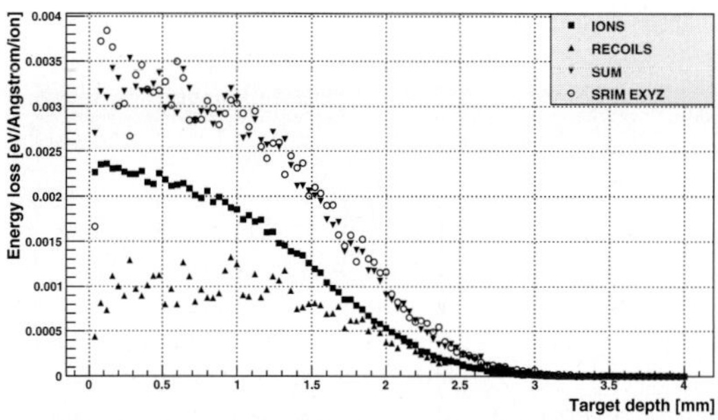

Fig. 5. Average ionization energy loss for 100 keV Sulfur ions (■) and those created in the cascade secondary nuclear recoils (▲) as a function of target depth, obtained from TRIM histrogramming. Total energy loss (▼) is compared with scaled results from the *EXYZ* TRIM output file analysis (○).

secondary recoils and by the sum of them. The latter is compared with scaled results from the *EXYZ* TRIM output file analysis. Both show the same character of the ionization energy loss along the target depth indicating here that more energy is lost at the beginning of the track (Tail) than at the end (Head). However, this distribution is along the target depth regardless of the ion range which, due to straggling, varies from ion to ion. This is likely not the case in an actual experiment where recoils are detected on an event-by-event basis and charge distribution along the track is measured with respect to its projected range. Thus for this work we study the energy loss and Head-Tail effect also using the *EXYZ* output file.

For Sulfur and Carbon ions the distributions of ionization charge were studied along: the track, projection onto a line between the first and last point of the track, and projection onto the axis of the initial ion motion. The latter were studied with and without diffusion. Diffused NIPs were redistributed in space according to the thermal diffusion formula: $\sigma = \sqrt{\frac{2k_BT}{e}}\sqrt{\frac{L}{E}}$ where: k_B is Boltzmann constant, T ambient temperature, e electron charge, E drift electric field and L drift distance. After numerical simplifications a diffusion formula was used of form: $\sigma[mm] = 0.72\sqrt{\frac{L[m]}{E[kV/cm]}}$ for three different drift lengths L: 0.1, 0.3 and 0.5 m and electric field in the DRIFT IIb detector E= 0.58 kV/cm.

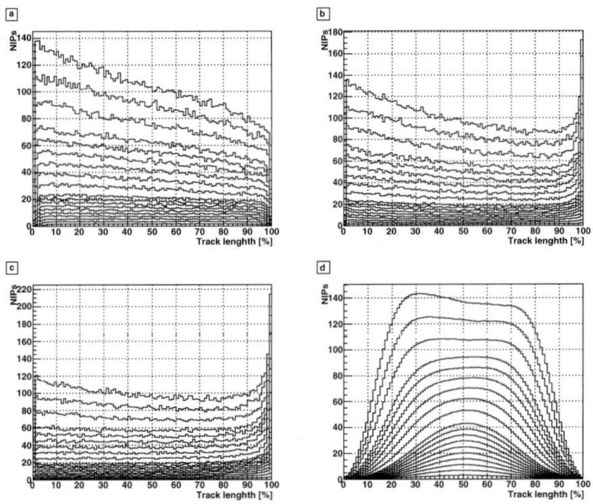

Fig. 6. Ionization charge distribution along Sulfur track (normalised to its range): (a) along the drift path, (b) projected onto the line between the beginning and end of the track, (c) projected onto the direction of the initial motion, (d) randomly diffused with maximum drift path of 50 cm and projected onto the direction of the initial motion. Displayed distributions are for the following ion energies: from 10 to 100 with a step of 10 keV then up to 250 with a step of 25 keV and up to 400 with a step of 50 keV.

Figs.[6] and [7] show NIP distributions along the Sulfur and Carbon tracks, respectively. For both, as expected, the ionization charge distributions along the line following the ion's motion (a) are not featured with a sharp rise at the end caused by the straggling. The negative slope of the distributions is very clear down to 40 keV and 20 keV for S and C, respectively from which at lower energies (where nuclear energy loss dominates) they become flat.

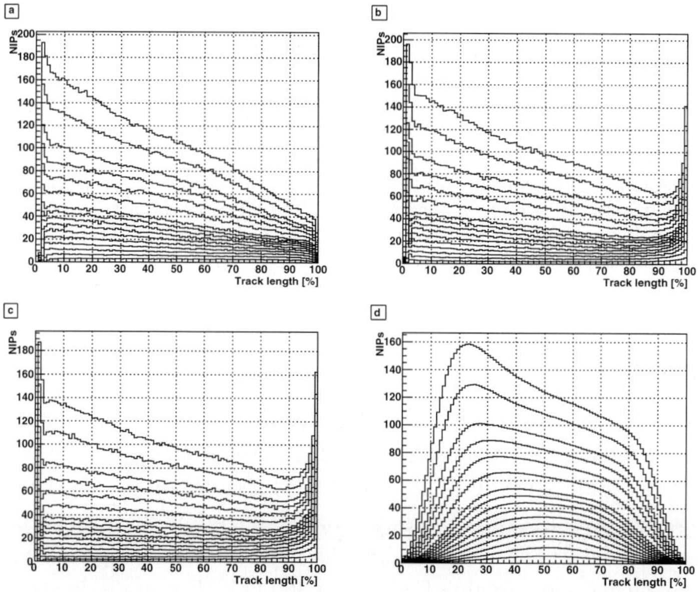

Fig. 7. The same as in Fig 6 for Carbon tracks.

Intermediate distributions of the charge projected onto a line between the track start and end point were studied. As one can see in (b) these still have negative slope but the sharp rise at the end is now clearly visible. The shape of distributions with projection onto the axis of the initial direction of motion is shown in (c). For Sulfur this is almost flat with a very sharp rise at the end, while for Carbon it remains with negative slope. An example of the impact of the diffusion caused by 0.5 m long drift, the maximum likely in DRIFT II, can be seen in Figs. 6d and 7d.

For diffused distributions the Head-Tail effect was studied by calculation of the mean position of the distribution as a function of ion energy and drift length. Fig. 8 shows the position of the mean along Carbon and Sulfur ion tracks with the head-tail effect dependent mainly on the ion energy. Points above 50% are for ion tracks with greater 'head' than 'tail' and those below for ions with the reversed effect.

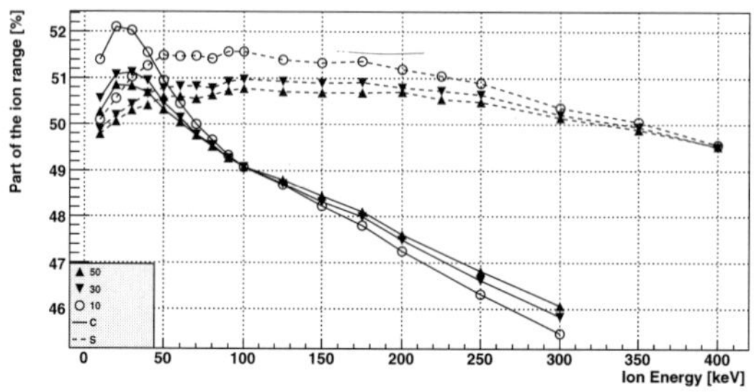

Fig. 8. Position of the mean of the charge distribution along Carbon (solid line) and Sulfur (dashed line) ion tracks showing the Head-Tail effect as a function of energy and for different maximum drift distances: 10 (○), 30(▼) and 50 (▲) cm. Points above 50 % indicate that more ionization charge is created at the end (Head) while those below that level, at the beginning (Tail) of the track.

4. Conclusions

In this work preliminary results of the Head-Tail effect of Carbon and Sulfur ions in low pressure CS_2 gas used in DRIFT detectors have been presented. Results were obtained using SRIM and TRIM simulation packages together with W-values available from existing data above ion energies of 20 keV. For lower ion energies W-values have been extrapolated using a polynomial fit to the existing data ranging from 20 to 300 keV. Results indicate the existence of the Head-Tail effect resulting respectively for Carbon and Sulfur greater Head for energies below 70 and 350 keV, above which Tail starts to dominate. As shown Head-Tail effect depends on the W-value which below 20 keV may be greater than values used in this work. This could impact significantly on the number of ionisation charge at the Head of the track and could cause the Tail to become dominant at lower energy.

References

1. D. P. Snowden-Ifft, *Results from DRIFT-IIa*, presentation at 7th *UCLA Symposium on Sources and Detection of Dark Matter and Dark Energy*, Los Angeles 2006.
2. J. Ziegler, SRIM 2006: *The Stopping and Range of Ions in Matter.*
3. D. P. Snowden-Ifft *et* al., *Nucl. Instr. Meth.* **A 498**, 155 (2003).
4. ESTAR: stopping power and range tables for electrons, http://physics.nist.gov/PhysRefData/Star/Text/ESTAR.html
5. G. L. Cano *et* al., *Phys. Rev.* **169**, 277 (1968).
6. S. Burgos *et* al., *Nucl. Instr. Meth.* **A** Available online 13 October 2007, In Press; arXiv:0707.1758

DIRECT SEARCH FOR SPIN-DEPENDENT WIMPS WITH CaF_2 DETECTOR

R. HAZAMA*, I. OGAWA, K. MUKAIDA, S. AJIMURA, Y. HIRANO,
K. ICHIHARA, Y. ISHIKAWA, M. ITAMURA, A. KATSUKI, K. KISHIMOTO,
K. MATSUOKA, H. MIYAWAKI, T. NITTA, H. SAKAI, S. SHIOMI, N. SUZUKI,
Y. TANAKA, Y. TATEWAKI, S. TOMII, S. UMEHARA, D. YOKOYAMA,
S. YOSHIDA, T. KISHIMOTO

Department of Physics and Laboratory of Nuclear Studies, Osaka University,
Toyonaka, Osaka 560-0043, Japan
**Present address: Graduate School of Engineering, Hiroshima University,*
Higashi-Hiroshima, Hiroshima 739-8527, Japan
E-mail: hazama@hiroshima-u.ac.jp

A CaF_2 scintillator complex system (ELEGANT VI) is developed to search for the axial-vector coupled dark matter (WIMPs) and study the neutrino-less double beta decay of ^{48}Ca. Active light guides of pure CaF_2 crystals which are on both sides of the central CaF_2(Eu) crystal act as a 4π active shield, combined with surrounding CsI(Tl) scintillators. The whole system is in operation at the underground laboratory (Oto Cosmo Observatory) located in Nara. In this article recent several improvements and our current status of the investigation are described, especially putting emphasis on the radioactive impurities of CaF_2 crystal and development of flash scaler/trigger system with FPGA (Field Programmable Gate Array) to do a single photon counting and a noise-cut via photon time distribution.

Keywords: WIMPs; CaF_2; Dark Matter search; Scintillator.

1. Introduction

The nature of dark matter, and the associated "missing mass problem", is one of the fundamental cosmological issues of modern astrophysics and its direct existence was recently boosted by several observations.[1] Direct detection of WIMPs(Weakly Interacting Massive Particles) dark matter by elastic nuclear recoil was suggested by Goodman and Witten in 1985.[2] Since then, various experiments, especially in conjunction with a progress of low level (rare) nuclear spectroscopy such as double beta decays, devoted their efforts to search this most promising candidate of particle dark matter (the Lighest Super-Particles(LSP) predicted by SUSY GUT theory like Neutralino), which will constitute most fraction of matter components of missing mass of the universe. Because of the low event rate (the cross section will be an order of 10^{-5} (for spin-dependent(SD)) ~ 1 (for spin-independent(SI)) pb) and the tiny energy deposition with only several keV, this is very challenging experiment. The scintillator approach has a great advantage to be able to utilize large detector mass which allow the use of the annual modulation signature, the ratio cost/mass is reasonably low (besides usually no cooling needed), and quite low radioactive backgrounds is achievable as well. In addition there is a wide choice of target nuclei. Scintillators include non zero spin target nuclei such as ^{23}Na, ^{127}I, ^{19}F and ^{129}Xe. Among WIMPs, vector-coupled (SI) one has been extensively studied so far, owing to the fact that coherence of scattering amplitude of all nucleon in the nucleus is expected, while axial-vector coupled (SD) one couple only to nuclei with non zero spin and several notable advantages of this SD interaction channel have been emphasized.[3]

2. Experimental setup

2.1. *CaF_2 scintillator complex and ELEGANT VI system*

As a target-nucleus, fluorine is suggested in particular to have the best properties for investigation of WIMP-nucleon SD interactions and to be favored due to the fact that the search for proton-sensitive SD candidates is in contrast somewhat neglected.[4] As target-detectors, several fluoride scintillating compounds are available such as CaF_2, CsF and BaF_2. In this comparison, light output, mass fraction of ^{19}F and hygroscopicity are criteria. The characteristics of these scnitillators are summarized including new scintillators under R&D efforts in the Table 1 in the previous publication.[5] Thus, CaF_2(Eu) crystal is most favorable scintillator among fluoride scintillating compounds.

One unit of the CaF_2 scintillator-complex consist of the combination of two kinds of CaF_2 crystals. One is central crystal made of Europium doped CaF_2(Eu) scintillator and the others are side crystals made of pure CaF_2 scintillators. The CaF_2(Eu) crystal with dimension of 45 mm cube has light guides of pure CaF_2 crystal with a size of 45 $\times$45 $\times$ 75mm^3, which are also sensitive to gamma rays. It is noticed that the side crystals made of pure CaF_2 act as not only light guide but also active shield against the background from outside materials, especially from Photomultiplier tube(PMT)s. The scintillation is viewed by two PMTs through light guides(CaF_2(pure)) of both sides of the crystal. Scintillation light from pure CaF_2 is dominantly in ultra violet(UV) region. On the other hand, scintillation light from central crystal CaF_2(Eu) is mainly normal visible light. The transmittance of CaF_2(Eu) is limited only for normal visible light, while pure CaF_2 transmit UV region in addtion to normal visible light. Thus central CaF_2(Eu) acts as the filter from the scintillation light from pure CaF_2. PMT is Hamamatsu H3178 assembly with the quartz window modification, which is essential to observe UV region of the scintillation light. Optical grease of KF96H (Shin'etsu silicon Co. Ltd.), which is transparent to UV light, was used for the connection between the PMT and the scintillation crystal. The refractive index is 1.44, 1.47, and 1.40 for CaF_2(Eu), CaF_2(pure) and the silicon oil, respectively, then we do not need to care reflection at the boundary. Taking advantage of the difference of the scintillation light between Eu doped CaF_2 and pure CaF_2, the difference of the pulse height between the both sides of PMTs are enhanced and the fired position in the scintillator-complexes are clearly identified.

The whole detector system (ELEGANT VI) consists of 25 CaF_2 complex modules and 40 CsI(Tl) scintillators with a size of $6.5 \times 6.5 \times 25\text{cm}^3$ for each crystal to achive 4π active shield. They are chosen to make active shield as hermetic as possible since no container is necessary due to non-deliquescence. CaF_2 detectors are divided into relatively small segments. True event is confined in a small region in space, on the other hand background events are supposed to be spread in space. Therefore small segmentation is effective to reduce background. We have installed two independent passive shielding systems. The first is a conventional shield consisting of an outer 10 cm layer of low activity lead and an inner 5 cm layer of OFHC copper. Inside these passive shields, there is an air-tight box to purge out radioactive radon gas in the air. The second is a 0.6 mm layer of Cd sheet and a 15 mm layer of LiH- loaded paraffin so as to shielding against the neutrons. The whole system is shielded by a 30cm thick boron-loaded water tank moderator. The whole system is in operation at the underground laboratory (Oto Cosmo Observatory) located Nara, Japan with 1400 m.w.e..

2.2. *System reorganization*

In order to reduce the background from PMTs further, the active light guide (pure CaF_2) of 75 mm length will be replaced with that of 200 mm length and some PMTs of CsI are changed with K-free PMTs. After the event selections by the 4π active shields, the remaining backgrounds mainly come from internal radioactive impurities. Delayed coincidence methods were used to measure directly the contamination inside the CaF_2(Eu) crystals. Our detector system has 'delayed-α trigger' to record a sequential second signal coming within 500 μs after the main (first) trigger. We can utilize ^{214}Bi $\rightarrow^{214}$Po ($T_{1/2}$=164μs) $\rightarrow^{210}$Pb for the uranium chain (U). In the thorium series (Th), we can use the time stamp of each events for the analysis with setting the software time window to be 0.5 or 1.0 sec for ^{220}Rn $\rightarrow^{216}$Po ($T_{1/2}$=145ms) $\rightarrow^{208}$Pb. The obtained U and Th contents for each crystals are shown in Fig. 1. As can be seen, two crystals (#1&#25) were highly contaminated, then these two crystals were removed from our ELEGANT VI system, we selected best nine low radioactive crystals and replaced central nine CaF_2(Eu) crystals by these low radioactive crystals, which are denoted by filled circle in Fig. 1.

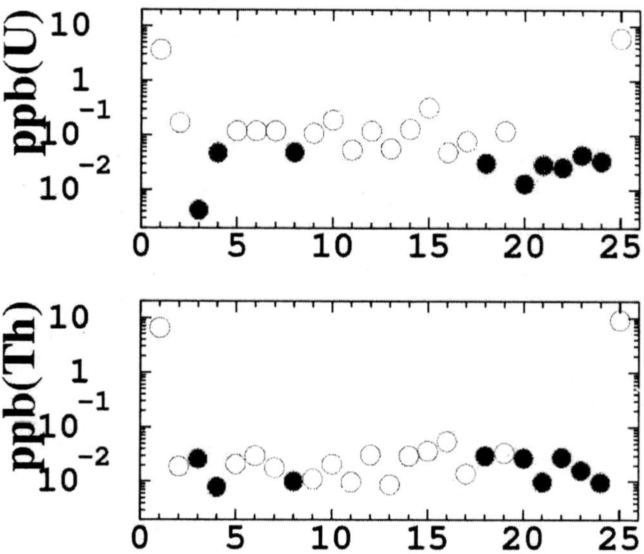

Fig. 1. Measured values of radioactive contamination for each crystals denoted by numbers(ID) from #1 to #25. Upper is for U-chain and Lower is for Th-chain impurities in the unit of ppb.

Reduction of internal radioactivities is a critical issue for low-background WIMPs search experiment. Thus, as for a relative comparison, the contamination levels of internal U/Th-chain inside of several CaF_2 scintillation crystals are summarized in Table 1.[a]

[a]It is noted that recent KIMS experiment with CsI(Tl) crystal scintillators (total mass of 34.8 kg) obtained ~0.75 ppt and ~0.38 ppt for U and Th mass concentration, respectively.[6]

Table 1. Contamination levels of U-chain and Th-chain for several CaF_2 crytals.

	Mass	U-chain (mBq/kg)	Th-chain (mBq/kg)	Ref.
ELEGANT VI CaF_2(Eu:0.17%)	6.7 kg	1.2(average) 0.052(best)	0.11(average) 0.036(best)	present work[7]
CANDLES III CaF_2(pure)	300 kg	0.039(average) < 0.02(best)	0.026(average) < 0.01(best)	8
DAMA Bicron-1	370 g	0.1	0.05	9
CaF_2(Eu:0.5%) Bicron-2	370 g	1.2	0.5	10
Tokyo-Kamioka CaF_2(Eu:0.5%)	310 g	~ 0.13*	~ 0.09*	11

Note: *They have not measured their actual U/Th contamination at all. They assumed their purity as the level of our CANDLES's undoped crytals such as 3.1ppt(3.8μBq/kg) and 23ppt(92μMq/kg), but this is just their assumption and there are no evidence. In order to estimate their actual crystal's U/Th impurities, the same technique which was developed for our ELEGANT VI was appiled for their spectrum. By looking at their energy region around 1 MeV, where there is indeed prominent alpha peaks. Their origin is attributed to decays of ^{238}U and ^{232}Th and their daughters. The observed spectrum has been fitted with the hypothesis of a mixture of U and Th chains, with ^{226}Ra out of equilibrium. Their simulation of internal U and Th backgrounds is indeed underestimated and almost all background in the low energy region will be dominated by internal U and Th backgrounds. Eu contribution is also about one sixth due to about ten years of underground storage (half) and one third of mass fraction.

2.3. *Flash scaler and low energy signal*

It is very important to treat properly the detector signals in the low energy region. A signal of one photoelectron is generated by an LED mounted on the crystal. It is found that one photoelectron energy of the CaF_2 scintillator-complex corresponds to 0.6 keV. The low energy signal like a few keV from CaF_2 scintillators looks like a bundle of many one photon signals in a long decay time(~ 1 μs). We developed a flash scaler (FS) which measures time distribution of one photon signals. The dwell time of FS is set to be 100 ns, which is provided by external 10 MHz clock and each channel has 8 bits scaler. The signals from the PMTs are amplified, fed to fast 300 MHz discriminator in order to avoid double pulse counting,[12] and recorded by FS. It is noted that the linearity between FS and ADC is verified until about 30 keV. The trigger is formed in the FPGA chip on the homemade UMEM (Universal MEmory Module) in a VME crate, which was originally developed for BNL E930 experiment. For low energy events, the trigger is started by single photoelectron and it is required to have at least one photoelectron for both Left and Right PMT, in addition to three photoelectrons in four μs to form an event trigger.

Instead of measuring charges from PMTs by ADCs, direct counting of the number of photoelectrons enables us to be free from the fluctuation on the pulse height of one photoelectron. With the advantage of statistically estimating the number of events occured at central CaF_2(Eu), left CaF_2(pure) and right CaF_2(pure) by utilizing three simple binomial function, we can fit a roll-off-ratio spectrum exactly without no uncertainty, which is shown in Fig. 2. Thus further separation of events occurred at CaF_2(Eu) from CaF_2(pure) can be achieved, compared with the use of a conventional ADC.[13] Preliminary test shows about a factor of 5 improvement for the background reduction, which is also shown in Fig. 2.

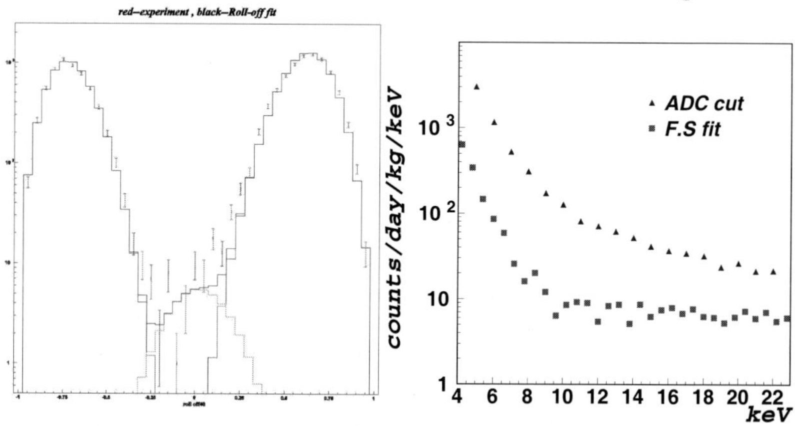

Fig. 2. LEFT: A typical roll-off-ratio spectrum generated from the flash scaler. In this case, total number of photoelectrons is 40. The solid line shows the fitting spectrum from three binomial functions. RIGHT: Energy spectrum of CaF_2(Eu) for #12 crystal, after roll-off cut by using conventional ADC's (triangle) and developed flash scaler (square).

3. Analysis

We analyzed the data after the reconstruction of ELEGANT VI system with introduction of FS and new trigger system, which was carried out in 2003. The current total analyzed live time after these improvements is 199.3 days (since 2004). Thus, we obtained 462.4 kg·d WIMP search data with the central eight crystals [b] , which were chosen as low radioactive crystals based on the GEANT simulation using each measured U/Th impurities in Fig. 1 to estimate the effect of U/Th chain in the low energy region.[5] The energy calibration was done by using ^{241}Am and ^{137}Cs every ~3 weeks and in between the gain variation was checked by using LEDs.

[b]Unfortunately, one of central module (#14) cannot be utilized for the analysis owing to a malfunction of its PMTs.

Off-line analysis to suppress the background and select the local events occurred at the CaF_2(Eu) crystal are done by setting the following conditions; i) no events in CsI(Tl) veto detectors, ii) only one CaF_2 scintillator complex records the event, iii) no following event recorded by 'delayed-α trigger', iv) roll-off cut, v) Time cut, vi) ADC cut. We extracted the number of events occurred at CaF_2(Eu) crystal by fitting the roll-off ratio spectra of ADC, not of FS with the binomial functions, yet. The PMT noise, which was also identified without the crystal, is seen by both PMTs which have very fast timing characteristics and are observed as a peak around zero asymmetry of time difference distribution of photoelectrons (T_{diff}) obtained by FS. A similar noise was suggested to be a spark in the dynode structure.[14] The continuum event orginated from accidental coincidence of dark noise is also seen. Probability distribution ($P(T)$) of the asymmetry between the signals from two PMTs was checked for the reference distribution using

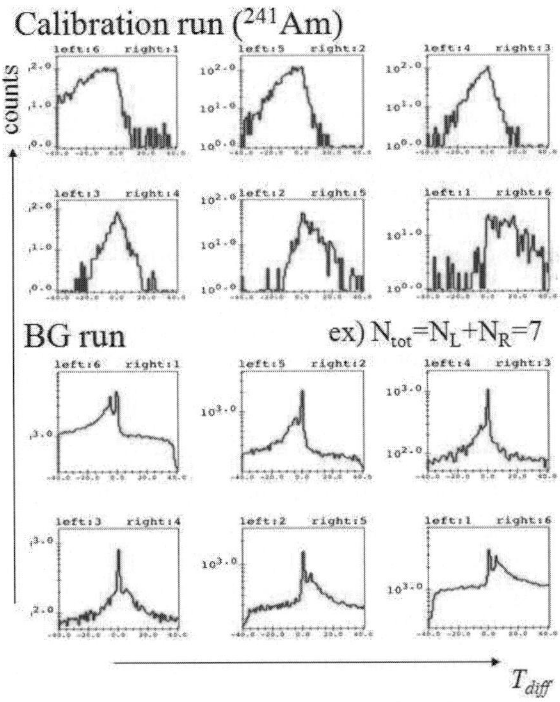

Fig. 3. Time difference distribution of photoelectrons (T_{diff}) obtained by FS for the reference distribution using ^{241}Am source (upper plot) and for actutal physics run (background: BG) (lower plot). This is the case that total number of photoelectrons is seven. Thus, there are only six combinations of ($N_L(N_R)$, $N_R(N_L)$)=(6,1), (5,2), and (4,3).

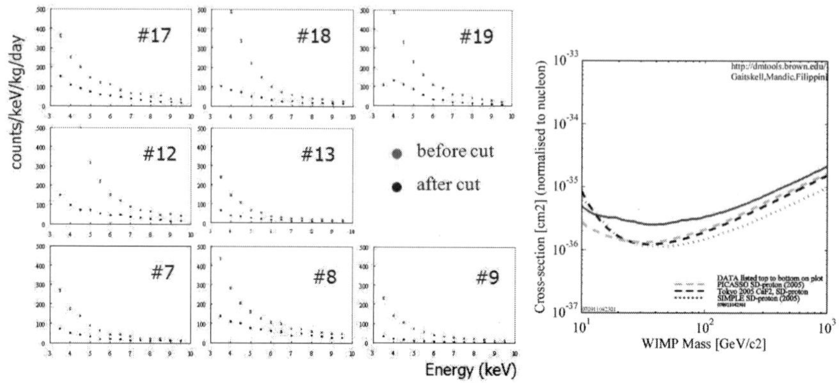

Fig. 4. LEFT: The energy spectra before and after cuts (Time, ADC, and roll-off by using ADC, not FS's binomial) for the central 8 CaF_2(Eu) crystals. RIGHT: Our present preliminary result (thick) for spin-dependent WIMP-nucleon elastic scattering; the exclusion plot is at 90 % C.L. It is compared with PICASSO,[15] Tokyo,[11] and SIMPLE[16] result.

^{241}Am source and its fitted decay time constant (τ) was utilzed to optimize the S/N ratio to cut these non-physics events (Time cut), which is shown in Fig. 3.

$$P(T) = \frac{N_L N_R}{\tau(N_L + N_R)} exp \frac{N_R}{\tau} T (T < 0), \frac{N_L N_R}{\tau(N_L + N_L)} exp \frac{N_L}{\tau} T (T > 0) \quad (1)$$

where total number of photoelectrons are constant ($N_{total} = N_L + N_R$). Additionally the PMT noise, which also induces an abnormally big pulse height with a narrow pulse width, is rejected by utilizing two-dimensional plot of ADC vs. number of photoelectrons (ADC cut).

4. Preliminary Result

The obtained energy spectra by using these cuts are shown in Fig. 4 and the preliminary exclusion plot, which is envelope plot of eight assembled one, for the WIMP-proton cross section on SD WIMPs by present measurement are also shown in Fig. 4, compared with the ^{19}F results by PICASSO,[15] Tokyo-Kamioka,[11] and SIMPLE[16] experiment. We expect the best sensitivity on ^{19}F can be obtained by using our best crystal and our higher quenching factor of ^{19}F [c], especially by roll-off cut with FS's binomial fitting. Much further reduction (about a factor of 5) is expected by fitting roll-off ratio

[c]Our value is 0.11-0.20.[17]

with an exact binomial function and even though by using conventional ADC's data, the current sensitivity is almost comparable with the world best result. Besides, further study is in progress to identify the source for the low energy background by doing simulation and measurement.

Acknowledgments

The authors thank Prof. Ejiri, for valuable discussions, and Prof. H. Ohsumi, Prof. K. Fushimi, Dr. N. Kudomi, Dr. K. Kume, the ELEGANTS group of Osaka University for the help during the run. The present work is supported by the Grant-in-Aid of Scientific Research and Ministry of Education, Science and Culture, Japan. The work of R. H., H. M., S. U., S. Y. were supported by JSPS Research Fellowships for Young Scientists. The work of R. H. partly was supported by the Sasakawa Scientific Research Grant from The Japan Science Society.

References

1. D. Clowe *et. al.*, astro-ph/0608407; M.J. Jee *et. al.*, *Astrophys. J.* **661**, 728 (2007); M.E. Peskin, *J. Phys. Soc. Jap.* **76**, 111017 (2007).
2. M. W. Goodman and E. Witten, *Phys. Rev.* **D31**, 3059 (1985).
3. J.I. Collar *et. al.*, *Phys. Rev. Lett.* **85**, 3083 (1988); T. Falk, A. Ferstl, K.A. Olive, *Astropart. Phys.* **13**, 301 (2000); T.A. Girard and F. Giuliani, *Phys. Rev.* **D75**, 043512 (2007).
4. V.A. Bednyakov *Phys. Part. Nucl.* **38**, 326 (2007); V.A. Bednyakov and F. Simkovic *Phys. Rev.* **D72**, 035015 (2005); V.A. Bednyakov and F. Simkovic *Phys. Part. Nucl.* **37**, S106 (2006); P.C. Divari *et. al.*, *Phys. Rev.* **C61**, 054612 (2000).
5. R. Hazama *et. al.*, Proc. KEK-RCNP International School and Miniworkshop for Scintillating Crystals and their Applications in Particle and Nuclear Physics, KEK, Japan, Nov. 2003, KEK Proceedings2004-4 131 (2004).
6. H.S. Lee *et. al.*, *Nucl. Instrum. and Meth.* **A571**, 644 (2007).
7. I. Ogawa *et. al.*, *Nucl. Phys.* **A730**, 215 (2004).
8. S. Umehara *et. al.*, *J. Phys.Conf. Ser.* **39**, 356 (2006).
9. R. Bernabei *et. al.*, *Astropar. Phys.* **7**, 73 (1997).
10. P. Belli *et. al.*, *Nucl. Phys.* **B563**, 97 (1999).
11. Y. Shimizu *et. al.*, *Phys. Lett.* **B633**, 195 (2006).
12. H. Sakai, Master Thesis, Osaka University (2002).
13. K. Mukaida, Master Thesis, Osaka University (2003).
14. H.S. Lee *et. al.*, *Phys. Lett.* **B633**, 201 (2006); B. Ahmed *et. al.*, *Astropar. Phys.* **19**, 691 (2003).
15. M. Barnabe-Heider *et. al.*, *Phys. Lett.* **B624**, 186 (2005).
16. T.A. Girard *et. al.*, *Phys. Lett.* **B621**, 233 (2005).
17. R. Hazama *et. al.*, *Nucl. Instrum. and Meth.* **A482**, 297 (2002).

SEARCH FOR NEUTRALINO DARK MATTER WITH THE AMANDA NEUTRINO TELESCOPE AND PROSPECTS FOR IceCube

A. RIZZO* (for the IceCube Collaboration)

Interuniversity Institute for High Energies, Vrije Universiteit Brussel, Brussels, B-1050, Belgium

**E-mail: alfio.rizzo@vub.ac.be, http://w3.iihe.ac.be*

If non-baryonic dark matter exists in the form of neutralinos, a neutrino flux is expected from the decay of neutralino pair annihilation products inside heavy celestial bodies. Data taken with the AMANDA-II neutrino telescope located at the South Pole, have been used in a search for this indirect dark matter signal. Results are presented from searches for neutralinos accumulated in the Sun, using AMANDA-II data from 2001, and in the centre of the Earth, using AMANDA-II data from 2001 to 2003. Future perspectives, in view of higher statistics data samples acquired during recent years and by the combined AMANDA-IceCube detector, are also discussed.

Keywords: Dark Matter; Neutralino; Neutrino Telescopes; AMANDA/IceCube; South Pole.

1. Introduction

The existence of dark matter is inferred from astrophysical observations that probe gravitational potentials[1,10,37] . One possible scenario to explain the mismatch between the required mass, needed to supply the derived gravitational potential, and the observed mass is represented by the Cold Dark Matter (CDM), i.e. non-relativistic massive particles (GeV or heavier) produced in the Big Bang[35] . This CDM makes up 23% of the energy density of the Universe, according to the WMAP measurements of the temperature anisotropies in the Cosmic Microwave Background, in combination with data on the Hubble expansion and the density fluctuations in the Universe[34] . The thermally averaged cross section of the dark matter (DM) particles at the freeze-out temperature is typical for weak interactions and explains why the DM does not cluster strongly in the centre of galaxies, like the baryons do. Therefore, the DM particles are generically called WIMPs,

Weakly Interacting Massive Particles[14,30] . One of the most popular and widely studied WIMP candidates is the lightest supersymmetric neutralino $\tilde{\chi}_1^0$ (or simply χ). In the Minimal Supersymmetric extension of the Standard Model (MSSM)[22] , where the multiplicative quantum number R-parity is conserved, the neutralino is the mixture of the superpartners of the B and W_3 gauge bosons and the neutral Higgs bosons, H_1^0 and H_2^0. The attractiveness of this candidate, stems from the fact that it is electrically neutral, and thus neither absorbs nor emits light, stable, and can be destroyed via pair annihilation (it is a Majorana particle), so that it may have survived since the Big Bang[13] . Consequently, relic neutralinos in the galactic halo will pass through massive bodies like the Sun or the Earth, where they can lose energy by scattering off nuclei. Over time, the neutralinos concentrate near the centres of these celestial bodies and annihilate producing Standard Model particles. The products of these annihilations will, in general, decay and produce neutrinos. The latter will be able to escape and would potentially be visible in a high energy neutrino telescope at the surface of the Earth as an excess over the atmospheric neutrino flux[21] .

In this paper we present the results of searches with the AMANDA (Antarctic Muon And Neutrino Detector Array) telescope (Sec. 2) for neutralino dark matter accumulated in the Sun (with the 2001 dataset, Sec. 4) and in the Earth (with the 2001–2003 dataset, Sec. 5). We present also the expected sensitivity for the configuration of the current IceCube detector, the one cubic kilometer neutrino telescope currently under construction at the South Pole (Sec. 2), combined with AMANDA (Sec. 6), and the expected sensitivity for the complete geometry of the IceCube detector (IceCube-80, Sec. 7) as well. We discuss furthermore current improvements and preliminary results from ongoing analyses on higher statistics data samples accumulated during recent years (Sec. 8).

2. The AMANDA and IceCube Detectors

AMANDA-II, the final stage of the AMANDA detector, has been operating since January 2000 with 677 optical modules (OMs) attached to 19 strings[8] . Most of the OMs are located between 1500 m and 2000 m below the South Pole ice surface; each OM consists of an 8 inch photomultiplier tube (PMT) enclosed in a glass sphere. The strings are arranged in three approximately concentric circles of 60 m, 120 m and 200 m diameter respectively. The previous configuration of AMANDA called AMANDA-B10, the inner core of 302 OMs on 10 strings, operated between 1997 and 1999. Muons from charged-current neutrino interactions near the array are detected by the

Cherenkov light they produce when traversing the ice. The PMT signals are amplified and sent to a majority logic trigger. Since 2001, an additional lower multiplicity trigger (referred to as "string trigger") is operational, which exploits not only the temporal information but also the space topology of the hit pattern. A detailed description of the reconstruction techniques used in AMANDA is given in Ref. 5 .

IceCube builds on the successful deployment and operation of the AMANDA telescope. In addition to its larger volume (about a cubic kilometer), IceCube differs from AMANDA in a significant way since the signals are digitized inside each optical module, to minimize loss of information from degradation of analog signal sent over long distances[3] . The IceCube neutrino observatory, the completion of which is planned for 2011, will consist of 4800 Digital Optical Modules (DOMs) installed on 80 strings between 1450 m and 2450 m below the surface[3,6] . The strings are deployed in a triangular grid pattern with a characteristic spacing of 125 m enclosing an area of one kilometer square. The In-Ice array is complemented by the IceTop surface shower array that can be used for calibration, for tagging of certain backgrounds and for cosmic-ray studies. The current (2007) IceCube configuration consists of 22 strings (referred to as "IceCube-22").

3. Signal and Background Simulations

The simulation of the neutralino-induced neutrino signal was performed using the program `WIMPSIMP`[11] based on `DARKSUSY`[25] for a sample of neutralino masses (50, 100, 250, 500, 1000, 3000 and 5000 GeV/c^2). Two annihilation channels were considered in our analysis for each mass tested, $\chi\chi \to b\bar{b}$ (referred to as "soft" channel) and $\chi\chi \to W^+W^-$ (referred to as "hard" channel, replaced with $\tau^+\tau^-$ in case of $m_\chi < m_W$). The neutrino-nucleon interactions in the ice around the detector, producing the detectable muon flux, were simulated with `PYTHIA`[33] . The simulated zenith angular range for the Solar WIMP analysis (Sec. 4), was restricted from 90° to 113°, since the muons induced by neutralinos can be distinguished, from the dominant atmospheric muon background, when the Sun is seen below the horizon at the South Pole. The zenith angle for the Earth WIMP analysis (Sec. 5) was simulated close to 180°, taking into account that light neutralinos annihilate less closely to the centre of the Earth than heavy neutralinos. The background for this search is represented by upward-going atmospheric neutrinos and mis-reconstructed downward-going atmospheric muons. The atmospheric neutrino flux was simulated according to Ref. 31 with `ANIS`[24] between 10 GeV and 10 PeV for zenith angles between 70° and 180°. The

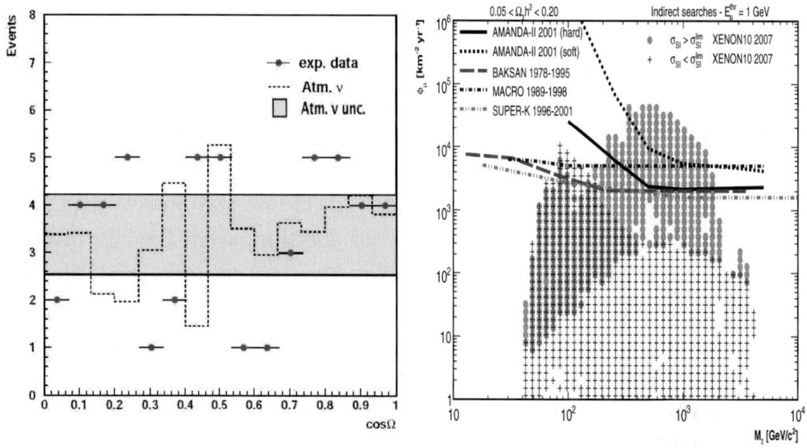

(a) (b)

Fig. 1. Figure 1(a) shows the distribution of the $\cos\Omega$, the space angle between the reconstructed tracks and the Sun's direction (m_χ=500 GeV/c^2). The data are represented by the dots and the atmospheric neutrino background by the dotted line. The shaded band represents the uncertainty in the expected atmospheric neutrino background. Figure 1(b) shows the upper limits (90% CL) on the muon flux as function of the neutralino mass, compared with other searches in the phase space of relevant neutralino models (+) and disfavoured models by XENON10[9] (•).

simulation of the atmospheric muons was made with CORSIKA[27] using the South Pole atmosphere parameters and protons as primaries. The sample was generated isotropically between zenith angles 0° and 85°, and with proton energies, E_p, between 600 GeV and 10^{11} GeV, with a differential energy distribution $\propto$ to $E_p^{-2.7}$.

4. To Catch a Neutralino from the Sun

The data set used contains 8.7×10^8 events collected in 143.7 days of effective live-time during 2001. The cuts on the variables used in this search, have been optimised independently for each mass in order to exploit the differences in the resulting muon energy spectra. The event selection was optimised in a "blind" way by explicitly excluding the time stamp of the event, and thus the Sun direction, in the development of the selection criteria (see Ref. 32 for details). Several quality criteria on the reconstructed track were adopted in order to select the events; which resulted in a rejection of 99.97% of the data (mostly downward-going muons), while the simulated background from atmospheric neutrinos was reduced to 30%. The simulated signal was reduced to between 46% and 54%, depending on the neutralino mass. The next step was a cut on a neural network (NN) output; the NN was optimised for the simulated signal and background sample (only atmospheric muons). The final event selection consisted in another series of cuts on the quality of the reconstructed track. In the search for a signal, a cone around the direction of the Sun and containing 90% of the expected signal was used. To estimate the expected background at the Sun position, the off-source data, in the declination band of the Sun, were used in order to further eliminate effects of uncertainties in the background simulation and in the total normalisation of the atmospheric neutrino flux (the relevant numbers are summarised in Refs. 4,32). Figure 1(a) shows the distribution of the cosine of the space angle Ω between the reconstructed tracks and the direction of the Sun (after un-blinding). This distribution is in agreement, both in shape and normalisation, with the expected atmospheric neutrino background, within the 25% overall systematics (the shaded band in the plot) on the neutrino flux[23] . Since no evidence of a statistically significant excess of events from the Sun direction was found, the upper limits on the muon flux at the Earth were calculated. Figure 1(b) shows the AMANDA limits on the muon flux from neutralino annihilation in the "hard" and "soft" channels in the Sun, as a function on the neutralino mass, compared to other indirect limits[7,16,19] and to theoretical predictions from MSSM models.

5. To Catch a Neutralino from the Earth

The data collected during the 2001–2003 period correspond to 688.0 days of live-time of the detector, with a total of 5.3×10^9 events. This search[18] besides the standard multiplicity trigger, makes use of the low energy string trigger, and it focuses on low mass WIMP models ($m_\chi \leq 250$ GeV/c^2) in order to achieve an improvement with respect to a previous analysis with AMANDA-B10[2] . The string trigger plays a key role in this new analysis.

The "blind" procedure for this search, consists in developing the analysis on only 20% of the data, and then deriving the results using the remaining 80% data sample. Because of an upgrade of the data acquisition system (DAQ) and change in the string trigger settings between 2001 and 2002, the optimisation was performed separately for 2001 and for the two combined years 2002–2003. The characteristics of the signal differs depending on the neutralino mass under examination. Therefore, the cuts are tuned for individual masses. In the first filtering step, upward-going events are selected. This reduces the data volume by two orders of magnitude. Then,

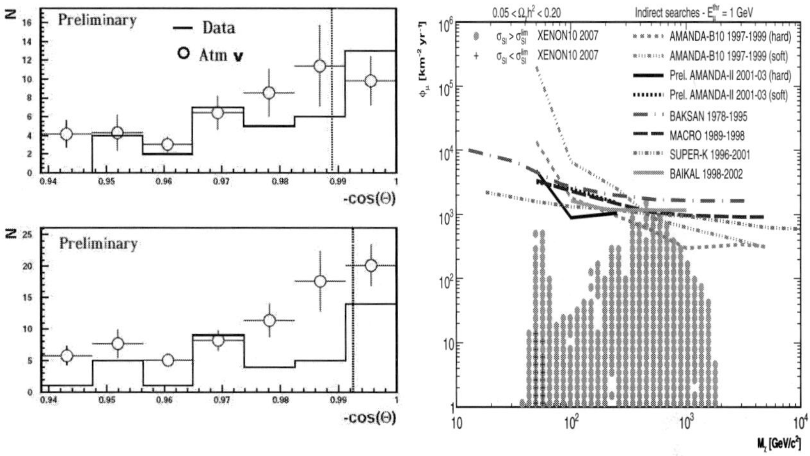

(a) (b)

Fig. 2. Figure 2(a) shows the distribution of the cosine of the zenith angle of the data, represented by the solid line, and atmospheric neutrino background, represented by the open circles. The mass is m_χ=100 GeV/c^2 in "soft" (top) and "hard" (down) channels. The dotted line indicates the location of the final cut. Figure 2(b) shows the upper limits (90% CL) on the muon flux as function of the neutralino mass, compared with other searches in the phase space of relevant neutralino models (+) and disfavoured models by XENON10[9] (•).

another selection on the zenith angle ($\theta > 140°$) is performed; following this, the amount of remaining data is only slightly more than 1% of the original triggered events. The sample, which is still dominated by mis-reconstructed downward-going atmospheric muons, needs to be reduced by a more refined method that consists of a series of sequential cuts (depending on the neutralino mass and annihilation channel) on the variables that form the selection criteria of the analysis. At the final stage the entire data (dominated by the atmospheric neutrinos) of the three years are combined to evaluate and to apply the final cut selection on the zenith angle. Figure 2(a) shows the distribution in cosine of the zenith angle of the 80% data sample and of the expected atmospheric neutrino background. The data and simulated atmospheric neutrino background are in reasonable agreement. Without significant excess of vertical tracks being observed, the upper limits on the muon flux from the centre of the Earth were calculated for each of the neutralino masses and annihilation channels. Figure 2(b) shows the AMANDA muon flux limits compared with other indirect searches and with theoretical predictions from MSSM models.

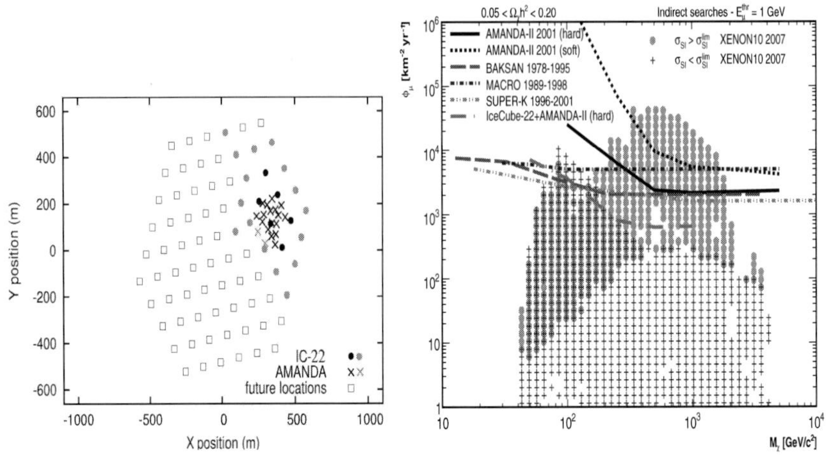

(a) (b)

Fig. 3. Figure 3(a) shows the top view of the 2007 IceCube. Strings are represented by the full circles in gray and black, including the 19 AMANDA strings represented by the crosses in gray and black. The inner strings, represented by the full circles and crosses in black, define the fiducial region, while the outer, represented by the full circles and crosses in gray, define the veto region. Uppermost OMs of the fiducial strings belong to the veto region. The open squares represent the future location of the IceCube strings. Figure 3(b) shows the 2007 expected sensitivity, for half a year of detector live-time, to muons from solar WIMP annihilation compared to other limits in the phase space of relevant neutralino models (+) and disfavoured models by XENON10[9] (•).

6. Expected Sensitivity for IceCube-22 Combined with AMANDA

The combined IceCube-22/AMANDA detector of 2007 consists of 41 strings. The sensitivity was estimated for half a year of detector live-time through MonteCarlo simulation[36] . The studied neutralino masses (m_χ = 50, 100, 250, 500 and 1000 GeV/c^2, "soft" and "hard" spectrum) were simulated with a new code called `WIMPSIM`[15] , which includes neutrino oscillations. In order to reject the atmospheric muon background, we searched for the "contained" events, i.e. neutrino events with the charged-current vertex inside a fiducial volume (see Fig. 3(a)), or for the events where first OM hit in the veto region came later than the OM hits in the fiducial region. The event selection started with a cut on the reconstructed zenith angle ($\theta \in [90°, 120°]$); a NN was trained and tested with the WIMP events and atmospheric muon background, and a cut on the NN output was applied. This was complemented by a cut on the hit multiplicity and

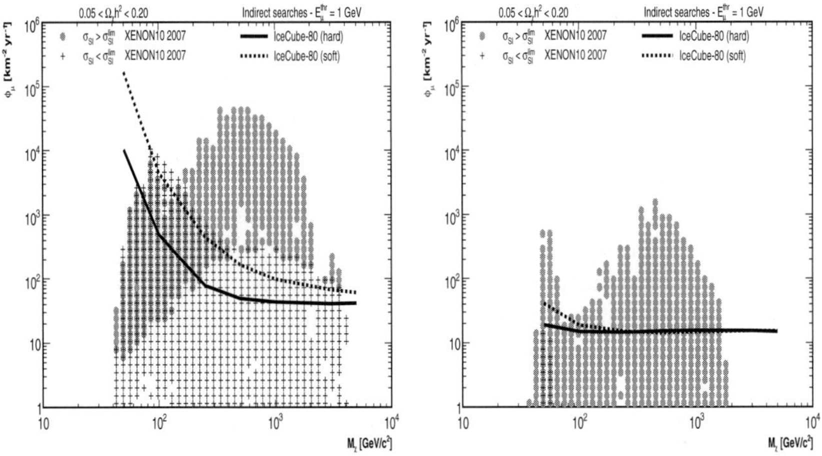

(a) (b)

Fig. 4. Figure 4(a) shows the predicted muon flux,with 5 years of detector live-time, for Solar WIMPs with IceCube-80. Figure 4(b) shows the predicted muon flux, with 10 years of detector live-time, for Earth WIMPs with IceCube-80.

on the reconstruction quality. After these cuts, only WIMP signal and atmospheric neutrino events remained; then events were selected in the Sun's direction within a search cone with the half opening angle between 3° and 10°. Combining all the information, the mean expected upper limit on the muon flux from neutralino annihilations in the Sun was calculated, which is a measure of the detector's WIMP sensitivity. Comparing these values with the AMANDA published limits for solar WIMP analysis[4] , we see (Fig. 3(b)) improvements of almost an order of magnitude for lower WIMP masses.

7. Expected Sensitivity for IceCube-80

The sensitivity of the complete geometry of the IceCube detector (IceCube-80) for neutrinos from WIMP annihilations in the Sun and in the Earth has also been studied[20] . In the analysis the simulation of noise was included, but the question of background rejection was not addressed. Hence these results can be viewed as a limit on the performance for perfect background rejection. The simulation of the signal ($m_\chi = 50, 100, 250, 500\ 1000$, 3000 and 5000 GeV/$c^2$, "soft" and "hard" spectrum) was performed with `WIMPSIMP`[11] and with `PYTHIA`[33] . Cuts on the quality of the reconstruction have been applied. Using an effective volume of 1 km^3 and a live-time of

10 years for the Earth and of 5 years for the Sun, the derived sensitivity on the muon flux for the Sun and the Earth (Fig. 4(a) and Fig. 4(b)) have been estimated.
A recent simulation, performed with the new simulation code `WIMPSIM`[15] and including the background rejection, has been done. The new limits, as expected, are more conservative but more realistic.

The role of IceCube for the searches for supersymmetric dark matter is expected to be complementary to that of the direct detection experiments. This is due to the fact that such experiments mainly constrain models with spin-independent neutralino-proton couplings whereas for neutralino capture in the Sun models with significant spin-dependent couplings may be very important[26] . Also, it should be stressed that direct and indirect searches probe different epochs of the history of the solar system and different parts of the neutralino velocity distribution.

8. Summary and Outlook

Figure 1(b) and Fig. 2(b) show the AMANDA upper limits (rescaled to a common muon threshold of 1 GeV) on the muon flux from neutralino annihilation in the Sun and in the Earth respectively, together with the results from other indirect searches[7,16,19] . The cosmologically relevant MSSM models are also shown, allowed (+) and disfavoured (•) by the direct search results from XENON10[9] . Compared to the previously published AMANDA results from searches for neutralinos in the Earth[2] , the analysis with 2001–2003 data benefits from the larger detector volume and the addition of the string trigger with its lower energy threshold. This makes it possible to improve the sensitivity especially for low neutralino masses; above 250 GeV/c^2 the effect is expected to be less pronounced. The new limits on the neutralino-induced muon flux are up to a factor 60 stronger than our earlier result. A similar improvement (with respect to Ref. 4) is expected for the solar neutralino analysis of 2001–2003 data, thanks to the increased detector exposure, improved reconstruction techniques and the string trigger. A preliminary analysis shows a factor 10–100 improvement of the effective volume at early analysis levels for low neutralino masses, mainly due to the inclusion of the string triggered events.

The neutralino searches will be continued with a larger set of AMANDA data from 2001–2006. Since 2007 AMANDA is embedded as a high-granularity subdetector of the IceCube neutrino telescope. This offers additional opportunities for the dark matter searches.

References

1. W. J. G. de Blok *et al.* *astro-ph/0103102.*
2. A. Achterberg *et al.* *Astropart. Phys.* **26**, 129 (2006).
3. A. Achterberg *et al.* *Astropart. Phys.* **26** 155 (2006).
4. M. Ackermann *et al.* *Astropart. Phys.* **24** 459 (2006).
5. J. Ahrens *et al.* *Nucl. Instrum. Meth.* **A 524** 169 (2004).
6. J. Ahrens *et al.* *IceCube Preliminary Design Document, The University of Wisconsin, Madison, Wisconsin* (2001).
7. M. Ambrosio *et al.* *Phys. Rev.* **D 60** 082002 (1999).
8. E. Andres *et al.* *Nature* **410** 441 (2001).
9. J. Angle *et al.* *arXiv:0706.0039.*
10. K. G. Begeman, A. H. Broeils and R. H. Sander *Monthly Notices of the Royal Astr. Society* **249** 523 (1991).
11. L. Bergström, J. Edsjö and P. Gondolo *Phys. Rev.* **D 58** 103519 (1998).
12. L. Bergström and A. Goobar Cosmology and Particle Astrophysics, *Praxing Publishing Ltd, Springer-Verlag* (2004).
13. L. Bergström *Rep. Prog. Phys.* **63** 793 (2000).
14. G. Bertone, D. Hooper and J. Silk *hep-ph/0404175.*
15. M. Blennow, J. Edsjö and T. Ohlsson *hep-ph/0709.3898.*
16. M. M. Boliev *et al.* Proc. of *Dark Matter in Astro- and Particle Physics, Dark '96*, (1997), eds: *Hans Volker Klapdor-Kleingrothaus and Y. Ramachers, World Scientific*, 711.
17. D. Chirkin and W. Rhode *hep-ph/0407075.*
18. A. Davour *Uppsala University* (2007), isbn = *978-91-554-6909-2.*
19. S. Desai *et al.* Phys. Rev. **D 70**, 083523 (2004).
20. J. Edsjö *AMANDA-IR 20000301* (2000).
21. J. Edsjö *Uppsala University* (1997).
22. J. Ellis *et al.* *Nucl. Phys.* **B 238**, 453 (1984).
23. Th. K. Gaisser and M. Honda *Ann. Rev. Nucl. Part. Sci.* **52**, 153 (2002).
24. A. Gazizov and M Kowalski *astro-ph/0406439.*
25. P. Gondolo *et al.* *Journ. of Cosm. and Astropart. Phys.* **0407**, 008 (2004).
26. F. Halzen and D. Hooper Phys. Rev. **D 73**, 123507 (2006).
27. D. Heck *et al.* *Forschungszentrum Karlsruhe Report FZKA, 6019* (1998). *http://ik1au1.fzk.de/ heck/corsika*
28. S. Hundertmark it AMANDA internal report (2000).
29. S. Hundertmark 'International Workshop on simulation and analysis methods for large neutrino telescope', Zeuthen, Germany, (1998).
30. G. Jungman, M. Kamionkowski and K. Griest Phys. Rept. **267**, 195 (1996).
31. P. Lipari *Astropart. Phys.* **1**, 195 (1993).
32. Y. Minaeva *Stockholm University* (2004).
33. T. Sjöstrand *Comp. Phys. Comm.* **82**, 74 (1994).
34. D. N. Spergel *et al.* *Astrophys. J. Suppl. Ser.* **170**, 377 (2007).
35. L. Wang *et al.* *Astrophysical Journal* **530**, 17 (2000).
36. G. Wikström (for the IceCube Collaboration), in Proc. of *30th Int. Cosmic Ray Conf. (ICRC 2007), arXiv:0711.0353.*
37. F. Zwicky *Helv. Phys. Acta.* **6**, 110 (1933).

PROSPECTS FOR THE SEARCH FOR DARK MATTER WITH GLAST

B. L. WINER
(for the LAT Dark Matter and New Physics Working Group)

Department of Physics, The Ohio State University,
Columbus, Ohio 43210, U.S.A.
E-mail: winer@mps.ohio-state.edu
www.physics.ohio-state.edu

Indirect detection of particle dark matter relies upon pair annihilation of Weakly Interaction Massive Particles (WIMPs), which is complementary to the well known techniques of direct detection (WIMP-nucleus scattering) and collider production (WIMP pair production). Pair annihilation of WIMPs results in the production of gamma-rays, neutrinos, and anti-matter. Of the various experiments sensitive to indirect detection of dark matter, the Gamma-ray Large Area Space Telescope (GLAST) may play the most crucial role in the next few years. After launch in 2008, The GLAST Large Area Telescope (LAT) will survey the gamma-ray sky in the energy range of 20 MeV-300 GeV. By eliminating charged particle background above 100 MeV, GLAST may be sensitive to as yet to be observed Milky Way dark matter subhalos, as well as WIMP pair annihilation spectral lines from the Milky Way halo. Discovery of gamma-ray signals from dark matter in the Milky Way would not only demonstrate the particle nature of dark matter; it would also open a new observational window on galactic dark matter substructure. Location of new dark matter sources by GLAST would dramatically alter the experimental landscape; ground based gamma ray telescopes could follow up on the new GLAST sources with precision measurements of the WIMP pair annihilation spectrum.

Keywords: Dark Matter, Gamma Rays, GLAST.

1. Introduction

Nonbaryonic dark matter is expected to make up ~25% of the total energy density of the universe. Empirical evidence for dark matter first came from the rotation curves of galaxies[1] where instead of falling velocities they observed constant velocities at large radial distances. More recently, weak lensing results from the collision of two galaxy clusters in the "bullet cluster" added more compelling evidence to the existence of dark matter.[2]

Beyond its existence however, almost nothing has been determined about the composition of the dark matter or its properties.

One of the currently favored models is one in which the dark matter is a weakly interacting massive particle (WIMP). Assuming supersymmetry exists, an excellent WIMP candidate would be the lightest supersymmetric particle, called the neutralino. This particle is theorized to be a Majorana particle and this property would allow for indirect detection by GLAST via pair annihilation to q-$\bar{q}$ pairs (continuum), $Z\gamma$, or $\gamma\gamma$ (lines). In these models the cross-section for the lines is suppressed and the cross-section for $\sigma(\chi\chi \rightarrow f_i\bar{f}_i) \propto m_f^2$ in the cold dark matter limit ($v \rightarrow 0$).[3] We will make use of the large energy range of GLAST to search for indirect evidence of dark matter and elucidate its particle properties.

2. GLAST Satellite

The Gamma-ray Large Area Space Telescope (GLAST) is the next generation of spaced-based gamma-ray observatory following in the footsteps of the Compton Gamma Ray Observatory that operated in the 1990's. GLAST is due to launch in the early summer of 2008 and will carry two different systems for gamma-ray detection. The Gamma-ray Burst Monitor (GBM) will cover a large portion of the visible sky and search for gamma-ray bursts in the energy range 5 keV – 25 MeV. The larger instrument, the Large Area Telescope (LAT), will cover the energy range 20 MeV – ~300 GeV. The field of view for the LAT covers approximately 20% of the sky and in normal operation it will give full sky coverage about every 3 hours. These studies presented in this paper utilize the LAT and the description below focuses solely on it.

The LAT detects gamma rays by pair production. Therefore, to reconstruct the direction and energy of a gamma-ray, the detector must provide a mechanism to induce pair production, measure the directions of the resulting electron-positron pair, and measure the energy of the pair. The detector must also provide good rejection against background sources. The LAT has three main detector subsystems along with a data acquisition and trigger system. The LAT is mounted to a satellite which provides power, telemetry control, and communications. The top and sides of the LAT are covered with a segmented Anti-Coincidence Detector (ACD). The ACD is constructed of scintillating tiles readout with wavelength shifting fibers connected to photo-detectors. The ACD provides the main rejection against cosmic-ray backgrounds because as charged particles pass through the ACD, they will deposit enough energy in the scintillator that a signal can be detected. It is important to note that the ACD is segmented so that only certain combinations of tiles can veto portions of the LAT. This avoids the "self-vetoing" problem that limited the operation of EGRET.

The main part of the LAT is a 4x4 grid of towers that each contains a tracking system and a calorimeter. The charged particle tracking system (TKR) is constructed of layers of silicon strips with alternating x-y orientation. Between the silicon layers are tungsten foil sheets that act as converters for the photons. There are a total of 36 layers of silicon strips. Software packages reconstruct hits in the silicon into tracks that represent the trajectories of charged particles. There is a total of approximately 90 square meters of silicon in the entire tracking system and 884,736 individual channels. This represents one of the biggest silicon tracking systems ever constructed and is only surpassed by the future detectors at the Large Hadron Collider. The calorimeter (CAL) is constructed from CsI(Tl) crystals readout by photo-diodes. The crystals are stacked up such that they form a hodoscope detector. Each tower in the LAT has 96 crystals arranged in 8 layers (4 aligned in the x-direction, 4 aligned in the y-direction) with 12 crystals per layer. The calorimeter is 8.6 radiation lengths deep. The ACD, TKR, and CAL are configured and controlled by a highly flexible and robust data acquisition (DAQ) system. The DAQ can be configured to have multiple hardware triggers based on signals from the three main subsystems. In addition, high level algorithms can be run on-board to (i) purify the sample of candidate gamma-rays and (ii) provide quick analysis of events to identify phenomena such as gamma-ray bursts.

The performance of the LAT greatly exceeds that of its predecessor, EGRET. The sensitivity of the LAT is about 50 times better than EGRET. At 10 GeV, the energy resolution is about 6% and the the angular resolution is about 0.1^o (68% containment). Additional information about the LAT instrument response can be found elsewhere.[4] GLAST contains no expendables and has a mission lifetime of at least 5 years. The LAT is being developed by at international collaboration of $\sim$ 250 scientists with the project managed by the Stanford Linear Accelerator Center.

3. Dark Matter Searches

The GLAST project has a very broad and rich scientific program that includes source identification, gamma-ray bursts, dark matter detection, the search for indications of quantum gravity, etc. This paper focuses on the indirect search for dark matter. We examine several possible detection channels and report on the expected sensitivity of each. Each of the these channels depends on several common factors. First, the source of gamma-rays is the annihilation of WIMPs where we use the lightest supersymmetric particle as a specific test case. The annihilation of the WIMPS can produce gamma-rays either directly or more commonly through the hadronization chain of quarks and the eventual decay of neutral pions. The primary decay products of the annihilation is predicted using tree-level final states that include $b\bar{b}$, $t\bar{t}$, $\tau^+\tau^-$, W^+, W^-, Z^0Z^0, $Z^0\gamma$, and $\gamma\gamma$, the later two being important detection of direct spectral lines (see below). The hadronization of the final states is carried out using the Lund Monte Carlo program Pythia 6.202.[5] The flux of gamma-rays is determined by assuming a density profile, such as Navarro-Frenk-White (NFW),[6] and integrating over the line of sight. Finally, the observed signal is determined using the detector response of the LAT that has been determined using a full GEANT simulation of the detector components. In addition, for each analysis there is an assumed integrated exposure by GLAST.

The backgrounds for these searches can be from diffuse Galactic gamma-rays, extragalactic gamma-rays, and charged particle background in the LAT. The rejection power of the LAT against charged particle backgrounds is assumed to be sufficiently high such that we do not include these in the present study. However, the backgrounds from diffuse emissions, galactic and extragalactic, must be included and represent at significant background to all analyses presented below. About 80% of the Milky-Way luminosity in gamma-rays is due to interactions of cosmic rays with gas in the interstellar medium. This background is modeled with GALPROP.[7] Two different parameter settings are used as models. One is referred to as the "conventional" model and the other as the "optimized" model. The optimized model attempts to re-tune parameters of the cosmic ray spectrum to better describe the high-energy excess of diffuse gamma-rays observed by EGRET. For recent reviews on cosmic ray propagation and diffuse gamma-ray emission see Ref. 8 and 9.

We study the sensitivity of several different observational approaches. Each approach has its own strength and weakness. The channels we consider are:

(1) Galactic Center:
Advantage – Good Statistics;
Disadvantage – Source Confusion
(2) Milky-Way Halo:
Advantage – Excellent Statistics;
Disadvantage – Galactic Diffuse Background

(3) Satellites and Subhalos:
Advantage – Small Background, Good Source ID;
Disadvantage – Low Statistics
(4) Spectral Lines:
Advantage – No astrophysical uncertainty, "Smoking Gun";
Disadvantage – Low Statistics

In the sections below we examine the current sensitivity of each of these approach. In addition, there are studies in other areas such as generic sources, extragalactic sources (e.g. dwarf galaxies), diffuse extragalactic from dark matter halos at all redshift. However, these will not be discussed.

3.1. *Galactic Center*

The Galactic center is expected to have a high density of dark matter and therefore should be a strong gamma-ray source. However, the Galactic center is also a very active region with objects capable of accelerating cosmic rays to high energy. It also contains very bright gamma-ray sources including the one recently discovered by H.E.S.S.,[10] MAGIC,[11] VERITAS,[12] and CANGAROO.[13] All these features present a substantial background to a dark matter search around the Galactic center. We estimate the sensitivity of GLAST by searching within 0.5 degrees of the galactic center for gamma-rays above an energy of 1 GeV. We scan the parameter space of $<\sigma v>$ vs WIMP mass assuming 5 years of LAT operation. The dark matter contribution is generated with DarkSusy v4.15[14] and a truncated NFW density profile as defined in.[15] A standard χ^2 analysis is performed to check that the results are consistent with EGRET observation. If it is compatible, a second χ^2 analysis is done to determine if GLAST can separate the WIMP contribution from the diffuse background. This is done for several final states of WIMP annihilation. Figure 1 shows the sensitivity for observing a WIMP contribution at a 3σ and 5σ confidence level. This analysis is sensitive to dark matter density profile at the Galactic center, which is not well known. A slightly steeper inner slope of the NFW profile would lead to an improved sensitivity by a factor of ten.

3.2. *Milky Way Halo*

Performing a search over the full Milky Way halo provides the advantage of large statistics. However, the sensitivity has a strong dependence on our systematic understand of the diffuse background over the full sky. This analysis examines the full sky, with the exception of within 10 degrees

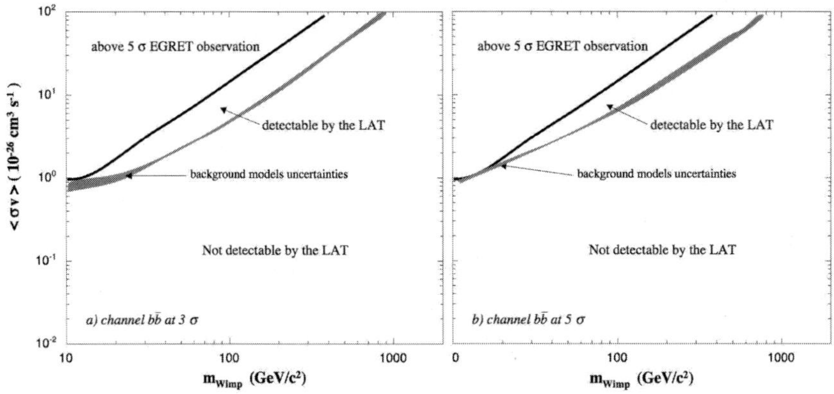

Fig. 1. The $< \sigma v >$ versus WIMP mass plane for the $b\bar{b}$ annihilation channel. The left plot is for 3σ significance and the right plot is for 5σ. The upper left of each plot represents the part of parameter space already excluded by EGRET data.

of the Galactic center, in a 3x3x30 (l, b, and log E) grid. We simulate the dark matter density profile with NFW. The dark matter energy distribution peaks at about 1 GeV for a WIMP mass of 50 GeV/c^2 and increases with increasing mass. The diffuse background contribution peaks at about 150 MeV. We create simulated data samples that represent 1 years worth of observation by the LAT and we perform a simultaneous fit to the spatial distribution and the energy. We create an ensemble of data sets at a variety of WIMP masses. At each mass we adjust the $< \sigma v >$ until 50% of our pseudo-experiments show a WIMP signal at 3σ or 5σ confidence. Figure 2 shows these values for a variety of WIMP masses and for two choices of the diffuse background. This represents only the statistical power of such a search. The final analysis in this channel will require a detailed study of the systemic uncertainties associated with the diffuse background. In this channel we also estimate our mass resolution for the cases where a signal can be observed. Our resolution for a WIMP mass of 100 GeV/c^2 (200 GeV/c^2) is 7% (11%).

3.3. *Satellites and Subhalos*

Because we expect structure in the universe to form hierarchically, the Milky Way halo should contain a large number of nearly isotropic subhalos with masses extending down to 10^6 $M_\odot$. These represent excellent targets for a dark matter search. To estimate our sensitivity, we use a satellite distribution generated by Taylor and Babul[16] each with a truncated NFW

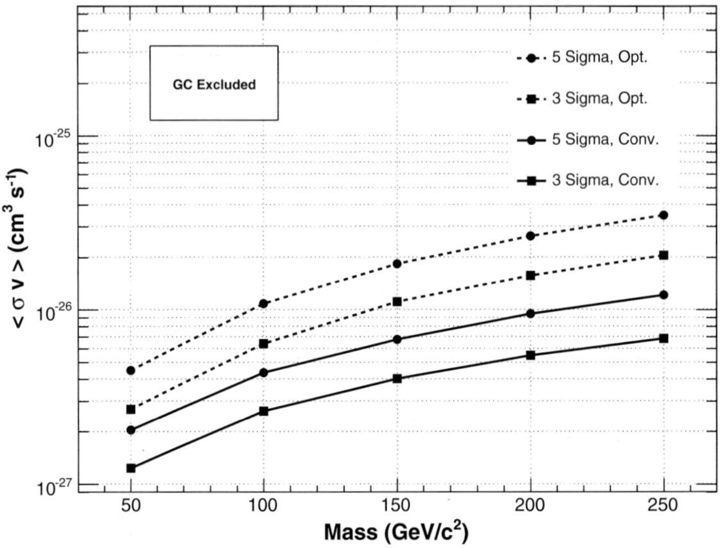

Fig. 2. This shows the sensitivity for the Milky Way halo analysis in terms of the $< \sigma v >$ required to obtain an observation of WIMP annihilation at either 3σ or 5σ significance versus WIMP mass. This represents one year of LAT data and two different assumptions of diffuse background are shown.

satellite structure. We then use a generic WIMP model assuming a WIMP mass of 100 GeV/c^2 and $< \sigma v >= 2.3 \times 10^{-26}$ cm^3 s^{-1}. The analysis assumes 1 and 5 years worth of LAT operation. The significance of the dark matter signal is estimated by calculating the number of gamma-rays $E > 1$ GeV within the satellites tidal radius divided by the square root of the total expected photons from both signal and background. Figure 3 shows the cumulative number of satellites with significance of at least 5σ. The figure shows the results assuming the two different models of diffuse background.

3.4. *Spectral Lines*

In addition to the continuum of gamma-rays due to WIMP annihilation into hadrons, WIMP annihilation can also lead to the production of gamma-ray lines from direct final states of $\gamma\gamma$ and γZ. These spectral lines could show up in the energy spectrum from astrophysical sources. Observation of spectral lines would represent a "Smoking Gun" for indirect dark matter detection. This analysis has the advantage that the background can be

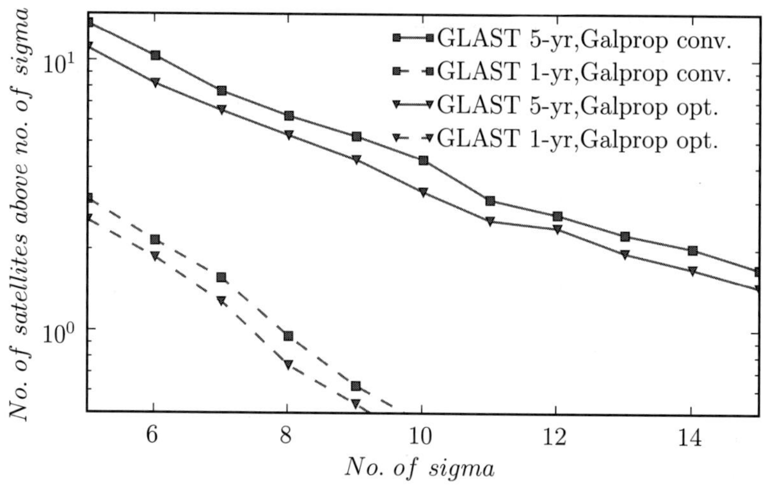

Fig. 3. This shows the estimated number of observable dark matter satellites for 1 and 5 years worth of LAT data. The background is both extragalactic diffuse and galactic diffuse (two choices).

determined directly from the data itself. However, in most supersymmetric models these decays are loop suppressed and therefore have branching ratios in the range of 10^{-2} to 10^{-4}. We estimate the sensitivity in terms of photons/cm^2/s/sr which is independent of specific source parameters, but not the diffuse model. The sensitivity also depends on the energy resolution of the LAT, which is determined from simulation. The LAT sensitivity to lines is shown in Figure 4. Shown are the 5σ confidence level limits for two cases: (i) where the WIMP line energy is unknown and (ii) where the WIMP line energy is known from other experimental sources such as collider based discoveries. In the case the WIMP energy is unknown, the sensitivity is slightly worse due to a penalty factor for searching in multiple energy windows.

4. Conclusions

The Gamma-ray Large Area Space Telescope will launch in 2008 and open up a new era in gamma-ray astroparticle physics. The Large Area Telescope, the primary detector on GLAST, will observe the sky over four orders of magnitude in energy. This will provide a new opportunity to study high energy phenomena in the cosmos including the possible elucidation of the nature of dark matter.

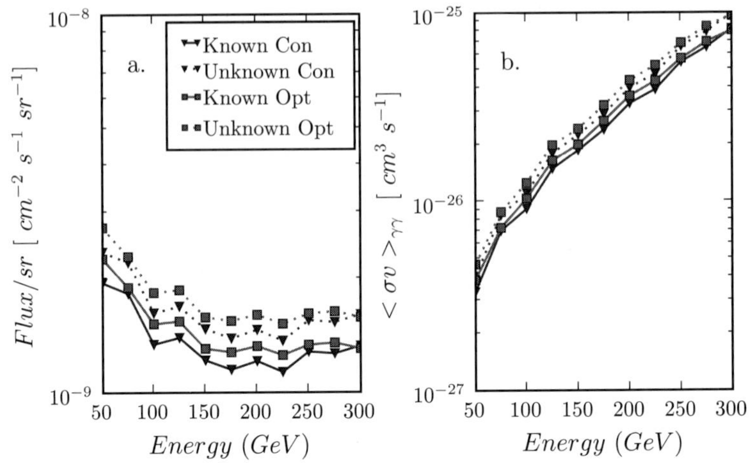

Fig. 4. The plots show the sensitivity for spectral lines from dark matter annihilation. The left side shows the sensitivity in terms of a source model independent quantity and the right in terms of $< \sigma v >$ and an assumed NFW density profile. Several contours are shown for different choices of diffuse galactic background and assumptions about known or unknown WIMP masses.

Acknowledgment

I want to thank my colleagues in the LAT Dark Matter and New Physics Working Group for the tremendous effort that lead to the results presented here.

References

1. Y.M. Georgelin and Y. P. Georgelin. Astronomy and Astrophysics, vol. 49, no. 1, May 1976, p. 57-79.
2. Clowe, D. et. al. astro-ph/0608407 (2006).
3. G. Jungman, M. Kamionkowski and K. Griest. hep-ph/9506380 (1996).
4. http://www-glast.slac.stanford.edu/software/IS/glast_lat_performance.htm
5. Pythia program package, see T. Sjöstrand, Comp. Phys. Comm. **82** 74 (1994).
6. J. F. Navarro, C. S. Frenk and S. D. White, Astrophys. J. **462** 563 (1996).
7. Galprop, http://galprop.stanford.edu/web_galprop/galprop_home.html
8. A. W. Strong, I. V. Moskalenko and V. S. Ptuskin, Ann. Rev. Nucl. Part. Sci. **57**, 285 (2007).
9. A. W. Strong, I. V. Moskalenko and O. Reimer, Astrophysics and Space Science Library vol 304, 279–310 (2004).
10. F. Aharonian et al.(HESS Collaboration), Astronomy and Astrophysics, 425,L13 (2004).

11. J. Albert et al., (MAGIC Collaboration), Astrophys. J. 638, L101, (2006).
12. K. Kosack et al., Astrophys. J. 608 L97 (2004).
13. K. Tsuchiya et al. (CANGAROO Collaboration), Astrophys. J. 606, L115 (2004).
14. P. Gondolo et al.,JCAP, **0407** 008 (2004).
15. E. Hayashi, J. F. Navarro, J. E. Taylor, J. Stadel and T. Quinn, Astrophys. J. **584** 541 (2003).
16. J. E. Taylor and A. Babul, Mon. Not. Roy. Astron. Soc. **364**, 535 (2005). J. E. Taylor and A. Babul, Mon. Not. Roy. Astron. Soc. **364**, 515 (2005).

CONSTRAINTS ON MACHOS WITH THE EROS-2 PROJECT

P. TISSERAND

Research School of Astronomy and Astrophysics,
Australian National University, Cotter Road
Weston Creek, ACT 2611, Australia
E-mail: tisserand@mso.anu.edu.au
http://www.mso.anu.edu.au

The EROS-2 collaboration has monitored during 6.7 years millions of stars in the Magellanic Clouds, in order to search for microlensing phenomena due to hypothetical compact dark objects in the Milky Way Halo ('machos'), constituting a component of Galactic dark matter. I will present the limit obtained by the EROS experiment on the macho content of the galactic halo, using only the brightest, well resolved stars ($\sim$ 7 million). The number of microlensing effects found is by far not numerous enough to account for the galactic dark matter. Our robust limit ($f < 0.07$ at 95% CL for 0.4 $M_\odot$) is now clearly in conflict with the MACHO experiment positive result ($0.05 < f < 0.42$ for 0.4 $M_\odot$). More generally, machos in the mass range $10^{-7} M_\odot < M < 5 M_\odot$ are ruled out as the primary occupants of the Milky Way Halo. Furthermore, a new constraint for higher machos masses (up to 150 solar masses) has been performed using the complete star sample. A preliminary limit is presented here.

Keywords: dark matter; Galactic Halo; microlensing.

1. Introduction

The rotation curves of spiral galaxies components lead to a dynamical mass estimate 5-10 times higher than that of the visible components. Since 1986 and Paczyński's proposal,[14] gravitational microlensing has been proven to be a powerful probe of galactic halo dark matter under the form of machos. The signature of these objects is a transient amplification of the light from extragalactic stars by the gravitational lens effect when they are close to their line of sight. Events are extremely rare and millions of stars, towards the Large and Small Magellanic Clouds (LMC and SMC), have thus been monitored photometrically since 1990 by several experiments (EROS, MA-

CHO, OGLE, MOA and SuperMACHO) for a period of years in order to obtain a useful detection rate.

Some microlensing candidates have been observed towards the two targets.[2,11] I present here the last result obtained towards the Magellanic Clouds by the EROS collaboration.[19] The full EROS2 data set (July 1996 till February 2003) have been produced and only a subsample of bright stars has been used to compile the new limit. The strategy of using only bright stars helps to discriminate against background events due to variable stars and allows a simple determination of the effects of source confusion (blending).

Section 2 presents the basics of the EROS-2 setup and general characteristics of the data. Section 3 describes the bright star sample selection and the analysis. The limit for mass machos $M < 15M_{\odot}$ is shown on Section 4 with a discussion of its significance. A new analysis that concentrate on longer timescale microlensing event have been recently performed. The preliminary limit up to 150 $M_{\odot}$ is presented in the last section.

2. The EROS2 Experimental Setup and Data Characteristics

The second phase of the EROS experiment (Experience de Recherche d'Objets Sombres) started in June 1996 with the aim of improving the sensitivity by at least an order of magnitude for higher mass objects ($M > 10^{-2}$). It was conducted until February 2003, using a 1 meter dedicated telescope, the MARLY, located at La Silla (Chile).

The observed fields correspond to 0.95 deg^2 each and they are twice as large in declination than in right ascension (1.38×0.69 deg^2). A total of 88 fields have been monitored toward the LMC and 10 toward the SMC. The positions of the fields are shown in Figure 1. For comparaison, the MACHO collaboration monitored 11.9 million stars over 13.4 deg^2, mostly in the central region of the LMC.[2]

3. Analysis and Bright Star Sample

The criteria used to select the subsample of bright star and the selection cuts applied on each light curve will be discuss in this section, more details can be found on section 3 and 4 in Tisserand et al., 2007. The analysis aims at detecting luminosity excursions, due to microlensing, on otherwise constant light curves. The smallest reachable microlensing timescale, t_E , is determined by the sampling of the fields; in practice, except for a few fields,

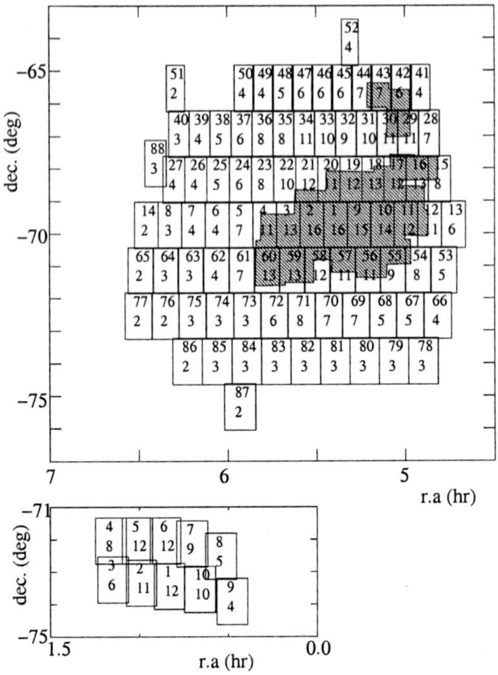

Fig. 1. Map of the EROS-2 LMC and SMC fields in equatorial coordinates. A total of 88 LMC and 10 SMC fields were monitored. The first number in each field is the field number and the second is the number of bright stars in the field in units of 10^4 used for the analysis. The two shaded regions (the larger one centered on the LMC bar) are the 13.4 deg^2 used by the MACHO collaboration to measure the optical depth (Alcock et al. 2000a).

the detection efficiency is very low below 2-3 days. The largest detectable timescale, about 800 days, is determined by the reduction in the detection efficiency when the timescale becomes so large that the baseline flux is not seen during the observing period.

3.1. *The Bright-Star Sample*

We have chosen to restrict our analysis to the Bright Sample of stars defined below. Besides the obvious advantage of ensuring a good photometric resolution, we do this to simplify the evaluation of the number of expected events predicted by a model. In crowded fields, this evaluation is complicated by "blending", i.e. the fact that photometry of a given object can be influenced by more than one star. These effects are rather small and simple to evaluate for the Bright-Star Sample (which should, strictly speaking, be called the Bright-Object Sample). Limiting the number of studied objects limits the sensitivity of the experiment so the magnitude cut must be a compromise between quantity and quality of objects. The efficiency calculations for unblended sources in the full star sample indicated that keeping only the 20% brightest stars reduces by only 50% the number of simulated microlensing events that pass our selection criteria. The photometric precision for the Bright-Star Sample turns out to be rather uniform, about 7% in average in both bands. The number of EROS-2 objects in the Bright-Star Sample is 6.05 million in the LMC and 0.90 million in the SMC.

3.2. *The Analysis*

The analysis consists of a series of selection criteria applied to the light curve of each object retained. It is guided using the simulation of microlensing described hereafter, which also serves to determine the efficiency of the selection procedure. The tuning of each cut and the calculation of the microlensing detection efficiency are done with simulated simple microlensing light curves : we surimpose artificial microlensing events on a representative sample of light curves, corresponding to 2% of the Bright-Star sample from each of the 98 monitored fields. The simulation takes into account the relative variation of photometric errors. Simulated light curves were then fed into the analysis chain to find the fraction that are recovered by our detection algorithm. The detection efficiency in a given t_E bin, $\epsilon(t_E)$, is then given by the ratio between the number of events passing all selection criteria in this bin and the number of microlensing events generated in the same bin.

3.3. *The Status of Published Magellanic Microlensing Candidates*

EROS has, in the past, used 11 candidates to place upper limits on the optical depth toward the LMC and to measure the depth toward the SMC. Six of the candidates are in the Bright-Star Sample considered here. Only EROS2-SMC-1 remains as a candidate microlensing event. The others have been eliminated either because continued observations of the same stars show further variability on the light curves, or because improved photometry, in some cases complemented by spectroscopy, lead to re-interpreting the candidates as variable stars or supernovae. See section 6 of Tisserand et al., 2007 for complete details.

4. Result on Low Machos Mass ($M < 15M_\odot$)

After the selection cuts have been applied, only one candidate microlensing event remains in the Bright-Star Sample, EROS2-SMC-1. It has been known since 1997 (Alcock et al. 1997c; Palanque-Delabrouille et al. 1998) and is identical to candidate MACHO-97-SMC-1. It has one of the two longest timescales (103 days) and the highest luminosity of all published microlensing candidates reported toward the Magellanic Clouds.

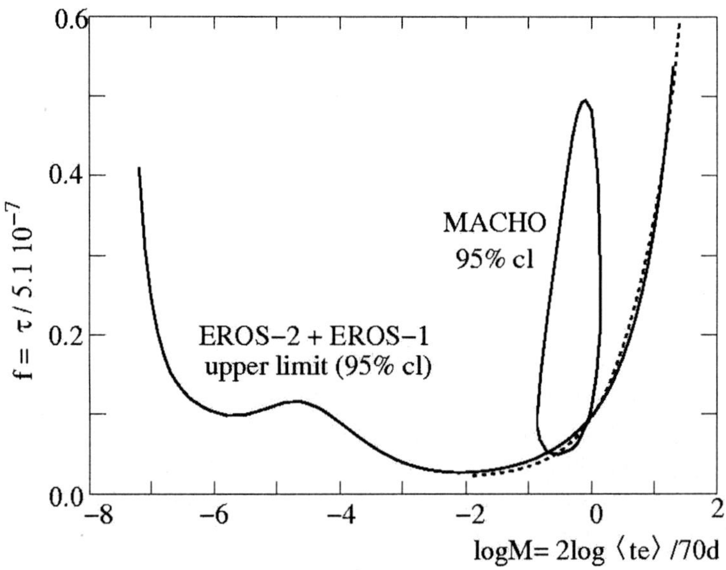

Fig. 2. The solid line shows the EROS 95% CL upper limit on $f = \tau_{lmc}/5.1 \times 10^{-7}$ based on no observed events in the EROS-2 LMC data and the EROS-1 data. The dashed line shows the EROS upper limit on lmc based on one observed SMC event in all EROS-2 and EROS-1 data assuming $\tau_{smc-halo} = 1.4\tau_{lmc}$. The MACHO 95% CL. curve is taken from Figure 12 (A, no lmc halo) of Alcock et al. (2000a).

The limits shown in Figure 2 rule out machos as the majority of Galactic dark matter over the range $0.6 \times 10^{-7}M_\odot < M < 15M_\odot$. The limits are $f < 0.04$ for $10^{-3}M_\odot < M < 10^{-1}M_\odot$ and $f < 0.1$ for $10^{-6}M_\odot < M < 1M_\odot$.

The observation of 17 events by MACHO with $\langle t_E \rangle \sim 40$ days suggested the existence of machos of $M \sim 0.4M_\odot$. Such objects could have been invisible (e.g. primordial black holes) or faint (e.g. cool white dwarfs). However, the latter are not seen at the corresponding level in multi-color surveys (Gates et al. 2004) and high proper-motion surveys (Reid 2005, and references therein).

At any rate, for a machos mass of $0.4M_\odot$, the EROS limit $\tau_{lmc}^{EROS} < 0.36 \times 10^{-7}$ at 95% CL. is significantly less than the MACHO result $\tau_{lmc}^{MACHO} = 1.2^{+0.4}_{-0.3} \times 10^{-7}$. However, there are considerable differences between the EROS and MACHO data sets that may help to resolve the conflict. Generally speaking, MACHO uses faint stars in dense fields (11×10^6 stars over 13.4 deg^2) while EROS-2 uses bright stars in sparse fields (7×10^6 stars over 84 deg^2). Of these bright EROS-2 stars, 2×10^6 are in MACHO fields.

The use of dense fields by the MACHO group suggests that the higher MACHO optical depth may be due, in part, to self-lensing in the inner parts of the LMC. This would contradict LMC models (Mancini et al. 2004) which suggest that only 1-3 MACHO events should be expected to be due to self-lensing. If it turns out that the self-lensing rate is higher than the model estimates in the MACHO fields but still negligible in the outer fields of the LMC, the disagreement between MACHO and this work is considerably reduced. Since only 1/3 of our Bright Sample stars are in MACHO fields, the EROS-2 95% CL upper limit on for the MACHO fields is multiplied by a factor 3, within 1 standard deviation of the MACHO result.

The use of faint stars by MACHO suggests two other possible explanations for the disagreement. The first would be contamination by variable stars that, in our Bright-Star Sample, are either not present or identified as such because of superior photometric precision. The low photometric precision for the faint LMC stars makes most of the events less convincing than the events on bright stars in the Galactic Bulge or the one EROS event in the SMC. Indeed, one of the MACHO A events, MACHO-LMC-23, has been identified as a variable star by a second variation in the EROS-2 data (see Tisserand and Milsztajn, 2005). Some of the MACHO candidates are very convincing microlensing candidates. In particular, MACHO-LMC 1, 5, 9, 14 and 21 are strong candidates based only on MACHO photometry while Bennett (2005) argued that MACHO-LMC-4, 13 and 15 are strong candidates because of high precision followup photometry. We might note, however, that of these candidates, MACHO-LMC-14 is most likely due to LMC self-lensing and MACHO-LMC-5 is due to lensing by a normal red-dwarf star in the disk of the Milky Way (Alcock et al. 2001a; Gould 2004).

The second possible explanation related to the use of faint source stars is the large blending effects that must be understood. Alcock et al. (2000a) suggest a 30% systematic error to reflect this uncertainty. The experience with the use of faint stars in the Galactic Bulge (see Hamadache et al., 2006) suggests that this uncertainty may be underestimated, though in principle the star distribution is better understood in the LMC than in the Bulge. At any rate, the extension of the limits presented here, either by EROS or by the OGLE-3 or SuperMACHO, will require the use of faint stars and a good understanding of blending.

5. Preliminary Result on Higher Macho Mass (5 to 200 $M_\odot$)

A new analysis that concentrate on longer timescale microlensing events has been performed. We use our full sample of LMC stars, 29×10^6 objects spread over 84 $\deg^2$ and monitored over 6.7 years. We find no events with Einstein radius crossing times greater than $t_E = 110$ days, whereas 15 (3) events would have been expected if the halo were comprised of objects of mass 10 $M_\odot$ (100 $M_\odot$).

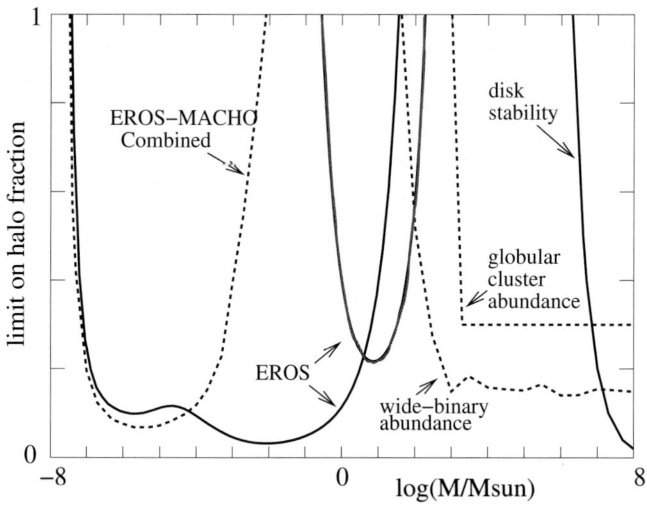

Fig. 3. Presentation of the microlensing limits on machos and the ones obtained from gravitational stability studies of bound system: the "globular cluster abundance" from Arras and Wasserman (1999), the wide–binary abundance from Yoo et al. (2004) and the solid line label "disk stability" summarizes the constraint from disk heating from Lacey and Ostriker (1985). Both 95 % C.L. limits on the halo fraction from the EROS expriment are in solid line. For $M > 10^{-7}$ $M_\odot$ the limit is everywhere better than $f < 0.2$ except for M between 1 and 1000 $M_\odot$.

This yields upper limits on the fraction, f, of the halo mass that could be comprised of machos (see Fig. 3). For $M_\odot < M < 30M_\odot$ we find $f < 0.3$ and for $0.2M_\odot < M < 150M_\odot$ we find $f < 1.0$. This rule out machos as the dominant halo component in the small gap between previous microlensing limits for $M < 15M_\odot$ and limits for $M > 43M_\odot$ based on the stability of wide-binary stellar systems.

References

1. C. Alcock *et al* (MACHO coll.) ApJ **491**, L11 (1997c).
2. C. Alcock *et al* (MACHO coll.) ApJ **542**, 281A (2000a).
3. C. Alcock *et al* (MACHO coll.) Nature, 414, 617 (2001a).
4. Arras, P. and Wasserman, I. 1999, MNRAS, 306, 257
5. E. Aubourg *et al* (EROS coll.), Nature 414, 617 (1993).
6. D. Bennett ApJ **633**, 906E (2005)..
7. E. Gates, G. Gyuk, H.C. Harris *et al* (SDSS coll.) ApJ **612**, 132 (2004).
8. A. Gould, ApJ **606**, 319 (2004).
9. C. Hamadache, L. Le Guillou, P. Tisserand, et al. (EROS coll.) A&A **454**, 185 (2006)
10. Lacey, G.C. and Ostriker, J.P. 1985, ApJ, 299, 633
11. T. Lasserre *et al* (EROS coll.) A&A **355L**, 39L (2000).
12. L. Mancini, S. Calchi Novati, Ph. Jetzer and G. Scarpetta A&A **427**, 61 (2004).
13. A. Milsztajn & T. Lasserre, Nucl. Phys. B Proc. Sup. 91, 413 (2001).
14. B. Paczyński, ApJ **304**, 1P (1986).
15. N. Palanque-Delabrouille *et al* (EROS coll.) A&A **332**, 1 (1998).
16. I.N. Reid Ann. Rev. of Astronomy and Astrophysic **43**, 247 (2005).
17. C. Renault *et al*, A&A **324**, L69 (1997).
18. P. Tisserand and A. Milsztajn, 2005, astro-ph/0501584, 5th Rencontres du Vietnam.
19. P. Tisserand *et al* (EROS coll.), A&A **469**, 387T (2007).
20. P. Tisserand, PhD thesis (2004), University of Nice-Sophia Antipolis.
21. A. Udalski, M. Szymanski, M. Kubiak, et al. (OGLE coll.) Acta Astron., 47, 431, 1997
22. A. Udalski, M. Szymanski, M. Kubiak, et al. (OGLE coll.) Acta Astron., 50, 307, 2000b
23. Yoo, J., Chanamé J. and Gould, A. 2004, ApJ, 601, 311
24. Z. Zatritsky, J. Harris, I. Thompson, and E. Grebel AJ **128**, 1606 (2004).

ANNIHILATING DARK MATTER

NICOLE F. BELL

School of Physics, The University of Melbourne,
Victoria 3010, Australia
E-mail: n.bell@unimelb.edu.au

We consider the annihilation of dark matter into Standard Model particles. We show that the least detectable annihilation products, namely neutrinos, can be used to set a conservative bound on the *total* annihilation cross section. This bound is much stronger than the unitarity bound for the most interesting mass range, and shows that dark matter halos cannot be significantly modified by annihilations. We also discuss the production of Galactic positrons via dark matter annihilation.

Keywords: Dark matter; annihilation; neutrinos; positrons.

1. Introduction

The self-annihilation cross-section is a fundamental property of dark matter. For a single dark matter species that was thermally populated in the early universe, the annihilation cross-section, $\langle \sigma v \rangle$, is completely specified by the relic density today, $\Omega_{CDM} \simeq 0.3$. This sets a natural scale for the annihilation cross-section at approximately $\langle \sigma v \rangle \sim 10^{-26}\ \text{cm}^3\ \text{s}^{-1}$. However, if dark-matter is a not a thermal relic very different annihilation cross sections are possible.

The self annihilation cross section also sets the annihilation rate in gravitationally-collapsed dark matter halos in the universe today.[1] In fact, models with large annihilation rates were invoked in Ref. 2 (KKT) as a means of altering the density profile of dark matter halos. This solves an apparent discrepancy between predicted (sharp cuspy) and observed (flat cored) halo profiles. (Note that a similar effect is obtained with large dark matter self-interactions rates.[3] While a large annihilation rate implies a large elastic scattering rate, the reverse is not true.) However, relative to the natural scale for a thermal relic, enormous annihilation rates are required to significantly soften halo cores. Irrespective of the cusp-core issue,

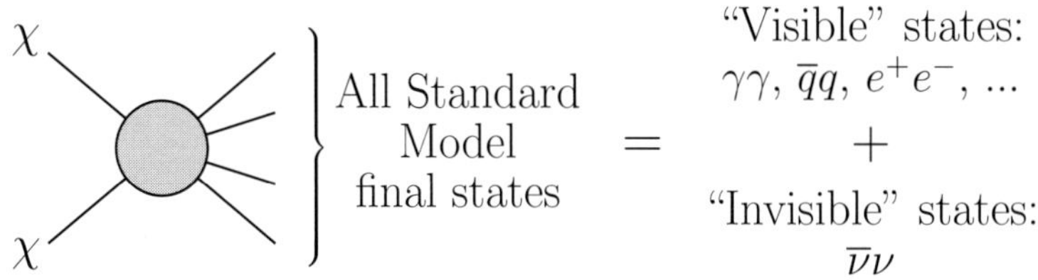

Fig. 1. Annihilation of dark matter into SM final states. Since all final states except neutrinos lead to the production of gamma rays, we can bound the total cross section from the limit on the neutrino signal, i.e., assuming Br("Invisible") = 100%.

we may interpret the KKT cross section as an upper bound, denoting the point at which annihilations would significantly distort halo profiles,

$$\langle \sigma_A v \rangle_{\rm KKT} \lesssim 3 \times 10^{-19} \frac{\rm cm^3}{\rm s} \left[\frac{m_\chi}{\rm GeV}\right] , \tag{1}$$

where m_χ is the dark matter mass.

Unitarity sets a general upper bound on $\langle \sigma_A v \rangle$, and can only be evaded in certain unusual cases[4–6] (e.g. non-pointlike dark matter candidates such as q-balls[6]). In the low-velocity limit where the cross section is assumed to be s-wave dominated, we have,[5]

$$\langle \sigma_A v \rangle \leq \frac{4\pi}{m_\chi^2 v} = 1.5 \times 10^{-13}\ {\rm cm^3\ s^{-1}} \left[\frac{\rm GeV}{m_\chi}\right]^2 \left[\frac{300\ {\rm km\ s^{-1}}}{v}\right] . \tag{2}$$

However, in the interesting GeV-TeV mass range the unitarity limit becomes extremely weak, allowing the possibility of models like KKT. Below, we set a strong and general bound on the dark matter annihilation rate by considering the appearance of annihilation products.

2. General Bounds on Dark Matter Annihilation

We assume that all annihilation products are Standard Model (SM) particles (i.e. we make the reasonable assumption that the dark matter is the lightest particle in the beyond-SM sector). We may then express the cross section in terms of branching ratios to "visible" and "invisible" final states such as gamma rays and neutrinos, respectively, as in Fig. 1. The KKT model requires stringent constraints on the branching ratio to "visible" final states" to avoid exceeding the cosmic diffuse gamma-ray background.[2]

Among SM final states, it is clear that all but neutrinos will inevitably produce gamma rays. Quarks and gluons hadronize, producing pions, where $\pi^0 \to \gamma\gamma$, and the decays of weak bosons and tau leptons also produce

π^0. The stable final state e^+e^- is not invisible, since it produces gamma rays either through electromagnetic radiative corrections[7] or energy loss processes,[8] while the final state $\mu^+\mu^-$ produces e^+e^- via decays. Given that limits on the $\bar{\nu}\nu$ final state will be weaker than the limits on all other final states, we can set a conservative upper limit on the *total* annihilation cross-section by setting the branching ratio to the "invisible" $\bar{\nu}\nu$ final state at $Br(\bar{\nu}\nu) = 100\%$.

The most direct approach to bound the $\chi\chi \to \bar{\nu}\nu$ cross section is to use the cosmic diffuse neutrino flux from dark matter annihilations in all halos in the universe as the signal.[9] (A complementary approach, with comparable sensitivity, is to consider the signal from annihilation in the Milky Way.[10]) The spectrum of diffuse annihilation products is given by[9]

$$\frac{d\Phi_\nu}{dE} = \frac{\langle\sigma_A v\rangle}{2} \frac{c}{4\pi H_0} \frac{\Omega_{DM}^2 \rho_{\rm crit}^2}{m_\chi^2} \int_0^{z_{\rm up}} dz \frac{\Delta^2(z)}{h(z)} \frac{dN_\nu(E')}{dE'}, \tag{3}$$

where $H_0 = 100\, h\,\mathrm{km\,s^{-1}\,Mpc^{-1}}$ is the Hubble parameter, Ω_{DM} is the dark matter density in units of the critical density $\rho_{\rm crit}$, we assume a flat universe, with $\Omega_{DM} = 0.3$, $\Omega_\Lambda = 0.7$, $h = 0.7$, and $h(z) = [(1+z)^3\Omega_{DM} + \Omega_\Lambda]^{1/2}$. We have assumed χ is its own antiparticle. The factor $\Delta^2(z)$ accounts for the increase in density due to the clustering of dark matter in halos, and we have conservatively adopted $\Delta^2 = 2 \times 10^5$ (see Ref. 9 for further details).

In order for the diffuse flux to be detectable, it must be larger than the atmospheric neutrino background. The atmospheric neutrino flux has been well measured by the Super-Kamiokande experiment,[11] with additional information provided by Fréjus[12] and AMANDA.[13] We choose a detection criteria such that the annihilation signal is 100% as large as the angle-averaged atmospheric neutrino ($\nu_\mu + \bar{\nu}_\mu$) background, integrated over a bin of width $\Delta \log_{10} E_\nu = 0.5$, just below $E_\nu = m_\chi$.

Our final bound on $\langle\sigma_A v\rangle$ is shown in Fig. 2. This bound is much stronger than the unitarity bound at the most interesting mass range, and shows that dark matter halos cannot be significantly modified by annihilations.

3. Dark Matter Annihilation and Galactic Positrons

The Galactic positrons,[14] as observed by their 0.511 MeV gamma-ray annihilation line, are difficult to account for with astrophysical sources. The emission region for this flux is consistent with a 2-d gaussian of full width at half maximum of ~ 8 degrees, located close to the Galactic center. This leads naturally to the proposal that the positrons are produced via the annihilation of dark matter particles into e^+e^- pairs at the Galactic center.[15]

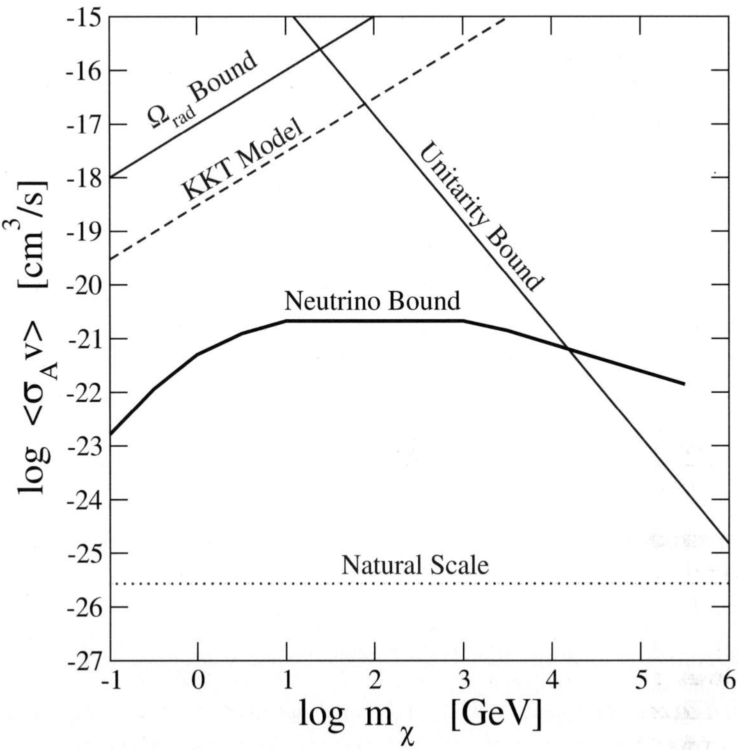

Fig. 2. Upper bounds on the dark matter total annihilation cross section in galaxy halos as a function of the dark matter mass.

However, the dark matter annihilation process $\chi\chi \rightarrow e^+e^-$ is necessarily accompanied by $\chi\chi \rightarrow e^+e^-\gamma$, arising due to electromagnetic radiative corrections.[7] Calculating the flux of gamma rays produced via this "internal bremsstrahlung" process, we find that the emission would violate COMTEL and EGRET constraints on the galactic diffuse gamma ray background, unless the dark matter mass is relatively low, less than about 20 MeV. Subsequently, Beacom and Yuksel[16] have refined this bound to $\lesssim 3$ MeV, by considering the inflight annihilation of the positrons. In fact these bounds apply to *any* positron production mechanism[7,16] and thus imply that no matter what the production process is, the Galactic positrons must be produced with quite low energies.

Acknowledgments

This article is based upon the results of Ref. 7 and Ref. 9. NFB thanks John Beacom, Gianfranco Bertone, and Greg Mack for enjoyable and productive collaborations.

References

1. G. Bertone, D. Hooper and J. Silk, Phys. Rept. **405**, 279 (2005).
2. M. Kaplinghat, L. Knox and M. S. Turner, Phys. Rev. Lett. **85**, 3335 (2000).
3. D. N. Spergel and P. J. Steinhardt, Phys. Rev. Lett. **84**, 3760 (2000).
4. K. Griest and M. Kamionkowski, Phys. Rev. Lett. **64**, 615 (1990).
5. L. Hui, Phys. Rev. Lett. **86**, 3467 (2001).
6. A. Kusenko and P. J. Steinhardt, Phys. Rev. Lett. **87**, 141301 (2001).
7. J. F. Beacom, N. F. Bell and G. Bertone, Phys. Rev. Lett. **94**, 171301 (2005).
8. E. A. Baltz and L. Wai, Phys. Rev. D **70**, 023512 (2004); D. P. Finkbeiner, astro-ph/0409027.
9. J. F. Beacom, N. F. Bell and G. D. Mack, Phys. Rev. Lett. *in press*, astro-ph/0608090.
10. H. Yuksel, S. Horiuchi, J. F. Beacom and S. Ando, arXiv:0707.0196.
11. Y. Ashie *et al.*, Phys. Rev. D **71**, 112005 (2005).
12. K. Daum *et al.*, Z. Phys. C **66**, 417 (1995).
13. J. Ahrens *et al.*, Phys. Rev. D **66**, 012005 (2002); K. Münich, in A. Achterberg *et al.*, astro-ph/0509330.
14. P. Jean *et al.*, Astron. Astrophys. **407** (2003) L55; J. Knodlseder *et al.*, Astron. Astrophys. **411**, L457 (2003); G. Weidenspointner *et al.*, astro-ph/0406178.
15. C. Boehm, D. Hooper, J. Silk, M. Casse, and J. Paul, Phys. Rev. Lett. **92** (2004) 101301.
16. J. F. Beacom and H. Yuksel, Phys. Rev. Lett. **97**, 071102 (2006).

DARK MATTER: SMOOTH OR COMPACT? LIMITS FROM GRAVITATIONAL MICROLENSING

GERAINT F. LEWIS

Institute of Astronomy, School of Physics,
University of Sydney, NSW 2006, Australia E-mail: gfl@physics.usyd.edu.au

Recent astrophysical clues point to dark matter being relativistically cold, resulting in the presence of substantial dark substructure within the halos of large galaxies. However, the gravitational lensing action of stars, coupled with a smooth dark matter component, can result in a signature which mimics the presence of dark matter subclumps. In this work, we examine the question of just how well such 'gravitational microlensing' can resolve the nature of dark matter, finding that the true resolution is limited by the scale of the source under consideration. However, multi-wavelength studies of gravitationally microlensed systems will provide both clues to the nature of dark matter and the details structure of cosmological sources.

Keywords: Cosmology: Dark Matter, Cosmology: Gravitational Lensing.

1. Introduction

Even though its gravitational effects dominate the cosmos, the true nature of dark matter remains one of the major mysteries of modern science, with ongoing terrestrial and heavenly searches to identify potential dark matter candidates. Substantial effort has been placed in understanding the cosmological growth of structure, with the adoption a cold (sub-relativistic) dark matter component predicting that, in the absence of the complex physics of gas hydrodynamics and star formation, dark matter structures in the universe should be scale-free, possessing self-similar structures (halos within halos) on a range of scales.

This picture appears to hold on the largest scales, but more recently interest has turned to the halos of large galaxies, like our own Milky Way, in the search for the small scale cosmic web. In this paper we review some recent discussion of these searches, focusing upon the issue of how gravitational microlensing my ultimately be faking the expecting signature in known gravitational lensed quasars. The paper looks at the addition of a smooth, as opposed to a compact, dark matter component, and focuses specifically on the issue of whether astrophysically compact matter can actually imitate the presence of smooth astrophysical dark matter.

2. Dark Matter Substructure

As noted previously, in a cold dark matter cosmos, structure form hierarchically with self-similar properties on a range of scales.[1] With this the halos of larger galaxies should be awash with smaller dark matter halos, and hot or warm dark matter would prevent the formation of substructure on the scale of galactic halos, rendering them smooth. Unfortunately, the thousands of expected halos are unseen and galaxies like the Milky Way possess only a handful of of dwarf companions.[6] The situation has been labeled a crisis, although it has been suggested that the majority of halos are truly dark, having ejected their gas during the earliest phases of galaxy formation; with this we must believe that we are surrounded by a population of galaxies that we cannot see directly.

While they emit no light, the gravitational pull of these smaller dark halos can be probed through the action of gravitational lensing. Our own local population of dark halos are too close to provide effective gravitational lenses, and so extragalactic gravitational lenses have been targeted, specifically those possessing close pairs of images.

For an overall smooth mass model, the simple laws of gravitational lensing predicts that such close pairs should possess virtually identical brightnesses [a]. However, for a growing number of observed gravitational lens systems, such close pairs possess quite discrepant image brightnesses. This situation led to the conclusion that the underlying lensing mass distribution was not smooth and the gravitationally lensed images were being demagnified by the presence of dark matter substruture on scales of $10^6 - 10^9 M_\odot$, precisely the range expected from cold dark matter structure growth.[2] While this appeared to be the astrophysical smoking gun for the detection of the temperature of dark matter, it was suggested that apparent under-bright images are due not to the presence of halo substructure, but are actually due to the gravitational lensing influence of stellar mass objects, known as gravitational microlensing, embedded within an overall smooth dark matter component.

[a] For an excellent introduction to gravitational lensing, the reader is referred to the work of Schneider, Ehlers and Falco (1992).[5]

3. Gravitational Microlensing

On small scales, the mass distribution in galaxies is granular, composed of stars, planets etc as well as, potentially, a component of compact dark matter, such as primordial black holes or dead stars. Each of these compact objects can act as an individual gravitational lens, subtly deflecting passing light rays, and a population of compact objects will produces a sea of small 'microimages'. A simple calculation of the deflection due to a solar mass object at cosmological distances reveals the resulting deflection is of order 10^{-6} arcsecs, which is also the typical separation of microimages, rendering them unresolvable to modern telescopes; hence, even if an image of a distant source is microlensed by stellar mass objects as its light passes through a lensing galaxy, the sea of microimages is seen as a single macroimage.

While the astrophysically small deflection appears to render microlensing unobservable, it has to be remembered that the relative configuration of source, microlenses and observer continually changes due to both bulk motions and internal velocity dispersions, and hence the lensing configuration also changes with time. With this the brightnesses and locations of the individual microimages will change with time, with the overall effect being that that brightness of the resultant macroimage will be seen to fluctuate. Moreover, small changes in the lensing configuration can result in the appearance or merging of bright image pairs, with the corresponding microlensing light curve display dramatic changes in brightness over a very short period.[4]

4. Ray-Tracing

With typical microlensing configurations, the light from a distant source will be influenced by the presence of many thousands on compact objects as it passes through a lensing galaxy. Such a situation presents a very daunting task to understand analytically, and the majority of headway in this area has been through the application of numerical techniques. The workhorse for the study of gravitational microlensing has been the 'ray-tracing' approach, first presented in a basic form by Young et al. (1981),[3] before being expanded by Kayser et al. (1986)[9] and implemented in earnest by Wambsganss (1990).[10]

At first glance, the technique appears extremely simple; trace a regular grid of rays though a field of lensing masses and projecting them onto the source plane. The resultant source plane distribution of rays is very inhomogeneous (see Figure 1) due to the influence of the lensing masses

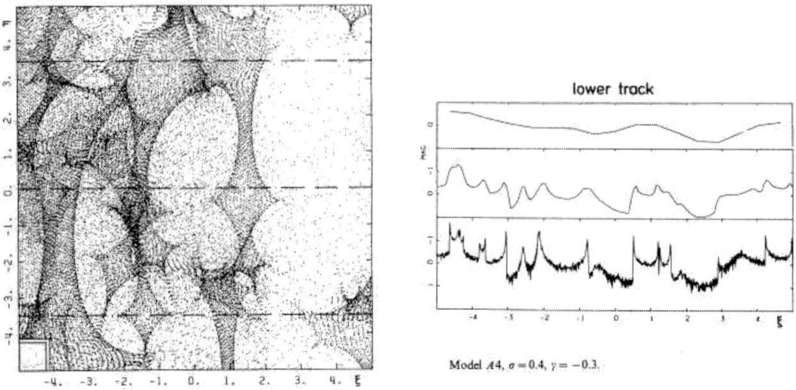

Fig. 1. An example of one of the early ray tracing experiments of Keyser et al (1986).[9] The left-hand panel presents the distribution of rays fired through a distribution of microlensing masses, whereas the right-hand panel presents the light curve of a source passing over the lower track. From the bottom to the top, the source size is increased and the degree in the microlensing light curve decreases.

and a comparison of the density of ray over the source, compared to the regular density of rays passing through the lens, gives a direct measure of the magnification. Hence, a cut across the source plane distribution of rays gives the measure of the light curve of a source which passes over that path (see the lower right-hand panel in Figure 1).

Importantly, it was realized the structure in the resulting light curve is dependent upon the size and shape of the source, with a larger source washing out some of the small scale structure seen in a microlensing magnification map (see upper right-hand panels in Figure 1), and hence the calculation of the light curve for a large source requires the convolution of the magnification map with the source profile. To undertake robust calculations with large populations of stars, many rays ($\sim 10^9 - 10^{10}$) need to be traced through the microlensing masses, requiring the application of novel techniques, such as tree-codes, and the use of supercomputers.

5. Microlensing Demagnification

In a seminal study, Schechter and Wambsganss (2002)[8] examined the influence of gravitational microlensing within systems of close image pairs. Undertaking a series of large ray tracing simulations, they showed that if the matter within the lensing galaxies were composed of purely compact matter, then demagnification by microlensing could not account for the observed discrepancy in brightnesses; this is a consequence of the magnification probability distributions resulting from microlensing in these systems being too narrow to give sufficiently different brightnesses.

While this may seem like a victory for those supporting the presence of halo-scale substructure, Schechter and Wambsganss (2002)[8] uncovered a quite non-intuitive result; within their simulations, they remove a fraction of the compact dark matter and replaced it with a smooth component. This significantly influences the resulting microlensing: although commonsense would suggest that the removal of microlenses would diminish the influence of microlensing and smooth out fluctuations, for one of the images within the pair, Schechter and Wambsganss (2002)[8] found the magnification probability distribution broadened substantially with the presence of smooth matter, possessing a tail of extreme demagnification even when the smooth matter component exceeded 95% of the total. This surprising result, coupled with the fact that it always to be a particular image (known the saddle-point) which is fainter, it was concluded that the detection of halo substructure in distant galaxy was not as robust a result as first thought.

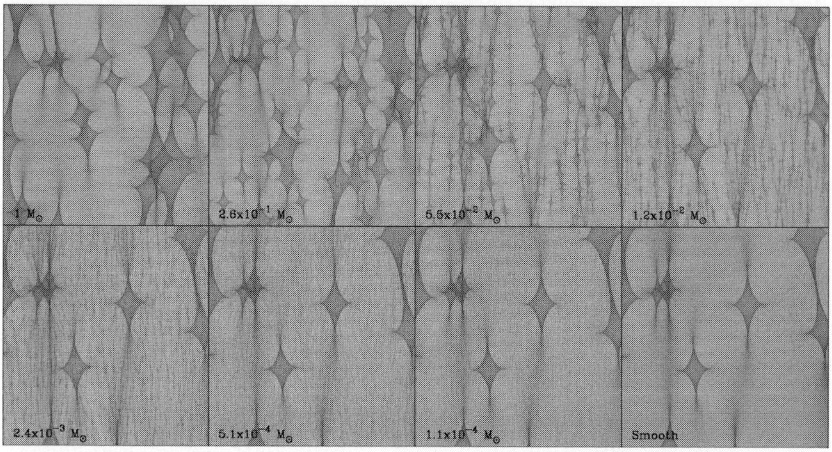

Fig. 2. This series of panels present the influence of a decreasing mass of dark matter. Throughout, the fraction of mass in stars of $1M_\odot$ and that in compact dark matter is equal, and in the top, left-hand panel the dark matter also has a mass of $1M_\odot$. Throughout the panels, the masses of the individual dark matter bodies is reduced, while the number is increased to ensure the surface density remains constant. In the lower right-hand panel, the dark matter component is replaced with smooth matter. Figure taken from Lewis and Gil-Merino (2006).[7]

6. But How Smooth is Smooth?

The results presented by Schechter and Wambsganss (2002)[8] dramatically show how a small number of stars embedded within a smooth dark matter component can spoof the demagnification expected from sub-halo substructure expected within the cold dark model. But does this result imply that dark matter is truly smooth on astrophysical scales, i.e. it is composed of elementary particles, or are we again being fooled? This question was the subject of research by Lewis and Gil-Merino (2006).[7] In this study, ray tracing was used to create an ensemble of gravitational microlensing magnification maps. The parameters for these maps ensured each possessed identical surface mass density, of which half was in solar mass compact objects. The remainder of the mass (assumed to be the dark matter component) was distributed as either smooth matter, or in compact objects with smaller and smaller masses.

Figure 2 presents a sample of the ensemble of magnification maps that were produced; in the upper left-hand panel, all matter is distributed in compact objects of mass $1M_\odot$. Progressing through the panels, the mass of the individual dark matter components is reduced (their mass is recorded in each panel), down to a minimum mass of $1.1 \times 10^{-4}M_\odot$. The final panel in this figure, in the lower right-hand corner, presents the magnification map for the case where the dark matter component is smoothly distributed. Clearly, as the individual masses of the dark matter component are reduced, the scale of structure apparent in the magnification map also decreases, such that at the smallest compact dark matter scale considered, the presence of the dark matter results in a very fine scale granulation of structure superimposed upon large scale magnification variations. This is immediately apparent when comparing the final two panels in the lower, right-hand corner; for the lowest compact mass and smooth dark matter component, the large scale magnification structure is identical, the difference only being in the small scale structure of the compact mass case.

Naively, this result runs contrary to the work of Schechter and Wambsganss (2002),[8] suggesting that the light curves, and hence magnification statistics, in the case with small scale compact dark matter and smooth dark matter should be identical, and hence their replacement with a smooth matter component should not have produced such a dramatic effect: this became the focus of Lewis and Gil-Merino (2006).[7]

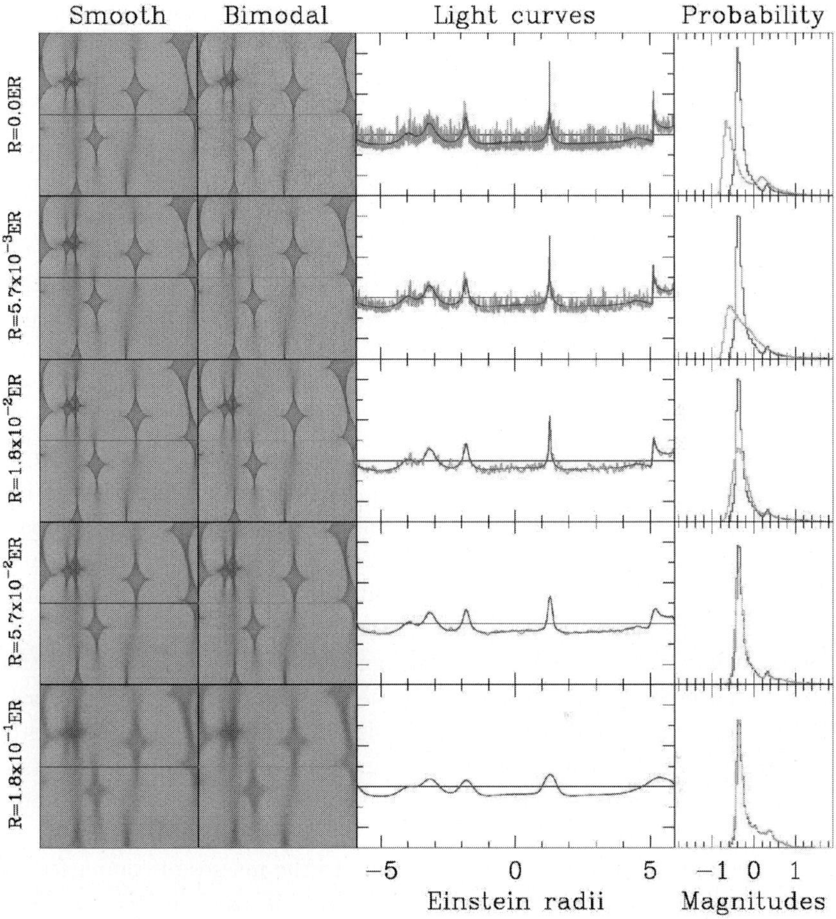

Fig. 3. The results of the microlensing experiments of Lewis and Gil-Merino (2006);[7] see text of Section 7 for detail.

7. Source Size Dependence

One important parameter defining the expected brightness variations for a microlensing scenario has so far been neglected, namely the size of the source under consideration. Lewis and Gil-Merino (2006) examined this for the magnification maps for the lowest mass compact dark matter and smooth distribution of dark matter, with the results presented in Figure 3; this figure presents a series of panels, from top to bottom. The first is the magnification map for the smooth dark matter component, whereas the second is for the case with a compact dark matter with individual masses of $1.1 \times 10^{-4} M_{\odot}$. As you move from top to bottom, these magnification maps are convolved with larger and larger sources (the size is given on the far left). In the next panel, the resultant light curves drawn from paths over the middle of the magnification map are presented, the red being for the smooth dark matter, while the green is for the compact dark matter. As expected, these paths possess generally similar overall structure, although clearly the compact dark matter introduces many small scale fluctuations into the light curve. It is, therefore, quite interesting to examine the magnification probability distributions for each case (presented in the right-hand panel of Figure 3; again red denotes the case for smooth mass, whereas the green is for the compact dark matter); the results are quite different, with the compact matter clearly introducing more substantial demagnifications. As we move down the figure, it's clear that the convolution with the larger source smooths out the magnification maps, and an examination of the resultant light curves show that the small scale structure in the compact dark matter case gets more and more washed out, and in the lowest panel the light curves in the two cases are indistinguishable. This situation is reflected in the magnification probability distributions, with that for the compact dark matter becoming indistinguishable from the smooth matter case at the largest source radii. An immediate consequence of this result is that below a particular source scale size, whether compact or smooth, dark matter will appear as being smooth, and hence there is a resolution to the mass of compact dark matter that can be probed in a gravitational microlensing experiment.

To judge the sensitivity of microlensing to the masses of compact dark matter, a further analysis was undertaken. For this, the magnification map for a compact dark matter component was continually convolved with a source of an increasing size, and the resulting cumulative magnification probability distribution was compared to that of an identical microlensing scenario where the dark matter was distributed smoothly, and were judged to be identical within a certain percentage tolerance. The results of this procedure for a large range of compact dark matter mass are presented in Figure 4, with the circles corresponding to a cumulative probability distribution of 10%, whereas the squares represent 5%; the solid line in this figure represent the Einstein radius (the scale size for gravitational microlensing) for the dark matter; for a solar mass star at cosmological distances, this corresponds to ~ 0.01pc. Clearly this figure reveals a significant trend in the resolution with mass scale, a trend which is roughly proportional to the

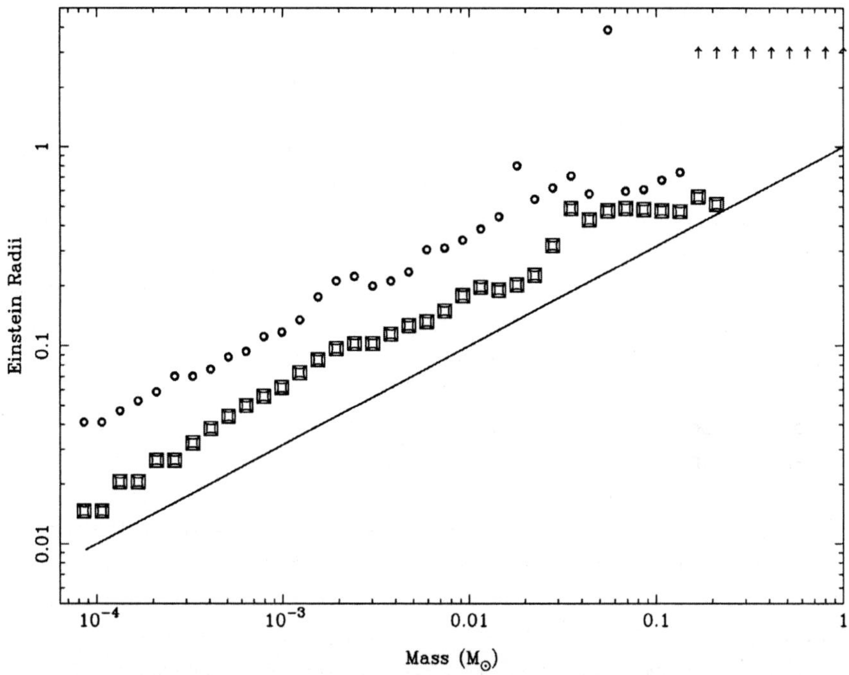

Fig. 4. The relation between the mass resolution due to source size in a gravitational lensing experiment as a function of the mass of the compact dark matter. Taken from Lewis and Gil-Merino (2006) .[7]

Einstein radius of the compact dark matter. Hence, the microlensing light curves of the optically emitting regions of quasars, which typically have a scale of ~ 0.1 Einstein radii, will be insensitive to the nature of dark matter below a mass scale of $\sim 0.01 M_{\odot}$.

8. So What's Next?

It's clear that gravitational microlensing can play an important role in the apparent brightness of multiply imaged quasars, but as we have seen, the size of the source provides a resolution scale below which all dark matter, either compact or smooth, is indistinguishable from smooth matter. Of course, for objects such as quasars, moving to higher energies focuses upon smaller source regions, although these, in themselves, possess significant temporal variability which can confuse any microlensing signature.

However, an alternative approach is to undertake a multi-wavelength, or better still spectroscopic, monitoring of a gravitationally lensed quasar.

The overall microlensing signature at all continuum wavelengths should be similar, although as we move to smaller wavelengths, and hence smaller source, the microlensing signature should possess a microlensing signature which is increasingly different from that of the smooth dark matter case, if the compact matter is present. Such a result, however, could only be achieved through a long-term monitoring program of microlensed quasars, although to-date, none have been planned. The next step in this theoretical analysis is to expand the parameter space explored, in terms of overall surface density and the proportion of dark matter present, with a goal of designing efficient survey strategies from which we can use microlensing to reveal the true nature of dark matter.

Acknowledgments

This research was supported under Australian Research Council's Discovery Projects funding scheme (project number DP0665574). GFL also thanks the School of Physics for their kind support of Dark2007.

References

1. S. D. M. White and M. J. Rees *M.N.R.A.S.* **183**, 341-358 (1978).
2. R. B. Metcalf and P. Madau *Astrophysical Journal* **563**, 9-20 (2001).
3. P. Young *Astrophysical Journal* **244**, 756-767 (1981).
4. K. Chang and S. Refsdal *Nature* **282**, 561-564 (1979).
5. P. Schneider, J. Ehlers and E. E. Falco "Gravitational Lenses", *Gravitational Lenses* (1992) **XIV, 560** pp. *Springer-Verlag, Berlin, Heidelberg, New York* also Astronomy and Astrophysics Library.
6. A. Klypin, A. V. Kravtsov, O. Valenzuela and F. Prada *Astrophysical Journal* **522**, 82-92 (1999) and *arXiv:astro-ph/9901240.*
7. G. F. Lewis and R. Gil-Merino *Astrophysical Journal* **645**, 835-840 (2006) and arXiv:astro-ph/0603623.
8. P. L. Schechter and J. Wambsganss *Astrophysical Journal* **580**, 685-695 (2002) and arXiv:astro-ph/0204425.
9. R. Kayser and S. Refsdal and R. Stabell *Astronomy & Astrophysics* **166**, 36-52 (1986).
10. J. Wambsganss and B. Paczynski and N. Katz *Astrophysical Journal* **352**, 407-412 (1990).

PART II

SUSY/SUGRA Phenomenology in Dark Matter, New Symmetries

LEPTONIC CP VIOLATION AND LEPTOGENESIS

M. N. REBELO*

Departamento de Física and Centro de Física Teórica de Partículas (CFTP)
Instituto Superior Técnico (IST), Av. Rovisco Pais, 1049-001 Lisboa, Portugal
CERN, Department of Physics, Theory Division, CH-1211 Genève 23, Switzerland
**Presently at CERN on sabbatical leave from IST.*
E-mail: margarida.rebelo@cern.ch and rebelo@ist.utl.pt

We review some recent results on the connection between CP violation at low energies and Leptogenesis in the framework of specific flavour structures for the fundamental leptonic mass matrices with zero textures.

Keywords: Leptonic CP violation; Neutrino masses; Leptogenesis.

1. Introduction

Neutrinos have masses which are much smaller than the other fermionic masses and there is large mixing in the leptonic sector. The Standard Model (SM) of electroweak interactions cannot accommodate the observed neutrino masses and leptonic mixing since in the Standard Model neutrinos are strictly massless: the absence of righthanded components for the neutrino fields does not allow one to write a Dirac mass term; the fact that the lefthanded components of the neutrino fields are part of a doublet of $SU(2)$ rules out the possibility of introducing Majorana mass terms since these would violate gauge symmetry; finally, in the SM, $B-L$ is exactly conserved, therefore Majorana mass terms cannot be generated neither radiatively in higher orders nor nonperturbatively. Therefore, neutrino masses require physics beyond the SM. At present, this is the only direct evidence for physics beyond the SM. The origin of neutrino masses remains an open question. It is part of a wider puzzle, the flavour puzzle, with questions such as whether or not there is a connection between quarks and leptons explaining the different patterns of flavour mixing in each sector and the different mass hierarchies. In the seesaw framework[1–5] the explanation of the observed smallness of neutrino masses is related to the existence of heavy neutrinos with masses that can be of the order of the unification scale and have profound implications for cosmology. Mixing in the leptonic sector leads to the possibility of leptonic CP violation both at low and at high energies. CP violation in the decay of heavy neutrinos may allow for the explanation of the observed baryon asymmetry of the Universe (BAU) through leptogenesis.[6] Neutrino physics may also be relevant to the understanding of dark matter and dark energy as well as galaxy-cluster formation. Recent detailed analyses of the present theoretical and experimental situation in neutrino physics and its future, can be found in Refs. 7 and 8.

In this work the possibility that BAU may be generated via leptogenesis through the decay of heavy neutrinos is discussed. Leptogenesis requires CP violation in the decays of heavy neutrinos. However, in general it is not possible to establish a connection between CP violation required for leptogenesis and low energy CP violation.[9,10] This connection can only be established in specific flavour models. The fact that in this framework the masses of the heavy neutrinos are so large that they cannot be produced at present colliders and would have decayed in the early Universe shows the relevance of flavour models in order to prove leptogenesis. In what follows it will be shown how the imposition of texture zeros in the neutrino Yukawa couplings may at the same time constrain physics at low energies and lead to predictions for leptogenesis.

2. Framework and Notation

The work described here is done in the seesaw framework, which provides an elegent way to explain the smallness of neutrino masses, when compared to the masses of the other fermions.

In the minimal seesaw framework, the SM is extended only through the inclusion of righthanded components for the neutrinos which are singlets of $SU(2) \times U(1)$. Frequently, one righthanded neutrino component per generation is introduced. This will be the case in what follows, unless otherwise stated. In fact, neutrino masses can be generated without requiring the number of righthanded and lefthanded neutrinos to be equal. Present observations are consistent with the introdution of two righthanded components only. In this case one of the three light neutrinos would be massless.

With one righthanded neutrino component per generation the number of fermionic degrees of freedom for neutrinos equals those of all other fermions in the theory. However neutrinos are the only known fermions which have zero electrical charge and this allows one to write Majorana mass terms for the singlet fermion fields. After spontaneous symmetry breakdown (SSB) the leptonic mass term is of the form:

$$\mathcal{L}_m = -[\overline{\nu_L^0} m_D \nu_R^0 + \frac{1}{2}\nu_R^{0T} C M_R \nu_R^0 + \overline{l_L^0} m_l l_R^0] + h.c. =$$
$$= -[\frac{1}{2} n_L^T C \mathcal{M}^* n_L + \overline{l_L^0} m_l l_R^0] + h.c. \qquad (1)$$

with the 6×6 matrix $\mathcal{M}$ given by:

$$\mathcal{M} = \begin{pmatrix} 0 & m_D \\ m_D^T & M_R \end{pmatrix} \qquad (2)$$

the upperscript 0 in the neutrino (ν) and charged lepton fields (l) is used to indicate that we are still in a weak basis (WB), i.e., the gauge currents are still diagonal. The charged current is given by:

$$\mathcal{L}_W = -\frac{g}{\sqrt{2}} W_\mu^+ \ \overline{l_L^0} \ \gamma^\mu \ \nu_L^0 + h.c. \qquad (3)$$

Since the Majorana mass term is gauge invariant there are no constraints on the scale of M_R. The seesaw limit consists of taking this scale to be much larger than the scale of the Dirac mass matrices m_D and m_l. The Dirac mass matrices are generated from Yukawa couplings after SSB and are therefore at most of the electroweak scale. As a result the spectrum of the neutrino masses splits into two sets, one consisting of very heavy neutrinos with masses of the order of that of the matrix M_R and the other set with masses obtained, to a very good approximation, from the diagonalisation of an effective Majorana mass matrix given by:

$$m_{eff} = -m_D \frac{1}{M_R} m_D^T \qquad (4)$$

This expression shows that the light neutrino masses are strongly suppressed with respect to the electroweak scale. There is no loss of generality in choosing a WB where m_l is real diagonal and positive. The diagonalization of $\mathcal{M}$ is performed via the unitary transformation:

$$V^T \mathcal{M}^* V = \mathcal{D} \qquad (5)$$

where $\mathcal{D} = \mathrm{diag}(m_1, m_2, m_3, M_1, M_2, M_3)$, with m_i and M_i denoting the physical masses of the light and heavy Majorana neutrinos, respectively. It is convenient to write V and $\mathcal{D}$ in the following block form:

$$V = \begin{pmatrix} K & G \\ S & T \end{pmatrix}; \qquad \mathcal{D} = \begin{pmatrix} d & 0 \\ 0 & D \end{pmatrix}. \qquad (6)$$

The neutrino weak-eigenstates are then related to the mass eigenstates by:

$$\nu_{i\,L}^0 = V_{i\alpha} \nu_{\alpha L} = (K, G) \begin{pmatrix} \nu_{iL} \\ N_{iL} \end{pmatrix} \quad \begin{pmatrix} i = 1,2,3 \\ \alpha = 1,2,...6 \end{pmatrix} \qquad (7)$$

and the leptonic charged current interactions are given by:

$$\mathcal{L}_W = -\frac{g}{\sqrt{2}} \left(\overline{l_{iL}} \gamma_\mu K_{ij} \nu_{jL} + \overline{l_{iL}} \gamma_\mu G_{ij} N_{jL} \right) W^\mu + h.c. \tag{8}$$

with K and G being the charged current couplings of charged leptons to the light neutrinos ν_j and to the heavy neutrinos N_j, respectively.

In the seesaw limit the matrix K coincides to an excellent approximation with the unitary matrix U_ν that diagonalises m_{eff} of Eq. (4):

$$-U_\nu^\dagger \, m_D \frac{1}{M_R} m_D^T \, U_\nu^* = d \tag{9}$$

and the matrix G verifies the exact relation:

$$G = m_D T^* D^{-1} \tag{10}$$

and is therefore very suppressed.

In a general framework, with $\mathcal{M}$ symmetric, without the zero block present in Eq. (2) the 3×6 physical matrix (K, G) of the 6×6 unitary matrix V would depend on six independent mixing angles and twelve independent CP violating phases.[11] This would be possible with a further extention of the SM including a Higgs triplet. The presence of the zero block reduces the number of independent CP violating phases to six.[12] In the seesaw framework massive neutrinos lead to the possibility of CP violation in the leptonic sector both at low and at high energies. CP violation at high energies manifests itself in the decays of heavy neutrinos and is sensitive to phases appearing in the matrix G.

3. Low Energy Leptonic Physics

The light neutrino masses are obtained from the diagonalisation of m_{eff} defined by Eq. (4) which is an effective Majorana mass matrix. The unitary matrix U_ν that diagonalises m_{eff} in the WB where the charged lepton masses are already diagonal real and positive is known as the Pontecorvo, Maki, Nakagawa, Sakata (PMNS) matrix,[13] and can be parametrised as:[14]

$$U_\nu = \begin{pmatrix} c_{12}c_{13} & s_{12}c_{13} & s_{13}e^{-i\delta} \\ -s_{12}c_{23} - c_{12}s_{23}s_{13}e^{i\delta} & c_{12}c_{23} - s_{12}s_{23}s_{13}e^{i\delta} & s_{23}c_{13} \\ s_{12}s_{23} - c_{12}c_{23}s_{13}e^{i\delta} & -c_{12}s_{23} - s_{12}c_{23}s_{13}e^{i\delta} & c_{23}c_{13} \end{pmatrix} \cdot P \tag{11}$$

with $P = \text{diag}\,(1, e^{i\alpha}, e^{i\beta})$, α and β are phases associated to the Majorana character of neutrinos. There are three CP violating phases in U_ν.

Experimentally it is not yet known whether any of the three CP violating phases of the leptonic sector is different from zero. The current experimental bounds on neutrino masses and leptonic mixing are:[14]

$$\Delta m^2_{21} = 8.0^{+0.4}_{-0.3} \times 10^{-5}\ \mathrm{eV}^2 \quad (12)$$
$$\sin^2(2\theta_{12}) = 0.86^{+0.03}_{-0.04} \quad (13)$$
$$|\Delta m^2_{32}| = (1.9\ \text{ to }\ 3.0) \times 10^{-3}\ \mathrm{eV}^2 \quad (14)$$
$$\sin^2(2\theta_{23}) > 0.92 \quad (15)$$
$$\sin^2\theta_{13} < 0.05 \quad (16)$$

with $\Delta m^2_{ij} \equiv m^2_j - m^2_i$. The angle θ_{23} may be maximal, meaning 45°, whilst θ_{12} is already known to deviate from this value. At the moment, there is only an experimental upper bound on the angle θ_{13}.

It is also not yet known whether the ordering of the light neutrino masses is normal, i.e, $m_1 < m_2 < m_3$ or inverted $m_3 < m_1 < m_2$. The scale of the neutrino masses is also not yet established. Direct kinematical limits from Mainz[15] and Troitsk[16] place an upper bound on m_β defined as:

$$m_\beta \equiv \sqrt{\sum_i |U_{ei}|^2 m_i^2} \quad (17)$$

given by $m_\beta \leq 2.3$ eV (Mainz), $m_\beta \leq 2.2$ eV (Troitsk). The forthcoming KATRIN experiment[17] is expected to be sensitive to $m_\beta > 0.2$ eV and to start taking data in 2010.[18]

It is possible to obtain information on the absolute scale of neutrino masses from the study of the cosmic microwave radiation spectrum together with the study of the large scale structure of the universe. For a flat universe, WMAP combined with other astronomical data leads to[19] $\sum_i m_i \leq 0.66$ eV (95% CL).

Neutrinoless double beta decay can also provide information on the absolute scale of the neutrino masses. In the present framework, in the absence of additional lepton number violating interactions, it provides a measurement of the effective Majorana mass given by:

$$m_{ee} = \left| m_1 U^2_{e1} + m_2 U^2_{e2} + m_3 U^2_{e3} \right| \quad (18)$$

The present upper limit is $m_{ee} \leq 0.9$ eV[20] from the Heidelberg-Moskow[21] and the IGEX[22] experiments. There is a claim of discovery of neutrinoless double beta decay by the Heidelberg-Moscow collaboration.[23] Interpreted in terms of a Majorana mass of the neutrino, this implies m_{ee} between 0.12 eV to 0.90 eV. This result awaits confirmation from other experiments and would constitute a major discovery.

It was shown that the strength of CP violation at low energies, observable for example through neutrino oscillations can be obtained from the following low energy WB invariant:[24]

$$Tr[h_{eff}, h_l]^3 = -6i\Delta_{21}\Delta_{32}\Delta_{31}\mathrm{Im}\{(h_{eff})_{12}(h_{eff})_{23}(h_{eff})_{31}\} \qquad (19)$$

where $h_{eff} = m_{eff}{m_{eff}}^{\dagger}$, $h_l = m_l m_l^{\dagger}$, and $\Delta_{21} = ({m_{\mu}}^2 - {m_e}^2)$ with analogous expressions for Δ_{31}, Δ_{32}. The righthand side of this equation is the computation of this invariant in the special WB where the charged lepton masses are real and diagonal. In the case of no CP violation of Dirac type in the leptonic sector this WB invariant vanishes; on the other hand, it is not sensitive to the presence of Majorana phases. This quantity can be computed in any WB and therefore is extremely useful for model building since it enables one to investigate whether a specific ansatz leads to Dirac type CP violation or not, without the need to go to the physical basis. It is also possible to write WB invariant conditions sensitive to the Majorana phases. The general procedure was outlined in Ref. 25 where it was applied to the quark sector. For three generations it was shown that the following four conditions are sufficient[24] to guarantee CP invariance:

$$\mathrm{Im\ tr}\left[h_l\,(m_{eff}\,m^*_{eff})\,(m_{eff}\,h^*_l\,m^*_{eff})\right] = 0 \qquad (20)$$

$$\mathrm{Im\ tr}\left[h_l\,(m_{eff}\,m^*_{eff})^2\,(m_{eff}\,h^*_l\,m^*_{eff})\right] = 0 \qquad (21)$$

$$\mathrm{Im\ tr}\left[h_l\,(m_{eff}\,m^*_{eff})^2\,(m_{eff}\,h^*_l\,m^*_{eff})\,(m_{eff}\,m^*_{eff})\right] = 0 \qquad (22)$$

$$\mathrm{Im\ det}\left[(m^*_{eff}\,h_l\,m_{eff}) + (h^*_l\,m^*_{eff}\,m_{eff})\right] = 0 \qquad (23)$$

provided that neutrino masses are nonzero and nondegenerate (see also Ref. 26). In Ref. 27 alternative WB invariant conditions necessary to guarantee CP invariance in the leptonic sector under less general circumstances are given.

4. Leptogenesis

The observed baryon asymmetry of the universe (BAU) is given by:[28]

$$\frac{n_B - n_{\overline{B}}}{n_\gamma} = (6.1^{+0.3}_{-0.2}) \times 10^{-10}. \qquad (24)$$

It is already established that this observation requires physics beyond the SM in order to be explained. One of the most plausibe explanations is Leptogenesis[6] where out-of-equilibrium L-violating decays of heavy Majorana neutrinos generate a lepton asymmetry which is partially converted through sphaleron processes[29] into a baryon asymmetry. The lepton number asymmetry ε_{N_j}, thus produced was computed by several authors.[30–34] Summing over all charged leptons one obtains for the asymmetry produced by the decay of the heavy Majorana neutrino N_j into the charged leptons $l_i^{\pm}$ ($i =$ e, μ, τ):

$$\begin{aligned}
\varepsilon_{N_j} &= \frac{g^2}{{M_W}^2} \sum_{k \neq j} \left[\text{Im} \left((m_D^\dagger m_D)_{jk} (m_D^\dagger m_D)_{jk} \right) \frac{1}{16\pi} \left(I(x_k) + \frac{\sqrt{x_k}}{1 - x_k} \right) \right] \frac{1}{(m_D^\dagger m_D)_{jj}} \\
&= \frac{g^2}{{M_W}^2} \sum_{k \neq j} \left[(M_k)^2 \text{Im} \left((G^\dagger G)_{jk} (G^\dagger G)_{jk} \right) \frac{1}{16\pi} \left(I(x_k) + \frac{\sqrt{x_k}}{1 - x_k} \right) \right] \frac{1}{(G^\dagger G)_{jj}}
\end{aligned} \tag{25}$$

where M_k denote the heavy neutrino masses, the variable x_k is defined as $x_k = \frac{{M_k}^2}{{M_j}^2}$ and $I(x_k) = \sqrt{x_k}\left(1 + (1 + x_k)\log(\frac{x_k}{1+x_k})\right)$. From Equation (25) it can be seen that, when one sums over all charged leptons, the lepton-number asymmetry is only sensitive to the CP-violating phases appearing in $m_D^\dagger m_D$ in the WB, where M_R is diagonal. Weak basis invariants relevant for leptogenesis were derived in:[9]

$$I_1 \equiv \text{ImTr}[h_D H_R M_R^* h_D^* M_R] = 0 \tag{26}$$
$$I_2 \equiv \text{ImTr}[h_D H_R^2 M_R^* h_D^* M_R] = 0 \tag{27}$$
$$I_3 \equiv \text{ImTr}[h_D H_R^2 M_R^* h_D^* M_R H_R] = 0 \tag{28}$$

with $h_D = m_D^\dagger m_D$ and $H_R = M_R^\dagger M_R$. These constitute a set of necessary and sufficient conditions in the case of three heavy neutrinos. See also.[33]

The simplest realisation of thermal leptogenesis consists of having hierarchical heavy neutrinos. In this case there is a lower bound for the mass of the lightest of the heavy neutrinos.[35,36] Depending on the cosmological scenario, the range for minimal M_1 varies from order 10^7 Gev to 10^9 Gev.[37,38] Furthermore, an upper bound on the light neutrino masses is obtained in order for leptogenesis to be viable. With the assumption that washout effects are not sensitive to the different flavours of charged leptons into which the heavy neutrino decays this bound is approximately 0.1 ev.[39–42] However, it was recently pointed out[43–51] that there are cases where flavour matters and the commonly used expressions for the lepton asymmetry, which depend on the total CP asymmetry and one single efficiency factor, may fail to reproduce the correct lepton asymmetry. In this cases, the calculation of the baryon asymmetry with hierarchical righthanded neutrinos must take into consideration flavour dependent washout processes. As a result, in this case, the previous upper limit on the light neutrino masses does not survive and leptogenesis can be made viable with neutrino masses reaching the cosmological bound of $\sum_i m_i \leq 0.66$ eV. The lower bound on M_1 does not move much with the inclusion of flavour effects. Flavour effects bring new sources of CP violation to leptogenesis and the possibility of having a common origin for CP violation at low energies and for leptogenesis.[52–55]

There are very interesting alternative scenarios to the minimal leptogenesis scenario briefly mentioned here. One was presented at this conference by A. Kusenko[56]

5. Implications from Zero Neutrino Yukawa Textures

The general seesaw framework contains a large number of free parameters. The introduction of zero textures and/or the reduction of the number of righthanded neutrinos to two, allows to reduce the number of parameters. In this work only zero textures imposed in the fundamental leptonic mass matrices are considered and, in particular, zero textures of the Dirac neutrino mass matrix, m_D in the WB where M_R and m_l are real and diagonal. Zero textures of the low energy effective neutrino mass matrix are also very interesting phenomenologically.[57] The physical meaning of the zero textures that appear in most of the leptonic mass ansätze was analysed in a recent work[58] where it is shown that some leptonic zero texture ansätze can be obtained from WB transformations and therefore have no physical meaning.

In general, zero textures reduce the number of CP violating phases, as a result some sets of zero textures imply the vanishing of certain CP-odd WB invariants.[59] This is an important fact since clearly zero textures are not WB invariant, therefore in a different WB the zeros may not be present making it difficult to recognise the ansatz. Furthermore, it was also shown[59] that starting from arbitrary leptonic mass matrices, the vanishing of certain CP-odd WB invariants, together with the assumption of no conspiracy among the parameters of the Dirac and Majorana mass terms, one is automatically lead to given sets of zero textures in a particular WB.

Frampton, Glashow and Yanagida have shown[60] that it is possible to uniquely relate the sign of the baryon number of the Universe to CP violation in neutrino oscillation experiments by imposing two zeros in m_D, in the seesaw framework with only two righthanded neutrino components. Two examples were given by these authors:

$$m_D = \begin{pmatrix} a & 0 \\ a' & b \\ 0 & b' \end{pmatrix} \quad \text{or} \quad m_D = \begin{pmatrix} a & 0 \\ 0 & b \\ a' & b' \end{pmatrix} \tag{29}$$

The two zeros in m_D eliminate two CP violating phases, so that only one

CP violating phase remains. This is the most economical extension of the standard model leading to leptogenesis and at the same time allowing for low energy CP violation. Imposing that the model accommodates the experimental facts at low energy strongly constrains its parameters.

In Ref. 61 minimal scenarios for leptogenesis and CP violation at low energies were analysed in some specific realisations of seesaw models with three righthanded neutrinos and four zero textures in m_D, where three of the zeros are in the upper triangular part of the matrix. This last particular feature was motivated by the fact that there is no loss of generality in parametrising m_D as:

$$m_D = U\, Y_\triangle \,, \tag{30}$$

with U a unitary matrix and $Y_\triangle$ a lower triangular matrix, i.e.:

$$Y_\triangle = \begin{pmatrix} y_{11} & 0 & 0 \\ y_{21}\, e^{i\,\phi_{21}} & y_{22} & 0 \\ y_{31}\, e^{i\,\phi_{31}} & y_{32}\, e^{i\,\phi_{32}} & y_{33} \end{pmatrix} , \tag{31}$$

where y_{ij} are real positive numbers. Choosing $U = 1$ reduces the number of parameters in m_D. Moreover, U cancels out in the combination $m_D^\dagger m_D$ relevant in the case of unflavoured leptogenesis, whilst it does not cancel in m_{eff}. Therefore choosing $U = 1$ allows for a connection between low energy CP violation and leptogenesis to be established since in this case the same phases affect both phenomena. The nonzero entries of m_D were written in terms of powers of a small parameter a la Frogatt Nielsen[62] and chosen in such a way as to accommodate the experimental data. Viable leptogenesis was found requiring the existence of low energy CP violating effects within the range of sensitivity of the future long baseline neutrino oscillation experiments under consideration.

In order to understand how the connection between CP violation required for leptogenesis and low energy physics is established in the presence of zeros in the matrix m_D, the following relation derived from Eq. (9) in the WB where M_R and m_l are real positive and diagonal is important:

$$m_D = iU_\nu \sqrt{d} R \sqrt{D}_R \tag{32}$$

with R an orthogonal complex matrix, $\sqrt{D}_R$ a diagonal real matrix verifying the relation $\sqrt{D}_R \sqrt{D}_R = D_R$ and $\sqrt{d}$ a real matrix with a maximum number of zeros such that $\sqrt{d}\sqrt{d}^T = d$. This is the well known Casas and Ibarra parametrisation.[63] From this equation it follows that:

$$m_D^\dagger m_D = \sqrt{D}_R R^\dagger \sqrt{d}^T \sqrt{d} R \sqrt{D}_R \tag{33}$$

Since the CP violating phases relevant for leptogenesis in the unflavoured case are those contained in $m_D^\dagger m_D$, it is clear that leptogenesis can occur even if there is no CP violation at low energies i.e. no Majorana- or Dirac-type CP phases at low energies.[10] Unflavoured leptogenesis requires the matrix R to be complex. In flavoured leptogenesis the separate lepton i family asymmetry generated from the decay of the kth heavy Majorana neutrino depends on the combination[45] $\mathrm{Im}\Big((m_D^\dagger m_D)_{kk'}(m_D^*)_{ik}(m_D)_{ik'}\Big)$ as well as on $\mathrm{Im}\Big((m_D^\dagger m_D)_{k'k}(m_D^*)_{ik}(m_D)_{ik'}\Big)$. The matrix U_ν does not cancel in each of these terms and it was shown that it is possible to have viable leptogenesis even in the case of real R, with CP violation in the PMNS matrix as the source of CP violation required for leptogenesis.

From Eq. (32) it is clear that one zero in $(m_D)_{ij}$ corresponds to having an orthogonality relation between the ith row of the matrix $U_\nu\sqrt{d}$ and the jth column of the matrix R:

$$(m_D)_{ij} = 0 \ : \qquad (U_\nu)_{ik}\sqrt{d}_{kl}R_{lj} = 0 \tag{34}$$

Ibarra and Ross[64] showed that, in the seesaw case with only two righthanded neutrinos, a single zero texture, has the special feature of fixing the matrix R, up to a reflection, without imposing any further restriction on light neutrino masses and mixing. The predictions from models with two zero textures in m_D were also analysed in detail in their work, including the constraints on leptogenesis and lepton flavour violating processes. The number of all different two texture zeros is fifteen. Two zeros imply two simultaneous conditions of the type given by Eq. (34). Compatibility of these two conditions implies restrictions on U_ν and $\sqrt{m_i}$. Only five of these cases turned out to be allowed experimentally, including the two cases of Eq. (29) in this reference.

All of these two zero texture ansätze satify the following WB invariant condition:[59]

$$I_1 \equiv \mathrm{tr}\left[m_D M_R^\dagger M_R m_D^\dagger, h_l\right]^3 = 0 \tag{35}$$

with $h_l = m_l m_l^\dagger$, as before. It was also shown[59] that for arbitrary complex leptonic mass matrices, assuming that there are no special relations among the entries of M_R and those of m_D this condition automatically leads to one of the two zero anzätze classified in Ref. 64. The assumption that M_R and m_D are not related to each other is quite natural, since m_D and M_R originate from different terms of the Lagrangian.

There are other CP-odd WB invariants which vanish for all of the two zero textures just mentioned, even if they arise in a basis where M_R is not diagonal. An example is the following WB invariant condition:[59]

$$I' \equiv \text{tr}\left[m_D m_D^\dagger, h_l\right]^3 = 0 \tag{36}$$

which is verified for any texture with two zeros in m_D in a WB where m_l is diagonal, while M_R is arbitrary.

The case of zero textures with three righthanded neutrinos was also considered in Ref 59. In this case the WB invariant I_1 always vanishes for three zero textures in m_D with two orthogonal rows, which implies that one row has no zeros. The case of three zeros corresponding to two orthogonal columns of m_D, which in this case implies that one column has no zeros leads to the vanishing of a new invariant I_2, defined by:

$$I_2 \equiv \text{tr}\left[M_R^\dagger M_R\ ,\ \ m_D^\dagger m_D\right]^3 \tag{37}$$

Four zero textures in the context of seesaw with three righthanded neutrinos are studied in detail in Ref. 65. It is shown that four is the maximum number of zeros in textures compatible with the observed leptonic mixing and with the additional requirement that none of the neutrino masses vanishes. It is also shown that such textures lead to important constraints both at low and high energies, and allow for a tight connection between leptogenesis and low energy parameters. It is possible in all cases to completely specify the matrix R in terms of light neutrino masses and the PMNS matrix. These relations are explicitly given in Ref. 65.

Acknowledgements

The author thanks the Organizers of the Sixth International Heidelberg Conference on Dark Matter in Astro and Particle Physics which took place in Sydney, Australia for the the warm hospitality and the stimulating scientifical environment provided. This work was partially supported by Fundação para a Ciência e a Tecnologia (FCT, Portugal) through the projects PDCT/FP/63914/2005, PDCT/FP/63912/2005 and CFTP-FCT UNIT 777 which are partially funded through POCTI (FEDER).

References

1. P. Minkowski, Phys. Lett. B **67**, 421 (1977).
2. T. Yanagida, *In Proceedings of the Workshop on the Baryon Number of the Universe and Unified Theories, Tsukuba, Japan, 13-14 Feb 1979*, edited by O. Sawada and A. Sugamoto

3. S. L. Glashow in "Quarks And Leptons. Proceedings, Summer Institute, Cargese, France, July 9-29, 1979," edited by M. Levy, J. L. Basdevant, D. Speiser, J. Weyers, R. Gastmans and M. Jacob *New York, Usa: Plenum (1980) 720 P.(Nato Advanced Study Institutes Series: Series B, Physics, 61).*
4. M. Gell-Mann, P. Ramond and R. Slansky in "Supergravity. Proceedings, Workshop At Stony Brook, 27-29 September 1979," edited by P. Van Nieuwenhuizen and D. Z. Freedman, *Amsterdam, Netherlands: North-holland (1979) 341p*
5. R. N. Mohapatra and G. Senjanovic, Phys. Rev. Lett. **44**, 912 (1980).
6. M. Fukugita and T. Yanagida, Phys. Lett. B **174**, 45 (1986).
7. R. N. Mohapatra *et al.*, Rept. Prog. Phys. **70**, 1757 (2007) [arXiv:hep-ph/0510213].
8. T. I. P. Group, arXiv:0710.4947 [hep-ph].
9. G. C. Branco, T. Morozumi, B. M. Nobre and M. N. Rebelo, Nucl. Phys. B **617**, 475 (2001) [arXiv:hep-ph/0107164].
10. M. N. Rebelo, Phys. Rev. D **67**, 013008 (2003) [arXiv:hep-ph/0207236].
11. G. C. Branco and L. Lavoura, Nucl. Phys. B **278**, 738 (1986).
12. T. Endoh, T. Morozumi, T. Onogi and A. Purwanto, Phys. Rev. D **64**, 013006 (2001) [Erratum-ibid. D **64**, 059904 (2001)] [arXiv:hep-ph/0012345].
13. B. Pontecorvo, Sov. Phys. JETP **7**, 172 (1958) [Zh. Eksp. Teor. Fiz. **34**, 247 (1957)]; Z. Maki, M. Nakagawa and S. Sakata, Prog. Theor. Phys. **28**, 870 (1962); B. Pontecorvo, Sov. Phys. JETP **26**, 984 (1968) [Zh. Eksp. Teor. Fiz. **53**, 1717 (1967)].
14. W. M. Yao *et al.* [Particle Data Group], J. Phys. G **33**, 1 (2006).
15. C. Kraus *et al.*, Eur. Phys. J. C **40**, 447 (2005) [arXiv:hep-ex/0412056].
16. V. M. Lobashev *et al.*, Phys. Lett. B **460**, 227 (1999).
17. A. Osipowicz *et al.* [KATRIN Collaboration], arXiv:hep-ex/0109033.
18. K. Valerius and K. Collaboration [KATRIN Collaboration], arXiv:0710.4906 [hep-ex].
19. D. N. Spergel *et al.* [WMAP Collaboration], Astrophys. J. Suppl. **170**, 377 (2007) [arXiv:astro-ph/0603449].
20. G. L. Fogli, E. Lisi, A. Marrone, A. Melchiorri, A. Palazzo, P. Serra and J. Silk, Phys. Rev. D **70**, 113003 (2004) [arXiv:hep-ph/0408045].
21. H. V. Klapdor-Kleingrothaus *et al.*, Eur. Phys. J. A **12**, 147 (2001) [arXiv:hep-ph/0103062].
22. C. E. Aalseth *et al.* [IGEX Collaboration], Phys. Rev. D **65**, 092007 (2002) [arXiv:hep-ex/0202026].
23. H. V. Klapdor-Kleingrothaus, I. V. Krivosheina, A. Dietz and O. Chkvorets, Phys. Lett. B **586**, 198 (2004) [arXiv:hep-ph/0404088].
24. G. C. Branco, L. Lavoura and M. N. Rebelo, Phys. Lett. B **180**, 264 (1986).
25. J. Bernabeu, G. C. Branco and M. Gronau, Phys. Lett. B **169**, 243 (1986).
26. H. K. Dreiner, J. S. Kim, O. Lebedev and M. Thormeier, Phys. Rev. D **76**, 015006 (2007) [arXiv:hep-ph/0703074].
27. G. C. Branco and M. N. Rebelo, New J. Phys. **7**, 86 (2005) [arXiv:hep-ph/0411196].

28. C. L. Bennett *et al.* [WMAP Collaboration], Astrophys. J. Suppl. **148**, 1 (2003) [arXiv:astro-ph/0302207].
29. V. A. Kuzmin, V. A. Rubakov and M. E. Shaposhnikov, Phys. Lett. B **155**, 36 (1985).
30. J. Liu and G. Segre, Phys. Rev. D **48**, 4609 (1993) [arXiv:hep-ph/9304241].
31. M. Flanz, E. A. Paschos and U. Sarkar, Phys. Lett. B **345**, 248 (1995) [Erratum-ibid. B **382**, 447 (1996)] [arXiv:hep-ph/9411366].
32. L. Covi, E. Roulet and F. Vissani, Phys. Lett. B **384**, 169 (1996) [arXiv:hep-ph/9605319].
33. A. Pilaftsis, Phys. Rev. D **56**, 5431 (1997) [arXiv:hep-ph/9707235].
34. W. Buchmuller and M. Plumacher, Phys. Lett. B **431**, 354 (1998) [arXiv:hep-ph/9710460].
35. S. Davidson and A. Ibarra, Phys. Lett. B **535**, 25 (2002) [arXiv:hep-ph/0202239].
36. K. Hamaguchi, H. Murayama and T. Yanagida, Phys. Rev. D **65**, 043512 (2002) [arXiv:hep-ph/0109030].
37. W. Buchmuller, P. Di Bari and M. Plumacher, Nucl. Phys. B **643**, 367 (2002) [arXiv:hep-ph/0205349].
38. G. F. Giudice, A. Notari, M. Raidal, A. Riotto and A. Strumia, Nucl. Phys. B **685**, 89 (2004) [arXiv:hep-ph/0310123].
39. W. Buchmuller, P. Di Bari and M. Plumacher, Nucl. Phys. B **665**, 445 (2003) [arXiv:hep-ph/0302092].
40. T. Hambye, Y. Lin, A. Notari, M. Papucci and A. Strumia, Nucl. Phys. B **695**, 169 (2004) [arXiv:hep-ph/0312203].
41. W. Buchmuller, P. Di Bari and M. Plumacher, Annals Phys. **315**, 305 (2005) [arXiv:hep-ph/0401240].
42. W. Buchmuller, P. Di Bari and M. Plumacher, New J. Phys. **6**, 105 (2004) [arXiv:hep-ph/0406014].
43. R. Barbieri, P. Creminelli, A. Strumia and N. Tetradis, Nucl. Phys. B **575**, 61 (2000) [arXiv:hep-ph/9911315].
44. T. Endoh, T. Morozumi and Z. h. Xiong, Prog. Theor. Phys. **111**, 123 (2004) [arXiv:hep-ph/0308276].
45. T. Fujihara, S. Kaneko, S. Kang, D. Kimura, T. Morozumi and M. Tanimoto, Phys. Rev. D **72**, 016006 (2005) [arXiv:hep-ph/0505076].
46. A. Pilaftsis and T. E. J. Underwood, Phys. Rev. D **72**, 113001 (2005) [arXiv:hep-ph/0506107].
47. O. Vives, Phys. Rev. D **73**, 073006 (2006) [arXiv:hep-ph/0512160].
48. A. Abada, S. Davidson, F. X. Josse-Michaux, M. Losada and A. Riotto, JCAP **0604**, 004 (2006) [arXiv:hep-ph/0601083].
49. E. Nardi, Y. Nir, E. Roulet and J. Racker, JHEP **0601**, 164 (2006) [arXiv:hep-ph/0601084].
50. A. Abada, S. Davidson, A. Ibarra, F. X. Josse-Michaux, M. Losada and A. Riotto, JHEP **0609**, 010 (2006) [arXiv:hep-ph/0605281].
51. S. Blanchet and P. Di Bari, JCAP **0703**, 018 (2007) [arXiv:hep-ph/0607330].
52. S. Pascoli, S. T. Petcov and A. Riotto, Phys. Rev. D **75**, 083511 (2007) [arXiv:hep-ph/0609125].

53. G. C. Branco, A. J. Buras, S. Jager, S. Uhlig and A. Weiler, JHEP **0709**, 004 (2007) [arXiv:hep-ph/0609067].
54. G. C. Branco, R. Gonzalez Felipe and F. R. Joaquim, Phys. Lett. B **645**, 432 (2007) [arXiv:hep-ph/0609297].
55. S. Uhlig, arXiv:hep-ph/0612262.
56. A. Kusenko, in these Proceedings.
57. P. H. Frampton, S. L. Glashow and D. Marfatia, Phys. Lett. B **536**, 79 (2002) [arXiv:hep-ph/0201008].
58. G. C. Branco, D. Emmanuel-Costa, R. G. Felipe and H. Serodio, arXiv:0711.1613 [hep-ph].
59. G. C. Branco, M. N. Rebelo and J. I. Silva-Marcos, Phys. Lett. B **633**, 345 (2006) [arXiv:hep-ph/0510412].
60. P. H. Frampton, S. L. Glashow and T. Yanagida, Phys. Lett. B **548**, 119 (2002) [arXiv:hep-ph/0208157].
61. G. C. Branco, R. Gonzalez Felipe, F. R. Joaquim, I. Masina, M. N. Rebelo and C. A. Savoy, Phys. Rev. D **67**, 073025 (2003) [arXiv:hep-ph/0211001].
62. C. D. Froggatt and H. B. Nielsen, Nucl. Phys. B **147**, 277 (1979).
63. J. A. Casas and A. Ibarra, Nucl. Phys. B **618**, 171 (2001) [arXiv:hep-ph/0103065].
64. A. Ibarra and G. G. Ross, Phys. Lett. B **591**, 285 (2004) [arXiv:hep-ph/0312138].
65. G. C. Branco, D. Emmanuel-Costa, M. N. Rebelo and Probir Roy "Four Zero Neutrino Yukawa Textures in the Minimal Seesaw Framework", CERN-PH-TH/2007-217

TOWARDS DARK ENERGY FROM STRING-THEORY

AXEL KRAUSE*

Arnold Sommerfeld Center for Theoretical Physics,
Department für Physik, Ludwig-Maximilians-Universität München,
Theresienstr. 37, 80333 München, Germany
**E-mail: axel.krause@physik.uni-muenchen.de*

We discuss the vacuum energy in string and M-theory with a focus on heterotic M-theory. In the latter theory a mechanism is described for maintaining zero vacuum energy *after* supersymmetry breaking. Higher-order corrections can be expected to give a sufficiently small amount of vacuum energy to possibly account for dark energy.

Keywords: M-theory, Dark Energy, Supersymmetry Breaking.

1. The Dark Energy Enigma

Cosmology underwent a revolution after it had been discovered in 1998 that our universe's current expansion accelerates, caused by some unknown, homogeneously distributed dark energy which dominates all other forms of energy or matter.[1,2] Under the assumption of a standard Friedmann-Robertson-Walker (FRW) cosmology, the dimming of distant type Ia supernovae, studies of angular anisotropies in the cosmic microwave background and studies of spatial correlations in the large-scale structure of galaxies have all led to this result. The price, on the other hand, for fitting the data without a dominating dark energy component is to adopt a spectrum of primordial density fluctuations which is not nearly scale-free, a Hubble constant which is globally lower than its locally measured value and a likely shift from the FRW to an inhomogeneous Lemaitre-Tolman-Bondi cosmology in order to explain the supernovae Ia Hubble diagram and the position of the b aryon acoustic oscillation peak in the autocorrelation function of galaxies.[3] It seems therefore less problematic to accept the existence of dark energy in combination with the ordinary FRW description.

In contrast to this rather firm observational evidence for dark energy, there is little consensus on the theory side on what the correct explanation for the nature of dark energy could be. Over the past years various proposals for its origin have been made which encompass

- *cosmological constant* Λ:
  leads to a time-independent energy-density $\rho_{DE} = \Lambda/8\pi G$, Λ being the cosmological constant, and results in string-theory from metastable non-perturbative vacua[4–6] or warped compactifications with branes.[7–17]
- *quintessence*:
  a dynamical scalar field rolls down an exponential potential and induces a time-dependent cosmological "constant" which evolves to small values at late times.[18–20] It opens up the possibility for tracker mechanisms,[21] which attempt to solve the coincidence problem, which arises from the near coincidence of the energy and the matter density in our present universe.
- *holographic dark energy*:
  to prevent a collapse into a black hole, the total energy of a region inside radius L should not exceed the mass of a black hole with same radius L. This leads to a bound on the quantum zero-point energy density within a spherical region of radius L.[22] Saturation of this bound gives a relation between the UV cut-off and the IR cut-off of a quantum field theory which implies an energy density of the right magnitude.[23–28] The actual equation of state depends, however, sensitively on the choice of the IR cut-off.[23,29]
- *ghost cosmology*:
  a ghost scalar ϕ has the wrong sign kinetic term and leads to an energy density unbounded from below. This renders the $\langle\phi\rangle = 0$ vacuum unstable. A non-minimal kinetic term could, however, allow the ghost to condense in a stable vacuum acquiring a non-zero constant velocity in field space.[30] This new vacuum breaks Lorentz invariance spontaneously and leads to an infrared modification of gravity. Depending on the choice of the non-minimal kinetic term, the solutions to the coupled ghost-gravity field equations describe transitions from early power-law FRW scale-factors, $a(t) \sim t^{(2n-1)/3n}$, including radiation and matter cosmologies, to late time dark matter or dark energy dominated cosmologies.[31]
- *modified gravity*:
  an accelerated expansion of the universe can either have its origin in a cosmological constant resp. scalar field added to the energy momentum tensor on the right hand side of the Einstein equa-

tions or in a modification of the geometrical part on its left hand side. The latter is the starting point for theories of modified gravity. For instance, one might add quadratic corrections[32] or inverse powers[33–35] of the Ricci-scalar R to the Einstein-Hilbert action or replace it by a general function $f(R)$ (see refs. 36,37 for reviews). Alternatively, brane-world constructions can also modify the geometrical part of the Einstein equation. In the DGP model[38] quantum corrections induce a brane Einstein-Hilbert term next to a bulk Einstein-Hilbert term. This leads to an IR modification of gravity at lengths beyond a cross-over scale and gives rise to a late time acceleration.

- *neutrino dark energy*:
  the apparent closeness of the dark energy and the neutrino mass scale motivates a linkage between them. Concretely, one extends the Standard Model by singlet right-handed neutrinos and allows their Majorana masses to vary with the acceleron, a dynamical scalar field. The acceleron provides the link between a quintessential dark energy and the neutrino masses.[39–41] See ref. 42 for a recent update.

2. Ubiquitous Finetuning Problem and Two-Step Strategy

In spite of the huge difference in the origins of the proposals for the nature of dark energy, they typically face one major problem. This is the required *tuning* or arbitrary choice of some parameters to generate the enormous smallness of the dark energy scale, $E_{DE} \simeq 1\text{meV}$, compared with any energy scale of fundamental physics such as the reduced Planck scale, $M_{Pl} = 2.4 \times 10^{18}\text{GeV}$ or the grand unification scale, $M_{GUT} = 2 \times 10^{16}\text{GeV}$. Supersymmetry in particle theory or supergravity models alleviates the discrepancy but does not nullify it. The problem persists in having a vacuum energy scale, *after* supersymmetry breaking, which is of order the supersymmetry breaking scale, $M_{SUSY} \approx 1\text{TeV}$, and thus still far too large to match the dark energy scale. This is the ubiquitous cosmological constant (CC) problem. Since all forms of energy gravitate, it is not enough to tune a particular sector of the theory to give an E_{DE} vacuum energy. Rather one has to ensure that at the same time none of the other sectors can develop energies surpassing E_{DE}. Moreover, in fundamental theories such as string-theory, low-energy parameters have their origin in dynamical scalar fields. Their values are thus determined dynamically and a fine-tuning is in principle unacceptable. Hence, the CC-problem cannot be glossed over in any proposal for dark energy and its dynamical solution is a necessary requisite for any successful explanation of dark energy.

The purpose of this article is to propose a *dynamical* way of generating the small dark energy scale. To this end, let us adopt the following natural 2-step approach to address the CC-problem

- First Step: find a mechanism which adjusts dynamically the vacuum energy *after* supersymmetry breaking to zero
$$V_{vac} = 0 \,. \tag{1}$$
This should be true at leading orders in some suitable expansion, like a $1/M_{Pl}$ expansion in effective field theories or an $l_s = \sqrt{\alpha'}$ expansion in string-theory.
- Second Step: higher-order perturbative and/or non-perturbative corrections to the theory are considered to lift the zero leading order vacuum energy to non-zero small positive values of the right magnitude
$$V_{vac} \simeq E_{DE}^4 \,. \tag{2}$$

Of course, generating the right magnitude for V_{vac} is not sufficient to explain dark energy. In addition, one has to check e.g. that also the equation of state parameter w lies sufficiently close to -1 to comply with observation. We won't discuss such additional checks further here, since already the generation of the right energy scale without tuning is a tremendous task and seems to be quite selective. The second step in this 2-step approach is further motivated by various numerical coincidences which relate the observed dark energy scale to expressions which could plausibly arise from subleading perturbative or non-perturbative corrections. For instance, the well-known relation

$$E_{DE}^4 \simeq M_{SUSY}^4 \times (M_{SUSY}/M_{Pl})^4 \,, \tag{3}$$

which gives the right size for E_{DE} might arise at subleading order in a perturbative expansion in M_{SUSY}/M_{Pl}. Also, non-perturbative instanton corrections might produce

$$E_{DE}^4 \simeq e^{-2/\alpha} M_{Pl}^4 \,, \tag{4}$$

where $\alpha \approx 1/137$ denotes the fine structure constant. A third numerical relation

$$E_{DE}^4 \simeq e^{-E_{Pl}/2M_{GUT}} E_{Pl}^4 \,, \tag{5}$$

where $E_{Pl} = 1.2 \times 10^{19}$GeV is the Planck energy, has been argued to arise in warped brane worlds[9,10] but might also have a non-perturbative origin.

With this 2-step approach in mind, let us in the sequel look at the prospects to realize it in supergravity, string-theory and finally M-theory.

3. Dark Energy and Supergravity

The best motivated theories beyond the Standard Model are based on supersymmetry, which we also adopt as one of the cornerstones in the following. Unbroken global supersymmetry has the attractive feature of enforcing a vanishing vacuum energy, as opposed to unbroken local supersymmetry. The latter being compatible with both zero and negative vacuum energies. The first question concerning supersymmetry is therefore whether it should be global or local. Here we note that a non-zero dark energy requires a curved spacetime, hence gravity and this implies local supersymmetry. Namely, in the presence of a non-zero homogeneous dark energy fluid, described by a diagonal energy-momentum tensor $T_{\mu\nu}^{DE} = \text{diag}(\rho, p, p, p)$, the Einstein field equations

$$R_{\mu\nu} - \frac{1}{2} R g_{\mu\nu} = \frac{8\pi G}{c^4} T_{\mu\nu}^{DE} \tag{6}$$

have non-vanishing diagonal components on the right hand side which imply via the geometrical left hand side a curved spacetime and no flat Minkowski spacetime. It is thus mandatory that we work in local supersymmetry, i.e. supergravity, which incorporates gravity.

If we then assume the phenomenologically favored $N = 1$ supergravity framework in four dimensions, there are two options *before* breaking supersymmetry. Supersymmetry is either compatible with an anti de Sitter (AdS) or Minkowski spacetime. Both spacetimes are maximally symmetric. In view of the fact that the generated vacuum energy after supersymmetry breaking should not deviate much from zero, one would prefer to start off at this stage with a Minkowski solution. The AdS option would introduce already at this stage a large negative vacuum energy whose compensation requires an extremely precise fine-tuning. However, in the Minkowski case, it is the solution itself which cannot be obtained without fine-tuning in supergravity. The $N = 1$ supersymmetry preserving vacua are characterized by a system of $N_c + 1$ equations, which involve the superpotential W

$$W = D_{\Phi_i} W = 0 \, , \qquad i = 1, \ldots, N_c \, . \tag{7}$$

But there are only N_c unknowns, the scalar components Φ_i of the chiral superfields. Since this system of equations is *overdetermined*, its solution is always *non-generic* and imposes fine-tuning on the physical parameters which enter the superpotential W and/or the Kähler potential K inside the Kähler covariant derivative $D_{\Phi_i}W = \partial_{\Phi_i}W + (\partial_{\Phi_i}K)W$. To deal with this fine-tuning there are two possibilities:

- one can accept the fine-tuning as unavoidable and a fundamental feature. In this case one might resort to a huge "*landscape of vacua*" combined with *anthropic* reasoning to try to make sense of a tiny, non-zero vacuum energy along the lines suggested in ref. 43 (for a recent discussion see ref. 44).
- alternatively one could go *beyond the effective* $N = 1$ *supergravity* in four dimensions and ask what string-theory has to offer in addition. After all supergravity captures only the massless spectrum of a string compactification and throws away all finite mass excitations.

Here, we decide to follow the latter route which brings us next to string-theory.

4. Heterotic vs Type IIB String-Theory

A priori there is a variety of five ten-dimensional string-theories which, together with eleven-dimensional supergravity, span the M-theory web. The five ten-dimensional perturbatively constructed string-theories represent special corners in this web at which the string couplings tends to zero. A generic point in this web, however, represents an eleven-dimensional theory, with the string coupling itself turning into the eleventh dimension when it grows.

To date, two regions have been identified in this M-theory web which are phenomenologically rich enough to admit a connection to the "real" cosmological and particle physics world. These are the ten-dimensional type IIB and the $E_8 \times E_8$ heterotic string theories. The latter is endowed with two E_8 gauge groups, one in a "hidden" sector and one in the "visible" sector. Both sectors interact only via (super)gravity. To bridge the gap from ten to four dimensions these theories have to be compactified on specific real 6-dimensional manifolds. For phenomenological reasons, and also for better technical control, one requires that this compactification preserves at least four supercharges in the effective four-dimensional supergravity theory. There is, however, an important distinction at this point. The type IIB theory has 32 supercharges in ten dimensions which is twice the amount of the $E_8 \times E_8$ theory. This difference has the consequence that supersymmetri c compactifications of the type IIB theory generically leave us with an

AdS spacetime in four dimensions with a large negative energy density

$$\text{Type IIB}: \qquad V_{vac} \ll 0\,, \tag{8}$$

whereas supersymmetric compactifications of the $E_8 \times E_8$ heterotic theory lead to a Minkowski solution[45] which has vanishing vacuum energy

$$\text{Heterotic } E_8 \times E_8: \qquad V_{vac} = 0\,. \tag{9}$$

As a consequence, to provide type IIB compactifications with a cosmologically relevant positive energy density, which can only be done by breaking supersymmetry, an additional "uplift" is required. This additional step, which in the simplest cases can be carried out by adding a supersymmetry-breaking anti-D-brane, adds positive energy density to the vacuum. In this approach one compensates the initially large negative AdS energy with the added positive "uplift" energy to end up with a desired but fine-tuned small vacuum energy. On the other hand, in the heterotic $E_8 \times E_8$ compactifications, we start off at zero vacuum energy without any need for an "uplift". Furthermore, supersymmetry breaking can only generate a *positive* or zero vacuum energy thanks to the theory's perfect square potential.[46] The challenge in the heterotic theory is thus to break supersymmetry and keep the vacuum energy small. This will be our main concern in the remainder .

5. Vacuum Energy after Supersymmetry Breaking in Heterotic String Compactifications

With the just given motivation, let us investigate the vacuum energy after supersymmetry breaking in the heterotic string. The aim is to implement the first step of our initially advocated 2-step approach, which is to break supersymmetry in such a way as to maintain at leading orders in α' the zero vacuum energy of the original supersymmetric theory. In fact such a mechanism had been proposed early on in the history of the heterotic string in ref. 46. The idea has been to consider a heterotic string compactification on a compact Calabi-Yau threefold $\mathbf{X}$ in the presence of 3-form Neveu-Schwarz flux H and a gaugino condensate in the hidden sector E_8. This type of compactification yields a *positive definite* potential

$$S_{pot} = -\frac{1}{2\kappa_{10}^2}\int_{\mathbf{R}^{1,3}\times\mathbf{X}} e^{-\phi}\Big(H - \frac{\alpha'}{16}e^{\phi/2}\mathrm{tr}\bar{\chi}\Gamma^{(3)}\chi\Big)^2\,, \tag{10}$$

with ϕ being the dilaton, κ_{10} the ten-dimensional gravitational coupling constant, χ the ten-dimensional hidden sector gaugino and $\Gamma^{(3)}$ an anti-symmetric three-index gamma matrix. The gaugino condensate $\langle \mathrm{tr}\bar{\chi}\Gamma^{(3)}\chi\rangle$

can only assume values proportional to the Calabi-Yau's holomorphic three-form Ω and its complex conjugate

$$\langle \mathrm{tr}\bar{\chi}\Gamma^{(3)}\chi\rangle \sim \Lambda^3\bar{\Omega} + \text{c.c.}\,, \tag{11}$$

where Λ^3 represents the gaugino condensate in the effective four-dimensional theory, see below. This perfect square potential seems to relax dynamically towards zero vacuum energy by balancing the condensate with a non-zero H-flux. The resulting alignment of Ω and its complex conjugate with H

$$H \sim \alpha' e^{\phi/2}(\Lambda^3\bar{\Omega} + \text{c.c.}) \tag{12}$$

fixes all complex structure moduli of the compactification. Furthermore, the 3-form flux H must be of Hodge type $H^{(0,3)}$ and $H^{(3,0)}$ which *breaks supersymmetry*. We would thus be tempted to conclude that the first step has successfully been implemented in heterotic string compactifications, obtaining a vanishing vacuum energy up to order $(\alpha')^2$ despite breaking supersymmetry.

This is indeed a very attractive dynamical mechanism were it not for a quantum effect which poses a serious obstruction. Soon after the the above mechanism had been proposed in ref. 46, it was realized by Rohm and Witten that the 3-form flux H had to be quantized to give a well-defined partition function.[47] In the heterotic theory the exact form part dB of the full 3-form flux, $H = dB + \frac{\alpha'}{4}(\omega_L - \omega_{YM})$, has to deliver quantized integer values when integrated over an arbitrary 3-cycle Σ_3

$$\frac{1}{2\pi\alpha'}\int_{\Sigma_3} dB = 2\pi N\,, \quad N \in \mathbf{Z}\,. \tag{13}$$

Moreover, the integrated Yang-Mills and Lorentz Chern-Simons terms

$$\frac{1}{2\pi\alpha'}\int_{\Sigma_3} \frac{\alpha'}{4}(\omega_L - \omega_{YM})\,, \tag{14}$$

which enter H to render it gauge invariant, are only well defined modulo integers.[47] Consequently, the *balancing equation*, which would result from minimizing the heterotic string perfect square potential, reads

$$N \overset{?}{=} \frac{\alpha_0}{8\pi}(\Lambda^3\bar{\Pi} + \bar{\Lambda}^3\Pi)\,, \tag{15}$$

after integration over some 3-cycle Σ_3. Here, $\alpha_0 = g_0^2/4\pi$ is the hidden sector gauge coupling, $\Pi = \int_{\Sigma_3}\Omega$ the period integral and

$$\Lambda^3 = \langle \mathrm{Tr}\lambda\lambda\rangle = 16\pi^2 M_{UV}^3 e^{-f_h/C_G} \tag{16}$$

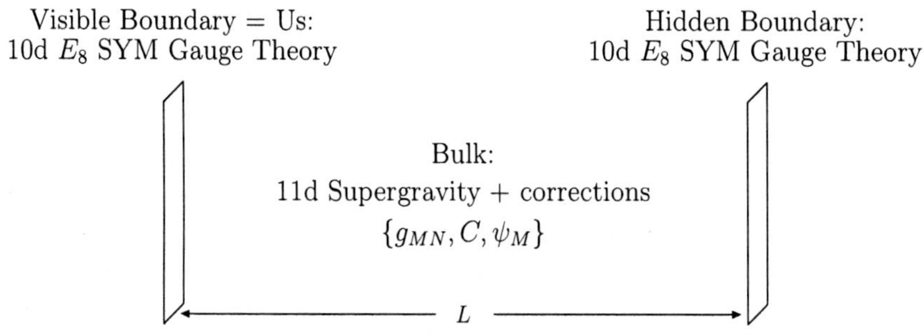

Fig. 1. The eleven-dimensional heterotic M-theory setup.

the gaugino condensate of the hidden sector four-dimensional gaugino λ with UV cut-off scale M_{UV}, hidden sector gauge kinetic function f_h and dual Coxeter number C_G related to the unbroken hidden sector gauge group G. The right hand side of the above balancing equation assumes exponentially small values due to the gaugino condensate. On the contrary, the left hand side assumes positive integers. It is therefore, in general, impossible to satisfy the heterotic string's balancing equation. In other words the quantization of H prohibits the dynamical relaxation of the vacuum energy to zero.

This obstruction disappears when we go from the ten-dimensional heterotic string-theory to the eleven-dimensional heterotic M-theory, as we will discuss next, following ref. 48.

6. Vacuum Energy after Supersymmetry Breaking in Heterotic M-Theory Compactifications

6.1. *Flux Compactification Geometry*

The essential new ingredient concerning the vacuum energy problem, when going from the ten-dimensional heterotic string to the eleven-dimensional heterotic M-theory, is the warped geometrical background which extends along the extra eleventh dimension. Note that when the string coupling g_s grows it turns into a geometrical entity, the size L of the extra eleventh dimension $L \sim g_s^{2/3}$ (see fig. 1). The bulk eleven-dimensional spacetime in heterotic M-theory is bounded by two ten-dimensional boundaries, located at $x^{11} = 0$ (visible boundary) and $x^{11} = L$ (hidden boundary), along the eleventh dimension $0 \leq x^{11} \leq L$ which represents the orbifold $\mathbf{S}^1/\mathbf{Z}_2$. The visible boundary is assumed to host us together with the Standard Model and its grand unified extension at higher energies. The hidden boundary hosts a mirror world which is only capable to interact with us via (super)gravitational interactions which propagate through the bulk. A direct exploration of the hidden boundary through, say, electromagnetic waves is forbidden by the fact that the electromagnetic $U(1)$ gauge theory is part of the Standard Model gauge theory which resides exclusively on the visible boundary but has no extension into the eleven-dimensional bulk. Heterotic M-theory incorporates therefore the brane-world picture as an essential ingredient.

The physics in the bulk, at wavelengths larger than the 11-dimensional Planck-length, is governed by 11-dimensional supergravity. Its 4-form field-strength G is sourced by the two $\mathbf{S}^1/\mathbf{Z}_2$ orbifold boundaries. They represent magnetic sources for G, much like a monopole represents a magnetic source for the ordinary electromagnetic gauge field-strength. The interesting consequence of this non-vanishing field-strength G is that, via the Einstein equations, they cause the spacetime to be curved. More specifically, since we require a compactification which conserves an $N = 1$ supersymmetry in the four large spacetime dimensions, it turns out that this spacetime has to be curved in a rather restricted way, namely by a warp-factor $e^{f(x^{11})}$ which depends only on the eleventh dimension x^{11}. Heterotic M-theory therefore operates on a spacetime which is described by a warped metric of the following type[49–51]

$$ds^2 = e^{-2f(x^{11})} g_{\mu\nu} dx^\mu dx^\nu + e^{2f(x^{11})} (g(\mathbf{X})_{lm} dy^l dy^m + (dx^{11})^2) \, . \quad (17)$$

The first part describes the 4-dimensional non-compact spacetime with metric $g_{\mu\nu}$, $\mu, \nu = 0, \ldots, 3$. The second and third part describe the Calabi-Yau metric $g(\mathbf{X})_{lm}$, $l, m = 1, \ldots, 6$ and the extension into the eleventh direction. The warp-factor is explicitly given by[49–51]

$$e^{f(x^{11})} = |1 - x^{11} Q_v|^{1/3} \, , \quad (18)$$

where

$$Q_v = -\frac{1}{8\pi V} \left(\frac{\kappa_{11}}{4\pi} \right)^{2/3} \int_{\mathbf{X}} J \wedge \left(\mathrm{tr} F \wedge F - \frac{1}{2} \mathrm{tr} R \wedge R \right) \quad (19)$$

represents the charge with which the visible boundary couples to the supergravity 3-form potential C, V denotes the unwarped Calabi-Yau volume, J the Kähler-form of the Calabi-Yau $\mathbf{X}$ and F resp. R the Yang-Mills resp. Riemannian curvature 2-forms on the visible boundary. From this geometry it is easy to see that the volume of **X** decreases along the eleventh direction and vanishes eventually at a critical length

$$L_c = 1/Q_v \, , \tag{20}$$

given simply by the inverse of the charge. The warp-factor on the hidden boundary, i.e. at $x^{11} = L$, which will play an important rôle later, can thus be expressed in geometrical terms as

$$e^{f(L)} = \left(|L_c - L|/L_c\right)^{1/3} . \tag{21}$$

It clearly vanishes when the size of the eleventh dimension $L \leq L_c$ reaches the critical length $L = L_c$.

With this warped background the question arises how does the warp-factor modify the heterotic perfect square potential? To answer this question, we must first clarify what becomes of the heterotic string perfect square potential when we go to heterotic M-theory. What happens is that the 3-form field-strength H gets lifted to the 4-form field-strength G, and the two E_8 super Yang-Mills gauge sectors of the $E_8 \times E_8$ gauge group become geometrically separated on the two boundaries (see fig. 1). The warping causes the Calabi-Yau volume to decrease from visible to hidden boundary. Since the gauge couplings on the boundaries are inversely proportional to the Calabi-Yau volumes on those sites, it follows that the gauge theory on the hidden boundary is naturally strongly coupled when the visible boundary gauge theory is weakly coupled. Gaugino condensation will therefore occur naturally at high energies on the hidden boundary, and the perfect square potential turns into[52] (for technical reasons one is working on the $\mathbf{S}^1$ covering space of the $\mathbf{S}^1/\mathbf{Z}_2$ orbifold)

$$S_{pot} = -\frac{1}{2\kappa_{11}^2} \int_{\mathbf{R}^{1,3} \times \mathbf{X} \times \mathbf{S}^1} \left(G - \frac{\sqrt{2}}{32\pi} \left(\frac{\kappa_{11}}{4\pi}\right)^{2/3} \mathrm{tr}\bar{\chi}\hat{\Gamma}^{(3)}\chi \wedge \delta_L \right)^2 . \tag{22}$$

The hat on $\hat{\Gamma}^{(3)}$ indicates that the 3-index gamma-matrix contains the vielbein of the warped metric, whereas δ_L denotes a Dirac-delta 1-form which localizes the gaugino condensate on the hidden boundary.

6.2. *Supersymmetry Breaking and Zero Vacuum Energy*

As in the heterotic string compactifications, we remain also with a perfect square potential in heterotic M-theory compactifications, which once more suggests that the system will relax dynamically towards zero potential energy. To see whether we can avoid the quantum obstruction which we faced for heterotic string compactifications, we have to analyze the balancing equation between G and the condensate. In contrast to the heterotic string, we have now a warped background entering the perfect square in a non-trivial way. Let us first analyze the condensate. Here one finds, after a careful analysis carried out in ref. 48, that the condensate which is located on the hidden boundary at $x^{11} = L$, acquires the following warp-factor dependence

$$\langle \mathrm{tr}\bar{\chi}\hat{\Gamma}^{(3)}\chi\rangle = 4\pi e^{-3f(L)}\alpha_0\big(\Lambda^3\bar{\Omega} + \bar{\Lambda}^3\Omega\big) \ . \tag{23}$$

Next, we have to take into account the quantization of the G flux. In M-theory this is given by the following quantization condition[53]

$$\frac{1}{\sqrt{2}}\left(\frac{4\pi}{\kappa_{11}}\right)^{2/3}\int_{\Sigma_4} G + \frac{\pi}{4}\int_{\Sigma_4} p_1(\mathbf{X}) = 2\pi N \ , \qquad N \in \mathbf{Z} \ . \tag{24}$$

The integration is performed over some arbitrary 4-cycle $\Sigma_4 \in H_4(\mathbf{X}, \mathbf{Z})$ and the second term includes the first Pontryagin class $p_1(\mathbf{X})$ of $\mathbf{X}$. The anticipated balancing of the G-flux with the condensate inside the perfect square tells us that G must have one index along the eleventh dimension and being proportional to Ω resp. its complex conjugate. The appropriate cycle over which we have to integrate to obtain a non-trivial result thus has to factorize like $\Sigma_4 = \Sigma_3 \times \mathbf{S}^1$, where Σ_3 is proportional to the Poincaré dual of Ω or $\bar{\Omega}$. Since this 4-cycle connects both boundaries, in principle, a further boundary contribution needs to be added to the above quantization condition.[54] However, with Σ_3 being proportional to the Poincaré dual of Ω or $\bar{\Omega}$, both this boundary contribution and the $p_1(\mathbf{X})$ term in the above G-flux quantization condition do not contribute.[54] We are thus left with the simple quantization rule

$$\frac{1}{\sqrt{2}}\left(\frac{4\pi}{\kappa_{11}}\right)^{2/3}\int_{\Sigma_3\times\mathbf{S}^1} G = 2\pi N \ , \qquad N \in \mathbf{Z} \ . \tag{25}$$

When we now apply this M-theory G-flux quantization to the balancing condition between G-flux and condensate at the minimum of the perfect square and integrate over the 4-cycle $\Sigma_3 \times \mathbf{S}^1$, we obtain the heterotic M-theory *balancing equation*

$$e^{3f(L)}N = \frac{\alpha_0}{8\pi}\left(\Lambda^3\bar{\Pi} + \bar{\Lambda}^3\Pi\right) \ , \tag{26}$$

where

$$0 \le e^{3f(L)} \le 1 \ . \tag{27}$$

Compared with the heterotic string balancing equation (15), the heterotic M-theory warped flux geometry led to the multiplication of the flux integer N by the cube of the *warp-factor*, being evaluated at the position of the hidden boundary. This has the following important consequences

- the quantum obstruction to the balancing of the condensate with the flux ceases to exist. The *continuous* warp-factor suppresses the quantized flux. Moreover, the size L of the $\mathbf{S}^1/\mathbf{Z}_2$ orbifold adjusts itself dynamically such as to set the perfect square potential to zero and satisfy eq. (26).
- this dynamical balancing implies a *stabilization of* L close to L_c, leading to $e^{3f(L)} \ll 1$. This is in nice agreement with phenomenological constraints which favor a stabilization around L_c.
- *supersymmetry is broken* since, as in the heterotic string case, G becomes of Hodge type $H^{(0,3)}$ and $H^{(3,0)}$ along $\mathbf{X}$.

To summarize, we find a *zero vacuum energy* after the system has evolved to the zero of its potential energy and supersymmetry has been broken. This accomplishes the first step of our 2-step approach proposed in the beginning. Let us now comment on the second step.

7. Dark Energy from Higher Order Corrections

The perfect square potential in heterotic M-theory compactifications included contributions at order $\kappa_{11}^{2/3} \sim 1/M_{11}^3$ and $(\kappa_{11}^{2/3})^2 \sim 1/M_{11}^6$, where M_{11} is the eleven-dimensional Planck mass. There is, however, no theorem which would protect the perfect square structure up to higher orders $(\kappa_{11}^{2/3})^n \sim 1/M_{11}^{3n}$, $n \geq 3$. These higher order contributions are suppressed by corresponding powers of the four-dimensional Planck mass in the effective four-dimensional theory and could thus generate a sufficiently small residual vacuum energy. One might therefore hope that perhaps a relation like eq. (3) might become derivable.

One should also add the known R^4 and their superpartner $C \wedge R^4$ corrections to the heterotic M-theory action. This leads to small corrections to the heterotic M-theory flux compactification geometry.[55] It is clear that this corrected background solution, being a supersymmetric solution to the field equations, must give itself a vanishing vacuum energy in the absence of supersymmetry breaking, as has been demonstrated in detail for the leading order warped background in ref. 51. So, it's again the task of the gaugino condensate and the induced G-flux to break supersymmetry and generate a non-zero vacuum energy. Since both the condensate and the

G-flux reside on the hidden boundary, where the warp-factor is smallest, higher order corrections to the perfect square potential which contribute to the vacuum energy can be expected to come with strong suppressions by high powers of the small warp-factor $e^{f(L)}$. Moreover, these higher order corrections are suppressed through additional powers of $1/M_{11}$. There has been recent progress in obtaining a heterotic M-theory action to all orders in $\kappa_{11}^{2/3}$.[56–58] This action would allow a quantitative study of the vacuum energy incorporating these higher order corrections to the vacuum energy and is presently under investigation.

8. Summary

We can therefore conclude that

- dynamical supersymmetry breaking is possible in heterotic M-theory compactifications with zero vacuum energy at leading orders in $\kappa_{11}^{2/3}$
- higher-order corrections can be expected to give a small correction to the vanishing leading order vacuum energy and could thus account for the observed dark energy
- the quantum obstruction of the heterotic string to obtain zero vacuum energy after supersymmetry breaking is avoided in heterotic M-theory through a dependence on the continuous warp-factor which is not available in the heterotic string
- the $\mathbf{S}^1/\mathbf{Z}_2$ size L, i.e. the dilaton in M-theory disguise, is stabilized close to the critical length L_c, an important fact for phenomenological and cosmological applications of heterotic M-theory[59–64]
- complex structure moduli are stabilized by G-flux which is generated as a dynamic response to the hidden sector gaugino condensate at the minimum of the perfect square potential

Even though it was not our focus here, let us finally comment on dark matter in the same framework. Moduli stabilization of heterotic M-theory compactifications favor a broken hidden E_8 gauge group.[61] This implies hidden matter which can interact with the visible matter only via bulk (super)gravitational interactions and has therefore been considered as dark matter candidates.[65] A complete analysis of the viability of this class of dark matter candidates still needs to be done but, if successful, this type of dark matter would nicely complement the outlined picture for the possible M-theory origin of dark energy.

References

1. A. G. Riess *et al.* [Supernova Search Team Collaboration], Astron. J. **116**, 1009 (1998) [arXiv:astro-ph/9805201].
2. S. Perlmutter *et al.* [Supernova Cosmology Project Collaboration], Astrophys. J. **517**, 565 (1999) [arXiv:astro-ph/9812133].
3. S. Sarkar, arXiv:0710.5307 [astro-ph].
4. G. Curio and A. Krause, Nucl. Phys. B **643**, 131 (2002) [arXiv:hep-th/0108220].
5. S. Kachru, R. Kallosh, A. Linde and S. P. Trivedi, Phys. Rev. D **68**, 046005 (2003) [arXiv:hep-th/0301240].
6. K. Intriligator, N. Seiberg and D. Shih, JHEP **0604**, 021 (2006) [arXiv:hep-th/0602239].
7. S. Kachru, M. B. Schulz and E. Silverstein, Phys. Rev. D **62**, 045021 (2000) [arXiv:hep-th/0001206].
8. S. H. H. Tye and I. Wasserman, Phys. Rev. Lett. **86**, 1682 (2001) [arXiv:hep-th/0006068].
9. A. Krause, Nucl. Phys. B **748**, 98 (2006) [arXiv:hep-th/0006226].
10. A. Krause, JHEP **0309**, 016 (2003) [arXiv:hep-th/0007233].
11. A. Kehagias and K. Tamvakis, Mod. Phys. Lett. A **17**, 1767 (2002) [arXiv:hep-th/0011006].
12. J. M. Cline and H. Firouzjahi, Phys. Lett. B **514**, 205 (2001) [arXiv:hep-ph/0012090].
13. N. Tetradis, Phys. Lett. B **509**, 307 (2001) [arXiv:hep-th/0012106].
14. H. P. Nilles, A. Papazoglou and G. Tasinato, Nucl. Phys. B **677**, 405 (2004) [arXiv:hep-th/0309042].
15. D. Kamani, arXiv:hep-th/0611339.
16. E. K. Park and P. S. Kwon, JHEP **0711**, 051 (2007) [arXiv:hep-th/0702171].
17. S. Das, D. Maity and S. SenGupta, arXiv:0711.1744 [hep-th].
18. C. Wetterich, Nucl. Phys. B **302**, 668 (1988).
19. B. Ratra and P. J. E. Peebles, Phys. Rev. D **37**, 3406 (1988).
20. C. Wetterich, Astron. Astrophys. **301**, 321 (1995) [arXiv:hep-th/9408025].
21. I. Zlatev, L. M. Wang and P. J. Steinhardt, Phys. Rev. Lett. **82**, 896 (1999) [arXiv:astro-ph/9807002].
22. A. G. Cohen, D. B. Kaplan and A. E. Nelson, Phys. Rev. Lett. **82**, 4971 (1999) [arXiv:hep-th/9803132].
23. M. Li, Phys. Lett. B **603**, 1 (2004) [arXiv:hep-th/0403127].
24. Y. g. Gong, Phys. Rev. D **70**, 064029 (2004) [arXiv:hep-th/0404030].
25. Q. G. Huang and M. Li, JCAP **0408**, 013 (2004) [arXiv:astro-ph/0404229].
26. K. Enqvist and M. S. Sloth, Phys. Rev. Lett. **93**, 221302 (2004) [arXiv:hep-th/0406019].
27. Y. g. Gong, B. Wang and Y. Z. Zhang, Phys. Rev. D **72**, 043510 (2005) [arXiv:hep-th/0412218].
28. Y. S. Myung, Phys. Lett. B **610**, 18 (2005) [arXiv:hep-th/0412224].
29. S. D. H. Hsu, Phys. Lett. B **594**, 13 (2004) [arXiv:hep-th/0403052].
30. N. Arkani-Hamed, H. C. Cheng, M. A. Luty and S. Mukohyama, JHEP **0405**, 074 (2004) [arXiv:hep-th/0312099].

31. A. Krause and S. P. Ng, Int. J. Mod. Phys. A **21**, 1091 (2006) [arXiv:hep-th/0409241].
32. A. A. Starobinsky, Phys. Lett. B **91**, 99 (1980).
33. S. Capozziello, S. Carloni and A. Troisi, arXiv:astro-ph/0303041.
34. S. Capozziello, V. F. Cardone, S. Carloni and A. Troisi, Int. J. Mod. Phys. D **12**, 1969 (2003) [arXiv:astro-ph/0307018].
35. S. M. Carroll, V. Duvvuri, M. Trodden and M. S. Turner, Phys. Rev. D **70**, 043528 (2004) [arXiv:astro-ph/0306438].
36. S. Nojiri and S. D. Odintsov, Int. J. Geom. Meth. Mod. Phys. **4**, 115 (2007) [arXiv:hep-th/0601213].
37. R. P. Woodard, Lect. Notes Phys. **720**, 403 (2007) [arXiv:astro-ph/0601672].
38. G. R. Dvali, G. Gabadadze and M. Porrati, Phys. Lett. B **485**, 208 (2000) [arXiv:hep-th/0005016].
39. P. Q. Hung, arXiv:hep-ph/0010126.
40. P. Gu, X. Wang and X. Zhang, Phys. Rev. D **68**, 087301 (2003) [arXiv:hep-ph/0307148].
41. R. Fardon, A. E. Nelson and N. Weiner, JCAP **0410**, 005 (2004) [arXiv:astro-ph/0309800].
42. J. R. Bhatt, P. H. Gu, U. Sarkar and S. K. Singh, arXiv:0711.2728 [hep-ph].
43. S. Weinberg, Phys. Rev. Lett. **59**, 2607 (1987).
44. R. Bousso, arXiv:0708.4231 [hep-th].
45. A. Strominger, Nucl. Phys. B **274**, 253 (1986).
46. M. Dine, R. Rohm, N. Seiberg and E. Witten, Phys. Lett. B **156**, 55 (1985).
47. R. Rohm and E. Witten, Annals Phys. **170**, 454 (1986).
48. A. Krause, Phys. Rev. Lett. **98**, 241601 (2007) [arXiv:hep-th/0701009].
49. G. Curio and A. Krause, Nucl. Phys. B **602**, 172 (2001) [arXiv:hep-th/0012152].
50. A. Krause, Fortsch. Phys. **49**, 163 (2001).
51. G. Curio and A. Krause, Nucl. Phys. B **693**, 195 (2004) [arXiv:hep-th/0308202].
52. P. Horava, Phys. Rev. D **54**, 7561 (1996) [arXiv:hep-th/9608019].
53. E. Witten, J. Geom. Phys. **22**, 1 (1997) [arXiv:hep-th/9609122].
54. A. Lukas, B. A. Ovrut and D. Waldram, Phys. Rev. D **57**, 7529 (1998) [arXiv:hep-th/9711197].
55. L. Anguelova and D. Vaman, Nucl. Phys. B **733**, 132 (2006) [arXiv:hep-th/0506191].
56. I. G. Moss, Phys. Lett. B **577**, 71 (2003) [arXiv:hep-th/0308159].
57. I. G. Moss, Nucl. Phys. B **729**, 179 (2005) [arXiv:hep-th/0403106].
58. I. G. Moss, Phys. Lett. B **637**, 93 (2006) [arXiv:hep-th/0508227].
59. E. Witten, Nucl. Phys. B **471**, 135 (1996) [arXiv:hep-th/9602070].
60. T. Banks and M. Dine, Nucl. Phys. B **479**, 173 (1996) [arXiv:hep-th/9605136].
61. M. Becker, G. Curio and A. Krause, Nucl. Phys. B **693**, 223 (2004) [arXiv:hep-th/0403027].
62. K. Becker, M. Becker and A. Krause, Nucl. Phys. B **715**, 349 (2005) [arXiv:hep-th/0501130].

63. K. Becker, M. Becker and A. Krause, Phys. Rev. D **74**, 045023 (2006) [arXiv:hep-th/0510066].
64. G. Curio and A. Krause, Phys. Rev. D **75**, 126003 (2007) [arXiv:hep-th/0606243].
65. A. Krause, in the proceedings of SUSY 2003, 11th Annual International Conference on Supersymmetry and the Unification of Fundamental Interactions, Tucson, Arizona, [arXiv:hep-th/0404001].

DARK MATTER AND DARK GAUGE FIELDS

D. V. AHLUWALIA

Department of Physics and Astronomy, Rutherford Building
University of Canterbury, Private Bag 4800
Christchurch 8020, New Zealand
E-mail: dharamvir.ahluwalia-khalilova@canterbury.ac.nz
www2.phys.canterbury.ac.nz/editorial/

CHENG-YANG LEE, D. SCHRITT, and T. F. WATSON

Department of Physics and Astronomy, Rutherford Building
University of Canterbury, Private Bag 4800
Christchurch 8020, New Zealand

Following the unexpected theoretical discovery of a mass dimension one fermionic quantum field of spin one half, we now present first results on two *local* versions. The Dirac and Majorana fields of the standard model of particle physics are supplemented by their natural counterparts in the dark matter sector. The possibility that a mass dimension transmuting symmetry may underlie a new standard model of particle physics is briefly suggested.

Keywords: Dirac, Majorana, Elko, Erebus, Nyx, Shakti, Dark Matter.

1. Introduction: Dark Matter and Its Darkness

The existence of dark matter hints towards new physics, and has often been thought to bring in new symmetries, such as super-symmetry. Yet, such directions can also take us far afield if we have not fully understood the technical and physical content of the well-known continuous and discrete symmetries of the standard model of particle physics.[1] It is to this possibility that this paper is devoted in the most conservative tradition of our craft.

At present, we do not know the spin of dark matter particles, nor do we have any inkling as to what type of interaction-inducing principle, if any, operates in the dark sector. A highly motivated candidate that answers these questions can suggest experiments that help make properties of dark matter concrete. On the positive side, it is now well established that dark matter interacts with the particles of the standard model predominantly via gravity. All other interactions seem to be highly, if not completely, suppressed.

Without calling upon any yet-unobserved symmetries, we invoke the minimal assumption that whatever dark matter is, it must transform in a well-defined manner from one inertial frame to another according to projective irreducible representations of the Poincaré algebra, supplemented by the discrete symmetries of spacetime reflections and charge conjugation. With this conservative ansatz we exploit recently gained insights[2,3] to construct two fermionic spin one half quantum fields. These carry mass dimension one, and satisfy the canonical locality requirement.

Defining darkness as the property that one set of fields carries limited or no interactions (except gravitational) with another set of fields, these new matter fields are dark with respect to the matter and gauge fields of the standard model (SM) of particle physics; and vice versa.

Identifying the new fields with dark matter (DM), we suggest a mass-dimension-transmuting symmetry principle that yields the DM sector Lagrangian density without any additional assumptions. This then becomes reminiscent of the super-symmetric paradigm but now the mass-dimension-transmuting symmetry does not take a fermion to a boson, but a SM fermion of mass dimension three-half to a DM fermion of mass dimension one.

2. Two New Quantum Fields

Here we present all the essential details for the construction of two spin one half quantum fields based upon the dual helicity eigenspinors of the charge conjugation operator (Elko)[a] introduced in references [2,3]. The breakthrough that this paper presents, and what constitutes the primary progress since the indicated 2005 publications, is that the quantum fields we now present are *local* in the canonical sense.

Background— To establish the notation and the working context, we here collect together certain facts and definitions. In case the reader feels a certain element of unfamiliarity, we urge her/him to consult Ref. [3,5] for pedagogic details.

[a]The acronym *Elko* for the eigenspinors of the charge conjugation operator originates from the German **E**igenspinoren des **L**adungs**k**onjugations**o**perators. The reader who wishes to study Elko similarities to and differences from the Majorana spinors may wish to consult references [3,4].

In the momentum space, dual helicity spinors have the generic form

$$\lambda(\mathbf{p}) = \begin{pmatrix} (\alpha\Theta)\,\phi_L^*(\mathbf{p}) \\ \phi_L(\mathbf{p}) \end{pmatrix} \tag{1}$$

where Θ is the Wigner time reversal operator for spin one half and reads

$$\Theta = \begin{pmatrix} 0 & -1 \\ 1 & 0 \end{pmatrix}. \tag{2}$$

The $\phi_L(\mathbf{p})$ is a massive left-handed (L) Weyl spinor, while $\alpha\Theta\phi_L^*(p)$ transforms as a right-handed (R) Weyl spinor associated with the same mass. If $\phi_L(\mathbf{p})$ carries a given helicity, then $\alpha\Theta\phi_L^*(p)$ is necessarily endowed with the opposite helicity. The charge conjugation operator C for the L$\oplus$R representation space is

$$C = \begin{pmatrix} \mathbb{O} & i\Theta \\ -i\Theta & \mathbb{O} \end{pmatrix} K \tag{3}$$

where K is the complex conjugation operator (see Ref. [3]). The $\lambda(\mathbf{p})$ become Elko satisfying, $C\lambda(\mathbf{p}) = \pm\lambda(\mathbf{p})$, if $\alpha = \pm i$. To construct a complete set of Elko we need the L$\oplus$R boost operator, and a complete dual-helicity basis for Elko at rest; i.e., $\lambda(\mathbf{0})$. The boost operator reads

$$\kappa = \begin{pmatrix} \exp\left(\frac{\boldsymbol{\sigma}}{2}\cdot\boldsymbol{\varphi}\right) & \mathbb{O} \\ \mathbb{O} & \exp\left(-\frac{\boldsymbol{\sigma}}{2}\cdot\boldsymbol{\varphi}\right) \end{pmatrix}. \tag{4}$$

In terms of the energy E and the momentum $\mathbf{p} = p\hat{\mathbf{p}}$ the boost parameter, $\boldsymbol{\varphi} = \varphi\hat{\mathbf{p}}$, is defined as $\cosh(\varphi) = E/m$, and $\sinh(\varphi) = p/m$, where m is the mass of the described particle. Since $(\boldsymbol{\sigma}\cdot\hat{\mathbf{p}})^2 = \mathbb{I}$, the boost operator is linear in $\mathbf{p}$[b]

$$\kappa = \sqrt{\frac{E+m}{2m}} \begin{pmatrix} \mathbb{I} + \frac{\boldsymbol{\sigma}\cdot\mathbf{p}}{E+m} & \mathbb{O} \\ \mathbb{O} & \mathbb{I} - \frac{\boldsymbol{\sigma}\cdot\mathbf{p}}{E+m} \end{pmatrix}. \tag{5}$$

Construction of Elko for the two new quantum fields— To construct the required Elko, we must first introduce a basis for the particle description

[b] It is this fact, when coupled with the observation that the Dirac spinors are eigenspinors of the L$\oplus$R parity operator, that yields the linearity of the Dirac operator $(i\gamma^\mu\partial_\mu - m)$. For a complementary discussion, the reader is referred to Sec. 5.5 of reference [6].

in its rest frame. Towards this end we shall take $\phi_L(\mathbf{0})$ to be eigenspinors of the helicity operator $(\boldsymbol{\sigma}/2)\cdot\hat{\mathbf{p}}$

$$\left[\frac{\boldsymbol{\sigma}}{2}\cdot\hat{\mathbf{p}}\right]\phi_L^{\pm}(\mathbf{0}) = \pm\frac{1}{2}\phi_L^{\pm}(\mathbf{0}). \tag{6}$$

Taking $\hat{\mathbf{p}} = (\sin\theta\cos\phi, \sin\theta\sin\phi, \cos\theta)$, we adopt a certain choice of phase factors so that[c]

$$\phi_L^{+}(\mathbf{0}) = \sqrt{m}\begin{pmatrix}\cos(\theta/2)e^{-i\phi/2}\\ \sin(\theta/2)e^{+i\phi/2}\end{pmatrix},\ \phi_L^{-}(\mathbf{0}) = \sqrt{m}\begin{pmatrix}-\sin(\theta/2)e^{-i\phi/2}\\ \cos(\theta/2)e^{+i\phi/2}\end{pmatrix}. \tag{7}$$

The complete basis for Elko that is now required for constructing the indicated quantum fields is

$$\xi_{\{-,+\}}(\mathbf{0}) = \lambda(\mathbf{0})\Big|_{\phi_L(\mathbf{0})\to\phi_L^{+}(\mathbf{0}),\ \alpha\to+i} \tag{8}$$

$$\xi_{\{+,-\}}(\mathbf{0}) = \lambda(\mathbf{0})\Big|_{\phi_L(\mathbf{0})\to\phi_L^{-}(\mathbf{0}),\ \alpha\to+i} \tag{9}$$

$$\zeta_{\{-,+\}}(\mathbf{0}) = \lambda(\mathbf{0})\Big|_{\phi_L(\mathbf{0})\to\phi_L^{-}(\mathbf{0}),\ \alpha\to-i} \tag{10}$$

$$\zeta_{\{+,-\}}(\mathbf{0}) = -\lambda(\mathbf{0})\Big|_{\phi_L(\mathbf{0})\to\phi_L^{+}(\mathbf{0}),\ \alpha\to-i} \tag{11}$$

with

$$\xi_{\{\mp,\pm\}}(\mathbf{p}) = \kappa\,\xi_{\{\mp,\pm\}}(\mathbf{0}),\quad \zeta_{\{\mp,\pm\}}(\mathbf{p}) = \kappa\,\zeta_{\{\mp,\pm\}}(\mathbf{0}). \tag{12}$$

The justification for the choice of phases and designations in equations (8-11) is far from trivial; but it is a straightforward generalisation of the reasoning found in Sec. 38 of reference [7] and Sec. 5.5 of reference [6].

If one works with Elko using the Dirac dual, $\overline{\eta}(\mathbf{p}) = \eta^\dagger(\mathbf{p})\gamma^0$, one finds[d] that these carry null norm.[9] One also encounters problems such as those found in Appendix P of reference [8]. In fact, the dual-helicity nature of the $\xi(\mathbf{p})$ and $\zeta(\mathbf{p})$ asks for the introduction of a new dual. We shall call it the *Elko dual.*[e] It is defined as

$$\stackrel{\neg}{\eta}_{\{\mp,\pm\}}(\mathbf{p}) := \mp i\,\eta^\dagger_{\{\pm,\mp\}}(\mathbf{p})\,\gamma^0 \tag{13}$$

---

[c]Please notice that the choice of phases here differs from that given in references [2,3].
[d]Here η stands symbolically for either ξ or ζ.
[e]The origin of this dual dates back to a preliminary work [10] on the subject.

where

$$\gamma^0 := \begin{pmatrix} \mathbb{O} & \mathbb{I} \\ \mathbb{I} & \mathbb{O} \end{pmatrix}. \tag{14}$$

With the Elko dual thus defined, we now have, by construction, the *orthonormality relations*

$$\overset{\neg}{\xi}_\beta(\mathbf{p})\,\xi_{\beta'}(\mathbf{p}) = +\,2m\,\delta_{\beta\beta'}\,, \quad \overset{\neg}{\xi}_\beta(\mathbf{p})\,\zeta_{\beta'}(\mathbf{p}) = 0 \tag{15}$$

$$\overset{\neg}{\zeta}_\beta(\mathbf{p})\,\zeta_{\beta'}(\mathbf{p}) = -\,2m\,\delta_{\beta\beta'}\,, \quad \overset{\neg}{\zeta}_\beta(\mathbf{p})\,\xi_{\beta'}(\mathbf{p}) = 0 \tag{16}$$

Here, β ranges over two possibilities: $\{+,-\}$ and $\{-,+\}$. The *completeness relation* is

$$\frac{1}{2m}\sum_\beta \left[\xi_\beta(\mathbf{p})\ \overset{\neg}{\xi}_\beta(\mathbf{p}) - \zeta_\beta(\mathbf{p})\ \overset{\neg}{\zeta}_\beta(\mathbf{p})\right] = \mathbb{I}. \tag{17}$$

The detailed structure underlying the completeness relations resides in the following *spin sums*

$$\xi_\beta(\mathbf{p})\ \overset{\neg}{\xi}_\beta(\mathbf{p}) = +m\,[\mathbb{I} + \mathcal{G}(\mathbf{p})] \tag{18}$$

$$\zeta_\beta(\mathbf{p})\ \overset{\neg}{\zeta}_\beta(\mathbf{p}) = -m\,[\mathbb{I} - \mathcal{G}(\mathbf{p})] \tag{19}$$

which together *define* $\mathcal{G}(\mathbf{p})$. Explicit calculation shows that $\mathcal{G}$ is an odd function of $\mathbf{p}$

$$\mathcal{G}(\mathbf{p}) = -\mathcal{G}(-\mathbf{p}). \tag{20}$$

It must be noted at this stage that the entire set of results is specific to the helicity basis. A careful examination is needed if one wishes to consider a basis in which $\phi_L(\mathbf{p})$ are not eigenspinors of the helicity operator but, say, some other operator of the type $(\boldsymbol{\sigma}/2)\cdot\hat{\mathbf{r}}$; where $\hat{\mathbf{r}}$ is not coincident with $\hat{\mathbf{p}}$. These two bases, as is evident on reflection, represent two *physically distinct* situations. Our preliminary results confirm that additional work is needed to extract further physical and mathematical content in such situations.

Two new quantum fields based on Elko: Erebus and Nyx— Following Greek mythology about primordial darkness we now introduce two fields which will later be identified with dark matter. We define Erebus as

$$e(x) = \int \frac{d^3p}{(2\pi)^3} \frac{1}{\sqrt{2mE(\mathbf{p})}} \sum_\beta \left[c_\beta(\mathbf{p})\xi_\beta(\mathbf{p}) e^{-ip_\mu x^\mu} + d^\dagger_\beta(\mathbf{p})\zeta_\beta(\mathbf{p}) e^{+ip_\mu x^\mu} \right].$$

It enjoys the same formal status as the Dirac field constructed from the Dirac spinors. We define Nyx as

$$n(x) = \int \frac{d^3p}{(2\pi)^3} \frac{1}{\sqrt{2mE(\mathbf{p})}} \sum_\beta \left[c_\beta(\mathbf{p})\xi_\beta(\mathbf{p}) e^{-ip_\mu x^\mu} + c^\dagger_\beta(\mathbf{p})\zeta_\beta(\mathbf{p}) e^{+ip_\mu x^\mu} \right].$$

It enjoys at the same formal status as the Majorana field constructed from the Dirac spinors. These formal similarities betray the true physical content of these new fields as we shall discover on deeper analysis. In a nutshell the difference is this: the mass dimension of Erebus and Nyx is one, not three half (see below). It is this aspect that renders them dark with respect to the Dirac and Majorana fields of the standard model of particle physics.[f] The creation, $c^\dagger_\beta$ and $d^\dagger_\beta$, and annihilation, c_β and d_β, operators that appear in the $e(x)$ and $n(x)$ fields satisfy the standard fermionic anti-commutation relations

$$\left\{ c_\beta(\mathbf{p}),\, c^\dagger_{\beta'}(\mathbf{p}') \right\} = (2\pi)^3\, \delta_{\beta\beta'}\, \delta^3(\mathbf{p}-\mathbf{p}') \tag{21}$$

$$\left\{ c_\beta(\mathbf{p}),\, c_{\beta'}(\mathbf{p}') \right\} = \left\{ c^\dagger_\beta(\mathbf{p}),\, c^\dagger_{\beta'}(\mathbf{p}') \right\} = 0 \tag{22}$$

with similar relations for the d's.

The Elko duals of Erebus and Nyx fields, $\stackrel{\neg}{e}(x)$ and $\stackrel{\neg}{n}(x)$, are obtained by the following substitutions in $e(x)$ and $n(x)$

$$\begin{gathered} \xi_\beta(\mathbf{p}) \rightarrow \stackrel{\neg}{\xi}_\beta(\mathbf{p}), \quad \zeta_\beta(\mathbf{p}) \rightarrow \stackrel{\neg}{\zeta}_\beta(\mathbf{p}) \\ e^{\pm ip_\mu x^\nu} \leftrightarrow e^{\mp ip_\mu x^\nu}, \quad c_\beta(\mathbf{p}) \leftrightarrow c^\dagger_\beta(\mathbf{p}), \quad d_\beta(\mathbf{p}) \leftrightarrow d^\dagger_\beta(\mathbf{p}). \end{gathered} \tag{23}$$

The propagator associated with the Erebus and Nyx fields follows from textbook methods, and is further elaborated in reference [3]. It entails evaluation of $\langle\ | T(e(x')\stackrel{\neg}{e}(x) | \ \rangle$, and $\langle\ | T(n(x')\stackrel{\neg}{n}(x) | \ \rangle$, where T is the fermionic time-ordering operator, and $|\ \rangle$ represents the vacuum state. The result for

[f]The reader may wish to consult reference [3] for the details of this argument.

Erebus as well as Nyx, in terms of spin sums, reads

$$S(x-x') = -\int \frac{d^3p}{(2\pi)^3} \frac{i}{2mE(\mathbf{p})} \sum_\beta \Big[\theta(t'-t)\,\xi_\beta(\mathbf{p})\overset{\neg}{\xi}_\beta(\mathbf{p}) e^{-ip_\mu(x'^\mu - x^\mu)} \\ -\theta(t-t')\,\zeta_\beta(\mathbf{p})\overset{\neg}{\zeta}_\beta(\mathbf{p}) e^{+ip_\mu(x'^\mu - x^\mu)}\Big] \tag{24}$$

Using the spins sums (18) and (19) yields (again, see reference 3 for additional details)

$$S(x-x') = \int \frac{d^4p}{(2\pi)^4} e^{ip_\mu(x^\mu - x'^\mu)} \frac{\mathbb{I} + \mathcal{G}(\mathbf{p})}{p_\mu p^\mu - m^2 + i\epsilon}. \tag{25}$$

Here, the limit $\epsilon \to 0^+$ is implicit. In view of (20), it is clear that, in the absence of a preferred direction, such as one arising from an external magnetic field, the second term in the above equation identically vanishes. As a result, $S(x-x')$ reduces to the Klein-Gordon propagator, modulo a 4×4 multiplicative identity matrix in the R$\oplus$L representation space. Consequently, both the Erebus and Nyx fields carry mass dimension one. This mass dimensionality forbids particles described by these fields to enter $SU(2)_L$ doublets of the SM. The fields $e(x)$ and $n(x)$ thus become first-principle candidates for dark matter. Following arguments similar to those presented in [2,3] the free Lagrangian densities associated with the Erebus and Nyx fields are

$$\mathcal{L}^{\text{Erebus}} = \partial^\mu \overset{\neg}{e}(x) \partial_\mu e(x) - m_e^2 \overset{\neg}{e}(x) e(x) \tag{26}$$

$$\mathcal{L}^{\text{Nyx}} = \partial^\mu \overset{\neg}{n}(x) \partial_\mu n(x) - m_n^2 \overset{\neg}{n}(x) n(x) \tag{27}$$

Here, we have made it explicit that the Erebus and Nyx fields need not carry the same mass.

3. A Pause

A brief pause in the development of our presentation now appears necessary. We physicists are trained to consider Lagrangian density, $\mathcal{L}$, almost as a God-given entity. Granted, one places some restrictions such as that it must be a Lorentz scalar, but beyond these well known caveats that $\mathcal{L}$ rarely is anything but a product of one's genius (such as was the case in Dirac's 1928 paper [11]) or a well-educated guess which often times runs along the lines, "for simplicity let's assume that $\mathcal{L}$ is linear in derivatives, ...". This is all wrong, in essence. Or, so is the lesson that stares us in the face of above-derived $\mathcal{L}$ for Erebus and Nyx. In fact, one could have as easily derived $\mathcal{L}^{\text{Dirac}}$ had one started not with the Elko, but with eigenspinors of the parity operator in the R⊕L representation space.[g]

In a nutshell: The kinematic structure as contained in $\mathcal{L}$ is determined entirely once one agrees that it must be based on a quantum field, and that this quantum field has certain properties under the symmetry operations inherent to the considered representation space. The details are obvious from the presented example of Erebus and Nyx.

4. Locality Structure of the Erebus and Nyx Fields

Having arrived at $\mathcal{L}$ for Erebus and Nyx in (26) and (27) one can immediately obtain the canonically conjugate momenta $\pi(x)$ for these fields. That done, a straightforward, though slightly lengthy calculation involving various Elko 'spin sums' shows that Erebus and Nyx are local. For Erebus we find

$$\{e(\mathbf{x},t), \pi(\mathbf{x}',t)\} = i\delta^3(\mathbf{x}-\mathbf{x}') \tag{28}$$

$$\{e(\mathbf{x},t), e(\mathbf{x}',t)\} = 0, \ \{\pi(\mathbf{x},t), \pi(\mathbf{x}',t)\} = 0 \tag{29}$$

An exactly similar set of equal-time anti-commutators exist for Nyx.

These results are in remarkable contrast, and constitute the breakthrough alluded to above, to the non-locality structure presented in references [2,3]. These results also provide a counter example to the expectations based on the 1966 work of Lee and Wick.[12] A resolution of this apparent discrepancy remains to be found.

5. Towards a Standard Model With Erebus and Nyx Fields

An intriguing fact that emerges from the above analysis is that the CPT structure of the spin one half R⊕L representation space is far richer than one would have expected. Such an analysis proceeds along similar lines as found in references [2,3] and yields, e.g., $(CPT)^2 = -\mathbb{I}$ for the Erebus and Nyx fields. The Dirac and Majorana fields are endowed with mass dimension and CPT properties that are in sharp contrast to those of the Erebus and Nyx fields. Therein may lie novel sources of violation of discrete symmetries,

[g]Please note that the usual parity operator in the R⊕L representation space can be easily constructed without reference to $\mathcal{L}^{\text{Dirac}}$, or the $\mathcal{L}$ for the Erebus and Nyx fields. If this is not obvious, kindly wait for one of the sequels to this paper.

such as parity and combined symmetry of charge conjugation and parity, and these may provide insights into such questions as raised by Saunders.[13]

In the present paper we do not further analyse these matters, but instead pose the question: does there exist a fundamental principle that one may invoke to extend the standard model of particle physics to systematically incorporate the Erebus and Nyx fields? We conjecture a symmetry operator $\mathcal{A}$ such that[h]

$$\mathcal{A}\mathcal{L}^{\text{Dirac}}\mathcal{A}^{-1} \rightarrow \mathcal{L}^{\text{Erebus}} \tag{30}$$

or, equivalently, $\mathcal{A}^{-1}\mathcal{L}^{\text{Erebus}}\mathcal{A} \rightarrow \mathcal{L}^{\text{Dirac}}$.[i] Clearly, such a symmetry, if it exists, transmutes the mass dimension of spin one half fermionic fields from 3/2 to 1 and from 1 to 3/2. The principle of local gauge interactions for the Erebus and Nyx fields in its $U(1)$ form is different from that of the SM: it corresponds to invariance of the Lagrangian density under $e(x) \rightarrow \exp[i\gamma^5\alpha(x)]\, e(x)$, and $n(x) \rightarrow \exp[i\gamma^5\alpha(x)]\, n(x)$. For the dark gauge fields we choose the name Shakti. This time we have invoked the mythology of the East.

Should $\mathcal{A}$ exist, and be realised by nature, the extended Lagrangian density would form a first-principle extension of the standard model to incorporate dark matter; i.e., if Erebus and Nyx indeed describe dark matter. The latter carries high plausibility given the natural darkness of the $e(x)$ and $n(x)$ fields with respect to the SM fields. We hasten to add that the Erebus and Nyx, just like their SM counterparts, are not, or need not, be self-referentially dark. It is prevented by Shakti.

6. Concluding Remarks

We have presented two new local quantum fields based on the dual helicity eigenspinors of the charge conjugation operator. These are fermionic, and carry mass dimension one. The latter result provides an *ab initio* origin of darkness of dark matter, and at the same time suggests the possible existence of a mass dimension transmuting symmetry which may be used to construct an extension of the standard model of particle physics to include the dark matter sector.

The SM matter and gauge fields and their counterparts in Erebus, Nyx, and Shakti may provide two self-referentially luminous sectors of the cosmos

---

[h] A related work has just appeared on the subject. We refer the reader to reference [14] for details.

[i] With a similar symmetry existing between the Majorana and Nyx fields.

which, although dark to each other, may be united by Higgs and gravity. Indeed, one may imagine whispering dark rivers and silvery dark moons in the night of the world inhabited by Erebus, Nyx, and Shakti — a world we see not, and yet a world to which owe our existence.

Acknowledgements

It is our great pleasure to thank all the organisers of "Dark 2007 – Sixth International Heidelberg Conference on Dark Matter in Astro and Particle Physics" for their warm hospitality.

On a personal note we are grateful to two dedicated powerhouses of physics, Hans Klapdor-Kleingrothaus and Irina Krivosheina, who by their relentless pursuit of knowledge and their scholarship have added to physics and its culture the spirit so well captured in Herman Hesse's *Das Glasperlenspiel* (The Glass Bead Game). This work was inspired by a talk Hans gave more than a decade ago at the Los Alamos National Laboratory and which led the senior author of the present manuscript to investigate Majorana's 1937 paper.[15–17] That in a series of papers Hans and his collaborators have provided a first glimpse into this beautiful realm attests to the genius of his single-minded pursuits and unique abilities. We give Hans, Irina and their team our very best wishes.

References

1. E. P. Wigner, Unitary representations of the inhomogeneous Lorentz group including reflections, in *Lectures of the Istanbul Summer School of Theoretical Physics (July 16 – August 4, 1962)*, (Gordon and Breach, New York, 1964).
2. D. V. Ahluwalia-Khalilova and D. Grumiller, *Phys. Rev. D* **72**, 067701 (2005) [arXiv:hep-th/0410192].
3. D. V. Ahluwalia-Khalilova and D. Grumiller, *JCAP* **0507**, 012 (2005) [arXiv:hep-th/0412080].
4. R. da Rocha and W. A. J. Rodrigues, *Mod. Phys. Lett. A* **21**, 65 (2006) [arXiv:math-ph/0506075].
5. L. H. Ryder, *Quantum Field Theory*, (Cambridge University Press, Cambridge, 1996).
6. S. Weinberg, *The quantum theory of fields. Vol. 1: Foundations*, (Cambridge University Press, Cambridge, 1995).
7. M. Srednicki, *Quantum field theory*, (Cambridge University Press, Cambridge, 2007).
8. I. J. R. Aitchison and A. J. G. Hey, *Gauge theories in particle physics: A practical introduction. Vol. 2: (IOP Publishers, Bristol, 2004)*.
9. D. V. Ahluwalia, *Int. J. Mod. Phys. A* **11**, 1855 (1996) [arXiv:hep-th/9409134].

10. D. V. Ahluwalia-Khalilova, *Extended set of Majorana spinors, a new dispersion relation, and a preferred frame*, arXiv:hep-ph/0305336. It is to be noted that many of the results presented in this preprint were later collaboratively developed with D. Grumiller, and published in reference [2,3].
11. P. A. M. Dirac, *Proc. Roy. Soc. Lond. A* **117**, 610 (1928).
12. T. D. Lee and G. C. Wick, Phys. Rev. **148**, 1385 (1966).
13. S. Saunders, "Mirroring as an A Priori Symmetry," *Philosophy of Science* (in press).
14. R. da Rocha and J. M. Hoff da Silva, *J. Math. Phys.* **48**, 123517 (2007) [arXiv:0711.1103 [math-ph]].
15. E. Majorana, *Nuovo Cim.* **14**, 171 (1937).
16. H. V. Klapdor-Kleingrothaus, A. Dietz, H. L. Harney and I. V. Krivosheina, *Mod. Phys. Lett. A* **16**, 2409 (2001) [arXiv:hep-ph/0201231].
17. H. V. Klapdor-Kleingrothaus and I. V. Krivosheina, *Mod. Phys. Lett. A* **21**, 1547 (2006).

STANDARD SUPERSYMMETRY FROM A PLANCK-SCALE STATISTICAL THEORY

ROLAND E. ALLEN*, ZORAWAR WADIASINGH, and SEIICHIROU YOKOO

Physics Department, Texas A&M University, College Station, Texas 77843, U.S.A.
** email: allen@tamu.edu*

We outline three new ideas in a program to obtain standard physics, including standard supersymmetry, from a Planck-scale statistical theory: (1) The initial spin 1/2 bosonic fields are transformed to spin 0 fields together with their auxiliary fields. (2) Time is defined by the progression of 3-geometries, just as originally proposed by DeWitt. (3) The initial (D-1)-dimensional "path integral" is converted from Euclidean to Lorentzian form by transformation of the fields in the integrand.

In earlier work it was shown that a fundamental statistical theory (at the Planck scale) can lead to many features of standard physics.[1–3] In some respects, however, the results had nonstandard features which appear to present difficulties. For example, the primitive supersymmetry of the earlier papers is quite different from the standard formulation of supersymmetry which works so admirably in both protecting the masses of Higgs fields from quadratic divergences and predicting coupling constant unification at high energy. Also, the fact that the theory was originally formulated in Euclidean time seems physically unsatisfactory for reasons mentioned below. Here we introduce some refinements in the theory which eliminate these two problems. The ideas in the following sections respectively grew out of discussions of the first author with Seiichirou Yokoo (on the transformation of spin 1/2 to spin 0 fields) and Zorawar Wadiasingh (on the transformation of the path integral from Euclidean to Lorentzian form).

1. Transformation of Original Spin 1/2 Fields Yields Standard Supersymmetry

In Refs. 2 and 3, the action for a fundamental bosonic field was found to have the form

$$S_b = \int d^4x \, \psi_b^\dagger \, i\sigma^\mu \partial_\mu \psi_b \tag{1}$$

at energies that are far below the Planck energy m_P (with $\hbar = c = 1$) and in a locally inertial coordinate system. This is the conventional form of the action for fermions, described by 2-component Weyl spinors, but it is highly unconventional for bosons, because a boson described by ψ_b would have spin 1/2. We can, however, transform from the original 2-component field ψ_b to two 1-component complex fields ϕ and F by writing

$$\psi_b(x) = \psi^+(x) + \psi^-(x) \tag{2}$$

$$\psi^+(\vec{x}, t) = \sum_{\vec{p},\omega} \phi(\vec{p}, \omega) \, u^+(\vec{p}) \, e^{i\vec{p}\cdot\vec{x}} e^{-i\omega t} \left(\omega + |\vec{p}|\right)^{1/2} \tag{3}$$

$$\psi^-(\vec{x}, t) = \sum_{\vec{p},\omega} F(\vec{p}, \omega) \, u^-(\vec{p}) \, e^{i\vec{p}\cdot\vec{x}} e^{-i\omega t} \left(\omega + |\vec{p}|\right)^{-1/2} \tag{4}$$

with

$$\vec{\sigma} \cdot \vec{p} \, u^+(\vec{p}) = + |\vec{p}| \, u^+(\vec{p}) \quad , \quad \vec{\sigma} \cdot \vec{p} \, u^-(\vec{p}) = - |\vec{p}| \, u^-(\vec{p}) \tag{5}$$

$$\phi(\vec{p}, \omega) = \int d^4x \, \phi(\vec{x}, t) \, e^{-i\vec{p}\cdot\vec{x}} e^{i\omega t} \, , \; F(\vec{p}, \omega) = \int d^4x \, F(\vec{x}, t) \, e^{-i\vec{p}\cdot\vec{x}} e^{i\omega t} \, .$$

Substitution then gives

$$S_b = V^{-1} \sum_{\vec{p},\omega} \left[\phi^*(\vec{p}, \omega) \left(\omega^2 - |\vec{p}|^2 \right) \phi(\vec{p}, \omega) + F^*(\vec{p}, \omega) \, F(\vec{p}, \omega) \right] \tag{6}$$

$$= \int d^4x \left[-\partial^\mu \phi^*(x) \, \partial_\mu \phi(x) + F^*(x) \, F(x) \right] \tag{7}$$

where $\partial^\mu = \eta^{\mu\nu} \partial_\nu$, $\eta^{\mu\nu} = diag(-1, 1, 1, 1)$, and V is a 4-dimensional normalization volume. This is, of course, precisely the action for a massless scalar boson field ϕ and its auxiliary field F.

With the fermionic action left in its original form, we now have the standard supersymmetric action for each pair of susy partners:

$$S_{fb} = \int d^4x \left[\psi_f^\dagger \, i\sigma^\mu \partial_\mu \psi_f - \partial^\mu \phi^*(x) \, \partial_\mu \phi(x) + F^*(x) \, F(x) \right] . \tag{8}$$

There is a major point that will be discussed at length elsewhere, in a more complete treatment of the present theory: The above transformation works only for $\omega + |\vec{p}| \geq 0$, since otherwise the sign of the integrand would be reversed. However, a stable vacuum already requires $\omega \geq 0$, so we must define time for would-be negative-frequency fields in such a way that this condition is satisfied.

2. Time is Defined by Progression of 3-Geometries in External Space

In our earlier work, the time coordinate x^0 was initially defined in exactly the same way as each spatial coordinate x^k, so x^0 was initially a Euclidean variable. For reasons given in the following section, however, this does not seem to be as physically reasonable as a picture in which time is Lorentzian when it is first defined. In this section, therefore, we move to a new picture in which the initial "path integral" Z_E still has the Euclidean form

$$Z_E = \int \mathcal{D}\left(\mathrm{Re}\,\phi\right) \mathcal{D}\left(\mathrm{Im}\,\phi\right) \, e^{-S} \; , \; S = \int d^{D-1}x \, \phi^* \left(\vec{x}\right) A \, \phi\left(\vec{x}\right) \qquad (9)$$

but there is *initially no time.* We are then confronted with the well-known situation in canonical quantum gravity,[4] where the "wavefunction of the universe" is a functional of only 3-geometries, with no time dependence. Roughly speaking, cosmological time is then defined by the cosmic scale factor R (except that there can be different branches for the state of the universe, corresponding to, e.g., expansion and contraction, as well as different initial conditions). More precisely, the progression of time is locally defined by the progression of local 3-geometries.

An analogy is a stationary state for a proton with coordinates $\vec{X}$ passing a hydrogen atom with coordinates $\vec{x}$. The time-independent Schrödinger equation can be written

$$\left(-\frac{\hbar^2}{2m_p}\nabla_p^2 + H_e\right) \Psi\left(\vec{X}\right) \psi\left(\vec{x}, \vec{X}\right) = E \, \Psi\left(\vec{X}\right) \psi\left(\vec{x}, \vec{X}\right) \qquad (10)$$

with Ψ required to satisfy

$$-\frac{\hbar^2}{2m_p}\nabla_p^2 \Psi\left(\vec{X}\right) = E \, \Psi\left(\vec{X}\right). \qquad (11)$$

Then the equation for ψ is

$$\left(-\frac{\hbar^2}{m_p}\Psi^{-1}\vec{\nabla}_p\Psi \cdot \vec{\nabla}_p - \frac{\hbar^2}{2m_p}\nabla_p^2 + H_e\right) \psi\left(\vec{x}, \vec{X}\right) = 0. \qquad (12)$$

The first term involves a local proton velocity

$$\vec{v}_p = \hbar \, \vec{\nabla}_p \theta / m_p, \qquad \Psi = \left|\Psi\right| e^{i\theta}. \qquad (13)$$

For a state in which the proton is moving rapidly, with

$$\Psi = \Psi_0 e^{i\vec{P}\cdot\vec{X}/\hbar}, \qquad (14)$$

and in which $\left(\hbar^2/2m_p\right)\nabla_p^2\psi$ is relatively small, we obtain

$$i\hbar\frac{\partial}{\partial t}\psi\left(\vec{x},t\right)=H_e\,\psi\left(\vec{x},t\right),\qquad\frac{\partial}{\partial t}\equiv\frac{\vec{P}}{m_p}\cdot\vec{\nabla}_p\,.\tag{15}$$

One then has an "internal time" defined within a stationary state.[5] Similarly, one can define time as a progression of 3-geometries, just as proposed 40 years ago by DeWitt, whose formulation of canonical quantum gravity (following the classical canonical decomposition of Arnowitt, Deser, and Misner, and the work of Dirac, Wheeler, and others) involves the local canonical momentum operator

$$\pi^{kl}\left(\vec{x}\right)=-\,i\frac{\delta}{\delta g_{kl}\left(\vec{x}\right)},\tag{16}$$

which corresponds to the proton momentum operator $-\,i\,\hbar\vec{\nabla}_p$ in the analogy above. After introducing the 3-dimensional metric tensor in the way described in Refs. 1-3, and the gravitational action in a way that will be described in a more complete treatment, we move from the original path-integral quantization to canonical quantization, with a state

$$\Psi_{\text{total}}=\Psi_{\text{gravity}}\left[g_{kl}\left(\vec{x}\right)\right]\;\Psi_{\text{otherfields}}\left[\phi_{\text{otherfields}}\left(\vec{x}\right),g_{kl}\left(\vec{x}\right)\right]\tag{17}$$

and time is defined essentially in the same way as in the analogy.

3. Transformation of 3-Dimensional "Path Integral" Changes Euclidean Factor $\mathrm{e}^{-\mathrm{S}}$ to Lorentzian Factor e^{iS}

A Euclidean path integral with the form of (9), but with time included, is formally transformed into a Lorentzian path integral

$$Z_L^D=\int\mathcal{D}\left(\mathrm{Re}\,\phi_L\right)\mathcal{D}\left(\mathrm{Im}\,\phi_L\right)e^{iS_L^D},\qquad S_L^D=\int d^Dx\,\mathcal{L}_L\tag{18}$$

through an inverse Wick rotation $x_E^0=t_E\rightarrow ix_L^0=it_L$. S_L^D has the usual form of a classical action, and it leads to the usual description of quantized fields via path-integral quantization. In other words, the standard equations of physics follow from S_L^D, and are therefore formulated in Lorentzian time. The Euclidean formulation, in either coordinate or momentum space, is ordinarily regarded as a mere mathematical tool which can simplify calculations and make them better defined.

Hawking, on the other hand, has suggested that Euclidean spacetime may actually be more fundamental than Lorentzian spacetime. In his well-known popular book, he says[6] "So maybe what we call imaginary time is really more basic, and what we call real is just an idea that we invent to help us describe what we think the universe is like." And in a more technical paper he states[7] "In fact one could take the attitude that quantum theory and indeed the whole of physics is really defined in the Euclidean region and that it is simply a consequence of our perception that we interpret it in the Lorentzian regime."

However, there is a fundamental problem with this point of view, because the factor $e^{iS_L^D}$ in the Lorentzian formulation results in interference effects, whereas the factor $e^{-S_E^D}$ in the Euclidean formulation does not. Also, a formal transformation from t_E to t_L mixes all of the supposedly more fundamental Euclidean times in the single Lorentzian time that we actually experience. Finally, it appears difficult to formulate a mathematically well-founded and physically well-motivated transformation of a general path integral from Euclidean to Lorentzian spacetime.

Here we adopt a very different point of view: (1) Nature is fundamentally statistical, essentially as proposed in Refs. 1-3, but the initial path integral (or partition function) does not contain the time as a fundamental coordinate. Instead time is defined by the local 3-space geometry (or more generally, (D-1)-space geometry). (2) It is, however, still necessary to transform from the Euclidean form (9), with e^{-S}, to the Lorentzian form (18), with e^{iS} (but also with no time coordinate, so that $D \to D-1$ in (18)), and this is our goal in the present section.

Consider a single complex scalar field ϕ with a 3-dimensional "Euclidean path integral"

$$Z_E = \int \mathcal{D}\left(\mathrm{Re}\,\phi\right) \mathcal{D}\left(\mathrm{Im}\,\phi\right)\, e^{-S}, \qquad S = \int d^3x\, \phi^*\left(\vec{x}\right) A\, \phi\left(\vec{x}\right). \tag{19}$$

In a discrete picture, the operator A is replaced by a matrix with elements $A\left(\vec{x}, \vec{x}'\right)$:

$$S = \sum_{x,x'} \phi^*\left(\vec{x}\right) A\left(\vec{x}, \vec{x}'\right) \phi\left(\vec{x}'\right). \tag{20}$$

A can be diagonalized to $A\left(\vec{k}, \vec{k}'\right) = a\left(\vec{k}\right) \delta_{\vec{k},\vec{k}'}$. Then

$$Z_E \equiv \left[\prod_{\vec{x}} \int_{-\infty}^{\infty} d\left(\mathrm{Re}\,\phi\left(\vec{x}\right)\right) \int_{-\infty}^{\infty} d\left(\mathrm{Im}\,\phi\left(\vec{x}\right)\right)\right] \exp\left(-\sum_{\vec{x},\vec{x}'} \phi^*\left(\vec{x}\right) A\left(\vec{x}, \vec{x}'\right) \phi\left(\vec{x}'\right)\right)$$

becomes[8]

$$Z_E = \left[\prod_{\vec{k}} \int_{-\infty}^{\infty} d\,\mathrm{Re}\,\phi\left(\vec{k}\right) \int_{-\infty}^{\infty} d\,\mathrm{Im}\,\phi\left(\vec{k}\right) \right] \exp\left(-\sum_{\vec{k}} \phi^*\left(\vec{k}\right) a\left(\vec{k}\right) \phi\left(\vec{k}\right) \right). \tag{21}$$

The Gaussian integrals over $\mathrm{Re}\,\phi\left(\vec{k}\right)$ and $\mathrm{Im}\,\phi\left(\vec{k}\right)$ may be evaluated as usual at each $\vec{k}$ to give

$$Z_E = \prod_{\vec{k}} \frac{\pi}{a\left(\vec{k}\right)} = \frac{\prod_{\vec{k}} \pi}{\det A}. \tag{22}$$

Here, and in the earlier papers, two representations of the path integral are taken to be physically equivalent if they give the same result for all operators A (including those which produce zero except for arbitrarily restricted regions of space and sets of fields). For example, we might define a path integral Z' with fields ϕ' and $\bar{\phi}'$ which are treated as independent and which each vary along the real axis. It is then appropriate to include the formal Jabobian, with a value of $1/2$, which would correspond to a transformation from $\mathrm{Re}\,\phi$ and $\mathrm{Im}\,\phi$ to $\phi' = \mathrm{Re}\,\phi + i\,\mathrm{Im}\,\phi$ and $\bar{\phi}' = i\left(\mathrm{Re}\,\phi - i\,\mathrm{Im}\,\phi\right)$. Since

$$\begin{aligned}
Z' &\equiv \left[\prod_{\vec{k}} \frac{1}{2} \int_{-\infty}^{\infty} d\left(\phi'\left(\vec{k}\right)\right) \int_{-\infty}^{\infty} d\left(\bar{\phi}'\left(\vec{k}\right)\right) \right] \exp\left(\sum_{\vec{k}} i\,\bar{\phi}'\left(\vec{k}\right) a\left(\vec{k}\right) \phi'\left(\vec{k}\right) \right) \\
&= \prod_{\vec{k}} \frac{1}{2} \int_{-\infty}^{\infty} d\left(\phi'\left(\vec{k}\right)\right) \int_{-\infty}^{\infty} d\left(\bar{\phi}'\left(\vec{k}\right)\right) \exp\left(i\,\bar{\phi}'\left(\vec{k}\right) a\left(\vec{k}\right) \phi'\left(\vec{k}\right) \right) \\
&= \prod_{\vec{k}} \frac{1}{2a\left(\vec{k}\right)} \int_{-\infty}^{\infty} d\left(a\left(\vec{k}\right) \phi'\left(\vec{k}\right)\right) 2\pi\,\delta\left(a\left(\vec{k}\right) \phi'\left(\vec{k}\right)\right) && (23) \\
&= \prod_{\vec{k}} \frac{\pi}{a\left(\vec{k}\right)} && (24) \\
&= Z_E && (25)
\end{aligned}$$

for any operator A, we regard Z_E and Z' as being physically equivalent.

Now let us define a "Lorentzian path integral" Z_L by

$$\begin{aligned}
Z_L &= \int \mathcal{D}\left(\mathrm{Re}\,\phi\right) \mathcal{D}\left(\mathrm{Im}\,\phi\right)\, e^{iS} && (26) \\
&\equiv \left[\prod_{\vec{x}} \frac{1}{i} \int_{-\infty}^{\infty} d\left(\mathrm{Re}\,\phi\left(\vec{x}\right)\right) \int_{-\infty}^{\infty} d\left(\mathrm{Im}\,\phi\left(\vec{x}\right)\right) \right] \exp\left(i \sum_{\vec{x},\vec{x}'} \phi^*\left(\vec{x}\right) A\left(\vec{x},\vec{x}'\right) \phi\left(\vec{x}'\right) \right).
\end{aligned}$$

Diagonalization of A gives

$$Z_L = \left[\prod_{\vec{k}} \frac{1}{i} \int_{-\infty}^{\infty} d\,\mathrm{Re}\,\phi\left(\vec{k}\right) \int_{-\infty}^{\infty} d\,\mathrm{Im}\,\phi\left(\vec{k}\right)\right] \exp\left(i \sum_{\vec{k}} \phi^*\left(\vec{k}\right) a\left(\vec{k}\right) \phi\left(\vec{k}\right)\right)$$

$$= \prod_{\vec{k}} \frac{1}{i} \frac{i\,\pi}{a\left(\vec{k}\right)} \tag{27}$$

$$= Z_E \,. \tag{28}$$

Then Z_E can be replaced by Z_L, which involves the original operator A and the original spatial coordinates $\vec{x}$, but a different form for the integrand. This replacement is possible because time is introduced only after Z is in Lorentzian form.

The transformation from Z_E to Z_L can be regarded as a transformation of the fields in the integrand, with the lines along which $\mathrm{Re}\,\phi$ and $\mathrm{Im}\,\phi$ are integrated each being rotated by 45° in the complex plane.[9]

4. Outline of Broad Program: From a Planck-Scale Statistical Theory to Standard Physics with Supersymmetry

The ideas above are part of a broad program to obtain standard physics, including supersymmetry, from a description at the Planck scale which is purely statistical. The major steps in the complete program are as follows:

(1) The fundamental statistical picture gives a $D-1$ "Euclidean action" for bosons only (and with no time yet):

$$Z_b^{D-1} = \int \mathcal{D}\left(\mathrm{Re}\,\phi\right) \mathcal{D}\left(\mathrm{Im}\,\phi\right) e^{-S_b}, \quad S_b = \int d^{D-1}x\, \mathcal{L}_b^{D-1} \,. \tag{29}$$

(2) Random fluctuations then give a "Euclidean action" with bosons, fermions, and a primitive supersymmetry:

$$Z_E^{D-1} = \int \mathcal{D}\left(\mathrm{Re}\,\phi\right) \mathcal{D}\left(\mathrm{Im}\,\phi\right) \mathcal{D}\left(\mathrm{Re}\,\psi\right) \mathcal{D}\left(\mathrm{Im}\,\psi\right) e^{-S}, \quad S = \int d^{D-1}x\, \mathcal{L}^{D-1} \,. \tag{30}$$

(3) Transformation of the integrand in the "path integral" changes the "Euclidean factor" e^{-S} to the "Lorentzian factor" e^{iS}:

$$Z_L^{D-1} = \int \mathcal{D}\left(\mathrm{Re}\,\phi\right) \mathcal{D}\left(\mathrm{Im}\,\phi\right) \mathcal{D}\left(\mathrm{Re}\,\psi\right) \mathcal{D}\left(\mathrm{Im}\,\psi\right) e^{iS}, \quad S = \int d^{D-1}x\, \mathcal{L}^{D-1} \,. \tag{31}$$

(4) The 3-dimensional gravitational metric tensor g_{kl} and $SO(N)$ gauge fields A_k (and their initial, primitive supersymmetric partners) result from

rotations of the vacuum state vector, in both 3-dimensional external space and $D-4$ dimensional internal space.

(5) Time is defined by the progression of 3-geometries in external space.

(6) The Einstein-Hilbert action for the gravitational field (as well as the cosmological constant), the Maxwell-Yang-Mills action for the gauge fields, and the analogous terms for the gaugino and gravitino fields are assumed to arise from a response of the vacuum that is analogous to the diamagnetic response of electrons.

(7) The gravitational field is approximately quantized via first a path-integral formulation and then the canonical formulation of Ref. 4.

(8) Heisenberg equations of motion are then obtained for all fields.

(9) Transformation of the initial spin 1/2 bosonic fields, followed by definition of standard gaugino and gravitino fields, gives standard supersymmetry.

(10) One finally obtains an effective action which is the same as that of standard physics with supersymmetry, except that particle masses, Yukawa couplings, and self-interactions are assumed to arise from supersymmetry breaking and radiative corrections.

A more complete treatment will be given in a much longer paper.

References

1. R. E. Allen, *Int. J. Mod. Phys. A* **12**, 2385 (1997), hep-th/9612041.
2. R. E. Allen, in *Proceedings of Beyond the Desert 2002*, edited by H. V. Klapdor-Kleingrothaus (IOP, London, 2003), hep-th/0008032.
3. R. E. Allen, in *Proceedings of Beyond the Desert 2003*, edited by H. V. Klapdor-Kleingrothaus (IOP, London, 2004), hep-th/0310039.
4. B. S. DeWitt, Phys. Rev. **160**, 1113 (1967).
5. J. F. Reading, 1979 (unpublished).
6. S. W. Hawking, *A Brief History of Time* (Bantam, London, 1988), p. 130.
7. S. W. Hawking, in *Euclidean Quantum Gravity*, edited by G. W. Gibbons and S. W. Hawking (World Scientific, Singapore, 1993), p. 74. This paper is reprinted from *Recent Developments in Gravitation*, Cargese Lectures, edited by M. Levy and S. Deser (Plenum, New York, 1978).
8. See, e.g., M. E. Peskin and D. V. Schroeder, *Introduction to Quantum Field Theory* (Addison-Wesley, Reading, Massachusetts, 1995), p. 285.
9. See, e.g., M. S. Swanson, *Path Integrals and Quantum Processes* (Academic, Boston, 1992), p. 24.

ON THE ORIGIN OF FAMILIES AND THEIR MASS MATRICES WITH THE APPROACH UNIFYING THE SPIN AND THE CHARGES, PREDICTION FOR THE DARK MATTER FAMILY

N.S. MANKOČ BORŠTNIK

Department of Physics, Faculty for Mathematics and Physics, University of Ljubljana
Jadranska 19, 1000 Ljubljana, Slovenia

The Approach unifying all the internal degrees of freedom—the spins and all the charges into only the spin—is offering a new way of understanding the properties of quarks and leptons, that is their charges and their couplings to the gauge fields, the appearance of families and their mass matrices. The starting Lagrange density for spinors in d $(= 1 + 13)$, which carry nothing but two kinds of spins (no charges) and interact with only the gravitational field through vielbeins and two kinds of spin connection fields—the gauge fields of the two kinds of the Clifford algebra objects manifests namely in $d = (1 + 3)$ the properties of fermions and bosons, postulated by the Standard model of the electroweak and colour interactions, with the Yukawa couplings included. In this talk a way of spontaneous breaking of the starting symmetry which leads to the properties of the observed fermions is presented and rough predictions for not yet measured fermions is made, with the dark matter candidates included.

Keywords: Unifying theory; the origin of families; dark matter.

1. Introduction

The Standard model of the electroweak and strong interactions (extended by assuming nonzero masses of neutrinos) fits with around 25 parameters and constraints all the existing experimental data. It leaves, however, unanswered many open questions, among which are questions about the origin of $U(1), SU(2)$ and $SU(3)$ charges, the families and the Yukawa couplings of quarks and leptons, as well as about the corresponding Higgs mechanism.

I assume in the Approach unifying spins and charges[1] a simple Lagrange density for spinors, which in $d = (1+13)$-dimensional space carry two kinds of spins and no charges and interact correspondingly with only the vielbeins and the two kinds of the spin connection fields. After appropriate breaks of the symmetry the starting action demonstrates the observed families of quarks and leptons coupled to the known gauge fields and carrying masses, determined by a part of the starting Lagrange density.

In this talk I demonstrate briefly how does spins and charges postulated by the Standard model appear in the Approach and what predictions can the Approach make for not yet observed families,[2] among which is also the dark matter candidate.

It is assumed that only a left handed Weyl spinor in (1+13)-dimensional space exists, carrying two kinds of spins and no charges: the ordinary spin determined by S^{ab} defined in terms of (the Dirac) γ^a operators and the spin determined by $\tilde{S}^{ab}$ defined in terms of $\tilde{\gamma}^a$, the second kind of the Clifford algebra objects[3,4]

$$\begin{aligned} S^{ab} &= \frac{1}{2}(\gamma^a\gamma^b - \gamma^b\gamma^a), \qquad \tilde{S}^{ab} = \frac{1}{2}(\tilde{\gamma}^a\tilde{\gamma}^b - \tilde{\gamma}^b\tilde{\gamma}^a), \\ &\{\gamma^a, \gamma^b\}_+ = 2\eta^{ab} = \{\tilde{\gamma}^a, \tilde{\gamma}^b\}_+, \\ &\{\tilde{\gamma}^a, \gamma^b\}_+ = 0 = \{S^{ab}, \tilde{S}^{cd}\}_-. \end{aligned} \tag{1}$$

At "physical energies" manifests (if appropriate breaks of the starting symmetry occur) the first kind of generators all the known charges while the corresponding gauge fields are the observed gauge fields. The first kind of generators is, together with the corresponding gauge fields, responsible also for that (diagonal) part of mass matrices of quarks and leptons, for which in the Standard model the Higgs field is needed. The second kind of generators is responsible for the appearance of families of quarks and leptons and accordingly for the Yukawa couplings of the Standard model of the electroweak and colour interactions.

The action[1] for a Weyl (massless) spinor in $d(= 1 + 13)$-dimensional space is as follows [a]

$$\begin{aligned} S &= \int d^dx\, \mathcal{L}, \\ \mathcal{L} &= \frac{1}{2}(E\bar{\psi}\gamma^a p_{0a}\psi) + h.c. = \frac{1}{2}(E\bar{\psi}\gamma^a f^\alpha{}_a p_{0\alpha}\psi) + h.c., \\ p_{0\alpha} &= p_\alpha - \frac{1}{2}S^{ab}\omega_{ab\alpha} - \frac{1}{2}\tilde{S}^{ab}\tilde{\omega}_{ab\alpha}. \end{aligned} \tag{2}$$

[a] Latin indices $a, b, .., m, n, .., s, t, ..$ denote a tangent space (a flat index), while Greek indices $\alpha, \beta, .., \mu, \nu, ..\sigma, \tau..$ denote an Einstein index (a curved index). Letters from the beginning of both the alphabets indicate a general index ($a, b, c, ..$ and $\alpha, \beta, \gamma, ..$), from the middle of both the alphabets the observed dimensions $0, 1, 2, 3$ ($m, n, ..$ and $\mu, \nu, ..$), indices from the bottom of the alphabets indicate the compactified dimensions ($s, t, ..$ and $\sigma, \tau, ..$). We assume the signature $\eta^{ab} = diag\{1, -1, -1, \cdots, -1\}$.

Here $f^{\alpha}{}_a$ are vielbeins (inverted to the gauge field of the generators of translations $e^a{}_\alpha$, $e^a{}_\alpha f^\alpha{}_b = \delta^a_b$, $e^a{}_\alpha f^\beta{}_a = \delta_\alpha{}^\beta$), with $E = \det(e^a{}_\alpha)$, while $\omega_{ab\alpha}$ and $\tilde{\omega}_{ab\alpha}$ are the two kinds of the spin connection fields, the gauge fields of S^{ab} and $\tilde{S}^{ab}$, respectively.

One Weyl spinor representation of S^{ab} in $d = (1+13)$ with the spin as the only internal degree of freedom manifests in four-dimensional "physical" space (if analyzed in terms of the subgroups $SO(1,3) \times U(1) \times SU(2) \times SU(3)$) as the ordinary ($SO(1,3)$) spinor with all the known charges of one family of the left handed weak charged and the right handed weak chargeless quarks and leptons of the Standard model. To manifest this we make a choice of $\tau^{Ai} = \sum_{s,t} c^{Ai}{}_{st} S^{st}$, with $c^{Ai}{}_{st}$ chosen in such a way that τ^{Ai} fulfil the commutation relations of the $SU(3), SU(2)$ and $U(1)$ groups: $\{\tau^{Ai}, \tau^{Bj}\}_- = i\delta^{AB} f^{Aijk} \tau^{Ak}$, with the structure constants f^{Aijk} of the corresponding groups, where the index A identifies the charge groups ($A = 1$ denotes $SU(2)$, $A = 2$ denotes one of the two $U(1)$ groups—the one following from $SO(1,7)$—$A = 3$ denotes the group $SU(3)$ and $A = 4$ the group $U(1)$ following from SO(6)) and index i identifies the generators within one charge group.[3,4]

We make a choice of the Cartan subalgebra set with $d/2 = 7$ elements in $d = 1+13$ for both kinds of generators: $S^{03}, S^{12}, S^{56}, S^{78}, S^{9\,10}, S^{11\,12}, S^{13\,14}$ and $\tilde{S}^{03}, \tilde{S}^{12}, \tilde{S}^{56}, \tilde{S}^{78}, \tilde{S}^{9\,10}, \tilde{S}^{11\,12}, \tilde{S}^{13\,14}$. Then we express the basis for one Weyl in $d = 1+13$ as products of nilpotents and projectors, which are binomials of γ^a[1]

$$\overset{ab}{(k)} := \frac{1}{2}(\gamma^a + \frac{\eta^{aa}}{ik}\gamma^b), \quad \overset{ab}{[k]} = \frac{1}{2}(1 + \frac{i}{k}\gamma^a\gamma^b), \tag{3}$$

respectively, which all are eigenvectors of S^{ab} and $\tilde{S}^{ab}$

$$S^{ab}\overset{ab}{(k)} := \frac{k}{2}\overset{ab}{(k)}, \quad S^{ab}\overset{ab}{[k]} := \frac{k}{2}\overset{ab}{[k]},$$
$$\tilde{S}^{ab}\overset{ab}{(k)} = \frac{k}{2}\overset{ab}{(k)}, \quad \tilde{S}^{ab}\overset{ab}{[k]} = -\frac{k}{2}\overset{ab}{[k]}\,. \tag{4}$$

We choose the starting vector to be an eigen vector of all the members of the Cartan set. In particular, the vector $\overset{03}{(+i)}\overset{12}{(+)}\overset{56}{(+)}\overset{78}{(+)}\overset{9\,10}{[-]}\overset{11\,12}{[+]}\overset{13\,14}{(-)}$ has the following eigenvalues of the Cartan subalgebra set S^{ab}: $(\frac{i}{2}, \frac{1}{2}, \frac{1}{2}, \frac{1}{2}, -\frac{1}{2}, \frac{1}{2}, -\frac{1}{2})$, respectively. With respect to the charge groups it represents a right handed weak chargeless u-quark with spin up and with the colour $(-1/2, 1/(2\sqrt{3}))(= (\tau^{33}, \tau^{38}))$.

Accordingly we may write one octet of the left handed and the right handed quarks of both spins and of one colour charge as presented in Table 1. Quarks of the other two colour charges and the colour chargeless leptons distinguish from this octet only in the part which determines the colour charge and τ^4 ($\tau^4 = 1/6$ for quarks and $\tau^4 = -1/2$ for leptons).

Taking into account the relations

$$\begin{aligned}
\gamma^a \overset{ab}{(k)} &= \eta^{aa} \overset{ab}{[-k]}, & \gamma^b \overset{ab}{(k)} &= -ik \overset{ab}{[-k]},\\
\gamma^a \overset{ab}{[k]} &= \overset{ab}{(-k)}, & \gamma^b \overset{ab}{[k]} &= -ik\eta^{aa} \overset{ab}{(-k)},\\
\tilde{\gamma^a} \overset{ab}{(k)} &= -i\eta^{aa} \overset{ab}{[k]}, & \tilde{\gamma^b} \overset{ab}{(k)} &= -k \overset{ab}{[k]},\\
\tilde{\gamma^a} \overset{ab}{[k]} &= \ i \overset{ab}{(k)}, & \tilde{\gamma^b} \overset{ab}{[k]} &= -k\eta^{aa} \overset{ab}{(k)} .
\end{aligned}$$

one easily sees that γ^a transform $\overset{ab}{(k)}$ into $\overset{ab}{[-k]}$, $\tilde{\gamma}^a$ transform $\overset{ab}{(k)}$ into $\overset{ab}{[k]}$, with unchanged value of S^{ab} (Eq.(4)). We shall use accordingly $\tilde{S}^{ab}$ to generate families.

Let us assume that a break from $SO(1,13)$ to $SO(1,7) \times SU(3) \times U(1)$ occurs at some scale at around 10^{17} GeV or higher and that at some lower scale at around 10^{13} GeV one further break occurs leading to the symmetry $SO(1,3) \times SU(2) \times U(1) \times SU(3) \times U(1)$. Then the starting action manifests as

$$\mathcal{L} = \bar{\psi}\,\gamma^m \Big\{ p_m - g^3 \sum_i \tau^{3i} A^{3i}_m - g^Y\, \tau^Y A^Y_m - g^1 \sum_{i=1,2,3} \tau^{1i} A^{1i}_m -$$

$$g^{Y'}\, Y' A^{Y'}_m - \frac{g^2}{\sqrt{2}} (\tau^{2+} A^{2+}_m + \tau^{2-} A^{2-}_m) \Big\} \psi - \sum_{s=7,8} \bar{\psi}\gamma^s p_{0s}\, \psi,$$

$$m, m' \in \{0,1,2,3\},\ s, s', t \in \{5,6,7,8\}, \tag{5}$$

with $Y = \frac{1}{2}(S^{56} - S^{78}) + \tau^4$, $Y' = \frac{1}{2}(S^{56} + S^{78}) - \tau^4 \tan^2\theta_2$, $\tau^{2\pm} = \frac{1}{2}(S^{58} + S^{67}) \pm i\frac{1}{2}(S^{57} - S^{68})$. The angle θ_2 determines mixing of the fields at very high energies, when $SO(4) \times U(1)$ breaks into $SU(2) \times U(1)$. First row manifests the starting action of the Standard model (before the electrowek break), second row with very heavy fields $A^{Y'}_m$ and $A^{2\pm}_m$ can at low energies be neglected. Third row determines mass matrices. It can be written as $\psi^\dagger\ \gamma^0 \{\overset{78}{(+)}\ p_{0+} + \overset{78}{(-)}\ p_{0-}\}\psi$, $p_{0\pm} = (p_{07} \mp i\ p_{08})$, while $p_{0s} = p_s - \frac{1}{2}S^{ab}\omega_{abs} - \frac{1}{2}\tilde{S}^{ab}\tilde{\omega}_{abs}$. Taking into account that $\tilde{S}^{ab} = \frac{i}{2}[\overset{ac}{\tilde{(k)}} + \overset{ac}{\tilde{(-k)}}][\overset{bc}{\tilde{(k)}} + \overset{bc}{\tilde{(-k)}}]$ for any c it follows that

$$-\sum_{(a,b)} \frac{1}{2} \overset{78}{(\pm)}\ \tilde{S}^{ab}\tilde{\omega}_{ab\pm} = - \sum_{(ac),(bd),\,k,l} \overset{78}{(\pm)}\overset{ac}{\tilde{(k)}}\overset{bd}{\tilde{(l)}}\ \tilde{A}^{kl}_{\pm}((ac),(bd)),$$

Table 1. The 8-plet of quarks—the members of $SO(1,7)$ subgroup, belonging to one Weyl left handed spinor representation of $SO(1,13)$. It contains the left handed weak charged quarks and the right handed weak chargeless quarks of a particular colour $(-1/2, 1/(2\sqrt{3}))$. Here $\Gamma^{(1,3)}, \Gamma^{(1,13)} = -1 = \Gamma^{(1,7)} \times \Gamma^{(6)}$ define the handedness in $(1+3)$, $(1+13)$ and $(1+7)$ parts of space, respectively, S^{12} defines the ordinary spin (which can also be read directly from the basic vector), τ^{13} defines the weak charge, τ^{2} defines the $U(1)$ charge from $SO(1,7)$, τ^{33} and τ^{38} define the colour charge and τ^{4} defines another $U(1)$ charge, which together with the first one defines $Y = \tau^4 + \tau^2 = \tau^4 + S^{56}$. The vectors are eigenvectors of all the members of the Cartan subalgebra set $(\{S^{03}, S^{12}, S^{56}, S^{78}, S^{910}, S^{1112}, S^{1314}\})$. The reader can find the whole Weyl representation in references, presented in .[1]

i		$\|{}^a\psi_i>$	$\Gamma^{(1,3)}$	S^{12}	$\Gamma^{(4)}$	τ^{13}	τ^{2}	τ^{33}	τ^{38}	τ^{41}	Y
		Octet, $\Gamma^{(1,7)} = 1,\ \Gamma^{(6)} = -1$, of quarks									
1	u_R^c	$\stackrel{03}{(+i)}\stackrel{12}{(+)} \mid \stackrel{56}{(+)}\stackrel{78}{(+)} \parallel \stackrel{9\,10}{[-]}\ \stackrel{11\,12}{[+]}\ \stackrel{13\,14}{(-)}$	1	$\frac{1}{2}$	1	0	$\frac{1}{2}$	$-\frac{1}{2}$	$\frac{1}{2\sqrt{3}}$	$\frac{1}{6}$	$\frac{2}{3}$
2	u_R^c	$\stackrel{03}{[-i]}\stackrel{12}{[-]} \mid \stackrel{56}{(+)}\stackrel{78}{(+)} \parallel \stackrel{9\,10}{[-]}\ \stackrel{11\,12}{[+]}\ \stackrel{13\,14}{(-)}$	1	$-\frac{1}{2}$	1	0	$\frac{1}{2}$	$-\frac{1}{2}$	$\frac{1}{2\sqrt{3}}$	$\frac{1}{6}$	$\frac{2}{3}$
3	d_R^c	$\stackrel{03}{(+i)}\stackrel{12}{(+)} \mid \stackrel{56}{[-]}\stackrel{78}{[-]} \parallel \stackrel{9\,10}{[-]}\ \stackrel{11\,12}{[+]}\ \stackrel{13\,14}{(-)}$	1	$\frac{1}{2}$	1	0	$-\frac{1}{2}$	$-\frac{1}{2}$	$\frac{1}{2\sqrt{3}}$	$\frac{1}{6}$	$-\frac{1}{3}$
4	d_R^c	$\stackrel{03}{[-i]}\stackrel{12}{[-]} \mid \stackrel{56}{[-]}\stackrel{78}{[-]} \parallel \stackrel{9\,10}{[-]}\ \stackrel{11\,12}{[+]}\ \stackrel{13\,14}{(-)}$	1	$-\frac{1}{2}$	1	0	$-\frac{1}{2}$	$-\frac{1}{2}$	$\frac{1}{2\sqrt{3}}$	$\frac{1}{6}$	$-\frac{1}{3}$
5	d_L^c	$\stackrel{03}{[-i]}\stackrel{12}{(+)} \mid \stackrel{56}{[-]}\stackrel{78}{(+)} \parallel \stackrel{9\,10}{[-]}\ \stackrel{11\,12}{[+]}\ \stackrel{13\,14}{(-)}$	-1	$\frac{1}{2}$	-1	$-\frac{1}{2}$	0	$-\frac{1}{2}$	$\frac{1}{2\sqrt{3}}$	$\frac{1}{6}$	$\frac{1}{6}$
6	d_L^c	$\stackrel{03}{(+i)}\stackrel{12}{[-]} \mid \stackrel{56}{[-]}\stackrel{78}{(+)} \parallel \stackrel{9\,10}{[-]}\ \stackrel{11\,12}{[+]}\ \stackrel{13\,14}{(-)}$	-1	$-\frac{1}{2}$	-1	$-\frac{1}{2}$	0	$-\frac{1}{2}$	$\frac{1}{2\sqrt{3}}$	$\frac{1}{6}$	$\frac{1}{6}$
7	u_L^c	$\stackrel{03}{[-i]}\stackrel{12}{(+)} \mid \stackrel{56}{(+)}\stackrel{78}{[-]} \parallel \stackrel{9\,10}{[-]}\ \stackrel{11\,12}{[+]}\ \stackrel{13\,14}{(-)}$	-1	$\frac{1}{2}$	-1	$\frac{1}{2}$	0	$-\frac{1}{2}$	$\frac{1}{2\sqrt{3}}$	$\frac{1}{6}$	$\frac{1}{6}$
8	u_L^c	$\stackrel{03}{(+i)}\stackrel{12}{[-]} \mid \stackrel{56}{(+)}\stackrel{78}{[-]} \parallel \stackrel{9\,10}{[-]}\ \stackrel{11\,12}{[+]}\ \stackrel{13\,14}{(-)}$	-1	$-\frac{1}{2}$	-1	$\frac{1}{2}$	0	$-\frac{1}{2}$	$\frac{1}{2\sqrt{3}}$	$\frac{1}{6}$	$\frac{1}{6}$

with the pair (a,b) in the sum before the equality sign running over all the indices which do not characterize the Cartan subalgebra, with $a,b = 0,\ldots,8$, while the two pairs (ac) and (bd) in the sum after the equality sigh denote only the Cartan subalgebra pairs (for $SO(1,7)$ we only have the pairs $(03),(12)$; $(03),(56)$;$(03),(78)$; $(12),(56)$; $(12),(78)$; $(56),(78)$); k and l run over four possible values so that $k = \pm i$, if $(ac) = (03)$ and $k = \pm 1$ in all other cases, while $l = \pm 1$.

2. Mass Matrices From Approach Unifying Spins and Charges

The mass term manifests after taking into account all the breaks of symmetries (from $SO(1,7) \times U(1)$ to the symmetry after the electroweak break $SO(1,3) \times U(1)$ in both sectors—S^{ab} and $\tilde{S}^{ab}$—in a equivalent way up to the point that at low energies only observable phenomena manifest) as the mass matrix[2] $\mathcal{L}_Y$

$$
\begin{aligned}
-\mathcal{L}_Y = \psi^+\gamma^0 \Big\{ & \overset{78}{(+)}\,(e\,QA_+ - \tilde{e}\tilde{Q}\tilde{A}_+ + \tilde{g}'\tilde{Q}'\tilde{Z}_+ + \tilde{g}^{Y'}\tilde{Y}'\tilde{A}^{Y'}_+ + \\
& \qquad\qquad \tilde{N}^{+3}_+\, A^{N^{+3}}_+ + \tilde{N}^{-3}_+\, A^{N_-}_+) + \\
& \overset{78}{(-)}\,(e\,QA_- + \tilde{e}\tilde{Q}\tilde{A}_- + \tilde{g}'\tilde{Q}'\tilde{Z}_- + \tilde{g}^{Y'}\tilde{Y}'\tilde{A}^{Y'}_+\tilde{N}^3_+\, A^{N_+}_- + \\
& \qquad\qquad \tilde{N}^3_-\, A^{N_-}_-) + \\
& \overset{78}{(+)} \sum_{\{(ac)(bd)\},k,l} \overset{ac}{(\tilde{k})}\overset{bd}{(\tilde{l})}\, \tilde{A}^{kl}_+((ac),(bd)) \;+ \\
& \overset{78}{(-)} \sum_{\{(ac)(bd)\},k,l} \overset{ac}{(\tilde{k})}\overset{bd}{(\tilde{l})}\, \tilde{A}^{kl}_-((ac),(bd)) \Big\}\psi,
\end{aligned} \tag{6}
$$

where $Q = S^{56}+\tau^4$, $Q' = -(\tau^4+\frac{1}{2}(S^{56}+S^{78}))\tan^2\theta_1+\frac{1}{2}(S^{56}-S^{78})$, $\tau^{1\pm} = \frac{1}{2}(\mathcal{S}^{58} - \mathcal{S}^{67}) \pm i\frac{1}{2}(\mathcal{S}^{57} + \mathcal{S}^{68})$, $\tilde{Q} = \tilde{S}^{56} + \tilde{\tau}^4$, $\tilde{Q}' = -(\tilde{\tau}^4 + \frac{1}{2}(\tilde{S}^{56} + \tilde{S}^{78}))\tan^2\tilde{\theta}_1 + \frac{1}{2}(\tilde{S}^{56} - \tilde{S}^{78})$, $\tilde{\tau}^{1\pm} = \frac{1}{2}(\tilde{S}^{58} - \tilde{S}^{67}) \pm i\frac{1}{2}(\tilde{S}^{57} + \tilde{S}^{68})$, $\tilde{Y}' = \frac{1}{2}(\tilde{S}^{56} + \tilde{S}^{78}) - \tilde{\tau}^4\tan^2\tilde{\theta}_2$, $\tilde{\tau}^{2\pm} = \frac{1}{2}(\tilde{S}^{58} + \tilde{S}^{67}) \pm i\frac{1}{2}(\tilde{S}^{57} - \tilde{S}^{68})$, $\tau^{41} := -\frac{1}{3}(\mathcal{S}^{9\,10} + \mathcal{S}^{11\,12} + \mathcal{S}^{13\,14})$, $\tilde{\tau}^{41} := -\frac{1}{3}(\tilde{S}^{9\,10} + \tilde{S}^{11\,12} + \tilde{S}^{13\,14})$.

The mass matrices of Eq.(6) manifest for quarks and leptons—under the above assumptions—as two times four times four matrices, one of the two four families are as heavy as the scale of the first break (at $\approx 10^{13}$ GeV). Less massive four families are presented[2] at Table 2. The terms $a_\pm$ are the diagonal terms to which S^{ab} and $\tilde{S}^{ab}$ contribute, $\tilde{A}^{13}_\pm = \tilde{A}_\pm\sin\tilde{\theta}_1 + \tilde{Z}_\pm\cos\tilde{\theta}_1$, $\tilde{A}^Y_\pm = \tilde{A}_\pm\cos\tilde{\theta}_1 - \tilde{Z}_\pm\sin\tilde{\theta}_1$, where $\tilde{A}_\pm$, $\tilde{Z}_\pm$ appear with $\tilde{Q}$ and

Table 2. The mass matrix for the lower four families of u-quarks (with the sign $-$) and d-quarks (with the sign $+$).

	I	II	III	IV
I	$a_\pm$	$\frac{\tilde{g}^m}{\sqrt{2}}\tilde{A}_\pm^{+N^+}$	$-\frac{\tilde{g}^1}{\sqrt{2}}\tilde{W}_\pm^+$	0
II	$\frac{\tilde{g}^m}{\sqrt{2}}\tilde{A}_\pm^{-N^+}$	$a_\pm$ $+\frac{\tilde{g}^m}{2}(\tilde{A}_\pm^{3N^-}+\tilde{A}_\pm^{3N^+})$	0	$-\frac{\tilde{g}^1}{\sqrt{2}}\tilde{W}_\pm^+$
III	$\frac{\tilde{g}^1}{\sqrt{2}}\tilde{W}_\pm^-$	0	$a_\pm$ $+\tilde{e}\tilde{A}_\pm+\tilde{g}'\tilde{Z}_\pm$	$\frac{\tilde{g}^m}{\sqrt{2}}\tilde{A}_\pm^{+N^+}$
IV	0	$\frac{\tilde{g}^1}{\sqrt{2}}\tilde{W}_\pm^-$	$\frac{\tilde{g}^m}{\sqrt{2}}\tilde{A}_\pm^{-N^+}$	$a_\pm+\tilde{e}\tilde{A}_\pm+\tilde{g}'\tilde{Z}_\pm$ $+\frac{1}{2}\tilde{g}^m(\tilde{A}_\pm^{3N^-}+\tilde{A}_\pm^{3N^+})$

$\tilde{Q}'$, respectively, with $\tilde{e}=\tilde{g}^Y\cos\tilde{\theta}_1, \tilde{g}'=\tilde{g}^1\cos\tilde{\theta}_1, \tan\tilde{\theta}_1=\frac{\tilde{g}^Y}{\tilde{g}^1}$. $\tilde{W}_\pm^\pm$ appear with $\tilde{\tau}^\pm$, while the fields $\tilde{A}_\pm^{\pm N^\pm}$ appear with $\tilde{\tau}^{\pm N^\pm}$, and $\tilde{A}_\pm^{3N^\pm}$ with $\tau^{3N^\pm}$, where $\tilde{\tau}^{N^{\pm i}}$ are the two $SU(2)$ generators of the $SO(1,3)$ group in the $\tilde{S}^{ab}$ sector.

The matrix elements in Table 2 depend (on the tree level) on unknown fields. We take them as free parameters and fit them using the Monte-Carlo program to the experimental data within the known accuracy (to the known masses and mixing matrix). We assume at this stage that mass matrices are real (neglecting the CP violating terms). Although it turns out that there are only two independent matrix elements for each type of quarks (or leptons), we still have too many free parameters to make any prediction. Assuming the values for the fourth family masses we get the mass matrix in MeV for the u-quarks

$$\begin{pmatrix} (9,22) & (-150,-83) & 0 & (-306,304) \\ (-150,-83) & (1211,1245) & (-306,304) & 0 \\ 0 & (-306,304) & (171600,176400) & (-150,-83) \\ (-306,304) & 0 & (-150,-83) & 200000 \end{pmatrix}$$

and for the d-quarks

$$\begin{pmatrix} (5,11) & (8.2,14.5) & 0 & (174,198) \\ (8.2,14.5) & (83,115) & (174,198) & 0 \\ 0 & (174,198) & (4260,4660) & (8.2,14.5) \\ (174,198) & 0 & (8.2,14.5) & 200000 \end{pmatrix},$$

corresponding to the following values for the masses of the quarks

$$m_{u_i}/GeV = (0.005, 1.220, 171., 215.),$$
$$m_{d_i}/GeV = (0.008, 0.100, 4.500, 285.)$$

and the mixing matrix for the quarks

$$\begin{pmatrix} -0.974 & -0.226 & -0.00412 & 0.00218 \\ 0.226 & -0.973 & -0.0421 & -0.000207 \\ 0.0055 & -0.0419 & 0.999 & 0.00294 \\ 0.00215 & 0.000414 & -0.00293 & 0.999 \end{pmatrix}.$$

By enlarging the masses of the fourth family of quarks, the fourth family (very) slowly decouples from the first three families.

3. Dark Matter Candidates

The data about the dark matter do not help very much to extract properties of the candidates, which the Approach unifying spins and charges is offering, although it looks like that the DAMA experiment might.

The break of $(SO(1,7) \times U(1)) \times SU(3)$ into $SO(1,3) \times (SO(4) \times U(1)) \times SU(3)$ splits eight families into $2\times$ four (with respect to the Yukawa interaction decoupled) families, while breaking of $(SO(4) \times U(1))$, first into $(SU(2) \times U(1))$ at some high scale and then further to $U(1)$ at the weak scale, leads to four massive families, whose mass matrix elements are dictated by the weak scale, and the additional four massive families, whose mass matrix elements are dictated by a much higher scale (with respect to the weak scale), which is the scale of breaking $SO(4) \times U(1)$ into $SU(2) \times U(1)$. I shall assume that the scale is around 10^{13} GeV, as estimated in the ref.[4]

While the fourth family could appear at around $200 GeV$ or higher, the lightest of the next four families—the fifth family— may have a mass at around 10^{13} GeV (the scale of breaking $SO(4) \times U(1)$ to $SU(2) \times U(1)$) or several orders of magnitude smaller. Since the mixing matrix turns out to have (due to the assumed way of breaking the symmetry) all the matrix elements between the lowest four families and the higher four ones equal to zero, the lowest family (u_d, d_5, ν_5, e_5) of the higher four families is stable.

The lowest— the fifth—family of quarks and leptons is accordingly the candidate to form the dark matter under the following conditions (see also[15,16]):

(i) The fifth family have to form heavy enough hadrons or any kind of neutral (with respect to the $U(1)$ and $SU(3)$ charge) clusters among themselves or with light families so that the corresponding average radius of a cluster and the abundance of the clusters allow their interaction with the ordinary matter to be small enough to be (up to now) observable only through their gravitational field and in the very last time also in the DAMA and other experiments. While clusters of light and heavy quarks have properties similar to the properties of light quarks clusters (up to a mass), have the heavy quarks clusters much smaller sizes. Using the nonrelativistic Bohr model with the $\frac{1}{r}$ (radial) dependence of the potential between two quarks $V = -\frac{3\alpha_c}{r}$ (or more then two quarks, if we neglect the details which would bring at most a correction of a factor of 10) to evaluate the (colour) binding energy and the size of a cluster, we find, for example for the mass equal to 10^4 GeV$/c^2$ the size equal to 10^{-4} fm and for the mass 10^{13} GeV the size 10^{-13} fm. The observed density of the dark matter does not put limitation on the type of clusters, allowing the clusters of heavy and light quarks, as well as clusters of only heavy quarks, as far as the heavy quarks are heavy enough. The DAMA experiment might distinguish between clusters of only heavy quarks and clusters of heavy and light quarks, since the number of measured events is proportional to the mass of the cluster to the minus third power (the size is proportional to the mass to minus the second power in the "Bohr-like" evaluations).

(ii) The hadron matter including the fifth family quarks had to have a chance during the evolution of the universe to be formed out of plasma, so that the today's ratio among the ordinary and the heavy hadron matter is around $6-7$.

(iii) The heavy hadron matter should have in the evolution of the universe, when clusters of galaxies were formed, its own way of forming clusters, although its density is (strongly) influenced by the gravitational field of the galaxies of the ordinary matter.

(iv) There are other limitations, some of them are discussed in.[15,16]

To say something about how heavy quarks clusters can be formed or about the surviving probability of the evolution of such clusters in the evolution (after the inflation) these mechanisms must be studied.

4. Concluding Remarks

In this talk I presented how does the Approach unifying spins and charges answer some of the open questions of the Standard model of the electroweak and colour interactions.

(i) We started with one Weyl spinor in $d = 1+13$, which carries two kinds of spins (no charges) and interacts correspondingly with vielbeins and gauge fields of the two kinds of the generators: S^{ab} and $\tilde{S}^{ab}$, corresponding to γ^a and $\tilde{\gamma}^a$—the two kinds of the Clifford algebra objects.

(ii) A simple starting Lagrange density for a Weyl spinor in $d = (1 + 13)$-dimensional space manifests in $d = (1+3)$-dimensional space the properties of all the quarks and the leptons of the Standard model (with the right handed neutrinos included) and the families.

(iii) There are S^{ab}, which determine in $d = (1+3)$ the spin and all the charges. One Weyl spinor representation includes (if analyzed with respect to the Standard model groups) the left handed weak charged quarks and leptons and the right handed weak chargeless quarks and leptons, coupled to all the corresponding gauge fields.

(iv) $\tilde{S}^{ab}$ generate an even number of families. It is a part of the starting action which manifests as Yukawa couplings of the Standard model $\psi^{\dagger}\gamma^0\gamma^s p_{0s}\psi$, $s = 7,8$, with $p_{0s} = -\frac{1}{2}S^{ab}\omega_{abs} - \frac{1}{2}\tilde{S}^{ab}\tilde{\omega}_{abs}$ contributing to diagonal and off diagonal elements of mass matrices.

(v) There are several possibilities of breaking the starting $SO(1,13)$ symmetry to the Standard model symmetry left after the electroweak break. We assume the break of $SO(1,13)$ to $SO(1,7) \times SO(6)$, then to $SO(1,7) \times SU(3) \times U(1)$, since $SO(1,7)$ manifests a left handed weak charged quarks and leptons and the right handed weak chargeless quarks and leptons. We assume several further breaks. One of possible breaks predicts the fourth family of quarks and leptons at the energies still allowed by the experimental data.[5] In this paper we present results only for the break from $(SO(1,7) \times U(1)) \times SU(3)$ to $(SO(1,3) \times SO(4) \times U(1)) \times SU(3)$ and further to $SO(1,3) \times (SU(2) \times U(1)) \times SU(3)$ and then to $SO(1,3) \times U(1) \times SU(3)$. We present masses of quarks and the corresponding mixing matrix for the „lightest" four families. Results depend on the masses of the fourth family. Letting the fourth family mass growing, the fourth family very slowly decouples from the first three. The mixing matrix predicts, for example, the changed values for $|V_{31}|/|V_{32}| = 0.128 - 0.149$, when four instead of three families at weak scale contribute to this value.

(vi) If the Approach is the right way beyond the Standard model, then there are more than just three families observed up to now. The stable fifth family seems a very promising candidate for forming the dark matter, if clusters of heavy (fifth family) quarks were formed in the evolution of the universe and survived up to today. It stays an open problem, how clusters of heavy quarks can be formed during the evolution of the universe, how does it happen that the today ratio between the ordinary hadron matter and the heavy hadron matter is the observed one (6 − 7), as well as how does the dark matter of these clusters behave when the galaxies of the ordinary matter (made out of light quarks) were formed to be spread as we measure now.

To decide whether or not the way of breaking symmetries presented in this paper is the right one, further studies going beyond the tree level are needed.

References

1. A. Borštnik Bračič, N. S. Mankoč Borštnik, hep-ph/0512062, *Phys Rev.* **D 74** (2006) 073013-16.
2. G. Bregar, M. Breskvar, D. Lukman, N. S. Mankoč Borštnik, *hep/ph-07082846.*
3. N. S. Mankoč Borštnik, H. B. Nielsen, *J. of Math. Phys.* **44**, 4817-4827 (2003), hep-th/0303224.
4. N. S. Mankoč Borštnik, H. B. Nielsen, *J. of Math. Phys.* **43**, 5782-5803 (2002), hep-th/0111257.
5. S. S. Bulanov, V. A. Novikov, L. B. Okun, A. N. Rozanov, M. I. Vysotsky, *Phys. Atom. Nucl.* **66**, 2169 (2003); *Yad. Fiz.* **66**, 2219 (2003), hep-ph/0301268v2.
6. N. S. Mankoč Borštnik, *Phys. Lett.* **B 292**, 25-29 (1992).
7. N. S. Mankoč Borštnik, *J. Math. Phys.* **34**, 3731-3745 (1993).
8. N. S. Mankoč Borštnik, *Modern Phys. Lett.* **A 10**, 587-595 (1995).
9. N. S. Mankoč Borštnik, *Int. J. Theor. Phys.* **40**, 315-337 (2001).
10. N. S. Mankoč Borštnik, H. B. Nielsen, *Phys. Rev. D* **62** (04010-14) (2000).
11. N. S. Mankoč Borštnik, H. B. Nielsen, *Phys. Lett.* **B 633**, 771-775 (2006), hep-th/0311037, hep-th/0509101.
12. A. Borštnik, N. S. Mankoč Borštnik, hep-ph/0401043, hep-ph/0401055, hep-ph/0301029.
13. N. S. Mankoč Borštnik, H. B. Nielsen, *Phys. Lett.* **B 633**, 771-775 (2006), hep-th/0311037, hep-th/0509101.
14. N. S. Mankoč Borštnik, H. B. Nielsen, *Phys. Lett.* **B 644**, 198-202 (2007).
15. Discussions on Dark matter properties by M. Y. Khlopov and N. S. Mankoc Borštnik in Proc. to the 10th International Workshop "What Comes Beyond the Standard Model", 17-27 of July, 2007, Volume II, *Ed. Norma Mankoč Borštnik, Holger Bech Nielsen, Colin Froggatt, Dragan Lukman,* DMFA Založništvo, Ljubljana December 2007.
16. G. Bregar, N. S. Mankoč Borštnik, "Discussions on Dark matter properties", Proceedings to the 10th International Workshop "What Comes Beyond the Standard Model", 17-27 of July, 2007, Volume II, *Ed. Norma Mankoč Borštnik, Holger Bech Nielsen, Colin Froggatt, Dragan Lukman,* DMFA Založništvo, Ljubljana December 2007.

CONSTRAINTS ON GAUSS-BONNET COSMOLOGIES

ISHWAREE P. NEUPANE

Department of Physics and Astronomy, University of Canterbury,
Private Bag 4800, Christchurch 8020, New Zealand
E-mail: ishwaree.neupane@canterbury.ac.nz

The modified Gauss-Bonnet theory can be motivated by a number of physical reasons, including: the uniqueness of a gravitational Lagrangian in four and higher dimensions and the leading order α' corrections in string theory and M theory. Such an effective theory of scalar-tensor gravity has been modeled in the recent past to explain both the initial cosmological singularity problem and the observationally supported cosmological perturbations. Here I present an overview of the recent advancements in the use of modified Gauss-Bonnet gravity to explain current observations, touching on key cosmological and astrophysical constraints applicable to theories of scalar-tensor gravity. The Gauss-Bonnet type modification of Einstein's theory admit nonsingular solutions for a wide range of scalar-curvature couplings. It also provides plausible explanation to some outstanding cosmological conundrums, including: the transition from matter dominance to dark energy and the late time cosmic acceleration. The focus is placed here to constrain such an effective theory of gravity against the recent cosmological and astrophysical observations.

Keywords: String theory and cosmology, Gauss-Bonnet gravity, dark energy.

1. Introduction

Einstein's general relativity has been very successful as a classical theory of gravitational interactions, especially, in a non-accelerating (or non-expanding) spacetime. In a cosmological background, the theory predicts spacetime singularities, so its modification is inevitable at high energy scales. Further the recently observed accelerating expansion of the universe[1] provides some insight to the possibility that general relativity together with ordinary matter and radiation, described by the standard model of particle physics, cannot fully explain the current observations. The question arises because the current observations[2] require in the fabric of the cosmos the existence of a dark energy component of magnitude about 73%, which does not 'clump' gravitationally. Another 23% of the mass-energy is in the form of mysterious non-baryonic dark matter, which 'clumps' gravitationally.

The bulk of the universe appears to be dark energy and dark matter. So far there is no fully consistent explanation of these energy components supported by a fundamental theory. The main focus of this meeting is obviously to update our knowledge on DARK matter and DARK energy searches and the physics behind these. The focus of my presentation will be on a possible resolution of dark energy problem within some string-inspired theories of scalar-tensor gravity.

2. Accelerating Universes and String-Inspired Models

The discovery that the expansion of the universe is currently accelerating is among the most significant of our times. Evidence in favour of this accelerated expansion (caused by putative dark energy) has led to a continued interest in scenarios that propose modifications to Einstein's general relativity. The proposals are of differing origins as well as motivations, some are based on theories of higher-dimensional gravity and others on consideration of one or more fundamental scalar fields and their interactions with higher-order curvature terms. Both these ideas are well motivated by supergravity and superstring theories, which incorporate Einstein's theory in a more general framework. There are several theoretical motivations to incorporate string theory into cosmological model building. Notably, gravitational interactions mediated by scalar fields, together with the standard graviton, are the best alternatives[3] to general relativity, which also provide a mathematically consistent framework to test the various observable predictions of higher dimensional theories of gravity, such as, string theory.

Typically the low energy limit of string theory or supergravity features scalar fields and their couplings to a unique combination of the three quadratic scalars R^2, $R_{\mu\nu}R^{\mu\nu}$ and $R_{\mu\nu\rho\lambda}R^{\mu\nu\rho\lambda}$, composed of the scalar curvature, the Ricci and Riemann tensors:

$$\mathcal{R}^2 \equiv R^2 - 4R_{\mu\nu}R^{\mu\nu} + R_{\mu\nu\rho\lambda}R^{\mu\nu\rho\lambda},$$

known as the Gauss-Bonnet term. This term arises, almost universally, in all versions of string theory as the leading order α' correction. An illustrative example is the following four-dimensional heterotic superstring model which describes the dynamics of graviton, dilaton S and the common (volume) modulus field T, arising from a compactification of 10D heterotic superstring theory on a symmetric 6D orbifold:[4]

$$\mathcal{L}_{\text{grav}} = \mathcal{L}_0 + \mathcal{L}_1, \tag{1}$$

where the string tree-level Lagrangian $\mathcal{L}_0$ is

$$\mathcal{L}_0 = \frac{R}{2\kappa^2} - \zeta \frac{2\Delta S \Delta \bar{S}}{(S+\bar{S})^2} - \gamma \frac{2\Delta T \Delta \bar{T}}{(T+\bar{T})^2} + \frac{1}{8}(\mathrm{Re}S)\mathcal{R}^2 + \frac{1}{8}(\mathrm{Im}S)\, R\tilde{R}, \quad (2)$$

while the modulus T dependent Lagrangian at the one loop level is

$$\mathcal{L}_1 = \Delta(T,\bar{T})\,\mathcal{R}^2 - i\Delta(T,\bar{T})\,R\tilde{R}, \quad (3)$$

where κ is the inverse Planck mass $M_P^{-1} = (8\pi G_N)^{1/2}$, G_N is Newton's constant, ζ and γ are numerical constants, $R\tilde{R} \equiv g^{-1/2}\epsilon^{\mu\nu\rho\lambda} R_{\mu\nu}{}^{\sigma\tau} R_{\rho\lambda\sigma\tau}$ (where $\epsilon^{\mu\nu\rho\lambda}$ is a totally anti-symmetric tensor) and $\Delta(T,\bar{T}) \propto \ln\left[(T+\bar{T})|\eta(iT)|^4\right]$. The Dedekind η-function is given by $\eta(iT) \equiv \mathrm{e}^{-\pi T/12}\prod_{n\geq 1}\left(1-\mathrm{e}^{-2n\pi T}\right)$. There can be additional terms in the four-dimensional effective Lagrangian, such as,

$$\mathcal{L}_{\mathrm{add}} = -V(S,T) - \cdots \quad (4)$$

which includes, within the context of string theory, some supersymmetry breaking non perturbative potentials coming from the dynamics of branes, fluxes and orientifold planes, as well as the back reaction effects from the localized sources. The potential usually consists of sum of exponential terms determined by the fluxes and the curvature terms;[5,6] this is related to the fact that upon dimensional reduction of a gravity theory, the potential is exponential in terms of canonically normalized scalar fields descending from the internal space metric and other modes.

To evaluate field equations obtained by varying a gravitational action, we consider approximately homogeneous and isotropic solutions given by the Friedmann-Robertson-Walker metric: $ds^2 = -dt^2 + a^2(t)\, d\mathbf{x}^2$, where $a(t)$ is the scale factor of the universe. $H \equiv \dot{a}/a$ defines the Hubble parameter and the dot denotes a derivative with respect to cosmic time t.

In a flat FRW background, the terms proportional to $R\tilde{R}$ give a trivial contribution. Defining $\mathrm{Re}S = e^{\varphi}/g_s^2$, $\mathrm{Re}T = e^{2\sigma}$, $\mathrm{Im}S \equiv \tau =$ const and $\mathrm{Im}T = 0$, the four-dimensional effective Lagrangian may be given by[4,7]

$$\mathcal{L}_{\mathrm{eff}} = \frac{R}{2\kappa^2} - \frac{\zeta}{2}(\nabla\varphi)^2 - \frac{\gamma}{2}(\nabla\sigma)^2 + \frac{1}{8}\left[\lambda f(\varphi) - \delta\,\xi(\sigma)\right]\mathcal{R}^2 - V(\varphi,\sigma), \quad (5)$$

where $\lambda \propto 1/g_s^2$, g_s is four-dimensional string coupling, τ is pseudoscalar axion and δ is a numerical constant. To leading order in string loop expansion, $f(\varphi) \propto \mathrm{e}^{\varphi}$ and $\xi(\sigma) = \ln 2 - \frac{\pi}{3}\mathrm{e}^{\sigma} + \sigma + 4\sum_{n=1}^{\infty}\ln(1-\mathrm{e}^{-2n\pi \mathrm{e}^{\sigma}})$. The latter implies that $d\xi/d\sigma \simeq -\mathrm{sgn}(\sigma)\frac{2\pi}{3}\sinh(\sigma) < 0$. Several authors have explored special features of the string-derived Lagrangian that might provide some characteristic features of the above model (see for example[7–16]).

In the discussion below we consider the simplest case of a single modulus, under the assumption that the compacification modulus σ would rapidly evolve along an instantaneous minimum determined by the condition $dV/d\sigma = 0$, such that $V(\varphi, \sigma) \approx \text{const} \times V(\varphi)$, while φ attains a constant value only at late times [a].

3. Modified Gauss-Bonnet Theory

As should be clear from the above discussion, the simplest version of scalar-tensor theories, which is perhaps sufficiently general for explaining the present evolution of our universe,[17] may be given by[18]

$$S_{\text{eff}} = \int d^4x\sqrt{-g}\left[\frac{R}{2\kappa^2} - \frac{\zeta}{2}(\nabla\varphi)^2 - V(\varphi) + \frac{\lambda}{8}f(\varphi)\mathcal{R}^2\right]. \tag{6}$$

Here $f(\varphi)$ is a function that, although computable in concrete string models, may be taken to be general for the present purpose. The most desirable property of the above type modification of Einstein's theory is that only the terms which are the second derivatives of the metric (or their product) appear in field equations – a feature perhaps most important in order to make a gravitational theory absence of (spin-2) ghosts[19] – thereby ensuring the uniqueness of their solutions.[20] Of course, one can supplement the above action with other higher derivative terms, such as those proportional to $(\nabla_\mu\varphi\nabla^\mu\varphi)^2$ and higher powers in R, $R_{\mu\nu}$ and $R_{\mu\nu\rho\lambda}$,[21,22] but in such cases it would only be possible to get special (asymptotic) solutions, so we limit ourselves to the above action.

Another important direction, which I will not review here, is the quest for a concrete construction of four-dimensional cosmology starting from some five-dimensional Gauss-Bonnet brane world models.[23–25] Of course, in spacetime dimensions $D \geq 5$, a pure Gauss-Bonnet term can lead to modification of Einstein field equation, even if $f(\varphi) = \text{const}$, and hence influence the four-dimensional cosmology defined on the 3-brane. Here we limit ourselves to the four-dimensional action and demand that $f(\varphi)$ is dynamical. In this case the GB term $\mathcal{R}^2$ is not topological, rather it can have an interesting dynamics, especially, on largest cosmological scales.

[a] This assumption may just be reversed and assume that the dilaton φ would evolve more rapidly as compared to σ; these all depend on an underlying model. Of course, the single field description in terms of φ (or σ) could underestimate the actual evolution of the universe at early epochs, like during inflation, because string compactifications invariably involve more than one scalar field, and the four-dimensional potential depends, in general, on all the moduli field of the compactification. Nevertheless, this simple approximation in a string-derived Lagrangian holds some validity as a post-inflation scenario.

4. Cosmological Perturbations and Stability Conditions

To explore the stability of an effective gravitational action, under large cosmological perturbations, one may consider the following perturbed metric about a flat Friedmann-Robertson-Walker (FRW) background:

$$ds^2 = -(1+2\varpi)dt^2 + 2a\partial_i\chi dx^i dt + a^2\left[(1+2\psi)\delta_{ij} + 2\partial_{ij}\eta + 2h_{ij}\right]dx^i\, dx^j, \tag{7}$$

where ϖ, χ, ψ, η denote scalar and h_{ij} denotes vector components of metric fluctuations, and $\partial_{ij} \equiv \Delta_i\Delta_j - (1/3)\delta_{ij}\Delta^2$. A remarkable property of the Gauss-Bonnet gravity is that the linearised action can be expressed (in the absence of matter fields) in the following explicit form:[26]

$$\delta^{(2)}\mathcal{S} \propto \int dta^3 \left[-A(t)\mathcal{R}\ddot{\mathcal{R}} + \frac{B(t)}{a^2}\,\mathcal{R}\Delta^2\mathcal{R}\right], \tag{8}$$

where $\mathcal{R}$ is a gauge invariant quantity:

$$\mathcal{R} \equiv \psi - \frac{H}{\dot\varphi}\,\delta\varphi, \tag{9}$$

so-called a comoving perturbation. For a linearized theory to be free of ghost and superluminal modes, the following conditions

$$A(t) > 0, \quad B(t) > 0, \tag{10}$$

known as stability conditions, should perhaps be satisfied. For quantum stability of (inflationary) solutions the speeds of propagation of scalar and tensor modes should also remain non-superluminal:

$$0 < c_{\mathcal{R}}^2 = 1 + \frac{\mu^2\left[4\epsilon(1-\mu) - \lambda\kappa^2\left(\ddot f - \dot f H\right)\right]}{(1-\mu)\left(2\zeta(1-\mu)\varphi'^2 + 3\mu^2\right)} \le 1, \tag{11}$$

$$0 < c_T^2 = \frac{1+\lambda\kappa^2\ddot f}{1+\lambda\kappa^2\dot f H} = \frac{1-\mu'+\epsilon\mu}{1-\mu} \le 1, \tag{12}$$

where $\epsilon \equiv \dot H/H^2 = H'/H$, $' \equiv d/d\ln a$ and $\mu \equiv -\lambda\kappa^2\dot f H = \Omega_f$. In fact, $A(t)$, $B(t) < 0$ implies a violation of unitarity, while $B(t) > A(t)$ implies the existence of a superluminal propagation or an ill-defined Cauchy problem. Moreover, in the case $A(t)B(t) < 0$, the system of equations could exhibit an exponential type of instability, leading to an imaginary c_T. Below we use the above relations for studying the stability of inflationary solutions.

5. Inflationary Constraints

One may put constrains on the strength of the coupling $f(\varphi)$ by considering observational limits on the spectral indices of scalar and tensor perturbations. For the present theory, and in the limit that $c_{\mathcal{R}}^2, c_T^2 \approx$ const, the spectral indices $n_{\mathcal{R}}$ and n_T are approximated by[26–28]

$$n_{\mathcal{R}} - 1 = 3 - \left| \frac{3 + \epsilon_1 + 2\epsilon_2 + 2\epsilon_3}{1 - \epsilon_1} \right|, \quad n_T = 3 - \left| \frac{3 - \epsilon_1 + 2\epsilon_4}{1 - \epsilon_1} \right|. \tag{13}$$

where $\epsilon_1 \equiv -\frac{\dot{H}}{H^2} = -\epsilon$, $\epsilon_2 = \frac{\ddot{\varphi}}{\dot{\varphi} H}$, $\epsilon_3 = \frac{\theta'}{2\theta}$, $\epsilon_4 \equiv -\frac{\mu'}{2(1-\mu)}$, $\theta \equiv \zeta + \frac{3\mu^2}{2(1-\mu)\varphi'^2}$ and $\mu \equiv -\lambda\kappa^2 \dot{f} H$. One more quantity of cosmological relevance is the tensor-to-scalar ratio, which, in the limit $|\epsilon_1| \ll 1$, is approximated by

$$r \approx 16 \frac{2\zeta x^2(1-\mu) + 3\mu^2}{(2-\mu)^2} \left(\frac{c_{\mathcal{R}}}{c_T} \right)^3. \tag{14}$$

The WMAP data put the constraints $0.94 < n_{\mathcal{R}} < 0.98$ and $r < 0.28$ for a single scalar field model.

6. Non-Singular Inflationary Solutions

6.1. *Absence of Scalar Potential*

Let us consider a non-singular inflationary solution obtainable by dropping the scalar potential. To quantify this, one sets $V(\varphi) = 0$. One then defines

$$\mathcal{F}(\varphi) \equiv -\lambda f(\varphi) H^2. \tag{15}$$

The magnitude of $\mathcal{F}$ should decrease with the expansion of the universe, so that all higher-order corrections to Einstein's theory become only subleading [b]. With $\mathcal{F} \equiv \mathcal{F}_0$, the explicit solution is given by

$$\frac{\dot{H}}{H^2} = -\mathcal{A} + \mathcal{B} \tanh \mathcal{B}(N + C), \tag{16}$$

where $N \equiv \ln a$, C is an integration constant and

$$\mathcal{A} \equiv \frac{5\mathcal{F}_0 + 1}{2\mathcal{F}_0}, \quad \mathcal{B} \equiv \sqrt{\mathcal{A}^2 - 6\mathcal{A} + 15} \tag{17}$$

The Hubble parameter is $H \propto \mathrm{e}^{-\mathcal{A}N} \cosh \mathcal{B}(N + C)$. The $\mathcal{F}_0 > 0$ solution, which allows $\dot{H} > 0$, supports a super-luminal expansion, see also.[29] It is possible to get a red-tilted scalar index ($n_{\mathcal{R}} < 1$) for $\mathcal{F}_0 > -2/3$.

---

[b] Particularly in the case $\mathcal{F} \simeq$ const $\equiv \mathcal{F}_0$, the coupled term $\lambda f(\varphi)\mathcal{R}^2$ is subleading to the Einstein term $R/2\kappa^2 = 3M_P^2(2H^2 + \dot{H})$ for $|\mathcal{F}_0| \ll 1$ or $\lambda \ll 1$.

6.2. *Inflating with an Exponential Potential*

Consider that $V(\varphi) \propto e^{-\beta(\varphi/\varphi_0)}$ and $f_{,\varphi} H^2 \propto \varphi'$; the latter choice is motivated by the fact that the coupling takes the form $f(\varphi) \propto e^{\beta(\varphi/\varphi_0)}$ in the limit $\varphi' \to$ const. The explicit solution for φ is

$$\varphi = \frac{2}{\beta}\varphi_0 \ln \frac{a/a_i}{\cosh \chi \ln(a/a_i)} + \text{const}, \quad \frac{\dot{H}}{H^2} = \frac{2\zeta\varphi'^2}{6+\zeta\varphi'^2} - \frac{\beta}{\varphi_0}\varphi'^2, \tag{18}$$

where $\chi \equiv \sqrt{(2\zeta\varphi_0^2 - 3\beta^2)/2\zeta\varphi_0^2}$, $\varphi' \equiv d\varphi/d(\ln a) = \dot{\varphi}/H$ and a_i is the initial value of scale factor $a(t)$. From this solution we can easily evaluate the indices $n_{\mathcal{R}}$ and n_T, using (13). The observation in Fig. 1 that $n_R \simeq 3$ at some early stage of inflation is not quite correct since in that region $c_{\mathcal{R}}^2$ and c_T^2 are varying considerably, for which there would be non-trivial corrections to the formulae (13). A result consistent with the WMAP data (e.g. $n_R \simeq 0.96$ and $n_T < 0.2$) can be obtained for $|\beta/\varphi_0| < \sqrt{\zeta}/4$.

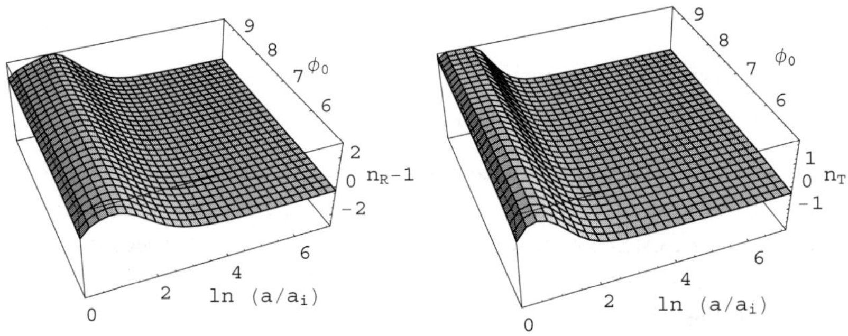

Fig. 1. The spectral indices n_R and n_T as the functions of $\phi_0 = 2\sqrt{\zeta}\varphi_0/\beta$ and $\ln(a/a_i)$.

7. Matter-Scalar Couplings

For constructing a realistic late time cosmology, one should consider the ordinary fields (matter and radiation) and also their possible couplings with the scalar field φ.

7.1. *Minimally Coupled Scalar Field*

In a flat FRW spacetime, the Gauss-Bonnet term $\mathcal{R}^2$ vanishes only at the stage of zero acceleration, and it flips its sign once the universe begins to accelerate. This effect can overturn the slope of the effective potential:

$$\Lambda(\varphi) \equiv V(\varphi) - \frac{\lambda}{8} f(\varphi)\mathcal{R}^2,$$

and push the universe transiently to a phantom era. Such an effect can be seen by considering the effective equation of state:

$$w_{\text{eff}} \equiv -1 - \frac{2\dot{H}}{3H^2} = \frac{p_{\text{tot}}}{\rho_{\text{tot}}} = w_m\Omega_m + w_r\Omega_r + w_\varphi\Omega_\varphi, \tag{19}$$

where m = matter and r = radiation. With the assumption that the ordinary matter is approximated by a non-relativistic perfect fluid (i.e. $w_m \simeq 0$ and $\Omega_r \ll 1$), we find $w_{\text{eff}} \simeq w_\varphi\Omega_\varphi$. Further, a simple calculation shows $\rho_\varphi + p_\varphi = \zeta\dot{\varphi}^2 + \lambda H^2(\ddot{f} - \dot{f}H) + 2\lambda H\dot{H}\dot{f}$.[18] To this relation, the stability conditions $1 > \kappa^2|\lambda\ddot{f}|$ and $|\ddot{f}| \geq |\dot{f}H|$ may be imposed, so as to keep the propagation speed of tensor and scalar modes non-superluminal under cosmological perturbations. Nevertheless, it is possible to get $p_\varphi + \rho_\varphi < 0$, or $w_\varphi < -1$, without making the cosmic expansion superluminal, or violating the condition $\dot{H} \leq 0$. This simple picture has obvious and intuitive appeal.

7.2. *Non-Minimally Coupled Scalar Field*

The constraints on the modified Gauss-Bonnet theory may arise by two different dynamics: one is the standard interaction effect between the scalar field φ and the Gauss-Bonnet term, while the other is the effect of non-minimal coupling between the scalar field φ and matter. The latter effect might perhaps be more significant than the former, especially, while applying the model into high density regions, or solar system experiments. To this reason, let us write the matter Lagrangian in a general form:

$$\mathcal{S}_{\text{matter}} = \mathcal{S}(A^2(\varphi)g_{\mu\nu}, \psi_m), \tag{20}$$

where $A(\varphi)$ measures the response of the geometry due to a time-variation of the field φ. Ordinary fields (matter and radiation) couple to $A^2(\varphi)g_{\mu\nu}$ rather than the Einstein metric $g_{\mu\nu}$ alone. Indeed, φ couples to the trace of the matter stress tensor, $g^{\mu\mu}_{(i)}T^{(i)}_{\mu\nu}$, so the radiation term (for which $w_r = 1/3$) does not contribute to the (Klein-Gordon) equation of motion for φ:

$$\dot{\rho}_\varphi + 3H\rho_\varphi\left(1 + w_\varphi\right) = -\dot{\varphi}(1 - 3w_i)\alpha_\varphi A(\varphi)\rho_m, \tag{21}$$

where $\rho_\phi \equiv \frac{\zeta}{2}\dot{\phi}^2 + V(\phi) - 3\lambda H^3\dot{f}$, $w_\phi \equiv p_\phi/\rho_\phi$ and $w_i \equiv p_i/\rho_i$. In order for current experimental limits on verification of the equivalence principle to be satisfied, the quantity

$$\alpha_{\varphi} \equiv \frac{d\ln A(\varphi)}{d\,(\kappa\varphi)}, \tag{22}$$

which measures the coupling of φ to background (baryonic and dark) matter, must be much smaller than unity, at least, on cosmological scales. The local GR constraints on α_φ and its derivatives imply that[34]

$$\alpha_{\varphi}^2 \leq 4 \cdot 10^{-5}, \qquad \beta_{\varphi} = \frac{d\alpha_{\varphi}}{d\varphi} > -4.5. \tag{23}$$

On large cosmological scales, where $\rho_m \lesssim \rho_\varphi$ and $V(\varphi) \sim 3H_0^2$, φ is expected to be sufficiently light, as for *quintessence*, $m_\varphi \equiv \sqrt{V_{\varphi\varphi}} \sim 10^{-33}\mathrm{eV}$, and the term on r.h.s. of eq. (21) may be safely ignored. In fact, in ref.,[35] smallness of α_φ was found to be linked to the smallness of the horizon-scale cosmological density fluctuation, $\delta\rho/\rho \sim 5 \times 10^{-5}$ (at the surface of last scattering). However, in high density regions, or within galactic distances, $\delta\rho/\rho \gg 10^{-5}$ and φ can be massive, like $m_\varphi \gtrsim 10^{-3}\mathrm{eV}$, in which case the observable deviations from Einstein's gravity are normally quenched on distances larger than a fraction of millimeter.

If $A(\varphi)$ is sufficiently flat near the current value of $\varphi = \varphi_0$, then the matter-scalar coupling can have only modest effects on cosmological scales. Especially, in the case that $A(\varphi) \propto e^{\nu(\varphi/M_P)}$, the above GR constraints may be satisfied only for a small ν ($\ll 1$). This restriction on the slope (or strength) of matter-scalar coupling may not apply to a gravitationally bound system, or in high density regions, where the field φ is not essentially light or weakly coupled to matter degrees of freedom (of the standard model). The latter argument is actually consistent with logics widely employed to various experiments designed to detect axion-like particles.

8. Late Time Cosmology

Making just one simplifying assumption that $\varphi \equiv \varphi_0 \ln[a(t)] + \mathrm{const}$, and then inverting the field equations following from (6), we find[7]

$$f(\varphi) = -f_0\,\mathrm{e}^{\beta(\varphi/\varphi_0)} - f_1, \quad V(\varphi) = \frac{2(\delta-1)}{3\lambda\,\kappa^4}\left(\frac{df(\varphi)}{d\varphi}\right)^{-1} \equiv V_0\,\mathrm{e}^{-\beta(\varphi/\varphi_0)}, \tag{24}$$

where $\beta = 1 + 3\delta$ and $\delta = \kappa^2\varphi_0^2/2$. These simplest choices for the potential and the scalar coupling admit the following simple solution[28]

$$a(t) = a_0\,t^{2/\beta}, \tag{25}$$

satisfying the relations:

$$\lambda f_0 = \frac{(\beta - 2\zeta\delta)\beta}{2\kappa^2(\beta+2)}, \quad V_0 = \frac{24(2-\beta) + 8\zeta\delta(10-\beta)}{(\beta+2)\beta^2\kappa^2}. \tag{26}$$

Acceleration requires $\beta < 2$ (if it is to be future eternal); thus, for the model to provide a solution to the dark energy problem, the strength of the GB coupling must grow with time, $f(\varphi) \propto e^{\beta(\varphi/\varphi_0)} \propto a(t)^\beta$. This is actually consistent with superstring models studied in.[4,8,18] One should, however, note that a growing $f(\varphi)$ does not necessarily mean that the term $f\mathcal{R}^2$ will dominate at late times the potential and/or the Einstein-Hilbert term. In the present universe $H_0 \sim 10^{-60} M_P$, which leads to $\kappa^2 V \simeq 10^{-120} M_P^2$, $R/6 \simeq H_0^2 \sim 10^{-120} M_P^2$ and $\kappa^2 f\mathcal{R}^2 \propto V^{-1} H_0^4 \equiv f_0\, e^{-120} M_P^2$. For $f_0 \ll 1$, $f\mathcal{R}^2$ is only subleading to V and R/κ^2. Terms higher powers in Ricci scalar (R^n with $n \geq 3$) contribute with H_0^{2n} and are thus subleading to $f(\varphi)\mathcal{R}^2$.

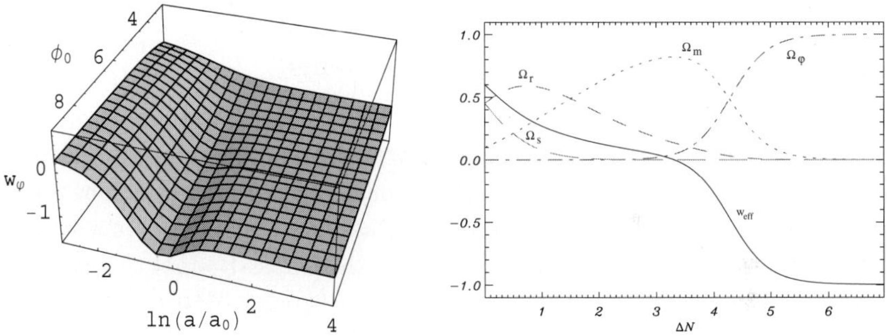

Fig. 2. (Left plot) The dark energy equation of state as a function of ϕ_0 and $\ln(a/a_0)$. (Right plot) The evolution of the fractional densities and $w_{\rm eff}$ with $\zeta = 1$, $(\beta/\varphi_0) = 3$, $\alpha_\varphi^2 = 10^{-5}$ (ordinary matter) and $\alpha_\varphi^2 = 10^{-2}$ (stiff-matter). The initial values are $\Omega_f^{\rm i} = 10^{-6}$, $\varphi_i' = 10^{-7}\, M_P$, $(V/H^2)_{\rm i} = 3\times 10^{-15}$, $\Omega_{\rm m} = 0.1$ (dots, brown), $\Omega_{\rm r} = 0.45$ (dashes, green), $\Omega_{\rm s}$ (long dashes, orange) and Ω_φ (dot-dash, red). Here $\Delta N = \ln(a/a_0) + C$; C may be chosen such that $N = 0$ corresponds to $\Omega_m \simeq 0.27$ and $\Omega_\varphi + \Omega_f \simeq 0.73$.[28]

Let me consider one specific model for which the dark energy equation of state can be less than -1 but transiently. This example is provided by the choice $V(\varphi) \propto e^{-\beta(\varphi/\varphi_0)}$ and $f_{,\varphi}H^2 \propto \varphi'$. In this case the explicit solution is given by (18). One takes $a_i = a_0 \equiv 1$, so that $a(t) < 1$ in the past. As shown on the left panel of Fig. 2, the dark energy equation of state $w_\varphi \equiv -1 - \frac{2\dot{H}}{3H^2}$ is less than -1, but only transiently, for $\phi_0 \equiv 2\sqrt{\zeta}\varphi_0/\beta \gtrsim 5$. This behaviour can be observed also in the presence of matter field.

The right panel of Fig. 2 represents a characteristic evolution for which the Gauss-Bonnet term never becomes dominant. In this plot the GB coupling $f(\varphi)$ has been chosen such that $\varphi' f_{,\varphi} H^2 \simeq$ const.

One may construct an explicit model by using the parametrization $f(\varphi) \equiv f_0\, e^{\alpha(\varphi/\varphi_0)}$ and $V(\varphi) \equiv V_0\, e^{-\beta(\varphi/\varphi_0)}$[30] and also replicate many observable properties of the universe from nucleosynthesis to the present epoch[28,31] (see also[32])). A possible drawback of this simple parametrization is, however, that, especially, for large slope parameters, like $\alpha > \beta \gtrsim \sqrt{3/2}(\varphi_0/M_P)$, the model may exhibit some kind of semi-classical instabilities associated with the linearized inhomogeneities or quantum fluctuations that grow explosively as the limit $c_T^2 < 0$ is approached or the tensor modes start to propagate faster than light's velocity.[33] The appearance of a superluminal mode, though not inevitable, could actually imply that one would have to invoke modifications of the simplest exponential parametrization or should allow only small slope parameters.

9. Time-Variation of Fundamental Constants

Scalar-tensor theories of gravity also entertain the result that some of the fundamental constants of nature may vary with time, including the Newton's constant, which are however tightly constrained by observations. On large cosmological scales, it is reasonable to assume that $A(\varphi) =$ const. In this case, the growth of matter fluctuations in the Gauss-Bonnet theory can be expressed in the following standard form:

$$\ddot{\delta} + 2\dot{\delta}H = 4\pi G_* \rho_m\, \delta, \tag{27}$$

where the normalized Newton's constant G_* may be given by[36]

$$G_* = G\left[1 + 3\,\Omega_f - \frac{\dot{\varphi}}{H}\left(\frac{\ddot{\varphi}}{\dot{\varphi}^2} + \frac{f_{\varphi\varphi}}{f_\varphi}\right)\Omega_f\right], \tag{28}$$

where $\Omega_f \equiv -\lambda\kappa^2\dot{\varphi}Hf_\varphi$. Unlike the slow roll relations $\ddot{\varphi}/\dot{\varphi}$, $\dot{\varphi} \ll 1$, the ratios like $\ddot{\varphi}/\dot{\varphi}^2$ and $\ddot{f}/\dot{f}$, which appear in the expression

$$\frac{f_{\varphi\varphi}}{f_\varphi} = \frac{d^2f/d\varphi^2}{df/d\varphi} = \frac{H}{\dot{\varphi}}\left(\frac{\ddot{f}}{\dot{\varphi}\dot{f}} - \frac{\ddot{\varphi}}{\dot{\varphi}^2}\right) \tag{29}$$

can be of order unity (in units $M_P = 1$). It is not improbable that $G_* \approx G$ for present value of the field, φ_0, and the coupling, $f(\varphi_0)$. In fact, almost every models of scalar-tensor gravity behave as Einstein's GR supplemented with a cosmological constant term Λ, if $\varphi' = \frac{\dot{\varphi}}{H} \ll M_P$ holds (at least) after the epoch of big bang nucleosynthesis. In the particular case that $f(\varphi) \propto e^{\beta(\varphi/\varphi_0)}$, we obtain

$$G_* = G\left[1 + \lambda f(\varphi) H^2 \left(\varphi'' + \epsilon\varphi' - 2\varphi'\right)\right]. \tag{30}$$

Thus one should satisfy, at least, one of the following conditions: (i) $|\lambda| \ll 1$, (ii) $|f(\varphi)|H^2 \ll 1$, or (iii) $|\varphi'| = |\dot{\varphi}/H| \ll M_P$, in order to get $G_* \simeq G$ at present. For a specific model studied in,[35] a safe upper bound is found to be $|\varphi_0'| < 0.84\, M_P$. Nevertheless, within solar system and laboratories distance, there exists a more stronger bound that $(dG_*/dt)/G_* < 0.01\, H_0$ (where H_0 is the Hubble expansion rate at present). This last condition translates to the constraint $|G_{\rm now} - G_{\rm nucleo}|/G_{\rm now}(t_{\rm now} - t_{\rm nucleo}) < 10^{-12}{\rm yr}^{-1}$. The quantity dG_*/dt is actually suppressed (as compared to G_*) by a factor of $\dot{\varphi}/H$, so it is necessary to satisfy $\varphi' \ll M_P$, at least, on large cosmological scales. Another opportunity for the model to overcome local gravity constraints coming from GR is to have a coupling $f(\varphi_0)$ which is nearly at its minimum. This is very much the approach one takes in a standard scalar-tensor theory.

10. Further Constraints

The growth of matter perturbations and the integrated Sachs-Wolfe (ISW) effect are the other effective ways of constraining the model under consideration.[36] In the case $\lambda f(\varphi)\mathcal{R}^2$ is subdominant to $V(\varphi)$ (thus $|\Omega_f| \ll 1$), the matter growth factor may be approximated by

$$\left(\frac{\dot{\delta}}{\delta}\right)_{\rm EGB} \approx \left(\frac{\dot{\delta}}{\delta}\right)\left[1 - \left(1 + \frac{H'}{H}\right)(1 + 0.75\,\Omega_m)\,\Omega_f\right]. \tag{31}$$

In an accelerating spacetime, so $1 + \epsilon \geq 0$, the Gauss-Bonnet coupling decreases the matter growth factor (as compared to the standard ΛCDM), for $\Omega_f > 0$. In view of the observational uncertainly in the growth rate of large scale structures,[37] $f_{\rm struc} \equiv \left(\dot{\delta}/\delta\right) = 0.51 \pm 0.1$, the Gauss-Bonnet energy density fraction Ω_f should perhaps not exceed 20%.[36] This last result may apply only to largest cosmological scales, and it is, by no means, applicable to gravitationally bound systems, such as, our solar system.

Under the post-Newtonian approximation:

$$ds^2 = -(1 + 2\Phi)(cdt)^2 + (1 - 2\Psi)\delta_{ij}dx^i\,dx^j \tag{32}$$

where $\Phi, \Psi \sim \mathcal{O}(GM/rc^2)$, the solar system constraints appear to be stronger than astrophysical constraints, mainly, due of a small fractional anisotropic stress[38] (see also[39]):

$$\hat{\gamma} - 1 \equiv \frac{\Psi - \Phi}{\Phi} \approx -2\Omega_f \left(1 - \frac{\dot{\varphi}}{H}\left(\frac{\ddot{\varphi}}{\dot{\varphi}^2} + \frac{f_{\varphi\varphi}}{f_\varphi}\right)\right) < 4 \times 10^{-5}. \quad (33)$$

When applied to solar system distances, this above expression demands that $\Omega_f \lesssim 10^{-5}$. There remains the possibility that the classical tests of Newtonian gravity, which typically deal with small perturbations in fixed (or time-independent) backgrounds are almost unaffected by the GB type modification of Einstein's theory.

11. Conclusions

The important ingredient of the present approach to dark energy cosmology is the treatment of gravitational coupling between the dynamical scalar field φ and the quadratic Riemann invariant of the Gauss-Bonnet form, which give rise to nonsingular cosmologies for a wide range of scalar-curvature couplings. The model also provide plausible explanation to some outstanding cosmological conundrums, including: the transition from matter dominance to dark energy and the late time cosmic acceleration. It is also worth noticing that the scalar-curvature coupling can easily trigger onset of a late dark energy domination. Despite these promising signs, it remains to be checked whether the Gauss-Bonnet modification of Einstein's theory will lead to genuine contact between observations and string theory.

String theory is known to be free from ghosts and superluminal modes, at least, in a flat ten-dimensional Minkowski background. This is perhaps not essentially the case in a four-dimensional FRW background. The effective string actions in four dimensions may well exhibit some unwarranted features, such as, short scale instabilities and superluminal propagation of tensor or scalar models, under an inhomogeneous (cosmological) perturbation. The model discussed here is perhaps not an exception, since, at least, for the potential and the Gauss-Bonnet coupling in simplest exponential forms, one can see that the tensor or scalar modes propagate at a speed faster than light at some stage, especially, for large slope parameters. The appearance of a superluminal mode is generic, and perhaps also acceptable, if such an effect is only transient.

In the present proposal for explaining a crossing of cosmological (dark energy) equation of state, $w_{DE} = -1$, and a superluminal propagation of scalar or tensor models, a number of important, physically falsifiable predictions can be made. These include a transient violation of Lorentz symmetry and the weak equivalence principle, associated with the microscopic effects of the coupling between φ and background matter in high density regions.

Ninety years after Einstein's proposition of general relativity with a cosmological constant, a modified cosmological scenario with its natural generalization is close to experimental test and possibly an outlet compatible with present experimental data. The coming generation of cosmological experiments, including Dark Energy Survey, will probably rule out the great majority of string-derived models, as well as exclude those class of scalar-tensor theories which give rise to unphysical states. *Time will tell.*

Acknowledgments

This research is supported by the Foundation for Research, Science and Technology (NZ) under Research Grant No. E5229.

References

1. S. Perlmutter *et al.* [Supernova Cosmology Project Collaboration], Astrophys. J. **517**, 565 (1999); A. G. Riess, *et al.* [Supernove Search Team Collaboration], Astrophys. J. **560**, 49 (2001).
2. D. N. Spergel *et al.* [WMAP Collaboration], Astrophys. J. Suppl. **148**, 175 (2003).
3. G. Veneziano, arXiv:hep-th/0002094.
4. I. Antoniadis, E. Gava and K. S. Narain, Nucl. Phys. B **383**, 93 (1992) [arXiv:hep-th/9204030]; I. Antoniadis, J. Rizos and K. Tamvakis, Nucl. Phys. B **415**, 497 (1994) [arXiv:hep-th/9305025].
5. I. P. Neupane, Phys. Rev. Lett. **98**, 061301 (2007) [arXiv:hep-th/0609086].
6. S. Panda, M. Sami and S. Tsujikawa, Phys. Rev. D **76**, 103512 (2007) [arXiv:0707.2848 [hep-th]].
7. I. P. Neupane, Class. Quant. Grav. **23**, 7493 (2006) [arXiv:hep-th/0602097]; arXiv:hep-th/0605265.
8. N. Kaloper, R. Madden and K. A. Olive, Nucl. Phys. B **452**, 677 (1995) [arXiv:hep-th/9506027].
9. R. Easther and K. I. Maeda, Phys. Rev. D **54**, 7252 (1996) [arXiv:hep-th/9605173].
10. S. Kalyana Rama, Phys. Lett. B **408**, 91 (1997) [arXiv:hep-th/9701154].
11. S. Kawai, M. A. Sakagami and J. Soda, Phys. Lett. B **437**, 284 (1998) [arXiv:gr-qc/9802033].
12. P. Kanti, J. Rizos and K. Tamvakis, Phys. Rev. D **59**, 083512 (1999) [arXiv:gr-qc/9806085].
13. D. A. Easson and R. H. Brandenberger, JHEP **9909**, 003 (1999) [arXiv:hep-th/9905175].
14. S. Tsujikawa, Phys. Lett. B **526**, 179 (2002) [arXiv:gr-qc/0110124].
15. Y. Ezawa, H. Iwasaki, M. Ohmori, S. Ueda, N. Yamada and T. Yano, Class. Quant. Grav. **20**, 4933 (2003) [arXiv:gr-qc/0306065].
16. Y. S. Piao, B. Feng and X. M. Zhang, Phys. Rev. D **69**, 103520 (2004) [arXiv:hep-th/0310206].

17. S. Nojiri, S. D. Odintsov and M. Sasaki, Phys. Rev. D **71**, 123509 (2005) [arXiv:hep-th/0504052].
18. B. M. N. Carter and I. P. Neupane, Phys. Lett. B **638**, 94 (2006) [arXiv:hep-th/0510109]; B. M. N. Carter and I. P. Neupane, JCAP **0606**, 004 (2006) [hep-th/0512262].
19. Y. M. Cho, I. P. Neupane and P. S. Wesson, Nucl. Phys. B **621**, 388 (2002) [arXiv:hep-th/0104227].
20. N. Dadhich, arXiv:hep-th/0509126.
21. K. I. Maeda and N. Ohta, Phys. Rev. D **71**, 063520 (2005) [arXiv:hep-th/0411093].
22. M. Sami, A. Toporensky, P. V. Tretjakov and S. Tsujikawa, Phys. Lett. B **619**, 193 (2005) [hep-th/0504154]; G. Calcagni, S. Tsujikawa and M. Sami, Class. Quant. Grav. **22**, 3977 (2005) [hep-th/0505193]; E. Elizalde, S. Jhingan, S. Nojiri, S. D. Odintsov, M. Sami and I. Thongkool, arXiv:0705.1211 [hep-th].
23. N. E. Mavromatos and J. Rizos, Phys. Rev. D **62**, 124004 (2000) [arXiv:hep-th/0008074]; I. P. Neupane, JHEP **0009**, 040 (2000) [arXiv:hep-th/0008190];
24. I. P. Neupane Phys. Lett. B **512**, 137 (2001) [arXiv:hep-th/0104226]; Class. Quant. Grav. **19** (2002) 5507 [hep-th/0106100].
25. J. E. Lidsey and N. J. Nunes, Phys. Rev. D **67**, 103510 (2003) [astro-ph/0303168]; S. Tsujikawa, M. Sami and R. Maartens, Phys. Rev. D **70**, 063525 (2004) [astro-ph/0406078].
26. J. C. Hwang and H. Noh, Phys. Rev. D **54**, 1460 (1996); C. Cartier, J. C. Hwang and E. J. Copeland, Phys. Rev. D **64**, 103504 (2001) [astro-ph/0106197].
27. Z. K. Guo, N. Ohta and S. Tsujikawa, Phys. Rev. D **75**, 023520 (2007) [hep-th/0610336].
28. B. M. Leith and I. P. Neupane, JCAP **0705**, 019 (2007) [hep-th/0702002].
29. S. Kanno and J. Soda, Phys. Rev. D **74**, 063505 (2006) [hep-th/0604192].
30. T. Koivisto and D. F. Mota, Phys. Lett. B **644**, 104 (2007) [astro-ph/0606078].
31. T. Koivisto and D. F. Mota, Phys. Rev. D **75**, 023518 (2007) [hep-th/0609155].
32. S. Nojiri, S. D. Odintsov and M. Sami, Phys. Rev. D **74**, 046004 (2006) [hep-th/0605039]; S. Tsujikawa and M. Sami, JCAP **0701**, 006 (2007) [hep-th/0608178].
33. G. Calcagni, B. de Carlos and A. De Felice, Nucl. Phys. B **752**, 404 (2006) [hep-th/0604201].
34. G. Esposito-Farese, AIP Conf. Proc. **736**, 35 (2004) [arXiv:gr-qc/0409081].
35. T. Damour, F. Piazza and G. Veneziano, Phys. Rev. D **66**, 046007 (2002).
36. L. Amendola, C. Charmousis and S. C. Davis, JCAP **0612**, 020 (2006) [hep-th/0506137].
37. D. J. Eisenstein *et al.* [SDSS Collaboration], Astrophys. J. **633**, 560 (2005).
38. L. Amendola, C. Charmousis and S. C. Davis, JCAP **0710**, 004 (2007) [arXiv:0704.0175 [astro-ph]].
39. T. P. Sotiriou and E. Barausse, Phys. Rev. D **75**, 084007 (2007).

PART III

Structure Formation and Gamma-Ray Sources

THE GALAXY PROFILE OF CLUSTERS IN THE COSMOS FIELD

J. BERIAN JAMES* and JOHN A. PEACOCK

Institute for Astronomy, University of Edinburgh,
Blackford Hill, Edinburgh EH9 3HJ, United Kingdom
** E-mail: jbj@roe.ac.uk*
http://www.roe.ac.uk/~jbj

for the COSMOS Collaboration

http://cosmos.astro.caltech.edu

We present the initial findings of a study into the assembly of galaxy clusters throughout cosmic time with the Cosmic Evolution Survey (COSMOS). Using cluster locations extracted from the X-ray temperature maps of the COSMOS-XMM survey paired with photometric redshifts, we examine the cluster-galaxy cross-correlation function in its projected form on scales beneath 10 Mpc/h for clusters in the mass range $10^{12} \leq M/M_{\odot} \leq 10^{14}$ and extending to a redshift $z \sim 1.25$. We find a strong positive trend in the amplitude of the cross-correlation function with increasing cluster mass and a weaker dependence on redshift, at least partially explicable by a systematic effect arising from the method of locating clusters.

Keywords: cosmology: large-scale structure.

1. Introduction

The present model of cosmological structure formation relies on a hierarchy of scale into which astronomical objects are ordered. Stars, galaxies, groups of galaxies, galaxy clusters and so on are built from the same initial perturbations in the underling dark matter density field. Denser-than-average regions of dark matter collapses and virialises to form objects known as 'haloes', within which galaxies form as baryonic gas condenses into stars.

The relationship between the luminous baryonic galaxies and the dark matter haloes is therefore a critical one. Understanding which properties of a halo influence the formation of the galaxies within it, and how this influence changes over time, is tantamount to understanding how galaxies themselves are constructed from the density perturbations at the very beginning of the Universe. To address this challenge, we will measure the clustering of galaxies within haloes and examine the variations in clustering with the associated halo and galaxy properties. Though haloes are by their nature not directly observable, their spatial locations and physical characteristics may be inferred from the galaxy distribution at large. In particular, rich clusters of galaxies mark the sites of massive virialised haloes, in which the intra-cluster gas is sufficiently hot to radiate at X-ray temperatures; clusters therefore act as an effective proxy for haloes. By observing with both optical and X-ray telescopes, the positions of the galaxies and the haloes can be determined independently.

In this work, we study the galaxy number density profile of clusters using observations from the Cosmic Evolution Survey (COSMOS)[1] and COSMOS-XMM[2] projects, which have yielded galaxy and cluster catalogues among other multi-wavelength data across a two-degree square field of sky. The details of the survey and observations are given in Section 2. To measure the clustering of galaxies within halos, we use the cluster-galaxy cross-correlation statistic as a function of radius, described in Section 3. In this short address, we present the initial results of the measurement, as well as a decomposition of the cross-correlation function by cluster mass and redshift; these results are shown in Section 4. The conclusions and implications of these findings are discussed within the context of halo-based structure formation in Section 5.

2. COSMOS: The Cosmic Evolution Survey

COSMOS is a multi-wavelength observational project to survey a two-square degree patch of near-equatorial sky (shown in Figure 1) for sources out to high-redshift. The initial programme was an HST Treasury Project using the Advanced Camera for Surveys (ACS); the largest survey carried out with the ACS, it utilised 10% (640 orbits) of observing time over a two year period.[3] Since that time, complementary observations of the field have been made through major commitments from the VLA radio telescope ,[4] ESO's VLT in Chile ,[5] ESA's XMM X-ray satellite ,[2] Japan's 8-meter Subaru telescope in Hawaii[6] and the Spitzer space telescope .[7]

For the measurements made in this work, spatial data comprising galaxy and cluster locations are required. Additional cluster properties, in particular mass, and galaxy properties, in particular colour and luminosity, give insight into the evolution of clusters throughout cosmic time. The following sections describe briefly how these data are derived from observations.

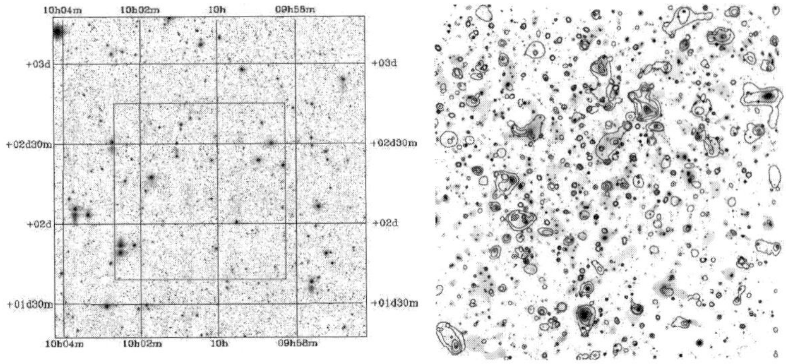

Fig. 1. Images of the COSMOS field, showing optical galaxies (left) and clusters from X-ray temperature maps (right). *Left:* The COSMOS field of two square degrees centred on $\alpha = 10:00:28.6$, $\delta = +02:12:21.0$. The background image is from the SDSS archive; the inner square demarcates the COSMOS region of interest: coverage is complete within this square and incomplete, though extensive, outwith (Credit: IRSA COSMOS Archive at `http://irsa.ipac.caltech.edu/data/COSMOS/`). *Right:* Cluster concentration in the XMM-COSMOS field *via* a wavelet reconstruction at energies of 0.5-2 keV. Grey colour corresponds to the surface brightness of 10^{-7} counts s^{-1} pixel^{-1}. A surface brightness level exceeding 10^{-5} counts s^{-1} pixel^{-1} is shown in black. Contours indicate the location and strength of galaxy structures, identified in the photo-z galaxy catalogue.[8]

2.1. *Galaxy catalogue*

The galaxy catalogue[9] is constructed from nine pointings of the 8.3m Subaru telescope with the Suprime-Cam instrument. Observations take place in a (non-standard) set of broad-band filters, the specifications of which are given in Table 1. The raw images are subject to bias subtraction, background subtraction and flat fielding before being run through the canonical source extractor, SExtractor.[11] Difficulties are encountered during source extraction around bright (foreground) objects; closely separated sources must be deblended and wrongly calibrating the strength of this source separation results in spurious over- or under-abundances of galaxy detections.

To account for these issues, difficult regions of the image, typically

Filter Name	Central Wavelength (Å)	Filter Width (Å)	Seeing Range ($''$)	Depth[1]	Offset from Vega System[2]
B_J	4459.7	897	0.4-0.9	27.3	-0.131
g^+	4779.6	1265	0.7-2.1	27.0	-0.117
V_J	5483.8	946	0.5-1.6	26.6	-0.004
r^+	6295.1	1382	0.4-1.0	26.8	0.125
i^+	7640.8	1497	0.4-0.9	26.2	0.379
z^+	9036.9	856	0.5-1.1	25.2	0.547

around the edges of bright foreground objects, are removed from the measurement *via* a catalogue mask that acts as a binary gate on the inclusion of detected objects within individual cells comprising a pixellation of the field. Such a mask is constructed in the first instance by hand, placing circular exclusion zones about obvious bright sources and image artifacts. This basic mask is made more rigorous by following two further procedures: i) to conservatively avoid all regions of contamination, the basic mask is grown by extending masked regions by four seconds of arc in all directions; ii) excluding regions in which the fraction of objects detected through the deblending procedure approaches unity, which is unlikely to occur naturally.

Applying this mask to the field yields a robust catalogue of angular galaxy positions as well as photometry in each of the bands given in Table 1. Distances to each galaxy can be determined from these bands, giving a catalogue of photometric redshifts[12] as well as myriad further information such as galaxy luminosities, colours and star formation rates; these will be studied in future work.

2.2. *Cluster catalogue*

The catalogue of clusters is extracted from X-ray temperature maps produced by observations with the XMM/Newton satellite[8] *via* a scale-wise wavelet reconstruction of the image in which compact sources (such as active nuclei) and highly diffuse background emission are excluded; significantly, this allows the determination of cluster masses *via* the X-ray scaling relations. The spatial distribution of clusters with the COSMOS field is shown in Figure 1.

However, the scale-wise shapelet reconstruction only yields angular position; the line-of-sight distance to the cluster must be determined independently. By cross-correlating the angular positions with the galaxy catalogue, a best-guess photometric redshift for the cluster centre is obtained.

3. The Galaxy-Cluster Cross-Correlation Function

The so-called **two-point correlation function** ξ is a basic and powerful descriptive tool. It is a measure of the *excess* probability of finding objects at separation r from one another; for a given distribution of points of mean density n, the average probability that a galaxy will have a neighbour at distance r is

$$dP = ndV\,[1 + \xi(r)]\,, \tag{1}$$

by which relation the two-point correlation function is implicitly defined.

For the Gaussian density field of the primordial Universe, this function is a Fourier transform pair to the power spectrum of density fluctuations; both functions have been calculated regularly for the observed galaxy distribution as new data become available. The canonical form of the two-point correlation function is a power law,[13]

$$\xi(r) = \left(\frac{r}{r_0}\right)^{\gamma}, \tag{2}$$

with the free parameters of scale length and gradient measured to be $r_0 =$ 4.7 Mpc and $\gamma = 1.8$.

The generalisation of the two-point correlation function to distributions containing two distinct populations of objects, *e.g.* galaxies and galaxy clusters, is the **two-point cross-correlation function**, *viz*

$$dP = n_c n_g \left[1 + \xi_{cg}(r)\right] dV_1 dV_2, \tag{3}$$

where n_g and n_c are the mean number densities of the galaxy and cluster populations, respectively. Quite often, and as in the case of galaxies and clusters, it is helpful to think of one object as an 'anchor', and that the function is measuring the distribution of one type of object about the other, *e.g.* the distribution of galaxies about the average cluster.

An important consequence of observations taking place in redshift- (velocity-) space is that the radial axis must be treated differently to the projection across the sky. The uncertainty in the photometric redshifts is quite high, but acts only in the radial direction; by decomposing the correlation function into line-of-sight and across-sky components, and projecting along the former, this effect of this uncertainty is minimised.

Given a pair of galaxies with position vectors $\mathbf{v}_1$ and $\mathbf{v}_2$, define first the redshift space separation $\mathbf{s} = \mathbf{v}_1 - \mathbf{v}_2$, and the line-of-sight $\mathbf{l} = \frac{1}{2}(\mathbf{v}_1 + \mathbf{v}_2)$. The components of the separation between the galaxies that are parallel (π) and perpendicular (r_p) to the line-of-sight are

$$\pi = \frac{\mathbf{s} \cdot \mathbf{l}}{|\mathbf{l}|} \tag{4}$$

$$r_p^2 = \mathbf{s} \cdot \mathbf{s} - \pi^2. \tag{5}$$

When using observations of the galaxy distribution, *e.g.* with apparent magnitude limited surveys, the density of galaxies as a function of redshift will not be uniform. If the probability that galaxy i is included in the catalogue is denoted ϕ_i, then the expected number of pairs at separation

(π, r_p) in a distribution of n_d galaxies will be

$$DD(\pi, r_p) = n_d \left[1 + \xi_v(\pi, r_p)\right] \sum_i d\pi dA_i \phi_i, \tag{6}$$

where dA_i is the area subtended by a ring dr_p about galaxy i. This is exactly the quantity that will be calculated from the raw data—it is simply a joint histogram in the space of the decomposed coördinates. To deal with the sum one must ask how many galaxies would be expected about object i if the clustering were of Poisson form. Hence, another quantity must be calculated: the histogram of n_r randomly clustered 'mock galaxies' about each real cluster *viz*

$$DR(\pi, r_p) = n_r \sum_i d\pi dA_i \phi_i. \tag{7}$$

The nomenclature is not arbitrary: DD refers to counts of separations between data points and data points, while DR is between data points and random points. The correlation function can now be extracted,

$$\xi_{cg}(\pi, r_p) = \frac{n_r}{n_d} \frac{DD}{DR} - 1. \tag{8}$$

To recover the clustering signal free from distortion, the **projected two-point correlation function** is used. The redshift space function is integrated through the line-of-sight axis, giving a measurement of the clustering strength as a function of separation in a direction perpendicular to radial:

$$w_p(r_p) = 2 \int_0^\infty \xi_{cg}(\sigma, \pi) d\pi \tag{9}$$

This is the particular quantity that will be used for measurement of the cluster-galaxy cross-correlation signal from the COSMOS data.

4. Results

In this section we present two results from our initial study: i) Figure 2 shows the projected cross-correlation function from the entire COSMOS galaxy and cluster catalogues; and ii) Figure 3 shows the same function for subsets of the data ordered by cluster mass and cluster redshift. For the latter, we split the clusters into groups by mass and redshift (three groups each, for a total of nine subsets) and measure the projected cross-correlation function again using all galaxies in the sample, but only the clusters in each group. To provide a comparison of the strength of clustering, we compare the amplitude of the cross-correlation function at a separation of 0.5 Mpc/h.

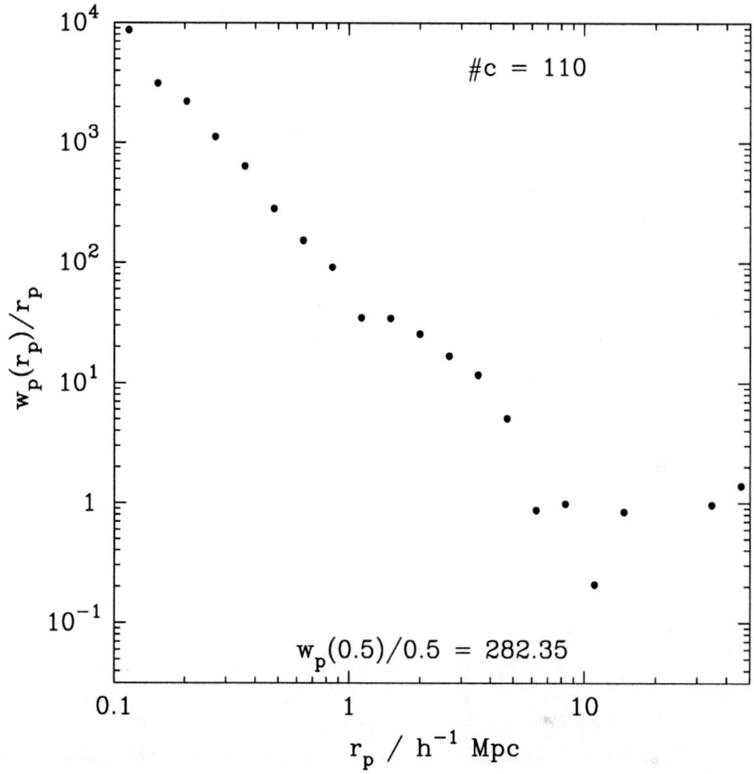

Fig. 2. The normalised galaxy-cluster cross-correlation amplitude as a function of projected separation r_p, in comoving Megaparsecs. The function obeys the expected power law relation with an interesting feature occuring at $r_p \sim 1$ Mpc/h, perhaps indicating the transition from the 1-halo to 2-halo term.

5. Discussion and Conclusions

In all cases the power-law form of the function is unsurprising: relative (to Poisson) clustering of galaxies is strongest at the centre of haloes, declining smoothly toward large radii. It is important to distinguish between two kinds of cluster-galaxy pairs: those in which the galaxy resides within the cluster with which it is paired form the '1-halo' contribution to the correlation function, while a galaxy (at larger separations) that is in a distinct clusters contributes to the '2-halo' term.

The transition from the 1-halo to 2-halo regime depends on the mass of the halo. In a sample like COSMOS, containing a range of cluster masses, the most massive clusters will dominate an average measurement of correla-

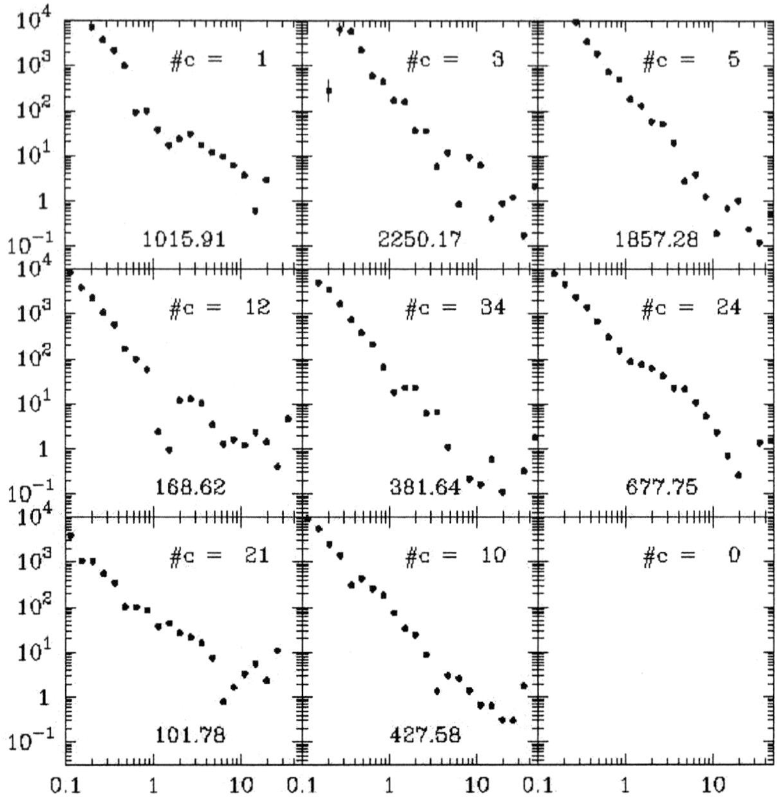

Fig. 3. The galaxy-cluster cross-correlation amplitude broken down by redshift, over the range $0 < z < 1$, and cluster mass, over the range $10^{12} \leq M/M_{\odot} \leq 10^{14}$. The spacing of the redshift bins is linear, *i.e.* in steps of $z+ = 0.\dot{3}$ going from left to right, while the spacing of the mass bins is logarithm, with each panel reporting from a factor of ten higher mass scale going from bottom to top. The figure at the top of each panel is the number of clusters to be found in this range of mass and redshift; for each bin, the cross-correlation measurement is made between just these clusters, but all available galaxies; the absence of clusters is the bottom-right hand bin (at high-redshift and lower-mass) is significant, suggesting an observational selection effect in the size of clusters as a function of redshift. The numeral at the bottom of each panel is the amplitude of the clustering at $r_p = 0.5$ Mpc/h; positive trends with both mass and redshift are apparent.

tion. A value of ~ 1 Mpc$/h$ is expected to be the transition scale for clusters of $\sim 10^{14} M_{\odot}$ in mass. There is a significant feature in the cross-correlation measurement at this scale, but it is not immediately clear what physical insight might be drawn from it, indicating that a study of the change in the transition point with mass will prove valuable.

As a step toward that end, a pronounced trend in clustering amplitude with cluster mass is observed, indicating that more massive haloes contain more strongly clustered galaxies. A similar, though substantially smaller trend with redshift also emerges, potentially indicating evolution in the cross-correlation function. However, the cluster mass and redshift are not independent: because the survey is magnitude limited, at larger distances only systematically brighter (*i.e.* more massive) clusters will be observable. Disentangling this relationship is possible through the use of a statistic such as Spearman's rank correlation coëfficient.

A forthcoming publication [14] will provide a thorough analysis of the results presented here, along with a comparison to halo-based theory in the form of analytic prediction of ξ_{cg}.

Acknowledgments

COSMOS is a collaborative enterprise and we have benefited from the assistance and discussion of many researchers within it. In particular, the galaxy and cluster catalogues are the result of enormous effort on the part of Peter Capak and Alexis Finoguenov, to whom we are most grateful; for the photometric redshift determination we owe thanks to Bahram Mobasher and Olivier Ilbert. Henry Joy McCracken and Gigi Guzzo have contributed directly to our effort to produce quality masks for the catalogues; Nick Scoville has coördinated the enterprise and provided useful feedback on this research. During the DARK 2007 conference, Arash Bodaghee provided interesting insights into the correlation between mass and redshift, of which the authors have been pleased to learn.

References

1. N. Scoville *et al., ApJS* **172** (2007), 1-8 (September 2007).
2. G. Hasinger *et al., ApJS* **172** (2007), 29-37 (September 2007).
3. N. Scoville *et al., ApJS* **172** (2007), 38-45 (September 2007).
4. E. Schinnerer *et al., ApJS* **172** (2007), 46-69 (September 2007).
5. S. J. Lilly *et al., ApJS* **172** (2007), 70-85 (September 2007).
6. Y. Taniguchi *et al., ApJS* **172** (2007), 9-28 (September 2007).
7. D. B. Sanders *et al., ApJS* **172** (2007), 86-98 (September 2007).
8. A. Finoguenov *et al., ApJS* **172** (2007), 182-195 (September 2007).
9. P. Capak *et al., ApJS* **172** (2007), 99–116 (September 2007).
10. A. U. Landolt, *AJ* **104** (1992), 340-371 (July 1992).
11. E. Bertin, E. and S. Arnouts, *A&AS* **117** (1996), 393-404 (June 1996).
12. B. Mobasher *et al., ApJS* **172** (2007), 117-131 (September 2007).
13. H. Totsuji and T. Kihara, *PASJ* **21**, 221 (1969).
14. J. B. James, *in preparation* (2007), *arXiv:astro-ph/0612344*.

SHEDDING LIGHT ON DARK MATTER: SEEING THE INVISIBLE WITH THE BULLET CLUSTER 1E0657-56

M. BRADAČ (for the Bullet cluster collaboration)

Department of Physics, University of California, Santa Barbara, CA 93106, USA
E-mail: marusa@physics.ucsb.edu, www.physics.ucsb.edu

The cluster of galaxies 1E0657-56 has been the subject of intense ongoing research in the last few years. This system is remarkably well-suited to addressing outstanding issues in both cosmology and fundamental physics. It is one of the hottest and most luminous X-ray clusters known and is unique in being a major supersonic cluster merger occurring nearly in the plane of the sky, earning it the nickname "the Bullet Cluster". Here I will present our measurements of the composition of this system, show the evidence for existence of dark matter, and describe limits that can be placed on the intrinsic properties of dark matter particles. In addition, I will explain how this cluster offers a serious challenge to MOdified Newtonian Dynamics (MOND) theories.

Keywords: cosmology: dark matter – gravitational lensing – galaxies: clusters: individual: 1E0657-56.

1. Introduction

The currently accepted cold dark matter model makes very precise predictions for the properties of galaxies and galaxy clusters that can be tested in great detail. There seem to be two crises; namely the amount of substructure in galaxies, and cuspy mass profiles, seen in simulations do not match the observations. These crises can be solved by invoking for example warm and/or self interacting dark matter; consequently they make forecasts about the properties of galaxy clusters (see e.g. 1 and references therein). In self-interacting dark matter scenarios, cores of dark matter halos are heated by collisions, and at least initially have lower central densities and shallower density profiles than in CDM. In warm dark matter scenarios the cluster profiles are also predicted to be less steep. The merger dynamics is also affected by the dark matter microphysics. In addition clusters provide a critical test for alternative gravity theories, that have been proposed to eliminate the need of dark matter altogether. By using gravitational lensing to measure the mass profiles of (merging) clusters, we are in a position to distinguish between different forms of dark matter (and gravity) from astrophysical observations. This will give a measurement of the interaction properties in advance of the expected direct detection of the dark matter particles. Finally, it is difficult to measure the properties of dark matter in galaxy clusters due to the presence and strong influence of baryons. However, by combining all observations we will not only learn about dark matter, but also be able to place meaningful constraints on the formation and evolution of clusters, further testing the ΛCDM paradigm.

The most striking example of such investigations to date has been 1E0657−56. In this system, the positions of dark matter halos and the dominant baryonic component (i.e. hot gas) are well separated, leading us to infer the clear presence and domination of a dark matter component (see Fig. 1 and 2,3). A union of the strong lensing data (information from highly distorted arcs) and weak lensing data (weakly distorted background galaxies) for the cluster mass reconstruction has been demonstrated to be very successful in providing a high-fidelity, high signal-to-noise mass reconstruction over a large area.[2] This allowed us to clearly detect and begin to study the properties of dark matter.

2. The Bullet Cluster 1E0657-56

The cluster of galaxies 1E0657−56 is one of the hottest, most X-ray luminous clusters known. Since its discovery,[4] it has been the subject of intense and ongoing research (see e.g. 5–9). In particular, Chandra X-ray observations by 6 revealed the cluster to be a supersonic merger in the plane of the sky with a textbook example of a bow shock, making this cluster a unique case in which to study hydrodynamical properties of interacting systems. The optical images show that the cluster has two distinct components, and the X-ray analysis reveals that the lower mass sub-cluster's gas has recently exited the core of the main cluster with a relative velocity of $4500^{+1100}_{-800}\mathrm{kms}^{-1}$. Detailed simulations of this system[10] revealed that the lower sub-cluster itself (galaxies and dark matter component) is likely moving with lower velocity of $\sim 2700\mathrm{kms}^{-1}$. Although this relative velocity still appears to be unusually large, an analysis of cosmological simulations demonstrates that it is well within the predicted range of the currently favored cosmological model.[11]

Due to its unique geometry and physical state, this cluster is the best known system in which to test the dark matter hypothesis.[8] The observed offsets between the gravitational lensing mass peaks (presented in Figs. 1

and 2) and the X-ray gas component give the most direct evidence for the presence of dark matter yet available.

Fig. 1. The color composite of the cluster 1E0657−56. Overlaid in *blue* shade is the surface mass density map from the weak lensing mass reconstruction. The X-ray emitting plasma is shown in red. Credit: Chandra Press Release; X-ray NASA/CXC/CfA Optical: NASA/STScI; Magellan/U.Arizona.[3]

3. Dark Matter Properties

The clear offset between the peaks of the major baryonic component (hot gas) and the total mass distribution (obtained from gravitational lensing) gives a strong evidence for the existence of dark matter. In addition, we see (see e.g. Fig. 2) that the total mass peak is consistent with the centroids of the collisionless galaxies belonging to the cluster (which can be recognized by their orange-yellow color in Fig. 2).

This observation already gives us a hint that dark matter is collisionless; in addition with detailed simulations (see 12) we were able to place upper limit on self-interaction cross-section of dark matter per unit mass of dark matter particle, σ/m. We take advantage of new, higher-quality observa-

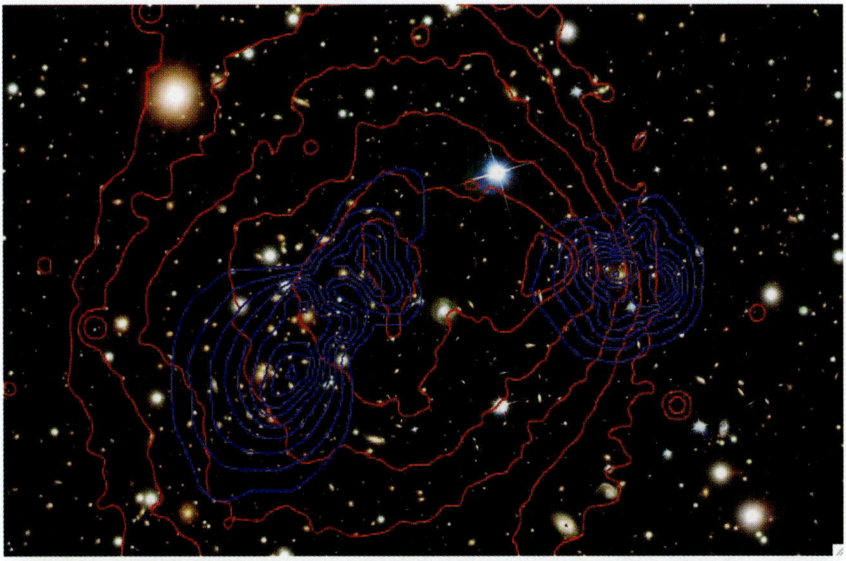

Fig. 2. The combined strong and weak lensing mass reconstruction (blue contours) of the cluster 1E0657−56. In contrast to the reconstruction using only weak lensing data (see Fig. 1), the addition of strong lensing data allows us for a much more detailed reconstruction in the centre of this cluster[2]. The X-ray emmision (shown in red contours) is clearly offset from the total mass distribution.

tional datasets by running N-body simulations of 1E0657-56 that include the effects of self-interacting dark matter, and comparing the results with observations of the hot gas (X-rays) and total matter distribution (strong and weak lensing). The addition of strong lensing is crucial here, as the results are sensitive to the central mass distribution in the cluster, the latter is difficult to obtain with weak lensing data only, since the signal-to-noise ratio of the weak lensing measurement is much lower (galaxy shapes are a noisy estimator of the underlying mass distribution) and consequently reconstructions need to be heavily smoothed.

This new method places an upper limit (68% confidence) of $\sigma/m < 1.25\text{cm}^2/\text{g} = 2.25\text{barn}/\text{GeV}$. If we make the assumption that the subcluster and the main cluster had equal mass-to-light ratios prior to the merger, we derive our most stringent constraint of $\sigma/m < 0.7\text{cm}^2/\text{g} = 1.3\text{barn}/\text{GeV}$, which comes from the consistency of the subcluster's observed mass-to-light ratio with the main cluster's, and with the universal cluster value, ruling out the possibility of a large fraction of dark matter particles being scattered away due to collisions. Our limit is a slight improvement over the previous result from analytic estimates, and rules out most of the $\sigma/m < 0.5 - 5\text{cm}^2/\text{g} = 1-10\text{barn/GeV}$ range invoked to explain inconsistencies between the standard collisionless cold dark matter model and observations.[13]

4. What Did We Learn

Massive and interacting clusters, while quite rare, are remarkably well-suited to addressing outstanding issues in both galaxy evolution and fundamental physics. However, in order to study the mass distribution, methods relying on hydrostatic (X-rays) or dynamical equilibrium are ill-suited for such systems.

We have applied a mass reconstruction method based on strong and weak gravitational lensing to the cluster 1E0657−56. We use deep, high resolution optical data to identify objects belonging to the same multiple-image systems. The same data are used to obtain weak lensing catalogs allowing us to obtain a strong+weak lensing mass map of the cluster core. Our main conclusions are the following:

(1) The majority of the mass is spatially coincident with the galaxies, which implies that the cluster mass must be dominated by a relatively collisionless form of dark matter. Combining these findings with detailed simulations we obtain upper limit on dark-matter self interaction cross section of $\sigma/m < 0.7\text{cm}^2/\text{g} = 1.3\text{barn/GeV}$.
(2) Using the combined strong and weak lensing mass reconstruction we derive a high-resolution, absolutely calibrated mass map. We detect the main cluster peak and a distinct mass concentration at the subcluster position, both clearly offset from the location of the X-ray gas in the system (at 10-σ and 6-σ significance for the main and the subcluster respectively - see Fig. 2).
(3) The high resolution data allow us to significantly detect the shapes of both the main mass component and the subcluster with no prior assumptions on their positions or profiles.

The rough limits on the dark-matter self interaction cross section from the pioneering work on 1E0657−56 can be substantially improved: by using more obviously merging systems. We are currently analysing the data from two such systems. In addition, these systems will allow us to derive constraints on the mass distribution of clusters from their very center to their outskirts, thereby providing a unique tool to study clusters and use them as dark matter laboratories.

Acknowledgments

I would like to thank the organizers for their kind invitation to this meeting. Support for this work was provided through Hubble Fellowship grant # HST-HF-01206.01 and through the grant # HST-GO-10200.05-A awarded by the Space Telescope Science Institute, which is operated by the Association of Universities for Research in Astronomy, Inc., for NASA, under contract NAS 5-26555.

References

1. J. P. Ostriker and P. Steinhardt, *Science* **300**, 1909(June 2003).
2. M. Bradač, D. Clowe, A. Gonzalez, P. Marshall, W. Forman, C. Jones, M. Markevitch, S. Randall, T. Schrabback and D. Zaritsky, *ApJ* **652**, 937(December 2006).
3. D. Clowe, P. Schneider, A. Aragón-Salamanca, M. Bremer, G. de Lucia, C. Halliday, P. Jablonka, B. Milvang-Jensen, R. Pelló, B. Poggianti, G. Rudnick, R. Saglia, L. Simard, S. White and D. Zaritsky, *A&A* **451**, 395(May 2006).
4. W. H. Tucker, H. Tananbaum and R. A. Remillard, *ApJ* **444**, 532(May 1995).
5. D. Mehlert, S. Seitz, R. P. Saglia, I. Appenzeller, R. Bender, K. J. Fricke, T. L. Hoffmann *et al.*, *A&A* **379**, 96(November 2001).
6. M. Markevitch, A. H. Gonzalez, L. David, A. Vikhlinin, S. Murray, W. Forman, C. Jones and W. Tucker, *ApJ* **567**, L27(March 2002).
7. R. Barrena, A. Biviano, M. Ramella, E. E. Falco and S. Seitz, *A&A* **386**, 816(May 2002).
8. D. Clowe, A. Gonzalez and M. Markevitch, *ApJ* **604**, p. 596(April 2004).
9. M. Markevitch, A. H. Gonzalez, D. Clowe, A. Vikhlinin, W. Forman, C. Jones, S. Murray and W. Tucker, *ApJ* **606**, p. 819(May 2004).
10. V. Springel and G. R. Farrar, *MNRAS* **380**, 911(September 2007).
11. E. Hayashi and S. D. M. White, *MNRAS* **370**, L38(July 2006).
12. S. Randall, M. Markevitch, D. Clowe, A. Gonzalez and M. Bradač, Constraints on the Self-Interaction Cross-section of Dark Matter from Numerical Simulations of the Merging Galaxy Cluster 1E0657−56, In preparation, (2007).
13. R. Davé, D. N. Spergel, P. J. Steinhardt and B. D. Wandelt, *ApJ* **547**, 574(February 2001).

COSMOLOGICAL WEAK LENSING AND DARK MATTER MAPPING WITH THE HUBBLE SPACE TELESCOPE

TIM SCHRABBACK[1,2], PATRICK SIMON[3,1], THOMAS ERBEN[1], PETER SCHNEIDER[1], JAN HARTLAP[1], CATHERINE HEYMANS[4], PHIL MARSHALL[5,6], HENDRIK HILDEBRANDT[1,2], CHRIS FASSNACHT[7], ERIC MORGANSON[5], MARUŠA BRADAČ[5,6], MARCO HETTERSCHEIDT[1], TIM EIFLER[1], JOAN-MARC MIRALLES[8,9,1], JÖRG DIETRICH[9], ROBERT FOSBURY[10], WOLFRAM FREUDLING[10], and NORBERT PIRZKAL[10,11]

[1]*Argelander-Institut für Astronomie, Universität Bonn, Auf dem Hügel 71, D-53121 Bonn, Germany;* [2]*Leiden Observatory, PO Box 9513, NL-2300 RA, Leiden, the Netherlands, Email: schrabback@strw.leidenuniv.nl;* [3]*Institute for Astronomy, University of Edinburgh, Royal Observatory, Blackford Hill, Edinburgh EH9-3HJ, UK;* [4]*University of British Columbia, Department of Physics and Astronomy, 6224 Agricultural Road, Vancouver, Canada;* [5]*Kavli Institute for Particle Astrophysics and Cosmology, P.O. Box 20450, MS29, Stanford, CA 94309, USA;* [6]*Physics department, University of California, Santa Barbara, CA 93106, USA;* [7]*Department of Physics, University of California, 1 Shields Avenue, Davis, CA 95616, USA;* [8]*Tècniques d'Avantguarda, Avda. Carlemany 75, Les Escaldes, AD-700 Principat d'Andorra;* [9]*European Southern Observatory, Karl-Schwarzschild-Str. 2, D-85741 Garching, Germany;* [10]*ST-ECF, European Southern Observatory, Karl-Schwarzschild Str. 2, 85741 Garching, Germany;* [11]*Space Telescope Science Institute, 3700 San Martin Drive, Baltimore, MD 21218, USA*

Cosmological weak lensing, the gravitational lensing effect of large-scale structure, provides a unique tool to study the large-scale matter distribution and constrain the power spectrum of matter fluctuations free of assumptions on the relation between luminous and dark matter. Due to the weakness of cosmological shear, it can only be detected statistically requiring accurate correction for systematic effects, most importantly the image point-spread-function. Here we present results from our cosmological weak lensing analysis of archival data from the Advanced Camera for Surveys on board the Hubble Space Telescope. Compared to ground-based observations the excellent resolution obtained from space provides a much higher number density of resolved galaxies which can be used to extract the shear signal. This enables us to locally suppress the shape noise induced by the intrinsic ellipticities of galaxies, which is particularly useful for dark matter mapping and constraining the small-scale power spectrum. We have already completed a pilot study using early data from the ACS Parallel Cosmic Shear Survey and the combined GEMS and GOODS mosaic of the *Chandra* Deep Field South (CDFS). From the low shear signal detected in the GEMS/GOODS data we conclude that this field is subject to strong sampling variance, with an exceptional under-density in its foreground. In a preliminary analysis of the HST/COSMOS Survey we detect residual systematics at small angular scales, whose origin is currently been investigated. Using only B-mode-free scales we find a preliminary estimate for the normalisation of the matter fluctuation power spectrum $\sigma_8 = 0.71 \pm 0.09$ (68% confidence) for a matter density $\Omega_m = 0.24$, where the error includes the uncertainties in the redshift distribution, the Hubble constant, and the shear calibration, as well as a Gaussian estimate for cosmic variance.

Keywords: Cosmology, gravitational lensing, large-scale structure.

1. Introduction

Gravitational lensing describes the deflection and distortion of light bundles in the gravitational potential of foreground matter concentrations. This effect can be described as a straightforward application of General Relativity. Hence, it provides a tool to study the total (mostly dark) foreground matter distribution free of assumptions on the relative distribution of luminous and dark matter. Some galaxy clusters are sufficiently massive to create prominent strong lensing features such as multiply-imaged background galaxies, with angular positions near the line-of-sight to the cluster (Fig. 1). However, the images of most galaxies on the sky are only weakly distorted by the projected tidal gravitational field of the typically less concentrated matter inhomogeneities of large-scale structure. Due to the intrinsic ellipticities of galaxies, these distortions cannot be detected from individual galaxies (for which a typical shear-signal-to-noise is of order $\sim 1/30$), but only from an ensemble of coherently lensed galaxies. Measuring the lensing effect of large-scale structure, which is termed *cosmic shear* or *cosmological weak lensing*, provides a powerful tool to study the power spectrum of the total matter fluctuations, so far yielding constraints on the matter content Ω_m and the power spectrum normalisation σ_8, see Ref. 1 for a compilation of recent values and Refs. 2–6 for reviews. Recently Ref. 7 presented a joint analysis from the CFHTLS-Wide, RCS, VIRMOS-DESCART, and GaBoDS surveys, amounting to a total sky area of $\sim 100\,\mathrm{deg}^2$, from which they find $\sigma_8(\Omega_m/0.24)^{0.59} = 0.84 \pm 0.05$. If accurate (photometric) redshifts are available for the source galaxies, the evolution of the shear signal can be constrained, which is a sensitive test to discriminate between different dark energy models due to its dependencies on structure growth and the involved distance ratios,[8,9] see Refs. 10–13 for first constraints on the dark energy equation-of-state parameter w.

Due to the weakness of cosmological shear, proper correction for sys-

Fig. 1. HST/ACS image of the central region of the galaxy cluster RXJ1347–1145. Besides the bright cluster members many background galaxies are visible, whose images are highly distorted by the tidal gravitational field of the cluster. In the case of cosmological weak lensing the distortions are typically much weaker and can only be detected statistically (Image credit: NASA/ESA/AIfA; Erben/Schrabback).

tematic effects, most importantly the image point-spread-function (PSF), is indispensable. The PSF dilutes the shear signal by isotropic blurring and creates spurious signal due to anisotropy, which can easily exceed the cosmological signal. A precise correction is particularly important in view of the next generation of cosmic shear surveys, which will provide the statistical power to measure cosmological shear with percent and sub-percent level accuracy aiming at precision measurements of dark energy properties. This will only be possible if the methods used to extract the shear signal by measuring galaxy shapes from noisy and PSF-distorted images have a comparable systematic accuracy. In order to assess the accuracy of existing methods and to provide a platform for further development, the Shear TEsting Programme[a] (STEP) was launched in 2004. In this project blind analyses of simulated images containing a hidden shear signal have been conducted.[14,15] These benchmarks show that several shape measurement techniques already reach an average relative systematic accuracy of a few percent, which is sufficient for current studies, but substantial development

[a]`http://www.physics.ubc.ca/~heymans/step.html`

is still required for future surveys. Important progress has also been made in the development of diagnostic tools for the detection of residual systematics, such as the galaxy-star cross-correlation[16] and the E-/B-mode decomposition,[17,18] where the curl B-mode cannot be created by lensing.

Most of the ongoing and planned weak lensing surveys use ground-based wide-field imaging data. These are hampered by image blurring (seeing) due to atmospheric turbulence which erases the shear information for unresolved galaxies limiting the number density of usable galaxies to $\lesssim 30/\mathrm{arcmin}^2$. The disturbing influence of Earth's atmosphere can be avoided by using space-based images, which provide a higher number density of resolved background sources and better control for systematics, on the cost of smaller sky coverage. The higher number density obtained from space helps to locally beat down the shape noise induced by the intrinsic ellipticities of the source galaxies. This is particularly important in order to constrain the small-scale power spectrum and reconstruct the projected matter distribution. In these proceedings we summarise results from our ongoing cosmological weak lensing analysis of archival data from the *Advanced Camera for Surveys* (ACS) on board the *Hubble Space Telescope* (HST), providing the currently largest set of data suitable for space-based cosmic shear measurements. Combining a relatively large field-of-view ($3\farcm3 \times 3\farcm3$) with high sensitivity (48% total throughput at 660 nm), good PSF-sampling ($0\farcs05$ per pixel) and space-based resolution ($\sim 0\farcs11$ at 800 nm), HST/ACS was the best instrument for weak lensing measurements at small angular scales until its unfortunate loss on January 27th, 2007.

2. Practical Concerns for Measuring Weak Lensing with ACS

Before being able to use the ACS data for weak lensing studies, it was necessary to develop several technical tools and carefully test our weak lensing pipeline:

Firstly, we developed an image reduction pipeline for ACS data, which fulfils the strict requirements weak lensing places on the data quality, e.g. regarding careful image registration and bad pixel masking.

Secondly, the detection of strong temporal variations of the ACS PSF (Fig. 2), which are interpreted as focus changes due to thermal breathing of the telescope, led to the development of a new PSF correction scheme. Classically, PSF variations across the field-of-view are modelled by interpolating measured stellar shape parameters using polynomial functions. However, due to the small field-of-view of ACS, high galactic latitude fields

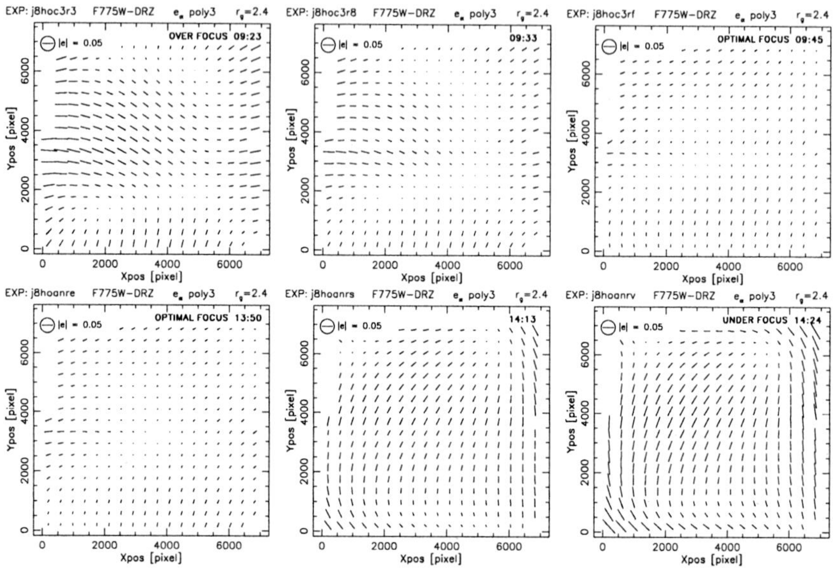

Fig. 2. Third-order polynomial fits to stellar ellipticities in the drizzled images of two series of subsequent ACS/WFC exposures, where the bars indicate the ellipticity of the PSF. The 400 second exposures were taken on 2002-08-28 (*upper* panels) and 2002-08-17 (*lower* panels), where the time indicated corresponds to the middle of the exposure (UT). The variations are interpreted as thermal breathing of the telescope. The *upper right* and *lower left* plots are near the optimal focus position, whereas the other exposures represent positive focus offsets (*upper left* panel) or negative focus offsets (*lower right* panel). Figure taken from Ref. 19.

contain only of order 10 stars, which are too few for a direct interpolation. In our correction scheme we fit the stars present in a galaxy field with PSF models derived from dense stellar fields. We determine the correction on the basis of single exposures in order to obtain an optimal time-dependence of the solution and exclude noise in the ellipticity measurement caused by resampling. We apply a principal component analysis (PCA) to the polynomial interpolation coefficients of the PSF anisotropy kernel[19,20] (a PSF-width-scaled version of the PSF ellipticity) computed for several hundred stellar field exposures of different ACS broad-band filters. From the PCA we find that indeed most of the PSF variation ($\sim$ 60% of the variance) occurs in a one-dimensional parameter space, which can be related to the focus position (Fig. 3). We estimate that our PSF correction scheme reduces the systematic contribution to the shear correlation functions due to PSF distortions to $< 2 \times 10^{-6}$ for galaxy fields containing at least 10 stars.[19]

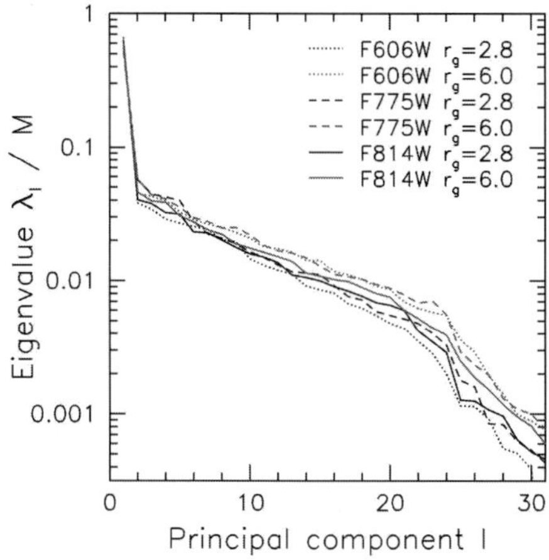

Fig. 3. Normalised eigenvalues λ_l/M for the first 30 principal components of the temporal variation of the ACS PSF polynomial coefficients. The normalised eigenvalue of a principal component gives the relative contribution of the model variation along the direction spanned by the corresponding principal vector, relative to the total variance in coefficient space. The different curves correspond to the ACS filters F606W, F775W, and F814W and two different PSF filter scales $r_g = (2.8, 6.0)$ pixels. In all cases about 60% of the total variation occurs along the direction of the dominant first principal component, which can be related to the focus position.

This corresponds to $\lesssim 5\%$ of the cosmological signal expected on scales of a single ACS field.

Thirdly, we carefully tested our weak lensing pipeline, which is based on the KSB-formalism,[20] using simulated images from the STEP Programme. After eliminating biases identified in the first round of simulations,[14] our pipeline performed among the most accurate methods with an average (relative) shear recovery accuracy of $\sim 2\%$ on the second simulation set.[15] It also achieves a similar average accuracy in a preliminary analysis of the third set of STEP simulations (Rhodes et al., in prep.), which features space-based weak lensing data, partially with ACS-like properties. In this test our updated pipeline also yields a good correction as a function of galaxy size and magnitude ($\sim \pm 5\%$ variation), which is important if one aims at measuring the shear signal as a function of source redshift. Concluding from STEP, the systematic accuracy of our pipeline should be well within the statistical noise of the ACS weak lensing data.

Finally, the continuous bombardment of ACS with cosmic rays creates defects in the CCDs reducing the charge-transfer efficiency (CTE). This leads to charge trails trailing behind objects in the read-out direction. In order to correct for this influence on galaxy shapes we apply a parametric correction model similar to the one proposed by Ref. 21.

3. Cosmological Weak Lensing Analysis of the ACS Archive

Our cosmological weak lensing analysis was done in two steps: In a pilot study[19] we first test the capabilities of ACS for cosmic shear measurements with early data from the ACS Parallel Survey (Sect. 3.1) and the combined GEMS and GOODS ACS mosaic of the *Chandra* Deep Field South (CDFS, Sect. 3.2). We are currently analysing two much larger data sets, the Extended ACS Parallel Survey (Sect. 3.1) and the HST/COSMOS Survey (Sect. 3.3).

3.1. *The Extended ACS Parallel Survey*

In the pilot study[19] we use a limited set of 59 independent ACS pointings from the ACS Parallel Cosmic Shear Survey. These parallel data are taken simultaneously with primary observations with another HST instrument. As the cameras point at positions separated by several arcminutes, parallel observations provide almost random pointings, which is helpful to measure a statistical quantity such as cosmological shear. Yet, due to the relatively small field-of-view of ACS, the shear correlation signal can only be measured at relatively small angular separations. In the parallel data of the pilot study we detect a low level of remaining systematics manifesting in a slight net alignment of the galaxies in the direction of the image y-axis. There are indications that this signal originates from the poor data quality caused by a lack of sufficient dithering of the data. We therefore developed several upgrades to our data reduction and analysis pipeline in order to reliably reject affected galaxies.

We are currently analysing a much larger data set consisting of 440 pointings or a total area of $1.22\,\mathrm{deg}^2$. Here we combine parallel data ($\sim 69\%$) with primary observations which resemble quasi-random pointings for cosmic shear, as they either target very near ($z < 0.07$) or distant ($z > 1.2$) objects with a low lensing efficiency. From this "Extended ACS Parallel Survey" we aim to constrain the small-scale power spectrum with high accuracy, which strongly depends on the largely unknown inter-play of baryons and dark matter. The analysis of this data set is done jointly with the HAGGLeS Project, which aims at the detection of galaxy-scale strong lenses in the ACS archive (Marshall et al., in prep.).

3.2. *GEMS/GOODS*

The combined GEMS and GOODS mosaic of the *Chandra* Deep Field South spans a total area of $\sim 28' \times 28'$ imaged with the ACS/WFC using the F606W filter. The exposure time for GEMS is 2.2ks, which we combine with a similarly deep stack for GOODS. In Ref. 19 we perform a number of diagnostic tests indicating that the remaining level of systematics is consistent with zero for the GEMS and GOODS data confirming the success of our PSF correction scheme. Combining our shear estimate of the GEMS and GOODS observations using 96 galaxies arcmin^{-2} with the photometric redshift catalogue of the GOODS-MUSIC sample,[22] we determine a *local single field estimate* for the mass power spectrum normalisation $\sigma_8 = 0.59^{+0.13}_{-0.17}\,(\text{stat}) \pm 0.07\,(\text{sys})$ (68% confidence assuming Gaussian sampling variance) at a fixed matter density $\Omega_\text{m} = 0.24$ for a ΛCDM cosmology, where we marginalise over the uncertainty of the Hubble constant and the redshift distribution. This estimate agrees only marginally with the WMAP-3 result[23] of $\sigma_8 = 0.761^{+0.049}_{-0.048}$ and is significantly below values found by recent ground-based surveys.[7] From this discrepancy we conclude that the CDFS is subject to strong sampling variance with a significant under-density of compact foreground structures. This is consistent with a recent study by Ref. 24, who find a strong deficiency of red galaxies in this field. Note that Ref. 25 originally derived a higher value $\sigma_8 = 0.79 \pm 0.15$ (for $\Omega_\text{m} = 0.24$) from the GEMS data. While we have been able to show that their and our shear measurements agree with a relative accuracy of $\sim 3\%$, we found that their σ_8 estimate is probably biased due to an under-estimation of the mean redshift of the survey.

Studies performing ray-tracing through N-body simulations indicate that non-linear evolution leads to significant non-Gaussian sampling variance,[26,27] which suggests that we actually under-estimate the statistical uncertainty. The frank conclusion is that the sky coverage of GEMS is too small to yield representative cosmological parameter constraints using cosmic shear. The line-of-sight in the CDFS seems to pass mostly through voids, whereas compact structures are rare. Following from the preliminary ray-tracing comparison by Hartlap et al. (in prep.), this seems to be a peculiar, but not extraordinary rare configuration, assuming a WMAP-3 cosmology.

Note that our σ_8 estimate from the CDFS was more discrepant with other weak lensing results at the time of the analysis. For example, early results from the CFHTLS yielded significantly higher σ_8 estimates of $\sigma_8 = 0.86 \pm 0.05$[12] and $\sigma_8 = 0.85 \pm 0.06$[11] computed for $\Omega_\text{m} = 0.3$, which corresponds to $\sigma_8 = 0.98 \pm 0.06$ and $\sigma_8 = 0.97 \pm 0.07$ for $\Omega_\text{m} = 0.24$. However, with the new analysis by Ref. 7, which makes use of a carefully calibrated redshift distribution and considers recent STEP results, the tension has been significantly reduced. It will be interesting to observe if this trend towards lower σ_8 estimates from cosmological weak lensing studies continues in the post-STEP era.

3.3. *COSMOS*

We also conducted a preliminary analysis of the HST/COSMOS Survey. Although this is also a single mosaic field, the approximately seven times larger sky area of 1.64 $\deg^2$ considerably reduces the influence of sampling variance compared to GEMS. From our analysis we have been able to confirm the main features of the projected dark matter maps found by Ref. 28. The maps show several mass peaks associated with galaxy clusters and groups, partially connected with filamentary structures (see Fig. 4). In addition, a large, relatively empty region is visible in the lower right (south-west) quadrant of the mosaic, with a diameter $\gtrsim 9h^{-1}$ Mpc at a typical lens redshift $z \sim 0.4$ (assuming a flat ΛCDM cosmology). We expect that this line-of-sight has similar statistical properties as the GEMS Survey, being a projection of mostly voids along the line-of-sight, but we did not test this quantitatively.

From the COSMOS data we have been able to measure the cosmic shear signal at a wide range of scales $0\overset{\prime}{.}1 \lesssim \theta \lesssim 80'$. At small scales we detect a significant B-mode signal (curl-component) indicating remaining uncorrected systematics (Fig. 5), where the affected scales are $\theta \lesssim 0\overset{\prime}{.}7$ for the shear two-point correlation function and $\theta \lesssim 5'$ for the dispersion of the aperture mass statistics.[30] The spurious B-mode signal might be related to residuals of the PSF correction, but this is still under investigation. Similar indications for remaining systematics have been reported by Ref. 31 for their analysis of the data. We perform a preliminary cosmological parameter estimation, in which we only use B-mode free scales ($\theta > 1'$ for the shear two-point correlation function), and a redshift distribution based on the public redshift catalogue by Ref. 32. From the COSMOS data we obtain an estimate for the power spectrum normalisation $\sigma_8 = 0.71 \pm 0.09$ (68% confidence) for $\Omega_{\rm m} = 0.24$, where the error includes the uncertainties in the redshift distribution, the Hubble constant, and the shear calibration, as well as a Gaussian estimate for cosmic variance. This result is in excellent agreement with the WMAP-3 result of $\sigma_8 = 0.761^{+0.049}_{-0.048}$. Yet, our estimate agrees only marginally with recent ground-based estimates, which favour slightly higher values of σ_8. However, the fact that no extraordinarily massive galaxy cluster is present in the field at redshifts with high lensing efficiency might suggest a σ_8 estimate rather at the low end of the expected range. Thus, we find it surprising that our measurement differs significantly from the estimate by Ref. 31 for the same field, who find $\sigma_8 = 0.90 \pm 0.08\,(\text{stat.}) \pm 0.10\,(\text{syst.})$ from a similar analysis using one redshift bin and $\sigma_8 = 0.96 \pm 0.04\,(\text{stat.})^{+0.06}_{-0.04}\,(\text{syst.})$ from a tomographic

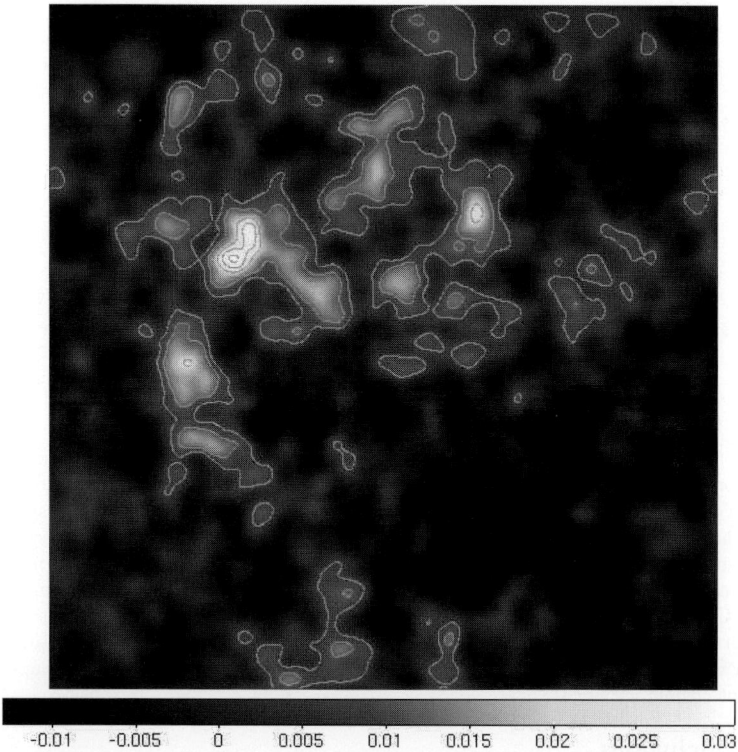

Fig. 4. Maximum likelihood reconstruction of the convergence κ (projected mass) in the COSMOS field from our shear catalogue applying noise propagation and Wiener filtering.[29] The absolute scale has been set such that the mean κ vanishes. The contours start at and are spaced in steps of $\kappa = 0.005$. Note that we use fewer galaxies due to our more stringent selection criteria (43/arcmin^2) compared to the independent analysis by Ref. 28, leading to a larger effective smoothing scale of our map.

analysis using three bins, both for $\Omega_{\rm m} = 0.24$. Note that differences within the statistical and systematic errors are not unexpected, given that independent data reduction and shear measurement pipelines have been applied, as well as different selection criteria. Yet, their tomographic results and our estimate are clearly not consistent. So far we have not been able to unambiguously identify the origin for this discrepancy. However, note that both catalogues still contain indications for systematic at small scales. Once these are eliminated by further improvements in the PSF correction and analysis scheme, the discrepancy might partially be resolved.

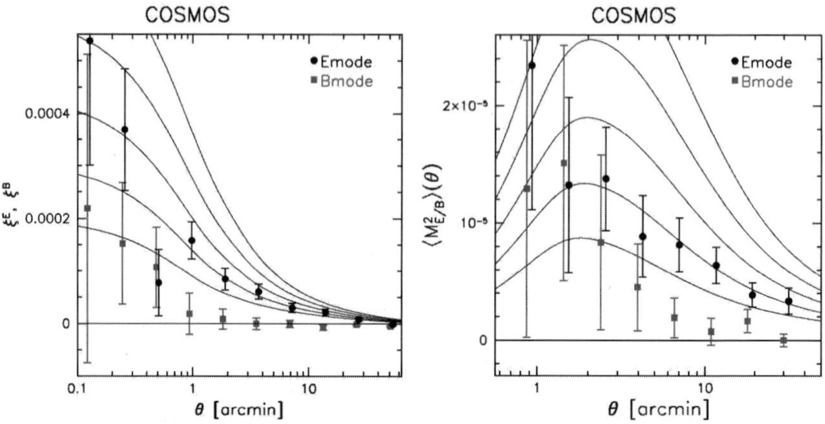

Fig. 5. E/B-mode decomposition of our COSMOS analysis using the shear correlation functions (left) and the dispersion of the aperture mass (right). While the E-mode signal can be created both by lensing and systematics, the B-mode must be due to systematics. It is therefore likely, that at small scales, where a significant B-mode is detected, also the E-mode signal will be affected by systematics. Thus, only the large, B-mode-free scales can be used for a cosmological parameter estimation. The curves indicate ΛCDM predictions for $\Omega_m = 0.3$ and $\sigma_8 = (0.6, 0.7, 0.8, 0.9, 1.0)$. For the decomposition using the correlation function an extrapolation of the shear signal to infinity is required, for which we assume $\sigma_8 = 0.7$. The error-bars include the uncertainty due to shape noise but not cosmic variance.

4. Conclusions

We found that the ACS PSF varies on short time scales, which is properly taken into account in our PSF correction scheme. From the analysis of the GEMS and GOODS data we conclude that the CDFS is an exceptionally under-dense field. Hence, although it is one of the best-studied fields in the sky, it cannot be considered to be fully representative for the cosmic mean. In our preliminary analysis of the HST/COSMOS data we recover the main structures in the dark matter maps by Ref. 28 with good consistency, which show that baryons trace the distribution of dark matter well on large scales. Our preliminary cosmological parameter estimation from COSMOS yields $\sigma_8 = 0.71 \pm 0.09$ consistent with WMAP3, but significantly below the value found by Ref. 31. However, note that our analysis is only preliminary as we still detect indications for remaining systematics, which we hope to eliminate with an upgraded analysis and correction scheme. A very interesting comparison will be possible with the Extended ACS Parallel Survey, which is expected to yield tight constraints on the cosmic shear signal at small angular scales, possibly providing interesting insights into the interplay of dark matter and baryons on scales of ~ 100 kpc.

Cosmological weak lensing studies with ACS, such as the one presented here, but also the work by Refs. 21,25,31,33,34 play a pathfinder role for dedicated future cosmic shear missions from space such as DUNE or JDEM/SNAP: Not only acting as proof of concept for envisaged future projects, they also provide first estimates for the small-scale signal, which can be used as input to estimate the accuracy of future studies. Furthermore, current projects with ACS yield substantial technical expertise, which will be very valuable for the planning, conduction, and analysis of future surveys. Following from the experience with ACS, future mission should, for example, aim to achieve maximal thermal stability to minimise temporal variations of the point-spread-function. Also, it will be crucial to use hardened CCD chips, which are subject to less CTE degradation. Concerning the analysis of future surveys, tools such as our PSF interpolation scheme, possibly upgraded using a principal component analysis, will be very valuable to achieve the systematic accuracy sought.

In general cosmological weak lensing is currently developing into a mature field of cosmology, which will yield precision constraints on the power spectrum, dark matter, and dark energy with upcoming surveys.

Acknowledgments

It is a pleasure to thank the organizers for the kind invitation and excellent organization of the conference. TS acknowledges support from the Studienstiftung des deutschen Volkes and the Deutsche Forschungsgemeinschaft through SFB/Transregio 33 "The Dark Universe".

References

1. M. Hetterscheidt, P. Simon, M. Schirmer, H. Hildebrandt, T. Schrabback, T. Erben and P. Schneider, *A&A* **468**, 859(June 2007).
2. M. Bartelmann and P. Schneider, *Phys. Rep.* **340**, 291(January 2001).
3. Y. Mellier, L. van Waerbeke, E. Bertin, I. Tereno and F. Bernardeau, Wide-field cosmic shear surveys, in *Astronomical Data Analysis II. Edited by Starck, Jean-Luc; Murtagh, Fionn D. Proceedings of the SPIE, Volume 4847, pp. 112-122*, December 2002.
4. H. Hoekstra, H. K. C. Yee and M. D. Gladders, *New Astronomy Review* **46**, 767(November 2002).
5. A. Refregier, *ARA&A* **41**, 645 (2003).
6. P. Schneider, C. S. Kochanek and J. Wambsganss, *Saas-Fee Advanced Course 33: Gravitational Lensing: Strong, Weak and Micro* (Springer Verlag, Berlin, 2006).
7. J. Benjamin, C. Heymans, E. Semboloni, L. Van Waerbeke, H. Hoekstra,

T. Erben, M. D. Gladders, M. Hetterscheidt, Y. Mellier and H. K. C. Yee, *MNRAS accepted (also astro-ph/0703570)* (March 2007).

8. J. Peacock, P. Schneider, G. Efstathiou, J. Ellis, B. Leibundgut, S. Lilly and Y. Mellier, *http://www.stecf.org/coordination/esa_eso/cosmology.php* (September 2006).
9. A. Albrecht, G. Bernstein, R. Cahn, W. L. Freedman, J. Hewitt, W. Hu, J. Huth, M. Kamionkowski, E. W. Kolb, L. Knox, J. C. Mather, S. Staggs and N. B. Suntzeff, *astro-ph/0609591* (September 2006).
10. M. Jarvis, B. Jain, G. Bernstein and D. Dolney, *ApJ* **644**, 71(June 2006).
11. H. Hoekstra, Y. Mellier, L. van Waerbeke, E. Semboloni, L. Fu, M. J. Hudson, L. C. Parker, I. Tereno and K. Benabed, *ApJ* **647**, 116(August 2006).
12. E. Semboloni, Y. Mellier, L. van Waerbeke, H. Hoekstra, I. Tereno, K. Benabed, S. D. J. Gwyn, L. Fu, M. J. Hudson, R. Maoli and L. C. Parker, *A&A* **452**, 51(June 2006).
13. T. D. Kitching, A. F. Heavens, A. N. Taylor, M. L. Brown, K. Meisenheimer, C. Wolf, M. E. Gray and D. J. Bacon, *MNRAS* , 138(April 2007).
14. C. Heymans, L. Van Waerbeke, D. Bacon, J. Berge, G. Bernstein, E. Bertin, S. Bridle, M. L. Brown, D. Clowe, H. Dahle, T. Erben, M. Gray, M. Hetterscheidt, H. Hoekstra, P. Hudelot, M. Jarvis, K. Kuijken, V. Margoniner, R. Massey, Y. Mellier, R. Nakajima, A. Refregier, J. Rhodes, T. Schrabback and D. Wittman, *MNRAS* **368**, 1323(May 2006).
15. R. Massey, C. Heymans, J. Bergé, G. Bernstein, S. Bridle, D. Clowe, H. Dahle, R. Ellis, T. Erben, M. Hetterscheidt, F. W. High, C. Hirata, H. Hoekstra, P. Hudelot, M. Jarvis, D. Johnston, K. Kuijken, V. Margoniner, R. Mandelbaum, Y. Mellier, R. Nakajima, S. Paulin-Henriksson, M. Peeples, C. Roat, A. Refregier, J. Rhodes, T. Schrabback, M. Schirmer, U. Seljak, E. Semboloni and L. van Waerbeke, *MNRAS* **376**, 13(March 2007).
16. D. J. Bacon, R. J. Massey, A. R. Refregier and R. S. Ellis, *MNRAS* **344**, 673(September 2003).
17. R. G. Crittenden, P. Natarajan, U.-L. Pen and T. Theuns, *ApJ* **568**, 20(March 2002).
18. P. Schneider and M. Kilbinger, *A&A* **462**, 841(February 2007).
19. T. Schrabback, T. Erben, P. Simon, J.-M. Miralles, P. Schneider, C. Heymans, T. Eifler, R. A. E. Fosbury, W. Freudling, M. Hetterscheidt, H. Hildebrandt and N. Pirzkal, *A&A* **468**, 823(June 2007).
20. N. Kaiser, G. Squires and T. Broadhurst, *ApJ* **449**, 460(August 1995).
21. J. D. Rhodes, R. J. Massey, J. Albert, N. Collins, R. S. Ellis, C. Heymans, J. P. Gardner, J.-P. Kneib, A. Koekemoer, A. Leauthaud, Y. Mellier, A. Refregier, J. E. Taylor and L. Van Waerbeke, *ApJS* **172**, 203(September 2007).
22. A. Grazian, A. Fontana, C. de Santis, M. Nonino, S. Salimbeni, E. Giallongo, S. Cristiani, S. Gallozzi and E. Vanzella, *A&A* **449**, 951(April 2006).
23. D. N. Spergel, R. Bean, O. Doré, M. R. Nolta, C. L. Bennett, J. Dunkley, G. Hinshaw, N. Jarosik, E. Komatsu, L. Page, H. V. Peiris, L. Verde, M. Halpern, R. S. Hill, A. Kogut, M. Limon, S. S. Meyer, N. Odegard, G. S. Tucker, J. L. Weiland, E. Wollack and E. L. Wright, *ApJS* **170**, 377(June 2007).

24. S. Phleps, C. Wolf, J. A. Peacock, K. Meisenheimer and E. van Kampen, *A&A* **468**, 113(June 2007).
25. C. Heymans, M. L. Brown, M. Barden, J. A. R. Caldwell, K. Jahnke, C. Y. Peng, H.-W. Rix, A. Taylor, S. V. W. Beckwith, E. F. Bell, A. Borch, B. Häußler, S. Jogee, D. H. McIntosh, K. Meisenheimer, S. F. Sánchez, R. Somerville, L. Wisotzki and C. Wolf, *MNRAS* **361**, 160(July 2005).
26. M. Kilbinger and P. Schneider, *A&A* **442**, 69(Octtober 2005).
27. E. Semboloni, L. van Waerbeke, C. Heymans, T. Hamana, S. Colombi, M. White and Y. Mellier, *MNRAS* **375**, L6(February 2007).
28. R. Massey, J. Rhodes, R. Ellis, N. Scoville, A. Leauthaud, A. Finoguenov, P. Capak, D. Bacon, H. Aussel, J.-P. Kneib, A. Koekemoer, H. McCracken, B. Mobasher, S. Pires, A. Refregier, S. Sasaki, J.-L. Starck, Y. Taniguchi, A. Taylor and J. Taylor, *Nature* **445**, 286(January 2007).
29. W. Hu and C. R. Keeton, *Phys. Rev. D* **66**, 063506(September 2002).
30. P. Schneider, L. van Waerbeke, B. Jain and G. Kruse, *MNRAS* **296**, 873(June 1998).
31. R. Massey, J. Rhodes, A. Leauthaud, P. Capak, R. Ellis, A. Koekemoer, A. Réfrégier, N. Scoville, J. E. Taylor, J. Albert, J. Bergé, C. Heymans, D. Johnston, J.-P. Kneib, Y. Mellier, B. Mobasher, E. Semboloni, P. Shopbell, L. Tasca and L. Van Waerbeke, *ApJS* **172**, 239(September 2007).
32. B. Mobasher, P. Capak, N. Z. Scoville, T. Dahlen, M. Salvato, H. Aussel, D. J. Thompson, R. Feldmann, L. Tasca, O. Lefevre, S. Lilly, C. M. Carollo, J. S. Kartaltepe, H. McCracken, J. Mould, A. Renzini, D. B. Sanders, P. L. Shopbell, Y. Taniguchi, M. Ajiki, Y. Shioya, T. Contini, M. Giavalisco, O. Ilbert, A. Iovino, V. Le Brun, V. Mainieri, M. Mignoli and M. Scodeggio, *ApJS* **172**, 117(September 2007).
33. A. Leauthaud, R. Massey, J.-P. Kneib, J. Rhodes, D. E. Johnston, P. Capak, C. Heymans, R. S. Ellis, A. M. Koekemoer, O. Le Fèvre, Y. Mellier, A. Réfrégier, A. C. Robin, N. Scoville, L. Tasca, J. E. Taylor and L. Van Waerbeke, *ApJS* **172**, 219(September 2007).
34. M. J. Jee, J. P. Blakeslee, M. Sirianni, A. R. Martel, R. L. White and H. C. Ford, *PASP accepted (also astro-ph/0710.5560)* **710**(Octtober 2007).

CAN MASSIVE DARK HALOES DESTROY THE DISKS OF DWARF GALAXIES?

B. FUCHS* and O. ESQUIVEL

Astronomisches Rechen-Institut am Zentrum für Astronomie der Universität Heidelberg, Mönchhofstraße 12 - 14, 69120 Heidelberg, Germany
* *E-mail: fuchs@ari.uni-heidelberg.de*

Recent high-resolution simulations together with theoretical studies of the dynamical evolution of galactic disks have shown that contrary to wide-held beliefs a 'live', dynamically responsive, dark halo surrounding a disk does not stabilize the disk against dynamical instabilities. We generalize Toomre's Q stability parameter for a disk-halo system and show that if a disk, which would be otherwise stable, is embedded in a halo, which is too massive and cold, the combined disk-halo system can become locally Jeans unstable. The good news is, on the other hand, that this will not happen in real dark haloes, which are in radial hydrostatic equilibrium. Even very low-mass disks are not prone to such dynamical instabilities.

Keywords: Galactic disks: stability; Dark halo.

1. Introduction

The classical paradigm is that at given total mass of the system a galactic disk is stabilized against local dynamical Jeans instabilities, if it is embedded in a dark halo. This can be seen for instance from Toomre's $Q_{\rm T}$ stability index for a stellar disk,

$$Q_{\rm T} = \frac{\kappa \sigma_{\rm d}}{3.36\, G \Sigma_{\rm d}}\,, \tag{1}$$

where κ denotes the epicyclic frequency of the orbits of the stars, $\sigma_{\rm d}$ is the radial velocity dispersion of the stars, and $\Sigma_{\rm d}$ the surface density of the disk. G denotes the constant of gravitation. If all other parameters are kept constant, but $\Sigma_{\rm d}$ is lowered, $Q_{\rm T}$ rises and the disk becomes more stable against Jeans instabilities. The physical reasoning is that the self-gravity of the disk, which has a destabilizing effect, is reduced by the surrounding halo. Similarly the onset of non-axisymmetric coherent large-scale instabilities of the entire disk such as the bar instability was thought to be damped by a surrounding halo. Ostriker & Peebles[16] showed in their classical numerical simulations of the dynamical evolution of a self-gravitating disk that the bar instability could be suppressed, if the disk was embedded in a halo potential. However, modern high-resolution simulations in which the surrounding halo is treated as a dynamically responsive system, have shown that actually the opposite is true. Athanassoula[1] showed that in her simulations of the bar instability the bar grows stronger, if the disk is embedded in a live dark halo rather than in a static halo potential. This was explained there and, particularly, in Athanassoula[2] as due to the effect of the live halo on the angular momentum exchange within the galaxy. First doubts about an entirely passive role of the halo were already raised by Toomre[18] . These findings were supported by theoretical studies of the swing amplification of shearing spiral density waves which is also enhanced, if the disk is embedded in a live dark halo instead of a static potential (Fuchs[9] , Fuchs & Athanassoula[10]).

Here we return to the severe local Jeans instability of a self-gravitating disk and investigate the effect of the presence of a live dark halo. In the next section we demonstrate how Toomre's concept of the $Q_{\rm T}$ parameter can be generalized in order to take into account the effect of such a halo. In the final section we discuss implications for realistic disk-halo systems (cf. also Esquivel & Fuchs[7]).

2. Modification of the $Q_{\rm T}$ Stability Index

We study the Jeans instability of an infinitesimally thin galactic disk using the model of a patch of the galactic disk developed by Toomre[17] , Goldreich & Lynden-Bell[11] , and Julian & Toomre[14] (cf. also Fuchs[8]). The patch is assumed to rotate around the galactic center and the differential rotation of the stars is approximated as a linear shear flow. The surface density is assumed to be constant over the patch. Polar coordinates are approximated by pseudo Cartesian coordinates (x, y) with x pointing in the radial direction and y in the direction of rotation, respectively. Toomre[17] (cf. also Lin[15]) has calculated the dynamical response of the disk to a small 'ring-like' perturbation of the gravitational disk potential of the form

$$\Phi_{\rm k} e^{i(\omega t + kx)} \tag{2}$$

by solving the linearized Boltzmann equation which describes the evolution of the phase space distribution of the disk stars. The induced density perturbation can be expressed as

$$\Sigma_{\rm k} e^{i(\omega t + kx)} = -\frac{\Sigma_{\rm d}}{\sigma_{\rm d}^2}\Bigg[1 - \exp\left(-\frac{k^2\sigma_{\rm d}^2}{\kappa^2}\right) \times \left(\mathrm{I}_0\left(\frac{k^2\sigma_{\rm d}^2}{\kappa^2}\right) + \sum_{n=1}^{\infty} \mathrm{I}_{\rm n}\left(\frac{k^2\sigma_{\rm d}^2}{\kappa^2}\right)\frac{2(\frac{\omega}{\kappa})^2}{(\frac{\omega}{\kappa})^2 - n^2}\right)\Bigg]\Phi_{\rm k} e^{i(\omega t + kx)}, \tag{3}$$

where the $\mathrm{I}_{\rm n}$ denote modified Bessel functions and $\Sigma_{\rm d}$ the background surface density of the disk, respectively. In deriving eq. (3) a Schwarzschild velocity distribution of the stars with the radial velocity dispersion $\sigma_{\rm d}$ has been adopted. The disk is assumed to be self-gravitating, so that the density – potential pair has to fulfill the Poisson equation implying

$$\Phi_k = -\frac{2\pi G}{|k|}\Sigma_k. \tag{4}$$

Equations (3) and (4) can be combined to a dispersion relation $\omega(k)$ for the 'ring'-like perturbations of the disk. In the limit $\omega = 0$ eq. (3) reduces to

$$\Sigma_{\rm k} e^{ikx} = -\frac{\Sigma_{\rm d}}{\sigma_{\rm d}^2}\left[1 - \exp\left(-\frac{k^2\sigma_{\rm d}^2}{\kappa^2}\right)\cdot \mathrm{I}_0\left(\frac{k^2\sigma_{\rm d}^2}{\kappa^2}\right)\right]\Phi_{\rm k} e^{ikx}. \tag{5}$$

This defines a line in a space spanned by $Q_{\rm T}$ and the wavelength of the perturbation $\lambda = 2\pi/k$ expressed in units of $\lambda_{\rm crit} = 4\pi^2 G\Sigma_d/\kappa^2$ which, as can be seen by a Taylor expansion of eq. (3), separates neutrally stable ($\omega^2 \geq 0$) from exponentially unstable ($\omega^2 < 0$) perturbations. The criterion that ensures that all perturbations are neutrally stable is the famous Toomre criterion

$$Q_{\rm T} \geq 1\,. \tag{6}$$

The model of a local patch of a galactic disk has been extended by Fuchs[9] by embedding it into a dark halo. All density gradients in the halo are neglected as in the disk so that the halo density distribution is assumed to be homogeneous. The dark matter particles follow straight-line orbits with an isotropic velocity distribution modelled by a Gaussian distribution. We can directly apply the results of Fuchs[9] . As is shown there the dark matter halo responds to the potential perturbation $\Phi_{\rm d}$ in the disk and develops potential perturbations $\Phi_{\rm h}$ which have the same radial structure $\exp(ikx)$ as in the disk. Equations (26) and (28) of Fuchs[9] have been obtained as solutions of the linearized Boltzmann equation for the dark matter particles and give the Fourier coefficients of the potential perturbation in the halo at the midplane of the disk. In the case $k_{\rm y} = 0$, $k_{||} = k$, and $\omega = 0$, which we consider here, there is only a contribution from non-resonant dark matter particles,

$$\Phi_{\rm hk} = \frac{4G\rho_{\rm h}}{\sigma_{\rm h}^2}\Phi_{\rm dk}\int_{-\infty}^{\infty} dk_z \frac{k}{(k^2+k_z^2)^2} = \frac{2\pi G\rho_{\rm h}}{\sigma_{\rm h}^2}\frac{1}{k^2}\Phi_{\rm dk}\,, \tag{7}$$

where $\rho_{\rm h}$ and $\sigma_{\rm h}$ denote the density of the dark halo and the velocity dispersion of the dark matter particles, respectively.

This induced perturbation of the gravitational potential of the dark halo has to be taken into account on the rhs of eq. (5),

$$\Phi_{\rm k} \rightarrow \Phi_{\rm dk} + \Phi_{\rm hk} \propto \Phi_{\rm dk}\,, \tag{8}$$

which means that the halo supports the perturbation of the disk and the density perturbation in the disk is stronger than in an isolated disk. Combining eqns. (5), (7), and (8) leads to an implicit equation that describes the line of neutrally stable ($\omega = 0$) perturbations in the space spanned by $Q_{\rm T}$ and $\lambda/\lambda_{\rm crit}$ as in the case of an isolated disk, but now modified by the extra halo term according to eqns. (7) and (8). This can be cast into dimensionless form as

$$\begin{aligned}\alpha Q_{\rm T}^2 = &\left[1 - \exp\left(-\frac{\alpha Q_{\rm T}^2}{(\lambda/\lambda_{\rm crit})^2}\right)\cdot {\rm I}_0\left(\frac{\alpha Q_{\rm T}^2}{(\lambda/\lambda_{\rm crit})^2}\right)\right]\frac{\lambda}{\lambda_{\rm crit}}\\ &\times\left(1+\beta\left(\frac{\lambda}{\lambda_{\rm crit}}\right)^2\right)\end{aligned} \tag{9}$$

with the parameters $\alpha = (3.36/2\pi)^2 = 0.286$ and $\beta = (2\pi G\rho_h/\sigma_h^2)\times(2\pi G\Sigma_d)^2/\kappa^4$.

In Fig. 1 we illustrate solutions of eq. (9) for various values of β. The case $\beta = 0$ reproduces Toomre's classical[17] result. The unstable perturbations ($\omega^2 < 0$) are located in the parameter space below the line. Thus for $Q_{\rm T} \geq 1$ all perturbations are neutrally stable ($\omega^2 \geq 0$). This is no longer the case, if finite values of β are considered. The graphs of the solutions shown in Fig. 1 always turn upwards at large wavelengths λ. Thus at large enough wavelengths[a] all perturbations of the disk – halo system become unstable and grow exponentially. This behaviour is related to the Jeans collapse of the halo component. Its Jeans length is given by $\lambda_{\rm J} = \sqrt{\pi\sigma_{\rm h}^2/G\rho_{\rm h}}$ or

$$\frac{\lambda_{\rm J}}{\lambda_{\rm crit}} = \frac{1}{\sqrt{2\beta}}\,. \tag{10}$$

[a]The roots of eq. (9) are $\lambda = 0$ and, if $\beta \leq 0.25$, $\lambda/\lambda_{\rm crit} = (1\pm\sqrt{1-4\beta})/2\beta$, respectively.

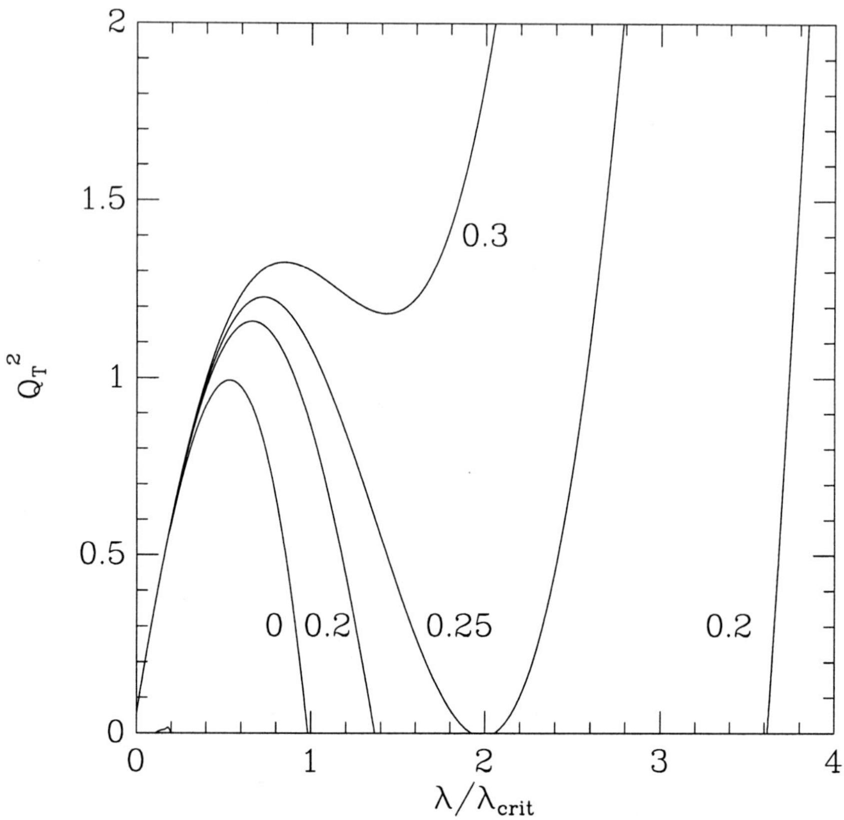

Fig. 1. Separation of stable from unstable perturbations of a self-gravitating disk embedded in a live dark halo. $Q_{\rm T}$ denotes the usual Toomre stability index and λ is the wavelength of the perturbation measured in units of $\lambda_{\rm crit}$. Unstable perturbations are located in the parameter space below the dividing lines. Lines are shown for values of the β-parameter, which describes the dynamical responsiveness of the dark halo, $\beta = 0$, 0.2, 0.25, and 0.3, respectively.

In real haloes the Jeans length will be of the order of the size of the halo or even larger, because otherwise the haloes would have collapsed to smaller sizes. Thus the inferred instability of the disk – halo system on large scales seems not to occur in real galaxies. As can be seen from Fig. 1 stability on small scales can be ensured by $Q_{\rm T}$ indices at thresholds which are slightly larger than in isolated disks.

3. Discussion and Conclusions

As a first application of the stability criterion derived here we test the stability of the Milky Way disk and the surrounding dark halo in the vicinity of the Sun. The local disk and halo parameters listed in Table 1 imply $Q_T = 2.8$ and $\beta = 0.0057$, respectively. If we include in our estimate the cold interstellar gas with a local surface density of $4 M_\odot/\mathrm{pc}^2$ (Dame[5]) and assume a velocity dispersion of the interstellar gas of $\sigma_g = 5$ km/s, which leads to a reduced mass weighted effective velocity dispersion of the combined stellar and gaseous disks, the parameter values change to $Q_T = 2.2$ and $\beta = 0.0078$, respectively. Equation (10) implies that $\lambda_J = 8\,\lambda_{crit} = 39$ kpc. Since the instability sets in nominally at $\lambda = 127\,\lambda_{crit}$ the Milky Way disk and halo system seems to be very stable.

Table 1. Local parameters of the Milky Way

Σ_d	$38\ M_\odot/pc^2$	(Holmberg & Flynn[12])
σ_d	40 km/s	(Jahreiß & Wielen[13])
κ	$\sqrt{2}\cdot 220$ km/s/8.5 kpc	(flat rotation curve)
ρ_h	$0.01\ M_\odot/pc^3$	(Bahcall & Soneira[3])
σ_h	220 km/s/$\sqrt{2}$	(isothermal sphere)
λ_{crit}	4.8 kpc	
λ_J	39 kpc	($\beta = 0.0078$)

In order to explore in what range the β-parameter of spiral galaxies is to be expected in general, we consider the model of a Mestel disk with the surface density $\Sigma_d = \Sigma_0\, R^{-1}$ embedded in a singular isothermal sphere representing the dark halo with the density distribution $\rho_h = \rho_0\, R^{-2}$. The rotation curve of the model galaxy is given by

$$v_c^2(R) = v_d^2(R) + v_h^2(R) \tag{11}$$

with the disk contribution $v_d^2(R) = 2\pi G \Sigma_d R =$ const. and the halo contribution $v_h^2(R) = 4\pi G \rho_h R^2 =$ const. (Binney & Tremaine[4]). Assuming isotropic velocity dispersions of the dark matter particles it follows from the radial Jeans equation,

$$\frac{1}{\rho_h}\frac{d\rho_h \sigma_h^2}{dR} = -\frac{v_c^2}{R}, \tag{12}$$

that their velocity dispersion is given by

$$\sigma_h^2 = \frac{1}{2}(v_d^2 + v_h^2), \tag{13}$$

because the dark matter particles are bound by both the gravitational disk and halo potentials. We find then

$$\beta = \frac{v_{\rm h}^2}{R^2}\frac{1}{v_{\rm d}^2+v_{\rm h}^2}\frac{R^2}{4}\frac{v_{\rm d}^4}{(v_{\rm d}^2+v_{\rm h}^2)^2} = \frac{1}{4}\frac{v_{\rm h}^2 v_{\rm d}^4}{(v_{\rm d}^2+v_{\rm h}^2)^3}\,, \tag{14}$$

which implies the maximal value

$$\beta \leq \beta_{\rm max}(v_{\rm d}^2 = 2v_{\rm h}^2) = 0.037\,. \tag{15}$$

This means that in realistic halo models its density cannot be increased, on one hand, and the velocity dispersion of the halo particles lowered, on the other hand, indiscriminately because the halo has to stay in radial hydrostatic equilibrium. Equation (15) implies $\lambda_{\rm J} = 3.7\,\lambda_{\rm crit}$, whereas the instability sets nominally in at $\lambda = 26\,\lambda_{\rm crit}$. In order to ensure stability at smaller wave lengths the Toomre stability index must be larger than $Q_{\rm T} \geq 1.02$.

As a further concrete example we analyse the dynamics of the low surface brightness galaxy F568-1. De Blok et al.[6] have observed its rotation curve and constructed several concurring mass models for the galaxy, which all fit the rotation curve equally well. Using the parameters of the models with isothermal haloes and with either a 'realistic' M/L-ratio of the disk or a 'maximum-disk' we have solved numerically the radial Jeans equation (12) for the dark matter particles,

$$\sigma_{\rm h}^2(R) = \frac{1}{\rho_{\rm h}(R)}\int_R^\infty dR\,\frac{\rho_{\rm h}(R)v_{\rm c}^2}{R}\,. \tag{16}$$

The resulting β-parameters of both models are shown in Fig. 2. As can be seen from Fig. 2 the β-parameters are of the same small order of magnitude as in the Milky Way or the simple disk – halo model described above.

We conclude from this discussion that embedded galactic disks are not prone to Jeans instabilities, provided their Toomre stability index is a few percent higher than $Q_{\rm T} = 1$. From a practical point of view the destabilizing effect of the surrounding dark halo on the Jeans instability of the embedded galactic disks seems to be negligible.

Acknowledgments

O.E. gratefully acknowledges financial support by the International-Max-Planck-Research-School for Astronomy and Cosmic Physics at the University of Heidelberg.

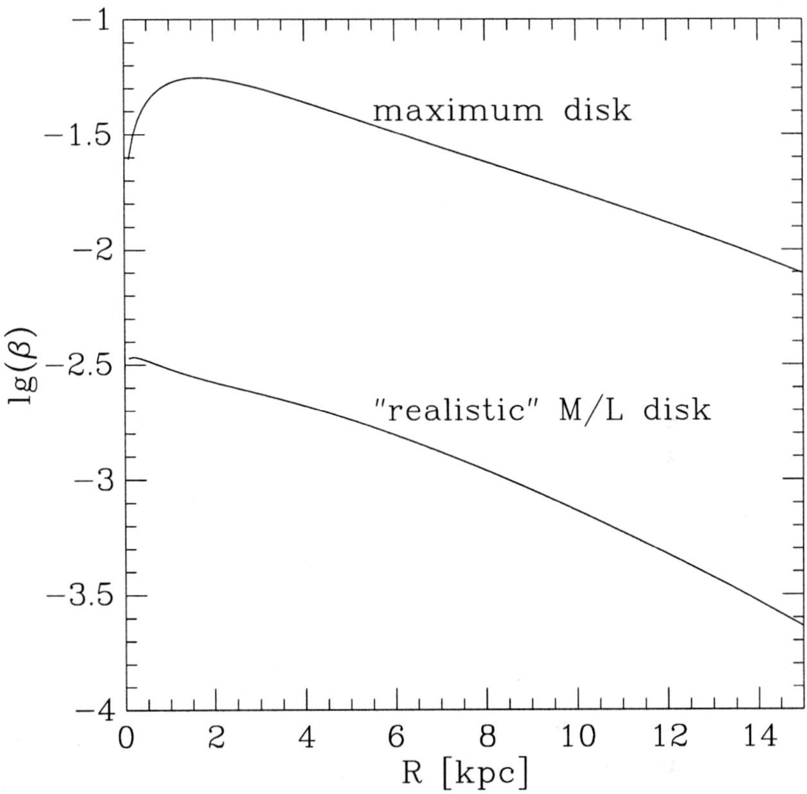

Fig. 2. β-parameters of mass models of the low surface brightness galaxy F568-1.

References

1. Athanassoula E., *ApJ* **569**, L83 (2002).
2. Athanassoula E., *MNRAS* **341**, 1179 (2003).
3. Bahcall J. N. & Soneira R. M., *ApJ* **44**, 73 (1980).
4. Binney J. & Tremaine S., *Galactic Dynamics* (Princeton: Princeton University Press, 1987).
5. Dame T. M., in: S.S. Holt and F. Verter (eds.), *Back to the Galaxy*, AIP Conf. Proc. Vol. 27, (Am. Inst. Phys., New York, 1993) p. 267.
6. de Blok W.J.G., McGaugh S.S. & Rubin V.C., *AJ* **122**, 2396 (2001).
7. Esquivel, O. & Fuchs B., *A&A* **468**, 803 (2007).
8. Fuchs B., *A&A* **368**, 107 (2001).
9. Fuchs B., *A&A* **419**, 941 (2004).
10. Fuchs B. & Athanassoula E., *A&A* **444**, 455 (2005).
11. Goldreich P. & Lynden-Bell D., *MNRAS* **130**, 125 (1965).

12. Holmberg J. & Flynn C., *MNRAS* **352**, 440 (2004).
13. Jahreiß H. & Wielen R., in: B. Battrick, M.A.C. Perryman & P.L. Bernacca (eds.), *HIPPARCOS '97* (ESA SP-402, Noordwijk, 1997) p. 675.
14. Julian W. H.& Toomre A., *ApJ* **146**, 810 (1966).
15. Lin C. C., *Galactic Astronomy*, Vol. 2, p. 1 (1970).
16. Ostriker P.J. & Peebles J. P. E., *ApJ* **186**, 467 (1973).
17. Toomre A., *ApJ* **139**, 1217 (1964).
18. Toomre A., *ARAA* **15**, 437 (1977).

NONEXTENSIVE STATISTICAL THEORY OF DARK MATTER AND PLASMA DENSITY DISTRIBUTIONS IN CLUSTERED STRUCTURES

M. P. LEUBNER

Institute for Astro- and Particle Physics, University of Innsbruck,
Innsbruck, A-6020, Austria
E-mail: manfred.leubner@uibk.ac.at

The classical Boltzmann-Gibbs-Shannon (BGS) extensive thermostatistics applies whenever gravitational and/or electromagnetic interactions are short ranged. However, clustered astrophysical structures are generally subject to nonlocal couplings making their behavior nonextensive. Moreover, lacking physical justification, the characteristics of observed dark matter (DM) and plasma density distributions galaxies and clusters were hitherto exclusively studied by empirical fitting relations. Based on a nonextensive generalization of the standard BGS statistics we review a recently introduced theory where the underlying tandem character of the entropy results fundamentally in a bifurcation of the corresponding density profiles. As natural consequence the radial structure of both, the kinetic DM and thermodynamic plasma distributions, manifest coexisting, are controlled by a single parameter quantifying the degree of correlations in the system. Significant consequences of nonextensive duality are illuminated and the density characteristics arising from long-range interactions with those available from standard BGS statistics are compared. The theoretically derived density profiles reproduce systematically the spatial dependences of DM and hot plasma distributions as observed or generated simulations. It is argued to favor the nonextensive approach, fundamentally derived from entropy generalization and accounting for nonlocal correlations, over phenomenological fitting procedures.

Keywords: Nonextensive statistics; Dark matter; Plasma; Density disrtibutions.

1. Introduction

Non-Gaussian distributions subject to a pronounced core and an extended tail are a persistent feature in a variety of different astrophysical environments, where the standard Boltzmann-Gibbs-Shannon (BGS) statistics does not apply. Those include the thermo-statistical properties of the interplanetary medium,[1–3] the probability distributions of the turbulent fluctuations in astrophysical plasmas,[4] systems subject to self-organized criticality[5,6] or gravitational equilibria of bound structures, we focus on here.

Dark matter (DM) and hot gas density distributions, as observed in relaxed galaxies or clusters or generated in numerical simulations, are usually modeled by simple empirical fitting functions. The phenomenological $\beta-$model[7] provides reasonable representations of the radial plasma density dependence where improvement was introduced via the double β-model to resolve the $\beta-$discrepancy.[8–10] Similarly, the radial density profiles of DM halos are analyzed with the aid of phenomenological fitting functions, thus lacking physical support as well.[11–17]

Physically, we regard the DM halo as an ensemble of self-gravitating, weakly interacting particles in dynamical equilibrium[18,19] and the hot gas component as an electromagnetic interacting high temperature plasma in thermodynamic equilibrium. Hence, astrophysical clusters experience long-range gravitational and/or electromagnetic interactions leading to correlations, such that the standard BGS statistics does not apply.

This situation motivates to introduce nonextensive statistics as physical environment for the analysis of DM and hot plasma density distributions,[20] which is fundamentally based on a generalized power-law entropy measure, accounting for couplings within the system considered. Hence, we retain long-range electromagnetic and gravitational interactions and generalize the concept of BGS thermo-statistics, applicable exclusively whenever a physical system is extensive, i.e. the entropy is additive. Historically, Renyi[21] introduced a generalization of the BGS entropy for statistical equilibrium from first principles, suitably extending the standard additivity to nonextensivity, later revived by Tsallis[22] achiving high popularity[23,24,26–30]).

2. Theory

Let us first illuminate the duality of nonextensive statistics considering as example the pseudo-additive entropy of two sub-systems A and B such that

$$S_\kappa(A+B) = S_\kappa(A) + S_\kappa(B) + \frac{1}{\kappa} S_\kappa(A) S_\kappa(B) \tag{1}$$

where S_κ denotes the entropy as depending on the entropic index κ, a parameter specifying the degree of nonextensivity in the system and accounting for long-range interactions. For $\kappa = \infty$ the last term on the right hand side cancels, leaving the additive entropy of standard BGS statistics. Hence, nonlocality is introduced through the last term on the right hand side of Eq. (1) assuming positive or negative definite values. Nonextensivity generates duality since positive κ-values correspond to less organized states of increased entropy whereas negative κ-values provide a higher organized state of decreased entropy,[4,20] as compared to the BGS state. Hence, κ can be interpreted as a bifurcation parameter measuring the two statistical realizations of ordering or disordering through correlations.

The duality of equilibria in nonextensive statistics is manifest in two families, the nonextensive thermodynamic equilibria and the equilibria of kinetic states, where both are related only via the sign of the coupling parameter κ.[29] Positive κ-values correspond to the stationary states of thermodynamics with finite positive heat capacity and negative κ-values to kinetic stationary states subject to negative heat capacity,[27] a typical property of self-gravitating systems.[18] The limiting BGS state for $\kappa = \infty$ is characterized by self-duality. Consequently we have to assign negative κ values to the DM and positive values to the hot plasma component.

Consistent with Eq. (1) the generalized entropy $S(\kappa)$ characterizing systems subject to long-range interactions reads[3,22]

$$S_\kappa = \kappa k_B \left(\sum p_i^{1-1/\kappa} - 1\right) \tag{2}$$

where p_i is the probability of the ith microstate, and k_B is Boltzmann's constant. The transformation[3] $\kappa = 1/(1-q)$ links the κ-formalism, usually applied in astrophysical plasma modeling, to the Tsallis context, commonly denoted as 'q-statistics'. The κ-context is conveniently symmetric in the interval $-\infty \leq \kappa \leq \infty$ and $\kappa = \infty$ recovers the classical BGS entropy $S_B = -k_B \sum p_i \ln p_i$ of statistical independence.

We mimic in the following gravitationally clustered structures by a spherical symmetric, self-gravitating and collisionless N-body system. Extremizing the generalized entropy (2) under conservation of mass and energy in a gravitational potential Ψ yields the energy distribution as

$$f^{\pm}(E_r) = B^{\pm} \left[1 + \frac{1}{\kappa} \frac{v^2/2 - \Psi}{\sigma^2}\right]^{-\kappa} \tag{3}$$

The superscripts refer to the positive or negative intervals of κ, accounting for less (+) and higher (-) organized states according to increased or decreased entropy,[20] σ represents the mean energy or variance of the distribution and $B^{\pm}$ are κ-dependent normalization constants.[3]

After incorporating the proper sign of κ into Eq. (3) the integrations over all velocities provide the density dependence of a system subject to long range interactions in a gravitational potential as

$$\rho^{\pm} = \rho_0 \left[1 - \frac{1}{\kappa} \frac{\Psi}{\sigma^2}\right]^{3/2-\kappa} \tag{4}$$

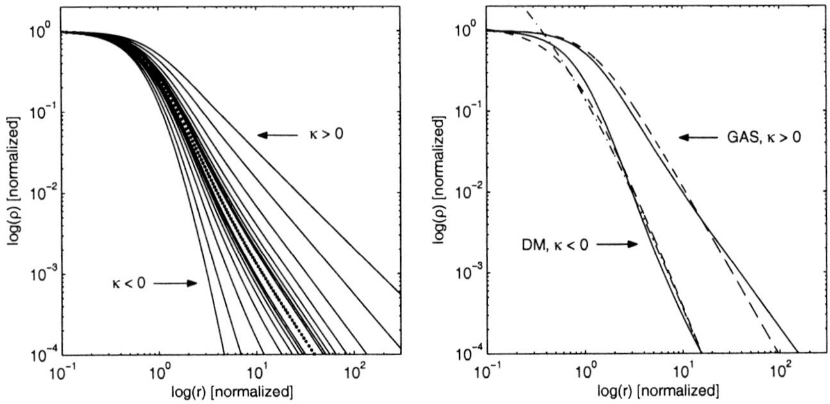

Fig. 1. Left: Normalized nonextensive family of density profiles for $\kappa = 3...10$. For increasing κ both sets of curves converge to the central, limiting isothermal sphere solution ($\kappa = \infty$, dots), ρ^- from inside and ρ^+ from outside. Right: Schematic comparison of a normalized nonextensive DM density profile ($\kappa = -7$, $\sigma = 1$, solid) with best fitting Burkert(dashed) and NFW (dashed-dotted) profiles. A radial nonextensive gas distribution ($\kappa = 7$, $\sigma = 1$) is compared with a single β–model (dashed).

Both, Eqs. (3) and (4) turn in the BGS limit for $\kappa \to \infty$ into the corresponding exponential distributions of the isothermal, self-gravitating sphere of gas, in this case identical to the phase-space density distribution of a system of uncorrelated particles.

Finally, we combine Poisson's equation $\Delta\Psi = -4\pi G\rho_t$ (the minus sign appears as consequence of the relative particle energy[31]) with Eq. (4) to find a second order nonlinear differential equation for the radial density variation of a spherically symmetric gas and DM distribution as

$$\frac{d^2\rho}{dr^2} + \frac{2}{r}\frac{d\rho}{dr} + (\frac{1}{3/2-\kappa} - 1)\frac{1}{\rho}(\frac{d\rho}{dr})^2 - \frac{4\pi G(3/2-\kappa)}{\kappa\sigma^2}\rho_t\rho(\frac{\rho}{\rho_0})^{-1/(3/2-\kappa} = 0 \qquad (5)$$

Here $\rho_t = \rho_g + \rho_{DM}$ denotes the total density of the gas (g) and DM components and we may neglect the gas contribution when dealing with DM density profiles only. The solutions for positive κ represent the thermodynamic state of the gas whereas the negative definite counterpart must be associated to the self-gravitating DM distribution. The limit $\kappa \to \infty$ recovers the equation defining the conventional isothermal sphere.[31]

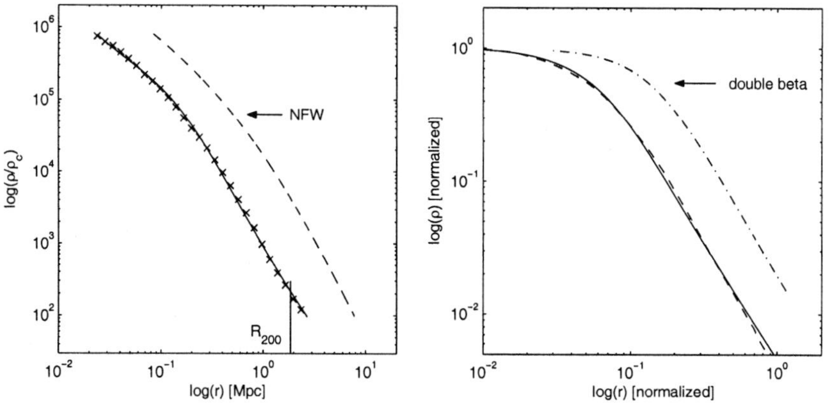

Fig. 2. Left: Normalized radial dark matter density profile obtained from simulations (crosses) and a fit of the nonextensive theory (solid) ($\kappa = -15$, $\sigma = 0.12$. The best fitting NFW profile is shifted to the right for comparison (dashed); R_{200} indicates the virial radius. Right: Normalized radial hot plasma density profile obtained from hydro-simulations (dashed) and a fit of nonextensive theory (solid) ($\kappa = 6.5$, $\sigma = 0.086$). Shifted to the right also the best fitting double β−model is provided (dashed-dotted).

Keeping the generally space dependent velocity dispersion $\sigma(r)$ provides a link between σ, κ and ρ as[32] $\kappa = 2\pi G\rho/(\sigma\nabla^2\sigma + (\nabla\sigma)^2)$. This results in a generalization of Eq. (5) where the derivatives of $\sigma(r)$ enter, a quantity to be known from other sources. Consequently also the spatial variation of the degree of nonextensivity $\kappa(r)$ is provided, a problem requiring further analyses.

3. Discussion and Conclusion

As natural consequence of nonextensive statistics the isothermal sphere profile bifurcates into two distribution families controlled by the sign and value of the coupling parameter κ and governed by the nonlinear differential equation (5). In a unique context, the self-gravitating DM component, a lower entropy state due to gravitational interaction, resides besides the second branch, a thermodynamic gas component of higher entropy.

Computed from Eq. (5) Fig. 1, left panel illuminates schematically the radial density profile characteristics, as depending on the parameter κ for both, DM below and plasma distributions above the limiting BGS solution of statistical independence, where increasing κ values correspond to a decoupling within the system. In Fig. 1, right panel, we compare one negative (DM) solution of Eq. (5) with the singular Navarro-Frenk-White (NFW)[13] and Burkert[11] model as well as one, positive (gas) solution with a single $\beta-$model.[7] The nonextensive solutions resemble not only the physical requirement $d\rho/dr = 0$ in the origin but generates, in particular due to the pronounced tail profiles, excellent representations of simulations and observation. Moreover, nonextensive theory provides naturally a context able to solve the $\beta-$discrepancy[10] since the entire nonextensive profile can be shown to be identical to a double $\beta-$model.[8] Finally, Fig. 2 presents a comparison of the nonextensive theory with results obtained in N-body and hydro-simulations. Fig. 2, left panel, shows a simulated DM profile[34,35] (crosses) as compared to the nonextensive theoretical curve (solid line) for a best fitting negative value of κ along with an empirical NFW-profile for comparison. The hot gas density distribution, generated in the corresponding hydrodynamic simulations,[36] is compared to the nonextensive solution to Eq. (5) for positive definte κ in the right panel of Fig. 2. Highly significant both plots demonstrate that the fundamental context of entropy generalization, accounting for long-range interactions, provides a physical and interpretable approach to the problem of density distributions in clustered large-scale structures, further confirmed by observations as well.[33,35]

The context of nonextensive statistics generates a bifurcation of the density distribution of the standard isothermal sphere solution, into two distinct power-law profiles. The self-gravitating DM component as lower entropy state with respect to the BGS context resides besides the thermodynamic plasma component of higher entropy. The transition is controlled only by the sign of the nonextensive parameter κ, whereas the density profile shape is determined by both, the values of κ and σ, representing the velocity dispersion. The dual nature of nonextensive theory provides a unique solution to the problem of DM and gas density distributions of clustered matter from fundamental physics, where both parameters admit physical interpretation. It is proposed to favor the physical family of nonextensive distributions over empirical models when fitting observed or simulated density profiles of astrophysical structures.

Acknowledgments

The author thanks E. van Kampen, T. Kronberger and M. Mair for their assistance. This work was supported by the Austrian Wissenschaftsfonds, grant number P20131-N16.

References

1. M. P. Leubner and N. Schupfer, *J. Geophys. Res.* **106**, 12993 (2001).
2. M. P. Leubner, *Phys. Plasmas* **11**, 1308 (2004).
3. M. P. Leubner, *Astrophys. J.* **404**, 469 (2004).
4. M. P. Leubner and Z. Voros, *Astrophys. J.* **618**, 547 (2005).
5. P. Bak, C. Tang and K. Wiesenfeld, *Phys. Rev. A* **38**, 364 (1988).
6. P. Bak., *How nature works*, New York, Copernicus (1996).
7. A. Cavaliere and R. Fusco-Femiano, *Astron. Astrophys.* **49**, 137 (1976).
8. Y-J. Xue and X-P. Wu, *MNRAS* **318**, 715 (2000).
9. N. Ota and K. Mitsuda, *Astron. Astrophys.* **428**, 757 (2004).
10. N. A. Bahcall and L. M. Lubin, *Astrophys. J.* **426**, 513 (1994).
11. A. Burkert, *Astrophys. J.* **447**, L25 (1995).
12. H. Zhao, *MNRAS* **278**, 488 (1996).
13. J. F. Navarro, C. S. Frenk and S. D. M. White, *Astrophys. J.* **490**, 493 (1997).
14. T. Fukushige and J. Makino, *Astrophys. J.* **477**, L9 (1997).
15. B. Moore, T. Quinn, F. Governato, J. Stadel and G. Lake, *MNRAS* **310**, 1147 (1999).
16. P. Salucci and A. Burkert, *Astrophys. J.* **537**, L9 (2000).
17. M. Ricotti, *MNRAS* **344**, 1237 (2003).
18. C. Firmani, E. D. Onghia, V. Avila-Reese, G. Chincarini and X. Hernandez, *MNRAS* **315**, L29 (2000).
19. D. N. Spergel and P. J. Steinhard, *Phys. Rev. Lett.* **84**, 3760 (2000).
20. M. P. Leubner, *Astrophys. J.* **632**, L1 (2005).
21. A. Renyi, *Acta Math. Hungaria* **6**, 285 (1955).
22. C. Tsallis, *J. Stat. Phys.* **52**, 479 (1988).
23. A. R. Plastino, A. Plastino and C. Tsallis, *J. Phys. A: Math. Gen.* **27**, 5707 (1994).
24. C. Tsallis, *Physica A* **221**, 277 (1995).
25. G. Kaniadakis, A. Lavagno and P. Quarati, *Phys. Lett. B* **369**, 308 (1996).
26. R. Silva, A. R. Plastino and J. A. S. Lima, *Phys. Lett. A* **249**, 401 (1998).
27. M. P. Almeida, *Physica A* **300**, 424 (2001).
28. J. S. Andrade, M. P. Almeida, A. A. Moreira and G. A. Farias, *Phys. Rev. E* **65**, 036121 (2002).
29. I. V. Karlin, M. Grmela and A. N. Gorban, *Phys. Rev. E* **65**, 036128 (2002).
30. Kaniadakis G., *Phys. Rev. E* **72**, 036108 (2005).
31. J. Binney and S. Tremaine, *Galactic Dynamics*, Princeton Univ. Press., Princeton, 1994.
32. J. Du, *Astrophys. Space Sci.* **312**, 47 (2007).
33. E. Pointecouteau, M. Arnaud and G. W. Pratt, *Astron. Astrophys.* **435**, 1 (2005).
34. E. van Kampen, R. Jimenez and J. A. Peacock, *MNRAS*, **310**, 43 (1999).
35. T. Kronberger, M. P. Leubner and E. van Kampen, *Astron. Astrophys.*, **453**, 21 (2006).
36. S. Schindler, W. Kapferer, W. Domainko, E. van Kampen, T. Kronberger, S. Kimeswenger, M. Ruffert, O. Mangete and D.Breitschwerdt, *Astron. Astrophys.* **435**, L25 (2005).

DARK MATTER HALO PROFILES IN SCALE-FREE COSMOLOGIES

S. R. KNOLLMANN

Astrophysikalisches Institut Potsdam,
An der Sternwarte 16, D-14482 Potsdam, Germany
**E-mail: sknollmann@aip.de*

C. POWER

Centre for Astrophycis & Supercomputing,
Swinburne University of Technology
PO Box 218, Hawthorn, Victoria 3122, Australia

A. KNEBE

Astrophysikalisches Institut Potsdam,
An der Sternwarte 16, D-14482 Potsdam, Germany

We performed a set of cosmological N-body simulations of scale-free models (i.e. $P(k) \propto k^n$) and investigated the central logarithmic slope of relaxed dark matter halo density profiles α and its dependence on the spectral index n of the linear matter power spectrum $P(k)$. The simulations are started and stopped according to a set of clear, reproducible and physically motivated criteria, allowing to compare haloes across models with different spectral indices. In each simulation we identify samples of well resolved haloes in dynamical equilibrium and we analyse their mass profiles. Using a "generalised" Navarro, Frenk & White profile in which the central logarithmic slope α is allowed to vary while the r^{-3} asymptotic form at large radii is preserved, we obtain best-fit central slopes for haloes in the different models. We find a strong correlation between α and n, such that α becomes shallower as n becomes steeper. When normalising the mass profiles by r_{-2}, the radius at which the logarithmic slope of the density profile is -2, we find that the differences between the models are no longer present. This becomes clearly visible when plotting the maximum slope $\gamma = 3(1 - \rho/\bar{\rho})$ as a function of r/r_{-2}: We observe a similar behaviour for haloes forming in different n models. This traces back to the concentration of the haloes, in steep-n cosmologies we find smaller concentration than in shallow-n cosmologies. Also, we do not find evidence for a convergence to a unique central asymptotic slope on the scale we can resolve.

Keywords: methods: N-body simulations – cosmology: theory – galaxies: haloes – dark matter.

1. Introduction

The currently favoured model of cosmological structure formation asserts that the Universe is dominated by some form of non-baryonic Cold Dark Matter (hereafter CDM), and it makes a number of generic predictions about how the dark matter clusters on small scales. In the seminal study of Navarro, Frenk & White[1] (hereafter NFW) the concept of a "universal" mass density profile for dark matter haloes has been introduced. This NFW profile in its extended form is written as

$$\rho(r) = \frac{\rho_c}{(r/r_s)^{\alpha}\,(1 + r/r_s)^{3-\alpha}}\,, \tag{1}$$

where r_s is a scale radius, ρ_c is a characteristic density and α is the central asymptotic logarithmic slope[a]. The NFW profile is "universal" in the sense that it describes the ensemble averaged mass profile of dark matter haloes in dynamical equilibrium, independent of virial mass, cosmological parameters and initial power spectrum.

Numerous studies have been undertaken to verify whether or not the NFW profile provides an adequate description of the mass profile of dark matter haloes. In the original work by NFW,[1] their halo sample was comprised of objects which contained of the order of 10^4 particles and the profiles are hence resolved down to 10% of the virial radius.[2] Further studies investigated haloes containing about two orders of magnitude more particles and confirmed the finding of NFW that CDM haloes are cuspy.[3–8] The exact value of α has been a matter of debate, but is expected to be in the range of $\alpha \sim 1$ to 1.5.[1,9] The highest resolved simulation to-date, which contained 130 million particles in a cluster sized dark matter halo, supports the argument for a central cusp, finding[10] a logarithmic inner slope of about $\alpha \sim 1.2$.

It is an interesting question, where this central slope originates from. May the halo structure depend on the primordial power spectrum $P(k)$, or in particular, the spectral index n of $P(k)$? One of the key predictions of the NFW papers is that the shape of the "universal" profile should hold in any hierarchical cosmology, including scale-free ones. Subsequent simulations confirmed this claim,[11] whereas a number of analytical studies[12–16] have claimed that the density slope should depend on n and are backed up but other numerical simulations.[17–22]

Especially interesting are the findings of Ricotti,[19] claiming that the central logarithmic slope depends explicitly on the effective spectral index

[a]A choice of $\alpha = 1$ corresponds to the original NFW profile.

n. It should be noted that a subsequent, comparable, work[23] could not confirm this. However, this topic has recently been revisited,[22] and the authors continue to argue in favour of the dependence of α on n. We are trying[24] to explicitly check this claims by performing a set of scale-free simulations, which we consider a "clean" approach as opposed to the previous work[19,22] where differently sized boxes have been used in a ΛCDM model to select different parts of the power spectrum $P(k)$, thereby having a different effective slope n for each box.

2. Simulations

The power spectrum in scale-free cosmologies follows a power law of the form

$$P\left(k\right) = a^2 A P_{\mathrm{WN}} \left(\frac{k}{k_{\mathrm{Ny}}}\right)^n , \tag{2}$$

where A is the amplitude scaling factor and

$$P_{\mathrm{WN}} = \left(\frac{L}{N}\right)^3 \quad \text{and} \quad k_{\mathrm{Ny}} = \frac{N\pi}{L}$$

denote the power of a random distribution of N^3 particles put in a box with size L and the Nyquist frequency, respectively. It should be noted that the simple scaling with the expansion factor a is only valid for an Einstein–de Sitter universe, which is the cosmology of our choice. We use $a = 1$ at the beginning of each simulation and then by specifying the amplitude scaling factor A the initial particle distribution can be generated.

In this work, we used a set of simulations containing 512^3 particles each and varied the spectral index n, using n=-0.50, -1.50, -2.25, -2.50 and -2.75.

2.1. *Starting and Stopping*

In our work[24] we give a detailed description of how to choose suitable starting values and when to stop the simulations to be able to cross-compare objects. The method we employ is three-fold. Firstly, we require all simulations to start with the same integrated power in the box (this gives us the only needed parameter A for setting up the simulation). Then we note, as has been argued[25] previously, that there exists a maximal expansion until which we can evolve our simulations. Furthermore we employ the Press&Schechter[26] theory to predict the mass growth within the simulation and stop once the same characteristic mass M_* has been reached in all simulations; this criterion has been used previously.[1]

2.2. *Performed Simulations and Halo Selection*

To perform the simulations, we used the parallel N-body code Gadget2[27] with 512^3 particles. We simulated five different models with varying spectral index n, the parameters of the simulations are given in columns 2–4 of table 1.

Table 1. Used parameters and inferred halos from the simulations.

model	n	A	$a_{\rm final}$	$N_{\rm halo}$	$N_{\rm vir}$	$N_{\rm vir,100}$
512-0.50	−0.50	0.0358	826.0	1928	1295	43
512-1.50	−1.50	0.0215	161.5	1229	761	55
512-2.25	−2.25	0.0109	43.9	704	384	34
512-2.50	−2.50	0.0076	27.3	479	232	20
512-2.75	−2.75	0.0046	(16.7)	350	113	17

The final output of each simulation has been analyzed with an MPI version of the AMIGA Halo Finder[b] (AHF, successor of MHF[28]) for identifying haloes and computing their integral and radial properties. Haloes are defined such that the virial mass is $M_{\rm vir} = 4\pi\rho_{\rm crit}\Delta_{\rm vir}r_{\rm vir}^3/3$, where the overdensity criterion $\Delta_{\rm vir} = 178$ for an Einstein-de Sitter cosmology.

The halo catalogues have been post-processed to only include objects which are resolved with more than 10^4 particles within the virial radius and are considered to be relaxed by two distinct proxies. We employed the virial ratio $-2T/U$, taking into account the mass-dependence of this quantity by performing a linear regression f on the binned logarithmic virial ratio, and the displacement of the centre of mass of all material within the virial radius with respect to the centre of potential of the halo, normalised to the virial radius of the halo. We reject all haloes for which $\Delta_r > 0.07$[29] and $f(N_{\rm vir}) - 0.05 < \log(-2T/U) < f(N_{\rm vir}) + 0.05$. We furthermore define a subset of those haloes fulfilling our selection criteria which contains only those objects which are resolved with 10^5 particles within the virial radius. The according numbers of objects which were then analyzed are given in columns 5–7 of table 1.

[b]AMIGA is freely available from download at `http://www.aip.de/People/AKnebe/AMIGA/`

3. Analysis

3.1. *Inner Profile Slope*

To each of the objects in our sample we fitted a set of analytic profiles of the form

$$\rho\left(r\right) = \frac{\rho_c}{\left(\frac{r}{r_s}\right)^{\alpha}\left(1+\frac{r}{r_s}\right)^{3-\alpha}} \tag{3}$$

were r_s and ρ_c are the fitted parameters and between the fits we varied the inner slope α in the range of $\alpha = 0.3, \ldots, 1.8$, obtaining a best-fit for each choice of inner slope. We then compute for each halo and each α the quality of the fit using the following formula:

$$Q_j = \frac{\sum_{\text{bins}}\left(\frac{X_{ij}-\rho_{\text{fit}}(r_i)}{X_{ij}}\right)^2}{\text{dof}}; \tag{4}$$

which is essential the relative deviation of the fit from the measured profile per degree of freedom. By then taking the average of this quality in each simulation and for each choice of α we can readily produce the plot shown in figure 1. Each line corresponds to one model and shows how the median quality of the fit changes depending on the choice of α. It can clearly be

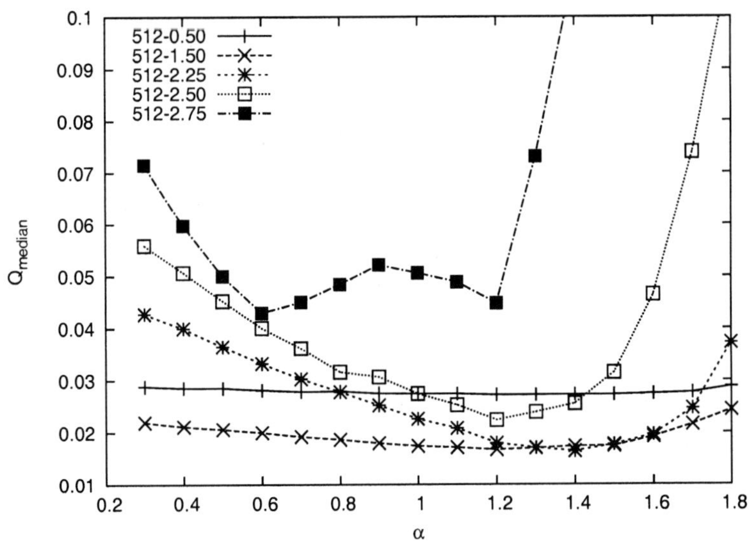

Fig. 1. The median of the quality in the different simulation boxes.

seen, that the different curve have their minimun at different values for α, we find $\alpha \simeq 1.6(1.4, 1.4, 1.2, 1.0)$ for the $n = -0.5(-1.5, -2.25, -2.50, -2.75)$ runs. Hence we find that the shallower the spectral index n, the steeper the preferred central slope α.

3.2. *The Maximum Slope*

It is now interesting to consider another quantity which does not rely on a given analytic profile and can be deduced in a non-parametric way. The maximum asymptotic slope,

$$\gamma(r) = 3\left(1 - \frac{\rho(r)}{\bar{\rho}(r)}\right) \tag{5}$$

uses the local mass density ρ and the enclosed mass density $\bar{\rho}(r)$ to derive an upper limit to the logarithmic slope of the density profile at radius r. It is assumed here, that the density is spherical symmetric and given interior to r by a power-law of the form $\rho(r) \propto r^{-\gamma}$.

Additionally, we choose to normalize the radius r to a certain scale radius r_{-2} which is defined as being the radius at which the density profile has a logarithmic slope of -2. In the case of a NFW-profile, this corresponds indeed to the scale-radius r_s of the analytic profile. In figure 2 the maximum

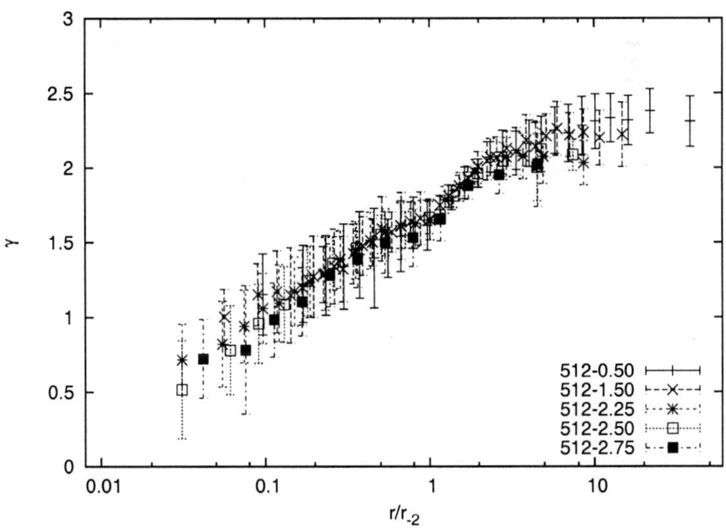

Fig. 2. The maximum slope γ is shown for all haloes containing more than 10^5 particles. The symbols correspond to the average slope in each simulation.

slope is plotted in these units. We can see, that by using this normalization the *average* behaviour is the same for all different cosmologies we simulated.

3.3. *The Concentration*

Having defined a scale-radius r_{-2}, we can readily calculate the concentration c for each halo, which is given by

$$c = \frac{r_{\rm vir}}{r_{-2}}. \tag{6}$$

We can see clearly from figure 3 that this quantity depends strongly on the spectral index n of the cosmology under investigation: the steeper the spectral index, the lower the average concentration of a halo in this cosmology.

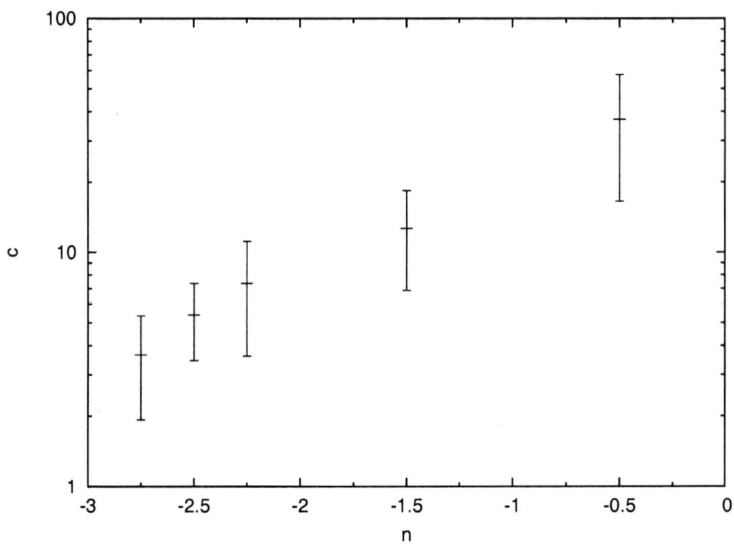

Fig. 3. Here we show the average concentration in each simulation calculated from all halos with more than 42000 particles.

4. Conclusions

We have shown that the inferred inner profile slope α of dark matter haloes, as given by the extended NFW profile, indeed shows a dependence on the spectral slope n of the power spectrum of initial density perturbations, namely for $n_1 > n_2$ we find $\alpha_1 > \alpha_2$. This would argue in favour of a non-universality of the profile shape. However, when suitably normalizing the profiles, we have shown that the maximum slope γ, which is inferred in a non-parametric way (without imposing a specific profile shape) behaves similar in all cosmological models.

We would like to point out, that the dependence of α on n can readily be explained by a varying concentration $c = r_{\rm vir}/r_{-2}$: The higher the concentration, the lesser resolved the profile part within r_{-2} becomes. That in turn means we will to be able to resolve a flattening of the profile which then strongly favours a steeping α in a χ^2-fit of an extended NFW profile. We have shown this dependence of c on n explicitly in figure 3. So in fact, the physical quantity which needs to be understood, is the concentration and the notation of a central asymptotic density profile slope α is a misleading one, as we do not find evidence for a convergence to a unique value and hence analytic profiles which assume a certain inner slope are always strongly depending on the resolution of the profile *within* r_{-2}.

References

1. J. F. Navarro, C. S. Frenk and S. D. M. White, *ApJ* **490**, 493(December 1997).
2. C. Power, J. F. Navarro, A. Jenkins, C. S. Frenk, S. D. M. White, V. Springel, J. Stadel and T. Quinn, *MNRAS* **338**, 14(January 2003).
3. T. Fukushige and J. Makino, *ApJ* **477**, L9+(March 1997).
4. B. Moore, F. Governato, T. Quinn, J. Stadel and G. Lake, *ApJ* **499**, L5+(May 1998).
5. Y. P. Jing and Y. Suto, *ApJ* **529**, L69(February 2000).
6. Y. P. Jing, *ApJ* **535**, 30(May 2000).
7. T. Fukushige and J. Makino, *ApJ* **557**, 533(August 2001).
8. Y. P. Jing and Y. Suto, *ApJ* **574**, 538(August 2002).
9. B. Moore, T. Quinn, F. Governato, J. Stadel and G. Lake, *MNRAS* **310**, 1147(December 1999).
10. J. Diemand, M. Zemp, B. Moore, J. Stadel and M. Carollo, *MNRAS* **364**, 665(December 2005).
11. S. Cole and C. Lacey, *MNRAS* **281**, 716(July 1996).
12. Y. Hoffman and J. Shaham, *ApJ* **297**, 16(Octtober 1985).
13. D. Syer and S. D. M. White, *MNRAS* **293**, 337(February 1998).
14. K. Subramanian, R. Cen and J. P. Ostriker, *ApJ* **538**, 528(August 2000).
15. K. M. Huffenberger and U. Seljak, *MNRAS* **340**, 1199(April 2003).
16. E. Salvador-Solé, A. Manrique, G. González-Casado and S. H. Hansen, *ApJ* **666**, 181(September 2007).
17. M. M. Crone, A. E. Evrard and D. O. Richstone, *ApJ* **434**, 402(Octtober 1994).
18. V. R. Eke, J. F. Navarro and M. Steinmetz, *ApJ* **554**, 114(June 2001).
19. M. Ricotti, *MNRAS* **344**, 1237(Octtober 2003).

20. R. Cen, F. Dong, P. Bode and J. P. Ostriker, *ArXiv Astrophysics e-prints* (March 2004).
21. D. Reed, F. Governato, L. Verde, J. Gardner, T. Quinn, J. Stadel, D. Merritt and G. Lake, *MNRAS* **357**, 82(February 2005).
22. M. Ricotti, A. Pontzen and M. Viel, *ApJ* **663**, L53(July 2007).
23. P. Colín, A. Klypin, O. Valenzuela and S. Gottlöber, *ApJ* **612**, 50(September 2004).
24. S. R. Knollmann, C. Power and A. Knebe, *MNRAS submitted* (2007).
25. G. Efstathiou, C. S. Frenk, S. D. M. White and M. Davis, *MNRAS* **235**, 715(December 1988).
26. W. H. Press and P. Schechter, *ApJ* **187**, 425(February 1974).
27. V. Springel, *MNRAS* **364**, 1105(December 2005).
28. S. P. D. Gill, A. Knebe and B. K. Gibson, *MNRAS* **351**, 399(June 2004).
29. A. F. Neto, L. Gao, P. Bett, S. Cole, J. F. Navarro, C. S. Frenk, S. D. M. White, V. Springel and A. Jenkins, *ArXiv e-prints* **706**(June 2007).

EFFECTS OF PRIMORDIAL MAGNETIC FIELD ON CMB AND MATTER POWER SPECTRUM

D. G. YAMAZAKI[a*], K. ICHIKI[b], T. KAJINO[a,c] and G. J. MATHEWS[d]

[a]*National Astronomical Observatory of Japan, Mitaka, Tokyo 181-8588, Japan*
** E-mail: yamazaki@th.nao.ac.jp*

[b]*Research Center for the Early Universe, School of Science, the University of Tokyo, 7-3-1 Hongo, Bunkyo-ku 113-0033, Japan*

[c]*Department of Astronomy, Graduate School of Science, University of Tokyo 7-3-1 Hongo, Bunkyo-ku, Tokyo 113-0033, Japan*

[d]*Center for Astrophysics, Department of Physics, University of Notre Dame, Notre Dame, IN 46556, U.S.A.*

The existence of a primordial magnetic field (PMF) would affect both the temperature and polarization anisotropies of the cosmic microwave background (CMB) and the formation of the large scale structure(LSS). It also provides a plausible explanation for the disparity between observations and theoretical fits to the CMB power spectrum and the LSS. Here we report on calculations of not only the numerical power spectrum of the PMF, but also the correlations between the PMF power spectrum and the primary curvature perturbations.

Keywords: Primordial Magnetic Field, CMB, Large Scale Structure.

1. Introduction

Magnetic fields in clusters of galaxies have been observed with a strength of $0.1 - 1.0\ \mu$ G. The existence of a primordial magnetic field (PMF) of order 1 nG whose field lines collapse as structure forms is one possible explanation for such magnetic fields to galactic clusters. The origin and detection of the PMF is, hence, a subject of considerable interest in modern cosmology. Moreover, a PMF could influence a variety of phenomena in the early universe[1] such as the cosmic microwave background (CMB),[2–5] and the matter power spectrum.[6] In the present work, in order to clarify the role of the PMF in CMB precisely, we must take into consideration the correlation between the CMB fluctuations induced by the PMF and those due to the primordial curvature and tensor perturbations. Also, we numerically evaluate the stochastic PMF power spectrum of the stochastic PMF and therefore avoid recourse to analytic approximations. We use the isocurvature initial condition for estimations of the CMB with the PMF effect in this paper. Throughout this letter we fix the best fit WMAP 3rd cosmological parameters of the ΛCDM model.[7] We use natural units $c = \hbar = 1$.

2. Primordial Magnetic Field

Since the trajectories of plasma particles are bent by Lorentz forces in a magnetic field, photons are indirectly influenced by the magnetic field through Thomson scattering. The energy density of the magnetic field can be treated as a first order perturbation upon a flat Friedmann-Robertson-Walker (FRW) background metric. In the linear approximation, the magnetic field evolves as a stiff source. Therefore, we can discard all back reactions from the magnetohydrodynamic (MHD) fluid onto the field itself. The conductivity of the primordial plasma is very large, so that the magnetic field is gfrozen-inh.[8] Furthermore, we can neglect the electric field, i.e. $E \sim 0$, and can decouple the time evolution of the magnetic field from its spatial dependence, i.e. $\mathbf{B}(\tau, \mathbf{x}) = \mathbf{B}(\mathbf{x})/a^2$ for very large scales , where a is the scale factor. We assume that the PMF is statistically homogeneous, isotropic and random. For such a magnetic field, the power spectrum can be taken as a power-law $S(k) =< B(k)B^*(k) >\propto k^{n_B}$[8] where n_B is the power-law spectral index of the PMF. The index n_B can be either negative or positive depending upon the physical processes of the field creation. Although possible origins of a PMF have been studied by many authors, there is no consensus how the PMF correlates with the primordial density fluctuations. In order to study the PMF effect in a more general manner, we here consider possible correlations between the PMF and the primordial density and tensor fluctuations. Therefore, we introduce a coefficient "s" to parameterize the correlation between the PMF and the primary power spectra,[6,9]

$$(2l+1)^2 C_l^{(\mathrm{X})} = \frac{4}{\pi}\int dkk^2[\Theta_{\mathrm{p}}^{(\mathrm{X})}(k)\Theta_{\mathrm{p}}^{(\mathrm{X})*}(k) + \Theta_{\mathrm{PMF}}^{(\mathrm{X})}(k)\Theta_{\mathrm{PMF}}^{(\mathrm{X})*}(k)$$
$$+s^{(\mathrm{X})}\{\Theta_{\mathrm{p}}^{(\mathrm{X})}(k)\Theta_{\mathrm{PMF}}^{(\mathrm{X})*}(k) + \Theta_{\mathrm{PMF}}^{(\mathrm{X})}(k)\Theta_{\mathrm{p}}^{(\mathrm{X})*}(k)\}], \quad (1)$$

where Θ is a transfer function, and X = S or T for scalar or tensor mode. The last term is the correlation between the primary and the PMF power spectra. Among many possible sets of two correlation coefficients that satisfy $-1 \leq s^{(X)} \geq 1$, the two cases of $s^{(S)} = s^{(T)} = \pm 1$ are expected to show the maximum absolute effects of the PMF, which represent the effective range of the resultant CMB anisotropies with the PMF. In this paper we have implemented a numerical method[6,10,11] to evaluate the PMF power spectrum.

3. Results and Discussions

We have explored effects of the PMF on the CMB and the matter power spectrum for the allowed PMF parameters ($B_\lambda < 10\ n$G and $n_B < -2.4$) deduced in our previous work.[4] In Fig. 1 we depict the CMB temperature and polarization anisotropies in the presence of the PMF taking into account the correlations. Since we obtain the temperature and polarization anisotropies of the CMB with the PMF from the isocurvature initial condition, the phase of CMB perturbation with the PMF is different $\pi/2$

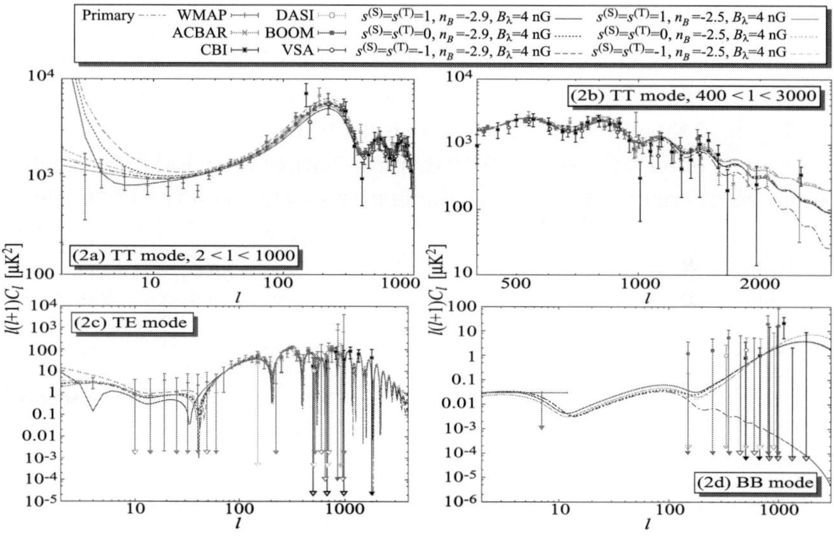

Fig. 1. CMB temperature and polarization anisotropies from the PMF. Panels (2a), (2b), (2c), and (2d) show TT($2 < \ell < 1000$), TT($400 < \ell < 3000$), TE($2 < \ell < 4000$), and BB($2 < \ell < 3000$) modes, respectively. Curves in all Panels are theoretical lines as indicated in the legend box on the figure. Lines in (2c) are plotted in the absolute value. Downward arrows for the error bars of Panels (2c) and (2d) indicate that the data points are positive and the lower error negative.

from those without the PMF (by the adiabatic initial condition) and odd peaks of the scalar mode of the CMB perturbations raise up for the positive correlation between the PMF and the energy density perturbations, or suppress for the negative one. These are compared with the observed power spectra (WMAP,[12,13] ACBAR,[14] CBI,[15,16] DASI,[17] BOOMERANG,[18] and VSA[19]). we see that the power spectral index of the PMF, n_B, is more effectively constrained from CMB observations for lower ℓ than those for higher ℓ. Furthermore,there is no discrepancy between observations and theories of the CMB polarization for models with the PMF for higher ℓ [(2c) and (2d) of Fig.1]. This shows that the PMF is one of the most effective candidates for explaining the CMB polarizations for higher ℓ. The power of CMB emission depend on frequency, and such difference expect to be due to foreground effects (e.g. the Sunyaev-Zel'dovich effect) , on the other hands, effects of the PMF is frequency-independent since the PMF affect the primary CMB as background. Therefore the degeneracy between the PMF and others foreground effects for higher ℓ is weaker, and we can simultaneously constrain the PMF and foreground effects individually by more than two different frequencies observation data set.

Figure. 2 shows that the models with the PMF for higher n_B give better fits to the observations(2dF[20]) than those without the PMF for smaller scale. Furthermore, we find that the matter spectra with the PMF effects

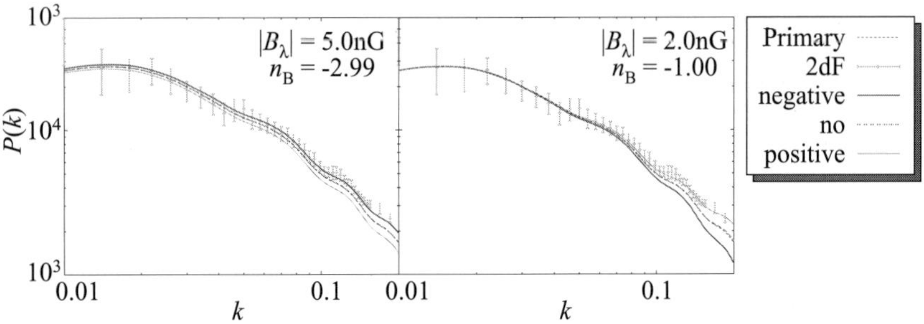

Fig. 2. Matter power spectra from the PMF with 2dF.[20] Curves in all Panels are theoretical Curves as indicated in the legend box on the figure. Theoretical Curves of the primary is computed from WMAP 3rd[12,13] cosmological parameters for ΛCDM model. The PMF pressure dominates the matter spectrum for $n_B < -1.5$ while the PMF tension dominates the matter spectrum for $n_B > -1.5$.[6] Therefore the PMF effect on the matter spectrum for the negative or positive correlation at $n_B = -2.99$ is difference from it at $n_B = -1.0$.

are sensitive to correlation factor $s^{(S)}$ and n_B for smaller scale. Thus we find that the strong degeneracy of PMF parameters is broken by the different effects of the PMF on the CMB power spectrum for lower and higher ℓ, and the matter power spectrum for smaller scale. We can also constrain more precisely PMF parameters from all modes of CMB temperature and polarization anisotropies, and the matter power spectrum.

4. Summary

The scalar mode from the PMF can be a main source for lower ℓ, while the vector mode can dominate for higher ℓ in the CMB temperature and polarization anisotropies. The matter spectra with the PMF effects are also sensitive to correlation factor $s^{(\mathrm{S})}$ and n_{B} for smaller scale. Thus the strong degeneracy of the PMF parameters,[3,4] which are the power law index n_{B} and amplitude B_λ of the PMF, and the correlation factors $s^{(\mathrm{X})}$, is broken by the different effects of the PMF on the CMB power spectrum for lower and higher ℓ and the matter power spectrum for smaller scale. If we combine our study, the surveys of the LSS, and future plans to observe the CMB anisotropies and polarizations for higher multipoles l, e.g. via the *Planck Surveyor*, we will be able to constrain the PMF more accurately, and explain the evolution and generation of the magnetic field on galaxy cluster scales along with the formation of the LSS.

References

1. D. Grasso, and H. R. Rubinstein, *Phys. Rept.* **348**, 163–266 (2001).
2. D. G. Yamazaki, et al., *Nuclear Physics A* **758**, 791–794 (2005a).
3. D. G. Yamazaki, et al., *Astrophys. J.* **625**, L1–L4 (2005b).
4. D. G. Yamazaki, et al., *Astrophys. J.* **646**, 719 (2006a).
5. D. G. Yamazaki, et al., *PoS(NIC-IX).* p. 194 (2006b).
6. D. G. Yamazaki, et al., *Phys. Rev.* **D 74**, 123518 (2006c).
7. D. N. Spergel, et al., (2006), `astro-ph/0603449`.
8. A. Mack, et al., *Phys. Rev.* **D 65**, 123004 (2002).
9. M. Giovannini, *Class. Quant. Grav.* **23**, 4991–5026 (2006).
10. D. G. Yamazaki, et al., submitted *Phys. Rev.* **D**
11. D. G. Yamazaki, submitted *Phys. Rev.* **D**
12. G. Hinshaw, et al., (2006), `astro-ph/0603451`.
13. L. Page, et al., (2006), `astro-ph/0603450`.
14. C.-L. Kuo, et al., (2006), `astro-ph/0611198`.
15. A. C. S. Readhead, et al., *Astrophys. J.* **609**, 498–512 (2004).
16. J. L. Sievers, et al., (2005), `astro-ph/0509203`.
17. E. M. Leitch, et al., *Astrophys. J.* **624**, 10–20 (2005).
18. W. C. Jones, et al., *Astrophys. J.* **647**, 823 (2006).
19. C. Dickinson, et al., *Mon. Not. Roy. Astron. Soc.* **353**, 732 (2004).
20. S. Cole, et al. [The 2dFGRS Collaboration], Mon. Not. Roy. Astron. Soc. **362**, 505 (2005)

REISSNER–NORDSTRÖM EXPANSION

EMIL M. PRODANOV*, ROSSEN I. IVANOV†
School of Mathematical Sciences, Dublin Institute of Technology,
Kevin Street, Dublin 8, Ireland
*E-mails: *Emil.Prodanov@dit.ie; †Rossen.Ivanov@dit.ie*

and V.G. GUEORGUIEV
University of California, Merced, USA
E-mail: vesselin@mailaps.org

We propose a classical mechanism for the cosmic expansion during the radiation-dominated era, assuming the Universe as a two-component gas. The first component is the ultra-relativistic "standard" fraction described by an equation of state of an ideal quantum gas of massless particles. The second component consist of superheavy charged particles and their interaction with the "standard" fraction drives the expansion. This interaction is described by the Reissner–Nordström metric purely geometrically — the superheavy charged particles are modeled as zero-dimensional naked singularities which exhibit gravitational repulsion. The radius of a repulsive sphere, surrounding a naked singularity of charge Q, is inversely proportional to the energy of an incoming particle or the temperature. The expansion mechanism is based on the "growing" of the repulsive spheres of the superheavy particles with the drop of the temperature — this drives apart all neutral particles and particles of specific charge q/m such that $\text{sign}(Q)q/m \geq -1$. The Reissner–Nordström expansion mechanism naturally ends at Recombination. We model the Universe during the Reissner–Nordström expansion as a van der Waals gas and determine the equation of state.

In 1971, Hawking[1] suggested that there may be a very large number of gravitationally collapsed charged objects of very low masses, formed as a result of fluctuations in the early Universe. A mass of 10^{14} kg of these objects could be accumulated at the centre of a star like the Sun. Hawking treats these objects classically and his arguments for doing so are as follows:[1] gravitational collapse is a *classical* process and microscopic black holes can form when their Schwarzschild radius is greater than the Planck length $(Gh/c^3)^{-1/2} \sim 10^{-35}$ m (at Planck lengths quantum gravitational effects do not permit purely classical treatment). This allows the existence of collapsed objects of masses from 10^{-8} kg and above and charges up to ± 30 electron units.[1] Additionally, a sufficient concentration of electromagnetic radiation causes a gravitational collapse — even though the Schwarzschild radius of the formed black hole is smaller than the photon's Compton wavelength which is infinite. Therefore, when elementary particles collapse to form a black hole, it is not the *rest* Compton wavelength hc/mc^2 that is to be considered — one should instead consider the *modified* Compton wavelength hc/E, where $E \sim kT \gg mc^2$ is the typical energy of an ultra-relativistic particle that went to form the black hole.[1] Microscopic black holes with Schwarzschild radius greater than the modified Compton wavelength hc/E, can form classically and independently on competing quantum processes. Hawking suggests that these charged collapsed object may have velocities in the range 50 – 10000 km/s and would behave in many respects like ordinary atomic nuclei.[1] When these objects travel through matter, they induce ionization and excitation and would produce bubble chamber tracks similar to those of atomic nuclei with the same charge. The charged collapsed objects survive annihilation and, at low velocities (less than few thousand km/s), they may form electronic or protonic atoms:[1] the positively charged collapsed objects would capture electrons and thus mimic super-heavy isotopes of known chemical elements, while negatively charged collapsed objects would capture protons and disguise themselves as the missing zeroth entry in the Mendeleev table. Such ultra-heavy charged massive particles (CHAMPS) were also studied by de Rujula, Glashow and Sarid[2] and considered as dark matter candidates. Dark Electric Matter Objects (DAEMONS) of masses just above 10^{-8} kg and charges of around ± 10 electron units have been studied in the Ioffe Institute and positive results in their detection have been reported[3] — observations of scintillations in ZnS(Ag) which are excited by electrons and nucleons ejected as the relic elementary Planckian daemon captures a nucleus of Zn (or S). The DAMA (DArk MAtter) collaboration also report positive results[4] in the detection of such particles using 100 kg of highly radiopure NaI(Tl) detector.

These superheavy charged particles can serve as driving force for the expansion of the Universe during the radiation-dominated epoch in a classical particle-scale model, which we recently proposed.[5] Along with this type of particles, within our model, magnetic monopoles can also play the same role for the expansion of the Universe: it has been suggested[6] that ultra-heavy magnetic monopoles were created so copiously in the early Universe that they outweighed everything else in the Universe by a factor of 10^{12}.

The classical mechanism of the cosmic expansion relies on the assumption that the Universe is a two-component gas. One of the fractions is that of ultra-relativistic "standard" particles of typical mass m and charge q with equation of state of an ideal quantum gas of massless particles. The other component consists of the superheavy charged particles of masses M (of around 10^{-8} kg and above) and charges Q (of around ± 10 electron charges and above) — exactly as those described earlier.

For an elementary particle such as the electron, the charge-to-mass ratio is $q/m \sim 10^{21}$ (in geometrized units $c = 1 = G$), while for the superheavy charged particles, $M \lesssim Q$. In view of this, the general-relativistic treatment of elementary particles or charged collapsed objects of very low masses also necessitates consideration from Reissner–Nordström (or Kerr–Newman) viewpoint — for as long as their charge-to-mass ratio remains above unity. We also treat the superheavy charged particles classically (in line with Hawking's arguments outlined earlier). That is, the superheavy charged particles are modelled as Reissner–Nordström naked singularities and the expansion mechanism is based on their gravito-electric repulsion. Instead of the Schwarzschild radius, the *characteristic length* that is to be considered now and compared to the *modified* Compton length,[1] will be the radius of the van der Waals-like impenetrable sphere that surrounds a naked singularity (see Cohen *et al.*[7] for a very thorough analysis of the radial motion of test particles in a Reissner–Nordström field). As shown in Prodanov *et al.*,[5] for temperatures below 10^{31} K, the radius of the impenetrable sphere of a superheavy charged particle of mass 10^{-8} kg and charge ± 10 electron units is greater than the *modified* Compton wavelength of the superheavy charged particle itself. The "standard" particles of the expanding Universe are therefore too far from the superheavy charged particles for quantum interactions to occur between the two fractions.

Consider a "standard" particle of specific charge q/m, and a superheavy charged particle of charge Q, such that $\mathrm{sign}(Q)q/m \geq -1$, with the "standard" particle approaching the superheavy charged particle from infinity. The pseudo-Newtonian potential of the field of the naked singularity is given by[5] $U(r) = -(mM)/r + (qQ)/r + [(m/2)(-M^2 + Q^2)]/r^2$ and the field is characterized by three regions.[5,7] The region between $r = 0$ and $r = r_0(T)$ is *impenetrable* due to the condition for reality of the kinetic energy of the incoming test particle . The turning radius $r_0(T)$ is given by[5,7] $r_0(T) = Q(q + m)/(kT)$ for very high temperatures. This can be thought of as the radius of an "impenetrable" sphere, surrounding the naked singularity, that grows with the drop of the temperature: the higher the energy (or the temperature), the deeper an incoming particle will penetrate into the gravitationally repulsive field of the naked singularity. The region between the turning radius $r_0(T)$ and the critical radius $r_c \geq r_0(T)$ is *repulsive.* At this critical radius, repulsion and attraction interchange:[5] $r_c = M(Q^2/M^2 - 1)/[1 - (qQ)/(mM)]$ and the region above r_c is *attractive.* As the temperature drops, the superheavy charged particles "grow" (incoming particles have lower and lower energies and turn back farther and farther from the naked singularity). When the temperature gets sufficiently low, the radius $r_0(T)$ of the impenetrable sphere of a superheavy charged particle grows to r_c (but not beyond r_c, as in the region $r > r_c$ the incoming probe is attracted and cannot turn back). This means that incoming particles have such low energies that they turn back immediately after they encounter the gravitational repulsion. Incoming particles of charge q such that $qQ > Mm$ do not even experience attraction — the repulsive region for such particles extends to infinity (the gravitational attraction will not be sufficiently strong to overcome the electrical repulsion) and for them $r_0(T)$ has no upper limit. The interaction between the two fractions of the Universe results[5] in power law expansion with scale factor $a(\tau) \sim \sqrt{\tau}$, corresponding to the expansion during the radiation-dominated era.

In our picture, the Universe has local Reissner–Nordström geometry, but globally, the geometry is that of Robertson–Walker. Namely, we confine our attention to the local spherical neighbourhood of a single naked singularity and consider the Universe as multiple copies (fluid) of such neighbourhoods. It is plausible to assume that these naked singularities are densely packed spheres that fill the entire Universe. Thus, the volume V of the Universe, at any moment during the Reissner–Nordström expansion, would be of the order of the number N of these particles, times the "volume" of the repulsive sphere of a superheavy charged particle: $V \sim Nr_0^3(T)$. Therefore, the number density of the superheavy charged particles is of the order of $r_0^{-3}(T)$. At Recombination, the free ions and electrons combine to form neutral atoms ($q = 0$) and this naturally ends the Reissner–Nordström expansion mechanism — a neutral "normal" particle will now be too far from an "usual" particle to feel the gravitational repulsion (the density of the Universe will be sufficiently low).

The standard treatment of a Robertson–Walker Universe uses the isotropicity and the homogeneity for modeling the energy-momentum sources as a perfect fluid. This applies for matter known observationally to be very smoothly distributed. On smaller scales, such as stars or even galaxies, this is a poor description. In our picture we model the Universe as a van der Waals gas and we use small scale density and pressure variables. Such modeling is possible in the light of the deep analogies between the physical picture behind the Reissner–Nordström expansion and the classical van der Waals molecular model: atoms are surrounded by imaginary hard spheres and the molecular interaction is strongly repulsive in close proximity, mildly attractive at intermediate range, and negligible at longer distances. The laws of ideal gas must then be corrected to accommodate for such interaction: the pressure should increase due to the additional repulsion and the available volume should decrease as atoms are no longer entities with zero own volumes. Therefore, at Recombination, the radius $r_0(T)$ of a superheavy charged particle will be of the order of $R_c = N^{-1/3} t_{recomb}$. We request that, once $r_0(T)$ becomes equal to R_c, i.e. when the charged particles recombine, then the potential of the interaction between a naked singularity and a particle of charge q, such that $qQ > mM$, becomes zero. For particles of charge q, such that $\text{sign}(Q)q/m \geq -1$ and also $qQ \leq mM$, the potential of the interaction becomes negligible earlier: when $r_0(T)$ reaches r_c.

With N superheavy particles, the effective space left for the motion of the "standard" fraction consisting of n particles, is reduced by a factor of N times the volume of the repulsive sphere of a superheavy particle. Then the "van der Waals" equation $p + (N^2\alpha)/V^2 = nkT[1 + (N\beta)/V]/V$ has: $\alpha = 2\pi \int_{r_0(T)}^{R} U(r)\, r^2\, dr = \pi m(M^2 - Q^2)[R - r_0(T)] + \pi m M[1 - (qm)/(QM)][R^2 - r_0^2(T)]$ and $\beta = 2\pi \int_0^{r_0(T)} r^2 dr = 2\pi r_0^3(T)/3$. Here $R = r_c$ if $\text{sign}(Q)q/m \geq -1$ and also $qQ \leq mM$, or $R = R_c$ if $qQ > mM$. The equation of state is $p = \eta\rho^{4/3} - \alpha/\beta^2$, where $\eta = \text{const}$. The second term depends on the temperature via α and β and becomes irrelevant towards the end, as $\alpha \to 0$ when $r_0(T) \to R$.

References

1. S. Hawking, Mon. Not. R. Astr. Soc. **152**, 75–78 (1971).
2. A. de Rujula, S.L. Glashow, and U. Sarid, Nucl. Phys. **B 333**, 173 (1990).
3. E.M. Drobyshevski, *Detection and Investigation of the Properties of Dark Electric Matter Objects: the First Results and Prospects*, astro-ph/0402367.
4. R. Bernabei, P. Belli, F. Montecchia, F. Nozzoli, F. Cappella, A. d'Angelo, A. Incicchitti, D. Prosperi, R. Cerulli, C.J. Dai, H.L. He, H.H. Kuang, J.M. Ma, and Z.P. Ye, *Recent DAMA Results*, in *Astroparticle, Particle and Space Physics, Detectors and Medical Physics Applications*, World Scientific Publishing 158 (2006).

5. E.M. Prodanov, R.I. Ivanov, and V.G. Gueorguiev, *Reissner–Nordström Expansion*, Astroparticle Physics **27** (150–154) 2007, hep-th/0703005;
E.M. Prodanov, R.I. Ivanov, and V.G. Gueorguiev, *Equation of State for a van der Waals Universe during Reissner–Nordström Expansion*, forthcoming.
6. J. Preskill, Phys. Rev. Lett. **43 (19)**, 1365 (1979).
7. J.M. Cohen and R. Gautreau, Phys. Rev. **D 19 (8)**, 2273–2279 (1979).

GALAXY-GALAXY LENS RECONSTRUCTION: EINSTEIN RINGS AND LENSED QSOS

B. J. BREWER

Institute of Astronomy, School of Physics, The University of Sydney NSW, 2006, Australia
E-mail: brewer@physics.usyd.edu.au

Gravitational lensing remains one of the most direct methods we have for the detection of dark matter. As observational abilities improve, it will become possible to map the projected mass distributions of lensing galaxies in detail. Since the equilibrium mass profile of an isolated halo depends on the properties of dark matter, additional constraints on dark matter physics will become possible. However, there are still some unresolved issues about how best to extract the mass profiles from observed images. These relate to taking into account prior information, and whether extended images provide any advantage over point images in terms of information content.

Keywords: Gravitational Lensing; Dark Matter Haloes.

1. Introduction

Gravitational lensing can be used as a powerful astrophysical tool for probing the density profiles of galaxies, and is one of the few ways in which dark matter or an equivalent effect can be detected. In addition, it often magnifies source objects by one to two orders of magnitude, particularly in those lens systems that we happen to discover, which just happen to be the brightest. This allows us to use the intervening gravitational lens as a kind of natural telescope, magnifying the source so that we can observe more detail than we would have been able to without the lens. The basic problem to be solved is one of parameter estimation: we describe the source by a set of parameters s_i (describing a non-negative surface brightness distribution over the source plane), and the lens by another set of parameters L_i (describing a non-negative projected mass density profile over the lens plane). We then seek to estimate these by fitting them to data D. Usually, the data are not good enough to uniquely determine the source and lens model parameters, so we explore the posterior distribution over the space of possible sources and lenses, which is given by Bayes' theorem:

$$p(s_i, L_i|D) \propto p(s_i, L_i)p(D|s_i, L_i) \quad (1)$$

This distribution describes our knowledge about the possible source and lens profiles after taking into account the data of an image [1].

An issue about lens modelling has been whether there is any advantage in searching for extended *Einstein Rings* as opposed to lensed quasars, since they may provide stronger constraints on the lens. In the following section we use a single system - RXS J1131-1231 - which is both a quadruply imaged quasar *and* an extended optical Einstein ring - to investigate this question.

2. Point vs. Extended Images: A Case Study: RXS J1131-1231

The discovery of this system was reported by [2]. It consists of a quadruply lensed quasar at redshift $z = 0.658$ lensed by a galaxy at $z = 0.295$. At the time of its discovery, it was the closest known gravitationally lensed quasar, with some evidence for an extended optical Einstein ring - the image of the quasar host galaxy. Initial simple modelling suggested that the source was magnified by a factor of ~ 50. Thus, subsequent observations with the ACS aboard the Hubble Space Telescope [3] at a wavelength of 555 nm should allow the recovery of the morphology of the quasar's host galaxy down to a resolution of about 0.01 arc seconds, and possibly tighter constraints on the lens model.

3. Assumptions

The particular lens model we assumed for this system was a pseudo-isothermal elliptical potential, primarily for computational speed but also because it is fairly general and realistic. This has five parameters: strength b, ellipticity q, orientation angle θ, and two parameters (x_c, y_c) for the central position. Although any Bayesian modelling can only explore a particular slice through the full hypothesis space we might have in our minds, using a simplified analytical model is often sufficient to illuminate the general properties of the true lens mass distribution.

For this system, we also chose to include an external shear component. All Einstein rings can be expected to reside in an environment where the external shear due to neighbouring galaxies is nonzero (Kochanek, private communication). So we also have two parameters for the external shear:

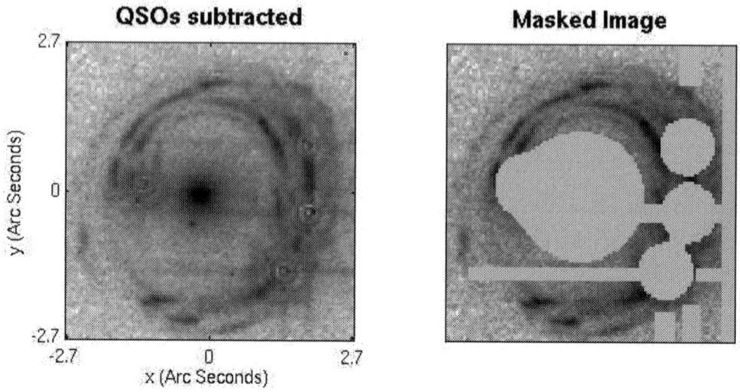

Fig. 1. On the left is the raw extended emission, after subtraction of the four quasar images (one on the left hand side of the image, three on the right). Residuals from this subtraction, as well as diffraction spikes, cause large systematic errors over the regions of the image that have been blanked out in the image on the right.

γ, its magnitude, and its orientation angle θ_{shear}. [3] have analysed the flux ratios of the quasar images and found that most of these ratios are consistent with a model of this type (elliptical potential plus external shear). In principle, we could adopt ever less restrictive parameterisations for the lens, to hunt for substructures in its density profile. This possibility will be discussed in section 5.

In terms of all of these parameters, the deflection angle formula, relating a point (x, y) in the image plane to a corresponding point (x_s, y_s) in the image plane, is

$$x_s = x - \alpha_x(x, y) \tag{2}$$

$$y_x = y - \alpha_y(x, y) \tag{3}$$

where the deflection angles α are given by the gradient of the potential

$$\phi(x, y) = \frac{1}{2}\gamma(x'^2 - y'^2) + b\sqrt{qx''^2 + y''^2/q} \tag{4}$$

and (x', y') are the ray coordinates in the rotated coordinate system whose origin is $(0, 0)$ and is aligned with the external shear (i.e. rotated by an angle θ_{shear}), and (x'', y'') are the ray coordinates in the rotated and translated coordinate system centred at (x_c, y_c) and oriented at an angle θ.

The physical interpretation of each of these parameters suggests a plausible prior range for their values. To represent this knowledge we used the following prior distributions (Table 1). Since these are broad distributions, and the data are good, the influence of these choices is negligible.

Table 1. Prior probability densities for the lens model parameters, and also the extra noise parameter σ.

Parameter	Prior Distribution
b	Normal, mean 1.8, SD 0.5, $b > 1$
q	Normal, mean 0.9, SD 0.2, $0 < q < 1$
x_c	Normal, mean -0.22, SD 0.5
y_c	Normal, mean -0.07, SD 0.5
θ	Uniform, between 0 and π
σ	Improper Uniform, $\sigma > 0$
θ_{shear}	Uniform, between 0 and 2π
γ	Exponential, mean 0.1

3.1. *How the Quasar Images Constrain the Lens*

The four quasar images can constrain the lens model because we require that the four image positions lens back to the same point in the source plane. Actually, with an upper limit of 10^{-3} pc for the continuum source size [4], this requirement is too strong - we can really only insist that they lens back to within $\sim$ 0.15 microarcseconds of each other. The magnifications of the images can also provide some information, although microlensing and absorption effects can limit the usefulness of including the magnifications as constraints.

We wrote a Markov Chain Monte Carlo code to explore the prior volume of the lens parameters, subject to the constraint that the four images lens back to within $\sim$ 0.1 microarcseconds of each other. Small astrometric uncertainties of 0.02" were also taken into account[a]. The results are shown in Table 2. Far from uniquely determining the lens model, the quasar images have only managed to give moderately strong constraints on the overall strength of the lens and the angle of orientation of the external shear. For all other parameters, the marginal distributions are very wide, in some cases nearly as wide as the priors, so the quasars have provided only a small amount of information about them.

4. Simple Modelling of the Extended Source

To take into account the extended images, we modelled the source as the sum of four elliptical Gaussians, with unknown brightness, orientation, cen-

[a]These correspond to ± half a pixel in the image. This is significantly larger than the uncertainties that would be obtained by subsampling the PSF and fitting to the observed QSO images, but uncertainties found in that way would be conditional on the TinyTim PSF being exactly correct, which is unlikely.

Table 2. Inferred lens parameters from the QSO astrometry alone. Throughout this paper, the units for b,x_c and y_c are arcseconds; θ and θ_{shear} are in radians.

Parameter	Value
b	1.776 ± 0.023
q	0.938 ± 0.034
x_c	-0.10 ± 0.10
y_c	-0.06 ± 0.02
θ	0.45 ± 0.42
γ	0.116 ± 0.034
θ_{shear}	1.24 ± 0.10

tral position and ellipticity for each. Similarly, the lensing galaxy light profile was modelled as a single elliptical Gaussianlike function in the image plane, with unknown slope. We used a simple Metropolis algorithm [1] to derive estimates and uncertainties on the lens parameters. These results are summarised in table 3 as an estimate for each lens parameter.

A typical simple model from the sample is shown in Figure 2. The scales on the axes are the same as those in Figure 1, so an idea of the magnification can be obtained visually. The four images have used the same colourmap, allowing direct comparisons of the surface brightnesses. This method favours a highly complex source (since the Gaussians do not overlap to the extent that they become indistinguishable), lensed by a slightly elliptical lens whose centre is located close to the centre (a small but statistically signif-

Table 3. Inferred lens parameters from the case where the extended source is modelled as elliptical gaussians. Note that the constraints are much tighter than those from the QSO astrometry.

Parameter	Value
b	1.792 ± 0.002
q	0.973 ± 0.004
x_c	-0.192 ± 0.004
y_c	-0.062 ± 0.004
θ	0.43 ± 0.08
γ	0.129 ± 0.003
θ_{shear}	1.24 ± 0.01
σ	0.206 ± 0.002

icant distance of 0.03 arcseconds) of the observed lensing galaxy. The flux contribution of the lens galaxy to the image is small compared to that from the source, but not negligible. This implies that simultaneous fitting of the lensed image and the lens galaxy light profile is required because the flux from the lens galaxy is diffuse and overlaps the image.

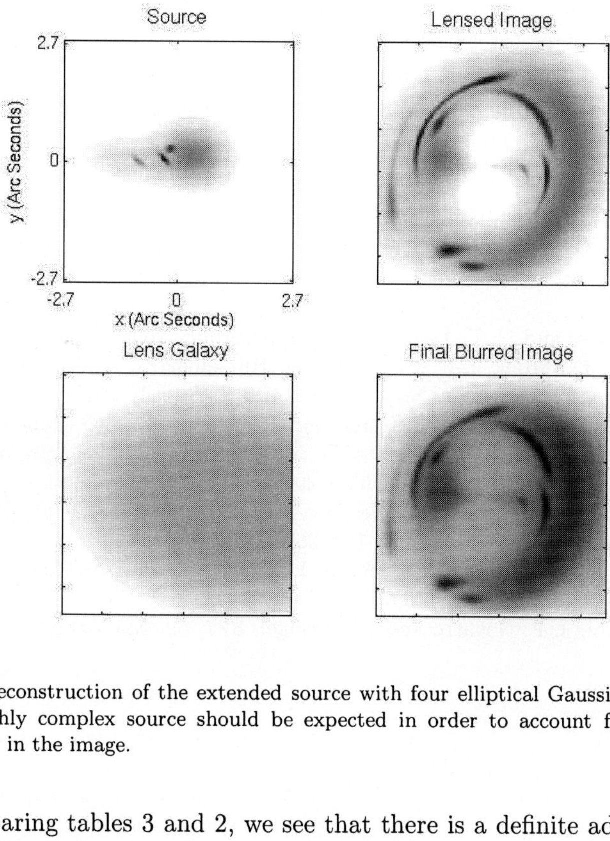

Fig. 2. Reconstruction of the extended source with four elliptical Gaussians, showing that a highly complex source should be expected in order to account for all of the complexity in the image.

Comparing tables 3 and 2, we see that there is a definite advantage to using the extended images in this case. Therefore it is not always true that there is little to be gained by modelling extended images [5] - it depends on their quality.

5. "Nonparametric" Lens Modelling

The ultimate goal of lens reconstruction techniques is to be able to reconstruct detailed mass maps that do not rely on a particular choice of parametric lens model, such as an elliptical potential or NFW profile. One approach that is popular is based on the PixeLens code [6], where the lens mass distribution is pixellated. It could be argued that this gives the lens model *too much* freedom, and an intermediate approach is needed.

An intermediate approach (providing more prior information than a pixellated model, but less than a single parametric model) would build up the lens mass distribution from small numbers of basis functions - sums of parametric mass blobs. This has been done in the cases of cluster lenses with point images (or extended images that are reduced to points for the analysis) [7] - but the ultimate goal would be to combine this approach for the lens modelling with information from multiple point *and* extended images to produce accurate maps of single lensing galaxy substructures. This could potentially shed light on the missing satellite problem [8] and the nature of dark matter.

References

1. B. J. Brewer and G. F. Lewis, **637**, 608 (February 2006).
2. D. Sluse, J. Surdej, J.-F. Claeskens, D. Hutsemékers, C. Jean, F. Courbin, T. Nakos, M. Billeres and S. V. Khmil, **406**, L43 (July 2003).
3. J.-F. Claeskens, D. Sluse, P. Riaud and J. Surdej, **451**, 865 (June 2006).
4. V. N. Shalyapin, L. J. Goicoechea, D. Alcalde, E. Mediavilla, J. A. Muñoz and R. Gil-Merino, **579**, 127 (November 2002).
5. I. Ferreras, P. Saha and S. Burles, *ArXiv e-prints* **710** (October 2007).
6. P. Saha and L. L. R. Williams, **127** 2604 (May 2004).
7. E. Jullo, J.-P. Kneib, M. Limousin, Á. Elíasdóttir, P. Marshall and T. Verdugo, *ArXiv e-prints* **706** (May 2007).
8. E. Hayashi, J. F. Navarro and V. Springel, **377**, 50 (May 2007).

DARK HALOES OF FOSSIL GALAXY GROUPS AND CLUSTERS: OBSERVATIONS AND SIMULATIONS

H. G. KHOSROSHAHI

Astrophysics Research Institute, Liverpool John Moores University,
Birkenhead, CH41 1LD, UK
E-mail: hgk@astro.livjm.ac.uk

A. DARIUSH[1], T. J. PONMAN[1], F. PEARCE[2]

[1]*School of Physics and Astronomy, University of Birmingham,*
Birmingham, B15 2TT, UK
[2]*School of Physics and Astronomy, University of Nottingham,*
Nottingham, NG7 2RD, UK

We present results from our recent studies of a class of galaxy groups and clusters known as "Fossils" which are dominated by a giant elliptical galaxy and are arguably the end product of mergers of galaxies within groups. Fossils are the archetypal relaxed systems thus seen as simple laboratories to study formation and evolution of galaxies and haloes in the absence of recent major mergers.

Multi-wavelength study of the largest sample of fossil systems in the observations complemented by cosmological simulations were undertaken to understand their origin and evolution. We study the dark matter distribution in fossils using their X-ray derived mass profiles. Fossils show higher halo concentration, for a given mass of the halo, compared to non-fossil groups and clusters indicating an early formation epoch for fossils. The study of their mass evolution in the Millennium simulations, including hot gas and semi-analytic galaxies, shows that their haloes form earlier than those of non-fossil systems. We estimate the space density of fossils in the simulations which agrees well with the observations.

Keywords: Galaxies, Clusters, Elliptical Galaxies, Intergalactic Medium, Dark Matter.

1. Introduction

Galaxy groups are key systems in advancing our understanding of structure formation and evolution. They contain the majority of galaxies in the universe, and are precursors to the most massive structures, i.e. clusters, giving them cosmological importance. They show departures from the scaling relations obeyed by galaxy clusters indicating that groups are not simply scaled-down versions of clusters. It is argued that galaxy groups are rapidly evolving, and many are not virialised.[1] Thus identifying and studying a sample of well-characterised galaxy groups can help us understand the origin of the some of the observed diversities in groups. Galaxy groups also play key role in understating the formation of luminous elliptical galaxies such as the brightest cluster galaxies (BCGs) arguably as a result of higher efficiency of galaxy-galaxy merger in low velocity environment of groups compared to rich clusters.

1.1. *Fossil Groups*

In the class of galaxy groups known as "fossil groups", the group is dominated optically by a single luminous elliptical galaxy at the centre of an extended luminous X-ray emission similar to that seen in bright X-ray groups (Fig 1). The X-ray is emitted as a result of gravitational shock heating during the collapse, and formation, of the system and is a powerful probe of the dark matter distribution in groups and clusters of galaxies. The X-ray emission in fossils is regular and symmetric, indicating the absence of recent group merging thus they are seen as simple laboratories to study formation and evolution of galaxies and haloes in the absence of recent mergers. The absence of $L^\star$ galaxies in fossils is argued to be the consequence of multiple mergers of galaxies within the group itself which could also mean that fossils are old galaxy systems as the time-scale of such mergers are usually over around 4 Gyr.[2] Thus fossils are the best candidates for virialised groups.

Observationally a galaxy group is classified as a fossil if it has an X-ray luminosity of $L_{X,bol} \geq 10^{42} h_{50}^{-2}$ erg s^{-1} spatially extended to few 100 kpc, and the dominant galaxy is at least 2 magnitudes brighter (in R-band) than the second ranked galaxy within half the projected virial radius of the group.[4] The X-ray criterion guarantees the existence of a group size galaxy halo while the optical criterion assures that the $M^\star$ galaxies are absent within the given radius which corresponds to the radius for orbital decay by dynamical friction.[5] No upper limit is placed on the X-ray luminosity or temperature, and recently a fossil galaxy cluster was found.[3]

2. Sample Selection and Observations

Our study makes use of a flux-limited sample of fossils found in the catalogue of spatially extended X-ray sources compiled by Wide Angle ROSAT

Fig. 1. Fossil cluster RXJ1416.4+2315. Cluster size X-ray emission observed with XMM-Newton surrounding a single giant elliptical galaxy. Other cluster members are at least 2 magnitude fainter than the central giant elliptical.[3]

Pointed Survey (WARPS) project. Details of the fossil identification and sample selection is given in Jones et al (2003)[4] which also represents the largest statistical sample of fossil groups studied to date. In addition, the nearest known fossil group NGC 6482 and the first discovered fossil group, RX J1340.5+4017, are included in our sample.[6]

All the systems were observed with NASA's Chandra X-ray telescope and some with ESA's XMM-Newton. In addition, the sample was observed using the observational facilities of Issac-Newton Group of Telescopes (ING) and Kitt-peak National Observatory (KPNO). The K_s-band observations of the sample using UIST/UKIRT were also performed to study the stellar properties of the central galaxy.[7]

3. Analysis and Results

Our analysis includes modelling of the X-ray surface brightness distribution of the sample to understand their morphology and the distribution of the X-ray emitting hot gas as well as the spectral analysis which allows us measure the IGM temperature profile. These measurements, together with spherical symmetry and hydrostatic equilibrium assumptions, gives the total gravitational mass distribution. The dark matter contribution is then calculated by subtracting the gas mass profile and the contribution from the central galaxy using the optical and near-IR observations. Having the temperature of the hot gas, T_X, total mass, M, and X-ray luminosity, L_X, from the X-ray analysis and the total optical light of the galaxies together with the velocity dispersion of the galaxies in each fossil group we study the scaling relations in fossils and comparison non-fossil groups.[6]

We confirm that, for a given optical luminosity of the group, fossils are more X-ray luminous than non-fossil groups. Fossils, however, fall comfortably on the conventional $L_X - T_X$ relation of galaxy groups and clusters, suggesting that their X-ray luminosity and their gas temperature are both boosted as a result of their early formation. This is supported by other scaling relations including the $L_X - \sigma$ and $T_X - \sigma$ relations in which fossils show higher X-ray luminosity and temperature for a given group velocity dispersion.

Dark matter haloes with an early formation epoch tend to be more concentrated.[8] Numerical studies also predict some mass dependency for the halo concentration, resulting from the fact that lower mass haloes generally form earlier.[9,10] In general observational results seem to agree well with the numerical predictions.[11,12]

We find that the haloes of fossil groups are more concentrated than those of non-fossil systems, for a given mass, which suggests that fossils have an early formation epoch. This is shown in Fig 2a, where the halo concentration parameter, c_{200}, for the fossils was obtained and compared with the observations of non-fossils[11] and numerical studies.[9,10]

Moreover, the $M_X - T_X$ relation suggests that fossils are hotter (Fig 2b), for a given total gravitational mass, both consistent with an early formation epoch for fossils. The entropy of the gas in low mass fossils appears to be systematically lower than that in normal groups, which may explain why the properties of fossils are more consistent with an extension of cluster properties.

4. Fossils in the Millennium Simulations

Observations show that fossil groups are outliers in some of the scaling relations (Fig 2) providing strong constraints for the theoretical models and simulations. In addition the simulations have the unique advantage that the formation and evolution of the individual haloes can be studied. The largest available cosmological simulation, the Millennium Simulations, were used to study fossils taking advantage of large statistics and the large halo mass range.

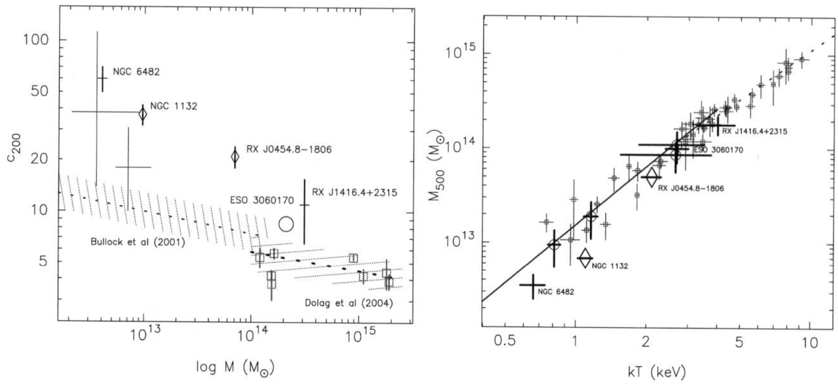

Fig. 2. a) $M - c_{200}$ relation: Comparison between the mass concentration in fossils and non-fossil groups and clusters. Three fossils with resolved temperature profile and two isolated Over-luminous Elliptical Galaxies OLEGs (diamonds)[13] are compared with non-fossil clusters (open squares).[11] The expected values of the dark matter concentration and its variation with the halo mass from the numerical studies[9,10] are also presented. b) $M - T$ relation for fossils: Bold crosses show fossils. Additional circle indicates the isothermality assumption in deriving the mass. Such an assumption results in over-estimation of the total mass. Two diamonds represent OLEGs.[13] Fossil groups appear to have higher temperatures, for their masses, in comparison to the non-fossil systems shown with open squares.[14] The thick solid line (blue) is the best orthogonal fit to the galaxy systems with $kT \leq 4$. The dashed line (blue) is the same for galaxy systems with $kT \geq 3$.

In order to extract fossil groups in the Millennium simulation, according to the observational selection criteria, we require a simulation suite that includes the baryonic physics of hot gas and galaxies, as well as a high resolution dark matter framework and a sufficient spatial volume to limit the effects of cosmic variance. For this study we use the dark matter Millennium Simulation,[15] a 10-billion particle model of a comoving volume of side $500h^{-1}$ Mpc, on top of which a publicly available semi-analytic galaxy model[16] has been constructed. For the hot gas we have repeated the Millennium simulation with a lower resolution simulation including gas physics utilising the same volume, phases and amplitudes as the original dark-matter-only model. This run accurately reproduces the structural framework of the Millennium Simulation. Below we summarize the main characteristics of the above simulations.

4.1. *Dark Matter Simulation*

The Millennium Simulation is based on a Cold Dark Matter cosmological model of structure formation, with a Dark Energy field Λ. The basic assumptions are those of an inflationary universe, dominated by dark matter particles, leading to a bottom-up hierarchy of structure formation, via collapsing and merging of small dense haloes at high redshifts, into the large virialised systems such as groups and clusters that contain the galaxies that we observe today. The simulation was performed using the publicly available parallel TreePM code Gadget2, achieving a 3D dynamic range of 10^5 by evolving 2160^3 particles of individual mass $8.6 \times 10^8 h^{-1}$ $M_\odot$.

4.2. *Semi-Analytical Galaxies*

Using the dark matter haloes of the original simulation,[15] Croton et al[16] have simulated the growth of galaxies, and their central supermassive black holes, by self-consistently implementing semi-analytic models of galaxies on the outputs of the Millennium Simulation. The semi-analytic catalogue contains 9 million galaxies at $z = 0$ down to a limiting absolute magnitude of $M_R - 5 \log h = -16.6$. In the semi-analytic formulation, galaxies initially form within small dark matter haloes. As the simulation evolves, such a halo may fall into a larger halo. The semi-analytic galaxy within this halo then becomes a satellite galaxy within the main halo and follows the track of its original dark matter halo (now a subhalo) until the mass of the subhalo drops below $1.72 \times 10^{10} h^{-1}$ $M_\odot$, which corresponds to a 20-particle limit in the Millennium Simulation. At this point the galaxy is assumed to spiral into the centre, on some fraction of the dynamical friction timescale, where it merges with the central galaxy of the larger halo.[16]

4.3. *Gas Simulations*

The Millennium Gas Simulations are a suite of hydrodynamical models, utilising the same volume, and values of initial perturbation amplitudes and phases as the parent dark-matter-only Millennium Simulation. We included a simple preheating scheme that is tuned to match the observed X-ray properties of clusters at the present day as we only utilise the hydrodynamical properties of the groups at $z = 0$, where the observational and simulation results are well matched.

Each of the Millennium Gas Simulations consists of 5×10^8 particles of each species, resulting in a dark matter mass of $1.422 \times 10^{10} h^{-1}$ $M_\odot$ per particle and a gas mass of $3.12 \times 10^9 h^{-1}$ $M_\odot$ per particle.

Control groups are necessary for our study in order to allow us to compare the properties of X-ray fossil and non-fossil groups similar to the observations. We define two control group samples, based on the magnitude difference of the two brightest members of the group (within $0.5R_{200}$ of the centre of the dark matter halo): (i) $0.8 \leq \Delta m_{12} \leq 1.0$, and (ii) $0.1 \leq \Delta m_{12} \leq 0.3$.

5. Results from the Simulations

Fig 3a shows a trace back of the mass build-up of the dark matter haloes as a function of expansion factor and redshift for both the X-ray fossils and the control groups 1 and 2. The plot suggest an earlier formation epoch for fossil groups in comparison to control groups. At any epoch fossils have larger fraction of their final mass compared to non-fossil haloes.[17]

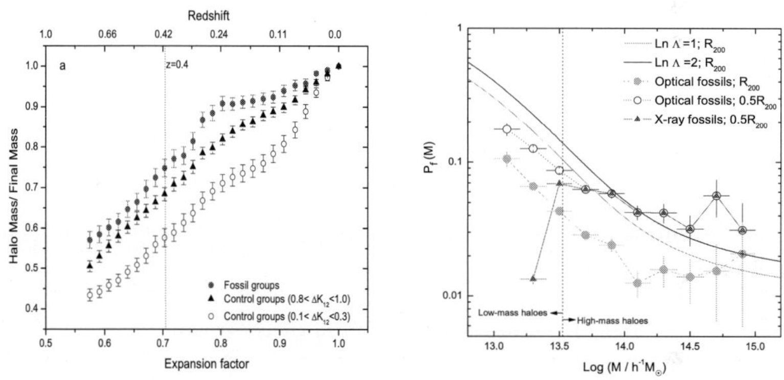

Fig. 3. a) Left: Tracing back the mass build-up of the dark matter haloes as a function of expansion factor and redshift for both the X-ray fossils (filled circles) and the control groups (open circles and filled triangles). b) Right: The probability of finding fossils in groups and clusters. The probability is shown for the fossils in the Millennium Simulations (data points) and analytical models (curves).

In order to understand the effect of the X-ray criteria in the fossil studies we identified 'optical fossils' which are the groups in which there exist a 2 magnitude luminosity gap only without taking any notice of their X-ray luminosity. Fig 3b shows the probability, $P_f(M)$, that a dark matter halo of mass M contains an X-ray fossil group (*filled triangles*), optical fossil group within $0.5R_{200}$ (*open circles*), or optical fossil group within R_{200} (*filled circles*) from the Millennium simulation. The fossil incidence rate from the analytical study of[18] for two values of $\mathrm{Ln}\Lambda = 1$ (*green line*) and $\mathrm{Ln}\Lambda = 2$ (*blue line*) is also plotted. The vertical dotted-line corresponds to halo mass $\sim 3.34 \times 10^{13}\,h^{-1}\,\mathrm{M}_{\odot}$.

5.1. *The Space Density of X-Ray Fossil Groups*

The integrated space density of X-ray fossil groups has been studied for small samples, each of three to five X-ray fossil systems, at different limiting luminosities.[4,19,20] We estimate the space density by systematically counting the fossil groups in the whole 500 h^{-1} Mpc survey volume of the Millennium Simulation at $z = 0$. For comparison with previous studies, we select and count X-ray fossil groups for three limiting X-ray luminosities ranging from 0.25–5$\times 10^{42}$ h^{-2}erg s^{-1}.

Our analysis show that for X-ray luminosities exceeding 2.5-5$\times 10^{42}$ h^{-2}erg s^{-1}, the space density of fossils in the Millennium simulation agrees within the errors with those estimated observationally.[?,19] At the lowest X-ray luminosities, the density from the Millennium simulations appear to be lower than observed, though the observational values have large uncertainties due to the small number of X-ray fossil groups and the effects of cosmic variance. A recent study[21] shows that one of the fossils in the sample of Jones et al (2003)[4] does not satisfy the fossil criterion of $\Delta m_{12} \geq 2$, which reduces the observational space density. Certainly the number of X-ray fossils found is heavily dependent on the X-ray luminosity threshold chosen and may be influenced by the scatter in X-ray group properties near this lower limit.

Table 1. Space densities of fossil galaxy groups.

L_X $10^{42}h^{-2}$ erg s^{-1}	Number of fossils	Density $10^{-7}h^3$ Mpc^{-3}		Present study $10^{-7}h^3$ Mpc^{-3}
> 0.25	5	320^{+216}_{-144}	J03[4]	104 ± 3
> 2.5	3	$16^{+15.2}_{-8.8}$	J03	22.4 ± 1.3
> 2.5	4	$36.8^{+47.2}_{-18.4}$	V99[19]	22.4 ± 1.3
> 2.5	3	~ 160	R00[20]	22.4 ± 1.3
> 5.0	4	$19.2^{+24.8}_{-9.6}$	V99	12.8 ± 1.0

6. Conclusions

We studied the X-ray properties of fossil galaxy groups as the extreme relaxed objects, in contrast to the bullet cluster, and found observational evidences that these systems have a relatively early formation epoch and hence can be used as cosmological tools to understand the IGM, dark matter halo and galaxy properties of old galaxy systems. The strongest support for the early formation of fossils came from their higher halo concentration compared to non-fossils as well as having hotter IGM for a given halo mass.

We studied the history of the mass assembly of fossil groups, selected using the usual observational criteria at $z = 0$, from a redshift $z = 0.8$ to the present day, within the Millennium simulation. We found that the space density of X-ray fossil groups is in close agreement with the observed space density of fossils with $L_X > 2.5 \times 10^{42}\ h^{-2}$erg s^{-1}. Although for low luminosity fossils simulations predict roughly 1/3 of the observed fossil space density, there are several potential factors that could lead to this difference. As well as significant uncertainties in the observational studies, the X-ray properties of haloes in the real Universe show far greater scatter than those seen in the preheating simulation used here. Given the X-ray luminosity threshold in the definition of an X-ray fossil, scatter in L_X will alter the X-ray fossil number density, since the number density of haloes is a steep function of mass. This will be worth revisiting in the future, when better observational estimates are available.

The probability of finding optical fossils with mass M, i.e, $P_f(M;\text{optical})$ is a decreasing function of: (a) group dark matter halo mass and, (b) the fraction of the virial radius within which the first and second brightest galaxies are being found. Conversely, as dark matter halo mass become small, the probability $P_f(M;\text{X-ray})$ for X-ray fossils decreases. Both high-mass and low-mass X-ray fossil groups are found to have assembled $\sim 90\%$ of their final masses by a redshift of $z = 0.24$. The corresponding mass fraction is about $\approx 70-80\%$ for two different sets of high-mass control samples, and $\approx 85\%$ for low-mass control samples, where groups fulfill the same X-ray luminosity criterion ($\gtrsim 3.34 \times 10^{14}\,h^{-1}\mathrm{M}_{\odot}$) but have the optical luminosity gaps corresponding $0.1 \leq \Delta m_{12} \leq 0.3$ and $0.8 \leq \Delta m_{12} \leq 1.0$ magnitudes.

References

1. J. Rasmussen, T. J. Ponman, J. S. Mulchaey, T. A. Miles, S. Raychaudhury 2006, MNRAS, **373**, 653.
2. Jones L. R., Ponman T. J., Forbes D.A., 2000, MNRAS, **312**, 139.
3. Khosroshahi H. G., Maughan B. J, Ponman T. J. & Jones L. R., 2006, MNRAS, **369**, 1211.
4. Jones L. R., Ponman T. J., Horton A., Babul A., Ebeling H., Burke D. J., 2003, MNRAS, **343**, 627.
5. Binney K., Tremaine S., 1987, Galactic Dynamics. Princeton Univ. Press, Princeton, NJ.
6. Khosroshahi H. G., Ponman T. J. and Jones L. R., 2007, MNRAS, **377**, 595.
7. Khosroshahi H. G., Ponman T. J. and Jones L. R., 2006, MNRAS Letters, **372**, 68.
8. Navarro J. F., Frenk C. S., White S. D. M., 1995, MNRAS, **275**, 720.

9. Bullock J. S. et al, 2001, **321**, 559.
10. Dolag K. et al, 2004, A&A, **416**, 853.
11. Pratt G. W., Arnaud M., 2005, A&A, **429**, 791.
12. Vikhlinin A., Kravtsov A., Forman W., Jones C., Markevitch M., Murray S. S., Van Speybroeck L., 2006, ApJ, **640**, 691.
13. Yoshioka T. et al, 2004, Adv. in Space Res, **34**, 2525.
14. Finoguenov A., Reiprich T. H., Bohringer, H., 2001, A&A, **368**, 749.
15. Springel, V. et al., 2005, Natur, **435**, 629.
16. Croton, D.J. et al., 2006, MNRAS, **365**, 11.
17. Dariush A., Khosroshahi H. G., Ponman T. J., Pearce F. Raychaudhury S., Hartley W., 2007, MNRAS, in press (astro-ph/0708.2702).
18. Milosavljević, M., Miller, C.J., Furlanetto, S.R. & Cooray, A., 2006, ApJ, **637**, L9.
19. Vikhlinin, A. et al 1999, ApJ, **520L**.
20. Romer A.K. et al., 2000, ApJS, **126**, 209.
21. Jeltema, T.E., Mulchaey, J.S., Lubin, L.M., Rosati, P., & Böhringer H., 2006, ApJ, **649**, 649.

SMALL SCALE STRUCTURE IN DARK MATTER MODELS AND CONSEQUENCES FOR GALAXY FORMATION

CHRIS POWER

Centre for Astrophysics and Supercomputing,
Swinburne University of Technology, Hawthorn,
Victoria 3122, Australia
E-mail: cpower@astro.swin.edu.au

A defining prediction of the Cold Dark Matter (CDM) model of cosmological structure formation is that the number density of low-mass dark matter haloes – or *small scale structure* – is large and increasing with decreasing mass. Using cosmological N-body simulations, we have studied *truncated* dark matter models in which low-mass halo formation is suppressed below a mass cut-off scale M_{cut}, which we vary between $5 \times 10^9 h^{-1} M_\odot$ and $10^{11} h^{-1} M_\odot$. We assess the impact of the mass cut-off on dark matter halo clustering, mass assembly and angular momentum growth on galaxy and group-mass scales compared to a fiducial ΛCDM model. As we would expect, the number density of haloes with masses $M \geqslant M_{cut}$ in truncated dark matter models is suppressed relative to the ΛCDM model. This leads to a reduction in the number density of low-mass haloes around galaxy- and group-mass haloes, and consequently a reduction in the frequency of minor mergers experienced by these systems. However, we find no significant differences between the angular momenta of galaxy- and group-mass haloes in truncated models when compared with those in the ΛCDM model. We consider the implications of this for our understanding of the growth of halo angular momentum and as a potential tests of the CDM paradigm.

Keywords: Cosmology; Dark Matter; N-body Simulations; Galaxy Formation.

1. Introduction

One of the key questions facing fundamental physics and cosmology at the turn of the 21st century concerns the nature of the dark matter. There is compelling observational evidence to suggest that approximately 80% of the matter content of the Universe is in the form of some exotic, non-baryonic dark matter.[1] Moreover, the clustering of this dark matter is believed to play a crucial role in galaxy formation and evolution.[2,3]

Theories of cosmological structure formation assert that this dark matter is *cold* – that is, the velocity distribution of the dark matter was non-relativistic at the time of decoupling, moments after the Big Bang. The coldness of this Cold Dark Matter (hereafter CDM) leads to two important predictions. The first is that the central densities of CDM haloes can be arbitrarily large[4] – that is, CDM haloes are *cuspy*. The second is that the CDM halo mass function – the comoving number density of haloes of mass M – increases with decreasing mass as a power-law $M^{-\alpha}$ ($\alpha \sim 2.0$)[5,6] and may extend down to mass scales of $10^{-6} M_{\odot}$.[7]

These two predictions lead to an expectation that the number density of low-mass CDM haloes – *small-scale structure* – should be large in any finite volume. High resolution cosmological N-body simulations of galaxy-mass CDM haloes show that the normalised mass function of substructure is indistinguishable from that measured in galaxy cluster mass haloes, and that it continues to increase as the mass resolution of the simulation increases.[8,9] In contrast, simulations of halo formation in dark matter models in which small scale power is suppressed, such as Warm Dark Matter, show that the abundance of substructure is suppressed whereas mass profiles retain their cuspiness.[10–13] Therefore it can be argued that it is the abundance of small-scale structure rather than central density cusps that is the defining characteristic of the CDM model.

While the CDM model may predict that the abundance of low-mass haloes is large, we do not observe a corresponding abundance of low-mass galaxies. This is because galaxy formation is predicted to be efficiently suppressed in the shallow potential wells of low-mass haloes.[14] A galaxy forms in a dark matter halo when the Jeans mass of gas is comparable to the halo mass and the halo is sufficiently massive to support efficient cooling. This occurs at virial temperatures $T_{\rm vir} \geqslant 10^4$K, which implies that haloes with virial masses of $\sim 10^9 h^{-1} {\rm M}_{\odot}$ are unable to support the formation of a galaxy. Furthermore, it is believed that cosmological reionisation (at $z \gtrsim 6$) was important in suppressing the cooling of gas onto haloes,[15] and so the population of satellite galaxies that we observe around our Galaxy at the present day are the small subset that formed their stars prior to reionisation.[16]

If the abundance of low-mass haloes is the defining prediction of the CDM model and if Nature has conspired to leave them dark, what astronomical observations can we make that will allow us to differentiate between the CDM model and dark matter models in which small scale structure is suppressed? One might expect that gravitational lensing would provide the means to address this question. Indeed, it has been speculated that anomalous flux ratios measured in gravitational lenses arise because of small perturbations introduced by small scale structure.[17] However, it has been argued that these anomalous flux ratios arise because of structure along the line of sight to the lens rather than in the lens itself,[18] and that systems in which this effect has been measured can be well fitted by smooth lens models.[19] Alternatively, one might expect that encounters between low-mass substructures and galactic discs will act as a source of dynamical heating of the discs, but recent studies suggest that such encounters are unimportant.[20,21]

Small scale structure may not affect the structure of a galaxy disc that is already in place, but could it play a role in setting the structure of the disc as it forms? The disc forms when gas condenses and cools in the deep potential well of its dark matter halo. If the dark matter and gas were initially well mixed, they will have been subject to the same gravitational tidal field and will have similar initial angular momenta. As the gas cools and condenses into a disc, its initial angular momentum is conserved and this is imprinted on the scale length of the disc.[22] The initial growth of halo angular momentum is driven by tidal torques arising from gravitational interaction with the large scale matter distribution,[23] but it has been argued strongly that minor mergers are the primary determinants of halo angular momentum at late times.[24] For this reason it is interesting to examine how halo angular momentum is affected in dark matter models in which the formation of low-mass haloes is suppressed, and whether this might have observationally measurable consequences for the scale lengths of galactic discs.

In the following sections we describe cosmological N-body simulations of dark matter models in which the formation of small scale structure is suppressed. We show that, by suppressing small scale power at early times, we reduce the number density and clustering of low-mass haloes at late times. This has important consequences for the frequency with which galaxies and groups experience minor mergers. We investigate how this affects the growth of halo angular momenta and how it is distributed within the halo at late times. Finally we discuss the implications of our results for our understanding of the generation of halo angular momentum and the formation of galactic discs.

2. Truncated Dark Matter Models

We have run a sequence of cosmological N-body simulations that follow the formation and evolution of dark matter haloes in a box of side $20h^{-1}$Mpc containing 256^3 equal-mass particles from a starting redshift of z=100 to z=0. For each run we have assumed a flat cosmology with a dark energy term ($\Omega_0 = 0.24, \Omega_\Lambda = 0.76$), a Hubble parameter of $H_0 = 73\mathrm{kms}^{-1}\mathrm{Mpc}^{-1}$, a power spectrum normalisation of $\sigma_8 = 0.74$ and a primordial spectral index of n_{spec}=0.95.

The runs differ in the spatial scale below which small-scale power is suppressed in the initial conditions. We use the same realisation of the ΛCDM power spectrum in each simulation but we introduce a *truncation* or sharp cut-off in the initial power spectrum at a spatial scale that depends on the mass scale below which we wish to suppress halo formation. Physically such a suppression arises because the dark matter free streams, thus acting as a damping mechanism that washes out primordial density perturbations and introduces a cut-off in the linear matter power spectrum. This gives a free streaming scale λ_{fs} that depends on the mass of the dark matter particle, which increases as the particle mass decreases. Provided λ_{fs} is small compared to the spatial scales we are interested in simulating, then the power spectrum will differ little from the ΛCDM power spectrum.[7] As λ_{fs} becomes important, however, it is essential to determine how the power spectrum changes. This has been calculated in detail for collisionless warm dark matter models.[25]

However, we do not wish to make any assumptions about the dark matter model other than that the dark matter is collisionless and low-mass halo formation is suppressed. For this reason, we follow adopt a simple truncation of the initial power spectrum, suppressing power at wavenumbers $k \geqslant k_{\mathrm{cut}}$. We choose k_{cut} by identifying a mass scale $\mathrm{M}_{\mathrm{cut}}$ and estimating the comoving length scale R_{cut},

$$R_{\mathrm{cut}} = \left(\frac{3\,\mathrm{M}_{\mathrm{cut}}}{4\pi}\frac{1}{\overline{\rho}}\right)^{1/3} \tag{1}$$

where $\overline{\rho}$ is the mean density of the Universe. We consider four truncated models with $\mathrm{M}_{\mathrm{cut}} = 5 \times 10^9, 10^{10}, 5 \times 10^{10}$ and $10^{11}h^{-1}\mathrm{M}_\odot$ (hereafter the TruncA, TruncB, TruncC and TruncD models respectively), as well as the fiducial ΛCDM model.

2.1. *Dark Matter Halo Abundance and Spatial Clustering*

As our starting point, we compare and contrast the mass function and spatial clustering of dark matter haloes in the ΛCDM and truncated models respectively. In Figure 1, we show the cumulative mass function $N(> M)$ for the different runs at z=0, 1, 2 and 3 (top right to bottom left panels). The cumulative mass function measures the comoving number density of haloes with masses in excess of M. For comparison we show the analytic mass function of Sheth & Tormen (1999) appropriate for our adopted ΛCDM cosmology (thin dotted curves), corrected for the effects of finite volume of our simulation boxes.[6]

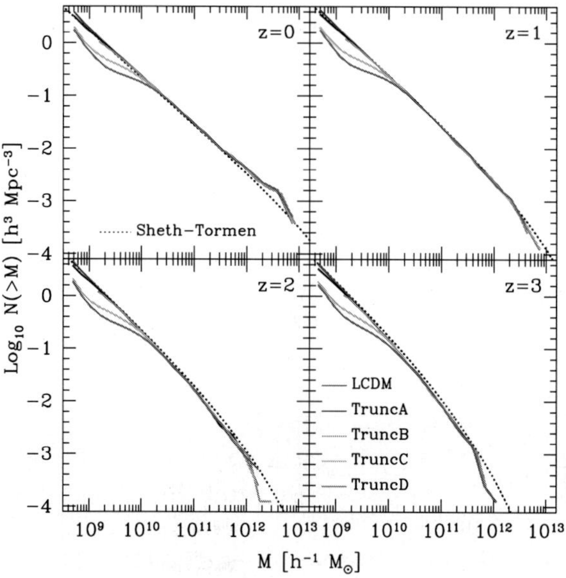

Fig. 1. **Evolution of the Halo Mass Function with Redshift.** The heavy solid curves in each panel show the mass function measured from the simulations at z=0, 1, 2 and 3 – the fiducial ΛCDM run (red), TruncA with $M_{cut} = 5 \times 10^9 h^{-1} M_\odot$ (blue), TruncB with $M_{cut} = 10^{10} h^{-1} M_\odot$ (green), TruncC with $M_{cut} = 5 \times 10^{10} h^{-1} M_\odot$ (cyan), and TruncD with $M_{cut} = 10^{11} h^{-1} M_\odot$ (magenta). The thin dashed curves correspond to the predicted mass function of Sheth & Tormen (1999), computed using the input initial power spectra for the respective simulations.

There is good agreement – to better than 10% – between the mass functions obtained in the runs for halo masses $M \gtrsim M_{cut}$. This can be understood by considering the mass variance of the linear density field $\sigma(M)$, which regulates when a particular mass scale collapses and virialises; $\sigma(M)$ is not particularly sensitive to the form of the power spectrum on small scales, and so it should be insensitive to suppression of small scale power at early times.

Differences between the ΛCDM and truncated runs become apparent for

$M \lesssim \mathrm{M_{cut}}$. Interestingly, the mass M at which $N(>M)$ begins to deviate significantly (by more than 10%) from $N(>M)_{\mathrm{CDM}}$ is approximately a factor of 2 to 2.5 smaller than $\mathrm{M_{cut}}$. The relationship between $\mathrm{M_{cut}}$, below which we suppress structure, and M, that we measure as the virial mass of a halo, is non-trivial, but we would expect M to be smaller that $\mathrm{M_{cut}}$ in general.

We note also that the mass function reveals an upturn at a mass scale that increases with $\mathrm{M_{cut}}$ at low masses (e.g. $M \sim 2 \times 10^9 h^{-1} \mathrm{M_\odot}$ in the TruncD model). This is the signature of the spurious unphysical haloes that afflict simulations of structure formation in which small scale power is suppressed.[26] We have taken care to account for these spurious haloes in our analysis.

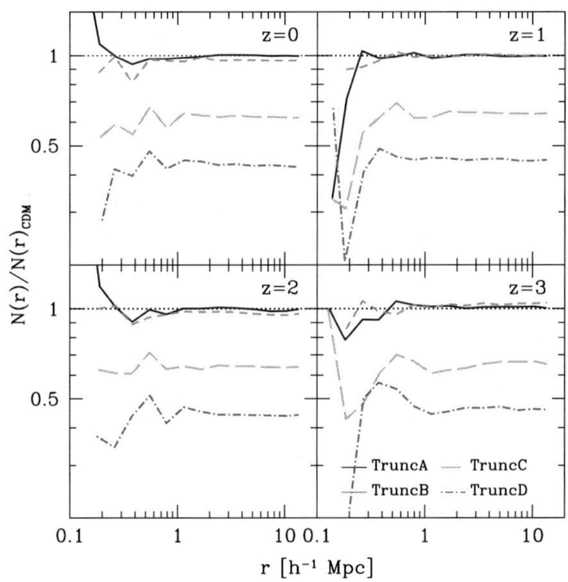

Fig. 2. **Evolution of Spatial Clustering with Redshift.** We examine how the 2-point clustering of pairs of haloes with virial masses in excess of $M_{\mathrm{vir}} \geqslant 10^9 h^{-1} M_\odot$ varies with redshift.

In Figure 2 we investigate how the clustering strength of haloes differs between the different dark matter models. We compute the separation between a pair of haloes i and j as $r = |\vec{r}_i - \vec{r}_j|$ and calculate $N(r)$, the number of haloes at a separation of r. In Figure 2, we show how the ratio $N(r)$ with respect to $N(r)$ in the fiducial ΛCDM model varies with radius for haloes in the TruncA (solid curves), TruncB (short dashed curves), TruncC (long dashed curves) and TruncD (dotted-dashed curves) models at z=0, 1, 2 and 3. We consider halo pairs where the primary mass is $M \geqslant 10^{11}h^{-1}\mathrm{M}_\odot$ and the secondary mass is $M \geqslant 3 \times 10^9 h^{-1}\mathrm{M}_\odot$.

This reveals that the clustering strength of low-mass haloes around high mass haloes decreases with increaing $\mathrm{M}_{\mathrm{cut}}$, although the dependence on $\mathrm{M}_{\mathrm{cut}}$ does not appear to be straightforward. In the TruncA and TruncB runs, we find that $N(r)/N(r)_{\mathrm{CDM}}$ is close to unity out to $r \simeq 10h^{-1}$ Mpc, never deviating by more than 10% to within $\sim 500h^{-1}$kpc at all redshifts. For the TruncC and TruncD runs, the suppression in clustering strength is quite marked – by $\sim$ 40% for the TruncC run and $\sim$ 50% for the TruncD run. Large deviations at small radii reflect the small numbers of very close pairs.

2.2. *Merging Rates*

Suppressing small scale power at early times leads to a reduction in the number density and clustering of low-mass haloes at $z \lesssim 3$. This implies that the number of neighbouring low-mass haloes and therefore the number of likely minor mergers a typical halo will experience during a given period should decline with increasing $\mathrm{M}_{\mathrm{cut}}$. In Figure 3 we show the merger rate $\Delta N/\Delta z/\Delta t$ varies with redshift. Here N represents the number of mergers in which the mass ratio is greater than 6%, which represents the lowest mass ratio merger that we can resolve in our smallest mass bin ($5 \times 10^{10}h^{-1}\mathrm{M}_\odot$ to $10^{11}h^{-1}\mathrm{M}_\odot$).

Differences between runs are immediately apparent and in the sense that we expect – as $\mathrm{M}_{\mathrm{cut}}$ increases, the merging rate decreases. For haloes with masses between $10^{11} \leqslant \mathrm{M}_{\mathrm{vir}}/\mathrm{h}^{-1}\mathrm{M}_\odot \leqslant 5 \times 10^{11}$, we find that the average merger rate in the TruncC (TruncD) model is a factor of $\sim 3(1.5)$ smaller than that in the fiducial ΛCDM model, and this is approximately constant with redshift. The difference is less pronounced for haloes with masses between $5 \times 10^{11} \leqslant \mathrm{M}_{\mathrm{vir}}/\mathrm{h}^{-1}\mathrm{M}_\odot \leqslant 10^{12}$, and for haloes with masses in excess of $10^{12}h^{-1}\mathrm{M}_\odot$ there is no discernible difference in the merging rates with redshift.

2.3. *The Growth of Halo Angular Momentum and Spin*

The results so far have revealed that galaxy- and group-mass haloes in our truncated runs are less strongly clustered than in the ΛCDM model,

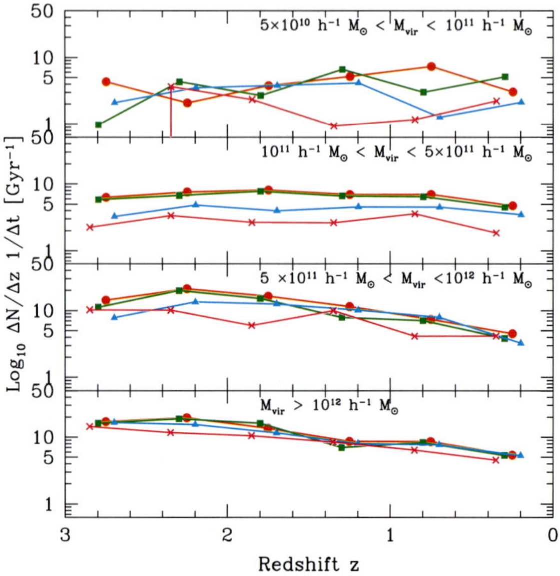

Fig. 3. **Impact on Merger Rate.** For each halo at z=0, we follow the main branch of its merger tree to higher redshifts and determine the number of mergers with mass ratios in excess of 6% experienced by the halo between z_0 and $z_1 > z_0$. From this we compute the merger rate per unit redshift per unit time. Within each of the mass bins we compute the average merger rate for haloes in the fiducial ΛCDM run (red filled circles), TruncB ($M_{cut} = 10^{10}h^{-1}M_\odot$; green filled squares), TruncC ($M_{cut} = 5 \times 10^{10}h^{-1}M_\odot$; cyan filled triangles) and TruncD ($M_{cut} = 10^{11}h^{-1}M_\odot$; magenta crosses).

and this leads to a reduction in the frequency with which galaxy-mass haloes ($M \sim 10^{11}h^{-1}M_\odot$) experience minor mergers. How is halo angular momentum and the related spin parameter affected?

The spin parameter quantifies the degree to which a halo is supported by rotation rather than velocity dispersion. We adopt the following definition of spin,

$$\lambda = \frac{J|E|^{1/2}}{GM^{5/2}}. \tag{2}$$

Here J is the total angular momentum of material of material contained within the virial radius of the halo, E is the total energy (gravitational plus kinetic) of this material, and M is the mass of this material.

In Figure 4 we show how the median spin of the galaxy- and group-mass haloes evolve with redshift. Filled circles, squares, triangles and crosses represent the median spin of the halo populations in the ΛCDM, TruncB,

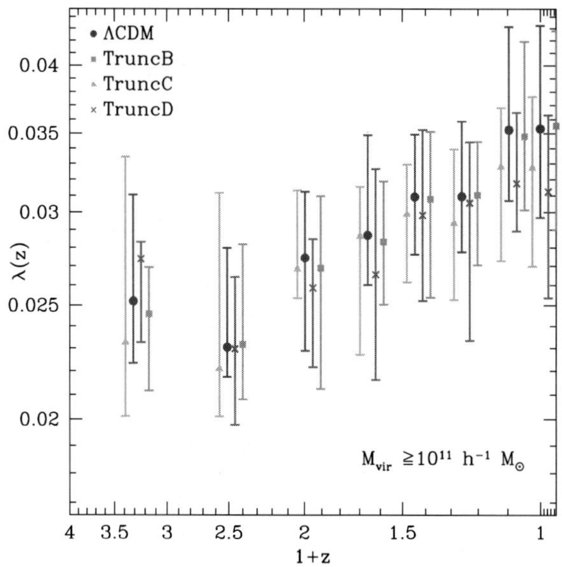

Fig. 4. **Variation of Median λ with Redshift.** We show how the median spin parameter λ_{med} varies with expansion factor. Lower and upper error bars represent the 45th and 55th percentiles. The filled circles, squares, triangles and crossed correspond to the fiducial ΛCDM, TruncB, TruncC and TruncD runs respectively.

TruncC and TruncD runs, and error bars indicate the 25th and 75th percentiles of the distribution. This figure is particularly interesting because it reveals that the median spin of the galaxy- and group-mass population does not differ significantly between models. It increases with decreasing redshift in proportion to $(1+z)^{-0.3}$ and if we include haloes with masses below M_{cut}, then we find that their spin grows at approximately the same rate.

The median spin of the halo population may evolve with redshift, but how does the spin of the main progenitor of a halo evolve over time? In Figure 5, we show how λ and the specific angular momentum J/M vary with redshift z for a selection of quiescent and violent merging histories. For each halo we determine the most significant merger δ_{max} that it has experienced since z=1, where we define δ_{max} as the ratio of the mass of the most massive halo accreted between z_0 and z z_0 (M^{acc}) by a halo's main progenitor and the virial mass of the main progenitor at z ($M_{vir}(z)$). This gives a distribution of δ_{max} and we identify haloes in the upper (lower) 20%

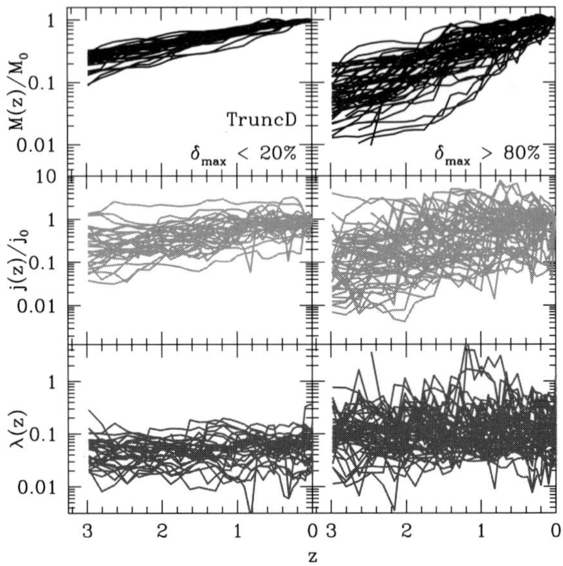

Fig. 5. **Variation of λ and J/M with Redshift for Relaxed and Unrelaxed Haloes.** We use the merging histories of haloes to identify two samples of haloes, one with a quiescent merging history ($\delta_{\rm max} \lesssim 0.2$ since z=3.0; left hand panels) and one with a violent merging history ($\delta_{\rm max} \gtrsim 0.8$ over the same period; right hand panels). The upper, middle and lower panels show the growth of halo virial mass (normalised to the virial mass at z=0), specific angular momentum J/M (normalised to the value at z=0) and the dimensionless spin parameter $\lambda = J|E|^{1/2}/GM^{5/2}$ as a function of expansion factor a.

of the distribution as systems with violent (quiescent) merging histories. We focus on the extremes in the TruncD run, but we note that the result does not depend sensitively on the model we consider.

This figure reveals a number of interesing points. The first is that the spin for a given halo it is a very noisy quantity, but when averaged over haloes in a similar dynamical state there is no clear correlation between spin and redshift. The second is that there is a clear offset between median spins in the "quiescent" and "violent" samples – haloes with violent merging histories tend to have higher spins (by factors of $\sim$ 3-4) than haloes with quiescent histories. The final point is that the specific angular momentum j for a given halo increases with redshift, although we would expect this given that the halo is growing and specific angular momentum correlates strongly with mass. However, what is interesting is that the rate of growth of j/j_0

is higher for haloes with violent mass accretion histories – as $\sim (1+z)^2$ compared to $\sim (1+z)$ for haloes in the quiescent sample.

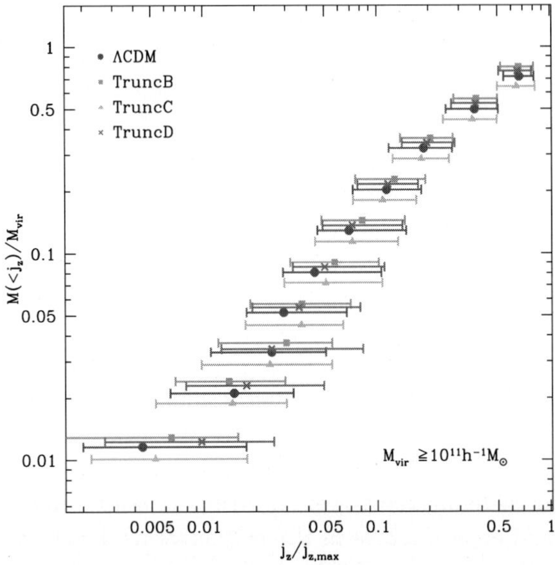

Fig. 6. **Specific Angular Momentum Profiles.** We use the method of Bullock et al. (2002) to determine the fraction of halo mass that has specific angular momentum of j_z or less. We consider only haloes that satisfy $M \geqslant 10^{11}h^{-1}M_\odot$. Note that we scale our profiles by j_{max}, the maximum specific angular momentum that we measure in our data; this is distinct from the j_{max} used in Bullock et al. (2002), who estimate j_{max} by fitting their universal angular momentum profile.

We conclude by examining the specific angular momentum profiles of galaxy- and group-mass haloes in our respective models. The specific angular momentum profile characterises the fraction of material within the virial radius that has specific angular momentum of j or less. Figure 6 shows the average of the z-component of the specific angular momentum profile $M(<j_z)$ of haloes in each of our models. For ease of comparison, we have applied small offsets to the data points from the truncated runs.

We observe that the profiles gently curve towards shallower logarithmic slopes with increasing j_z – $M(<j_z) \sim j_z^{1.2}$ for the lowest angular momentum material and $M(<j_z) \sim j_z^{1/2}$ for the highest angular momentum material. However, the r.m.s. scatter is large for a given $M(<j_z)$ in all our models and we find no statistically significant differences between models.

3. Summary

We have investigated how the angular momentum and spin of galaxy- and group-mass haloes at late times are affected by suppressing small scale power at early times. As expected, this leads to a reduction in the number density and clustering of low-mass haloes, and consequently a suppression in the frequency with which galaxy-mass haloes experience minor mergers. However, we do not find any effects in the angular momentum content and spin of our galaxy- and group-mass haloes. We have considered the redshift evolution of the median spin of the halo population, the redshift evolution of the spin of the main progenitors of haloes with quiescent and violent merging histories, and specific angular momentum profiles of haloes at $z=0$, and in all instances we find no statistically significant differences between the truncated models and the fiducial ΛCDM model.

This is an interesting result. Previous studies[24] have argued that it is minor mergers rather than tidal torquing by the surrounding matter distribution[23] that drives the growth of halo angular momentum and spin at late times. Minor mergers are sensitive to the shape of the power spectrum at small scales, while tidal torques are sensitive to the large scale gravitational field and therefore the power spectrum at large scales.[27] For these reasons we would expect haloes in the respective models to be subject to similar large scale gravitational fields and therefore similar tidal torquing. The apparent absence of any significant dependence of halo spin or angular momentum on M_{cut} suggests that the tidal field continues to be important for angular momentum growth at later times[a]. However, it also suggests that galactic disc scale length is unlikely to be sensitive to the abundance of small scale structure, which might have provided a interesting if indirect test of the CDM paradigm. Clearly more thought is required if we are to shed light on the dark matter.

References

1. D. N. Spergel, R. Bean, O. Doré, M. R. Nolta, C. L. Bennett, J. Dunkley, G. Hinshaw, N. Jarosik, E. Komatsu, L. Page, H. V. Peiris, L. Verde, M. Halpern, R. S. Hill, A. Kogut, M. Limon, S. S. Meyer, N. Odegard, G. S. Tucker, J. L. Weiland, E. Wollack and E. L. Wright, *Astrophysical Journal Supplement* **170**, 377 (2007).
2. S. D. M. White and M. J. Rees, *Monthly Notices of the Royal Astronomical Society* **183**, 341 (1978).

[a] Of course, major mergers may also play a role, but these are comparatively rare events and will tend to act as a stochastic source of angular momentum.

3. S. D. M. White and C. S. Frenk, *Astrophysical Journal* **379**, 52 (1991).
4. S. Tremaine and J. E. Gunn, *Physical Review Letters* **42**, 407 (1979).
5. D. S. Reed, R. Bower, C. S. Frenk, A. Jenkins and T. Theuns, *Monthly Notices of the Royal Astronomical Society* **374**, 2 (2007).
6. Z. Lukic, K. Heitmann, S. Habib, S. Bashinsky and P. M. Ricker, *ArXiv Astrophysics e-prints* (2007).
7. A. M. Green, S. Hofmann and D. J. Schwarz, *Monthly Notices of the Royal Astronomical Society* **353**, L23 (2004).
8. A. Klypin, A. V. Kravtsov, O. Valenzuela and F. Prada, *Astrophysical Journal* **522**, 82 (1999).
9. L. Gao, S. D. M. White, A. Jenkins, F. Stoehr and V. Springel, *Monthly Notices of the Royal Astronomical Society* **355**, 819 (2004).
10. P. Colín, V. Avila-Reese and O. Valenzuela, *Astrophysical Journal* **542**, 622 (2000).
11. V. Avila-Reese, P. Colín, O. Valenzuela, E. D'Onghia and C. Firmani, *Astrophysical Journal* **559**, 516 (2001).
12. A. Knebe, J. E. G. Devriendt, A. Mahmood and J. Silk, *Monthly Notices of the Royal Astronomical Society* **329**, 813 (2002).
13. P. Colin, O. Valenzuela and V. Avila-Reese, *ArXiv e-prints* **709** (2007).
14. G. Efstathiou, *Monthly Notices of the Royal Astronomical Society* **256**, 43P (1992).
15. A. Benson, C. Lacey, C. Baugh, S. Cole and C. Frenk, *Monthly Notices of the Royal Astronomical Society* **333**, p. 156 (2002).
16. A. Benson, C. Frenk, C. Lacey, C. Baugh and S. Cole, *Monthly Notices of the Royal Astronomical Society* **322**, p. 177 (2002).
17. S. Mao and P. Schneider, *Monthly Notices of the Royal Astronomical Society* **295**, p. 587 (1998).
18. J. Chen, A. Kravtsov and C. Keeton, *Astrophysical Journal* **592**, p. 24 (2003).
19. N. Evans and H. Witt, *Monthly Notices of the Royal Astronomical Society* **345**, p. 1351 (2003).
20. A. Font, J. Navarro, J. Stadel and T. Quinn, *Astrophysical Journal* **563** (2001).
21. S. Kazantzidis, J. Bullock, A. Zentner, A. Kravtsov and L. Moustakas, *ArXiv Astrophysics e-prints* **708** (2007).
22. S. Fall and G. Efstathiou, *Monthly Notices of the Royal Astronomical Society* **193**, p. 189 (1980).
23. S. White, *Astrophysical Journal* **286**, p. 38 (1984).
24. M. Vitvitska, A. Klypin, A. Kravtsov, R. Wechsler, J. Primack and J. Bullock, *Astrophysical Journal* **581**, p. 799 (2002).
25. P. Bode, J. Ostriker and N. Turok, *Astrophysical Journal* **556**, p. 93 (2001).
26. J. Wang and S. White, *Monthly Notices of the Royal Astronomical Society* **380**, p. 93 (2007).
27. C. Power and A. Knebe, *Monthly Notices of the Royal Astronomical Society* **370**, p. 691 (2006).

HIGHLIGHTS FROM INTEGRAL & A SYNTHETIC VIEW OF γ-RAY SOURCES

A. BODAGHEE

INTEGRAL Science Data Centre, University of Geneva Observatory, Versoix, 1290, Switzerland
E-mail: arash.bodaghee@obs.unige.ch
obswww.unige.ch

Selected scientific highlights are presented from the first 5 years of observations of the gamma-ray sky by ESA's INTEGRAL space telescope. Its unprecedented angular resolution and sensitivity at high energies ($\gtrsim 20\,\mathrm{keV}$) has allowed INTEGRAL to detect around 500 objects, many of which are new. Sources that have been classified are predominantly represented by active galactic nuclei (AGN) and X-ray binaries (XRBs) whereby a compact object (a supermassive black hole in AGN, usually a neutron star in XRBs) accretes matter from a large disk (AGN) or from a stellar companion (XRBs, often mediated by a disk). Together with unclassified sources, they account for nearly all of the diffuse Galactic background emission. Furthermore, INTEGRAL has created an all-sky map of the 511 keV distribution helping to identify potential dark matter sites. The distribution of Al-26 follows massive star-forming regions and reflects the rotation of the Galaxy. Gamma-ray bursts (GRBs) are detected in the wide field of view (FOV) at a rate of 1 per month, but INTEGRAL's design also enables it to detect GRBs outside its FOV. Previously rare, XRBs with supergiant companions are an emerging class. This underscores INTEGRAL's ability to peer through the dust that enshrouds these sources and which made them invisible to previous X-ray surveys. Their increasing numbers (as well as those of other classes) offer larger samples on which to perform statistical analyses. A synthetic view of populations of γ-ray sources is instrumental for highlighting signatures of stellar and galactic evolution. In addition, it permits a speculation on the nature of the roughly 100 sources that remain unclassified.

Keywords: INTEGRAL; gamma-ray sources; compact objects; stellar evolution; absorption; AGN; GRB; positronium.

1. Introduction

X-rays and γ-rays occupy nearly half of the total electromagnetic spectrum ($\lambda \lesssim 100\,\text{Å}$), yet astronomers in the 1950s and '60s did not expect to detect high-energy sources. Theoretically, the energy flux of a source in the

γ-rays follows $f_\nu \propto \nu^{-1}$ which gives a photon flux proportional to ν^{-2}. Extrapolation of Solar values to nearby stars led to the conclusion that there would not be enough of a signal given the state of technology. It was a pleasant surprise when a series of X-ray observations of the Moon aboard rocket flights revealed Sco X-1 and the bright X-ray background.[1]

1.1. *The INTEGRAL Space Telescope*

Since then, space telescopes such as Uhuru, Einstein, EXOSAT, and more recently CGRO, ASCA, Beppo-SAX, and ROSAT, have found $\sim 100{,}000$ sources with emission above 0.1 eV. High-energy astrophysics is currently experiencing its Golden Age with several space-based X/γ-ray missions operating simultaneously: Chandra, INTEGRAL, RXTE, Suzaku, Swift, and XMM-Newton.

INTEGRAL stands for the INTErnational Gamma-Ray Astrophysics Laboratory,[2] which is an X/γ-ray space telescope of the European Space Agency. The main instruments aboard INTEGRAL are the imager (IBIS) and the spectrometer (SPI) both operating in the ~20 keV–10 MeV range. IBIS provides a $20° \times 20°$ FOV with spatial, spectral, and timing resolutions of $\lesssim$ 1 arcmin, 1 keV at 40 keV, and 1 ms, respectively. The instrument is optimized for surveying the soft γ-ray sky, revealing faint sources in crowded fields, and recording fine timing characteristics. The germanium detectors that comprise SPI provide a spectral resolution of 2.2 keV at 1.33 MeV and a FOV with a radius of $\sim 16°$. Its angular resolution of $\lesssim 2.5°$ makes it well suited for the detection of diffuse emission. Its all-sky monitor has a timing resolution of 0.1 ms (without spatial or spectral information) allowing it to detect GRBs outside the FOV. An X-ray imager (JEM-X) and an optical camera (OMC) complement IBIS and SPI.

The 4 instruments are co-aligned so that INTEGRAL is able to observe the Universe in a spectrum that stretches from the optical band to the γ-rays. With an apogee of ~154,000 km, INTEGRAL spends ~90% of its 3-day orbit at altitudes far above the Van Allen radiation belts. This orbit was chosen so that deep, uninterrupted exposures(~200 ks) can be made with minimal contamination from the background, and to prevent instrumental damage. Data from the space telescope are transferred via ground-based receptors to the INTEGRAL Science Data Centre (ISDC) which provides a near-real time initial scientific analysis of incoming data before archiving and distributing it to the community.[3]

Gamma-rays are notoriously difficult to focus or reflect. They will simply pass through traditional lenses and mirrors, or be absorbed by material that tries to deviate them. Using the latter fact, one could place a semi-transparent mask between a γ-ray source and a detector, and deduce the properties of the source based on its shadow. This is the basic idea behind the coded-mask technique employed by all high-energy instruments on INTEGRAL.

A coded-aperture mask is an array of elements that are transparent or opaque to photons of a given energy range. The shape and size of the cells and their distribution are arbitrary, but some patterns are more effective than others. The mask is placed a certain height above a detector of similar dimensions (or smaller). Light from each source in the instrument's field of view (FOV) illuminates the coded-mask pattern onto the detector plane in a specific way. The detector records the $x-y$ coordinates on the plane, the arrival time, and the energy of every photon that it receives. An image of the sky can then be generated via deconvolution.

One of the advantages of coded-mask imaging is that source and background information are recorded simultaneously. In any given direction, the set of detector pixels that are opaque to the source defines a background that is measured under identical conditions. Another benefit is a much wider FOV for coded-mask instruments ($\sim$ degrees) than for classical telescopes ($\sim$ arcmins).

1.2. *γ-Ray Sources*

Various types of objects emit in the γ-ray domain. In most of them, the high-energy emission stems from an accelerated charge. Usually, this involves matter falling into a deep gravitational well, as in the case of an accreting compact object (CO). Electrons can also be accelerated by strong magnetic fields (i.e. cyclotron/synchrotron emission), the electrostatic fields of ions (i.e. Bremsstrahlung), or by energetic photons (Compton scattering). Violent shocks between clouds, electron-positron annihilation, and the radioactive decay of nucleosynthesis products can also produce energetic photons.

In the Milky Way, γ-ray emitters generally involve a CO consuming its companion (i.e. donor) star. Such objects are called X-ray binaries (XRBs). The CO is typically a neutron star, but a black hole candidate or white dwarf are also possible. The mass of the companion is used to distinguish the 2 main classes of XRB. When a CO is paired with a star that has $M \lesssim 1\,M_\odot$, the system is classified as a low mass X-ray binary (LMXB). If the CO has a companion of $M \gtrsim 10\,M_\odot$, the system is a high-mass X-ray binary (HMXB).

Besides XRBs, the Milky Way hosts other types of γ-ray sources such as cataclysmic variables, molecular clouds, supernova remnants, radio pulsars, pulsar wind nebulae, anomalous X-ray pulsars, soft-gamma repeaters, and others. Often, additional processes (other than accretion onto a CO) are invoked to explain a portion of the high-energy emission. These sources are categorized henceforth as "Miscellaneous."

Beyond the Milky Way and the Magellanic Clouds, the primary sources of γ radiation are active galactic nuclei (AGN) and γ-ray bursts (GRBs). These sources are among the most luminous objects in the Universe.

An AGN is a galaxy whose core is a suppermassive black hole ($M \sim 10^8\, M_\odot$). Stars, gas, and dust are consumed by the black hole after being spun and heated in a disk. The black hole and its disk are embedded in a torus of molecular gas. The disk, which is sometimes visible in the IR/optical/UV/X-rays, generate magnetic fields that trigger material outflows. Matter in the inner disk can also be ejected through the jets. In the radio, these jets are seen to extend to lobes located several Mpc away from the galaxy.

A GRB is a prodigious release of energy from a random direction that, for a short time ($\sim$0.1–100 s) can be as luminous as all other objects in the Universe combined. Shocks between expanding shells ejected from hypernovae and merging black holes are believed to produce the emission. Being unique and ephemeral events, they are clearly different from the other classes of sources discussed thus far.

Even though INTEGRAL can easily spot γ-ray sources, a broad spectral coverage remains the most reliable way of classifying a high-energy source into one of the above mentioned categories. Therefore, astrophysicists studying high-energy sources routinely complement their γ-ray data with observations taken from lower energies such as the X-rays, UV/optical, infrared, and radio bands. Given the variable nature of γ-ray emitters, follow-up observations of a newly-discovered source are ideally performed immediately after their discovery. INTEGRAL results have triggered $\sim$300 Astronomer's Telegrams (or 25% of the total) and nearly as many publications.

2. Galactic Surveys & IGR Sources

Its wide FOV in addition to unprecedented angular, spectral, and timing resolution in its energy range, have enabled INTEGRAL to detect 500 objects in 5 years of operations. In most cases, γ-ray sources are detected by stacking images into a mosaic thereby raising the effective exposure time.

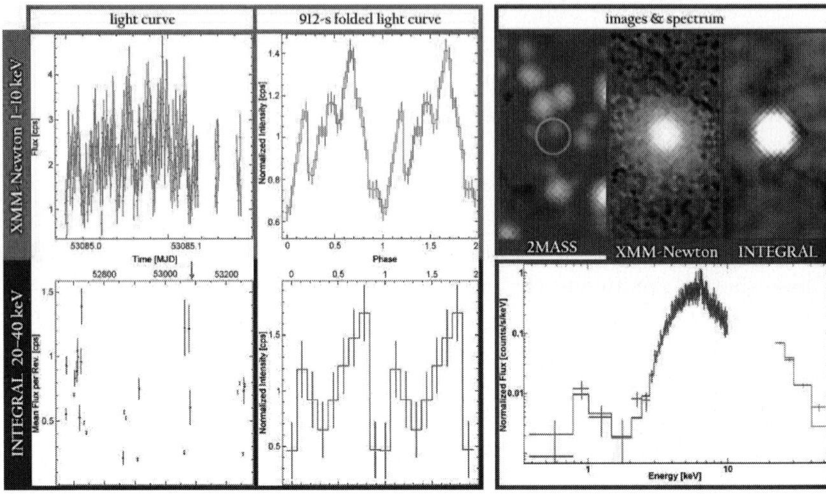

Fig. 1. Multi-wavelength imaging, timing, and spectral analyses of the absorbed HMXB pulsar IGR J16393−4643.

INTEGRAL devotes a quarter of its available time to scans of the Galactic Plane,[4,5] the Galactic Center.[6–9] Other interesting regions such as the Crux Arm,[10] Cassiopeia,[11] and the Large Magellanic Cloud[12] are regular targets of observing campaigns. Full-sky surveys have become possible[13,14] thanks to a recent increase in time dedicated to filling in less exposed regions of the sky.

Many of the sources detected in these surveys are new. Their identifying prefix IGR stands for INTEGRAL Gamma-Ray source. A comprehensive list of IGRs and all related publications can be found at the following URL: `http://isdc.unige.ch/~rodrigue/html/igrsources.html`. The first IGR source, IGR J16318−4848, presents the highest column density ever seen in a Galactic object ($N_{\rm H} \sim 10^{24}\,{\rm cm}^{-2}$).[15,16] Such pronounced absorption can only be attributed to matter local to the source since it is 2 orders of magnitude greater than the column density expected along the line of sight (which is also rather high).

Soon after this discovery, dozens of other such objects emerged to form a growing class of supergiant HMXBs.[17–19] They were missed by previous X-ray surveys due to their strong intrinsic and extrinsic absorption, and due to their location among luminous sources in the crowded Galactic Plane and spiral arms. Photons emitted above 20 keV are energetic enough to penetrate obscuring matter along the line of sight. By operating in this en-

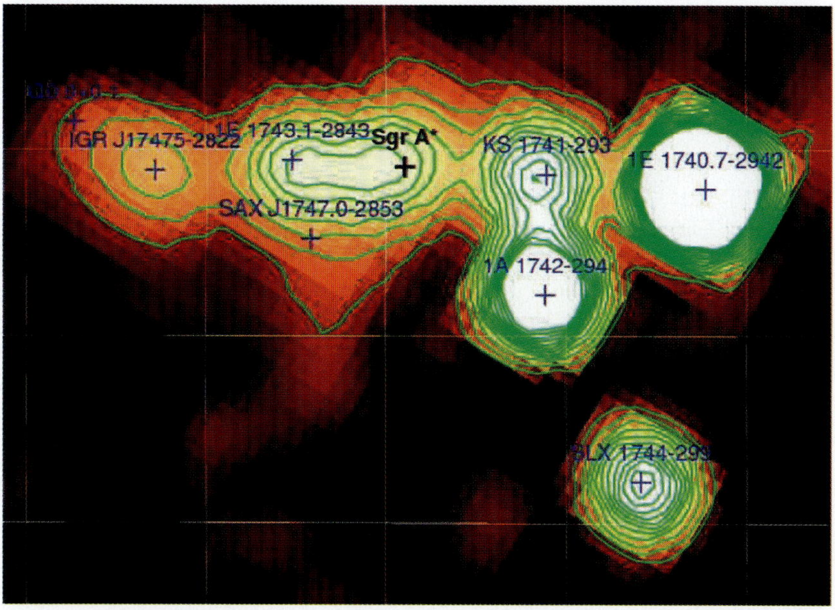

Fig. 2. INTEGRAL-IBIS mosaic image of the Galactic Center in the 20–40 keV range. This deep exposure accumulates 7 Ms of data with contours denoting significance levels of 6 to more than 75 σ (Bélanger et al., 2006).

ergy range, INTEGRAL is able to unveil objects that are otherwise hidden behind both the shell of material around the source, and the dust and gas in their direction. Given the large mass of the donor star in these systems, they are relatively young. An increase in their ranks permits their use as independent probes of stellar and Galactic evolution.[8,20,21]

Coherent pulsations were discovered in several absorbed IGRs.[22–26] Figure 1 presents the timing and spectral behavior of one such source, IGR J16393−4643. The combined characteristics point to a persistent emitter where the CO is a neutron star with a spin period of ~ 1 ks, and where the companion star is a supergiant of spectral type O or B. The pulsation is caused by a misalignment between the neutron star's axis of rotation and its magnetic field which channels material that is to be accreted along the field lines to the poles. In order to exhibit such strong attenuation of the spectrum $\lesssim 5$ keV (i.e. absorption), the neutron star must be deeply embedded in and accreting from the stellar wind. This is corroborated by the persistent emission, the presence of iron lines at 6.4 and 7.1 keV, and by a soft excess emission at 1 keV.

INTEGRAL has also helped to shed light on a related group of HMXBs:

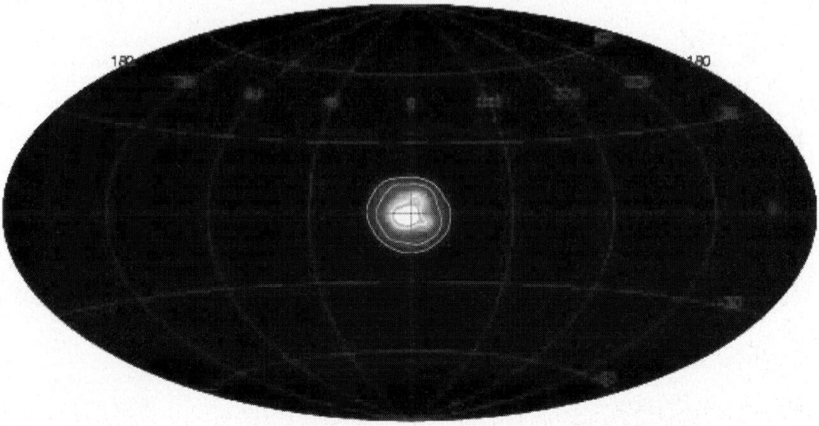

Fig. 3. All-sky distribution of the 511 keV line from e^-–e^+ annihilation as measured by INTEGRAL-SPI. The effective exposure time at the Galactic Center is 5 Ms (Weidenspointner et al., 2006).

the emerging class of Supergiant Fast X-ray Transients (SFXTs).[27] Orbits of the neutrons star in these systems are larger than in the persistent emitters described above. The sporadic (and in at least 1 system episodic) nature of the high-energy emission in SFXTs is believed to occur whenever the neutron star passes through layered stellar winds[28] and/or from the accretion of stellar wind clumps.[29,30]

3. Galactic Center

Deep exposures of the Galactic Center illustrate well the unique abilities of INTEGRAL to unveil γ-ray sources (Fig. 2). Studying an environment that is heavily absorbed and densely populated with bright objects, IBIS revealed γ-ray emission from the vicinity of the Milky Way's central black hole, Sgr A*, as diffuse radiation stemming from a compact source.[9] Emission from this source, dubbed IGR J17456−2901, is Compton reflected off of a molecular cloud (Sgr B2 = IGR J17475−2822) located a few hundred parsecs from the Galactic Center. The characteristics of this scattered light suggest that the massive black hole at the heart of our Galaxy was more active in the recent past.[31]

4. Very High-Energy Sources

The γ-ray source associated with Sgr A*, IGR J17456−2901, has also been observed at very high energies (TeV) by HESS.[32] Several IGR sources are

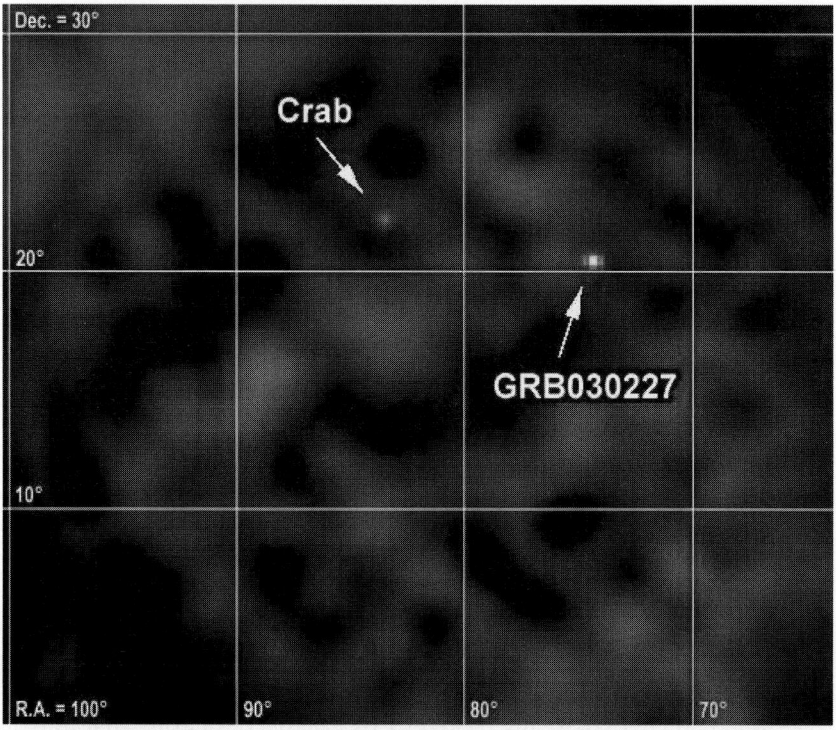

Fig. 4. Image of a GRB captured by INTEGRAL-SPI. The energy range is 20–200 keV for an exposure time of 18 s (Beckmann et al., 2004).

likely to represent the soft γ-ray counterparts to TeV sources. Most of them are shell-type supernova remnants (SNR) or pulsar wind nebulae (PWN). Synchrotron processes near the pulsar account for the soft γ-ray emission, whereas the TeV photons are explained by electron jets from the pulsar undergoing inverse Compton upscattering against the Cosmic Microwave Background.[33–35]

5. Extragalactic Sources

INTEGRAL has helped fill gaps in the broad-band spectra of bright, well-known AGN such as 3C 273.[36–38] However, the majority of extragalactic sources detected by INTEGRAL are low luminosity Seyfert galaxies.[39–41] Over 150 AGN have been detected between 20 and 100 keV, with 30% of them emitting above 100 keV, and 10% above 150 keV.[42] For the most part,

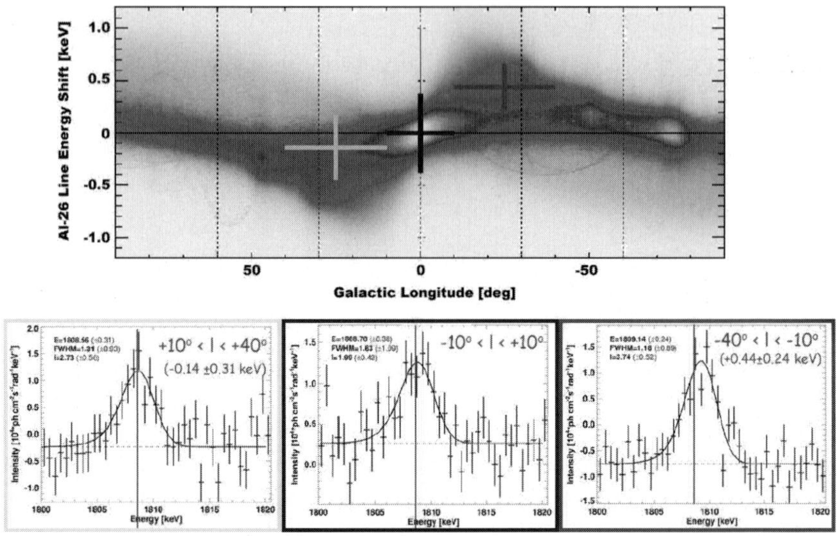

Fig. 5. INTEGRAL-SPI spectra and distribution of Galactic ^{26}Al. The spectra are based on 4 Ms of exposure time (Diehl et al., 2006).

they are situated relatively nearby with an average redshift of $z \sim 0.02$. Recently, IBIS resolved a hard X-ray extension to the Coma cluster.[43]

Around 50 GRBs were discovered by INTEGRAL equating to a rate of 1 per month. Capturing a GRB in the INTEGRAL FOV is a serendipitous discovery as the target of the observation is never the GRB itself. The large FOV of IBIS and SPI certainly enhances their chances of spotting a GRB. Plus, the walls of SPI act as an anti-coincidence shield enabling it to detect GRBs outside the FOV at a rate of 1 per week.

During an observation of the Crab, SPI caught a GRB whose intensity dwarfs that of the target (Fig. 4). Bear in mind that the Crab is one of the brightest sources in the γ-ray sky and it serves as the standard candle in high-energy astronomy (source fluxes are traditionally given in units of mCrabs for ease of comparison). The brightest GRB yet detected by INTEGRAL exhibits polarized emission[44] while the least luminous GRB has a redshift of only ~ 0.1 making it the closest one ever seen.[45]

6. Positron Annihilation

Emission at 511 keV from electron-positron annihilation is strongly detected by the Ge spectrometers of SPI.[46–48] Since the origin of the line is still uncertain, any information about the distribution of the emission could

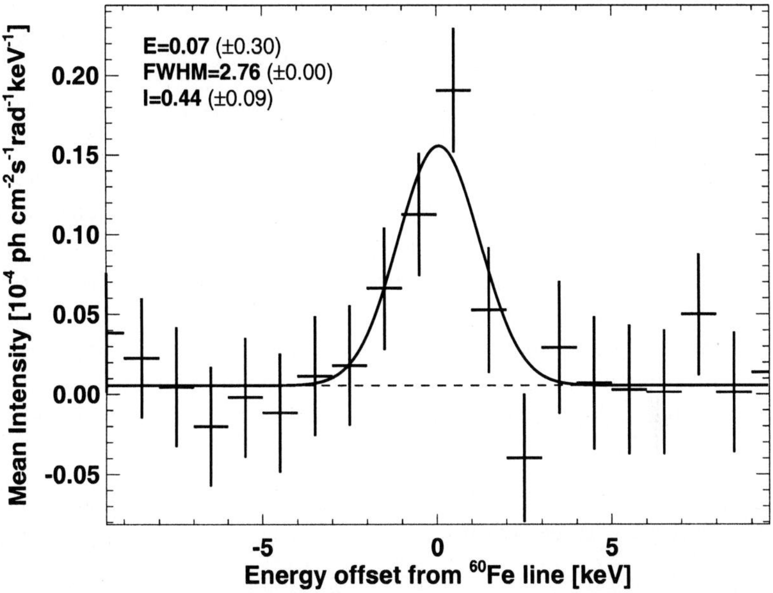

Fig. 6. Spectrum of the ^{60}Fe line in the Galaxy taken by INTEGRAL-SPI with 7 Ms of exposure time (Wang et al., 2007).

help solve the puzzle. It appears that the 511 keV distribution is mostly confined to the Galactic Center and Bulge. The emission is symmetrical around the Galactic Center with an extension of 8° FWHM. Furthermore, recent measurements have revealed some contributions from the disk and halo,[49] although the bulge/disk luminosity ratio is between 3 and 9.[48] These results place constraints on the main source of positrons and on the shape of the Milky Way's dark matter halo.[50] The portion attributable to the disk is compatible with β^+ decay from ^{26}Al and ^{44}Ti produced by core-collapse supernovae.[48] Bulge emission could originate from Sgr A*,[51] old stellar populations, LMXBs, novae, type Ia supernovae, cosmic rays interacting with the ISM, and/or light dark matter.[48,52]

7. Nucleosynthesis

INTEGRAL has observed other bright emission lines. Among the most prominent of these is the line at 1809 keV from the radioactive decay of ^{26}Al. Produced by stars in their terminal stages, and with a half-life of $\tau_{1/2} \sim 7\times10^5$ y, ^{26}Al serves as a key tracer of recent massive star formation.

Constraints from the line provide an independent estimate of the Galactic supernova (type Ib/c, II) rate at 1.9±1.1 per century.[53] This implies a star formation rate of ~7.5 per year. In addition, SPI detects shifts in the line corresponding to the rotation of the inner Galaxy, thus confirming a Galactic origin for the emission.

Emission lines at 1173 & 1333 keV constitute the first conclusive signal of ^{60}Fe decay.[54] Iron-60 is another diagnostic of massive stars since it is a byproduct of late-stage stellar nucleosynthesis with $\tau_{1/2} \sim 2 \times 10^6$ y. The ratio of ^{60}Fe to ^{26}Al is roughly 15% or well below theoretical predictions (40%).

In Cas A, lines at 67.8 and 78.4 keV attributed to ^{44}Ti are detected at the 3σ level.[55,56] The half-life is only 86 y so young, nearby supernova remnants such as Cas A are obvious targets of searches for this short-lived isotope.

8. Hard X-Ray Background

The peak of the diffuse cosmic X-ray background is situated around 30 keV or well within INTEGRAL's effective range. The intensity and origin of the emission are not well understood. It is believed that the bulk of the radiation can be attributed to the summed emission from all AGN in the Universe. Luminosity functions and cumulative spectra derived from IBIS surveys of AGN are consistent with those of the cosmic X-ray background if one assumes little change in the column density distribution of AGN since $z \sim 1.5$[41] or assuming an evolution of the AGN population.[57]

Known and unidentified point sources from IBIS surveys account for the near-totality of the diffuse Galactic background emission with only a negligible contribution from diffuse emission processes.[58] However, the energy spectra from these point sources drop off sharply at high energies, leaving open the possibility for diffuse processes and/or unresolved weak γ-ray sources to explain the harder ridge emission.[59]

9. A Synthetic View of γ-Ray Sources

The large numbers of objects detected by INTEGRAL make it possible to perform statistical analyses on the different classes of sources. Immediately apparent from Table 1 is the lack of LMXBs discovered by INTEGRAL. Whereas the other classes (HMXBs, AGN, and Misc.) are similarly represented in both new and previously-known sources, only 6 new LMXBs have been found. At the same time, half of all IGRs are unclassified.

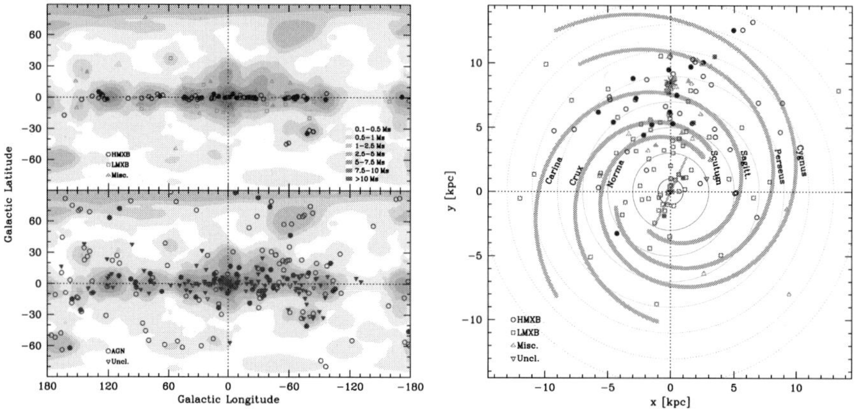

Fig. 7. Spatial distribution of the different classes of sources detected by INTEGRAL-IBIS. The image on the right presents sources whose distances are known overlaid on a 4-arm spiral model of the Galaxy (Russeil, 2003) with the Sun placed 8.5 kpc from the center (red star). Filled symbols represent new sources (≡ IGRs). The exposure map covers the first 500 revolutions.

Table 1. The number of sources from each of the major classes detected by INTEGRAL-IBIS are listed for new (≡ IGRs) and previously-known sources.

	HMXBs	LMXBs	AGN	Misc.	Uncl.	Total
IGRs	32 (15%)	6 (3%)	50 (23%)	15 (7%)	111 (52%)	214
previously-known	46 (16%)	76 (27%)	113 (40%)	32 (11%)	18 (6%)	285
Total	78 (16%)	82 (16%)	163 (33%)	47 (9%)	129 (26%)	499

Figure 7 presents the spatial distribution of sources detected by IBIS. Given the large mass of the companion, HMXBs are young systems ($\lesssim 10^7$ yr) so they are located close to regions of recent star formation, i.e. close to the Galactic Plane and in the outer disk. On the other hand, the small mass of the companion star in LMXBs means that they are old ($\sim 10^{10}$ yr) so they are found predominantly in the Galactic Bulge and inner disk, or they have had time to migrate off the plane. The unclassified sources are expected to include many AGN hidden among them, but they are distributed spatially rather like a Galactic population composed essentially of LMXBs.

On average, Galactic IGRs are 4 times more absorbed than sources detected by INTEGRAL that were previously known (Fig. 8). Kolmogorov-Smirnov tests yield a probability of $\ll 0.01\%$ that the distributions are

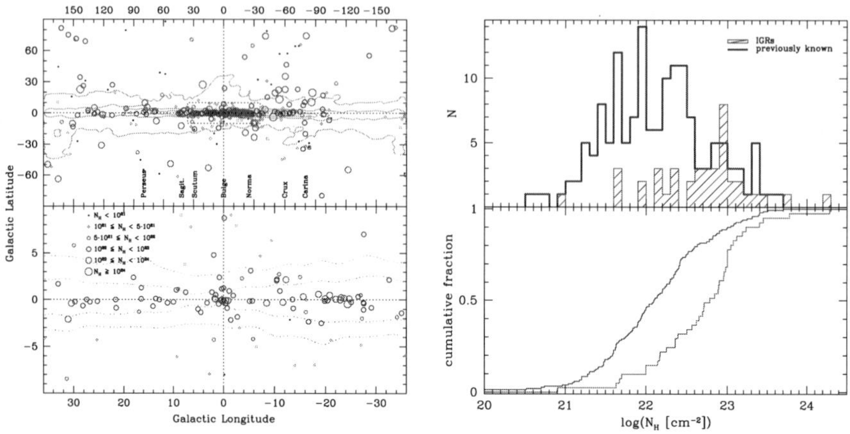

Fig. 8. *Left*: Spatial distribution of sources detected by INTEGRAL-IBIS where the symbol size is proportional to the reported column density. The lower panel focuses on the Galactic Bulge and excludes extragalactic sources. Dotted contours denote N_H expected along the line of sight (10^{21}–10^{22} cm^{-2}). *Right*: Distribution of reported column density and Kolmogorov-Smirnov test for Galactic sources that are new ($\equiv$ IGRs) or that were previously known.

statistically compatible. Curiously, the same is not true of extragalactic IGRs suggesting that Compton-thick AGN are rare. The map of reported column densities presented in Fig. 8 is akin to an X-ray of the Milky Way's body. It unveils a cluster of absorbed sources towards the Norma Arm, consistent with the view that these systems are young and hence should be near sites where massive stars were recently formed.

Pulse periods of IGRs are mostly in the 100–1000 s range. This is longer than the average pulsation period of sources detected by INTEGRAL that were previously known, but it is a typical value for the wind-fed pulsars of the type that IBIS is finding. An exceptional case is the accreting millisecond pulsar IGR J00291+5934 which rotates once every 1.7 ms.[60] In terms of orbital periods, IGRs display a double-peaked distribution owing to underlying populations of short-period cataclysmic variables and LMXBs, mixed with long orbital-period HMXBs. This bimodal distribution is consistent with the distribution of orbital periods from sources that were known prior to INTEGRAL.

10. Outlook

Results from the first 5 years of INTEGRAL cover a broad range of topics: compact objects; accretion/emission processes; dark matter searches;

positron annihilation; nucleosynthesis; stellar/galactic evolution; clusters of galaxies, etc. With a mission lifetime guaranteed for at least another 5 years, INTEGRAL will continue to provide key insights into the most energetic phenomena of the Cosmos.

Acknowledgments

AB thanks the conference organizers for extending an invitation to present this topic at Dark 2007, and for the chance to visit Oz. This publication uses observations obtained with the ESA science mission INTEGRAL. The INTEGRAL instrument and data centres were directly funded by ESA member states and the USA (NASA).

References

1. R. Giacconi *et al.*, *Physical Review Letters* **9**, 439 (1962).
2. C. Winkler *et al.*, *Astronomy & Astrophysics* **411**, L1 (2003).
3. T. J.-L. Courvoisier *et al.*, *Astronomy & Astrophysics* **411**, L53 (2003).
4. A. J. Bird *et al.*, *Astrophysical Journal Letters* **607**, L33 (2004).
5. A. J. Bird *et al.*, *Astrophysical Journal Letters* **636**, 765 (2006).
6. M. G. Revnivtsev *et al.*, *Astronomy Letters* **30**, 382 (2004).
7. G. Bélanger *et al.*, *Astrophysical Journal Letters* **601**, L163 (2004).
8. A. Lutovinov *et al.*, *Astronomy & Astrophysics* **433**, L41 (2005).
9. G. Bélanger *et al.*, *Astrophysical Journal* **636**, 275 (2006).
10. M. G. Revnivtsev *et al.*, *Astronomy Letters* **32**, 145 (2006).
11. P. R. den Hartog *et al.*, *Astronomy & Astrophysics* **451**, 587 (2006).
12. D. Götz *et al.*, *Astronomy & Astrophysics* **448**, 873 (2006).
13. A. J. Bird *et al.*, *Astrophysical Journal Supplement* **170**, 175 (2007).
14. R. Krivonos *et al.*, *ArXiv Astrophysics e-prints* (2007).
15. T. J.-L. Courvoisier *et al.*, *IAU Circular* **8063**, 3 (2003).
16. M. G. Revnivtsev *et al.*, *Astronomy Letters* **29**, 587 (2003).
17. R. Walter *et al.*, IGR J16318-4848 & Co.: a new population of hidden high-mass X-ray binaries in the Norma Arm of the Galaxy, in *The INTEGRAL Universe. Proceedings of the Fifth INTEGRAL Workshop*, 2004.
18. E. Kuulkers, An absorbed view of a new class of INTEGRAL sources, in *AIP Conf. Proc. 797: Interacting Binaries: Accretion, Evolution, and Outcomes*, 2005.
19. R. Walter *et al.*, *Astronomy & Astrophysics* **453**, 133 (2006).
20. A. J. Dean *et al.*, *Astronomy & Astrophysics* **443**, 485 (2005).
21. A. Bodaghee *et al.*, *Astronomy & Astrophysics* **467**, 585 (2007).
22. A. Lutovinov *et al.*, *Astronomy & Astrophysics* **430**, 997 (2005).
23. J. Rodriguez *et al.*, *Astronomy & Astrophysics* **432**, 235 (2005).
24. J. Rodriguez *et al.*, *Monthly Notices of the Royal Astronomical Society* **366**, 274 (2006).
25. J. A. Zurita Heras *et al.*, *Astronomy & Astrophysics* **448**, 261 (2006).

26. A. Bodaghee *et al.*, *Astronomy & Astrophysics* **447**, 1027 (2006).
27. I. Negueruela *et al.*, Supergiant Fast X-ray Transients: a new class of high mass X-ray binaries unveiled by INTEGRAL, in *The X-ray Universe 2005, ESA SP-604*, 2006.
28. L. Sidoli *et al.*, *ArXiv e-prints* **710** (2007).
29. J.-C. Leyder *et al.*, *Astronomy & Astrophysics* **465**, L35 (2007).
30. R. Walter and J. Zurita Heras, *ArXiv e-prints* **710** (2007).
31. M. G. Revnivtsev *et al.*, *Astronomy & Astrophysics* **425**, L49 (2004).
32. F. Aharonian *et al.*, *Astronomy & Astrophysics* **425**, L13 (2004).
33. A. Malizia *et al.*, *Astrophysical Journal Letters* **630**, L157 (2005).
34. P. Ubertini *et al.*, *Astrophysical Journal Letters* **629**, L109 (2005).
35. R. Landi *et al.*, *Monthly Notices of the Royal Astronomical Society* **380**, 926 (2007).
36. T. J.-L. Courvoisier *et al.*, *Astronomy & Astrophysics* **411**, L343 (2003).
37. M. Türler *et al.*, *Astronomy & Astrophysics* **451**, L1 (2006).
38. M. Chernyakova *et al.*, *Astronomy & Astrophysics* **465**, 147 (2007).
39. L. Bassani *et al.*, *Astrophysical Journal Letters* **636**, L65 (2006).
40. V. Beckmann *et al.*, *Astrophysical Journal* **638**, 642 (2006).
41. S. Sazonov *et al.*, *Astronomy & Astrophysics* **462**, 57 (2007).
42. A. Bazzano *et al.*, *Astrophysical Journal Letters* **649**, L9 (2006).
43. D. Eckert *et al.*, *Astronomy & Astrophysics* **470**, 835 (2007).
44. S. McGlynn *et al.*, *Astronomy & Astrophysics* **466**, 895 (2007).
45. S. Y. Sazonov *et al.*, *Nature* **430**, 646 (2004).
46. J. Knödlseder *et al.*, *Astronomy & Astrophysics* **441**, 513 (2005).
47. P. Jean *et al.*, *Astronomy & Astrophysics* **445**, 579 (2006).
48. G. Weidenspointner *et al.*, *Astronomy & Astrophysics* **450**, 1013 (2006).
49. G. Weidenspointner *et al.*, *ArXiv Astrophysics e-prints* (2007).
50. Y. Ascasibar *et al.*, *Monthly Notics of the Royal Astronomical Society* **368**, 1695 (2006).
51. K. S. Cheng *et al.*, *Astronomy & Astrophysics* **473**, 351 (2007).
52. M. Pospelov and A. Ritz, *Physics Letters B* **651**, 208 (2007).
53. R. Diehl *et al.*, *Nature* **439**, 45 (2006).
54. W. Wang *et al.*, *Astronomy & Astrophysics* **469**, 1005 (2007).
55. J. Vink, *Advances in Space Research* **35**, 976 (2005).
56. M. Renaud *et al.*, *New Astronomy Review* **50**, 540 (2006).
57. V. Beckmann *et al.*, *Astrophysical Journal* **652**, 126 (2006).
58. F. Lebrun *et al.*, *Nature* **428**, 293 (2004).
59. J. Knödlseder *et al.*, Soft gamma-ray galactic ridge emission as unveiled by SPI aboard INTEGRAL, in *American Institute of Physics Conference Series*, 2007.
60. D. K. Galloway *et al.*, *Astrophysical Journal Letters* **622**, L45 (2005).

PART IV

Dark Matter and Dark Energy

NON-LINEAR MATTER POWER SPECTRUM TO 1% ACCURACY BETWEEN DYNAMICAL DARK ENERGY COSMOLOGIES

M. J. FRANCIS*, G. F. LEWIS

Institute of Astronomy, University of Sydney
Sydney, NSW, Australia
** E-mail: mfrancis@physics.usyd.edu.au*

E. V. LINDER

Lawrence Berkeley National Laboratory
Berkeley, California, United States

We present a matching procedure for rapidly calculating the non-linear mass power spectrum in dynamical dark energy cosmologies to percent level accuracy in the range of scales between $0.1 < k < 3$. This procedure is verified by large N-body simulations and is valid at any redshift for cosmologies consistent with current observations. Such accuracy in the power spectrum is necessary for next generation cosmological mass probes. Our matching procedure reproduces the CMB distance to last scattering and delivers sub-percent level accuracy in the matter power spectra at $z = 0$ and $z \approx 3$. We discuss the physical implications for probing dark energy with surveys of large scale structure.

Keywords: methods:N-body simulations — methods: numerical — dark matter — dark energy — large-scale structure of Universe.

1. Introduction

The apparent acceleration of the Universe and the existence of an exotic energy component with negative pressure, dubbed 'dark energy', it implies is a startling discovery of modern cosmology. While distance measures, such as the observation of Supernova[9,18] and the Cosmic Microwave Background[21] are important probes of the expansion history of the Universe, and hence the nature of dark energy, they are largely unable to distinguish between models with dark energy or possible modifications to gravity. To break this and other parameter degeneracies, observations of cosmic structure provide an important piece of the dark energy puzzle.

The mass power spectrum is a key statistic of the large scale structure in the Universe and cosmological probes such as weak gravitational lensing, baryon acoustic oscillations, or galaxy/cluster abundances depend crucially on accurate knowledge of the mass power spectrum for physical interpretation of the data from large scale surveys.

Most commonly the mass power spectrum is approximated by the 'Halofit' form of,[20] and this is used to determine cosmological parameters from the data (or estimate future precision of parameter extraction). However, the[20]— formula is calibrated only on ΛCDM models, and for these has a precision of ~10%. This will be insufficient for future large structure surveys that aim to explore the acceleration of the cosmic expansion and the properties of dark energy responsible for it.[7] estimate that for weak gravitational lensing surveys, for example, 1% accuracy in knowledge of the mass power spectrum will be required.

N-body simulations provide a well-tested technique for calculating the dark matter power spectrum at the percent level.[4] However, carrying out simulations for every possible cosmological model is obviously impractical. If one could devise a mapping procedure that matched models with the same key physical quantities, ideally to a single class of cosmologies like ΛCDM, then this would greatly aid the study of the cosmological information carried by the distribution and growth of large scale structure.

2. Dark Energy and Cosmic Structure

The power spectrum of matter density perturbations in the Universe cannot be analytically calculated (at present) once the growth of the perturbations has become non-linear and proceeded far enough to create collapsed virialised objects with complex merger histories. The full, nonlinear mass power spectrum requires N-body simulations and this work studies the dark energy effects on the full, non-linear mass power spectrum to models with dynamical dark energy, utilizing the model independent, physically motivated[10] evolving equation of state $w(a) = w_0 + w_a(1 - a)$.

Previous work in this area,[8,13–17,25] has focused on constant equation of state models, or a limited set of evolving equation of state models. See[3] for more details of previous literature.

In studying the non-linear power spectrum of mass fluctuations, a natural place to start is with the linear power spectrum. The effects of dark energy on the linear mass power spectrum can be calculated through the relation

$$P(k,a) = \frac{D^2(a)}{D^2(a_i)} P(k,a_i) \tag{1}$$

(see e.g.[1]) with the growth factor $D(a)$ given by the formula (e.g.[5,12])

$$D'' + \frac{3}{2}\left[1 - \frac{w(a)}{1+X(a)}\right]\frac{D'}{a} - \frac{3}{2}\frac{X(a)}{1+X(a)}\frac{D}{a^2} = 0 \qquad (2)$$

with derivatives with respect to scale factor a, and where $X(a)$ is the ratio of the matter to dark energy densities, given by $X(a) = \frac{\Omega_m}{1-\Omega_m}e^{-3\int_a^1 d\ln a' w(a')}$, with Ω_m the dimensionless present matter density.

The non-linear power spectrum cannot be written in terms of a simple differential equation and requires the use of large volume, high resolution, N-body simulations. These are computationally expensive and therefore accurate semi-analytic fitting formulas derived from simulation results are a valuable tool. The most widely adopted current formula, sometimes called Halofit, was presented in.[20] This formula is motivated by the halo model of structure growth with free parameters in the function set by fitting to a large suite of simulations. All these simulations, however, were of cosmological constant, $w = -1$, cosmologies.[17] produced a fitting formula as a multipolynomial series for constant w models, intended to be used to modify the Smith et al result. This modification was estimated to be accurate to within a few percent in the range of cosmologies encompassed by the simulation grid.

Taking a different approach,[13] demonstrated that when the *linear* growth factors between different constant w models were matched at a high redshift point as well as at $z = 0$, by compensating with other cosmological parameters, the *non-linear* power spectrum from N-body simulations also matched to much better than a percent at those redshifts, as well as matching to one to two percent at any intermediate redshift. Additionally,[13] also found that the distance to the surface of last scattering, $d_{\rm lss}$, closely matched when their growth matching criteria was implemented, preserving CMB constraints. With this formalism the power spectrum for a dark energy model can be matched to, say, a ΛCDM cosmology. Hence one can either use an appropriately matched Halofit result or carry out a vastly reduced suite of only ΛCDM simulations to achieve the desired accuracy on the mass power spectrum.

This article concentrates on developing accurate matching of the non-linear mass power spectrum for dynamical dark energy models. We employ a somewhat different matching procedure from[13] explicitly matching the distance to CMB last scattering $d_{\rm lss}$ and the mass fluctuation amplitude σ_8 at the present and studying the effect on the growth. In this respect, our approach is essentially the converse of the approach of.[13] The geometric factor of the distance to CMB last scattering suffices to incorporate substantially the CMB constraints on the dark energy parameters. Since dark energy had a negligible density in the early universe (except in special, early dark energy models, e.g. see[2,11,24]), the physical size and nature of features in the CMB at the surface of last scattering is largely insensitive to the properties of dark energy. However, the angular size of such features is set through the angular diameter distance, which does depend on the properties of dark energy, since it relates to the expansion history of the universe $a(t)$. Therefore, dark energy models giving the same distance to the last scattering surface are largely degenerate with respect to the CMB (some differences relating to secondary anisotropies such as the ISW effect remain, see[6]). For a given dynamic dark energy model (w_0, w_a), there is a corresponding constant equation of state $w_{\rm eff}$, say, that gives the same $d_{\rm lss}$ as the dynamical model, holding all other cosmological parameters (such as the physical matter density $\Omega_m h^2$) fixed. This article examines the relation between the non-linear mass power spectra of the dynamical and the $w_{\rm eff}$ models. Once a tight correspondence is established, one can then either employ a constant w fitting formula such as from,[17] carry out only a suite of constant w simulations, or adjust the other cosmological parameters such that one chooses $w_{\rm eff} = -1$ and requires only ΛCDM simulations. We discuss these alternatives further in §6.

3. Simulation Details

The simulations were performed using the GADGET-2 N-body code,[22] modified to incorporate the background evolution $a(t)$ appropriate for dynamical dark energy cosmologies with $w(a) = w_0 + (1-a)w_a$. For more details and on the simulations and numerical convergence see[3]

All simulations in this paper used the best fit cosmological parameters from[21] of $\Omega_m = 0.234$, $h = 0.74$, $\Omega_b = 0.0407$ and $\sigma_8 = 0.76$ in a flat ΛCDM universe. For each set of simulations, a constant equation of state $w_{\rm eff}$ is selected and several values for the parameters w_0 and w_a that maintained the same $d_{\rm lss}$ were calculated. One consequence of this methodology is that these $w(a)$ models cross the value $w = -1$ at some point in cosmic history. Debate exists surrounding the physical validity of crossing between the phantom regime, defined as $w < -1$, and $w > -1$. This will eventually be settled by a microphysical theory for dark energy, rather than merely a phenomenological description. With this in mind we do not consider the issue of phantom energy and phantom crossing further.

We select three values, $w_{\rm eff} = -0.9, -1, -1.1$, as the foundations for our comparison of $w(a)$ cosmologies. This range is in accord with constraints on constant w from current cosmological data sets[19,21] and provides a reasonable variety for testing the matching procedure. For each constant w model, simulations were carried out for four more $w(a)$ models with matching distance to the LSS. The dark energy models used are summarized in Table 1.

Table 1. Distance Matched Models. Simulations were carried out for five models (including $w = w_{\rm eff}$) for each of three values of $w_{\rm eff}$, where all five models in a column had identical distances to CMB last scattering.

$\|w_{\rm eff} = -0.9\|$		$\|w_{\rm eff} = -1.0\|$		$\|w_{\rm eff} = -1.1\|$	
w_0	w_a	w_0	w_a	w_0	w_a
−1.1	0.620	−1.2	0.663	−1.3	0.707
−1.0	0.319	−1.1	0.341	−1.2	0.363
−0.8	−0.336	−0.9	−0.359	−1.0	−0.381
−0.7	−0.686	−0.8	−0.732	−0.9	−0.778

4. Results

The ratio of power measured in the simulation outputs at $z = 0$ relative to the central $w_{\rm eff}$ model for each of the sets of simulations was calculated. The trends were observed to be very similar regardless of the value of $w_{\rm eff}$ and hence for brevity, only the $w_{\rm eff} = -1$ models are displayed, in Fig. 1.

The most outstanding result is the excellent agreement between the

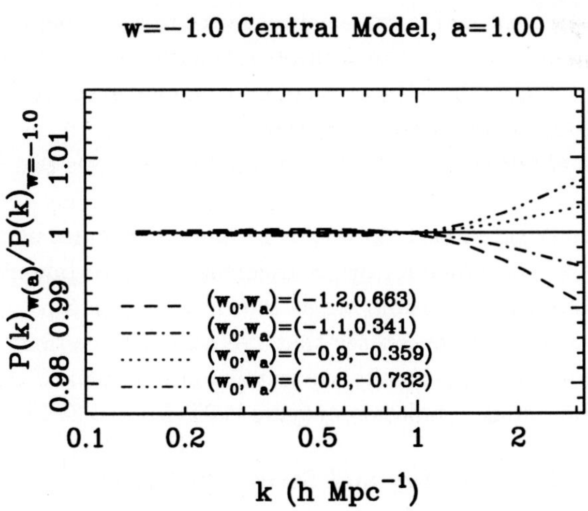

Fig. 1. Ratio of the non-linear mass power spectrum at z=0 relative to the $w = -1$ model for models with matched $d_{\rm lss}$.

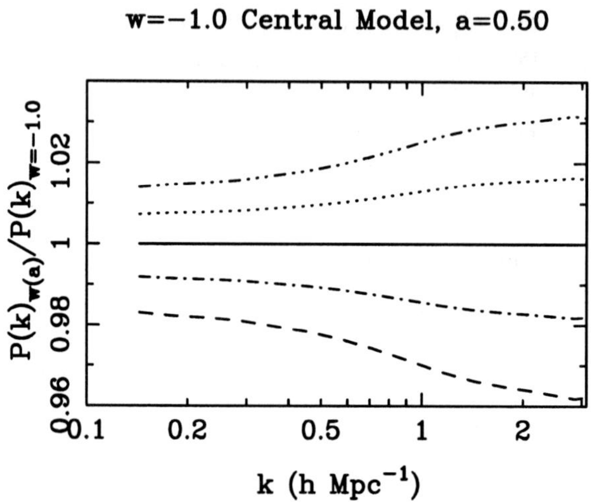

Fig. 2. As Fig. 1, but at $z = 1$

mapped power spectra, at the 0.1% level for $k < 1\,h\mathrm{Mpc}^{-1}$ and $\lesssim 1\%$ for $k < 3\,h\mathrm{Mpc}^{-1}$ ($\lesssim 0.5\%$ at the higher k for the less rapidly varying dark energy models).

For cosmological structure probes, we are interested not just in how well we can predict the power spectrum at $z = 0$ but across all redshifts. In the simulations performed, data was output at a number of times. One epoch of particular interest for cosmology is $z \simeq 1$ which is a 'sweet spot' for dark energy measurement and is extremely relevant to a number of forthcoming cosmological surveys. Hence an accurate estimation of power here, provided by the distance matching scheme, is of great importance for understanding possible constraints on dark energy cosmologies.

The ratio of power measured in the simulations boxes at $z = 1$ is shown in Figure 2.

We can do even better, however, by realizing that much of the difference in power can be accounted for by the difference in linear power. Scaling this out via Equation 1, the results are as shown in Figure 3.

From these figures we can see that the combined distance and growth matching procedure is generally accurate to better than 1%. The greatest deviation found in all simulation outputs is 2% for $k \approx 3\,h\mathrm{Mpc}^{-1}$ at $z = 1$.

5. Evolving Dark Energy and Structure Growth

The matching prescription used in this article produces a mapping between the matter power spectra of dark energy cosmologies accurate to $\lesssim 1\%$ over

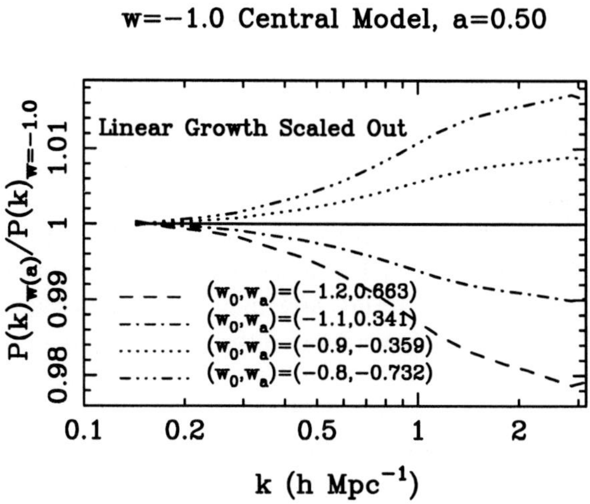

Fig. 3. As Fig. 2, but with the linear growth difference scaled out. Note the reduced y-axis scale relative to Fig. 2.

a wide range of wavemodes and a wide range of cosmic history. This agreement 1) indicates that simple physical quantities determine the nonlinear power spectrum over this range, leading to the prospect of understanding structure formation on a fundamental level even in dynamical dark energy cosmologies, 2) points the way to significant advances in computational efficiency by reducing the dimension of the grid of simulations necessary to produce accurate estimations of power spectra required for interpretation of cosmological probes such as weak gravitational lensing, baryon acoustic oscillation, and other large scale structure surveys, and 3) identifies a degeneracy that makes it difficult to distinguish between models lying on a particular subsurface of the cosmological parameter space.

To try to ameliorate the degeneracy, we note that an evolving equation of state does imprint a small but systematic effect on the non-linear matter power spectrum. The general trend shown by Figure 1 is that dark energy with a less negative value today but more negative value at high redshift (i.e. negative w_a) gives greater non-linear power at $k \gtrsim 1\,h\mathrm{Mpc}^{-1}$ than its d_{lss}-matched w_{eff} model. Similarly, more negative equations of state today with positive w_a possess less power in the same range. This deviation is however relatively small, remaining less than 2% out to $k = 3\,h\mathrm{Mpc}^{-1}$. Even so, this partial degeneracy is not too worrying since it can readily be broken by other cosmological dependencies (e.g. the geometric distance dependence entering with the mass power spectrum into the weak lensing shear power spectrum or the baryon acoustic scale) or by complementary cosmological probes.

Those models that show a greater non-linear growth, given that the models have the same σ_8 today, are those that in the early universe had a greater contribution of the matter density relative to the dark energy density; these correspond to the models with today $w_0 > w_{\rm eff}$ and in the recent universe had *lower* matter density relative to dark energy density. This suggests that non-linear growth is more sensitive to conditions (including the effects of dark energy) in the early, matter dominated universe than it is to conditions in the later, accelerated era of dark energy domination.

6. Conclusion

The mass power spectrum lies at the foundation of many cosmological observables, such as the weak lensing shear statistics of galaxies, the large scale structure clustering distribution (including baryon acoustic oscillations), and cluster abundances. To utilize any of these cosmological measurements as next generation probes of large scale structure, cosmology, or dark energy requires clear and accurate understanding of the mass power spectrum over the range of models under consideration, e.g. dynamical dark energy not just ΛCDM.

The main results of this article demonstrate that non-linear mass power spectra of dynamical dark energy models with smooth equation of state evolution can be determined to percent accuracy by calculating the power for the constant equation of state cosmology that gives a matching distance to the CMB last scattering surface. By varying other parameters as well, such as the matter density Ω_m and h keeping $\Omega_m h^2$ constant (see[13]), one can envision mapping a wide variety of dark energy models (at least those without rapidly varying $w(a)$ or small scale spatial inhomogeneity) to ΛCDM models, resulting in significant gains in computational efficiency.

Finding the distance matched models as described in this paper is a trivial task numerically requiring the integration of a differential equation and a one dimensional parameter search. This simple procedure however provides a mapping that is accurate to a percent. Of course the accuracy of the resultant power spectrum estimation is ultimately only as good as the accuracy in the model being mapped to. For that model we can then utilise fitting formulas, such as Halofit, for a rough, $\sim 10\%$ accuracy or more generally perform N-body simulations with the desired parameters. However, the distance matching scheme in this case allows a much reduced grid of simulations to be carried out while still maintaining a high degree of accuracy.

Acknowledgments

MF acknowledges the support of a Science Faculty UPA, thanks Chris Power and Jeremy Bailin for helpful discussions, advice and pieces of code, and thanks LBNL and SNAP for hospitality and support during much of the research presented in this article. We thank Martin White for pointing out the nice method of getting high resolution FFT's without needing large arrays and for other useful discussions. We thank Robert E. Smith for his helpful and insightful comments. This work has been supported by the Australian Research Council under grant DP 0665574 and in part by the Director, Office of Science, US Department of Energy under grant DE-AC02-05CH11231.

References

1. Coles P. & Lucchin, F., 2002 "Cosmology: The origin and evolution of cosmic structure", *Wiley & Sons.*
2. Doran, M. & Robbers, G., *JCAP* **0606**, 026 (2006).
3. Francis, M. J., Lewis, G. F., & Linder, E. V. *MNRAS* **380**, 1079 (2007).
4. Heitmann, K., Ricker, P., Warren, M., & Habib, S., *ApJS*, **160** 28 (2005).
5. Hu, W. *Phys. Rev. D* **66**, 083515 (2002).
6. Hu W. & Dodelson S., *ARA&A* **40**, 171 (2002).
7. Huterer, D., & Takada, M. *Astroparticle Physics* **23**, 369 (2005).
8. Klypin A, Macciò, A., Mainini, R., & Bonometto, S., *ApJ* **599**, 31 (2003).
9. Knop R. et al, *ApJ* **598**, 102 (2003).
10. Linder E., *Phys. Rev. Lett.* **90**, 091301 (2003).
11. Linder E., *Astroparticle Physics* **26**, 16 (2006).
12. Linder E. & Jenkins A., *MNRAS* **346**, 573 (2003).
13. Linder E. & White M., *Phys. Rev. D* **72**, 061304(R) (2005).
14. Ma, Z. 2007, *ApJ* 665, 887 (2007).
15. Ma, C.-P., Caldwell, R. R., Bode, P., & Wang, L. 1999, *ApJL* **521**, L1 (1999).
16. Macciò A. et al, *Phys. Rev. D* **69**, 123516 (2004).
17. Mcdonald P., Trac H. & Contaldi C., *MNRAS* **366**, 547 (2006).
18. Riess, A. et al, *ApJ* **607**, 665 (2004).
19. Seljak, U., Slosar, A., & McDonald P., *JCAP* **0610**, 014 (2006).
20. Smith, R.E., et al. *MNRAS* **341**, 1311 (2003).
21. Spergel, D. N., et al. *ApJS* **170**, 377 (2007).
22. Springel V., *MNRAS* **364**, 110 (2005).
23. Takada, M. *Phys. Rev. Lett.* **74**, 043505 (2006).
24. Wetterich, C. *Phys. Lett. B* **594**, 17 (2004).
25. White, M., & Vale, C., *Astroparticle Physics* **22**, 19 (2004).

GENERAL RELATIVITY (PLUS DARK MATTER AND DARK ENERGY) OR ALTERNATIVE THEORIES OF GRAVITY

A. F. ZAKHAROV

National Astronomical Observatories of Chinese Academy of Sciences, 20A Datun Road, Chaoyang District, Beijing, 100012, China

Institute of Theoretical and Experimental Physics, B. Cheremushkinskaya, 25, 117259, Moscow, Russia

Center of Advanced Mathematics and Physics, National University of Sciences and Technology, Rawalpindi, Pakistan

Bogoliubov Laboratory for Theoretical Physics, Joint Institute for Nuclear Research, 141980 Dubna, Russia
E-mail: zakharov@itep.ru

V. N. PERVUSHIN

Bogoliubov Laboratory for Theoretical Physics, Joint Institute for Nuclear Research, 141980 Dubna, Russia
E-mail: pervush@theor.jinr.ru

F. DE PAOLIS* and G. INGROSSO**

Department of Physics and INFN, University of Salento, CP 193, I-73100 Lecce, Italy
**E-mail: depaolis@le.infn.it, ** E-mail: ingrosso@le.infn.it*

A. A. NUCITA

XMM-Newton Science Operations Centre, ESAC, ESA, PO Box 50727, 28080 Madrid, Spain
E-mail: nucita@le.infn.it

One could call 2006 as the year of cosmology since in the year two US scientists were awarded by the Nobel prize for their studies of Cosmic Microwave Background (CMB) spectrum and anisotropy. Studies of CMB anisotropy done with the Soviet spacecraft Prognoz-9 by the Relikt-1 team are reminded. Problems of modern cosmology are outlined. We discuss conformal cosmology parameters from supernovae data in brief. Two approaches to solve the basic problems of cosmology, such as dark matter and dark energy, are discussed, the first (standard) possibility is to introduce new particles, fields etc, the second possibility is to try to change a gravity law to fit observational data. We discuss advantages and disadvantages of the second choice.

Keywords: General Relativity and Gravitation; Cosmology; Observational Cosmology; Cosmological tests; Supernovae; Cosmic Microwave Background Radiation.

1. Introduction

For a scientific community 2005 was the World year of physics due to publications of the famous Einstein's papers and the birth of modern (contemporary) physics in 1905. The next year (2006) was also remarkable for a physical community (it was the year of astrophysics and cosmology) since in 2006 the Nobel prize for CMB studies was presented to J. Mather and G. Smoot, moreover J. Mather and the COBE (COsmic Background Explorer) team was awarded by the Peter Gruber prize on cosmology in the same year at the General Assembly of the International Astronomical Union in Prague. No doubt, 2007 will be recognized as the very important year for astrophysical (or probably for scientific in general) community since a number of of discoveries were done even at present date. For Russian space science 2007 is the jubilee year since the first artificial satellite (Sputnik) was launched fifty years ago on October 4, 1957.

A structure of the paper is the following. In section 2 we remind results of CMB anisotropy studies done with the Soviet Relikt-1 mission. In section 3 we outline a current status of cosmological studies and parameters of conformal cosmological models in brief. In section 4 we discuss a success of alternative way to solve DM and DE problem (so-called, a geometrical approach or $f(R)$-models), where a gravity law is substituting with changing classical Lagrangian for gravity and severe constraints on parameters of so-called R^n theory from Solar system data will be given. Conclusions will be presented in section 5.

2. CMB anisotropy studies, Relikt-1 & COBE

Cosmic microwave background (CMB) existence was predicted in the framework of the the so called Big Bang cosmological model,[1–4] however, first estimates CMB temperature $T_{CMB} \sim 50^0$ K were rather rough since it was assumed that the age of the Universe is about 2–3$\times 10^9$ years (because of the Hubble constant estimate was very high $H = 500\ km/(s \cdot Mpc)$ due to errors in distance measurements at the time), but later Gamow re-estimated the temperature $T_{CMB} \sim 6^0$ K.[5] The CMB radiation was discovered by A. Penzias and R. Wilson[6] in an unexpected way (and they awarded by the Nobel prize for the discovery in 1978). The physical meaning the discovery was explained in the introductory paper[7] published in the same issue of the Astrophysical Journal as the paper[6] with a description of the discovery.

Because the Earth moves in respect to the CMB, a dipole temperature anisotropy of the level of is expected. A dipole temperature anisotropy of the level of $\Delta T/T = 10^{-3}$ was observed, it corresponds to a peculiar velocity 380 km/s of the Earth towards the constellation Virgo (first measurements of dipole anisotropy were not very precise,[8] but later the accuracy was significantly improved[10–12]). Moreover, based on results of balloon measurements it was claimed several times about features of quadrupole anisotropy,[12,13] however, the quadrupole term is too high in these experiments as further studies had indicated, or reversely the realistic level of quadrupole anisotropy is to low to be detected in the balloon experiments.

In 1983, in the Soviet Union the Relikt-1 experiment* was conducted aboard the Prognoz-9 spacecraft† in order to investigate CMB radiation from space for the first time in history. As many other Prognoz missions, the scientific payload was prepared by the Space Research Institute of the Soviet Academy of Sciences. Dr. Igor Strukov was the principal investigator.

The spacecraft Prognoz-9, had an 8 mm band radiometer with an extremely high sensitivity of 35 μK per second and it was launched into a high apogee orbit with a 400,000 km semi-axis. The high orbit was a great advantage of the mission since it allows to reduce of geomagnetic field impact on measurements. A disadvantage of the experiment was that the observations were conducted only in one spectral band, therefore, it was a freedom for a theoretical interpretation of the results, in contrast, multi-band measurements provide very small room for alternative explanations of anisotropy.

The radiometer scanned the entire celestial sphere for six months. Computer facilities (and therefore, data analyzing) were relatively slow in the time. Preliminary analysis of anisotropy studies indicate upper limits on

---

*In Russian literature CMB radiation is called as a relic radiation, therefore the name Relic ("Relikt" in Russian) was chosen.

†In Russian "Prognoz" means "forecast". Prognoz spacecrafts were designed originally for studies of Solar activity (Solar flares in particular) and its impact on geophysics, geomagnetic fields and space weather.[14] The spacecrafts were launched at high elliptical orbits with apogee about 200 000 km.

anisotropy,[15] but in this case even the negative results (upper limits) were extremely important to evaluate a sensitivity to design detectors for next missions (including COBE).

In 1986, at the Space Research Institute it was decided to prepare a next generation of space experiments to study the anisotropy of CMB, and start to develop the Relikt-2 project. The sensitivity of the detectors was planned to be in 20 times better than the Relikt-1 sensitivity. The Libris satellite was scheduled to carry the Relikt-2 payload and the spacecraft was planned to be located near the Lagrangian point L_2 (in the Sun – Earth system).[‡] Originally, it was a plan to launch the Libris spacecraft in 1993–1994, however, the project has not been realized, basically due to a lack of funds. To prepare Relikt-2, the team members re-analyzed Relikt-1 data and finally in beginning of 1991 they discovered signatures of the quadrupole anisotropy, but I. Strukov required to check the conclusions again and again. The discovery of anisotropy by the Relikt-1 spacecraft was first reported officially in January 1992 at the Moscow astrophysical seminar and the Relikt-1 team submitted papers in Soviet Astronomy Letters[16] and Monthly Notices of Royal Astronomical Society[17] (soon after the papers were published).

Relikt-1 results are described in an adequate way at the at the official NASA web-site http://lambda.gsfc.nasa.gov/product/relikt/[§] "... The Relikt Experiment Prognoz 9, launched on 1 July 1983 into a high-apogee (700,000 km) orbit, included the Relikt-1 experiment to investigate the anisotropy of the CMB at 37 GHz, using a Dicke-type modulation radiometer. During 1983 and 1984 some 15 million individual measurements were made (with 10% near the galactic plane providing some 5000 measurements per point). The entire sky was observed in 6 months. The angular resolution was 5.5 degrees, with a temperature resolution of 0.6 mK. The galactic microwave flux was measured and the CMB dipole observed. A quadrupole moment was found between 17 and 95 microKelvin rms, with 90% confidence level. A map of most of the sky at 37 GHz is available..." (also, references,[16,17] given at the NASA web-site).

The Nobel Prize in Physics for 2006 was awarded to John Mather and

---

[‡]The trajectories near the L_2 point are suitable for different space experiments, for example, the orbit was selected for the WMAP spacecraft (it is one of the most successful mission to study the CMB anisotropy and such an orbit was planned to be used for the future Millimetron cryogenic telescope (it scheduled to be launched around 2016)).

[§]See, also, an article "Cosmic microwave background radiation" in Wikipedia, the free encyclopedia (http://en.wikipedia.org/wiki/Main_Page).

George Smoot. The Royal Swedish Academy of Sciences had issued the Press Release, dated on 3 October 2006, on The Nobel Prize in Physics 2006.¶ The Royal Swedish Academy of Sciences has decided to award the Nobel Prize in Physics for 2006 jointly to John C. Mather (NASA Goddard Space Flight Center, Greenbelt, MD, USA), and George F. Smoot (University of California, Berkeley, CA, USA) "for their discovery of the blackbody form and anisotropy of the cosmic microwave background radiation". Really, J. Mather was the project scientist of the COBE mission and was responsible for measurements of CMB spectrum (he was the PI of the Far InfraRed Absolute Spectrometer (FIRAS) experiment) and measurements of CMB spectrum were rather successful. G. Smoot did a great contribution into measurements of the dipole anisotropy[11] and the quadrupole one,[18] since he was PI of the differential microwave radiometer (DMR) aboard the COBE satellite and found the quadrupole anisotropy with these facilities, but both anisotropies were discovered earlier by other people.

The discovery of anisotropy by the Relikt-1 spacecraft was first reported officially in January 1992 at the Moscow astrophysical seminar. Relikt-1 team submitted their paper in Soviet Astronomy Letters and Monthly Notices of Royal Astronomical Society on January 19, 1992 and on February 3, 1992, respectively.

On April 21, 1992, G. Smoot (the head of DMR experiment aboard the COBE mission) and his co-authors submitted a paper in Astrophysical Journal (Letters).[18] On April 22, 1992, Smoot reported at a press conference about the discovery of the CMB anisotropy with the COBE satellite. After that mass media reported this results as the main science success. In 1992, COBE results about discovery of the CMB anisotropy were reported elsewhere. However, no doubt, the COBE collaboration knew results of Relikt-1 team and even quoted their upper limit on the quadrupole anisotropy published at the paper.[15] Sometimes, results of both (COBE and Relikt-1) collaborations were presented at the same scientific meetings. In particular, in June 1992, at the conference on Particle Astrophysics in Blois (France) results on discoveries of CMB anisotropy were presented at a plenary session by representatives of both teams (Relikt-1 and COBE), namely M. V. Sazhin and G. Smoot. The Smoot's paper was published in the conference proceedings,[19] but unfortunately, eventually, Sazhin's contribution was not published there.

¶Information from the Nobel prize committee web-site http://www.kva.se.

However, summarizing, one could say that since papers of the Relikt-1 team were submitted on January 19, 1992 and February 3, 1992 in Soviet Astronomy and Monthly Notices of Royal Astronomy Society respectively, but COBE paper was submitted on April 21, 1992, one would conclude that the discovery one quadrupole anisotropy was done by the Relikt-1 team and published in papers.[16,17]

3. Precise cosmology & Conformal Cosmological Models

Since 1998 people claimed that cosmology started to be a precise science. Really, if the standard cosmological paradigm was chosen, now we can evaluate cosmological with a precision about 10 % (typically only statistical errors are indicated and it is assumed that systematical errors are negligible). However, even in the framework of the standard paradigm of the Big Bang cosmology, our understanding the best fit for cosmological model is changing. For example, in the beginning of 1998 it was found that the Universe is opened and $\Omega_m \sim 0.2 - 0.4$ (based on observations of galactic clusters[20] (immediately, mass media had reported that eventually scientists proved that the Universe would be expanding forever), but at the end of 1998, it was concluded that the Universe is flat $\Omega_m = 0.3, \Omega_\Lambda = 0.7$[22,23] (assuming that the SNeIa are standard candles).

We will remind basic relations for CC model parameters (see papers[24–30] for details) considering the General Relativity with an additional scalar field Q, as usually people did to introduce quinessence[31–34] (earlier, the approach was used for inflationary cosmology, see for example, paper[35] and references therein)

$$S_{CR} = \int d^4x\sqrt{-g}\left[-\frac{1}{6}R(g) + \frac{1}{2}\partial_\mu Q\partial^\mu Q - V(Q)\right], \tag{1}$$

where we used the natural units

$$M_{\text{Planck}}\sqrt{\frac{3}{8\pi}} = \hbar = c = 1, \tag{2}$$

therefore, we have the following expressions for density and pressure of the scalar field, p_Q and ρ_Q, respectively,[31–34]

$$p_Q(t) = \frac{1}{2}\dot{Q}^2 + V(Q),$$

$$\rho_Q(t) = \frac{1}{2}\dot{Q}^2 - V(Q), \tag{3}$$

and equation of state (EOS) such as $p_Q = w_Q \rho_Q$, where

$$w_Q = \frac{\frac{1}{2}\dot{Q}^2 - V(Q)}{\frac{1}{2}\dot{Q}^2 + V(Q)}, \tag{4}$$

($-1 \leq w_Q \leq 1$, for "natural" potentials $V(Q) \geq 0$). In contrast with quintessence model where one uses typically $\dot{Q}^2 \ll V(Q)$, below for CC model we will use an approximation $\dot{Q}^2 \gg V(Q)$ (for a standard representation of the potential $V(Q) = \frac{1}{2}mQ^2$, where m is a mass of the field, the approximation corresponds to a massless field model) and we have

$$w_Q = \frac{\frac{1}{2}\dot{Q}^2}{\frac{1}{2}\dot{Q}^2} = 1, \tag{5}$$

or on the other words, a rigid EOS for the scalar field $p_Q = \rho_Q$ ($p_{\rm rig} = \rho_{\rm rig}$, since for our future needs an origin of the EOS is not important, hereafter, we will call the component such as the rigid matter).

We suppose that a measurable interval is defined by the Weyl-like ratio

$$ds^2_{\rm meas} = \frac{ds^2_{\rm Einstein}}{ds^2_{\rm units}} = \frac{g_{\mu\nu}dx^\mu dx^\nu}{[g_{\mu\nu}dx^\mu dx^\nu]_{\rm units}} \tag{6}$$

of two Einstein intervals

$$ds^2_{\rm Einstein} = g_{\mu\nu}dx^\mu dx^\nu \tag{7}$$

one of which $ds^2_{\rm units}$ plays a role of measurement units determined by one of masses. The Weyl definition means that there is no device measuring the absolute value of any length. Nevertheless, this definition gives us a possibility to fit also the units of observations as two choices of observable variables. These choices correspond to two different cosmological models: the Standard Cosmology with observable masses $m_{\rm sc} = m$ and Einstein's intervals in homogeneous approximation

$$ds^2_{\rm sc} = dt^2 - a^2(dx^k)^2 = a^2(\eta)[(d\eta)^2 - (dx^k)^2] \tag{8}$$

and the Conformal Cosmology, where all masses

$$m_{\rm cc} = ma(\eta) \tag{9}$$

are scaled by the factor a and measurable intervals

$$ds^2_{\rm cc} = (d\eta)^2 - (dx^k)^2 \tag{10}$$

do not depend on the cosmological scale factor a.

In general case, both SC and CC are described by equation

$$(a')^2 = \rho(a), \tag{11}$$

where

$$\rho(a) = \rho_0 \sum_{J=-2,0,1,4} \Omega_J a^J \tag{12}$$

is the conformal energy density connected with the SC one by the transformation

$$\rho(a) = a^4 \, \rho_{SC}(a) \ , \tag{13}$$

factor $a^4 = a^3 a$ arises due to differences of length and mass units in SC and CC, Ω_J is partial energy density marked by index J running a set of values $J = -2$ (rigid), $J = 0$ (radiation), $J = 1$ (mass), and $J = 4$ (Λ-term) in correspondence with a type of matter field contributions; here $\sum_{J=-2,0,1,4} \Omega_J = 1$ is assumed.

In terms of the standard cosmological definitions of the redshift

$$1 + z \equiv \frac{1}{a(\eta)} \ . \tag{14}$$

the density parameter $\Omega_c(z) = \sum_{J=-2,0,1,4} \Omega_J a^J$ in Eq. (12) takes the form

$$\Omega_c(z) = \Omega_{\rm rig}(1+z)^2 + \Omega_{\rm rad} + \frac{\Omega_m}{(1+z)} + \frac{\Omega_\Lambda}{(1+z)^4} \ , \tag{15}$$

where $\Omega_{\rm rig} = \Omega_{-2}, \Omega_{\rm rad} = \Omega_0, \Omega_m = \Omega_1, \Omega_\Lambda = \Omega_2$.

Then the equation (11) takes the form

$$H_0 \frac{d\eta}{dz} = \frac{1}{(1+z)^2} \frac{1}{\sqrt{\Omega_c(z)}} \ , \tag{16}$$

and determines the dependence of the conformal time on the redshift factor. This equation is valid also for the conformal time - redshift relation in the SC where this conformal time is used for description of a light ray.

A light ray traces a null geodesic, i.e. a path for which the conformal interval $(ds^L)^2 = 0$ thus satisfying the equation $dr/d\eta = 1$. As a result we obtain for the coordinate distance as a function of the redshift

$$H_0 r(z) = \int_0^z \frac{dz'}{(1+z')^2} \frac{1}{\sqrt{\Omega_c(z')}}. \tag{17}$$

The equation (17) coincides with the similar relation between coordinate distance and redshift in SC.

In the comparison with the stationary space in SC and stationary masses in CC, a part of photons is lost. To restore the full luminosity in both SC and CC we should multiply the coordinate distance by the factor $(1+z)^2$.

This factor comes from the evolution of the angular size of the light cone of emitted photons in SC, and from the increase of the angular size of the light cone of absorbed photons in CC. However, an observable length in SC contains an additional factor $(1+z)^{-1}$

$$\ell_{\rm SC} = a\int \frac{dt}{a} = \frac{r}{1+z}\,. \tag{18}$$

in comparison with measurable distances in CC

$$\ell_{\rm CC} = \int \frac{dt}{a} = \frac{r}{1+z}\,. \tag{19}$$

(that coincide with the coordinate ones).

Thus, we obtain the relations for observable lengths in SC and CC, respectively

$$\ell_{\rm SC}(z) = (1+z)r(z)\,, \tag{20}$$

$$\ell_{\rm CC}(z) = (1+z)^2 r(z)\,, \tag{21}$$

simply because of $\ell_{\rm SC}(z) = a\ell_{\rm CC}(z)$. It means that the observational data are described by different manners in SC and CC, since functions $H_0\ell(z)$ have different representations in SC and CC models therefore, observational data can be fitted with different parameters of the functions as we will demonstrate below.

3.1. *Magnitude-Redshift Relation*

Typically to test cosmological theories one should check a relation between an apparent magnitude and a redshift. In both SC and CC models it should be valid the effective magnitude-redshift relation:

$$\mu(z) \equiv m(z) - M = 5\log\left[H_0\ell(z)\right] + \mathcal{M}, \tag{22}$$

where $m(z)$ is an observed magnitude, M is the absolute magnitude, $\mathcal{M}$ is a constant with recent experimental data for distant SNe. Values of μ_i, z_i and σ_i could be taken from observations of a detected supernova with index i (σ_i^2 is a dispersion for the μ_i evaluation). Since we deal with observational data we should choose model parameters to satisfy an array of relations (22) by the best way because usually, a number of relations is much more than a number of model parameters and there are errors in both theory and observations (as usual we introduce indices for the relations corresponding to all objects). Typically, χ^2-criterium is used to solve the problem, namely, we calculate

$$\chi^2 = \sum_i \frac{(\mu_i^{\text{theor}} - \mu_i)^2}{\sigma_i^2}, \tag{23}$$

where μ_i^{theor} are calculated for given z_i with the assumed theoretical model and after that we can evaluate the best fit model parameters minimizing χ^2-function.

However, in principle, other alternatives for cosmology are not completely ruled out, for example, a conformal cosmological model was discussed in papers.[24–30] Using "gold" and "silver" 186 SNe Ia[36] we confirm in general and clarify previous conclusions about CC model parameters,[37] done earlier with analysis of smaller sample of SNe Ia data[24–30] that the pure flat rigid CC model could fit the data relatively well since $\Delta\chi^2 \approx 44.3$ (or less than 20 %) in respect of the standard cosmology flat model with $\Omega_m = 0.28$. Other pure flat CC models should be ruled out since their χ^2 values are too high. For the total sample, if we consider CC models with a "realistic" constraint $0.2 \leqslant \Omega_m \leqslant 0.3$ based on other astronomical or cosmological arguments except SNe Ia data, we conclude that the standard cosmology flat model with $\Omega_m = 0.28$ is still preferable in respect to the fits for the CC models but the preference is not very high (about 5 % in relative units of χ^2 value), so the CC models could be adopted as acceptable ones taking into account possible sources of errors in the sample and systematics.

In Fig. 1 we compare the SC and CC fits for the effective magnitude-redshift relation if we put "natural" constraints on $\Omega_m \geqslant 0$. In this case the χ^2 difference between two CC models ($\Delta\chi^2 \approx 16$) is not very high and a difference between this fit and the SC best fit for a flat model is about $\Delta\chi^2 \approx 10$ (or less than 5%), it means the CC fit is at an acceptable level. For references, we plotted also pure flat CC models, so that rigid, matter, lambda and radiation models are shown with thin dotted, short dashed, dot dash, dash dot dot dot lines, respectively. One can see that only pure flat rigid CC model has relatively low χ^2 values (and it could be accepted as a rough and relatively good fit for cosmological SNe Ia data), but other models should be definitely ruled out by the observational data.

Thus, for CC model fits calculated with SNe Ia data, in some sense, a rigid equation of state could substitute the Λ-term (or quintessence) in the Universe content. As it was mentioned in the paper[37] the rigid matter can be formed by a free massless scalar field. The best CC models provide

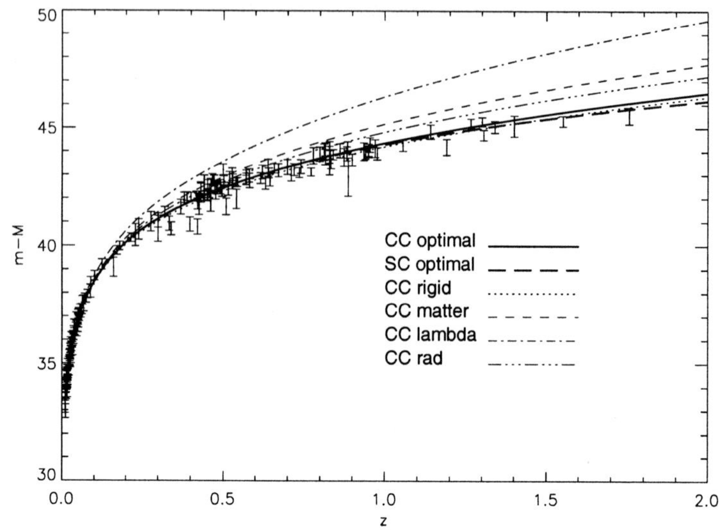

Fig. 1. $\mu(z)$-dependence for cosmological models in SC and CC. The data points include 186 SN Ia (the "gold" and "silver" sample) used by the cosmological supernova HST team.[36] For a reference we use the best fit for the flat standard cosmology model with $\Omega_m = 0.27, \Omega_\Lambda = 0.73$ (the thick dashed line), the best fit for CC is shown with the thick solid line. For this CC model we assume $\Omega_m \geq 0$.

almost the same quality fits of SNe Ia data as the best fit for the SC flat model, however the last (generally accepted) model is more preferable.

4. Standard cosmology vs. f(R) gravity

The standard cosmological model has problems to understand an origin of Dark Matter (DM) and dark energy (DE). Another approach was proposed and it was based on an assumption that gravity is different from standard general relativity it can be described by a modified Lagrangian[38] (however, there are tensions with Solar system constraints[39]). The model was successful to explain an acceleration of the Universe, but it has problems to fit Solar system data. Recently, it has been proposed a generalization of the approach in the framework of higher order theories of gravity – also referred to as $f(R)$ (or fourth order) theories – a modification of the gravity action with the form[40–56,60]

$$\mathcal{A} = \int d^4x\sqrt{-g}[f(R) + \mathcal{L}_m], \tag{24}$$

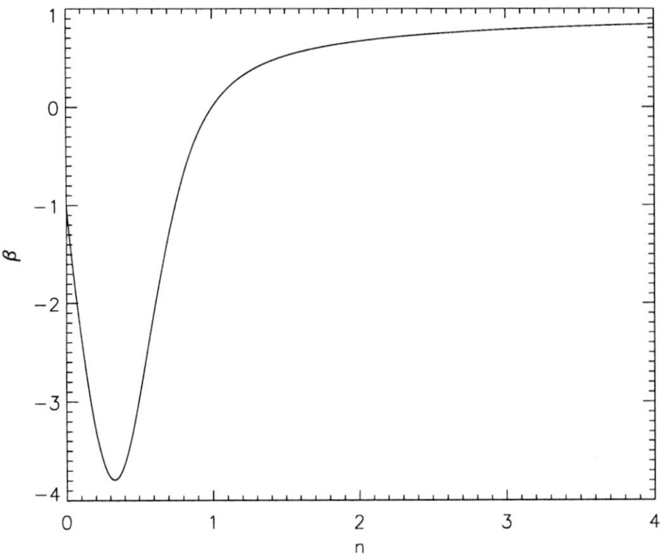

Fig. 2. The parameter β as a function of n for fourth order gravity.

where $f(R)$ is a generic function of the Ricci scalar curvature and $\mathcal{L}_m$ is the standard matter Lagrangian. For example, if $f(R) = R + 2\Lambda$ the theory coincides with General Relativity (GR) with the Λ term. In particular, it was considered a power law function $f(R)$ theories of the form $f(R) = f_0 R^n$. As a result, in the weak field limit, the gravitational potential is found to be

$$\Phi(r) = -\frac{Gm}{2r}\left[1 + \left(\frac{r}{r_c}\right)^{\beta}\right], \tag{25}$$

where

$$\beta = \frac{12n^2 - 7n - 1 - \sqrt{36n^4 + 12n^3 - 83n^2 + 50n + 1}}{6n^2 - 4n + 2}. \tag{26}$$

The dependence of the β parameter on the n power is shown in Fig. 2.[61] Of course, for $n \to \infty$ it follows $\beta \to 1$, while for $n = 1$ the parameter β reduces to zero and the Newtonian gravitational field is recovered. On the other hand, while β is a universal parameter, r_c in principle is an arbitrary parameter, depending on the considered system and its typical scale. Consider for example the Sun as the source of the gravitational field and the

Earth as the test particle. Since Earth velocity is $\simeq 30$ km s^{-1}, it has been found that the parameter r_c varies in the range $\simeq 1 - 10^4$ A.U.

The authors of the paper[60] fix r_c to be $r_c = GM_\odot/v^2$, $M_\odot$ being the solar mass and $v \sim 30$ km s^{-1}, in order to have no deviation from Newtonian potential for a circular Earth orbit with $r \simeq 1$ A.U., meaning that r_c has to be 1 A.U.

Once r_c and β has been fixed, authors of the paper[60] used them to study deviations from the standard Paczynski light curve for gravitational microlensing[57] and claimed that the implied deviation can be measured. It is clear that for gravitational microlensing one could detect observational differences between GR and an alternative theory (the fourth order gravity in particular), so that one should have different potentials at the scale R_E (the Einstein radius) of the gravitational microlensing . For the Galactic microlensing case R_E is about 1 A.U. This is a reason why the authors[60] have selected r_c at a level of astronomical units to obtain observable signatures for non-vanishing β.

Fourth order gravity theories were very successful to explain standard cosmological data such as SNe Ia fits, an acceleration of the Universe[40–56] rotation curves for galaxies[58] and it was suggested that the standard general relativity plus DM and DE may be distinguished from R^n approaches with gravitational microlensing,[60] but Solar system data (planetary orbital periods, in particular) put severe constraints on parameters of these theories.[61]

4.1. *Planetary Constraints*

We now discuss some observational consequences of fourth order gravity which, after fixing the core radius $r_c \simeq 1$ A.U., still depends on the parameter β in the weak field approximation.

A constraint on the fourth order gravity theory can be obtained from the motion of the solar system planets. Let us consider as a toy model a planet moving on circular orbit (of radius r) around the Sun. From Eq. (25), the planet acceleration $a = -\partial\Phi(r)/\partial r$ is given by

$$a = -\frac{Gm}{2r^2}\left[1 + \left(\frac{r}{r_c}\right)^\beta - \beta\left(\frac{r}{r_c}\right)^\beta\right] . \qquad (27)$$

Accordingly, the planetary circular velocity v can be evaluated and, in turn, the orbital period P is given by

$$P = P_K\sqrt{2}\left[1+\left(\frac{r}{r_c}\right)^\beta - \beta\left(\frac{r}{r_c}\right)^\beta\right]^{-1/2}, \qquad (28)$$

where $P_K = [4\pi^2 r^3/(Gm)]^{1/2}$ is the usual Keplerian period. In order to compare the orbital period predicted by the fourth order theory with the Solar System observations, let us define the quantity

$$\frac{\Delta P}{P_K} = \frac{|P - P_K|}{P_K} = |f(\beta, r_c) - 1| \qquad (29)$$

being $f(\beta, r_c)$ the factor appearing on the right hand side of equation (28) and multiplying the usual Keplerian period.

There is a question about a possibility to satisfy the planetary period condition – vanishing the Eq. (29) – with β parameter which is significantly different from zero. Vanishing the right hand side of Eq. (29) we obtain the relation

$$\ln r = \ln r_c - \frac{\ln(1-\beta)}{\beta}, \qquad (30)$$

so that Eq. (30) should be satisfied for all the planetary radii. This is obviously impossible since the fourth order theory defines β as a parameter, while the specific system under consideration (the Solar system in our case) allows us to specify the r_c parameter. Hence the right hand side of Eq. (30) is fixed for the Solar system, implying that it is impossible to satisfy Eq. (30) even with two (or more) different planetary radii.

Just for illustration we presented the function $f(\beta, r_c)$ as a dependence on β parameter for fixed r_c and planetary radii r (see Figs. in paper[61]). As one can see from Eq. (30) (and Figs. in paper[61] as well) for each planetary radius $r > r_c$ there is $\beta \in (0, 1)$ satisfying Eq. (30), but the β value depends on fixed r_c and r, so that they are different for a fixed r_c and different r. Moreover, if we have at least one radius $r \leq r_c$, there is no solution of Eq. (30). Both cases imply that the β parameter should be around zero.

As one can note, only for β approaching zero it is expected to recover the value of the Keplerian period. In the above mentioned figure, the calculation has been performed for the Earth orbit (i.e. $r = 1$ A.U.).

Current observations allow also to evaluate the distances between the Sun and the planets of the Solar System with a great accuracy. In particular, differences in the heliocentric distances do not exceed 10 km for Jupiter and amount to 180, 410, 1200 and 14000 km for Saturn, Uranus, Neptune and Pluto, respectively.[62] Errors in the semi-major axes of the inner planets are even smaller (see e.g. Table 2 in the paper[63]) so that the relative error in the orbital period determination is extremely low. As an example, the orbital period of Earth is $T = 365.256363051$ days with an error of $\Delta T = 5.0 \times 10^{-10}$ days, corresponding to a relative error of $\Delta T/T$ less than 10^{-12}. These values can be used in order to constrain the possible values of both the parameters β and r_c introduced by the fourth order gravity theory. This can be done by requiring that $\Delta P/P_K \lesssim \Delta T/T$ so that, in the case of Earth, $|f(\beta, r_c) - 1| \lesssim 10^{-12}$ which can be solved with respect to β once the r_c parameter has been fixed to some value. For $r_c = 1$ AU and $r_c = 10^4$ AU (i.e. the two limiting cases considered in the paper[60]) we find the allowed upper limits on the β parameter to be 4.0×10^{-12} and 3.9×10^{-13}, respectively (since $\Delta P/P_K = \Delta\beta[-1 + \ln(r/r_c)]/4$).

A more precise analysis which takes into account the planetary semi-major axes and eccentricities leads to variations of at most a few percent, since the planetary orbits are nearly circular. Therefore, in spite of the fact that orbital periods of planets are not generally used to test alternative theories of gravity (since it is taken for granted that the weak field approximation of these theories gives the Newtonian limit), we found that these data are important to constrain parameters of the fourth order gravity theory.

4.2. *Discussion*

GR and Newtonian theory (as its weak field limit) were verified by a very precise way at different scales. There are observational data which constrain parameters of alternative theories as well. As a result, the parameter β of fourth order gravity should be very close to zero (it means that the gravitational theory should be very close to GR). In particular, the β parameter values considered for microlensing,[60] for rotation curves[40] and cosmological SN type Ia[55] are ruled out by solar system data.

No doubt that one could also derive further constraints on the fourth order gravity theory by analyzing other physical phenomena such as Shapiro time delay, frequency shift of radio photons,[64] laser ranging for distant objects in the solar system, deviations of trajectories of celestial bodies from ellipses, parabolas and hyperbolas and so on. But our aim was only to show that only $\beta \simeq 0$ values are not in contradiction with solar system data in spite of the fact that there are a lot of speculations to fit observational data with β values significantly different from zero.

5. Conclusions

In conclusion we note that, unfortunately, even well-informed Russian (and other) authors did not cite Relikt-1 results in papers on cosmology, where the COBE anisotropy result was quoted as the only experiment discovered the phenomenon. It means the Nobel prize winner (1978) P. L. Kapitza[‖] wrote quite correctly about this kind of problems in 1946: "... Our main national defect is an underestimation of our powers and overestimation of foreign ones. So, an extra modesty is much more defective than an extra self-confidence... Very often a cause of unused innovations is that usually we underestimate our own discoveries and overestimate foreign ones..."

For CC model fits calculated with SNe Ia data, in some sense, a rigid equation of state could substitute the Λ-term (or quintessence) in the Universe content. The rigid matter can be formed by a free massless scalar field.[37]

Solar system data put severe limits on parameters of alternative theories of gravity,[61] thus, the Solar system limits a gravity law not only at scales about several A.U., but also at cosmological distances, in this class of alternative theories of gravity.

Acknowledgments

AFZ is grateful to the National Natural Science Foundation of China (NNSFC) (Grant # 10233050) and National Basic Research Program of China (2006CB806300) for a partial financial support of the work and the Pakistan Higher Education Commission for a partial support through a start-up grant and a travel support.

AFZ acknowledges also organizers of the DARK-2007 Conference in Sydney, especially prof. H.-V. Klapdor-Kleingrothaus and Dr. I. Krivosheina for their kind invitation and attention to this contribution.

References

1. G. Gamow, *Phys. Rev.* **70**, 572 (1946).
2. G. Gamow, *Phys. Rev.* **74**, 505 (1948).
3. G. Gamow, *Nature* **162**, 680 (1948).

[‖]He was awarded by the Nobel prize for his experimental discovery of superfluidity and received the prize in the same year as Penzias and Wilson got. He was the favorite Rutherford's student and worked for a long time in UK, therefore, he knew Western and Russian mentalities very well.

4. R. A. Alpher, *Phys. Rev.* **74**, 1577 (1948).
5. G. Gamow, *Vistas in Astronomy* **2**, 1726 (1956).
6. A. A. Penzias, R. W. Wilson, *Astrophys. J.* **142**, 419 (1965).
7. R. H. Dicke, P. J. E. Peebles, P. G. Roll, D. T. Wilkinson, *Astrophys. J.* **142**, 414 (1965).
8. E. K. Conklin, *Nature*, **222**, 971 (1969).
9. P. S. Henry, *Nature*, **231**, 516 (1971).
10. B. E. Corey, D. T. Wilkinson, *Bull. Amer. Astron. Soc.*, **8**, 351 (1976).
11. G.F. Smoot, M. V. Gorenstein, R. A. Muller, *Phys. Rev. Lett.*, **39**, 898 (1977).
12. R. Fabbri, I. Giudi, F. Melchiorri, V. Natali, *Phys. Rev. Lett.*, **44**, 1563 (1977).
13. S. P. Boughn, E. S. Cheng, D. T. Wilkinson, *Astrophys. J.*, **243**, L113 (1981).
14. Problems of Solar activity and space system "Prognoz", eds. S. N. Vernov et al., (Nauka, Moscow, 1977, in Russian).
15. A. A. Klypin, M. V. Sazhin, I. A. Strukov, D. P. Skulachev, *Sov. Astron. Lett.* **13**, 104 (1987).
16. I. A. Strukov, A. A. Brukhanov, D. P. Skulachev, M. V. Sazhin, *Sov. Astron. Lett.* **18**, 153 (1992).
17. I. A. Strukov, A. A. Brukhanov, D. P. Skulachev, M. V. Sazhin, *Mon. Not. R. Astron. Soc.*, **258**, 37 (1992).
18. G.F. Smoot, C. L. Bennett, A. Kogut et al., *Astrophys. J.*, **396**, L1 (1992).
19. G.F. Smoot, in *Particle Astrophysics, Proc. of the 4th Rencontres de Blois*, Eds. G. Fontaine, J. Tran Thanh Van (Gif-sur-Yvette, Editions Frontieres, 1993) p. 409.
20. N.A. Bahcall, X. Fan, *Publ. Nat. Academy of Science*, **95**, 5956 (1998).
21. N.A. Bahcall, X. Fan, *Astrophys. J.* **504**, 1 (1998).
22. A.G. Riess et al., *Astron. J.* **116**, 1009 (1998).
23. S. Perlmutter et al., *Astrophys. J.* **517**, 565 (1999).
24. D. Behnke, D. B. Blaschke, V. N. Pervushin, D. V. Proskurin, *Phys. Lett. B* **530**, 20 (2002).
25. D. B. Blaschke, S. I. Vinitsky, A. A. Gusev, V. N. Pervushin, D. V. Proskurin, *Phys. Atom. Nucl.* **67**, 1050 (2004);
26. B. M. Barbashov, V. N. Pervushin, A. F. Zakharov, V. A. Zinchuk, *Int. J. Mod. Phys. A* **21**, 5957 (2006).
27. B. M. Barbashov, V. N. Pervushin, A. F. Zakharov, V. A. Zinchuk, *Physics Letters B* **633**, 458 (2006).
28. B. M. Barbashov, V. N. Pervushin, A. F. Zakharov, V. A. Zinchuk, *Int. J. Geom. Meth. Mod. Phys.* **4**, 171 (2007).
29. B. M. Barbashov, V. N. Pervushin, A. F. Zakharov, V. A. Zinchuk, *Phys. Atom. Nucl.* **70**, 191 (2007).
30. B. M. Barbashov, V. N. Pervushin, A. F. Zakharov, V. A. Zinchuk, "Quantum gravity as theory of superfluidity," in *A Century of Relativity Physics: ERE 2005; XXVIII Spanish Relativity Meeting*, edited by L. Mor-

nas and J. D. Alonso, AIP Conference Proceedings 841, American Institute of Physics, New York, 2006, p. 362.
31. R.R. Caldwell, R. Dave, P.J. Steinhardt, *Phys. Rev. Lett.* **80**, 1582 (1998).
32. L. A. Boyle, R. R. Caldwell, M. Kamionkowski, *Phys. Lett. B* **545**, 17 (2002).
33. T. Padmanabhan, Arxiv:astro-ph/0411044.
34. Y.-F. Cai, H. Li, Y.-S. Piao, X.-M. Zhang, *Phys. Lett. B* **646**, 141 (2007).
35. A. D. Linde, Arxiv:0705.0164v2[hep-th].
36. A.D. Riess, L.-G. Strolger, J. Tonry et al., *Astrophys. J.*, **607**, 665 (2004).
37. A. F. Zakharov, A. A. Zakharova, V. N. Pervushin, ArXiv:astro-ph/0611657.
38. S. M. Carroll, V. Duvvuri, M. Trodden, M. S. Turner, *Phys. Rev. D* **70**, 043528 (2004).
39. C.-G. Shao, R.-G. Cai, B. Wang, R.-K. Su, *Phys. Lett. B* **633**, 164 (2006).
40. S. Capozziello, R. Garattini, *Class. Quant. Grav.* **24**, 1627 (2007).
41. S. Capozziello, A. Stabile, A. Troisi, arXiv:0708.0723[gr-qc].
42. S. Capozziello, S. Nojiri, S. D. Odintsov, A. Troisi, *Phys. Lett. B* **639**, 135 (2006).
43. S. Capozziello, A. Troisi, V. F. Cardone, *New Astron. Rev.* **5**, 341 (2007).
44. S. Capozziello, V. F. Cardone, A. Troisi, *J. Cosm. Astropart.* **8**, 1 (2006).
45. S. Capozziello, S. Nojiri, S. D. Odintsov, *Phys. Lett. B* **634**, 93–100 (2006).
46. S. Capozziello, V. F. Cardone, E. Elizalde, S. Nojiri, S. D. Odintsov, *Phys. Rev. D* **73**, 043512 (2006).
47. S. Capozziello, A. Troisi, *Phys. Rev. D* **72**, 044022 (2005).
48. S. Capozziello, V. F. Cardone, E. Piedipalumbo, C. Rubano, *Class. Quant. Grav.* **23**, 1205 (2006).
49. S. Capozziello, V. F. Cardone, A. Troisi, *Phys. Rev. D* **71**, 043503 (2005).
50. S. Capozziello, V. F. Cardone, M. Francaviglia, *Gen. Rel. Grav.* **38**, 711 (2006).
51. S. Carloni, P. K. S. Dunsby, S. Capozziello, A. Troisi, *Class. Quant. Grav.* **22**, 4839 (2005).
52. S. Capozziello, A. Stabile, A. Troisi, ArXiv:0709.0891 [gr-qc].
53. S. Capozziello, M. Francaviglia, ArXiv:0706.1146 [astro-ph].
54. S. Capozziello, *Intern. J. Geom. Meth. Mod. Phys.* **4**, 53 (2007).
55. A. Borowiec, W. Godlowski, M. Szydlowski, *Int. J. Geom. Meth. Mod. Phys.* **4**, 183 (2007).
56. A. Borowiec, W. Godlowski, M. Szydlowski, *Phys. Rev. D* **74**, 043502 (2006).
57. B. Paczynski, *Astrophys. J.*, **506**, 1 (1986).
58. S. Capozziello, V. F. Cardone, A. Troisi, *Mon. Not. Roy. Astron. Soc.* **375** 1423 (2007).
59. C. F. Martins, P. Salucci, ArXiv:astro-ph/0703243.
60. S. Capozziello, V. F. Cardone, A. Troisi, *Phys. Rev. D* **73**, 104019 (2006).
61. A. F. Zakharov, A. A. Nucita, F. De Paolis, G. Ingrosso, *Phys. Rev. D* **74**, 107101 (2006).
62. E. V. Pitjeva, *Solar System Research* **39**, 176 (2005).
63. J. W. Moffat, *Class. Quantum Grav.* **23** 6767 (2006).
64. B. Bertotti, L. Iess and P. Tortora, *Nature* **425**, 374 (2003).

PART V

Galaxy Clusters, Low Mass Stellar Evolution

LOW MASS STELLAR EVOLUTION WITH WIMP CAPTURE AND ANNIHILATION

P. C. SCOTT* and J. EDSJÖ

Cosmology, Particle Astrophysics and String Theory, Physics, Stockholm University & High Energy Astrophysics and Cosmology Centre (HEAC), AlbaNova Univeristy Centre, SE-106 91 Stockholm, Sweden
**Presenting author; pat@physto.se*

M. FAIRBAIRN

PH-TH, CERN, Geneva, Switzerland & King's College London, WC2R 2LS, UK

Recent work has indicated that WIMP annihilation in stellar cores has the potential to contribute significantly to a star's total energy production. We report on progress in simulating the effects of WIMP capture and annihilation upon stellar structure and evolution near supermassive black holes, using the new DarkStars code. Preliminary results indicate that low-mass stars are the most influenced by WIMP annihilation, which could have consequences for upcoming observational programs.

Keywords: cosmology; stellar evolution; dark matter; WIMPs; galactic centre; WIMP burners.

Our current understanding of cosmology is that over 20% of the mass-energy in the universe is in the form of 'dark matter',[1–3] the composition of which remains unknown. One promising group of dark matter candidates are weakly interacting massive particles (WIMPs), attractive because the masses and couplings associated with the weak scale naturally lead to a relic abundance of dark matter consistent with present-day observations.

WIMPs should posses a small but non-zero weak scattering cross-section with standard model particles. This means that they could scatter off atomic nuclei in stars, become gravitationally captured and eventually congregate in stellar cores. WIMP accretion by stars has been studied extensively,[4–7] typically with a view to observing neutrinos produced by WIMP self-annihilation in the core of the Sun. Others have considered the influence of larger concentrations of WIMPs upon the structure of the stars themselves, focusing on conductive energy transport[8–12] and energy production by annihilations.[13–15]

We provide preliminary results from simulations of the influence of WIMP capture and annihilation upon main sequence stars, using the evolutionary code DarkStars. We have also presented results obtained with a simpler static stellar structure code,[16] using some results presented herein for comparison. DarkStars is built upon the stellar evolution package EZ,[17] derived from Eggleton's STARS code,[18–20] and a generalised version of the capture routines in DarkSUSY,[21] which are based upon the capture expressions of Gould.[7] The code includes a detailed treatment of conductive energy transport by WIMPs, using the expressions of Gould & Raffelt[22] with correction factors derived from their accompanying numerical solutions to the Boltzmann equation.[23] The WIMP radial distribution is allowed to deviate from a strictly isothermal Gaussian in a manner consistent with the treatment of the conductive energy transport. DarkStars, its theoretical underpinnings and application to a range of different stars will be described in full in a coming publication.[24]

We focus on conditions obtained near supermassive black holes, where the highest ambient dark matter densities are expected to be found.[25,26] We use the maximal values of both the spin-dependent[27] ($10^{-38}\,\mathrm{cm}^2$) and spin-independent[28] ($10^{-44}\,\mathrm{cm}^2$) WIMP-nucleon cross-sections currently allowed by direct and indirect detection experiments, and assume a 100 GeV WIMP mass. The annihilation cross-section is set to $3\times10^{-26}\,\mathrm{cm}^3\,\mathrm{s}^{-1}$, as demanded by relic abundance considerations.[3] For simplicity we currently work with dark matter halo parameters for the Sun, assuming a Gaussian WIMP velocity distribution of width $270\,\mathrm{km\,s}^{-1}$ and a stellar proper motion of $220\,\mathrm{km\,s}^{-1}$ relative to the halo. We assume a metallicity of $Z = 0.02$, which is in the vicinity of the Sun's.[29]

Given a large, constant ambient WIMP density, capture and annihilation quickly equilibrate inside stars, such that annihilation provides a roughly constant source of additional energy. Beacuse the WIMPs congregate very close to the stellar centre, this energy is produced in a much smaller region than that from nuclear burning. The initial, concentrated, rapid injection of energy raises the luminosity gradient in the core of a star, steepening the temperature gradient and prompting the creation of a central convection zone (or the rapid expansion of the existing one). The increased efficiency of energy transport causes the core to cool and expand, increasing the stellar radius and decreasing surface temperature and central density.

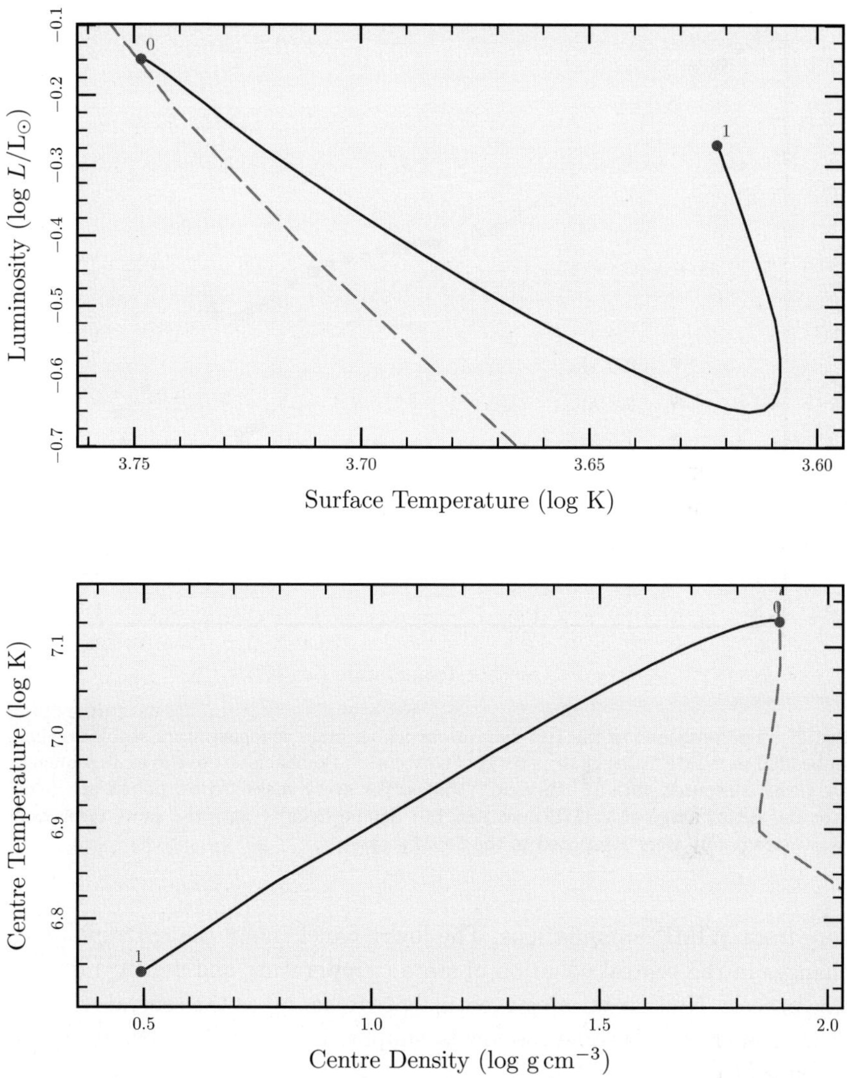

Fig. 1. The path followed in the HR (top) and central equation of state (bottom) diagrams by a $1.0\,M_\odot$ star, evolved for 100 Myr in a 10^{10} GeV cm^{-3} WIMP halo. The dashed lines indicate the zero age main sequence, which defines the boundary for hydrogen-burning in the lower plot (to the left of the line, hydrogen fusion cannot occur). The red points indicate the starting (0) and final (1) positions.

The top panel of Fig. 1 shows the evolution of a $1.0\,M_\odot$ WIMP burner in the HR diagram as it adjusts to the presence of the extra energy in its

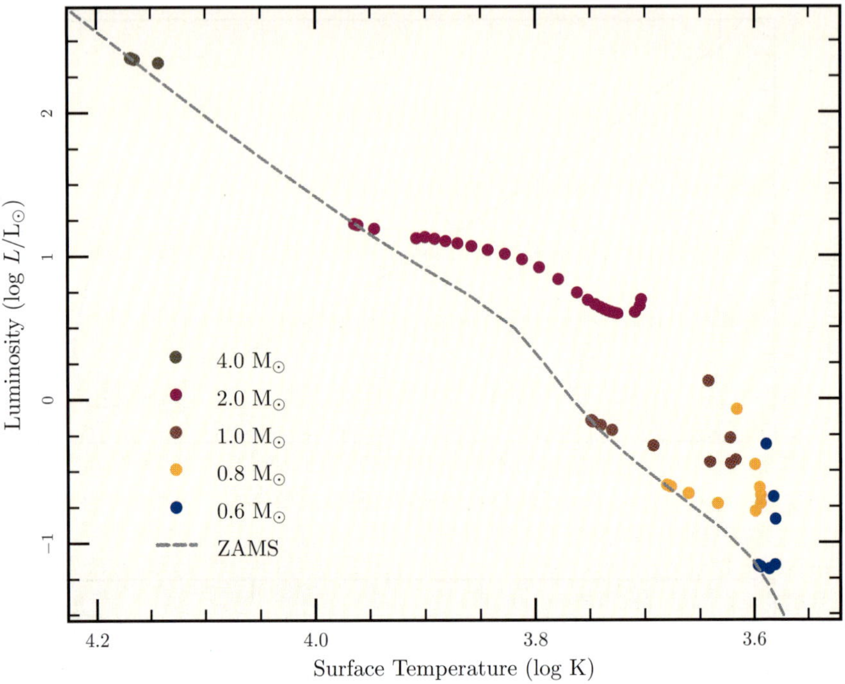

Fig. 2. The positions in the HR diagram at which main sequence stars stabilise when embedded in WIMP halos of $10^7 - 5\times10^{10}$ GeV cm^{-3}. Points closest to the main sequence are those computed with 10^7 GeV cm^{-3} halos. For every mass shown, points are given over the entire range of WIMP densities, but not necessarily with the same resolution (e.g. more points were computed in the 2.0 $M_\odot$ case).

core from WIMP annihilations. The lower panel shows the corresponding changes in the central equation of state (temperature and density) during this process. Nuclear burning eventually switches off as the central temperature and density become too low to support it, leaving WIMPs to power the star alone.

The surface cooling brought about by the star's expansion allows H^- ions to survive to increasing depths, with the resulting opacity increase causing the surface convection zone to expand. If enough WIMPs are present, the expanding surface and core convection zones will meet. This results in a fully convective star with a total luminosity which rises once again, as the further enhancement in energy transport allows the energy reaching the surface to outstrip that powering the overall expansion. Figure 2 shows the location in the HR diagram of main sequence stars of various masses in differing ambient dark matter densities from $10^7 - 5 \times 10^{10}\,\mathrm{GeV\,cm^{-3}}$, after having completely adjusted to the effects of the WIMPs. The change in direction of the 'tracks' approximately corresponds to the point at which nuclear burning turns off entirely, the star becomes fully convective and its luminosity undergoes some increase after the initial decrease.

The effects of WIMPs upon main sequence stars are most pronounced at low stellar masses, simply because the energy from nuclear burning scales as roughly the third or fourth power of M, whereas the WIMP capture rate is almost linear in M. Since WIMPs are in principle an eternal source of energy, WIMP burners will shine and occupy the same position in the HR diagram indefinitely. (After being evolved for a further 30 Gyr beyond what we show here, the star in Fig. 1 was virtually indistinguishable from itself at age 100 Myr.) This suggests that main sequence WIMP burners could be found by examining regions where stars cannot have formed recently, looking for populations where lower mass stars appear oddly younger than higher mass ones. Whilst the stars we describe are far too cool to solve the 'paradox of youth' reported at the centres of M31 and our own galaxy,[30,31] some of the explanations for this paradox demand the presence of a fainter, as-yet unobserved population of lower-mass stars. If main sequence WIMP burners exist anywhere, such stars would be prime examples. Since upcoming observations of the galactic centre will soon reach the sensitivity required to detect such a population, discovery of WIMP burners might be just around the corner.

References

1. L. Bergström, *Rep. Prog. Phys.* **63**, 793 (2000).
2. G. Bertone, D. Hooper and J. Silk, *Phys. Rep.* **405**, 279 (2005).
3. D.N. Spergel *et al.*, *ApJS* **170**, 377 (2007).
4. W.H. Press and D. N. Spergel, *ApJ* **296**, 679 (1985).
5. K. Griest and D. Seckel, *Nucl. Phys. B* **283**, 681 (1987).
6. A. Gould, *ApJ* **321**, 560 (1987).
7. A. Gould, *ApJ* **321**, 571 (1987).
8. D.N. Spergel and W.H. Press, *ApJ* **294**, 663 (1985).
9. A. Bouquet and P. Salati, *ApJ* **346**, 284 (1989).
10. P. Salati, *ApJ* **348**, 738 (1990).
11. D. Dearborn, G. Raffelt, P. Salati, J. Silk and A. Bouquet, *ApJ* **354**, 568 (1990).
12. A. Bottino, G. Fiorentini, N. Fornengo, B. Ricci, S. Scopel and F.L. Villante, *Phys. Rev. D* **66**, 053005 (2002).
13. P. Salati and J. Silk, *ApJ* **338**, 24 (1989).
14. I.V. Moskalenko and L.L. Wai, *ApJ* **659**, L29 (2007).

15. G. Bertone and M. Fairbairn (2007), arXiv:0709.1485.
16. M. Fairbairn, P. Scott and J. Edsjö (2007), arXiv:0710.3396.
17. B. Paxton, *PASP* **116**, 699 (2004).
18. P. P. Eggleton, *MNRAS* **151**, 351 (1971).
19. P. P. Eggleton, *MNRAS* **156**, 361 (1972).
20. O. R. Pols, C. A. Tout, P. P. Eggleton and Z. Han, *MNRAS* **274**, 964 (1995).
21. P. Gondolo, J. Edsjö, P. Ullio, L. Bergström, M. Schelke and E.A. Baltz, *Cosmology Astropart. Phys.* **7**, 8 (2004).
22. A. Gould and G. Raffelt, *ApJ* **352**, 654 (1990).
23. A. Gould and G. Raffelt, *ApJ* **352**, 669 (1990).
24. P. Scott, J. Edsjö and M. Fairbairn, in preparation.
25. D. Merritt, *Phys. Rev. Lett.* **92**, 201304 (2004).
26. G. Bertone and D. Merritt, *Phys. Rev. D* **72**, 103502 (2005).
27. J. Angle *et al.* (2007), arXiv:0706.0039.
28. S. Desai *et al.*, *Phys. Rev. D* **70**, 083523 (2004).
29. M. Asplund, N. Grevesse and A.J. Sauval, in *Cosmic Abundances as Records of Stellar Evolution and Nucleosynthesis*, eds. T.G. Barnes III and F.N. Bash, *ASP Conf. Ser.* **336**, 25 (2005).
30. A.M. Ghez *et al.*, *ApJ* **586**, L127 (2003).
31. P. Demarque and S. Virani, *A&A* **461**, 651 (2007).

THE WEDDING OF MODIFIED DYNAMICS AND NON-EXOTIC DARK MATTER IN GALAXY CLUSTERS

B. FAMAEY[1,*], G. W. ANGUS[2], G. GENTILE[3], H. Y. SHAN[4], H. S. ZHAO[2,4]
[1]*IAA, Université Libre de Bruxelles, Bvd du Triomphe, 1050 Bruxelles, Belgium*
[2]*SUPA, Univ. of St. Andrews, KY16 9SS Fife, UK*
[3]*Univ. of New Mexico, 800 Yale Blvd NE, Albuquerque, New Mexico 87131, USA*
[4]*National Astronomical Observatories, Beijing 100012, PRC*
** E-mail: bfamaey@ulb.ac.be*

We summarize the status of Modified Newtonian Dynamics (MOND) in galaxy clusters. The observed acceleration is typically larger than the acceleration threshold of MOND in the central regions, implying that some dark matter is necessary to explain the mass discrepancy there. A plausible resolution of this issue is that the unseen mass in MOND is in the form of ordinary neutrinos with masses just below the experimentally detectable limit. In particular, we show that the lensing mass reconstructions of the rich clusters 1E0657-56 (the bullet cluster) and Cl0024+17 (the ring) do not pose a new challenge to this scenario. However, the mass discrepancy for cool X-ray emitting groups in which neutrinos cannot cluster pose a more serious problem, meaning that dark baryons could present a more satisfactory solution to the problem of unseen mass in MOND clusters.

Keywords: gravitation; dark matter; galaxy clusters; gravitational lensing.

1. Introduction

Data on large scale structures point towards a Universe dominated by dark matter and dark energy.[1] Discovering the nature of these mysterious components of the Universe is, without a doubt, the major challenge of modern astrophysics, nay of physics as a whole. Nowadays, the dominant paradigm is that dark matter is actually made of non-baryonic weakly interacting massive particles, the so-called "cold dark matter" (CDM), and that the mysterious dark energy is well represented by a cosmological constant (Λ) in Einstein equations. The ΛCDM cosmological model has known a remarkable success in explaining and predicting diverse data sets corresponding to the Universe at its largest scales, including the CMB radiation, galaxy redshift surveys, distant supernovae data and absorption lines in the spectra of distant quasars. Nevertheless, a number of observations on galactic scales appear to be at variance with a number of CDM predictions. For instance, measurements of non-circular motions in the Milky Way have shown that there is actually very little room for dark matter inside the solar radius,[2] where CDM simulations predict a cuspy density profile. External galaxies have also been used to compare the predicted cuspy CDM density profiles with the observations, in particular rotation curves of dwarf and spiral galaxies show evidence for dark matter halos with a central constant density core[3] at odds with the CDM predictions. Another interesting problem faced by CDM on galactic scales is the overabundance of predicted satellite galaxies compared to the observed number in Milky Way-sized galaxies.[4] What is more, it is now well-documented that rotation curves suggest a correlation between the mass profiles of the baryonic matter (stars + gas) and dark matter.[5] Some rotation curves, like the one of NGC1560[6] even display obvious features (bumps or wiggles) that are also clearly visible in the stellar or gas distribution. A solution to all these problems, and especially the baryon-DM relation, could be a new specific interaction between baryons and some exotic dark matter made of, e.g., dipolar particles.[7,8] On the other hand, it could indicate that, on galaxy scales, the observed discrepancy rather reflects a breakdown of Newtonian dynamics in the ultra-weak field regime: this alternative explanation to solve the dark matter problem is known as the Modified Newtonian Dynamics (MOND[9]) paradigm, which postulates that for accelerations below $a_0 \approx 10^{-10}\mathrm{m\,s^{-2}}$ the effective gravitational attraction approaches $(g_N a_0)^{1/2}$ where g_N is the usual Newtonian gravitational field. Without resorting to galactic dark matter, this simple prescription is known to reproduce galaxy scaling relations in spirals and ellipticals (Tully-Fisher, Faber-Jackson, fundamental plane) as well as the details of the rotation curves of individual spiral galaxies[10] over five decades in mass. In particular, the recent kinematic analysis of tidal dwarf galaxies belonging to the NGC 5291 system,[11] showing a mass discrepancy unexpected in the CDM context, strongly argues in favour of MOND.[12,13] Moreover, the paradigm successfully predicts the local galactic escape speed from the solar neighbourhood,[14,15] the statistical bar frequency in spirals,[16] as well as the velocity dispersions of satellite galaxies around their hosts.[17,18] Recent developments in the theory of gravity have also added plausibility to the case for modification of gravity through the advent of Lorentz-covariant theories of gravity yielding a MOND behaviour in the appropriate limit.[19–21] Although rather fine-tuned and still being a far cry from a fundamental theory underpinning the MOND paradigm, these theories remarkably allow for new predictions regarding cosmology[22–25] and gravitational lensing.[26,27] Hereafter we notably investigate the weak-lensing properties of some galaxy clusters in MOND.

2. The Modified Dynamics in Galaxy Clusters

While having an amazing predictive power on galactic scales, the simple MOND prescription badly fails in galaxy clusters without an additional unseen component. Indeed, in rich clusters of galaxies, the observed acceleration is typically larger than a_0 in the central regions, meaning that the MOND prescription is not enough to explain the observed discrepancy between visible and dynamical mass there,[28–30] a conclusion that can be reached by computing the centripetal gravity as a function of radius in the cluster (and thus the corresponding enclosed MOND mass) from the density and temperature profiles of X-ray gas and from the assumed hydrostatic equilibrium of the cluster.

At very large radii, the discrepancy is about a factor of two, meaning that there should be as much dark matter (mainly in the central parts) as observed baryons in MOND clusters. The main characteristic of this MOND dark matter is thus that it should cluster at galaxy cluster scales but not at galaxy scales. An ideal candidate, whose free-streaming length is known to be high, is at the same time the only dark matter particle that we know for sure to exist, the neutrino. We know that ordinary neutrinos have mass[31] and that they have a number density comparable to photons, meaning that they indeed contribute to the mass budget of the Universe. However, in order to reach the densities needed to account for the MOND missing mass in galaxy clusters, they should have a mass at the limit of their experimental detection, i.e. 2 eV. This idea[29] has the great advantage of naturally reproducing most cluster scaling relations including the luminosity-temperature relation,[30] while accounting for the bulk of the missing mass in galaxy clusters. Moreover, in their modelling of the CMB anisotropies, Skordis et al.[22] showed that such a significant non-baryonic component (with $\Omega_n \simeq 0.15$) was actually helpful to prevent the MOND Universe from accelerating too much, keeping $\Omega = 1$ as a constraint on the amount of dark energy (although MOND might have the ability to drive late-time acceleration without resorting to dark energy[32]).

On the other hand, given that, in the global baryon inventory at low redshift, about 20% of the baryons are still missing, and that the observed baryons in clusters only account for 5 to 10% of those produced during Big Bang nucleosynthesis,[33–35] there is plenty of room for this dark matter to be baryonic in MOND, since there should be as much dark matter (mainly in the central parts) as observed baryons in MOND clusters. Knowing exactly how many baryons hide in the Warm-Hot Intergalactic Medium (WHIM) is thus imperative if one wants to exclude this hypothesis.

3. The Bullet Cluster 1E0657-56

Keeping in mind this known discrepancy between the observable and dynamical masses of galaxy clusters in MOND, it is then useful to ask which new challenge is posed to the MOND paradigm by the gravitational lensing map of the bullet cluster[36,37] (see M. Bradac's contribution to these proceedings). In this extremely interesting object, the collisionless component (galaxies and a hypothesised collisionless dark matter component) and the fluid-like X-ray emitting plasma have been spatially segregated due to the collision of the two progenitor galaxy clusters. However, the lensing convergence map is centered on the minor baryonic collisionless component (galaxies) rather than on the dominant baryonic X-ray emitting gas component: this was argued[36] to be the first direct empirical proof of the existence of dark matter, independently of the validity of General Relativity at galaxy cluster scales. However, while the linear relation between the matter density and the gravitational potential implies that the convergence parameter is a direct measurement of the projected surface density in General Relativity, this is not the case anymore in MOND due to the non-linearity of the modified Poisson equation. Actually, it has been shown that, in MOND, it is possible to have a non-zero convergence along a line of sight where there is zero projected matter.[38] However, in the specific case of the bullet cluster, solving the non-linear Poisson equation for the observed matter density in various line-of-sight configurations showed that the convergence map always tracks the dominant baryonic component:[39] this means that non-linear effects, being capable of counteracting this trend, turn out to be very small. The presence of large amounts of collisionless dark matter in this cluster is thus necessary in MOND.

However, by applying a simple potential-density approach, we[40] have been able to estimate the needed quantities of such collisionless dark matter in the bullet cluster, finding that the central densities around the galaxies were in accordance with the maximum density of 2 eV neutrinos, from the Tremaine-Gunn[41] limit for a 9 keV ($\sim 10^8$ K) cluster:

$$\rho_\nu^{\max} = 7 \times T(\mathrm{keV})^{3/2} \times 10^{-5} \mathrm{M}_\odot \mathrm{pc}^{-3} \tag{1}$$

$\sim 2 \times 10^{-3} \mathrm{M}_\odot/\mathrm{pc}^3$. However, a problem might exist from strong lensing data at the center of the collisionless component of the least massive cluster, a problem similar to the one discussed in section 5. We however conclude that the weak-lensing map of the bullet cluster in itself is not a new challenge to the "MOND+neutrinos" hypothesis, meaning that the amount of dark matter required is globally consistent with that suggested by the previous analyses[29] from hydrostatic equilibrium of X-ray emitting clusters. However, if it turns out that the MOND dark matter should rather be in baryonic form, then the bullet cluster provided the interesting constraint that it should be of collisionless nature (e.g. MACHO's or dense clumps of cold gas, but see also Mahdavi et al.[42] for a counter-example).

We finally note that possible non-trivial contributions from the vector field of relativistic MOND theories in non-stationary configurations[23–25] were neglected, which could only decrease the need for dark matter in this system (but not in other clusters close to a steady-state equilibrium), and that the high-speed encounter of the clusters making up the bullet could actually be a standard manifestation of MOND long-range interaction.[43]

4. The Ring in Cl0024+17

Recently, a comprehensive weak lensing mass reconstruction of the rich galaxy cluster Cl0024+17 at $z = 0.4$[44] has been argued to have revealed the first dark matter structure that is offset from both the gas and galaxies in the cluster. This structure is ringlike, located between $r \sim 60''$ and $r \sim 85''$. It was argued to be the result of a collision along the line-of-sight of two massive clusters 1-2 Gyr in the past. It has also been argued[44] that this offset was hard to explain in MOND.

Assuming that this ringlike structure is real and not caused by instrumental bias or spurious effects in the weak lensing analysis (due e.g. to the unification of strong and weak-lensing), and that cluster stars and galaxies do not make up a high fraction of the mass in the ring (which would be too faint to observe anyway), is this really hard to explain in MOND?

First of all, it has recently been shown[45] that, considering the boost of the gravitational field in MOND as the effect of some virtual dark matter (which makes it easier to compare with Newtonian and General Relativistic predictions), a peak in this virtual matter distribution *generically* appears close to the transition radius of MOND $r_t = (GM/a_0)^{1/2}$, especially when most of the mass of the system is well-contained inside this radius (which is the case for the cluster Cl0024+17). This means that the ring in Cl0024+17 could be the first manifestation of this pure MOND phenomenon. However, the sharpness of this virtual dark matter peak strongly depends on the choice of the μ-function, controlling the transition from the $1/r^2$ Newtonian regime to the $1/r$ MOND regime.[8] A sharp transition of the μ-function is needed to reproduce the ringlike structure observed in Cl0024+17, meaning that if the simple μ-function[2,46] recently used to fit many galaxy rotation curves is chosen, the ring cannot be adequately reproduced by this pure MOND phenomenon.

In this case, a collisional scenario would be needed in MOND too, in order to explain the feature as a peak of cluster dark matter. Indeed, as explained above, we already know that there is a mass discrepancy in MOND clusters, and we know that this dark matter must be in collisionless from (e.g., neutrinos or dense clumps of cold gas). So the results of the simulation with purely collisionless dark particles[44] would surely be very similar in MOND gravity. In case the missing mass in clusters is in baryonic form, we do not really have a quantitative limit on the density of MOND dark matter that would be allowed in the ring. But since we know that the "MOND + neutrinos" hypothesis works fine in other similar rich clusters, we can follow the approach of Angus et al.[40] and test this hypothesis in Cl0024+17. If the missing mass is in the form of dark baryons, this is an effective way to compare the dark density to what should be expected in similar clusters in MOND.

Let us note that this cluster was already studied[47] in the framework of MOND, however this was prior to the detection of the ringlike structure. The cluster was found to be marginally consistent with 2 eV neutrinos, using a Hernquist profile with a total mass of $3.5 \times 10^{14} \mathrm{M}_\odot$ and a core radius of 0.3 Mpc. In a latter version, a cored model was tried, including also the strong lensing data, and a model consistent with a neutrino mass of 4 eV was found. However, they assumed a simple spherical model without any line-of-sight structure, contrary to the spirit of the collision scenario invoked to explain the ringlike feature. Given the uncertainty of the density models, it is unclear if existing data for this system actually rule out the 2 eV neutrinos. We hereafter rather focus on the newly discovered ringlike structure to see if it presents a new challenge to the "MOND+neutrinos" hypothesis.

The main limit on the neutrino ability to collapse in clusters comes from the Tremaine-Gunn limit,[41] stating that the phase space density must be preserved during collapse. Assuming the same temperature for the neutrino fluid as for the baryons, the maximum density of a mixture of all neutrino types all having a 2 eV mass for a cluster of a given temperature T (in keV) is then given by Eq. (1). This means that for Cl0024+17 whose mean emission weighted temperature is $T = 4.25^{+0.40}_{-0.35}$ keV,[44] the Tremaine-Gunn limit for the density of neutrinos is $\rho_\nu^{\rm max} = 6.1^{+0.9}_{-0.7} \times 10^{-4} {\rm M}_\odot {\rm pc}^{-3}$.

A detailed simulation of Cl0024+17 would involve numerically solving the non-linear Poisson equation of MOND. However since observationally consistent relativistic MOND theories[19,20] always enhance the gravitational lensing, the surface density of the ring derived from General Relativity is always an upper limit to the actual density in MOND. Moreover, the gravity at the position of the ring is of the order of $\sim 2a_0$, meaning that MOND effects just start to be important (except for the peculiar mechanism discussed earlier in the case of a sharp transition[45]). This means that, as a first-order approximation, we can simply consider the density of the ring in General Relativity as an upper limit on the MOND density, and compare it to the Tremaine-Gunn limit. The convergence parameter is $\kappa = 0.69$ in the ring,[44] but the background is estimated[44] to contribute up to $\kappa = 0.65$, which would be the convergence if no ring was present, meaning that the convergence due to the ring itself is $\kappa_r = 0.04$. Adopting the effective distance $D_{\rm eff} = D_l D_{ls}/D_s = 0.9$ Gpc (where D_s, D_l, and D_{ls} are the distance from the observer to the source, from the observer to the lens, and from the lens to the source, respectively), we find that the MONDian upper limit of the surface density of the ring is $\Sigma = \kappa_r \times \Sigma_c = 70 {\rm M}_\odot {\rm pc}^{-2}$. Given that the ring is $25''$ wide, i.e. 0.15 Mpc wide for a distance of 1.2 Gpc, it is sensible to consider that its depth along the line-of-sight is of the same order of magnitude leading to $\rho = \Sigma/(0.15\ {\rm Mpc}) = 4.6 \times 10^{-4} {\rm M}_\odot {\rm pc}^{-3}$, i.e. significantly less (at more than 2σ) than the Tremaine-Gunn limit. We thus conclude that the ringlike structure in Cl0024+17, if real and not caused by spurious effects in the weak lensing analysis, does not pose a new challenge to MOND in galaxy clusters.

5. Low Temperature X-Ray Emitting Groups

While we have shown that the widely advertised lensing analysis of the clusters 1E0657-56 and Cl0024+17 do not pose any new challenges to the "MOND + neutrinos" hypothesis, we show hereafter that low-mass X-ray emitting groups do provide a much more serious problem. Indeed, Eq. (1) implies that 2 eV neutrinos would stop contributing significantly to the mass density in cooler clusters or groups, since their maximum density is proportional to $T^{3/2}$. The pure "MOND + neutrinos" hypothesis thus predicts that the MOND mass discrepancy should decrease with decreasing temperature. However, when analyzing the hydrostatic equilibrium of X-ray emitting groups with $0.6\,{\rm keV} < T < 2\,{\rm keV}$, in which neutrinos cannot cluster, one finds[48] a mass discrepancy that cannot be explained by neutrinos. This of course does not mean that 2 eV neutrinos cannot be present to alleviate the mass discrepancy in rich clusters, but it means that there is *more* MOND hidden mass than just neutrinos, especially in cool groups.

6. Conclusion

We thus conclude that, while having an amazing predictive power on galactic scales, the simple MOND prescription fails at present in galaxy clusters, where some dark matter is needed. If this dark matter is assumed to be in the form of 2 eV neutrinos (at the limit of experimental detection), then the bulk of the problem can be solved in rich clusters, including the bullet cluster and the ringlike feature observed in Cl0024+17. However, neutrinos cannot cluster in cool groups with $0.6\,\mathrm{keV} < T < 2\,\mathrm{keV}$, where a discrepancy is still observed. One solution could then be that dark matter in MOND is in the form of a 4th sterile neutrino with a mass around 6-10 eV. Another possibility is that the new fields that are invoked in relativistic versions of MOND might behave as a dark matter fluid in galaxy clusters.[49] However these explanations seem to be slightly acts of the last resort, whilst another, more elegant, possibility would be that the MOND cluster dark matter is simply in the form of cold gas clouds or MACHO's, since there are enough missing baryons at low redshift to account for all the MOND hidden mass in galaxy groups and clusters (except if more baryons are detected in WHIM in between). An interesting possibility is then that this baryonic dark matter is in the form of dense clumps of cold gas of only a Jupiter mass and a temperature of a few Kelvins,[50] which would behave in a collisionless way. In any case, one should understand why this MOND dark matter component vanishes for systems with $T < 0.6$ keV. As a final remark, it should be highlighted that this additional unseen component in MOND only appears in systems with an abundance of ionised gas and X-ray emission, whatever consequence this might have on the nature of this dark matter.

References

1. D. Spergel et al., 2007, *ApJS* **170**, 377.
2. B. Famaey, J. Binney, 2005, *MNRAS* **363**, 603.
3. G. Gentile et al. 2004, *MNRAS* **351**, 903.
4. B. Moore et al., 1999, *ApJ* **524**, L19.
5. F. Donato et al., 2004, *MNRAS* **353**, L17.
6. A. H. Broeils, 1992, *A&A* **256**, 19.

7. L. Blanchet, 2007, *CQGra* **24**, 3529.
8. B. Famaey et al., 2007, *Phys. Rev. D* **75**, 063002.
9. M. Milgrom, 1983, *ApJ* **270**, 365.
10. R. H. Sanders, S.S. McGaugh, 2002, *ARA&A* **40**, 263.
11. F. Bournaud et al., 2007, *Science* **316**, 1166.
12. G. Gentile et al., 2007, *A&A* **472**, L25.
13. M. Milgrom, 2007, *ApJ* **667**, L45.
14. B. Famaey et al. 2007, *MNRAS* **377**, L79.
15. X. Wu et al., 2007, *ApJ* **665**, L101.
16. O. Tiret, F. Combes, 2007, *A&A* **464**, 517.
17. G. W. Angus et al., 2007, *MNRAS*, in press, arXiv:0709.1966.
18. O. Tiret et al., 2007, *A&A*, in press, arXiv:0710.4070.
19. J. Bekenstein, 2004, *Phys. Rev. D* **70**, 083509.
20. T. G. Zlosnik et al., 2007, *Phys. Rev. D* **75**, 044017.
21. J. -P. Bruneton, G. Esposito-Farèse, 2007, arXiv:0705.4043.
22. C. Skordis et al., 2006, *Phys. Rev. Lett.* **96**, 011301.
23. S. Dodelson, M. Liguori, 2006, *Phys. Rev. Lett.* **97**, 231301.
24. T.G. Zlosnik et al., 2007, arXiv:0711.0520.
25. A. Halle, H. S. Zhao, 2007, arXiv:0711.0958.
26. H. S. Zhao et al., 2006, *MNRAS* **368**, 171.
27. D. Xu et al., 2007, arXiv:0710.4935.
28. A. Aguirre et al., 2001, *ApJ* **561**, 550.
29. R. H. Sanders, 2003, *MNRAS* **342**, 901.
30. R. H. Sanders, 2007, *MNRAS* **380**, 331.
31. Y. Fukuda et al., 1998, *Phys. Rev. Lett.* **81**, 1562.
32. L. M. Diaz-Rivera et al., 2006, *Phys. Rev. D* **73**, 083503.
33. M. Fukugita et al., 1998, *ApJ* **503**, 518.
34. J. Silk, 2006, arXiv:astro-ph/0603209.
35. S. S. McGaugh, 2007, arXiv:0707.3795.
36. D. Clowe et al., 2006, *ApJ* **648**, L109.
37. M. Bradac et al., 2006, *ApJ* **652**, 937.
38. G. W. Angus et al., 2006, *MNRAS* **371**, 138.
39. M. Feix et al., 2007, *A&A*, in press, arXiv:0707.0790.
40. G. W. Angus et al., 2007, *ApJ* **654**, L13.
41. S. Tremaine, J. E. Gunn, 1979, *Phys. Rev. D* **42**, 40710.
42. A. Mahdavi et al., 2007, *ApJ* **668**, 806.
43. G. W. Angus, S.S. McGaugh, 2007, *MNRAS*, in press, arXiv:0704.0381.
44. M. J. Jee et al., 2007, *ApJ* **661**, 728.
45. M. Milgrom, R.H. Sanders, 2007, arXiv:0709.2561.
46. H. S. Zhao, B. Famaey, 2006, *ApJ* **638**, L9.
47. R. Takahashi R., T. Chiba, 2007, *ApJ*, in press, arXiv:astro-ph/0701365.
48. G. W. Angus et al., 2007, arXiv:0709.0108.
49. H. S. Zhao, 2007, *ApJ*, in press, arXiv:0710.3616.
50. D. Pfenniger, F. Combes, 1994, *A&A* **285**, 94.

PART VI

Massive Black Holes

THE DARK MASS AT THE CENTER OF THE MILKY WAY: NIR AND X-RAY FLARE EMISSION FROM SGR A*

A. ECKART[1,2], L. MEYER[1,2], M. ZAMANINASAB[1,2], R. SCHÖDEL[1], K. MUZIC[1,2], D. KUNNERIATH[1,2], G. WITZEL[1], C. STRAUBMEIER[1], A. ZENSUS[2,1]

[1] *I.Physikalisches Institut, Universität zu Köln, Zülpicher Str.77, 50937 Köln, and*
[2] *Max-Planck-Institut für Radioastronomie, Auf dem Hügel 69, 53121 Bonn, Germany*
** E-mail: eckart@ph1.uni-koeln.de*

F. K. BAGANOFF[3], M. MORRIS[4], W. J. DUSCHL[5,8], M. DOVCIAK[6], V. KARAS[6], and J. MOULTALKA[7]

[3] *Center for Space Research, Massachusetts Institute of Technology, Cambridge, MA 02139-4307, USA* - [4] *Department of Physics and Astronomy, University of California Los Angeles, Los Angeles, CA 90095-1562, USA* - [5] *Institut für Theoretische Physik und Astrophysik, Christian-Albrechts-Universität zu Kiel Leibnizstr. 15 24118 Kiel, Germany* - [6] *Astronomical Institute, Academy of Sciences, Boční II, CZ-14131 Prague, Czech Republic* - [7] *Observatoire Midi-Pyrénées, 14, Avenue Edouard Belin, 31400 Toulouse, France* - [8] *Steward Observatory, The University of Arizona, 933 N. Cherry Ave. Tucson, AZ 85721, USA*

We report on recent near-infrared and X-ray observations of the SgrA* counterpart associated with the super massive 3–4$\times 10^6 M_\odot$ black hole at the Galactic Center. The goal is the investigation of the variable emission from SgrA*. The observations have been carried out using the NACO adaptive optics (AO) instrument at the European Southern Observatory's Very Large Telescope (July 2005, May 2007) and the ACIS-I instrument aboard the *Chandra X-ray Observatory* (July 2005). We report on a polarized NIR flare synchronous to a 8×10^{33} erg/s X-ray flare in July 2005, and a new flare in May 2007 that shows the highest sub-flare to flare contrast observed until now. The observations can be interpreted in a model involving a temporary disk with a short jet. In the disk component the quasi-periodic flux density variations can be explained due to spots on relativistic orbits around the central super massive black hole. The profile variations for the May 2007 flare are interpreted as a variation of the spot structure due to differential rotation within the disk.

Keywords: Black hole physics, accretion, accretion disks, Galactic Center.

1. Introduction

Over the last decade, the investigation of the dynamics of stars has provided compelling evidence for the existence of a super massive black hole (SMBH) at the center of the Milky Way. At a distance of only ∼8 kpc a SMBH of mass $\sim 4\times 10^6 M_\odot$ can convincingly be identified with the compact radio and infrared source Sagittarius A* (Sgr A*).[1–13] Due to its proximity Sgr A* provides us with a unique opportunity to understand the physics and possibly the evolution of super massive black holes at the nuclei of galaxies. However, Sgr A* is remarkably faint in all wavebands, challenging current theories of matter accretion and radiation surrounding SMBHs. The feeble emission ($10^{-9\ldots-11}$ of the Eddington rate) is due to a combination of a low accretion rate with a low radiation efficiency. An intense discussion among the theoretical community at present focuses on radiatively inefficient accretion flow and jet models. Additional strong evidence for a SMBH at the position of Sgr A* came from the observation of flare activity on hourly time scales both in the X-ray and NIR wavelength domain. The first successful simultaneous NIR/X-ray campaigns combined NACO and Chandra as well as mostly quasi-simultaneous mm-data from BIMA, SMA and VLA.[14,15] The NIR/X-ray variability is probably also linked to the variability at radio through sub-millimeter wavelengths showing that variations occur on time scales from hours to years.[16–20] The surprisingly low luminosity has motivated many theoretical and observational efforts to explain the processes that are at work in the immediate vicinity of Sgr A*. For a recent summary of accretion models and variable accretion of stellar winds onto Sgr A* see Ref. 21,22.

The temporal correlation between rapid variability of the near-infrared (NIR) and X-ray emission suggests that the emission showing 10^{33-34} erg/s flares arises from a compact source within a few ten Schwarzschild radii of the SMBH.[14,15] For Sgr A* we assume $R_s=2R_g=2\mathrm{GM}/c^2\sim 8$ μas, with R_s being one Schwarzschild radius and R_g the gravitational radius of the SMBH. By now for several simultaneous flare events the authors found no time lag larger than an upper limit of ≤10 minutes, mainly given by the required binning width of the X-ray data. The flaring state can be explained with a synchrotron self-Compton (SSC) model involving up-scattered submillimeter photons from a compact source component. Inverse Compton scattering of the THz-peaked flare spectrum by the relativistic electrons then accounts for the X-ray emission. This model allows for NIR flux density contributions from both the synchrotron and SSC mechanisms. Indications for red and variable NIR spectra[13,23,24] is indicative of a possible exponential cutoff of the NIR/MIR synchrotron spectrum.[15]

There is also evidence for a quasi-periodic modulation of the NIR and X-ray emission[25–31,39] (in the following we refer to this phenomenon as QPO: quasi periodic oscillation). The NIR flare emission is polarized with a well defined range over which the position angle of the polarized emission is changing ($60^o \pm 20^o$).[26–29] All these observations can be explained in a model of a temporary accretion disk harboring a bright orbiting spot possibly in conjunction with short jet and suggest a stable orientation of the source geometry over the past few years.

The millimeter/submillimeter wavelength polarization of Sgr A* is variable in both magnitude and position angle on timescales down to a few hours. Marrone et al. (Ref. 32) present simultaneous observations made with the Submillimeter Array polarimeter at 230 and 350 GHz with sufficient sensitivity to determine the polarization and rotation measure at each band. From their measurements they deduce an accretion rate that does not vary by more than 25% and - depending on the equipartition constraints and the magnetic field configuration - amounts to 2×10^{-5} to 2×10^{-7} $M_\odot$ yr^{-1}. The mean intrinsic position angle is $167°\pm7°$ with variations of $\sim31°$ that must originate in the sub-millimeter photosphere of SgrA*.

Here, we present new polarimetric NIR observations from May 2007 and X-ray measurements from 2005 that were obtained using *Chandra* and that were taken in parallel with the NIR polarization measurements reported by Eckart et al. (Ref. 26). In Section 2 we summarize the observations and the data reduction. The observational results and modeling of the data are presented in Section 3 and a more general discussion of available infrared and X-ray variability data on Sgr A* is given in Section 4. In Section 5 we summarize and draw some conclusions.

2. Observations and Data Reduction

As part of a large observing campaign Sgr A* was observed in May 2007 and July 2005, using the VLT[a]. In July 2005 we carried out simultaneously X-ray observations using the Chandra observatory. In the following we merely describe the data (depicted in Fig. 2). Details of the observations and data reduction are given in Eckart et al. (2008, submitted to A&A, see also Ref. 26).

[a]Based on observations at the Very Large Telescope (VLT) of the European Southern Observatory (ESO) on Paranal in Chile; Program: 075.B-0093.

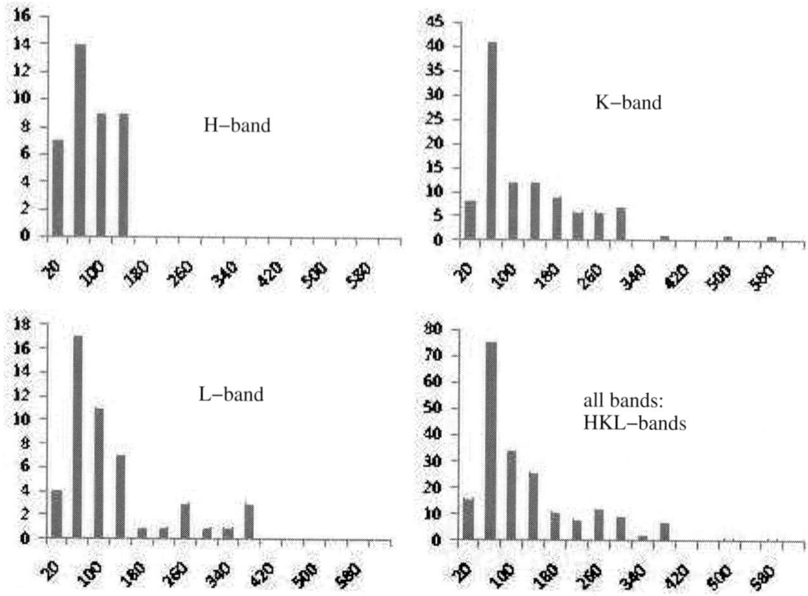

Fig. 1. Statistics of the distribution of ESO imaging experiments on SgrA* using NACO covering 2002 until summer 2007. The width of the bins is 40 minutes.

In general the imaging data volume obtained at the VLT using NACO is ideally suited to study the variability of SgrA* at infrared wavelengths in great detail. In Fig. 1 we show a statistics of all VLT NACO observing runs on SgrA* covering 2002 until summer 2007. Starting in fall 2006 the data is not complete. The length of the run was determined over time intervals that did not show gaps larger than 30 minutes. If runs were separated by gaps larger than 30 minute the data were considered to belong to an independent observing run. For all observing runs in excess of 40 minutes duration we obtained 32 (H-band), 96 (K-band), 45 (L-band) and 186 (all bands) runs. In the bins covering shorter runs also contain observations of a few minutes only. The shorter observating are predominantly used to derive the proper motions of stars in the central stellar cluster. The observations with durations of 100 minutes or longer are those that are most valuable for variability studies.

2.1. *Comparison of the NIR/X-ray flare events*

The *Chandra* X-ray data fully cover the observed polarized NIR flare that we observed at the VLT in July 2005. The X-ray data show a 8×10^{33} erg/s

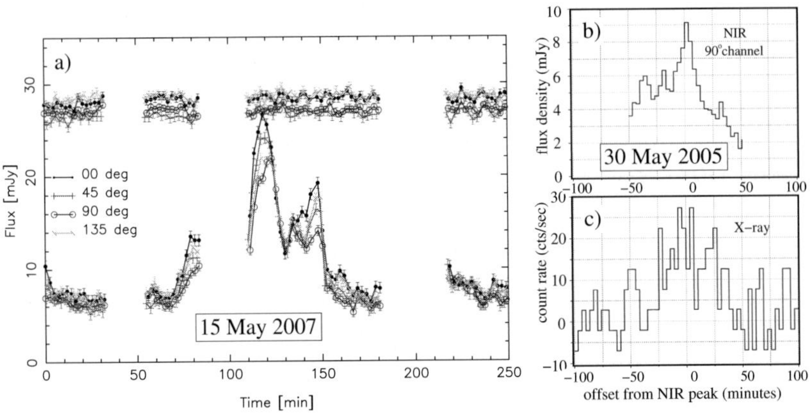

Fig. 2. **a):** The flux of the observed flare of Sgr A* on 15 May 2007 in different channels as a function of time - each channel depicted in different color. The light curve of a constant star S2, is shown in the same plot and shifted by a few mJy for a better view. **b) and c):**The NIR (top) and X-ray (bottom) data for the SgrA* flare observed on 30 May 2005. To highlight both the flare and the sub-flare structure we plot the flux in the NIR 90° polarization channel. Both data sets are shown in 207 bins. The X-ray data are corrected for the intermediate quiescent emission. The peak of the NIR fare occurred at 02:56:00 UT $\pm$3 minutes. To within about $\pm$7 minutes the X-ray peak time occurred at the same time.

flare that is about 3 times as bright as the quiescent emission from SgrA*. In Fig. 2 we show the X-ray data using a 207 and a 628 second bin size. The cross-correlation of the X-ray data with the flux densities in the individual NIR polarization channels show that the flare event observed in the two wavelength bands is simultaneous to within less than 10 minutes. The two sub-peaks in the cross-correlation function correspond to two apparent sub-peaks in the X-ray light curve that can, however, not be taken as significant given the SNR of $\sim$3 cts/s per integration bin. In the X-ray domain there is no clear indication for a QPO sub-flare structure as observed in the NIR. The NIR sub-flare contrast defined as the sub-flare height divided by the height of the overall underlying flare flux density ranges between 0.3 and 0.9.

3. Modeling Results

3.1. *Relativistic disk modeling of the variable flare emission*

We interpret our polarized infrared flare events that show indications for quasi-periodic variability via the emission of spots on relativistic orbits around the central SMBH in a temporary disk.[26–29] The model calculations are based on the KY-code by Dovciak, Karas, & Yaqoob (Ref. 33) and are usually done for a single spot orbiting close to the corresponding last stable orbit. The possibility to explore effects of strong gravity via time-resolved polarimetrical observations of X-rays was originally proposed by Connors & Stark (Ref. 34).

3.2. *A multi component disk model*

The observed NIR/X-ray properties of the SgrA* light curves raise a number of questions: Can we expect a sub-flare structure in the X-ray domain using a synchrotron self-Compton model? What is the approximate flux distribution within a temporary accretion disk around Sgr A*? This is also closely related to more general questions of how the observed light curve properties vary if the life time of the spot, shearing and synchrotron cooling time scales are considered. In the following we describe an extended SSC model that includes a disk structure and allow us to calculate light curves in the NIR and X-ray domain in order to discuss the questions posed above. A detailed description of the used SSC model is given in Ref. 15.

3.2.1. *The SSC disk model*

In order to explain the time dependent flare properties we assume that the sub-flare and disk component can be described by a number of individual synchrotron and SSC emitting source components. Combining the light amplification curve for individual orbiting spots and the simple SSC model described above, we can obtain zero order time dependent flare characteristics from the NIR to the X-ray domain.

As a starting point we used synchrotron models that represent a high flux density, i.e. flaring, and a low flux density state. Greenhough et al. (Ref. 35) outline the importance of scaling properties of the transport processes operating within accretion disks. Pessah et al. (Ref. 36) present a scaling law between magnetic stress in units of the gas pressure and the vertical disk cell size in units of the pressure scale height implying that the magnetic field and source component size follow a power law relation. Therefore we assume that the essential quantities of the SSC models, i.e. the turnover flux density S_m, frequency ν_m and the source size θ of the individual source components are distributed as power laws with the boundary values taken from the high and low flux density state models. The corresponding exponents of the distribution are α_{S_m}, α_{ν_m} and α_θ:

$$N(S) \propto S_m^{\alpha_S} \quad , \quad N(\nu) \propto \nu_m^{\alpha_\nu} \quad , \quad N(\theta) \propto \theta^{\alpha_\theta} \quad .$$

For example if $\alpha_S = 0$ the flux densities of the source components cover the full range between the minimum and maximum values. For $\alpha_S > 0$ and $\alpha_S < 0$ there is an increasing preference towards larger and lower flux density values, respectively. Similarly this is true for α_{ν_m} and α_θ.

The ISCO around a non-rotating black hole with spin parameter a=0 is $6R_g$. Assuming the co-rotating case, that radius will shrink for higher spin parameters. For a rotating black hole with a=0.5 the radius is $\sim$4.4R_g. Model calculations have shown[27,28] that for Sgr A* spin parameters $a \geq 0.5$ and source components orbiting at radii larger than the ISCO are very likely. It was furthermore shown that the disk is small with an outer disk radius extending not much further than $2R_s$ beyond the ISCO. With source components sizes of the order of 1.5 R_g (Ref. 27,28) we can safely assume that the disk is well sampled using a total of 10 Gaussian shaped disk sections with random values of S_m, ν_m and θ taken from the described power law distributions in order to model the entire accretion disk. The brightest of these sections will then represent the orbiting spot and the rest will account for the underlying disk. This setup will, of course, also allow for several bright spots. As a simple - but still general - model we assumed the source components to be equally spaced along the circumference of a constant orbit. While orbiting, the flux density of each component will follow the achromatic magnification curves that can be calculated as a function of the spin parameter a, inclination i and orbital radius. In addition, we infer a Gaussian shaped weightening function with a FWHM of about 3 orbital periods, which resembles the observed flare lengths quite well.

3.2.2. *Results of the modeling*

An important result of the simulations is that the observed total NIR and X-ray flux densities can successfully be modeled simultaneously with the observed sub-flare contrast. In addition, the best fits to the NIR and X-ray flux densities lie within or close to regions of high NIR flux density weighted magnetic field strength. This demonstrates that the combination of the SSC modeling and the idea of a temporary accretion disk can realistically describe the observed NIR polarized flares that occur synchronous with the 2-8 keV X-ray flares. We also find that the exponential α_S of the assumed power law distribution for the synchrotron peak flux S_m results in best model results for values around α_S=-1±1. A value of α_S=0 (which is included) represents senarii in which source components cover the entire range of flux densities with an equal probability for each value rather being biased towards similarly faint or bright components. This provides high sub-flare contrast values. An exponent of α_S=-1 favors lower flux density values. In the SSC model high contrast is provided by the SSC contribution to the NIR spectral range, also allowing for χ^2-fits at lower flux density weighted magnetic field strengths around 30 G rather than 60 G as for the synchrotron model.

A comparison of the July 2005 X-ray data with noise sections added from the 207 s X-ray light curve to the noise free modelling results demonstrates that at the given SNR and data sampling QPOs in the X-ray data are difficult to determine, even if they have a modulation contrast similar to that observed in the NIR (see Fig. 3). Bright spots may on average have smaller sizes or lower cutoff frequencies. An increase of SSC X-ray flux density due to an increase of THz peak synchrotron flux may be compensated by this effect. Hence the sub-flare contrast may be much lower in the X-ray compared to the NIR domain.

In Fig. 4 we show the modeling results for the May 2007 NIR and the July 2004 simultaneous NIR/X-ray data on SgrA* using our time dependent flare emission model. For the 2007 data we implemented a double hot spot model in the KY code that was briefly described in Section 3.1. For the 2004 data we involved a model consisting of 7 components at increasing distances from the SMBH starting at the inner last stable orbit. The components line up on opposite sides of the SMBH close to the flare center in time, thus providing a maximum amount of Doppler amplification before and afterwards. This gives rise to the two NIR flare events labeled III and IV (Ref. 15). Motivated by the fact that the May 2007 data shows evidence for a spot evolution due to differential rotation within the relativistic disk, we assumed that this may result in an increase of the source size for the individual spots as well. Therefore, starting at the center of the flare event, we assumed in both cases a 30% increase of the source component sizes over 30 to 40 minutes, i.e. about two orbital time scales. This results in a sharp decrease of the SSC X-ray flux density and therefore in a very good representation of the 2004 measurements (see X-ray flares labeled $\phi 3$ and $\phi 4$ in Fig. 4 and Ref. 15). Based on these time dependent model assumptions we would have expected a similarly strong evolution of the X-ray flare light curve for the May 2007 NIR observations as shown in the bottom right panel of Fig. 4.

Such a scenario may also explain the 2006 July 17 Keck NIR/X-ray light curves reported by Hornstein et al. (Ref. 23). The authors measured an NIR flare without a detectable X-ray counterpart. It was delayed by about 45 minutes from a significant X-ray flare, during which no NIR data was taken. Assuming that the X-ray flare was accompanied by an unobserved NIR flare as well, this event may have been very similar in structure to our July 2004 flare.

4. Alternative Models

Explaining SgrA* with alternative solutions for a MBH becomes increasingly difficult (see discussion in appendix of Ref. 26). Stellar orbits near SgrA* make a universal Fermion ball solution for compact galactic nuclei highly unlikely and especially the fact that SgrA* appears to be a strongly variable and mass accreting object, represents a problem for the stability constrains that boson or fermion balls have. It is, for instance, quite a delicate process to form a boson star and preventing it from collapsing to a super massive black hole despite of further accretion of matter, a non spherically symmetric arrangement of forces as in the case of a jet or matter being in orbit around the center but well within the boson star. Such a massive boson star scenario could already be excluded for the nucleus of MCG-6-30-15 (Ref. 40). In the case of a stationary boson star the orbital velocity close to the ∼3 R_S radius LSO is already ∼3 times lower than that of a Schwarzschild MBH (Ref. 40) and relativistic effects are severely diminished and further reduced at even smaller radii. If the indicated quasi-periodicity that is indicated especially by the new infrared polarization data is due to orbital motion of a spot within a temporal accretion disk then a stationary boson star can be excluded as an alternative solution for SgrA*, since in this case one expects the orbital periods to be larger.

5. Summary and Conclusion

We have measured a significant X-ray flare that occurred synchronous to a NIR flare with polarized sub-flares. This confirms the previous finding[14,15,37] that there exists a class of X-ray flares that show simultaneous NIR emission with time lags of less than 10 minutes. In addition there are lower energy flare events that are bright in the infrared and are not detected in the X-ray domain.[23] In the relativistic disk model the May 2007 polarimetric NIR measurements of a flare event with the highest sub-fare to flare contrast observed until now, provides direct evidence for a spot evolution during the flare. This fact provides further strong support for the

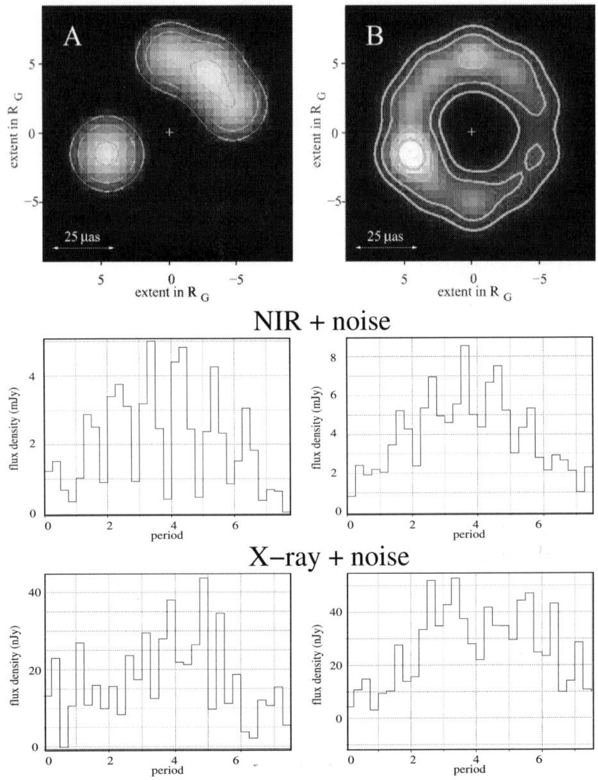

Fig. 3. As results from model calculations, we show for two cases representative flux density distributions and NIR/X-ray light model curves with noise. The flux density distributions are shown along the last stable orbit perimeter of the super massive black hole associated with Sgr A* (upper panel). Here, no truncation at or just within the last stable orbit has been applied. We show the results for the synchrotron (A, left panels) and synchrotron self Compton model (B, right panels). The contour lines are at 12, 25, 50, and 75% of the peak of the flux density distribution. The NIR and X-ray light curves are representative for the median values calculated in Eckart et al. (2008, subm. to A&A). For comparison we added sections of the 207 s light curve from 2005 scaled to the peak values and bin size of the X-ray model light curves. For the NIR we added 0.4 mJy of random Gaussian noise. The bin size of the model data corresponds to 207 s for a 14 min period. The position of Sgr A* is indicated by a white cross.

interpretation of the NIR polarimetry data within a relativistic disk model. Combined with the assumption of spot expansion due to differential rotation, the combined SSC disk model can explain the July 2004 flare (Ref. 15) and possibly also the 2006 July 17 flare reported by Ref. 23.

The combination of relativistic amplification curves with a simple SSC mechanism allows us zero order interpretations in a time dependent flare

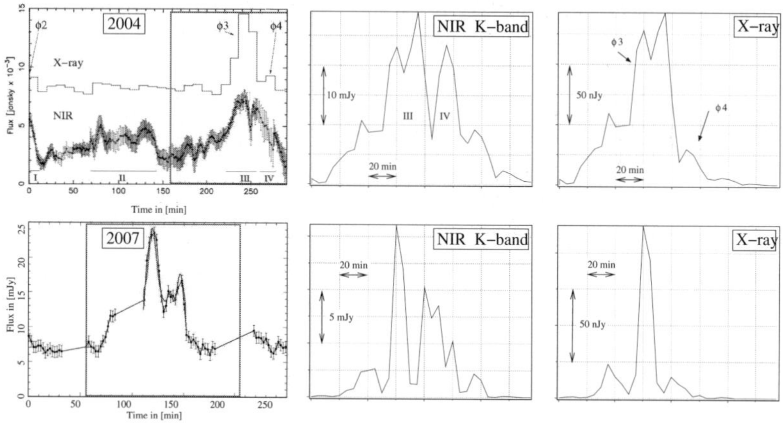

Fig. 4. Application of the time dependent flare emission model presented in Section 3.2.1 to the data obtained in May 2007 (bottom) and July 2004 (top). In the panels on the left we show the available NIR and X-ray data with the modeled portion indicated by a red line. In the middle and on the right we show the NIR K-band and X-ray light curve derived from a multi component disk model. In both cases, starting at the center of the flare event, we assumed a 30% increase of the source component sizes over about 40 minutes i.e. two orbital time scales. With the additional assumption of a flux decrease of ~1 magnitude due to synchrotron losses the model provides a very good qualitative and quantitative representation of the 2004 measurements (see Eckart et al. 2006a). For the 2007 NIR data there are no simultaneous X-ray data available but the K-band light curve indicates an evolution of the source components. The lower right panel therefore represents the light curve we would have expected.

emission model. We find that the temporary accretion disk around Sgr A* can well be represented by a multi component model with source properties that are bracketed by those of a simple flare and a quiescent model. We have used a ($\gamma_e \sim 10^3$) synchrotron model in which the source component spectral indices are compatible with the constant value of $\alpha = 0.6 \pm 0.2$ reported by Ref. 23. A steeper spectral index of $\alpha = 1.3$ allows for direct synchrotron and SSC contributions in the NIR. In both cases the component flux densities can be represented by a power spectrum $N(S) \propto S_m^{\alpha_S}$ with an exponent α_S close to -1. The multicomponent model explains the quasi-periodic sub-flare structure at infrared wavelengths and shows that with adequate sensitivity and time resolution they should be detectable in the X-ray domain as well.

Eckart et al. (2008, subm. to A&A) present a model in which a combination of a temporary accretion disk occurs in combination with a short jet. Such a source structure may explain most of the observed properties of

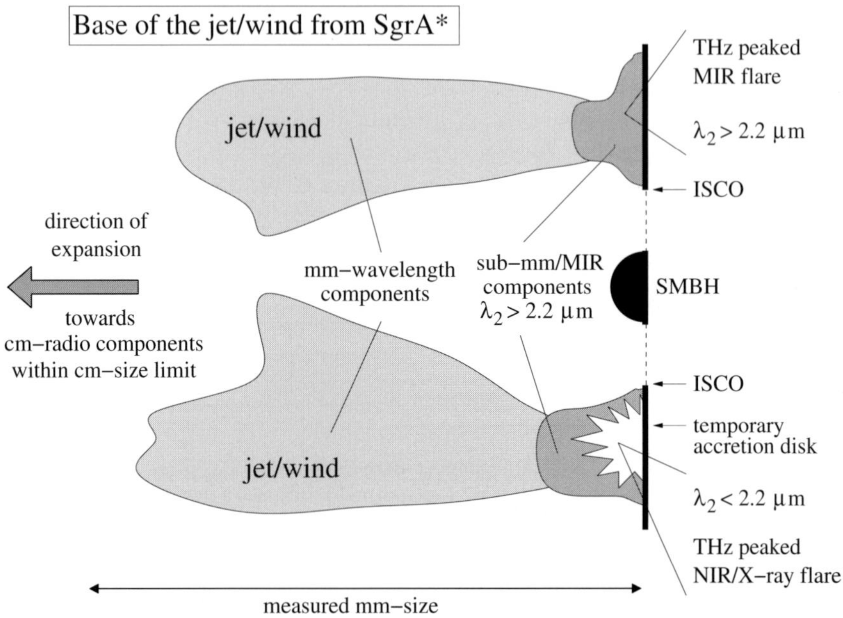

Fig. 5. A possible source structure for the accretion disk around the SMBH associated with Sgr A*. In this sketch the disk is shown as a vertical thick line to the right. Extending to the left, we show one side above the disk. Higher energy flare emission (lower part) is responsible for the observed NIR/X-ray flare emission. Lower energy flare emission (upper part) may substantially contribute to long wavelength infrared emission. In addition to the expansion towards and beyond the the mm-source size, radial and azimuthal expansion within the disk may occur. Here λ_2 is the wavelength corresponding to the upper synchrotron cutoff frequency ν_2.

Sgr A*. Such a configuration is sketched in Fig.5. In this figure the disk is seen edge-on. Details of expected jet geometries are discussed by Markoff, Bower & Falcke (Ref. 38).

Simultaneous NIR K- and L-band measurements in combination with X-ray observations should lead to a set of light curves that should allow us to prove the proposed model and to discriminate between the individual higher and lower energy flare events. Simultaneous X-ray measurements are important to clearly distinguish between high and low energy events. To do so, it is required to separate the thermal non-variable bremsstrahlung and the non-thermal variable part of the Sgr A* X-ray flux density. This capability is provided by the ACIS-I instrument aboard the *Chandra X-ray Observatory* and is essential to have, especially in the case of weak X-ray flare events in which the X-ray flare intensity is of the order of the extended bremsstrahlung component associated with SgrA* - or even below. These can clearly be identified in combination with infrared data.

Acknowledgements

Part of this work was supported by the German *Deutsche Forschungsgemeinschaft, DFG* via grant SFB 494. L. Meyer, K. Muzic, M. Zamaninasab, and D. Kunneriath are members of the International Max Planck Research School (IMPRS) for Radio and Infrared Astronomy at the MPIfR and the Universities of Bonn and Cologne.

References

1. Eckart, A. & Genzel, R. 1996, Nature 383, 415.
2. Genzel, R., Eckart, A., Ott, T., Eisenhauer, F., 1997, MNRAS 291, 219.
3. Genzel, R., Pichon, C., Eckart, A., Gerhard, O. E., Ott, T., 2000, MNRAS 317, 348.
4. Ghez, A., Klein, B.L., Morris, M. & Becklin, E.E. 1998, ApJ, 509, 678.
5. Ghez, A., Morris, M., Becklin, E.E., Tanner, A. & Kremenek, T. 2000, Nature 407, 349.
6. Ghez, A.M., Wright, S.A., Matthews, K., et al. 2004a, ApJ 601, 159.
7. Ghez, A.M., Hornstein, S.D., Bouchez, A., Le Mignant, D., Lu, J., Matthews, K., Morris, M., Wizinowich, P., Becklin, E.E., 2004b, AAS 205, 2406.
8. Ghez, A.M., Salim, S., Hornstein, S. D., Tanner, A., Lu, J. R., Morris, M., Becklin, E. E., Duchêne, G.,2005, ApJ 620, 744.
9. Eckart, A., Genzel, R., Ott, T. and Schödel, R. 2002, MNRAS, 331, 917.
10. Schödel, R., Ott, T., Genzel, R., Hofmann, R., Lehnert, M., Eckart, A., Mouawad, N., Alexander, T., 2002, Natur 419, 694.
11. Schödel, R., Ott, T., Genzel, R., Eckart, A., Mouawad, N., Alexander, T., 2003 ApJ 596, 1015.
12. Eisenhauer, F.; Schdel, R.; Genzel, R.; Ott, T.; Tecza, M.; Abuter, R.; Eckart, A.; Alexander, T., 2003, ApJ 597, L121.
13. Eisenhauer, F., Genzel, R., Alexander, T., Abuter, R., Paumard, T., Ott, T., Gilbert, A., Gillessen, S., 2005, ApJ 628, 246.
14. Eckart, A.; Baganoff, F. K.; Morris, M.; Bautz, M. W.; Brandt, W. N.; Garmire, G. P.; Genzel, R.; Ott, T.; Ricker, G. R.; Straubmeier, C.; Viehmann, T.; Schödel, R.; Bower, G. C.; Goldston, J. E., 2004, A&A 427, 1.
15. Eckart, A.; Baganoff, F. K.; Schödel, R.; Morris, M.; Genzel, R.; Bower, G. C.; Marrone, D.; et al. 2006a, A&A 450, 535.
16. Bower, G.C., Falcke, H., Sault, R.J. and Backer, D.C., 2002, ApJ, 571, 843.
17. Herrnstein, R.M., Zhao, J.-H., Bower, G.C., & Goss, W.M., 2004, AJ, 127, 3399.
18. Zhao, J.-H., Young, K.H., Herrnstein, R.M., Ho, P.T.P., Tsutsumi, T., Lo, K.Y., Goss, W.M. & Bower, G.C., 2003, ApJL, 586, L29.

19. Zhao, J.-H., Herrnstein, R.M., Bower, G.C., Goss, W.M., & Liu, S.M., 2004, ApJL, 603, L85.
20. Mauerhan, J.C.; Morris, M.; Walter, F.; Baganoff, F.K., 2005, ApJ 623 L25.
21. Yuan, F., 2006, JPhCS 54, 427.
22. Cuadra, J.; Nayakshin, S., 2006, JPhCS 54, 436.
23. Hornstein, S. D.; Matthews, K.; Ghez, A. M.; Lu, J. R.; Morris, M.; Becklin, E. E.; Rafelski, M.; Baganoff, F. K., 2007, astro-ph:0706.1782.
24. Gillessen, S., et al., 2006, JPhCS 54, 411.
25. Genzel, R., Schoedel, R., Ott, T., et al. 2003, Nature, 425, 934.
26. Eckart, A.; Schödel, R.; Meyer, L.; Trippe, S.; Ott, T.; Genzel, R., 2006b, A&A 455, 1.
27. Meyer, L., Eckart, A., Schödel, R., Duschl, W. J., Muciz, K., Dovciak, M., Karas, V., 2006a, A&A 460, 15.
28. Meyer, L., Schödel, R., Eckart, A., Karas, V., Dovciak, M., Duschl, W. J., 2006b, A&A 458, L25.
29. Meyer, L., Schödel, R., Eckart, A., Duschl, W. J., Karas, V., Dovciak, M., 2007, A&A 473, 707.
30. Belanger, G.; Terrier, R.; de Jager, O. C.; Goldwurm, A.; Melia, F., 2006, JPhCS 54, 420.
31. Aschenbach, B., Grosso, N., Porquet, D. and Predehl, P., 2004, A&A 417, 71.
32. Marrone, D.P.; Moran, J.M.; Zhao, J.-H.; Rao, R. 2007, ApJ 654, L57.
33. Dovciak, M., Karas, V., Yaqoob, T. 2004, ApJS, 153, 205.
34. Connors, P. A.; Stark R. F., 1977, Nature, 269, 128.
35. Greenhough, J., Chapman, S.C., Chaty, S., Dendy, R.O., and Rowlands G., 2001, A&A, arXiv:astro-ph/0107074.
36. Pessah, M.E., Chan, C.-K., Psaltis, D., 2007, eprint:0705.0352.
37. Yusef-Zadeh, F., Roberts, D., Wardle, M., Heinke, C. O., Bower, G. C., 2006, ApJ 650, 189.
38. Markoff, S.; Bower, G.C.; Falcke, H., 2007, to be published in MNRAS.
39. Karas, V., Dovciak, M., Eckart, A., Meyer, L., 'Flares from spiral waves by lensing and time-delay amplification?', Proc. of the Workshop on the Black Holes and Neutron Stars, eds. S. Hledik and Z. Stuchlik, 19-21 September 2007 (Silesian University, Opava), astro-ph0709.3836.
40. Lu, Y., & Torres, D.F., 2003, Int. Journal of Modern Physics D, Vol.12, No. 1, pp. 63-77.

STERILE NEUTRINOS AND THE RAPID FORMATION OF SUPERMASSIVE BLACK HOLES

M. C. RICHTER, G. B. TUPPER and R. D. VIOLLIER*

Centre for Theoretical Physics and Astrophysics, Department of Physics, University of Cape Town, Private Bag, Rondebosch 7701, South Africa,
raoul.viollier@uct.ac.za

Massive sterile neutrinos that are embedded in a renormalizable seesaw-like extension of the standard model of particle physics are a most attractive dark matter candidate. Indeed, lowering the seesaw mass-scale from the grand unification scale to well below the weak interaction symmetry breaking scale, the lowest-mass right-handed neutrino may acquire a mass around 10 keV/c^2 and thus become a quasi-stable dark matter particle due to its small Yukawa coupling. Moreover, this low mass-scale seesaw mechanism may explain the observed oscillation properties of the left-handed active neutrinos as well as the baryon asymmetry through ordinary mixing. Here we show that self-gravitating degenerate balls, made of such sterile neutrinos, can resolve the longstanding mystery of the formation of supermassive black holes at redshift larger than six, less than a Gyr after the big bang.

1. Introduction

Supermassive black holes of $\sim 3 \times 10^9 M_\odot$[1] are already present at redshift $z = 6.42$ or 850 Myr after the big bang, as is evident from the recently discovered[2] quasar SDSS J114816.64+525150.3. Unlike the galaxies, which are thought to assemble through hierarchical mergers of smaller galaxies,[3] the supermassive black holes form anti-hierarchically, i.e. the larger black holes prior to the smaller ones. This is supported by the observation[4] that the number of brighter quasars peaks at $z \sim 2$, while that of lower-luminosity active galactic nuclei peaks at $z \sim 0.6$. There is only a narrow window of about 485 Myr between the appearance of the first stars at $z \sim 11$, or 365 Myr after the big bang when reionization of the universe begins,[5] and the appearance of the first quasars[2] at $z = 6.42$ when reionization ends, during which at least some of the most massive black holes must have formed.

*Author to whom any correspondence should be addressed.

It has been known for more than three decades[6] that the growth of a 3 $M_{\odot}$ black hole of stellar origin, formed at reionization $\sim$ 0.4 Gyr after the big bang and accreting baryonic matter at the Eddington limit with a radiative efficiency of $\epsilon = 0.13$, would reach a mass of $3 \times 10^9 M_{\odot}$ only $\sim$ 1.6 Gyr after the big bang, as its mass doubling time is $\sim$ 40 Myr. This seems to preclude the existence of a $3 \times 10^9 M_{\odot}$ black hole at $z = 6.42$ or 0.85 Gyr after the big bang. A shorter accretion time can, in the absence of frequent black hole mergers, only be achieved by lowering the radiative efficiency or increasing either the accretion rate or the seed black hole mass.[4] While the former two possibilities are not really attractive in the case of baryonic matter, increasing the seed black hole mass by several orders of magnitude is not attractive either, because, in spite of the ongoing intensive search, the evidence for intermediate mass black holes, with masses between 20 $M_{\odot}$ and $10^6 M_{\odot}$, is rather weak and controversial.[7] Regardless of whether the supermassive black holes were produced through the accretion of baryonic matter onto stellar-mass black holes, or through binary mergers starting from stellar-mass black holes, these would inevitably have left a trail of observable intermediate-mass black holes, which ran out of baryonic matter or black hole supplies sometime and somewhere in this universe.

Thus apart from the easily detectable upper limit of $M_{\rm max} \sim 3 \times 10^9 M_{\odot}$, there also seems to be a lower mass limit of $M_{\rm min} \sim 10^6 M_{\odot}$ of the supermassive black holes. In this context, it is important to note that the mass of the black hole at the center of M87[8] is equal to that of the earliest quasar[2] observed at $z = 6.42$. As these are both archetypical examples of the most massive black holes at vastly different epochs of the universe, one may infer that $M_{\rm max}$ has not increased significantly from $z = 6.42$ to $z = 0$, or from 0.85 Gyr to 13.7 Gyr after the big bang. Indeed, although Eddington-limited baryonic matter accretion is essential for seeing the quasars, this is a transient phenomenon with an estimated total lifetime of a few tens of Myr,[4] that presumably contributes very little to the final mass of the most massive black holes today.

In summary, it seems that a consistent theory of the formation of the supermassive black holes should be able to explain both observational facts, the nearly time-independent upper and lower mass limits of the supermassive black holes, as well as the early formation of the most massive black holes around $z = 6.42$. As both the baryonic matter accretion, as well as the black hole merger scenarios, do not seem to provide us with a satisfactory description of these observational facts, we are led to ask the pertinent question, whether an alternative scenario, based on the accretion of mainly dark matter, may do better. Galaxies are, indeed, dominated by dark matter, and as part of this dark matter is concentrated in the galactic centers, it may very well have contributed to the formation of the supermassive black holes. Of course, in order to make definite predictions, one needs to rely on a well-defined and consistent dark matter candidate. Thus, in section 2, we discuss the physical and cosmological properties of our sterile neutrino dark matter candidate, while in section 3, we explore an astrophysical consequence of this dark matter scenario, namely the formation of self-gravitating supermassive degenerate sterile neutrino balls, with masses between $10^6 M_\odot$ and $3 \times 10^9 M_\odot$. We then discuss how these preformed neutrino balls convert efficiently and anti-hierarchically into supermassive black holes, using stellar-mass seed black holes as catalysts. In section 4, we describe the dynamics of the accretion of degenerate sterile neutrinos onto a black hole in a simple nonrelativistic Thomas-Fermi field theory, based on the Lane-Emden equation, while our conclusions are presented in section 5.

2. Physics and Cosmology of Sterile Neutrinos

Three decades ago, Minkowski[9] introduced a set of three right-handed sterile neutrinos in a renormalizable seesaw-like extension of the minimal standard model of particle physics, in order to allow for non-zero Dirac and Majorana neutrino masses. Apart from the standard model Lagrangian $\mathcal{L}_{SM}$ with three left-handed massless neutrinos, the new minimal standard model Lagrangian $\mathcal{L}_{\nu MSM}$ includes the kinetic terms of the three right-handed neutrinos, Yukawa coupling (or Dirac mass) terms and Majorana mass terms for the three right-handed neutrinos, i.e.

$$\mathcal{L}_{\nu MSM} = \mathcal{L}_{SM} + \overline{N}_I i\gamma_\mu \partial^\mu N_I - F_{\alpha I}\overline{L}_\alpha N_I \tilde{\Phi} - \frac{M_I}{2}\overline{N_I}^c N_I + h.c. \qquad (1)$$

Here, $\tilde{\Phi}$ are the components of the adjoint Higgs doublet and L_α the lepton doublets and N_I the sterile neutrino singlets. Thus, compared to the original standard model, the new Lagrangian $\mathcal{L}_{\nu MSM}$ has 18 additional parameters: 3 Majorana masses, associated with the right-handed neutrinos N_I, and 15 new Yukawa couplings in the leptonic sector (i.e. 3 Dirac masses, 6 mixing angles and 6 CP-violating phases). These parameters can be chosen such as to fit the solar, atmospheric, accelerator and reactor neutrino oscillation experiments, as well as the observed baryon asymmetry, with two nearly degenerate sterile neutrinos in the mass range between 1 GeV/c^2 and 20 GeV/c^2, and a third one with mass around 10 keV/c^2.[9] While the two heavy sterile neutrinos are unstable and thus may only be observed at particle accelerators, the third one is a most promising quasi-stable dark matter candidate, that can only be detected astrophysically through its radiative decay into an active neutrino. This light sterile neutrino interacts with the standard model particles only through its tiny mixing with the active neutrinos. The bulk part of the light sterile neutrinos is produced $\sim 2.3\ \mu$s after the big bang, at temperatures

$$T \sim 328 \left(\frac{mc^2}{15 \text{ keV}}\right)^{1/3} \text{MeV}/k \ , \tag{2}$$

well ahead of the quark-gluon and chiral restoration phase transitions, through incoherent resonant[10] and non-resonant[11] scattering of the active neutrinos. For a wide range of the parameters of the sterile neutrino, these are generated out of thermal equilibrium, yielding a sterile neutrino mass fraction of the total mass-energy of this universe which is consistent with that of nonbaryonic dark matter. In order to fix our ideas, we now choose the initial lepton asymmetry to be

$$L_{\nu_e} = \frac{n_{\nu_e} - n_{\bar{\nu}_e}}{n_\gamma} = 10^{-2} \ , \tag{3}$$

the mixing angle of the sterile neutrino ν_s with the active neutrino ν_e to be given by $\sin^2 \vartheta = 10^{-13}$, and the sterile neutrino to have a mass $m = 15$ keV$/c^2$.With these parameters, the mass fraction of the sterile neutrinos produced in the early universe is indeed $\Omega_{\nu_s} \sim 0.24$,[10] i.e. equal to that of the nonbaryonic dark matter, derived from the WMAP data.[5] The choice of $m = 15$ keV$/c^2$ is consistent with the constraints from the flux power spectrum of the Lyα forest at $z \sim 4$ contained in the Sloan Digital Sky Survey (SDSS), i.e. $m > 14$ keV$/c^2$[12] and $m \gtrsim 9$ keV$/c^2$.[13]

The same tiny mixing angle, which prevents the thermal equilibration and thus the overproduction of the sterile neutrino in the early universe, also renders the sterile neutrino quasi-stable,[14] with a lifetime much larger than the age of the universe. In fact, with our choice of the parameters, $\sin^2 \vartheta = 10^{-13}$ and $m = 15$ keV$/c^2$, the sterile neutrino decays with a lifetime of

$$\tau(\nu_s \to \nu_e \nu \bar{\nu}) = \frac{192\pi^3}{G_F^2 m^5 \sin^2 \vartheta} = 1.21 \times 10^{19} \text{ yr} \ , \tag{4}$$

predominantly into a ν_e and a neutrino-antineutrino pair, ν_i and $\bar{\nu}_i$, carrying the flavours $i = e, \mu$ or τ.[14] These standard neutrinos are all virtually unobservable because their energy is too small. But there is also a subdominant radiative decay mode, with a branching ratio

$$\frac{\tau\left(\nu_s \to \nu_e \nu \bar{\nu}\right)}{\tau\left(\nu_s \to \nu_e \gamma\right)} = \frac{2\alpha}{8\pi} = 0.784 \times 10^{-2} \tag{5}$$

into a potentially observable photon and a ν_e.[14] However, due to the smallness of its partial decay width of

$$[\tau(\nu_s \rightarrow \nu_e \gamma)]^{-1} = 0.649 \times 10^{-21} \mathrm{yr}^{-1} \ , \tag{6}$$

these photons of energy $mc^2/2$, which may be the "smoking gun" of the sterile neutrino, are difficult to observe as well. In fact, for the chosen model parameters, a sterile neutrino dark matter concentration of mass M has a luminosity of merely

$$L_X = \frac{Mc^2}{2\tau(\nu_s \rightarrow \nu_e \gamma)} = 1.84 \times 10^{25} (M/M_\odot) \ \mathrm{erg/s} \ , \tag{7}$$

in photons of $mc^2/2 = 7.5$ keV energy. Thus the best places to look for these photons are the diffuse extragalactic X-ray background, as well as the X-rays emitted by large galaxy clusters, low-surface-brightness and dwarf galaxies that are dominated by nonbaryonic dark matter.[13]For an initial lepton asymmetry $L_{\nu_e} \sim 10^{-10}$, which is of the order of the baryon asymmetry[5]

$$B = \frac{n_b}{n_\gamma} \sim 6 \times 10^{-10} \ , \tag{8}$$

mainly non-resonant neutrino scattering contributes to the production of sterile neutrino dark matter. These sterile neutrinos inherit a nearly thermal energy spectrum from the active neutrinos,[11] which allows them to play the role of warm dark matter in the large-scale structure of the universe, the clusters of galaxies and the galactic halos. They may also erase the undesirable excessive substructure on the galactic scales.

The initial lepton asymmetry does not need to be of the same order of magnitude as the baryon asymmetry. However, for a larger initial lepton asymmetry, like $L_{\nu_e} \sim 10^{-2}$, there is, in addition to non-resonant neutrino scattering, also resonant or matter-enhanced neutrino scattering contributing to the production of dark matter.[10] The latter yields cool sterile neutrinos that have a distorted quasi-degenerate spectrum, with an average energy of about two-thirds of that of the warm sterile neutrinos, due to the resonant Mikheyev-Smirnov-Wolfenstein (MSW) oscillations.[15]

A relatively large initial lepton asymmetry of $L_{\nu_e} \sim 10^{-1}$ to 10^{-2} is also what seems to be required, to bring the observed light-element abundances in line with the number of three active neutrinos at nucleosynthesis.[16] Sterile neutrinos may as well be responsible for the pulsar kicks of up to ~ 1600 km/s, which magnetars acquire in supernova explosions.[17] Thus our sterile neutrino meets all the constraints which an acceptable dark matter particle must fulfil,[10,12,13] but it has several remarkable additional properties that make it a rather unique candidate for dark matter.

3. The Symbiotic Black Hole Formation Scenario

For our model parameters $m = 15\ \text{keV}/c^2$, $\sin^2\vartheta = 10^{-13}$ and $L_{\nu_e} = 10^{-2}$, cool (or resonant) dark matter dominates over warm (or non-resonant) dark matter by a factor of about three.[10] The cool sterile neutrinos become nonrelativistic $\sim$ 22 min after the big bang, well after nucleosynthesis, and they begin, together with the warm sterile neutrinos and baryonic matter, to dominate the expansion of the universe $\sim$ 79 kyr after the big bang, well ahead of recombination. Thus the primordial density fluctuations of sterile neutrino dark matter have enough time to grow nonlinear and form degenerate sterile neutrino balls,[18] through a process called gravitational cooling,[19] prior to the appearance of the first quasars, 850 Myr after the big bang. This collapse process may start ahead of reionization, perhaps as early as $\sim$ 320 Myr after the big bang. Initially, the free-falling sterile neutrino dark matter, dominating baryonic matter by about a factor of six,[5] drags the baryonic matter along towards the center of the collapse. The baryonic gas will get heated, reionized and may evaporate, but the free fall of the neutrinos will not be inhibited by the Eddington radiation limit. Eventually, the quasi-degenerate sterile neutrino dark matter hits, $\sim$ 640 Myr after the big bang, the degeneracy pressure, bouncing off a number of times, while ejecting a fraction of the dark matter at every bounce. The neutrino ball finally settles in a condensate of degenerate sterile neutrino matter at the center of the collapsed object, as has been shown in calculations based on time-dependent Thomas-Fermi mean field theory.[19]

The smallest mass that may collapse is the mass contained within the free-streaming length at matter-radiation equality, $\sim$ 79 kyr after the big bang. For $m = 15\ \text{keV}/c^2$, this free-streaming mass is $M_{\text{warm}} \sim 7\times10^6 M_\odot$ in the case of warm (or non-resonant) sterile neutrinos, and $M_{\text{cool}} \sim 2\times10^6 M_\odot$ in the case of the dominant cool (or resonant) sterile neutrinos.[10] As part of the neutrino dark matter is ejected during the collapse process, the minimal mass of a degenerate sterile neutrino ball may be somewhat smaller than M_{cool}, perhaps $M_{\text{min}} \sim 10^6 M_\odot$, consistent with the lower mass limit of the observed supermassive black holes.

The maximal mass that a self-gravitating degenerate neutrino ball can support gravitationally, is the Oppenheimer-Volkoff (OV) limit[20]

$$M_{\text{max}} = 0.5430\, M_{\text{Pl}}^3\, m^{-2}\, g^{-1/2}\ . \tag{9}$$

For $m = 15\ \text{keV}/c^2$ and spin degeneracy factor $g = 2$, this mass $M_{\text{max}} =$

$2.789 \times 10^9 M_\odot$ is consistent with that of the most massive black holes observed in our universe,[12] As such a supermassive degenerate neutrino ball has a radius of only 4.45 Schwarzschild radii,[20] it is almost a black hole. Thus the maximal mass scale of these objects may be linked to the existence of a sterile neutrino of $\sim$ 15 keV/c^2 mass, in a similar fashion as the maximal mass scale of the neutron stars is linked to the effective mass of the neutron.[21]

Since the gravitational potential in the interior of a neutrino ball is nearly harmonic, these objects, in particular those near the upper mass limit, are ideal breeding grounds for stars of mass $M \gtrsim 25 M_\odot$.[22] As soon as such a central massive star is formed from a collapsing molecular hydrogen cloud that was attracted to the neutrino ball, it may be kicked out through close binary encounters with intruding stars on a time scale of $\sim$ 250 Myr. However, before that happens, a star of 25 $M_\odot$ will evaporate large portions of its hydrogen and helium envelope and become a Wolf-Rayet star. And about 3 Myr after its formation, the star undergoes a core collapse supernova explosion of type Ic, leaving a black hole of 3 to 4 $M_\odot$ at the center of the neutrino ball. Some of these most massive supernova explosions, occurring in high-mass neutrino balls, may be observable as long-duration γ-ray bursts.[23] As in contrast to pulsars, black holes do presumably not acquire "black hole kicks" during a supernova explosion, the velocity of the stellar-mass black hole will be small compared to the escape velocity from the center of the neutrino ball. For a ball of $3 \times 10^6 M_\odot$ mass and 25 light-days radius,[18] consisting of degenerate sterile neutrinos of 15 keV/c^2 mass and degeneracy factor $g = 2$, the escape velocity from the center is 1700 km/s, while for a ball of the same sterile neutrinos at the OV-limit, with 2.8×10^9 $M_\odot$ mass and 1.4 light-days radius, the escape velocity from the center is the velocity of light.[20]

The supernova explosion of the massive star, giving birth to a stellar-mass black hole at the center, sparks the rapid growth of the black hole through nearly radiationless, and therefore, Eddington-unlimited accretion of mainly degenerate sterile neutrino dark matter from the surrounding neutrino ball, until the supplies dry up. In this symbiotic scenario,[22] the anti-hierarchical formation of the bright quasi-stellar objects and low-luminosity active galactic nuclei may be understood in terms of the escape velocity from the center of a $3 \times 10^6 M_\odot$ neutrino ball, which is 176 times smaller than that of a $3 \times 10^9 M_\odot$ neutrino ball. The low-mass neutrino ball may, therefore, have a much harder time capturing a molecular hydrogen cloud that is able to produce a massive star. In particular, a low-mass neutrino ball may experience a large number of unsuccessful attempts, leading to ordinary low-mass stars, or neutron stars after supernova explosion, prior to delivering the expected stellar-mass black hole. These undesirable stellar-mass objects will eventually be ejected from the neutrino ball through pulsar kicks or close binary encounters with intruder stars from the surrounding star cluster, thus clearing the scene for the next attempt at forming this stellar-mass black hole. The randomness of this process may very well delay the formation of a stellar-mass black hole at the center of a low-mass neutrino ball by several Gyr, while a neutrino ball at the top of the mass scale may easily deliver the stellar-mass black hole on its first attempt in less than 10 Myr. Of course, these sketchy ideas will have to be tested in realistic numerical simulations. However, if this scenario is correct, some low-mass neutrino balls may still be around at some galactic centers. For instance, a $10^6 M_\odot$ neutrino ball would reveal itself through its X-ray emission of 2×10^{31} erg/s at 7.5 keV, for our model parameters.

4. Accretion of a Neutrino Halo into a Black Hole

A particle that is initially at rest at the surface of a $3 \times 10^6 M_\odot$ neutrino ball reaches the center in the free-fall time $\tau_F \sim 35$ yr. This is also the time frame in which the accretion process onto the central black hole reaches a steady-state flow. In the steady-state approximation, the flow is governed by Bernoulli's eq.[22]

$$\phi(r) + \frac{1}{2}(u^2(r) + v_F^2(r)) = \phi(r_H) = \text{const} \ . \tag{10}$$

Here $u(r)$ is the flow velocity of the infalling degenerate sterile neutrino fluid, $v_F(r)$ its Fermi velocity, $\phi(r)$ the gravitational potential and r_H the radius of the halo. Assuming that the flow makes the gravitational potential ϕ extremal for all values of the radius r, with respect to variations that satisfy the constraint of mass conservation,

$$\rho u = \frac{m^4 \, g \, v_F^3 \, u}{6\pi^2\hbar^3} = \text{const} \ , \tag{11}$$

we obtain

$$u^2(r) = \frac{1}{3}v_F^2(r) = c_S^2(r) \ , \tag{12}$$

which means that the inflow of the sterile neutrinos is trans-sonic, i.e. it flows at the local velocity of sound $c_S(r)$. Thus Bernoulli's eq. is now

$$\frac{2}{3}v_F^2(r) = \phi(r_H) - \phi(r) = GM_\odot \frac{v(x)}{bx} \ , \tag{13}$$

which defines the quantity $v(x)$. Using Poisson's eq., one can readily verify that $v(x)$ fulfils the Lane-Emden eq.

$$\frac{1}{x}\frac{d^2v(x)}{dx^2} = -\left(\frac{v(x)}{x}\right)^{3/2} , \tag{14}$$

provided the length scale is

$$b = \frac{4}{3}\left(\frac{3\pi\hbar^3}{4\sqrt{2}gm^4G^{3/2}M_\odot^{1/2}}\right)^{2/3} . \tag{15}$$

Thus for $m = 15$ keV/c^2 and $g = 2$ we have $b = 2.587$ lyr. The total mass enclosed within a radius $r = bx$ is[24]

$$M(x) = M_\odot\left[v(x) - xv'(x)\right] . \tag{16}$$

Various solutions $v(x)$ of the Lane-Emden eq.(14), all having the total mass $M = 2.714\ M_\odot$, are shown in Fig. 1. There are three distinct classes of

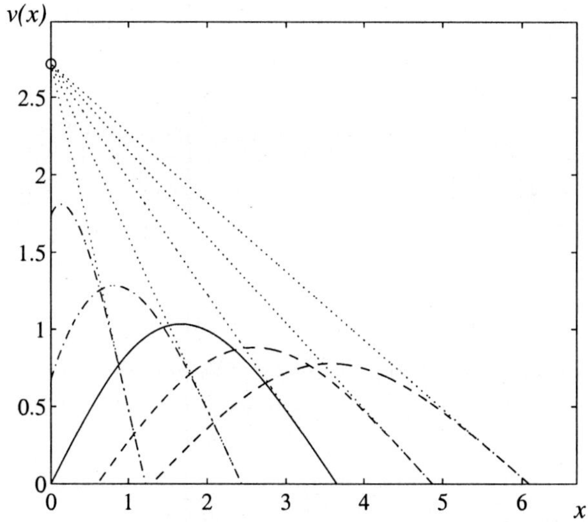

Fig. 1. Various solutions of the Lane-Emden equation, all having total mass $M = M_P + M_H = 2.714\ M_\odot$. The solid, dashed and dash-dotted lines represent the $E-$, $F-$ and $M-$ solutions, respectively.

solutions: The M-solutions exhibit a central point mass $M_P = M_\odot v(0)$, surrounded by a self-gravitating degenerate sterile neutrino halo. The F-solutions describe shells of self-gravitating degenerate neutrino matter that are gravitationally unstable. The E-solution, with $v(0) = 0$ and $v'(0) = 1$, stands for a pure neutrino ball with $M_P = M_\odot v(0) = 0$. Our focus is on the M-solutions of the Lane-Emden eq.(14) because these describe, in the steady-state approximation, the various stages of the accretion history of a black hole surrounded by a degenerate sterile neutrino halo. For instance, decreasing the halo radius $r_H = bx_H$, while keeping the total mass M fixed, causes the central point mass M_P to increase, as seen in Fig.1. The solutions of the Lane-Emden eq.(14) with $M_P = M_\odot v(0) > 0$, for arbitrary mass $\tilde{M}$ can be obtained noting that, if $v(x)$ is a solution, $\tilde{v}(x) = A^3 v(Ax)$ with $A > 0$ is a solution as well. Thus all the masses and radii scale as $\tilde{M} = A^3 M$ and $\tilde{r} = r/A$.[24] The mass accretion rate into a sphere, containing the mass M_C within a radius r_C from the center, is thus given by

$$\frac{dM_C}{dt} = 4\pi r_C^2 \rho(r_C) u(r_C) \tag{17}$$

or

$$\frac{dM_C}{dt} = \frac{\sqrt{3} g m^4 G^2 M_\odot^2}{2\pi\hbar^3} \left[v(x_C)\right]^2 = \frac{f(\mu)}{\tau M_\odot} M_C^2 \ , \tag{18}$$

where we have introduced the universal time-scale

$$\tau = \frac{2\pi\hbar^3}{\sqrt{3} g m^4 G^2 M_\odot} = 1.488 \times 10^7 \ \frac{2}{g} \left(\frac{15 \text{ keV}}{mc^2}\right)^4 \text{ yr} \ . \tag{19}$$

The shut-off parameter

$$f(\mu) = \frac{v\left(x_C\right)^2}{\left[v(x_C) - x_C v'(x_C)\right]^2} \tag{20}$$

is a function of the mass ratio $\mu = M_C/M$. We now choose $r_C = bx_C$ to be the radius at which the escape velocity reaches the velocity of light, i.e. x_C is given by

$$\frac{1}{x_\odot} = \frac{v(x_C)}{x_C} - v'(x_H) \ , \tag{21}$$

where $r_\odot = bx_\odot$ is the Schwarzschild radius of the sun with

$$x_\odot = \frac{2GM_\odot}{bc^2} = 1.207 \times 10^{-13} \left(\frac{g}{2}\right)^{2/3} \left(\frac{mc^2}{15 \text{ keV}}\right)^{8/3} \ . \tag{22}$$

As for $M \ll M_{\max}$ and $\mu < 1$, we may approximate $v(x_C) \sim v(0)$ and $f(\mu) \sim 1$, eq.(18) agrees well with standard Bondi accretion theory.[25]

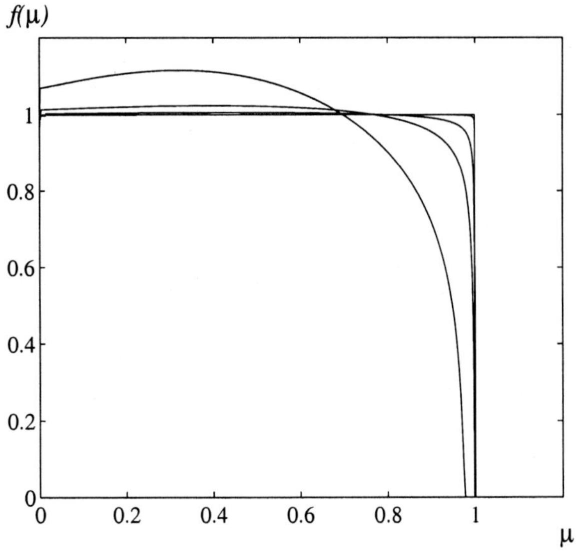

Fig. 2. Various shut-off parameters $f(\mu)$ as a function of the mass ratio $\mu = M_C/M$. The neutrino ball masses vary from $M = 10^6 M_\odot$ for the box function, through $M = 10^7, 10^8$ to $10^9\ M_\odot$ for the curve with the largest peak value.

Integrating eq.(18) using this approximation, the growth of the black hole is given by

$$M_C(t) \sim \frac{M_C(0)}{1 - t/\tau_A} \ , \tag{23}$$

yielding an accretion time scale of $\tau_A = \tau M_\odot / M_C(0)$. During the accretion process, both M_C and x_C grow, while x_H shrinks as a function of time, eventually causing x_C and x_H to converge and $v(x_C)$ to vanish. The shut-off parameter $f(\mu)$, shown in Fig.2 as a function of the mass ratio $\mu = M_C/M$, is for a neutrino ball of mass $M = 10^6\ M_\odot$ a simple Heavyside function. As M increases towards $M_{\max}$, this curve starts deviating from the simple box form, thus signalling the breakdown of our non-relativistic theory. Since, for $M \ll M_{\max}$, the black hole growth is approximately given by the Bondi formula, $\dot{M}_C \propto M_C^2$, we expect the mass growth curves to match the Bondi solution closely, with the attractive feature that the solution of eq.(18) eventually brings the growth to a halt. The mass ratio μ is shown in Fig.3 as a function of time in units of the accretion time scale τ_A. For low-mass central black holes, these curves are indistinguishable from the Bondi solution, represented by the dotted curve. Here we start with $\mu(0) = 0.1$ at

$t = 0$ to illustrate the differences for halo masses between 10^6 and $10^9 M_\odot$. The universal time scale is $\tau = 14.88$ Myr, and the accretion time scales for a $3\times10^6 M_\odot$ neutrino ball onto 3 and 4 $M_\odot$ seed black holes are, therefore, $\tau_A = 4.96$ Myr and $\tau_A = 3.72$ Myr, respectively.

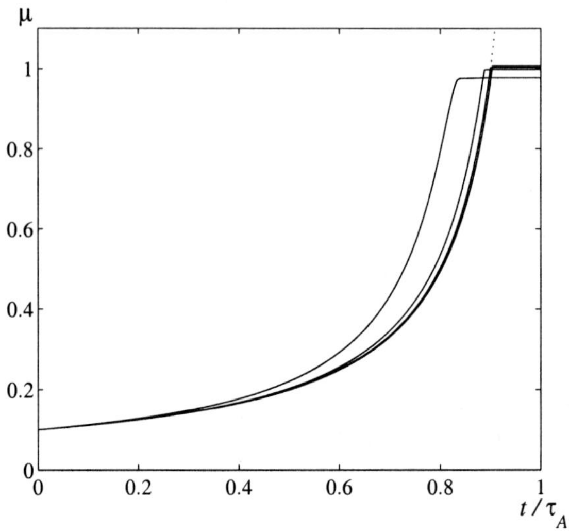

Fig. 3. The growth of a central black hole as a function of time, in units of the accretion time τ_A, for halos of between 10^6 and $10^9 M_\odot$, as in Fig. 2. For $\mu < 0.1$, the growth curves match the Bondi growth curve too closely to distinguish.

5. Conclusions

In summary, the neutrino balls are almost entirely swallowed by the seed black holes in an accretion time scale $\tau_A < 5$ Myr, thus converting these rapidly into supermassive black holes with negligible residual sterile neutrino halos. The most massive black holes may, therefore, form between 650 Myr and 840 Myr after the big bang. Although the nonrelativistic Thomas-Fermi mean field theory breaks down for $M \sim M_{\rm max}$ and $r_H \sim r_C$, we expect these basic results to persist in a relativistic theory of the accretion process.

There are three main features which distinguish the accretion of degenerate sterile neutrino dark matter from that of baryonic matter, playing a decisive role in the rapid growth of a stellar-mass black hole to the supermassive scale. Firstly, in contrast to the clumping of baryonic matter, neither the neutrino ball formation nor the neutrino halo accretion onto a black hole is inhibited by the Eddington radiation limit. Secondly, the matter densities of the degenerate sterile neutrino balls are much larger than those of any known form of baryonic matter having the same total mass, leading to much faster growth of the black holes through neutrino dark matter. Thirdly, the preformed degenerate sterile neutrino balls have, for $m \sim 15$ keV$/c^2$ and $g = 2$, masses in the same range as the observed supermassive black holes, which sets a natural limit to the growth of the black holes.

We, therefore, conclude that supermassive neutrino balls, with stellar-mass black holes at their center, indeed offer an intriguing symbiotic scenario, in which baryonic matter conspires with degenerate sterile neutrino dark matter, to form these galactic supermassive black holes, with masses between $10^6\ M_\odot$ and $3 \times 10^9\ M_\odot$, rapidly and efficiently.

Acknowledgements

This research is supported by the Foundation for Fundamental Research and the National Research Foundation of South Africa.

References

1. M. Schmidt, *Nature* **197** (1963) 1040;
E.E. Salpeter, *Astrophys. J.* **140** (1964) 796;
Y.B. Zel'dovich, *Dokl. Akad. Nauk SSSR* **155** (1964) 67;
D. Lynden-Bell, *Nature* **223** (1969) 690;
J. Kormendy and D. Richstone, *Ann. Rev. Astron. Astrophys.* **33** (1995) 581;
J. Kormendy and L.C. Ho, astro-ph/0003268;
L.C. Ho and J. Kormendy astro-ph/0003267.
2. C.J. Willott, R.J. McLure and M.J. Jarvis, *Astrophys. J.* **587** (2003) L15;
F. Walter *et al.*, *Nature* **424** (2003) 406.
3. R. Ellis, *Nature* **395** (1998) A3-8.
4. F. Combes, astro-ph/0505463.
5. D.N. Spergel *et al.*, astro-ph/0603449.
6. J.G. Hills, *Nature* **254** (1975) 295;
M.J. Rees, *Ann. Rev. Astro. Astrophys.* **22** (1984) 471.
7. M.C. Miller and E.J.M. Colbert, *Int. J. Mod. Phys.* **D13** (2004) 1.
8. F. Macchetto *et al.*, *Astrophys. J.* **489** (1997) 579.
9. P. Minkowski, *Phys. Lett.* **B67** (1977) 421;
T. Asaka, S. Blanchet and M. Shaposhnikov, *Phys. Lett.* **B631** (2005) 151;
T. Asaka and M. Shaposhnikov, *Phys. Lett.* **B620** (2005) 17;
M. Shaposhnikov and I. Tkachev, hep-ph/0604236.
10. X. Shi and G.M. Fuller, *Phys. Rev. Lett.* **82** (1999) 2832;
K. Abazajian, G.M. Fuller and M. Patel, *Phys. Rev.* **D64** (2001) 023501;

K.N. Abazajian and G.M. Fuller, *Phys. Rev.* **D66** (2002) 023526;
K. Abazajian and S.M. Koushiappas, astro-ph/0605271.

11. S. Dodelson and L.M. Widrow, *Phys. Rev. Lett.* **72** (1994) 17;
R. Barbieri and A.D. Dolgov, *Phys. Lett.* **B237** (1990) 440;
K. Kainulainen, *Phys. Lett.* **B244** (1990) 191.
12. U. Seljak, A. Makarov, P. McDonald and H. Trac, astro-ph/0602430.
13. M. Viel, J. Lesgourgues, M.G. Haehnelt, S. Matarrese and A. Riotto, astro-ph/0605706;
A. Boyarski, A. Neronov, O. Ruchayskiy, M. Shaposhnikov and I. Tkachev, astro-ph/0603660.
14. P.B. Pal and L. Wolfenstein, *Phys. Rev.* **D25** (1982) 766.
15. S.P. Mikheyev and A. Yu. Smirnov, *Sov. J. Nucl. Phys.* **42** (1985) 913;
L. Wolfenstein, *Phys. Rev.* **D17** (1978) 2369.
16. G. Steigman *Int. J. Mod. Phys.* **E15** (2006) 1.
17. A. Kusenko and G. Segré, *Phys. Lett.* **B396** (1997) 197;
A. Kusenko and G. Segré, *Phys. Rev.* **D59** (1999) 061302.
18. M.A. Markov, *Phys. Lett.* **10** (1964) 122;
R. Ruffini, *Lett. Nuovo Cim.* **29** (1980) 161;
R.D. Viollier, D. Trautmann and G.B. Tupper, *Phys. Lett.* **B306** (1993) 79;
F. Munyaneza, D. Tsiklauri and R.D. Viollier, *Astrophys. J.* **509** (1998) L105;
F. Munyaneza, D. Tsiklauri and R.D. Viollier, *Astrophys. J.* **526** (1999) 744.
19. N. Bilić, R.J. Lindebaum, G.B. Tupper and R.D. Viollier, *Phys. Lett.* **B515** (2001) 105.
20. J.R. Oppenheimer and G.M. Volkoff, *Phys. Rev.* **55** (1939) 374;
N. Bilić, F. Munyaneza and R.D. Viollier, *Phys. Rev.* **D59** (1999) 024003.
21. W. Baade and F. Zwicky, *Phys. Rev.* **45** (1934) 138;
A. Hewish, S.J. Bell, J.D.H. Pilkington, P.F. Scott, and R.A. Collins, *Nature* **217** (1968) 709.
22. M.C. Richter, G.B. Tupper and R.D. Viollier *Journal of Cosmology and Astroparticle Physics* **12** (2006) 015.
23. R.A. Chevalier and Z. Li, *Astrophys. J.* **520** (1999) L29;
P.A. Mazzali *et al.*, astro-ph/0603567.
24. R.D. Viollier, F.R. Leimgruber and D. Trautmann, *Phys. Lett.* **B297** (1992) 132.
25. H. Bondi, *Mon. Not. Roy. Astron. Soc.* **112** (1952) 195.

PART VII

Search for Hot and Warm Dark Matter

THE KARLSRUHE TRITIUM EXPERIMENT ON THE NEUTRINO MASS (KATRIN)

ERNST OTTEN

Institute of Physics, Staudinger Weg 7, University of Mainz, D-55099 Mainz, GERMANY, E-mail: otten@uni-mainz.de

A preview is given on a next generation β-spectroscopy experiment under development and construction; it aims at lowering the $m^2(\nu_e)$-uncertainty by another factor of 100 reaching a sensitivity limit $m(\nu_e) < 0.2\,\mathrm{eV}$.

1. Introduction

The question whether neutrinos have a finite rest mass or are lacking mass as well as charge, like the photon, came up already when W. Pauli[1] postulated in 1930 the existence of this particle in order to explain the continuous spectrum of β-decay. With Fermi's fundamental theory of weak interactions[2] a quantitative analysis of β-spectra with respect to the electron neutrino mass $m(\nu_e)=m(\bar{\nu}_e)$ became possible (We do not discriminate between the masses of particles and antiparticles assuming CPT-symmetry). Since the sensitivity of the β-spectrum to the neutrino mass maximizes at small neutrino energy, measurements have to be focussed onto the β-endpoint region. At given energy resolving power of the β-spectrometer $E/\Delta E$ a low β-endpoint energy is favoured, of course, as found in the case of tritium decay for example with $E_0 \sim 18.6\,\mathrm{keV}$, $T_{1/2}=12.3\mathrm{a}$. The long history of neutrino mass search in tritium β-decay counts about a dozen experiments started by Curran et al. in the late forties; it yielded an upper limit of $m(\nu_e)<1\,\mathrm{keV}$.[3] units of eV). The most recent limit of

$$m(\nu_e) < 2\,eV \quad (90\%C.L.) \tag{1}$$

has been published by the particle data group[4] based on the latest generation of experiments performed at the University of Mainz[5] and at the Institute of Nuclear Research at Troitzk near Moscow.[6] Considering that the observable in such a missing mass search is m^2 rather than m, the limit has been improved by more than 5 orders of magnitude meantime.

2. The Physics Potential of the KATRIN Experiment

From a workshop in Bad Liebenzell in 2001 emerged a letter of intent for the KATRIN-Experiment[7] in close collaboration of group members from the foregoing neutrino mass experiments at Los Alamos (now at University of Washington, Seattle), Mainz, and Troitsk with Forschungszentrum Karlsruhe. A design report[8] was approved in 2004. Construction of the experiment is under way and expected to terminate in 2010. The experiment aims at an improvement of the sensitivity limit by an order of magnitude down to $m(\nu_e) < 0.2\,\mathrm{eV}$. The true challenge becomes clear by pointing at the experimental observable $m^2(\nu_e)$ whose uncertainties have then to be lowered by two orders of magnitude.

Improving tritium β-spectroscopy by a factor of 100 requires evidently brute force, based on proved experimental concepts. It was decided, therefore, to build a MAC-E-Filter with a diameter of 10 m, corresponding to a 100 times larger analysing plane as compared to the pilot instruments at Mainz and Troitsk. Accordingly one gains a factor of 100 in quality factor which we may define as the product of accepted cross section of the source times resolving power $E/\Delta E$ for the emitted β-particles. Figure 1 shows the spectrometer tank on its way to Forschungszentrum Karlsruhe, figure 2 a scheme of the whole, 70 m long set-up.

Fig. 1. The KATRIN-spectrometer passes the village Leopoldshafen on its way from river Rhine to Forschungszentrum Karlsruhe on November 9, 2006.

A decay rate of order 10^{11} Bq is aspired from a source with diameter $\sim$8cm. This allows for an energy resolution $\Delta E \sim 1.5\,\mathrm{eV}$. The choice be-

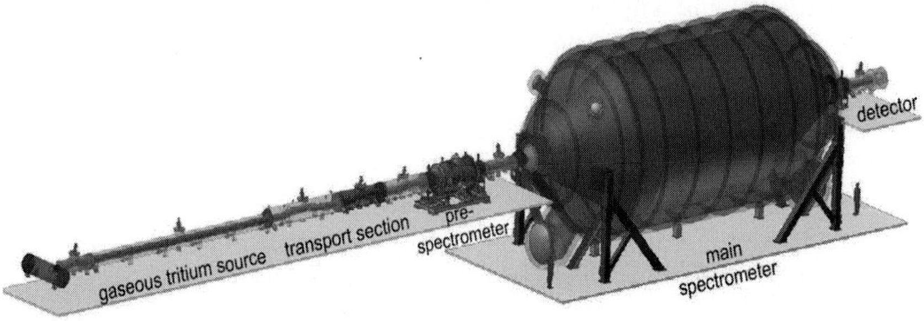

Fig. 2. Scheme of the 70 m long KATRIN setup consisting of windowless gaseous T2-source, differential pumping and cryo-trapping section, small pre-spectrometer and large main spectrometer, segmented pIN-diode detector.

tween a gaseous and a frozen T_2-source was not simple. The latter could have been built with modest effort; but the aspired activity requires a thickness of 500 monolayers at which it would charge up to a voltage of 10 V across the frozen film (see[9]); this would ruin the resolution. Since a way out of this problem has yet to be found, preference was given to a windowless gaseous T_2-source (WGTS) in spite of its extraordinary demands in size and cryo technique, which are required to handle the flux of 10^{19} T_2-molecules/s safely. T_2 is injected in the middle of a 10 m long source tube kept at a temperature of 27 K by a 2-phase liquid Neon bath. The integral column density of the source has to be stabilized within 0.1%. For background reasons, the T_2-flux entering the spectrometer should not exceed 10^5/s. This will be achieved by differential pumping sections (DPS), followed by cryo pumping sections (CPS) which trap residual T_2 on Argon frost at 4 K. Each system reduces the cross flow by 10^7. The T_2-gas collected by the DPS-pumps will be purified and recycled. A pre-spectrometer will transmit but the uppermost end of the β-spectrum into the main spectrometer in order to reduce the rate of background producing ionisation events there-in.

The β-particles will be guided from the source through the spectrometer to the detector within a magnetic flux tube of 190 Tcm^2, which is provided by a series of superconducting solenoids. This tight transverse confinement by the Lorentz force applies also to the 10^{11}/s daughter ions, emerging from β-decay in the source tube, as well as to the 10^{12}/s electron ion pairs produced therein by the β-flux. The question, how this plasma is getting neutralized or at which potential it might charge up, has been raised and treated but recently.[10] One can show that the high longitudinal conductance of the plasma will stabilize the potential along a magnetic field line to that value which this field line meets at the point where it crosses a rear wall. This gives a handle to control the plasma potential. Meanwhile the Troitsk group has performed a first experiment on the problem.[11] They have mixed ^{83m}Kr to their gaseous T_2 and searched for a broadening of the L3-conversion line at 30.47 keV which might be due to an inhomogeneous source potential. Their data fit is compatible with a possible broadening of 0.2 eV, which would not affect their results but suggests further investigation at KATRIN.

The sensitivity limit of KATRIN on $m^2(\nu_e)$ has been simulated (see below) on the basis of a background rate of 10^{-2} cts/s, observed at Mainz and Troitsk. Whether this small number can also be reached at the so much larger KATRIN-instrument – or even be lowered – has yet to be proven. On one hand the large dimensions are helpful since they improve straight adiabatic motion due to reduced field gradients. On the other hand the central flux tube faces a 100 times larger electrode surface at the filter potential from which secondary electrons might sneak in. This background source can be cut back by two orders of magnitude by a repelling grid in front of the electrode.[12] Such grids will also be installed at KATRIN.

Simulated spectra for 1 year of data taking at KATRIN are shown in figure 3. Due to the gain in signal to background ratio (S/B) the region of optimal mass sensitivity around S/B=2 has moved much closer to the endpoint and one recognizes prima vista a marked mass effect already for $m(\nu_e)=0.5$ eV. But one also notices that the typical third power rise of the integral spectrum is delayed by several eV. This is mainly due to the width of the spectrum of ro-vibrational excitations of the daughter molecule. At KATRIN this effect will be felt for the first time and to the extent that at the upper end of this spectrum at $E_0 - E \sim 4$ eV, which falls in the region of optimal mass sensitivity of KATRIN, the mass dependent signal will be weakened by a factor of 2 with respect to the case of a sharp recoil energy. This evokes once more the old dream of an atomic tritium source, and one should start hard thinking whether and how this could be turned true in a phase II of KATRIN.

Figure 4 shows simulations of the total uncertainty of the observable $m^2(\nu_e)$ and corresponding upper mass limits which are expected from the KATRIN-Experiment after 3 years of data taking at background rates of 10^{-2} cts/s and 10^{-3} cts/s, respectively. They are plotted as function of the width of the spectral interval, as measured with equidistant or optimized distribution of settings for filter potential as well as for measuring time.

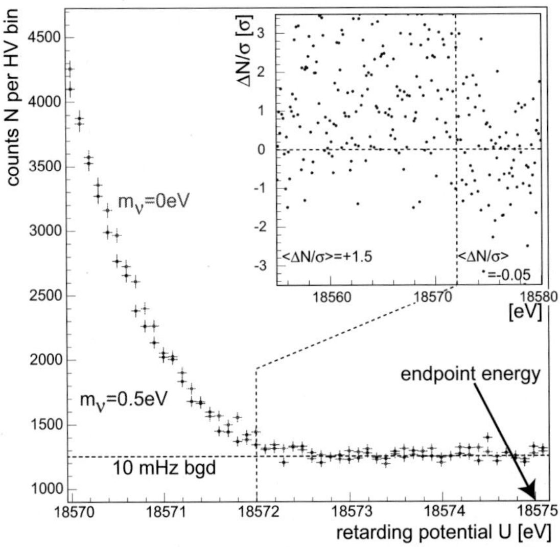

Fig. 3. Simulated β-spectrum near the endpoint after 1 year of KATRIN-running. Below 18571 eV the simulation for a mass of 0.5 eV (lower symbols) clearly separates from that for zero mass (upper symbols).

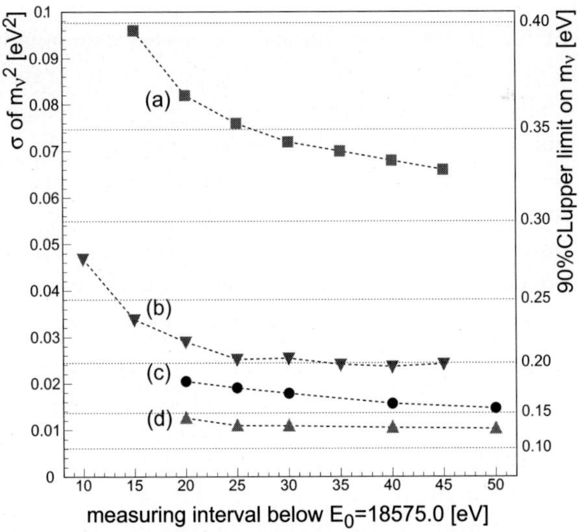

Fig. 4. Simulations of total mass squared uncertainty expected at KATRIN after 3 years of running, calculated in dependence of the fit interval under following conditions. Spectrometer diameter =7 m as originally proposed:[7] (a); final 10 m design:[8] (b, c, d); background = 10^{-2} counts/s: (a, b, c); background = 10^{-3} counts/s: (d); equidistant measuring point distribution: (a, b); measuring point distribution optimized according to local mass sensitivity: (c, d).

The dependence on the interval length is pretty flat in particular for the lower choice of background. For the reference value $B=10^{-2}$ cts/s one expects to reach a total uncertainty somewhat below 0.02 eV^2, corresponding to an upper mass limit of somewhat better than 0.2 eV at 90% C.L. This goal is realistic only in case that also the systematic uncertainties can be cut back by 2 orders of magnitude as compared to present. Fortunately, the improved signal to noise ratio is very helpful also in this regard, as it reduces the decisive correlated uncertainties between energy in the source, endpoint and mass squared through schifting the region of maximum mass sensitivity closer to the endpoint and throug shortening the measured spectral interval. In addition, relevant experimental parameters like potentials of source and filter, resolution function, source thickness, etc. will have to be controlled with substantially improved precision. Also energy loss spectrum and total cross section of scattering on source molecules must be re-measured at higher accuracy. These matters are discussed in detail in the KATRIN Design Report.[8]

3. Final Remarks

If research on radioactive decay still got to play a role in fundamental physics then it has to face tremendously increased demands in sensitivity and precision of experiments. Hence these have to be transformed from table top to really large scale undertakings which require high investments and the commitment of large collaborations for more than a decade in R&D, building and running them. The search for the absolute neutrino mass from single and double β-decay is just undergoing this transition. The present limits on the eV-scale have been obtained from medium sized experiments which could still be handled by single groups. Their potential is fairly exhausted by now, but they have also paved the way towards the coming generation of large scale experiments.

KATRIN, the forthcoming electrostatic filter for β-spectroscopy of tritium, will break the 1 eV limit shortly after getting started and then will approach its sensitivity limit of 0.2 eV within a couple of years of running. There it will meet the mass scale which is addressed by the fairly realistic claim of evidence of neutrino less double β-decay of ^{76}Ge, derived from the Heidelberg/Moscow Experiment.[13] If KATRIN doesn't find a finite mass signal its new, refined mass limit will have exhausted pretty much the space open for degenerate neutrino masses. Any substantial progress in terrestrial neutrino mass determination will be valuable input and crosscheck for our understanding of the role of neutrinos in astrophysics.

Regarding β-spectroscopy of tritium, KATRIN will probably have no competition, since the experiment requires enormous efforts and alternatives are yet missing. This could lead to a somewhat uneasy situation, in particular, if a finite but small mass signal happens to appear. How to satisfy then the principle of checking a new result independently? Here calorimetric detection of rhenium β-decay has a chance to enter the scene. Present results encourage further developing its technology and upgrading its scale. Envisaging a sensitivity competitive to KATRIN is not unrealistic, although it might still be a long way to go (see the MARE proposal[14]).

References

1. Pauli 1930 in a letter to a *"Local meeting on radioactivity at Tübingen"*, Germany and *"Rapports du Septiéme Conseil de Physique Solvay"*, Brussels, 1933 *Paris: Gauthier-Villars* (1934).
2. Fermi E. *Z. Physik* **88** (1934) 11.
3. Curran S.C., Angus J. and Cockroft A. *Phys. Rev.* **76** (1949) 853.
4. Yao W.M. et al. (Particle Data Group) *J. Phys.* **G33** (2006) 1 (URL: http://pdg.lbl.gov)
5. Kraus Ch.et al., *Eur. Phys. J.* **C40** (2005) 447.
6. Lobashev V.M. et al. *Phys. Lett.* **B460** (1999) 227.
7. KATRIN Collaboration (A. Osipowicz et al.) *"KATRIN: A Next Generation Tritium Beta Decay Experiment With Sub-eV Sensitivity for the Electron Neutrino Mass. Letter of Intent", hep-ex/*0109033.
8. Angrick J. (for the KATRIN-Collaboration) (Bonn, Fulda, Karlsruhe, Mainz, Moscow, Münster, Prague, Seattle, Swansea) (2004) "KATRIN Design Report" Wissenschaftliche Berichte Forschungszentrum Karlsruhe **7090** *http : //www − ik.fzk.de/katrin.*
9. Bornschein B., Bonn J., Bornschein L., Otten E.W. and Weinheimer Ch. *J. Low Temperature Phys.* **131** (2003) 69.
10. Nastoyashchii A.F., Titov N.A., Morozov I.N., Glück F. and Otten E.W. *American Nuclear Society: Fusion Science and Technology* **48** (2005) 743.
11. Lobashev V.M. Proc. 11th Int. "Workshop on Neutrino Telescopes", Venice, Italy 22-25 Feb 2005, *ed* M. Baldo-Ceolin, *Padova, Papergraf* (2005) 507.
12. Müller B. Diploma thesis (2002) Mainz University, to be published; Flatt B. PHD Thesis (2004) Mainz University, to be published.
13. Klapdor-Kleingrothaus H.V., Dietz A., Harney H.L. and Krivosheina I.V. *Mod. Phys. Lett.* **A16** (2001) 2409; Klapdor-Kleingrothaus H.V., Krivosheina I.V., Dietz A. and Chkvorets O., *Physics Letters* **B586** (2004) 198; Klapdor-Kleingrothaus H.V. and Krivosheina I.V. *Mod. Phys. Lett.* **A21** (2006) 1547.
14. A. Monfardini et al. in Proc. of 11th Intern. Workshop on Low Temperature Detectors (LTD-11), Tokyo, Japan, 31 Jul - 5 Aug 2005; *Prog. Part. Nucl. Phys.* **57** (2006) pp. 68-70, *Nucl. Instrum. Meth.* **A559** (2006) pp. 346-348; e-Print: hep-ex/0509038.

HOT DARK MATTER AND NEUTRINOLESS DOUBLE BETA DECAY: WORLD STATUS OF THE FIELD

HANS V. KLAPDOR-KLEINGROTHAUS *

Heidelberg, Germany,
prof.klapdor-kleingrothaus@hotmail.de, http : //www.klapdor − k.de

A final independent analysis of the data of the ^{76}Ge experiment in Gran Sasso (HEIDELBERG-MOSCOW experiment) which is since 15 years the worldwide by far most sensitive double beta experiment, is presented for the main set of data taken in the period Nov. 1995 - May 2003.[4,5] In this period the time structure of all events have been registered. Two different methods of pulse shape analysis[4,5,7–9,15,17–19] lead to selection of potential $0\nu\beta\beta$ events at $Q_{\beta\beta}$ *with almost no* γ*-background* in a wide range around $Q_{\beta\beta}$. The background achieved in the energy region of the $Q_{\beta\beta}$ value for double beta decay is $5{\cdot}10^{-3}$ events $\mathrm{kg^{-1}yr^{-1}keV^{-1}}$. A line at $Q_{\beta\beta}$ is observed at a confidence level of more than 6σ. The identification of this line as signal of neutrinoless double beta decay is the most reasonable explanation. This means total lepton number violation and a Majorana nature of the neutrino as fundamental consequences for particle physics. The half-life for this process is determined to be $\left(2.23\ ^{+0.44}_{-0.31}(stat)\right) \times 10^{25}$ y. Assuming vanishing right-handed weak currents or other more exotic contributions to the $0\nu\beta\beta$ decay amplitude, a value of $\langle m\rangle = \left(0.32\ ^{+0.03}_{-0.03}\right)$ eV is derived. Normalizing the nuclear matrix element for $2\nu\beta\beta$ decay to the experimental value, the lower limit for $\langle m_\nu\rangle$ could be lower, down to ~ 0.22 eV. With this value neutrinos contribute at least 4.7% to the dark matter in the Universe.

Keywords: Total lepton number violation; Majorana nature of the neutrino, Neutrino mass and mixing; Beta-decay; Double beta decay; HEIDELBERG-MOSCOW experiment; High purity Ge detectors, hot dark matter in the Universe.

1. Introduction

Nuclear double beta decay provides an extraordinarily broad potential to search for beyond standard model physics.[1–3,27] Its occurrence has

*Spokesman of HEIDELBERG-MOSCOW, GENIUS-TF and HDMS Collaborations

enormeous consequences: it means that *total lepton number is not conserved*, and second that the *neutrino is a Majorana particle.* Furthermore it can provide, *under some assumptions*, an absolute scale of the neutrino mass, and yields sharp restrictions for SUSY models, leptoquarks, compositeness, left-right symmetric models, test of special relativity and equivalence principle in the neutrino sector and others.[1–3]

Among the many existing efforts in search for this process, since 15 years now the HEIDELBERG-MOSCOW experiment[13] which operated 11 kg of enriched ^{76}Ge in the period August 1990 - November 2003 in the Gran Sasso underground laboratory is the by far most sensitive double beta experiment, and probably will remain this for the next decade.

The *first* and up to now *only* evidence for $0\nu\beta\beta$ decay has been reported from this experiment.[3–6,11,12,19] From the analysis of the full spectrum *without any background subtraction* a line was observed at $Q_{\beta\beta}$ on a 4.2σ c.l. [a]. In the period Nov. 1995 - May 2003 which delivered *the main set of data*, the time structures of all events have been registered by using a 250 mHz flash ADC. In this paper we concentrate on an analysis of *these* data, and try to select $0\nu\beta\beta$ event candidates from the surrounding γ -background by two different and completely independent methods *of pulse shape analysis.*[7–9,15,17–19]

For a proof of observation of neutrinoless double beta decay a method is required, which fulfills the following criteria:

(1) select $0\nu\beta\beta$ events at $Q_{\beta\beta}$;
(2) reduce strongly surrounding γ events.

In this paper we show that both of our pulse shape analysis methods, neuronal net method[4–6,15] and method based on calculated pulse-shape[17,19] libraries, fulfill these criteria, and lead to consistent results. We show that we have now *two projections of events with almost no background*, which both prove the existence of a line at $Q_{\beta\beta}$ which we interpret as $0\nu\beta\beta$ signal. The signal seen in *each* of the pulse shape analyzed spectra has a confidence level of $\sim$6σ.

---

[a] An independent analysis, of the data taken *until 2000*, which confirmed our earlier result published in,[6] has been performed in.[36] Another analysis of part of the data,[40] which was performed by the Kurchatov part of the collaboration - which left the collaboration in 2001 - has been shown to be *wrong* in,[4] page 201, and,[5] p.382, 385. Some earlier criticism has been *ruled out* already in,[11,37] i.e. it was already history before the higher statistics data were presented in.[4,5]

2. Structure of Double Beta Events and the Corresponding Pulse Shapes

2.1. *Monte Carlo Simulated Tracks and the Pulse Shapes*

The half-life for the neutrinoless decay mode is given by[28,29]

$$[T_{1/2}^{0\nu}(0_i^+ \rightarrow 0_f^+)]^{-1} = \qquad (1)$$

$$C_{mm}\frac{\langle m\rangle^2}{m_e^2} + C_{\eta\eta}\langle\eta\rangle^2 + C_{\lambda\lambda}\langle\lambda\rangle^2 + C_{m\eta}\langle\eta\rangle\frac{\langle m\rangle}{m_e} + C_{m\lambda}\langle\lambda\rangle\frac{\langle m_\nu\rangle}{m_e} + C_{\eta\lambda}\langle\eta\rangle\langle\lambda\rangle,$$

$\langle m\rangle = |m_{ee}^{(1)}| + e^{i\phi_2}|m_{ee}^{(2)}| + e^{i\phi_3}|m_{ee}^{(3)}|$, where $m_{ee}^{(i)} \equiv |m_{ee}^{(i)}|\exp(i\phi_i)$ $(i = 1,2,3)$ are the contributions to the effective mass $\langle m\rangle$ from individual mass eigenstates, with ϕ_i denoting relative Majorana phases connected with CP violation, and $C_{mm}, C_{\eta\eta}, ...$ denote nuclear matrix elements squared. Ignoring contributions from right-handed weak currents on the right-hand side of the above equation, only the first term remains.

The tracks of $0\nu\beta\beta$ events in a Ge detector (see Fig. 6 from[7]) and their resulting electrical pulses depend in principle on the spectral-angular correlations of the emitted electrons. These dependencies have been studied recently for the case of ^{76}Ge.[7–9,17,18]

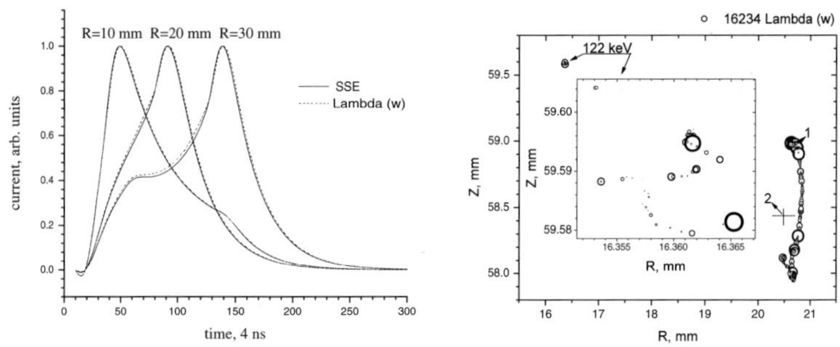

Fig. 1. Right: Typical calculated event for $0\nu\beta\beta$ decay without photon emission (bremstrahlung) in a Ge detector. Left: Comparison of the calculated electrical pulses from a large size $0\nu\beta\beta$ event with a corresponding 'zero range' pulse (SSE) (see[17,18]).

2.2. *Zero Range Library*

Since our Monte Carlo calculations (see[7–9]) show, that most $\beta\beta$ -events have a small size, a library of pulses has been built up for pulses, in which *all energy* of the event is assumed to be *deposited in one point* in the detector *(zero range approximation)*, for the four enriched ^{76}Ge detectors of the HEIDELBERG-MOSCOW experiment,[17,18] which measured the pulse shapes of events in the period Nov. 1995 - May 2003, throughout the detector volume. This is found to be a good approximation for ~90% of the $\beta\beta$ events (see[17,18]). An example is shown in Fig.1 where a relatively large size microscopically calculated $0\nu\beta\beta$ event pulse shape (here the final pulse is a sum of typically a few hundred subpulses), is compared with the result assuming a point-like event. The pulses *measured* in the HEIDELBERG-MOSCOW experiment will later in this paper be compared to pulses of this library.

The method of calculation of the pulses of the zero range library follows the standard procedure (see[20,22]) and has been described in.[17] The library of pulses was calculated as function of location (R,Z) of the event in the detectors. For optimization of the library, γ -source measurements have been performed with a collimator in the period July to September 2004 in Gran Sasso, in which shapes of pulses were measured as function of radius and height of the detectors (see[17]).

2.3. *The Neuronal Net Method*

A general description of the neuronal method used is given in.[15] It was applied in a first step already in.[4,5] This neuronal method (which we call N or 'total' neuronal net here), was trained with one γ -double escape line (DE), and a second one (called R) trained with *two* γ lines. For the following discussion we have to explain some notations. The 'total' projection N by the neuronal method can be subdivided into events seen *only* by method N (called NN here), events seen by methods N *and* R (called NR), seen by methods N,R and the method (called H) developed in,[14] (called HNR). So the total projection by N can be written as sum of subprojections N=HNR+NN+NR+HN+HR. In Ref.[17] we show the degree of suppression obtained by N (and also by the method described in section 2.2) for a normal γ line in contrast to a double escape γ line. In Fig. 31 of[5] we show the projection obtained by HNR+NN (which we call 'global' neuronal net here) which contains *most* of the N-events. We will show below that the selectivity of events around $Q_{\beta\beta}$ increases *drastically*, if we use only the subclass NN of the projection.

3. Analysis of the HEIDELBERG-MOSCOW Experiment with Neuronal Net and Calculated Libraries

The nature of invention of suitable PSA methods is not simple. We will see, that the experimental data exhibit a greater swing of potential than indicated by theory.[7,8,17,18] This will be exploited here. We discuss now the pulse shape analysis (PSA) of the spectra measured in the period Nov. 1995 - May 2003 (see[4,5]) by the two PSA methods.

The situation after application of the neuronal net optimized to select single site events from the 1592 keV line, as shown in Ref.,[17] was that shown in Figs. 30, 31 of.[5] A gamma line consisting of practically *only* MSE events, as the 2016.7 keV line from ^{214}Bi which is as E0 transition, produced by two subsequent γ -transitions, seems to be *stronger* reduced than a normal γ -line (similar to the 2505.7 keV line, see discussion below). On the other hand it is obvious that the suppression obtained is not sufficient to clearly separate a signal at $Q_{\beta\beta}$ from the surrounding background γ -lines in particular for low statistics lines, (see Figs. 31 and 30 in[5]). The reason is that normal γ-lines according to Monte Carlo simulations always contain ~30% of SSE (see, e.g[5,7,8,16,23]).

We thus *have to find a way to suppress the* γ *-lines stronger.* We achieve this, by accepting in the cuts only a *subclass* of events, namely when accepting only the subclass NN of the neuronal net selection (see Fig. 2) and.[17] In the library method, applying a cut with *lower* χ^2 than optimal for selection of 1592 keV SSE γ -events (see Fig.13 in[17]) also leads to a sharp suppression of γ-lines. Accepting only measured pulses which are fitted by a library event with small χ^2, ($\chi^2 < 0.4, 0.4, 0.7, 0.35$ for detectors 2,3,4,5, respectively) [b], the result is a *very* drastic reduction of *any* γ - lines in the measured spectrum (see Fig. 3).

While for the 'global' neuronal net cut (more precisely HNR+NN, see section 2.3) anywhere in the spectrum the γ -lines are suppressed, but far from completely suppressed (see Fig. 30, 31 in[5]), *we now obtain an almost complete suppression* when applying the neuronal subset (NN), and the low-χ^2 cut selection (see Table 1). Fig. 4, shows the results of the cuts obtained in the range 2000-2100 keV[c] in the full detector volumes *and* those obtained when cutting 1 or 2 mm from the detector borders. This is the *typical length* which could be passed by electrons emitted *on the surface* (e.g. conversion electrons). By such cut we should loose on the other hand almost *no* $\beta\beta$ - events, since for most $0\nu\beta\beta$ events produced in this border zone, not the full energy will be deposited in the detector and thus they will not show up

---

[b]We define chisquared here as a sum of squared deviations between the time structures of measured pulse and library pulse: $\chi^2 = \sum_{n=t05-f}^{t05-l} \frac{(i_{exp}^n - i_{lib}^n)^2}{w5} \cdot 10^3$. Here i_{exp}^n and i_{lib}^n are the (normalized) experimental and library currents for the (time) channel n (one time step is 4 ns per channel of the flash-ADC sampling frequency of 250 MHz). $w5$ is the width at the 5% level of the amplitude (see Fig. 2). χ^2_{min} gives the degree of agreement of a measured pulse with the library pulse 'closest' in shape (time structure) to the measured pulse. For a detailed description see.[17]

[c]The non-integral numbers of counts in the spectra in Figs. 4,5 are simply a binning effect (see[4]).

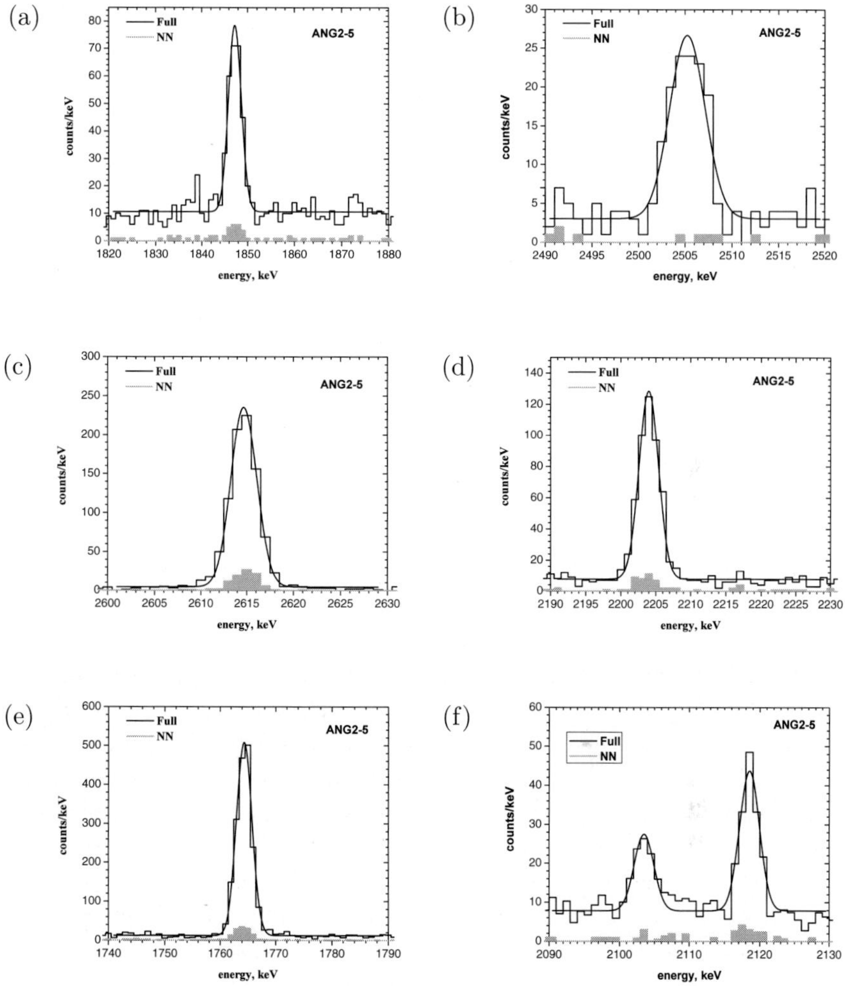

Fig. 2. Full lines, and lines from the neuronal net (NN) pulse shape selected spectrum measured Nov. 1995 - May 2003 (see also Fig. 9 and 33 respectively in[4,5]) for detectors 2,3,4,5 in the energy regions of some strong background lines. Even some fine details of the composition of γ -lines are 'seen' by this selection of NN type events. E.g. the 2505.7 keV line (see[4,5]) originating from summation of the subsequent 1173.2 and 1332.5 keV γ - lines from ^{60}Co - naturally expected be a *multiple site* event is *erased to 100%*, as it is also the case when applying the zero range library (see Fig. 3 and also Fig. 4 in[17]). The neuronal net also reduces the 2016.7 keV line from ^{214}Bi which is as E0 transition also produced by two subsequent transitions (see[4,5]), stronger than normal full energy (FE) γ -peaks (see Fig. 31 in[5]).

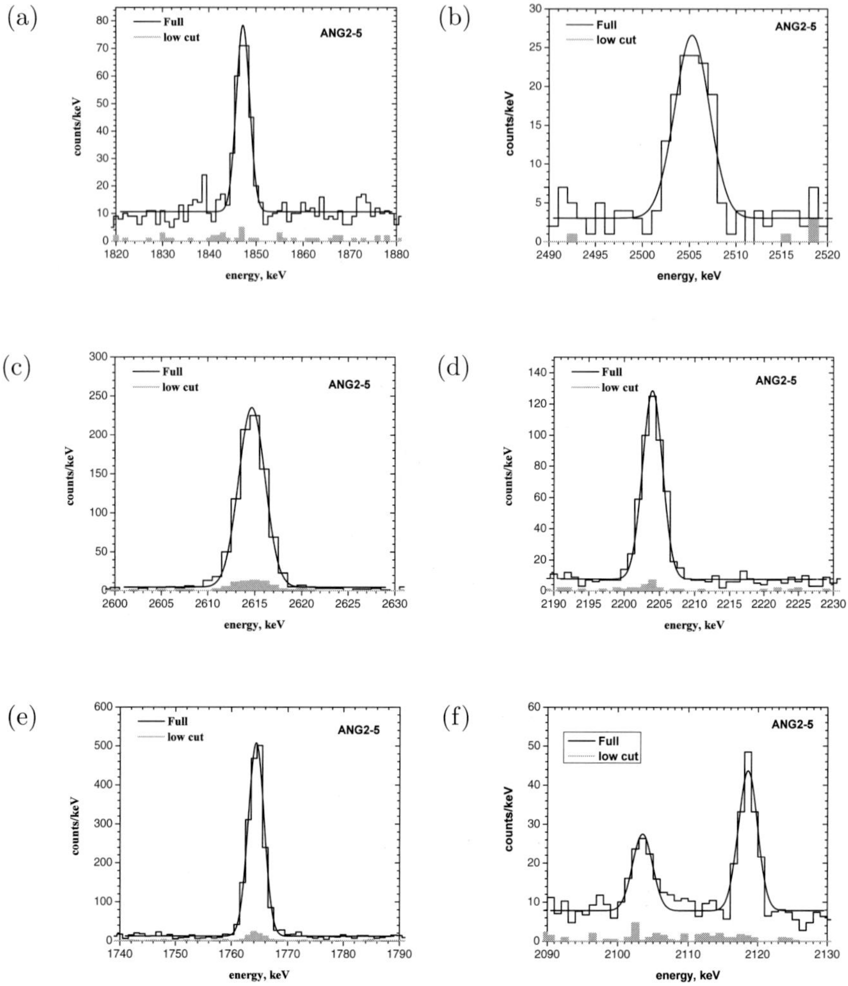

Fig. 3. Lines from the full spectrum measured Nov. 1995 - May 2003 (see also Figs. 15 and 16 of,[5] and Fig.3 of[4]) – note that in the captions of these figures it should read 'from Nov. 1995 to May 2003' and the pulse shape selected spectrum on the basis of the zero range library with small χ^2 cut (see text) for detectors 2,3,4,5 in the energy regions of some strong background lines. The γ -lines are drastically reduced. Even some fine details of the composition of γ -lines are 'seen' by this selection of small site events. E.g. the 2505.7 keV line (see[4,5]) originating from summation of the subsequent 1173.2 and 1332.5 keV γ -lines from ^{60}Co - naturally expected be a *multiple site* event is *erased to 100%*, as it also the case when applying the neuronal net (see Fig. 2).

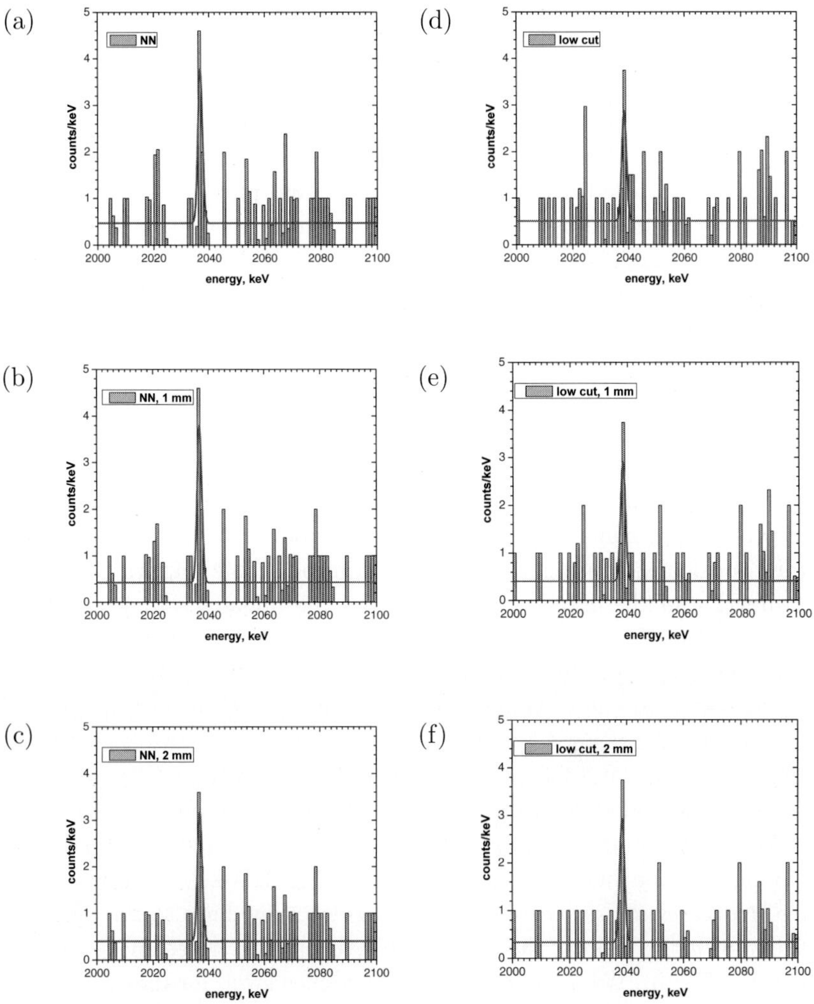

Fig. 4. Left column: The pulse shape selected spectra (subsection NN by neuronal net, see also Fig. 33 from[5]) measured with detectors 2,3,4,5 from 1995 to 2003 in the energy range of (2000-2100) keV. Shown are the events in the full detector (a) and cutting boundary areas of 1(2) mm - (b) and (c). Signals near $Q_{\beta\beta}$ are found: 7.05$\pm$1.11, 7.18$\pm$1.03, and 5.91$\pm$0.97 events (a, b, c, respectively), corresponding to 6.4, 7.0 and 6.1σ confidence levels. Right column: The pulse shape selected spectrum after applying the zero range library with 'low' cuts (d), (e), (f) measured with detectors 2,3,4,5 from 1995 to 2003 in the energy range of (2000-2100) keV. Shown are the events in the full detector (d) and cutting boundary areas of 1(2) mm -(e) and (f). Signals are found near $Q_{\beta\beta}$: 5.05$\pm$1.23, 5.33$\pm$1.08, and 5.54$\pm$0.96 events (d, e, f, respectively), corresponding to 4.1, 4.9 and 5.8σ levels.

in the line at $Q_{\beta\beta}$. According to Monte Carlo calculations about 4% of all $0\nu\beta\beta$ events produced homogeneously in the detector, are lost in this way in the border range (see[8]).

The surrounding γ -lines are now practically completely erased (compare to Fig.5 left). This includes the structure at $\sim$ 2030 keV. The background level is reduced to $\sim 5{\cdot}10^{-3}$ kg y keV. Only a tiny part is still seen of the 2053 keV ^{214}Bi line. The indication of a structure seen at $\sim$ 2024 keV in the full spectrum (Figs. 17,18 in[5]) which is still visible in Figs. 5 left, and 4 without border cut, seems to be a surface impurity vanishing with the 2 mm cut completely. What is remaining for the cut with the pulse library of section 2.2 is a signal at 2038.4 $\pm$ 0.3(stat) $\pm$ 1.2(syst) keV on a 4.1, 4.9, 5.8σ level for the 0, 1, 2 mm border cuts. The estimated systematic error comprises ballistic deficit, energy non-linearity, etc. It is important to be noted, that the result is *not* dependent on details of the applied library – it is obtained for the library in which the impurity distribution in the detectors is approximated by a constant distribution[17] and *also* when the library is calculated using the realistic variation of the impurities[21] along the direction of the detector axis, given by the manufacturers (see Table 2 in[17]), and further, if we change the detector parameters within limits, still allowing to reproduce the location of the events measured with the collimated source. The χ^2 values of the cuts change slightly, but the selected events remain the same.

From the cut with the neuronal net (NN - see section 2.3) we obtain (Fig.4 left) a signal at 2036.7 $\pm$ 0.2(stat) $\pm$1.2(syst) keV on a 6.4, 7.0, 6.1 σ level for the 0, 1, 2 mm border cut. The fits in Fig. 4 are made, *conservatively* assuming only one line in the energy window 2000-2100 keV, and *treating all other structures as constant* background. *If* we allow lines at the positions of the known ^{214}Bi lines, the confidence levels in Fig. 4 become somewhat larger (see Fig. 5). We obtained similar confidence levels for the signal when we analyzed a slightly smaller energy window, 2000-2060 keV (see[19]).

The spectra shown in Fig. 4 should be compared to the full spectra measured, shown in Figs. 5(left) and Fig. 4 and 19 in.[4,5] It is visible, that after applying the (NN) method, or the pulse shape library, all γ -lines are drastically reduced and only a line at $Q_{\beta\beta}$ survives.

If we *sum* the complementary spectra shown in Fig. 4, a line at 2037.4 $\pm$ 0.5(stat) $\pm$ 1.2(syst) keV is seen at a 5.4, 6.4, 6.7σ level (0, 1, 2 mm border cut) (see Fig. 9 in[19]). That the c.l. is smaller than would be expected from the summed information from the two different PSA approaches, has its

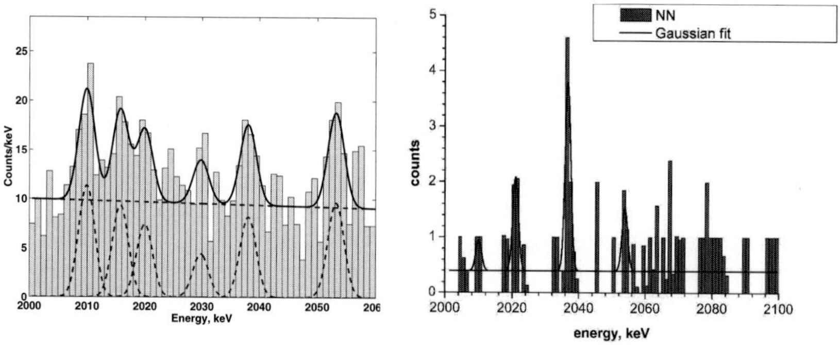

Fig. 5. Right: The pulse shape selected spectrum (selection by the neuronal net-(NN) method, see also Fig. 33 from[5] and text below) measured with detectors 2,3,4,5 from Nov. 1995 to May 2003 in the energy range of (2000-2100) keV. The fit assumes a constant background and three lines at the position of known Bi lines and a signal at $Q_{\beta\beta}$. The signal at $Q_{\beta\beta}$ consists of 7.27±1.05 events (6.9σ). Left: The full spectrum measured in the same period (from[5]), in the energy range 2000-2060 keV.

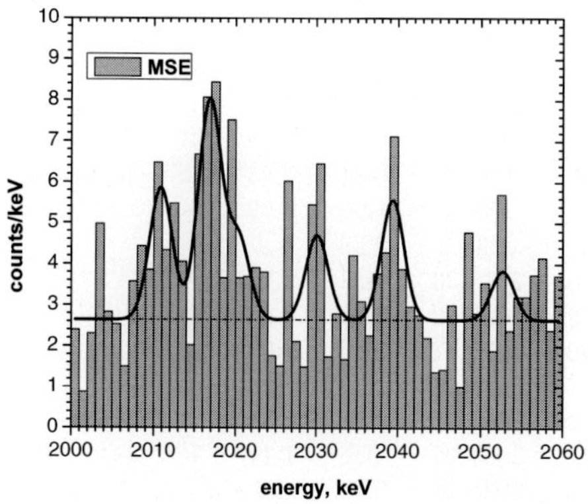

Fig. 6. The part of the spectrum shown in Fig.5 left, identified as multiple site events by the 'global' neuronal net. The ^{214}Bi line at 2016.7 keV is now dominant (compare Fig. 19 in[5]), consistent with the fact, that as a E0-transition it is a summation line of two cascading lines, and thus mainly consisting of MSE. A γ line is indicated at 2039.3 keV, which cannot be separated from the $0\nu\beta\beta$ signal (see Figs.4, 5, and Fig.9 in[17]) with the energy resolution of a Ge detector.

reason in the larger width of the line resulting from the slight difference in energy in the lines from both methods, which again is coming from the ballistic deficit (see below).

Let us now look into the *individual* events and discuss them. First we note, that the events found by the zero range library with the low χ^2, and by the neuronal net subset (NN) are found to be *complementary* (see Table1).

How can we understand these findings? *Parts* of the full line at $Q_{\beta\beta}$ in Fig.5 (left) can be *projected out of the full data, practically without background,* by the *subset* NN (defined in section 2.3) of the neuronal net given in[4,5] *and also* by the low-χ^2 cut of the pulse shape analysis (PSA) described in this paper and earlier in.[17,19] *All* events projected out by these two procedures are also selected by the full ('total', (see section 2.3)) neuronal net of[4,5,15] (see Table1). *They contain no multiple site events (except for one event).* So *all* results presented here and given earlier[4,5] are fully consistent.

Table 1. **All** *global* neuronal net accepted events (plus one MSE event) in the energy region of 2035-2043 keV for detectors 2,3,4,5.[4,5] In 'Index' column: 0 -*global* neuronal net selection NN+HNR; 1- Events selected by subclass of neuronal net method (NN); 2 - selected by low-χ^2 (see text) library cuts.

Det.	run	event	type	Date	Energy	χ^2	R cm	Index: 0	Index: 1	Index: 2
5	3357	2948	HNR	24.09.02	2034.39	0.33	3.45	+		+
3	3383	180	HNR	18.10.02	2035.14	0.77	1.84	+		
3	1627	172	HNR	15.05.99	2036.0	0.49	2.51	+		
3	749	295	NN	08.07.97	2036.04	1.68	2.7	+	+	
5	1067	421	NN	20.02.98	2036.31	0.98	2.65	+	+	
4	2613	47	NN	23.07.01	2036.37	1.72	1.99	+	+	
4	1762	97	HNR	28.08.99	2036.43	3.38	2.75	+		
3	3463	27	NN	07.01.03	2036.49	6.84	2.7	+	+	
5	2390	199	NN	27.01.01	2036.57	0.81	3.82	+	+	
2	2094	42	NR	20.05.00	2036.91	0.3	1.31			+
3	3385	354	HNR	21.10.02	2037.22	0.8	1.35	+		
5	1787	359	HNR	17.09.99	2037.22	0.31	1.91	+		+
5	2744	430	NN	21.11.01	2037.35	1.18	3.63	+	+	
3	108	112	NN	23.03.96	2037.58	0.81	2.45	+	+	
4	1271	251	HNR	01.08.98	2038.0	1.96	3.52	+		
3	2216	526	MSE	10.09.00	2038.3	0.38	1.1			+
4	1465	46	HNR	03.01.99	2038.33	5.9	2.75	+		
3	49	226	HNR	28.12.95	2038.53	0.26	2.27	+		+
4	1592	144	HNR	18.04.99	2038.57	0.53	1.99	+		+
2	735	131	HNR	28.06.97	2038.91	2.31	3	+		
4	2133	138	NR	24.06.00	2038.92	0.56	1.64			+
5	670	486	NN	09.05.97	2038.92	0.7	1.17	+	+	
3	3501	142	HNR	07.02.03	2038.97	0.74	2.7	+		
5	1036	73	HNR	26.01.98	2039.92	5.88	0.92	+		
3	1513	342	HNR	13.02.99	2039.93	0.55	2.27	+		
5	1650	206	HNR	03.06.99	2040.82	0.25	3.69	+		+
4	3449	74	HNR	22.12.02	2041.0	0.64	3.69	+		+
4	1378	490	HNR	27.10.98	2041.53	0.55	2.05	+		+
4	1125	296	HNR	05.04.98	2042.56	0.7	2.23	+		
4	3476	259	HNR	18.01.03	2042.75	0.78	1.35	+		
4	1833	74	HNR	22.10.99	2042.97	1.76	3.69	+		

In the neuronal net cut, no events of the types R, HN and HR are seen in the range 2036-2043 keV. From Table 1 we see that in addition to HNR events only three events are observed by the low cut of the PSA method - from which two are of NR type - i.e. which are *also* seen by the total neuronal net, and one is of MSE type. In total, by the two projections, which show each a line near $Q_{\beta\beta}$ with almost no background, we find 9+8=17 events in the energy range 2036-2043 keV according to Table 1. From these, 11±1.8 events from the *fits* in Figs. 4, 5 and Fig. 9 from[19] remain in the line. This means we can identify the major part of the 19.6 ± 5.4 events determined from the fit of the line in the *full* spectrum (Fig.5 left and Fig.19 and Table 7 in[5]), as $0\nu\beta\beta$ *candidate events.*

Considering only the NN events from Table 1 and the neuronal-net selected HNR events found by the low-cut PSA, these are together 14 events, consistent with the 12.4±3.7 events obtained by the neuronal net cut HNR+NN (see Fig. 31 in[5]).

That the two projections select different events, we may understand in the following way: the projections may select events in different parts of the volume of the detector. Table 1 shows that the neuronal (NN) mainly selects events in the outer part of the detector, while the zero range library mainly selects in the inner parts. This would easily explain the slightly different energies of the lines observed by both methods - by the ballistic deficit (see below). The reason may lie in tiny deviations of the calculated electric fields from the real ones for the outer part of the detectors for the library method, and in limitations of the training process with the 1592 keV line for the neuronal method for the inner part (the source was positioned outside the detectors!).

It is important to stress, that *if we relax a bit the sharp cut conditions*, for example admit in the library cut also the still good chisquared values of the events selected by the NN neuronal net selected events, or admit *all* events found by the 'global' neuronal net (see Fig. 31 in[5]), *then all selected events* are seen by *each* of these methods. The softer library cuts then are, however, not any more able to sufficiently strongly suppress the gamma background events, which then are also partly accepted by the cuts (see Fig. 31 in,[5] which shows the neuronal net cut including in addition to NN events also HNR events - and see Table 1).

We should keep in mind that the selection methods have to look for differences between gamma events and double beta events - which could be *extremely small.* In that sense the selection methods are *very* powerful, but such effect of slight inhomogeneities of the methods over the detector volume, as we may find it, when strongly sharpening the cuts, may also not be surprising. It may be a question, whether the precision of the simulations of the electric field in the detectors can be pushed to such level at all, that such inhomogeneities fully disappear.

That on the other hand both selections together select *all* $0\nu\beta\beta$ candidates, seems to be supported by the fact, that if we select (by the 'global' neuronal net) *only* multiple site events, we get a signal of 10.3 $\pm$ 3.3 events at 2039.3 $\pm$ 0.63(stat) keV (Fig. 6), which together with the 11$\pm$1.8 events observed by fit of the *sum* of the NN and low χ^2 cut spectra (Fig. 9 from[19]) is consistent with the 19.6 $\pm$ 5.4 events observed in the full spectrum (Fig. 5, and Fig. 19 in[5]). The line at 2039.3 keV cannot be resolved from the $0\nu\beta\beta$ line by the energy resolution of the detector.

Concluding, by the *two* projection procedures, the *major part* of the line near $Q_{\beta\beta}$ in the full spectrum is selected *as single site events*, while at the same time the surrounding γ -lines are *drastically reduced.* The selected events are interpreted as events of $0\nu\beta\beta$ decay.[d] Fig.7 shows these pulses, and the reproduction by the zero range approximation of single site events of section 2.2. The radii at which the events are observed in the detector (determined by the zero range approximation) are given in Table1.

The energy of the line observed, with 2037.4 $\pm$ 0.5(stat) $\pm$ 1.2(syst) keV seems to be slightly below the 'best' value reported for $Q_{\beta\beta}$ [33] of 2039.006$\pm$0.050 keV. Other measurements report $Q_{\beta\beta}$ =2040.71$\pm$0.52 keV,[30] 2038.56$\pm$0.32 keV[31] and 2038.668$\pm$2.142 keV.[32]

Part of this difference, and the difference in energy of the (NN) and low-cut lines in Fig. 4, seems to be due to the effect of the ballistic deficit (see[23,24]). Since in axial Ge detectors different radial location of events correspond to large variation in the charge collection times, and the latter may come close to the differentiation time of the electronic circuit, this leads if uncorrected, to a dependence of the measured energy from the radial location of the individual event and to a broadening of the observed line. We have investigated this effect. E.g. for detector ANG2 for single site events from the 2614 keV line from ^{228}Th a shift in the energy positions of around 7 channels ($\sim$ 2.4 keV) to lower energy is visible for the relatively slow pulses that start from the peripheral parts of the crystal with respect to the fastest pulses with initial radii of around 1.5 cm (Fig.8). For the other detectors this difference amounts to $\sim$ 1-1.6 keV. For energies around $Q_{\beta\beta}$, due to the linear dependence of the energy shift on the event energy, a difference of $\sim$ 1.8 keV is expected for detector ANG2 and of $\sim$ 0.8-1.3 keV for the others (see[19]).

---

[d]Furthermore, a γ line at 2039+1022=3061 keV is *not* found in the measured spectrum (see[5]), so that the line at $Q_{\beta\beta}$ can not be interpreted as a double escape line of any γ-line.

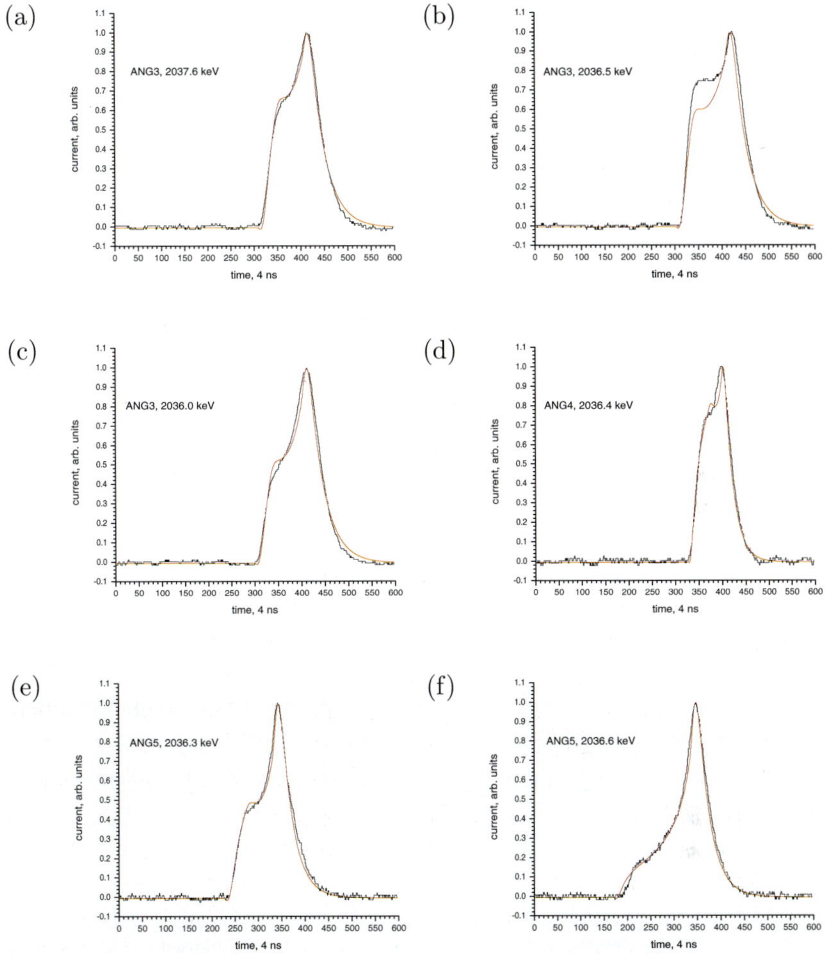

Fig. 7. Time structure *of those* events (black) measured in the HEIDELBERG-MOSCOW experiment in the energy region 2036-2042 keV by the four enriched ^{76}Ge detectors (ANG2, ANG3, ANG4, ANG5), during the period Nov. 1995 - May 2003 selected by the neuronal net (NN selection) (Figs. (a)-(h)) and by the low-cut PSA method (Figs. (i)-(r)) as single-site events (see Table 1, and Fig. 9 in[4]). The red lines show best reproduction of the events *by the zero range* pulse shape library, as described in this paper. The χ^2 for the fit by the zero range library and the radii determined from the latter are: NN selection: Det.3: (a)- 0.81(2.45 cm), (b)- 6.84(2.7 cm) and (c)- 1.68((2.7 cm); Det.4: (d)- 1.72(1.99 cm); Det.5: (e)- 0.98(2.65 cm), (f)- 0.81(3.82 cm), (g)- 1.18(3.63 cm), (h)- 0.7(1.17 cm). Low-cut selection: Det.2: (i) - 0.3(1.31 cm); Det.3: (j) - 0.38((1.1 cm) and (k) - 0.26(2.27 cm); Det.4: (l) - 0.53(1.99 cm), (m) - 0.56(1.64 cm), (n) - 0.64(3.69 cm), and (o) - 0.55(2.05 cm); Det.5: (p) - 0.31(1.91 cm) and (q) - 0.25(3.69 cm), (see also Table 1).

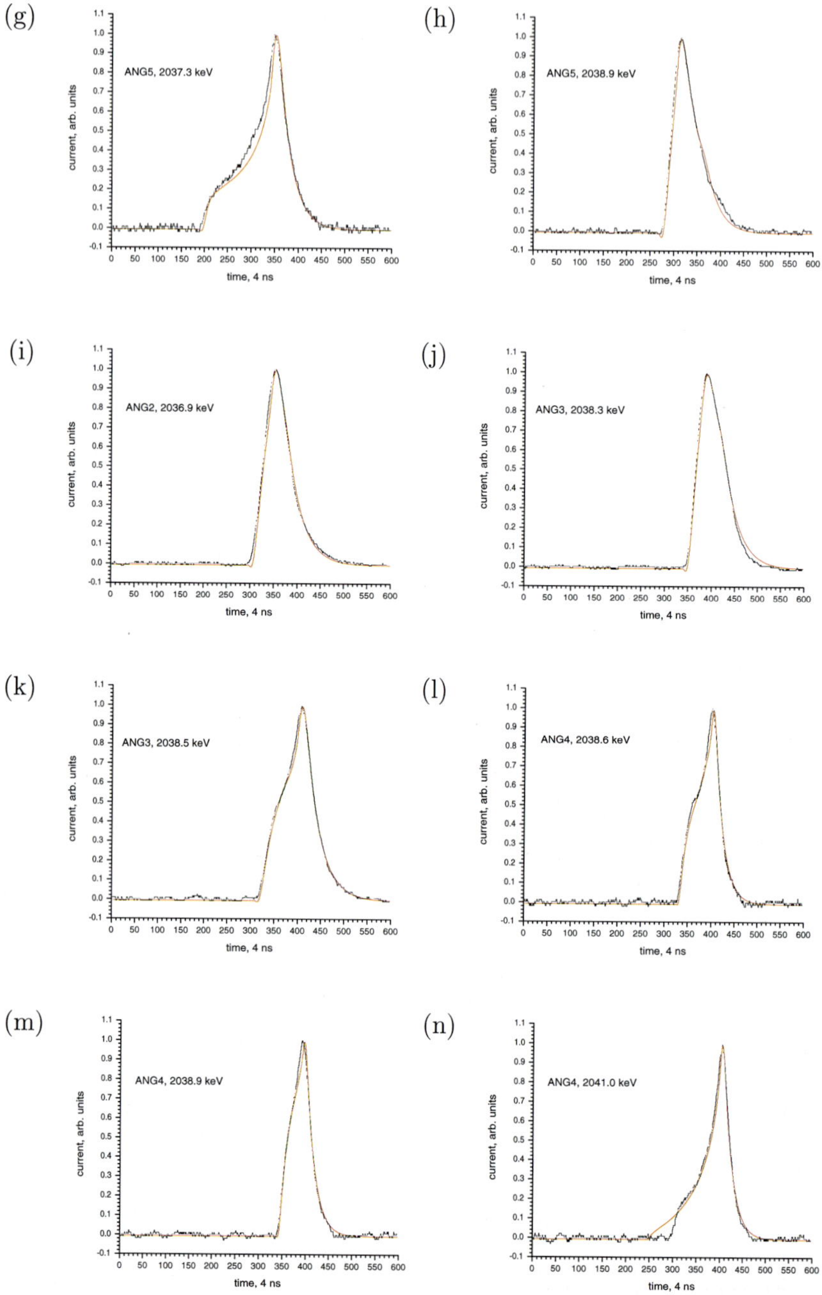

Fig. 7 (Continued)

(o) (p) (q)

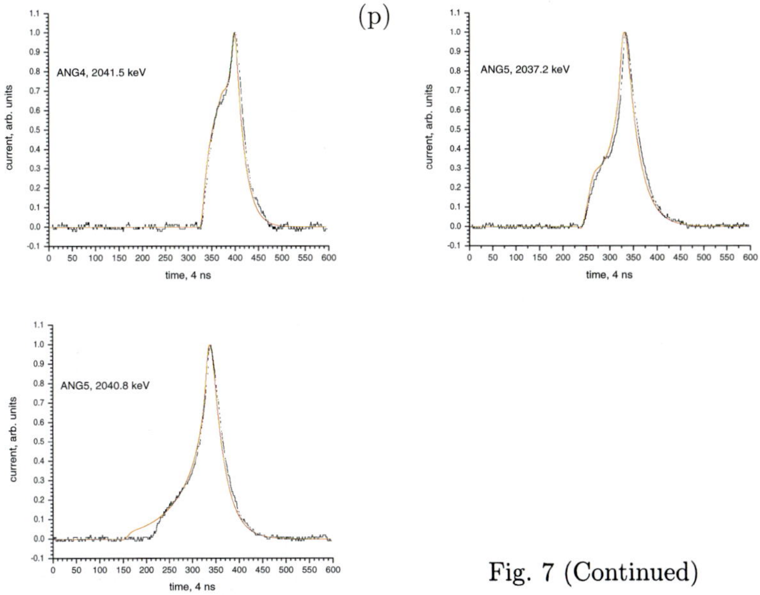

Fig. 7 (Continued)

Finally, Figs. 9 (a-e) show for completeness *all* events in the energy range 2036-2042 keV around $Q_{\beta\beta}$, which are *not* accepted by the neuronal NN selection or the zero range library as single site events (SSE). These events are identified as MSE by the (global) neuronal net and the zero range library, or not accepted by the NN neuronal net. Three of the events are identified as H events. Figs.7 and 9 (a-e) *together* thus show *all* events measured in the range 2036-2042 keV by detectors 2,3,4,5 in the period 1995-2003.

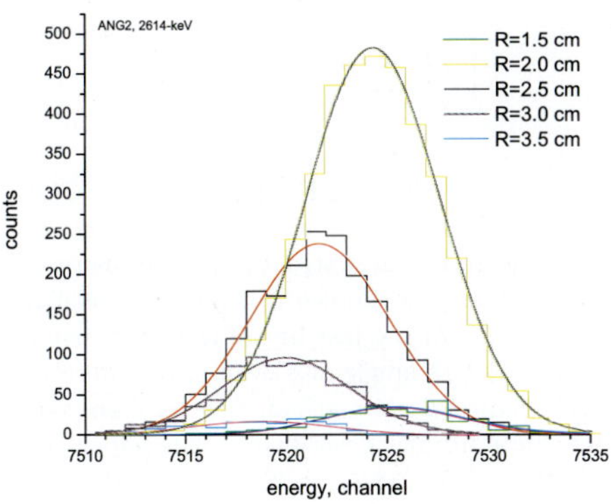

Fig. 8. To the ballistic deficit. Events of the γ -line 2614 keV (only events which are accepted by the zero range library as single site events, are accepted here) measured with a ^{228}Th source for detector Nr. 2. Shown is the measured energy for events at different radial locations.

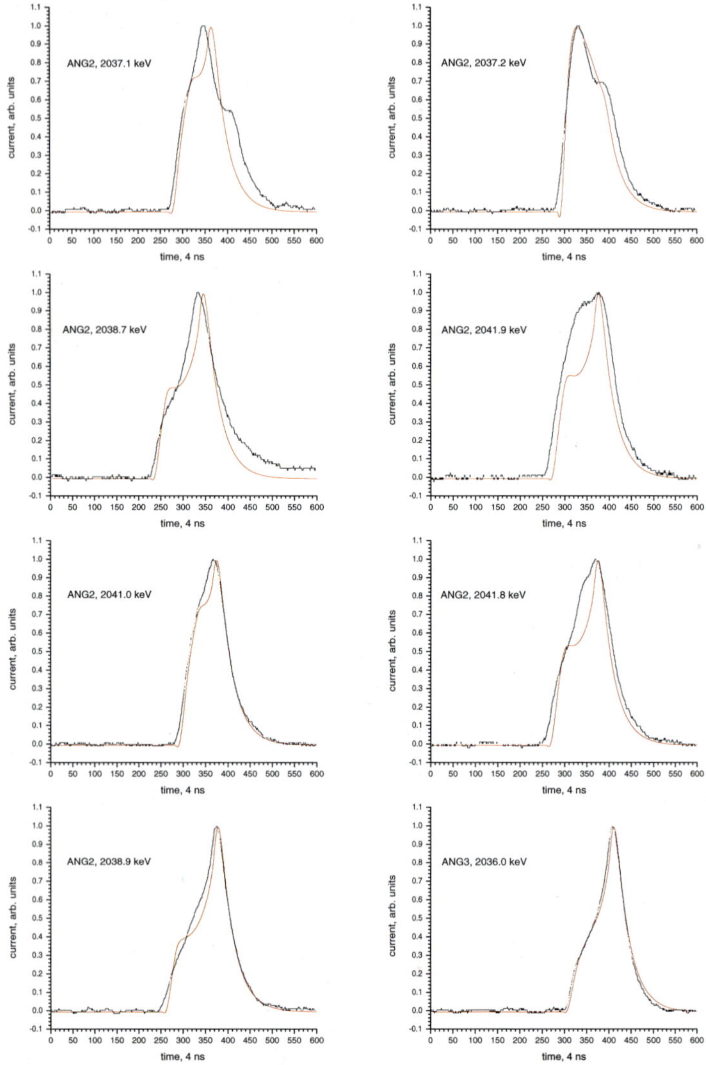

Fig. 9.a Time structure *of all* events (black) measured in the HEIDELBERG-MOSCOW experiment in the energy region 2036-2042 keV which were identified as MSE by the neuronal net (N) and the zero range library, or were not accepted by the NN neuronal net selection. In the first three lines of the figure events are shown which were identified, by the 'global' neuronal net[4,5] *as multiple site events.* In the last line of the figure events are shown, identified by the 'global' *neuronal net*[4,5] as single site events, but were *not accepted* by the low-cut approximation and NN neuronal net method. The red lines show *best* fits of the events *by the zero range* pulse shape library, described in this paper. The determined χ^2 of the events are Det2: 19.3, 9.44, 8.54, 33.23, 1.78, 13.02, 2.31, Det3: 0.49. I.e. they are not accepted as $0\nu\beta\beta$ SSE events.

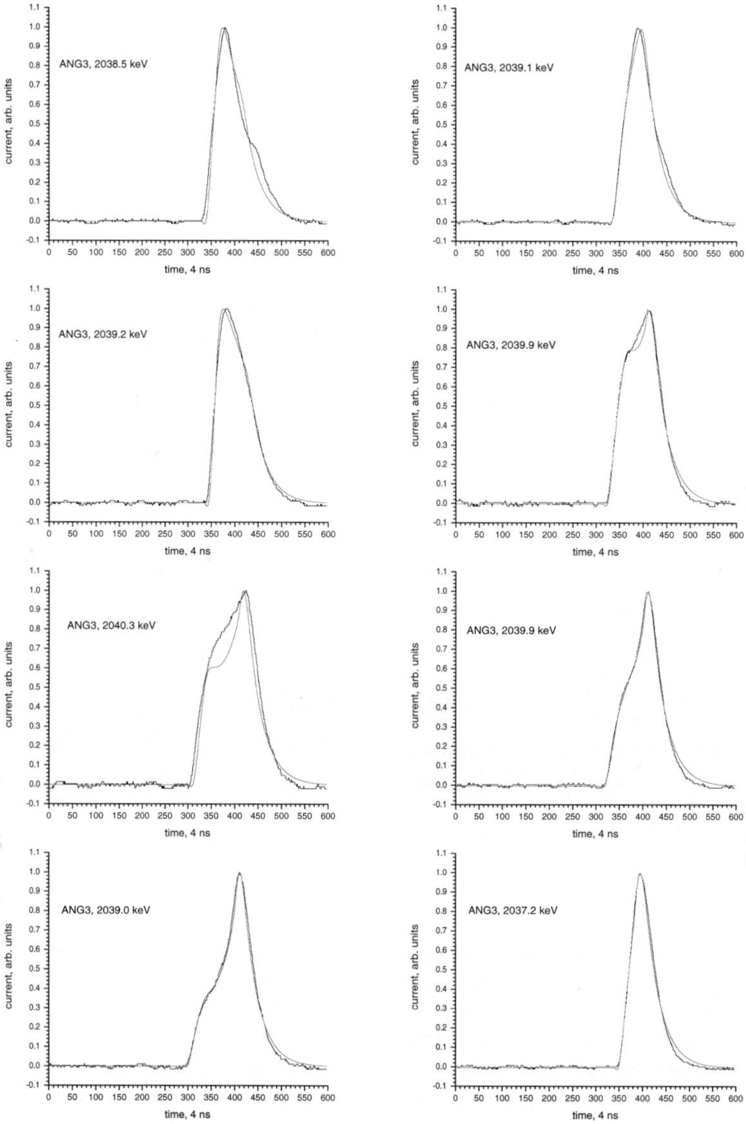

Fig. 9.b: The first five events (from the left to the right) were identified, by the 'global' neuronal net[4,5] *as multiple site events.* The last three events in the figure were identified by the global *neuronal net*[4,5] as single site events, but were not accepted by the low-cut approximation and NN neuronal net method. The red lines show best fits of the events *by the zero range* pulse shape library, described in this paper. The determined χ^2 of the events are (Det3: 5.0, 1.42, 1.82, 1.04, 9.63, 0.55, 0.74, 0.8). I.e. they are not accepted as $0\nu\beta\beta$ SSE events.

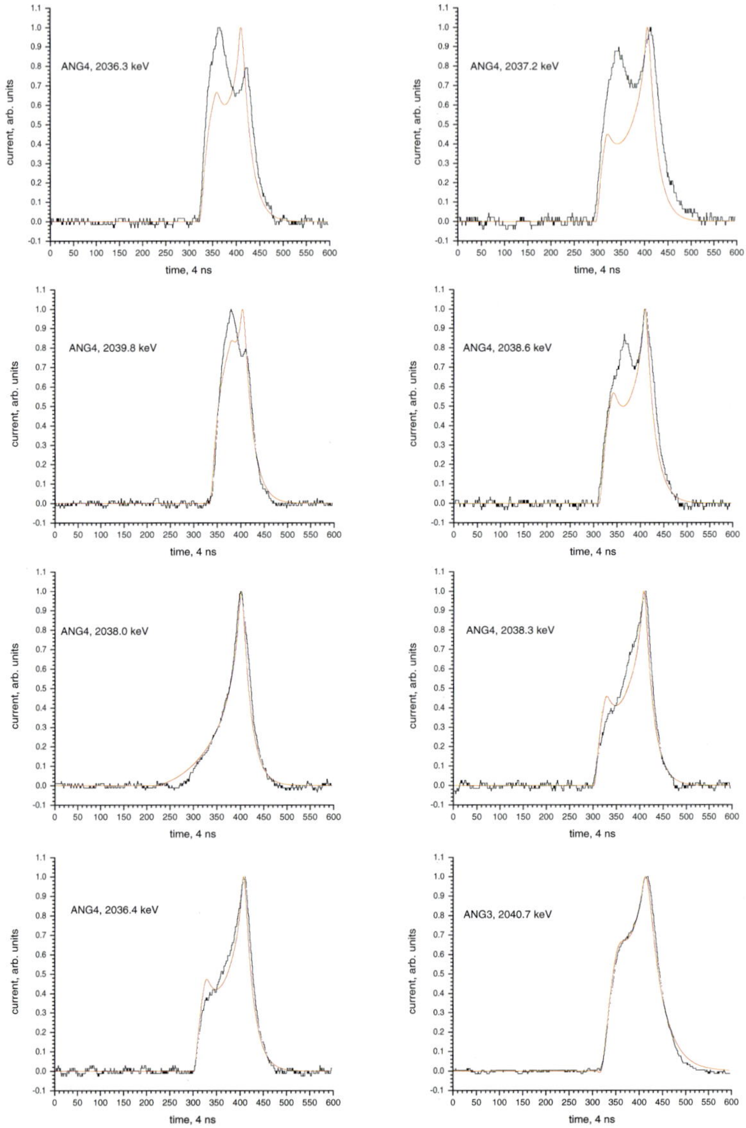

Fig. 9.c: The first four events do not pass the global neuronal net conditions for SSE (MSE or H (last event)). The rest of the events shown are identified by the *neuronal net*[4,5] as single site events, but were not accepted by the low-cut approximation and NN neuronal net method. The red lines show best fits of the events *by the zero range* pulse shape library, described in this paper. The determined χ^2 for the events of the events are Det4: 37.63, 50.03, 9.74, 20.99, 1.96, 3.38; Det3: 0.62. I.e. they are not accepted as $0\nu\beta\beta$ SSE events by both methods.

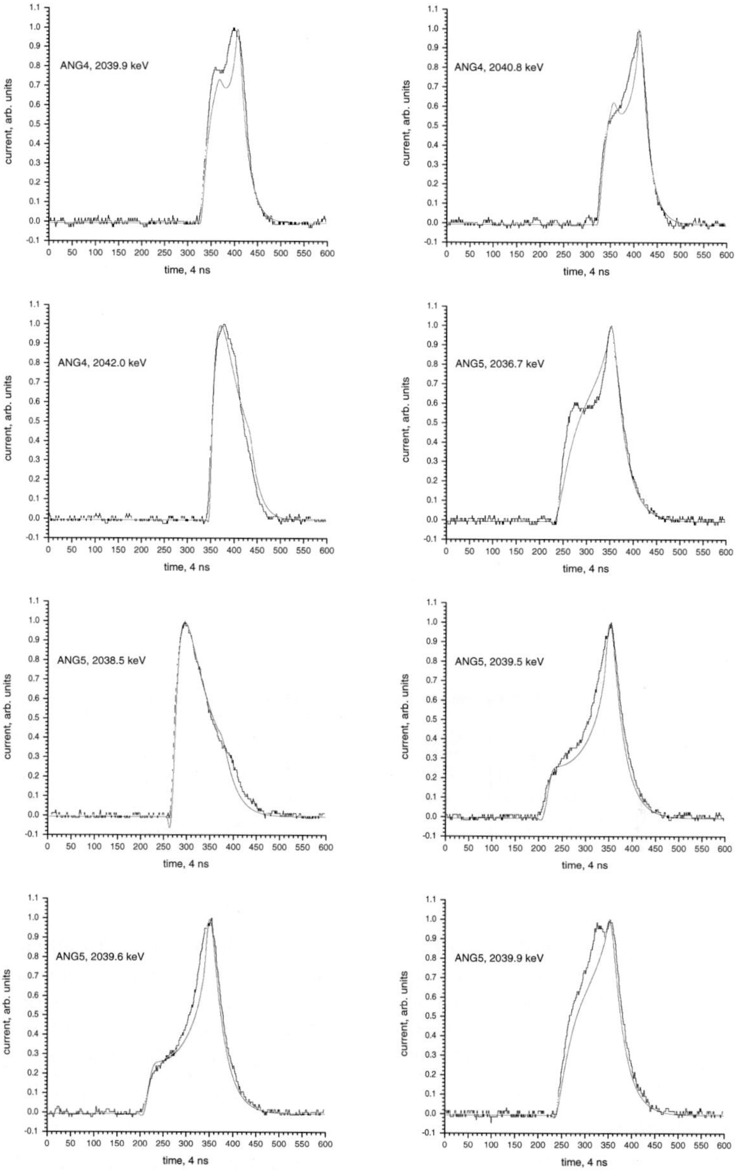

Fig. 9.d: All events shown were identified, by the global neuronal net[4,5] *as multiple site events*. The red lines show best fits of the events *by the zero range* pulse shape library, described in this paper. The determined χ^2 for the events of the events are Det4: 10.0, 3.85, 5.19; Det5: 6.37, 1.63, 4.46, 4.58, 8.7. I.e. they are not accepted as $0\nu\beta\beta$ SSE events by both methods.

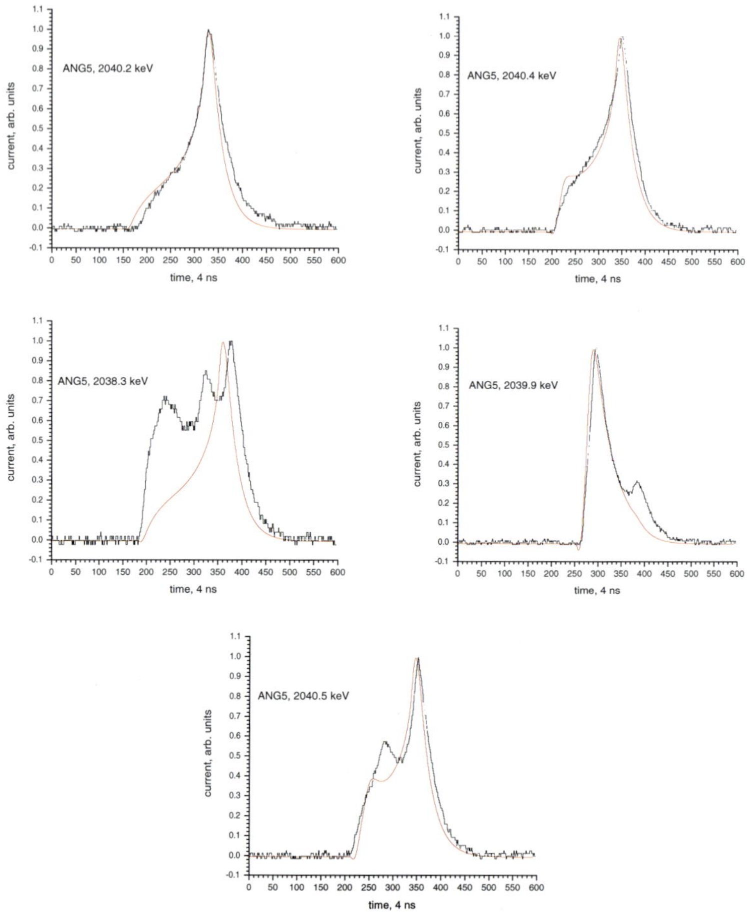

Fig. 9.e: The first three events were not accepted by the global neuronal net. The first is MSE, the second (upper line, right) is H, the third is H. The last two events are HNR and HN. The red lines show best fits of the events *by the zero range* pulse shape library, described in this paper. The determined χ^2 for the events of the all events are (Det5: 2.46, 1.91, 86.55, 5.88, 5.7). I.e. they are not accepted as $0\nu\beta\beta$ SSE events.

4. Conclusion

We gave a description of the present status of evidence for the observation of $0\nu\beta\beta$ decay. Two independent methods of pulse shape analysis have been used for the analysis of the data taken in the period November 1995 - May 2003 in the HEIDELBERG-MOSCOW experiment. The first one uses a neuronal net for selection of $0\nu\beta\beta$ event candidates and has been partly described and applied already in.[4,5,15,19] Some details are given here for the first time. The second method developed recently uses a pulse shape library of point-like events calculated from the field distributions in the detectors as function of radius R and height Z of the detectors. These calculations have been described in Ref.,[17] where it also has been shown microscopically, by Monte Carlo calculations, that the pulse shapes of such library events provide a good approximation of the pulse shapes of realistic $0\nu\beta\beta$ events and their spatial energy distribution.

By each of the methods, part of the line found at $Q_{\beta\beta}$ in the *full* spectrum is clearly *projected out with almost no* background from any surrounding γ rays, on a $\sim 6\sigma$ confidence level. It is found, that *both sets of events* are *contained* in the selection of the *global* neuronal net used in Refs.[4,5] We consider this identification of events at $Q_{\beta\beta}$ and practically full suppression of the surrounding γ-background as decisive proof for the observation of the $0\nu\beta\beta$ process. This proves that total lepton number is violated and, that the neutrino is a Majorana particle. This is the central message from double beta decay for particle physics.

Starting from an intensity of the line in the pulse shape selected spectrum (sum of NN (Fig. 4 -left) and low-cut spectra (Fig. 4 - right), see Fig. 9 of[19]) of 11±1.8 events (average of intensity of the line near $Q_{\beta\beta}$ in the sum spectra in the 2000-2060 keV range for 0, 1, 2 mm cuts, see[19]) leads to a half-life for $0\nu\beta\beta$ decay of $T^{0\nu}_{1/2} = (2.23\ ^{+0.44}_{-0.31}) \times 10^{25}$ y. This agrees with the central value of the half-life determined earlier from the *full* spectrum (see Fig. 9 of[4,5]) within a 1.7σ error of the latter. From the half-life one can derive *some* information on the effective neutrino mass $\langle m \rangle$ and the right-handed weak current parameters $\langle \eta \rangle$, $\langle \lambda \rangle$. In general we *cannot* determine the individual contributions of $\langle m \rangle, \langle \eta \rangle, \langle \lambda \rangle$ from a single $\beta\beta$ experiment, but only *upper limits* (see eq. (1)). *Only* under the *assumption*, that only *one* of the terms contributes to the decay process, and ignoring potential other processes connected with SUSY theories, leptoquarks, compositeness, etc. (see[1,2])), we can fix $\langle m \rangle = (0.32\ ^{+0.03}_{-0.03})$ eV, *or* $\langle \eta \rangle = (3.05\ ^{+0.26}_{-0.25}) \times 10^{-9}$, *or* $\langle \lambda \rangle = (6.92\ ^{+0.58}_{-0.56}) \times 10^{-7}$. *Thus we can fix the effective neutrino mass only, when we assume* $\langle \eta \rangle, \langle \lambda \rangle = 0$. In that sense it is highly premature to compare, as often done, such number with numbers deduced from e.g. WMAP, SDSS, or other cosmological experiments, or to use it as a landmark for future tritium experiments - other than as an upper limit. This fact also strongly relativates the popular efforts to calculate the nuclear matrix elements entering into eq. (1). They are not very relevant at the moment for the extraction of *fundamental* physics. We use in derivation of the limits given above the nuclear matrix elements of.[26,29] Since the corresponding matrix element for $2\nu\beta\beta$ decay underestimates this decay by ~30% (see[4,5,25]) these calculations may also underestimate the $0\nu\beta\beta$ matrix element, and consequently the effective neutrino mass (assuming $\eta = \lambda = 0$) could be lower, down to ~0.22 eV.

The value of $\langle m \rangle$=0.22 eV corresponds to a mass eigenvalue of the (degenerate) neutrinos of 0.2 - 0.8 eV.[2] This value which has to be taken as an upper limit is consistent with the value deduced from cosmological experiments. The most careful analysis of the SDSS data, by Tegmark et al.,[38] yields $\sum m_i \leq 1.75$ eV. The result of $0\nu\beta\beta$ decay defines under the assumption $\eta = \lambda = 0$ a *lower* limit for the contribution of neutrinos to hot dark matter of 4.7%.

It might be of some interest, to consider the *potential of present and presently planned future experiments:* (1) to confirm the present result, (2) to deliver additional information on the individual contributions of $\langle m \rangle, \langle \eta \rangle, \langle \lambda \rangle$, and others, to the $0\nu\beta\beta$ amplitude. For such discussion we refer to.[2,7–9,19,34,35]

It might be finally asked *what else* could be done in principle to prove observation of double beta decay - after having proved that there is a line at $Q_{\beta\beta}$ - after having shown that this line is not a double escape peak (which should consist of single site events) of an unknown gamma line (see[4,5]) and after having shown that -with two completely independent methods - only these events are selected at $Q_{\beta\beta}$, while at the same time gamma events around $Q_{\beta\beta}$ are suppressed almost completely.

The answer to this question in our opinion is : *Practically nothing else* could be done. An idea could be to look for angular correlations of the emitted electrons for the double beta candidate events. *However*, even *if* some future experiment *could* measure these angular correlations, probably no clear conclusion could be made (see the discussion of the expected angular correlation, for different combinations of $\langle m \rangle, \langle \eta \rangle, \langle \lambda \rangle$ in[7,8,35]). It might be added that one of the most promising of the projected experiments (CUORE) has even unfortunately no chance to perform a pulse shape analysis as it has been performed for the HEIDELBERG-MOSCOW experiment in the present paper, since the signal produced in the bolometer cannot differentiate between a $\beta\beta$ and γ -event. Also the recently proposed SNO$^+$ experiment[39] which hopefully will develop a 5σ sensitivity for a signal at $\langle m_\nu \rangle$=0.05 eV until 2015, has a chance to deduce a signal for double beta decay, as result of its very low energy resolution, only by observation of a deformation of a spectrum of relatively high background. For a detailed discussion of future experiments we refer to Ref.[2]

Concluding, the search for double beta decay may have reached some point, after which further progress will require huge efforts and *different* types of experiments than considered at present (see[2,7,8,34,35]).

Acknowledgements

The author would like to thank the Deutsche Forschungsgemeinschaft (DFG) for the important support of this work (436 RUS 113/670/0-2). He is deeply indebted to Dr. Irina Krivosheina for the decisive contribution she made in the performance and evaluation of the HEIDELBERG-MOSCOW double beta decay experiment.

References

1. H.V. Klapdor-Kleingrothaus, "60 Years of Double Beta Decay - From Nuclear Physics to Beyond the Standard Model", *World Scientific, Singapore* (2001) 1281 pages.
2. H.V. Klapdor-Kleingrothaus, "70 Years of Double Beta Decay - From Nuclear Physics to Beyond the Standard Model", *World Scientific, Singapore* (2008).
3. H.V. Klapdor-Kleingrothaus, *Int. Journ. of Mod. Phys.* **D13** (2004) 2107.
4. H.V. Klapdor-Kleingrothaus, I.V. Krivosheina, A. Dietz et al., *Phys. Lett.* **B586** (2004) 198-212.
5. H.V. Klapdor-Kleingrothaus, A. Dietz, I.V. Krivosheina et al., *Nucl. Instr. Methods* **A522** (2004) 371-406.
6. H.V. Klapdor-Kleingrothaus et al. *Mod. Phys. Lett.* **A16** (2001) 2409-2420, hep-ph/0201231; *Found. Phys.* **31** (2002) 1181-1123 (Corrigenda: **33** (2003) 678-684); H.V. Klapdor-Kleingrothaus, A. Dietz and I.V. Krivosheina *Part. Nucl. Lett.* **110** (2002) 57-79.
7. H.V. Klapdor-Kleingrothaus, I.V. Krivosheina, I.V. Titkova, *Phys. Lett.* **B632** (2006) 623-631.
8. H.V. Klapdor-Kleingrothaus, I.V. Krivosheina, I.V. Titkova, *Phys. Rev.* **D73** (2006) 013010.
9. H.V. Klapdor-Kleingrothaus, I.V. Krivosheina, I.V. Titkova, *Intern. J. of Modern Phys.* **A21** (2006) 1159-1188.
10. M. Günther et al., Heidelberg-Moscow Coll., *Phys. Rev.* **D55** (1997) 54.
11. H.V. Klapdor-Kleingrothaus, In Proc. of International Conf. "Neutrinos and Implications for Physics Beyond the Standard Model", Stony Brook, USA, 11-13 October, 2002, ed. R. Shrock, *World Scientific*, 2003, 367-382, and hep-ph/0303217, and in Proc. of Intern. Conf. "Beyond the Desert 2003", BEYOND03, Tegernsee, Germany, 9-14 June, 2003, *Springer, Heidelberg* (2004) ed. H.V. Klapdor-Kleingrothaus, pp. 307-364 and in Proc. of International Conf. "Neutrino Telescopes", February 2005, Venice, Italy, ed. M. Baldo-Ceolin., p. 215, hep-ph/0512263.
12. H.V. Klapdor-Kleingrothaus, *Nucl. Phys.* **B (Proc. Suppl.) 143** (2005) 229-232, Proc. of International Conf. "Neutrino 2004", 14-19 June, 2004,

Paris, France, and *Nucl. Phys.* **B (Proc. Suppl.) 145** (2005) 219, Proc. of International Worksh. "Neutrino Oscillation - NOW 2004", 11-17 September, 2004, Bari, Italy, eds. by G. Fogli et al.

13. H.V. Klapdor, *Proposal*, Internal Report, MPI-1987-V17, September 1987.
14. J. Hellmig, H.V. Klapdor-Kleingrothaus, *Nucl. Instrum. Meth.* **A 455** (2000) 638-644; and J. Hellmig, F. Petry, H.V. Klapdor-Kleingrothaus, *Patent* **DE19721323A**.
15. B. Majorovits, H.V. Klapdor-Kleingrothaus. *Eur. Phys. J.* **A6** (1999) 463 and B. Majorovits, Dissertation December 2000, MPI-Heidelberg.
16. B. Maier, *Dissertation* Nov. 1995, Univ. of Heidelberg.
17. H.V. Klapdor-Kleingrothaus, I.V. Krivosheina, V. Mironov, I.V. Titkova, *Phys. Lett.* **B636** (2006) 235-247.
18. H.V. Klapdor-Kleingrothaus, I.V. Krivosheina, I.V. Titkova, *Mod. Phys. Lett.* **A21** (2006) 1257-1278.
19. H.V. Klapdor-Kleingrothaus, I.V. Krivosheina, *Mod. Phys. Lett.* **A21** (2006) 1547-1566.
20. K. Vetter et al., *Nucl. Instr. Meth.* **A452** (2000) 105-114 and 223-238; Th. Kröll and D. Bazzacco, *Nucl. Instr. Meth.* **A463** (2001) 227-249.
21. *ORTEC company*, T.W. Raudorf, private communication, 2006.
22. T.W. Raudorf, M.O. Bedwell and T.J. Paulus, *IEE Transaction on Nucl. Science* **NS-29, No.1** (1982) 764-768.
23. G.F. Knoll, "Radiation Detection and Measurement", second ed. 1989, *John Wiley & Sons.*
24. F.S. Goulding and D.A. Landis, *IEEE Trans.Nucl. Sci.* **35** (1998) 119-124; B.W. Loo et al., *IEEE Trans. Nucl. Sci.* **35** (1998) 114-118; S.M. Hinshaw, D.A. Landis, *IEEE Trans. Nucl. Sci.* **37** (1990) 374-377; A. Georgiev and W. Gast, *IEEE Trans. on Nucl. Science* **40** (1993) 770-779.
25. Ch. Dörr, H.V. Klapdor-Kleingrothaus, *Nucl. Instr. Meth.* **A513** (2003) 596.
26. A. Staudt, K. Muto, H.V. Klapdor-Kleingrothaus, *Eur. Lett.* **13** (1990) 31.
27. K. Grotz, H.V. Klapdor, "Die Schwache Wechselwirkung in Kern-, Teilchen- und Astrophysik", *B.G. Teubner, Stuttgart* (1989), 458 pages, "The Weak Interaction in Nuclear, Particle and Astrophysics", *IOP Bristol* (1990), *Moscow, MIR* (1992) and *Shandong Science and Technology Press, Jinan, China* (1998).
28. K. Muto, H.V. Klapdor, in "Neutrinos", "Graduate Texts in Contemporary Physics", ed. H.V. Klapdor, *Berlin, Germany: Springer* (1988) 183-238.
29. K. Muto, E. Bender and H.V. Klapdor, *Z. Phys. A - Atomic Nuclei* **334** (1989) 187 and 177.
30. R.J. Ellis et al., *Nucl. Phys.* **A** 435 (1985) 34-42.
31. J.G. Hykawy et al., *Phys. Rev. Lett.* **67** (1991) 1708.
32. G. Audi, A.H. Wapstra, *Nucl. Phys.* **A595** (1995) 409-480.
33. G. Douysset et al., *Phys. Rev. Lett.* **86** (2001) 4259 - 4262, and I. Bergström et al., in Proc. of Intern. Conf. on Particle Physics Beyond the Standard Model, BEYOND'02, Oulu, Finland June 2002, IOP, 2003, ed. H.V. Klapdor-Kleingrothaus and in BEYOND 2003, Castle Ringberg, Germany, 10 - 14 June, 2003, *Springer, Heidelberg*, 2003, ed. *H.V. Klapdor-Kleingrothaus.*
34. H.V. Klapdor-Kleingrothaus *"A New View on the Future of Nuclear Double*

Beta Decay", in Proc. 6th Intern. Worksh. on the Identification of Dark Matter "idm2006", Rhodos, Greece, September 11-16, 2006, *eds. M. Axenides, G. Fanourakis, J. Vergados, World Scientific* (2007) pp. 423-432.

35. H.V. Klapdor-Kleingrothaus, in Proc. "2nd Scandanavian Neutrino Workshop (SNOW 2006)", Stockholm, Sweden, 2-6 May 2006, *eds* T. Ohlsson, M. Blennow et al., Stockholm, Roy. Swed. Acad. Sci. (2006), *Phys. Scripta* **T127** (2006) pp. 40-42.
36. K.Ya. Gromov, V.A. Bednyakov, V.I. Fominykh, V.G. Chumin, *J. Part. Nucl. Lett.* **3** (2006) pp. 157-164 and also these Proceedings.
37. H.V. Klapdor-Kleingrothaus et al., *Phys. Lett.* **B578** (2004) 54-62 and *Nucl. Instr. Meth.* **A511** (2003) 335; H.V. Klapdor-Kleingrothaus *hep-ph/0205228.*
38. SDSS Collaboration: M. Tegmark et al., *Phys. Rev.* **D69** (2004) pp. 103501-1–26, and *astro-ph/*0310723.
39. M.C. Chen (for the SNO^{++} collaboration) **Nucl. Phys.** *B 145* Proc Suppl. (2005) pp. 65-68.
40. [Kur05] A.M. Bakalyarov et al. (Moscow group of Heidelberg-Moscow experiment), *hep-ex/*0309016. in Proc. of 4th Intern. Conf. on Nonaccelerator New Physics (NANP 03), Dubna, Russia, 23-28 Jun 2003, *Phys. Part. Nucl. Lett.* **2** (2005) pp. 77-81, *Pisma Fiz. Elem. Chast. Atom. Yadra* **2** (2005) pp. 21-28.

STRUCTURE OF BACKGROUND IN THE HEIDELBERG – MOSCOW EXPERIMENT ON SEARCH FOR AND INVESTIGATION OF DOUBLE BETA DECAY OF ^{76}Ge

K. YA. GROMOV, V. A. BEDNYAKOV*, V. I. FOMINYKH and V. G. CHUMIN

Dzhelepov Laboratory of Nuclear Problems, Joint Institute for Nuclear Research, Dubna, Russia

Independent analysis of the spectra from the Heidelberg–Moscow experiment has been carried out. A direct comparison of the peak intensities in the spectra allows the conclusion that in the energy range up to 3200 keV the background is due to detection of γ-rays from the decay of trace impurities of anthropogenic and cosmogenic nuclides occurring in the apparatus between the HPGe detectors and the main (Pb and Cu) shielding. The ^{226}Ra, ^{214}Pb, and ^{214}Bi γ-rays observed in the spectra are shown to arise in the ^{226}Ra rather than ^{238}U decay chain. The estimates of the expected intensities of weak ^{214}Bi γ-rays in the spectrum region in the vicinity of $Q_{\beta\beta}(^{76}\mathrm{Ge})$=2039 keV do not contradict observation of the peak at 2039 keV as claimed by Klapdor-Kleingrothaus and coauthors.

1. Introduction

The double neutrinoless beta decay ($0\nu2\beta$-decay) of nuclei

$$A(Z,N) \to A(Z+2, N-2) + 2e^-,$$

which may proceed only with violation of the lepton number conservation law by two units ($\Delta L = 2$) is generally recognized to be a process of high scientific significance. Experimental observation of the $0\nu2\beta$-decay is of great interest as direct and unambiguous indication of going beyond the standard model of electroweak interactions where the lepton number is strictly conserved ($\Delta L = 0$). In addition, by the Schechter–Valle theorem[1] the probability of the $0\nu2\beta$-decay means that at least one of the neutrinos has a nonzero mass and moreover that the neutrino is a Majorana particle (i.e. has a Majorana nature, when the particle coincides with its antiparticle).

*E-mail: Vadim.Bednyakov@jinr.ru

Many collaborations in various underground laboratories have been trying to detect the neutrinoless beta decay using various isotopes potentially capable of decaying via the $0\nu2\beta$-mode, such as Ge, Se, Mo, Cd, Te, Xe. The currently best result has been obtained in the joint German–Russian Heidelberg–Moscow experiment with five large highly pure Ge detectors enriched with germanium-76, which are located deep underground in the Gran Sasso low-background laboratory (Italy). The limit for the $0\nu2\beta$-decay period of ^{76}Ge was established to be $T_{1/2}(0\nu2\beta) \geq 1.9 \times 10^{25}$ years (90 % C.L.).[2] Later the spectral region expected to embrace the energy of the ^{76}Ge double beta decay ($Q_{\beta\beta} = 2039.0$ keV) was thoroughly analyzed again under the supervision of Prof. Klapdor-Kleingrothaus, and the results of the analysis allowed the authors to state observation of a peak at 2039 keV in the spectrum[3,4] and.[5] In the same spectral region lines of energy 2011, 2017, 2022 and 2053 keV were detected and ascribed to the ^{214}Bi decay.[6]

Soon after Klapdor-Kleingrothaus et al. reported observation of the 2039-keV peak[3,4] there appeared papers by C.E. Aalseth et al.,[7] F.Feruglio et al.,[8] and Yu.G. Zdesenko et al.,[9] questioning observation of the ^{76}Ge $0\nu2\beta$-decay in.[3,4] It is stated in[7] and[9] that the ^{214}Bi γ-lines and the 2039-keV line observed by Klapdor-Kleingrothaus et al.[3,4] are spurious. In[8] this spectral region was analyzed by the same methods as in[3,4] and practically the same lines as in[3,4] were revealed. It was stressed, however, that more accurate data on the background structure in this spectral region were needed to draw a final conclusion. The well-grounded answer to this criticism was first given by Klapdor-Kleingrothaus in[10] and then the observation of this line was supported by the results of special additional investigations published by the authors within the years.[11–18] The results of the most complete analysis of the background spectrum of the Heidelberg–Moscow experiment obtained with careful Monte-Carlo simulation based on the GEANT4 package are published in.[19]

Yet, it was of interest to find out what conclusions about the background structure could be drawn from a direct comparison of the peak intensities observed in the background spectrum. To do this, Prof. Klapdor-Kleingrothaus kindly provided the authors of this paper with the files of three spectra measured in the Heidelberg–Moscow experiment. Spectrum I is a total spectrum of all 5 detectors (№ 1, 2, 3, 4, and 5) measured over the period from August 1990 to May 2000 in the energy range up to 2850 keV (it corresponds to Fig. 3 from[5]). Spectrum II is the total spectrum of all 5 detectors (№ 1, 2, 3, 4, and 5) measured over the period from August 1995 to May 2000 in the energy range up to 8350 keV. Spectrum III is the spectrum of detector № 4 measured over the period from August 1995 to May 2000 in the energy range up to 8350 keV.

The Heidelberg–Moscow experimental setup is described in detail, for example, in.[5,20] It should be noted that Spectrum III is of particular interest because detector № 4 has additional Cu shielding 27.5 cm thick installed between this detector and the lead shielding.

2. Identification of γ-Lines in the Background Spectrum

The results yielded by our analysis of Spectrum I (Fig. 3 from[5]), including γ-ray energies, intensities (areas of photopeaks), and identifications are presented in Table 1. To identify γ-rays we used Firestone's Tables of Isotopes (hereinafter referred to as TI),[6] tables of the Japanese Institute of Atomic Energy (JAERI-Data, Code 98-008 and JNDC(JPN)-170/L), and tables "Gamma-ray from radioactive decay, listed in order of increasing energy" by U.Rous, W.Westmeier, I.Warneke. Magdeburg-Darmstadt, Germany, 1978.

The following possible cases were considered:
– γ-rays arise from the decay of nuclides belonging to the radioactive ^{238}U (4n+2) or ^{232}Th (4n) families;
– γ-rays arise from the decay of long-lived anthropogenic nuclides;
– γ-rays arise from the decay of cosmogenic (n, γ or μ^-, γ) nuclides.

Column 3 of Table 1 shows nucleus (A, Z) to which the particular decay γ-transition is ascribed. Columns 4 and 5 show the energy of this γ-transition and the intensity, a_γ, of the γ-rays per decay of the nucleus (per decay in the radioactive chain) from the TI.[6] The last column presents the calculated ratios $\frac{S_{\gamma i}}{a_{\gamma i}} = N \cdot \mathcal{E}(E_{\gamma i})$ (see formula (2) below). We skip figures with used experimental spectra, because they are qualitatively the same as in.[2,19] Our conclusions given below rely on quantitative analysis of the intensities of the lines in these spectra.

(4n+2) chain decay nuclei. Energies of 25 γ-lines allow them to be ascribed to the decay of the ^{214}Pb ($T_{1/2} = 26.8$ min) and ^{214}Bi (19.7 min) nuclei from this chain. They are short-lived nuclei arising from the decay of a long-lived parent nucleus in the setup. Radiations of short-lived nuclides from the (4n+2) family observed in the background spectra are usually attributed to uranium contamination of the setup, the short-lived nuclides assumed to be in secular equilibrium with uranium. Yet, one overlooks the fact that in the uranium-238 family the short-lived nuclei ^{226}Ra ($1.6 \cdot 10^3$ y),

^{214}Pb (26.8 min) and ^{214}Bi (19.7 min) are preceded by the long-lived nuclei ^{234}U ($2.5 \cdot 10^5$ y) and ^{230}Th ($7.5 \cdot 10^4$ y). Therefore, accumulation of ^{226}Ra and daughter nuclei in the time after separation of impurity uranium from the ore should be ignored and the ^{226}Ra, ^{214}Pb and ^{214}Bi γ-lines in the spectra should be ascribed to radium-226 contamination of the setup. The

Table 1. Spectrum I analysis results. Gamma-rays of energies 230 to 2850 keV in the Heidelberg–Moscow experiment.

Experiment		Identification			
E_γ, keV	$S_\gamma \cdot 10^{-3}$	^AZ	E_γ, keV	a_γ	$N\mathcal{E}(E_\gamma) \cdot 10^{-3}$
186.0(2)	0.7(2)	^{226}Ra	186.21	0.033	21(6)
238.68(10)	5.51(15)	^{212}Pb	238.63	0.433	12.7(4)
241.83(14)	1.77(5)	^{214}Pb	241.98	0.075	23.6(7)
295.22(10)	3.16(22)	^{214}Pb	295.21	0.115	17.1(18)
299.7(3)	0.39(3)	^{212}Pb	300.09	0.033	11.9(9)
328.30(10)	~ 0.3	^{228}Ac	328.0	0.030	~ 10.2
337.8(2)	0.83(16)	^{228}Ac	338.3	~ 0.113	~ 7.4
339.5(3)	0.37(14)	^{228}Ac	341.0	0.0038	97
347.5(4)	0.16(10)	^{214}Bi	348.0	0.0068	200
351.90(10)	5.40(15)	^{214}Pb	351.92	0.358	15.1(4)
409.7	~ 0.17(14)	^{228}Ac	409.45	0.019	~ 8.9
427.86(14)	0.68(11)	^{125}Sb	427.88	0.30	2.3(4)
463.05(17)	0.53(14)	^{228}Ac	463.0	0.044	12(3)
		^{125}Sb	463.4	0.105	5.0(3)
511.11(13)	1.70(19)	^{208}Tl	510.77	0.082	21(2)
569.77(14)	0.59(6)	^{207}Bi	569.70	0.977	0.60(9)
579.90(10)	0.193(10)				
583.20(11)	1.72(9)	^{208}Tl	583.19	0.307	5.6(3)
600.50(14)	0.58(7)	^{125}Sb	600.60	0.175	3.3(4)
609.27(10)	4.8(2)	^{214}Bi	609.31	0.448	10.7(4)
622.80(10)	0.15(3)				
635.9(3)	0.36(7)	^{125}Sb	635.95	0.112	3.2(6)
661.66(10)	14.3(3)	^{137}Cs	661.66	0.851	16.8(3)
727.3(2)	0.31(10)	^{212}Bi	727.33	0.066	4.7(15)
768.11(10)	0.40(10)	^{214}Bi	768.36	0.048	8.3(21)
785.6(3)	0.15(5)	^{212}Bi	785.37	0.011	13.6(45)
795.45(16)	0.40(6)	^{228}Ac	795.00	0.0434	9.2(14)
~ 802	~ 0.3				
~ 806	~ 0.24	^{214}Bi	806.17	0.0112	~ 21
~ 810	~ 0.13	^{58}Co			
834.9(2)	0.46(11)	^{54}Mn			
840.6(2)	0.33(10)	^{228}Ac	840.4	0.0093	35(10)
860.66(10)	0.24(8)	^{208}Tl	860.56	0.045	5.3(18)
910.93(11)	1.78(12)	^{228}Ac	911.2	0.266	6.5(15)
934.01(10)	~ 0.37	^{214}Bi	934.04	0.0303	~ 12
964.82(10)	0.26(2)	^{228}Ac	964.77	0.051	5.1(4)
968.65(12)	0.78(6)	^{228}Ac	968.97	0.162	4.8(4)
999.7(3)	0.09(5)				

Table 1. *(Continued)*

Experiment		Identification			
E_γ, keV	$S_\gamma \cdot 10^{-3}$	^AZ	E_γ, keV	a_γ	$N\mathcal{E}(E_\gamma) \cdot 10^{-3}$
$\sim$ 1001.9	$\sim$ 0.05(2)	^{234}Pa	1001.0	0.0084	$\sim$ 6.0
1063.70(10)	0.44(7)	^{207}Bi	1063.66	0.745	0.59(9)
1093.5	0.10(5)	^{228}Ac	1095.68	0.0013	77(38)
1120.16(14)	1.08(11)	^{214}Bi	1120.29	0.148	7.3(7)
1124.3(3)	0.38(7)				
1147.3	0.11(5)				
1155.4(3)	0.17(6)	^{214}Bi	1155.19	0.0164	10.4(36)
1173.21(11)	3.42(14)	^{60}Co	1173.2	1.00	3.42(14)
1237.82(10)	0.37(6)*	^{214}Bi	1238.11	0.0586	6.3(10)
1281.6(1)	0.17(4)	^{214}Bi	1280.96	0.0144	11.8(28)
1332.46(12)	3.21(13)	^{60}Co	1332.47	1.00	3.21(13)
1363.4(5)	0.08(2)				
1377.5(2)	0.39(5)	^{214}Bi	1377.67	0.392	9.9(13)
1401.10(10)	$\sim$ 0.11	^{214}Bi	1401.50	0.0155	$\sim$ 7.1
1407.6(3)	0.22(5)	^{214}Bi	1408.0	0.028	$\sim$ 7.8
1460.48(12)	9.50(19)	^{40}K	1460.81	0.105	90.5(20)
$\sim$ 1509	$\sim$ 0.08	^{214}Bi	1509.23	0.021	$\sim$ 4
$\sim$ 1587	$\sim$ 0.05	^{228}Ac	1588.21	0.0327	$\sim$ 1.5
1620.4(2)	0.08(2)	^{212}Bi	1620.5	0.015	5.3(13)
$\sim$ 1630	$\sim$ 0.04	^{228}Ac	1630.6	0.016	$\sim$ 2.5
$\sim$ 1660	$\sim$ 0.10	^{214}Bi	1661.3	0.0114	$\sim$ 9
1729.3(2)	0.33(4)	^{214}Bi	1729.60	0.0288	11.5(14)
1764.4(2)	1.27(6)	^{214}Bi	1764.49	0.154	8.3(4)
1847.5(2)	0.19(4)	^{214}Bi	1847.42	0.0204	9.3(20)
2102.8(9)	0.07(3)	^{208}Tl	SEP 2614.5		
2118.3(3)	0.13(2)	^{214}Bi	2118.55	0.0114	11.4(18)
2204.2(2)	0.27(4)	^{214}Bi	2204.21	0.0486	5.6(8)
2292.1(3)	0.034(17)	(^{214}Bi)	2293.4	0.0030	11(6)
2447.8(2)	0.06(2)	^{214}Bi	2447.86	0.015	4.0(13)
2505.9(2)	0.10(2)	^{60}Co	$\Sigma 1172 + 1332$		
2614.5(1)	0.82(5)	^{208}Tl	2614.5	0.36	2.28(14)

presence of ^{238}U in the Heidelberg–Moscow setup could be estimated from the spectral γ-lines of ^{234m}Pa (1.17 min) which is in equilibrium with ^{238}U. The observed 1001.9 keV γ-decay can be ascribed to the ^{234m}Pa decay and identified as the highest-intensity transition in this decay. However, the intensity of this spectral line is very low ($S_\gamma \cong 0.09 \cdot 10^3$). No other evidence for the presence of ^{238}U in the setup was found. Thus, contamination of the setup by ^{238}U is not proved.

The number of detected decays of a nucleus followed by emission of γ-rays with energy E_γ (i.e. the photopeak area $S(E_\gamma)$) is given by the expression

$$S(E_\gamma) = N \cdot a(E_\gamma) \cdot \mathcal{E}(E_\gamma), \tag{1}$$

where N is number of decays of the nucleus within the time of the experiment, $a(E_\gamma)$ is the intensity of the γ-rays with energy E_γ per decay of the nucleus (per decay in the equilibrium chain), $\mathcal{E}(E_\gamma)$ is the detection efficiency for γ-rays with the energy E_γ at the photopeak. Column 6 of Table 1 presents the calculated quantities

$$N\mathcal{E}(E_{\gamma i}) = \frac{S_{\gamma i}}{a_{\gamma i}}. \tag{2}$$

We have no data which allows to determine N and $\mathcal{E}(E_\gamma)$ separately, but even their product allows some helpful conclusions. Dependence of $N\mathcal{E}(E_\gamma)$ on E_γ for the ^{226}Ra, ^{214}Pb and ^{214}Bi γ-lines is displayed in Fig. 1. Though

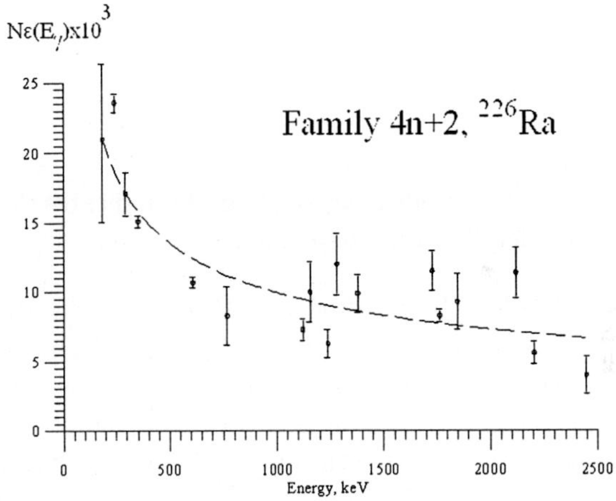

Fig. 1. Dependence of $N\mathcal{E}(E_\gamma)$ on E_γ for γ-lines from the ^{226}Ra chain decay

the errors in the values $N\mathcal{E}(E_\gamma)$ are large, these values may be assumed to increase monotonically as the energy decreases, which additionally confirms that the nuclei in question belong to one equilibrium chain, namely, the ^{226}Ra chain. The $N\mathcal{E}(E_\gamma)$ value for the 347.5-keV γ-transition which drops out of the smooth dependence probably indicates the presence of another, unidentified background nuclide emitting γ-rays of the same energy. In the equilibrium chain N is constant. Therefore, an increase in the region of low energies ($E_\gamma < 500$ keV) directly indicates that a considerable number of all ^{226}Ra, ^{214}Pb and ^{214}Bi nuclei decay in the close vicinity to the detectors (between the detectors and the passive shielding).

4n decay chain nuclei. The γ-ray energies (Table 1) allow about 20 γ-transitions to be ascribed to the decay of the nuclei from this chain: ^{228}Ac ($T_{1/2} = 6.1$ h), ^{212}Pb (10.6 h), ^{212}Bi (60.6 min) and ^{208}Tl (3.1 min). Monotonic dependence of $N\mathcal{E}(E_\gamma)$ on the γ-ray energy does not contradict (see Table 1) the fact that above nuclides are produced in one equilibrium decay chain. The parent long-lived nucleus of this chain may be ^{232}Th ($1.4 \cdot 10^{10}$ y) or ^{228}Ra (5.8 y). An increase in $N\mathcal{E}(E_\gamma)$ in the region $E_\gamma <$ 500 keV indicates that long-lived parent nuclei whose decay results in the above nuclei are localized in the vicinity of the detectors (between the passive shielding and the detectors). Spectrum II and Spectrum III show a peak at 3199 keV which we identify as a peak of a sum of pulses from the γ-ray cascade 584.2 keV + 2614.5 keV = 3198.7 keV. Formation of the sum peak also indicates that ^{208}Tl decay in the close vicinity of the detectors.

^{60}Co beta decay. The transitions γ1173 ($N\mathcal{E}(E_\gamma)$=3420) and γ1332 ($N\mathcal{E}(E_\gamma)$=3210) are ascribed to this decay. All three spectra show a 2505-keV peak of the sum of pulses from these transitions. The estimated average γ1173 and γ1332 detection efficiency of about 3% indicates that ^{60}Co nuclei decay in the close vicinity of the detectors.

Gamma-rays from the ^{125}Sb decay. Four γ-transitions (Table 1) can be ascribed on the basis of E_γ to the β^--decay of ^{125}Sb. The energy of one of them, 463.4 keV, coincides with 463.0 from the ^{228}Ac decay within the error. Taking $N\mathcal{E}(E_\gamma)$ for the ^{228}Ac 463.0-keV γ-line to be approximately $7 \cdot 10^3$, as for the neighboring γ-rays from the 4n chain, we get $N\mathcal{E}(E_\gamma) \sim 3 \cdot 10^3$ for the ^{125}Sb γ-line at 463.4 keV. Table 2 lists energies of the γ-rays ascribed

Table 2. ^{125}Sb γ-rays observed in Spectrum I

E_γ, keV	427.9	463.4	600.5	635.9
$N\mathcal{E}(E_\gamma)$	$2.3(4) \cdot 10^3$	$\sim 3 \cdot 10^3$	$3.3(4) \cdot 10^3$	$3.2(6) \cdot 10^3$

to the ^{125}Sb decay and $N\mathcal{E}(E_\gamma)$ for them. Thus, $N\mathcal{E}(E_\gamma)$ values can be considered as being not in conflict with ascribing the above γ-lines to the decay of the same ^{125}Sb nucleus.

Gamma-rays from the electron capture decay of nuclei. There is no doubt about identification of γ-rays at 1460.8 keV (β^+EC ^{40}K,

$N\mathcal{E}(E_\gamma)$=90500), 569.7 keV, and 1063.7 keV (EC ^{207}Bi, $N\mathcal{E}(E_\gamma)$=600 and 590). Dörr and Klapdor-Kleingrothaus[19] ascribe four more peaks to the nuclei undergoing electron capture decay. They assumed that the $E_\gamma \approx 810$ keV peak results from electron capture in the ^{58}Co ($T_{1/2} = 70.8$ d) nucleus, the $E_\gamma = 834.9$ keV arises from the EC in the ^{54}Mn ($T_{1/2} = 312$ d) nucleus, the $E_\gamma = 840.6$ keV peak is due to summation of pulses from $E_\gamma = 834.9$ keV and X-rays, and the γ1124.3 peak is caused by summation of 1115.2-keV γ-rays and X-rays arising from electron capture in the ^{65}Zn ($T_{1/2} = 244$ d) nucleus. The ^{58}Co, ^{54}Mn, and ^{65}Zn nuclei are relatively short-lived ones, and their radiation may manifest itself in the spectrum as their residual activity after being left to stand for 200 days before the beginning of data collection.[5] In the course of the experiment the intensity of the above γ-lines should considerably (in approximately 10 years) decrease. Under this condition identification of $E_\gamma \approx 810$ keV, $E_\gamma = 834.9$ keV, and $E_\gamma = 840.6$ keV does not contradict anything.

In Table 1 the 840.6-keV line is ascribed to the ^{228}Ac decay while the value $N\mathcal{E}(E_\gamma) = 35(10)\cdot 10^3$ is a few times larger than that of the neighboring ^{228}Ac lines. With the contribution to the 840.6-keV peak from the sum of $E_\gamma = 834.9$ keV and X-rays taken into account, $N\mathcal{E}(E_\gamma)$ approaches the average value at this energy for the 4n chain. Interpretation of the 1124.3-keV line as a sum of pulses from $E_\gamma = 1115.2$ keV and X-rays following EC in ^{65}Zn is a different matter. The fluorescence yield in this Z-region is $\omega_k = 0.47$. If ^{65}Zn nuclei as the other setup impurities are located outside the detectors, then the X-ray detection efficiency (solid angle "source-detector") should be < 1 and a peak at $E_\gamma = 1115.2$ keV should appear in the spectrum with intensity larger than the intensity of the peak of the sum with the X-rays. The peak $E_\gamma = 1115.2$ keV can completely disappear due to summing with X-rays and Auger electrons only in the case when ^{65}Zn is located inside the material of the detectors. This is scarcely probable. Therefore we believe the 1124.3-keV line has to be explained.

The analysis of Spectrum II and Spectrum III in the region up to 2850 keV yields results that generally agree within the errors with the results yielded by the analysis of Spectrum I (Table 1). Spectrum III measured with a copper shield (27.5 cm) between the main lead shielding and the detector allows the conclusion that the background from external sources (rocks in the underground laboratory) is negligibly small, i.e., that it is small impurities between the main shielding and the detectors which are responsible for the background in the Heidelberg–Moscow experiment. An appreciable (5–7 times) decrease in the intensity of the γ661-keV peak from the

^{137}Cs decay in Spectrum III may indicate that the ^{137}Cs nuclide is present in the lead shielding, though contamination of the assembly of detectors № 1, 2, 3, 5 with ^{137}Cs is not impossible either. Finally, a comparison of Spectra I and II with Spectrum III is in conflict with the assertion (Table 2 in[19]), that the ^{40}K and ^{210}Pb nuclides are present in the lead shielding materials.

It is difficult to analyze Spectra II and III in the region above 3500 keV because there are no reference energy values in this region. A wide peak at about 5240 keV manifests itself in both spectra. In[5,20] and[21] this peak is said to be due to detection of α-particles from the decay of ^{210}Po ($T_{1/2}$ = 138 d, $E_{\alpha 0} = 5305$ keV, $I_{\alpha 0} = 100$ %), produced in the β^--decay of ^{210}Pb ($T_{1/2} = 22,3$ y). It is assumed that ^{210}Pb is located on the inner surface of the detector shell. The energy of the α-particles decreases (by $\sim$ 70 keV) as they pass through the dead surface layer of the detectors.

Bakalyarov et al.[21] simulated the background spectrum of all five detectors in the region of 3500–8000 keV. They believe that the observed spectrum is satisfactorily explained by detection of α-particles from the decay of nuclides of the uranium and thorium equilibrium chains on the assumption that uranium and thorium nuclei are located, like ^{210}Pb nuclei, on the inner surface of the shell of the detectors. However, this conclusion does not agree with the above assertion that observation of ^{226}Ra, ^{214}Pb, and ^{214}Bi in the background spectrum is caused by contamination of the setup with ^{226}Ra rather than ^{238}U. If the setup is contaminated with a small amount of uranium, this may cause observation of radiation from ^{238}U ($4.5 \cdot 10^9$ y), ^{234}Th (24.1 d), ^{234m}Pa (1.2 min), and ^{234}U ($2.5 \cdot 10^5$ y) in the background spectrum. The peaks due to α-particles from the decay of ^{230}Th ($8 \cdot 10^4$ y, $E_{\alpha 0} = 4.77$ MeV, $I_{\alpha 0} = 76$ %) are ruled out in the spectrum. Therefore, we think that explanation of the events observed in the energy region above 3500 keV needs more careful consideration.

The results of identification of γ-rays in the background spectrum of the Heidelberg–Moscow experiment are summarized in Table 3.

The table presents A, Z, $T_{1/2}$ of the isotope whose γ-rays are observed in the background spectrum, and energies of the γ-rays (for the ^{226}Ra and ^{238}Th equilibrium chains there are the numbers of γ-transitions identified in the chain). Column 4 lists values $N\mathcal{E}(E_\gamma) = \dfrac{S(E_\gamma)}{a(E_\gamma)}$ (see formula (2)) which allow an idea about the number of detected decays. It should be borne in mind, of course, that the detection efficiency decreases with increasing E_γ and depends on the place where these nuclei decay. The last

Table 3. Results of identification of γ-rays in the background spectrum

^{A}Z	$T_{1/2}$, years	E_γ, keV	$N\,\mathcal{E}(E_\gamma)$	$Q_{\beta-}$ or Q_{EC}, keV
^{226}Ra→	$1.6 \cdot 10^3$	25-γ	2400–5000	^{214}Bi, β^- 3270
				^{210}Tl, β^- 5497
^{232}Th→	$1.4 \cdot 10^{10}$	21-γ	1300–2300	^{208}Tl, β^- 4992
^{40}K	$1.3 \cdot 10^9$	1460	90500	EC 1505
^{60}Co	5.3	1172, 1332	3300	β^- 2824
^{137}Cs	30.2	661.7	16800	EC 1220
^{125}Sb	2.7	422.9; 463.0; 600.6; 635.9	3000	β^- 767
^{207}Bi	38	569.7; 1063.7	600	EC 2405
^{54}Mn	0.85	834.9 (840.6)	460	EC 1377
^{58}Co	0.19	810	130	EC 2307

column of Table 3 lists energies of the β^--decay ($Q_{\beta-}$) or EC (Q_{EC}) of the corresponding isotopes. For the ^{226}Ra and ^{232}Th decay chains there are energies of the β-decay of the chain member isotopes for which $Q > 2000$ keV and whose decay may give rise to 2039-keV γ-rays of interest. For five nuclides Q_β or Q_{EC} is larger than 2000 keV. But comparison of the available TI data[6] on their γ-spectra with the intensities of the spectral γ-lines given in Table 1 allows the presence of the 2039-keV line to be ruled out.

However, it is not impossible that some γ-transitions from Table 1 proceed in reactions with background neutrons and μ-mesons. These reactions could be followed not only by γ-rays from the decay of the corresponding nuclide but also by high-energy γ-rays. In this connection it is of interest to trace the variation in the intensity of γ-rays over the exposure period (13 years) of the Heidelberg–Moscow experiment. It might be possible to observe the expected decrease in the intensity of ^{125}Sb and ^{60}Co γ-rays as well as disappearance of the γ-rays ascribed to the ^{54}Mn and ^{58}Co decay. A decrease in the intensity of the γ-rays from the decay of the nuclides belonging to the 4n chain may indicate that this chain begins with ^{228}Ra ($T_{1/2} = 5.8$ y) nucleus. An increase in the intensity of some lines may also be observed.

3. Spectral Region of 2000–2060 keV

C.E. Aalseth et al.[7] and Yu.G. Zdesenko et al.[9] state that weak γ-lines of ^{214}Bi, observed by Klapdor-Kleingrothaus et al.[3–5] in the spectral region 2000–2060 keV *cannot manifest themselves* in the spectrum. Accordingly, the 2039.0-keV line with the intensity close to the intensity of those weak Bi lines could not be observed either. In Klapdor-Kleingrothaus's reply[10] to the criticism,[7] and in[5] intensities of peaks in the region 2000–2060 keV are

Table 4. Experimental and calculated (expected) intensities of weak peaks from the ^{214}Bi decay. Row 1: experimental intensities of peaks according to;[10] Row 2: intensities of γ-rays per ^{214}Bi decay; Row 3: expected intensities of peaks (summation of pulses is ignored); Row 4: the same as in Row 3 but according to;[7] Row 5: intensities of peaks calculated in[10] with summation of peaks taken into account;

No.	E, keV	γ2204.2	γ2010.7	γ2016.7	γ2021.8	γ2052.9	γ2039.0
1.	$S_{\rm expt}$ from[10]	319(22)	37.8(102)	13.0(85)	16.7(88)	23.2(90)	12.1(83)
2.	a_γ% according to[6]	4.86	0.050(6)	$\equiv 0$	0.020(6)	0.078(11)	
3.	$S = 319 \cdot a_{\gamma i}/4.86$	319	3.3(4)	$\equiv 0$	1.3(4)	5.1(7)	
4.	S according to[7]	319	3.2	0.38	1.26	5.0	
5.	$S_{\gamma i} + S_\Sigma$ from[10]	319	12.2(6)	15.6(7)	1.2(1)	4.7(3)	

evaluated. They are presented in the first row of Table 4. Expected intensities of weak ^{214}Bi peaks can be found by using intensities of strong ^{214}Bi lines in the background spectrum and data on intensities of ^{214}Bi decay γ-rays from the TI[6] (row 2 in Table 4). The detection efficiency for γ-rays in the region 2000–2060 keV and the detection efficiency for 2204-keV γ-rays can be considered identical within errors, and the expected intensities of weak ^{214}Bi γ-rays can be calculated by using intensities of 2204-keV γ-rays in the background spectrum. The results of these calculation are presented in row 3 of Table 4. Row 4 presents the results of similar evaluations from[7] normalized to the intensity of the 2204-keV γ-peak equal to 319 events in[10] (row 1 of the table). The values in rows 3 and 4 are seen to coincide except the value for the 2016.7-keV peak. The values for $S_{\gamma 2016.7}$ in[7] is calculated incorrectly. According to the TI,[6] the intensity of 2016.7-keV γ-rays is identically equal to zero because it is a transition of the $E0$ type. In,[7] for the intensity of 2016.7-keV γ-rays they used the intensity of conversion electrons of this γ-transition.[6] The experimentally observed peak at 2016.7 keV should be attributed to summation of pulses from cascade γ-transitions of 1408 keV and 609 keV from the 0^+ 2016.7-keV level to the 0^+ ground state of ^{214}Po in the decay of ^{214}Bi. According to the ^{214}Bi decay scheme, a considerable pulse summation effect should also be responsible for formation of the 2010.7-keV peak (as a cascade 1401 keV + 609 keV = 2010 keV). Thus, reliability of observation of weak ^{214}Bi lines can be estimated only from the peaks at 2021.8 keV and 2052.9 keV, whose intensity, according to the ^{214}Bi decay scheme, suffers only an insignificant (below 1%) decrease due to summation in the spectrum.

Analyzing Spectrum I we obtained the intensity of the γ2204-keV peak equal to 290(30) events, which agrees with the 319(22) events found in.[5,10] Consequently, the intensity of the γ2021.8-keV and γ2052.9-keV peaks in Spectrum I is expected to be 1.3 and 5.1 events respectively. According to our estimations, the average background in the region 2000–2060 keV of Spectrum I is $10\frac{\text{events}}{\text{keV}}$ or $0.18\frac{\text{events}}{\text{keV}\cdot\text{kg}\cdot\text{year}}$, which agrees with background measurements in.[5,10,19,21] Considering the resolving power of the detectors (FWHM $\cong$ 3.5 keV), the expected intensities of the 2021.8-keV and 2052.9-keV peaks should exceed the average background by approximately 4% and 15% respectively. This is why the statement of C.E. Aalseth et al.[7] and Yu.G. Zdesenko et al.[9] that weak ^{214}Bi peaks could not be observed in[3,4] should be thought of as somewhat premature, at least for the γ2052.9-keV peak. Estimates of expected intensities of peaks from the ^{214}Bi decay in Table 4 are not in conflict with the fact that the peaks at 2021.8 keV and 2052.9 keV and the peak at 2039 keV reported in[3–5] are really observed in the Heidelberg–Moscow experiment.

Row 5 of Table 4 presents intensities of weak ^{214}Bi peaks calculated in[5,10] with allowance for summation of pulses from the cascade γ-transitions (simulation of the γ-spectrum with the fitted location of ^{214}Bi nuclei in the vicinity of the detectors). As expected, the intensities of the γ2021.8-keV and γ2052.9-keV peaks did not change in comparison with the calculations (rows 3 and 4), while intensities of the γ2010.7-keV and γ2016.7-keV peaks appreciably increased. In[5,10] they arrive at the conclusion that the calculated expected intensities of the peaks agree with the experimental values within double the experimental error. It is noteworthy that the calculated intensities of the γ2021.8-keV and γ2052.9-keV peaks are a few times smaller than the experimental ones while the calculated and experimental intensities of the γ2010.7-keV and γ2016.7-keV peaks are close in value.

We have considered possibilities of the 2039-keV peak being formed in the spectrum as a result of detecting γ-rays from the decay of radioactive nuclides which may be present in the setup as natural impurity or may be produced in the structural materials as a result of interaction with background neutrons or muons. In our consideration we used the entire list of γ-rays emitted by radioactive nuclides[22] and the TI.[6] We did not find any possibility for that. To illustrate, let us consider two cases of radioactive nuclear decay giving rise to γ-rays of energy close to 2039.0 keV.

Among all isotopes identified in the background spectrum only ^{234m}Pa ($T_{1/2}$ = 1.17 min) emits γ-rays of energy close to 2039.0 keV (E_γ =

2041.2 keV, $I_\gamma = 1.1 \cdot 10^{-4}$ % of decays).[23] The background spectrum of the Heidelberg–Moscow experiment[19] shows 1000-keV γ-rays ascribed to the ^{234m}Pa decay. The intensity of this γ-transition, the strongest in the ^{234m}Pa decay, is 0.84 % of decays.[6] The area of the γ1000-keV peak in Spectrum I is 90(50) events. Thus, the area of the γ2041.2 peak in Spectrum I is estimated at 0.020 events. Clearly, these γ-rays cannot make a noticeable contribution to the 2039-keV peak because, according to Table 4, its intensity is 12.1(83) events.

2037.2-keV γ-rays arise from the β^--decay of ^{77}Ge ($T_{1/2} = 11.3$ h). This nuclide may be produced in the reaction ^{76}Ge$(n,\gamma)^{77}$Ge. Yet, the ^{77}Ge decay gives rise to 1999.6-keV γ-rays with an intensity ten times as high as that of 2037.2-keV γ-rays. However, 1999.6-keV γ-rays were not observed in.[3–5] Note that γ-rays resulting from the β^--decay of nuclides occurring in the material of the detectors do not form peaks because summation of pulses from β-particles with pulses from γ-rays smear out any lines.

4. Conclusion

Thus, our analysis of the spectra from the Heidelberg–Moscow experiment kindly provided by Prof. Klapdor-Kleingrothaus allows the following conclusions.

Firstly, the Heidelberg–Moscow experiment shows the best suppression of the background among low-background experiments. There is no experimental evidence for a contribution to the background spectrum from external sources (rocks in the background laboratory). The γ-rays observed in the background spectrum in the region up to 3200 keV arise from the decay of trace amounts of anthropogenic or cosmogenic nuclides contaminating the setup between the detectors and the main shielding.

Secondly, contrary to previous opinions, γ-rays of ^{226}Ra and daughter nuclides are shown to arise in the radium-226 rather than uranium-238 equilibrium chain.

Thirdly, the expected intensities of weak ^{214}Bi γ-lines in the spectral region of 2000–2060 keV determined by a direct comparison to strong Bi lines in the spectrum do not contradict the fact that these lines, and the 2039.0-keV line caused by detection of two β-particles from the $0\nu2\beta$-decay of ^{76}Ge,[3] are really observed in this experiment. The statistics gained after the year 2000 and the pulse-shape analysis of events in the 2039.0-keV peak confirm this conclusion (see, for example,[11–18]).

The results obtained may help both to confirm observation of the

2039-keV line from the $0\nu2\beta$-decay of ^{76}Ge in the Heidelberg–Moscow experiment and to design new low-background experiments.

The authors are grateful to Prof. Klapdor-Kleingrothaus for providing background spectra of the Heidelberg–Moscow experiment and for helpful discussions. The investigation was supported by the RFBR (Projects 04-02-17144 and 02-02-04009).

References

1. Schechter J., Valle J. W. F. *Phys. Rev. D* **25**, 2951 (1982).
2. Klapdor-Kleingrothaus H. V. et al. *Eur. Phys. J. A* **12**, 147 (2001).
3. Klapdor-Kleingrothaus H. V., Dietz A., Harney H. L., Krivosheina I. V. *Mod. Phys. Lett. A* **16**, 2409 (2001).
4. Klapdor-Kleingrothaus H. V., Dietz A., Krivosheina I. V. *Part. Nucl. Lett.* **110**, 57 (2002).
5. Klapdor-Kleingrothaus H. V., Dietz A., Krivosheina I. V. *Found. Phys.* **32**, 1181 (2002); Erratum-ibid. **33**, 679 (2003).
6. Firestone R.B., Shirley V.S.. *Table of Isotopes* (8th edition). J.Wiley and Sons, (1998).
7. Aalseth C. E. et al. *Mod. Phys. Lett. A* **17**, 1475 (2002).
8. Feruglio F., Strumia A., Vissani F. *Nucl. Phys. B* **637**, 345 (2002); *Addendum-ibid. B* **659**, 359 (2003).
9. Zdesenko Yu. G. et al., *Phys. Lett. B* **546**, 206 (2002).
10. Klapdor-Kleingrothaus H.V. *hep-ph/0205228.*
11. Klapdor-Kleingrothaus H.V. *hep-ph/0302237.*
12. Klapdor-Kleingrothaus H.V. *Int. J. Mod. Phys. A* **18**, 4113 (2003).
13. Klapdor-Kleingrothaus H.V., Dietz A., Krivosheina I. V., Dorr C., Tomei C. *Nucl. Instrum. Meth. A* **510**, 281 (2003).
14. Klapdor-Kleingrothaus H.V. et al., *Nucl. Instrum. Meth. A* **511**, 335 (2003).
15. Klapdor-Kleingrothaus H.V., Dietz A., Krivosheina I.V., Dorr C., Tomei C. *Phys. Lett. B* **578**, 54 (2004).
16. Klapdor-Kleingrothaus H.V., Dietz A., Krivosheina I.V. et al., *Nucl. Instrum. Meth. A* **522**, 371 (2004).
17. Klapdor-Kleingrothaus H.V., Krivosheina I.V. et al., *hep-ph/0404062.*
18. Klapdor-Kleingrothaus H.V., Krivosheina I. V., Dietz A. et al., *Phys. Lett. B* **586**, 198 (2004).
19. Dörr C., Klapdor-Kleingrothaus H.V. *Nucl. Instr. Meth.* **A513**, 596 (2003).
20. Gunther M. et al. *Phys. Rev. D* **55**, 54 (1997).
21. Bakalyarov A. M., Balysh A. Y., Belyaev S. T., Lebedev V. I., Zhukov S. V. *Part. Nucl. Lett.* **2, Nr. 2**, 21 (2005).
22. Reus U., Westmeier W. Acompilation of gamma-rays origінaning from radioactive decay. *Inst. Nucl. Chemistry*, University of Marburg, Germany.
23. Brudanin V. B. et al. *Part. Nucl. Lett.* **1, Nr. 5**, 84 (2004).

DARK MATTER'S X-FILES

ALEXANDER KUSENKO

Department of Physics and Astronomy, University of California, Los Angeles, CA 90095-1547, USA

Sterile neutrinos with keV masses can constitute all or part of the cosmological dark matter. The electroweak-singlet fermions, which are usually introduced to explain the masses of active neutrinos, need not be heavier than the electroweak scale; if one of them has a keV-scale mass, it can be the dark-matter particle, and it can also explain the observed pulsar kicks. The relic sterile neutrinos can be produced by several different mechanisms. If they originate primarily from the Higgs decays at temperatures of the order of 100 GeV, the resulting dark matter is much "colder" than the warm dark matter produced in neutrino oscillations. The signature of this form of dark matter is the spectral line from the two-body decay, which can be detected by the X-ray telescopes. The same X-rays can have other observable manifestations, in particular, though their effects on the formation of the first stars.

1. Neutrino Masses and the Emergence of Sterile Neutrinos

Most discoveries in particle physics amount to either a measurement of some parameter related to a known particle, or a detection of some new degrees of freedom, new particles. The discovery of the neutrino mass[1] is both. Not only is it a measurement of the non-zero mass, but it also implies the existence of some additional, SU(2) singlet fermions, "right-handed" neutrinos. The corresponding particles can be made very heavy even for small masses of the active neutrinos (the seesaw mechanism[2]), but they can also be light, in which case they are called sterile neutrinos. The name *sterile neutrino* was coined by Bruno Pontecorvo, who hypothesized the existence of the right-handed neutrinos in a seminal paper,[3] in which he also considered vacuum neutrino oscillations in the laboratory and in astrophysics, the lepton number violation, the neutrinoless double beta decay, some rare processes, such as $\mu \to e\gamma$, and several other questions that have dominated the neutrino physics for the next four decades. Most models of the neutrino masses introduce sterile (or right-handed) neutrinos to generate the masses of the ordinary neutrinos via the seesaw mechanism.[2] The seesaw lagrangian

$$\mathcal{L} = \mathcal{L}_{\rm SM} + \bar{N}_a \left(i\gamma^\mu \partial_\mu\right) N_a - y_{\alpha a} H \, \bar{L}_\alpha N_a - \frac{M_a}{2} \, \bar{N}_a^c N_a + h.c. \, , \qquad (1)$$

where $\mathcal{L}_{\rm SM}$ is the lagrangian of the Standard Model, includes some number n of singlet neutrinos N_a ($a = 1, ..., n$) with Yukawa couplings $y_{\alpha a}$. Here H is the Higgs doublet and L_α ($\alpha = e, \mu, \tau$) are the lepton doublets. Theoretical considerations do not constrain the number n of sterile neutrinos. In particular, there is no constraint based on the anomaly cancellation because the sterile fermions do not couple to the gauge fields. The experimental limits exist only for the larger mixing angles.[4] To explain the neutrino masses inferred from the atmospheric and solar neutrino experiments, $n = 2$ singlets are sufficient,[5] but a greater number is required if the lagrangian (1) is to explain the r-process nucleosynthesis,[6] the pulsar kicks[7,8] and the strength of the supernova explosion,[9,10] as well as dark matter.[11–15] The same particle can play an important role in the formation of the first stars[18] and other astrophysical phenomena.[19]

The scale of the right-handed Majorana masses M_a is unknown; it can be much greater than the electroweak scale,[2] or it may be as low as a few eV.[16] The seesaw mechanism[2] can explain the smallness of the neutrino masses in the presence of the Yukawa couplings of order one if the Majorana masses M_a are much larger than the electroweak scale. Indeed, in this case the masses of the lightest neutrinos are suppressed by the ratios $\langle H \rangle / M_a$. However, the origin of the Yukawa couplings remains unknown, and there is no experimental evidence to suggest that these couplings must be of order 1. In fact, the Yukawa couplings of the charged leptons are much smaller than 1. For example, the Yukawa coupling of the electron is as small as (10^{-6}).

The Majorana mass can arise from the Higgs mechanism.[17] For example, let us consider the following modification of the lagrangian (1):

$$\mathcal{L} = \mathcal{L}_0 + \bar{N}_a \left(i\gamma^\mu \partial_\mu\right) N_a - y_{\alpha a} H \, \bar{L}_\alpha N_a - \frac{h_a}{2} \, S \, \bar{N}_a^c N_a + V(H, S) + h.c. \, , \quad (2)$$

where $\mathcal{L}_0$ includes the gauge and kinetic terms of the Standard Model, H is the Higgs doublet, S is the real boson which is SU(2)-singlet, L_α ($\alpha = e, \mu, \tau$) are the lepton doublets, and N_a ($a = 1, ..., n$) are the additional singlet neutrinos. After the symmetry breaking, the Higgs doublet and singlet fields each develop a VEV, $\langle H \rangle = v_0 = 247$ GeV, $\langle S \rangle = v_1$, and the singlet neutrinos acquire the Majorana masses $M_a = h_a v_1$. As discussed below, this model is suitable for generating dark matter in the form of sterile neutrinos.

2. Dark Matter in the Form of Sterile Neutrinos

Sterile neutrino is a dark matter candidate. Since the singlet fermions are introduced anyway to explain the observed neutrino masses, one can ask whether the same particles can be the dark matter. Because of the small Yukawa couplings, the keV sterile neutrinos are out of equilibrium at high temperatures. However, there are several ways in which the relic population of sterile neutrinos can be produced.

- Sterile neutrinos can be produced from neutrino oscillations as was proposed by Dodelson and Widrow (DW).[11] If the lepton asymmetry is negligible, this scenario appears to be in conflict with a combination of the X-ray bounds[21] and the Lyman-α bounds,[22,23] although it is possible to evade this constraint if the lepton asymmetry of the universe is greater than (10^{-3}).[13] On the other hand, observations of dwarf spheroids point to a non-negligible free-streaming length for dark matter,[24] which favors warm dark matter. It is also possible that the sterile neutrinos make up only a fraction of dark matter,[23] in which case they can still be responsible for the observed velocities of pulsars.[7,20]
- The bulk of sterile neutrinos could be produced from decays of S bosons at temperatures above the S boson mass, $T \sim 100$ GeV.[20] In this case, the Lyman-α bounds on the sterile neutrino mass are considerably weaker than in the DW case because the momenta of the sterile neutrinos are red-shifted as the universe cools down from $T \sim 100$ GeV.
- Sterile neutrinos can be produced from their coupling to the inflaton,[25] or the radion.[26]

It is important to note that only in the first case, the DW scenario, the dark matter abundance is directly related to the mixing angle. In contrast, if the relic population of sterile neutrinos arises from the Higgs decays, their abundance is determined by the coupling h in eq. (2), while the mixing angle is controlled by a different coupling y.

We also note that both models, with largangians (1) and (2), allow for some production of sterile neutrinos from oscillations, but in the case of the singlet Higgs decays (2) the bulk of the sterile dark matter could be produced at $T \sim 100$ GeV, regardless of the value of the mixing angle, which can be vanishingly small.

Indeed, if the couplings of S to H are large enough, while $h < 10^{-6}$, the S boson can be in equilibrium at temperatures above its mass, while the sterile neutrino with a small mixing angle can be out of equilibrium at all times. Since S is in thermal equilibrium at high temperatures, some amount of sterile neutrinos can be produced in decays $S \to NN$:

$$\Omega_{\nu_s} = 0.2 \left(\frac{33}{\xi}\right) \left(\frac{h}{1.4 \times 10^{-8}}\right)^3 \left(\frac{\langle S \rangle}{\tilde{m}_S}\right), \tag{3}$$

where ξ is the change in the number density of sterile neutrinos relative to T^3 due to the dilution taking place as the universe cools. For example, in the Standard Model, the reduction in the number of effective degrees of freedom that occurs during the cooling from the temperature $T \sim 100$ GeV to a temperature 1 keV causes the entropy increase and the dilution of any species out of equilibrium by factor $\xi \approx 33$.

At the same time, the sterile neutrino mass is determined by the VEV of S:

$$h\langle S \rangle \sim \text{keV} \quad \Longrightarrow \quad \langle S \rangle \sim \frac{\text{keV}}{h} \sim 10^2 \text{GeV} \tag{4}$$

Based on the required values of Ω_s and the mass, we conclude that the Higgs singlet should have a VEV at the electroweak scale. The dark matter abundance, as in eq. (3), was first computed for a model in which the S field served as the inflaton with a potential adjusted to have $\langle S \rangle \gg m_S$, and much smaller values of h, ξ were considered.[25] The alternative possibility,[20] $\langle S \rangle \sim m_S$, which some may find more natural, places the singlet Higgs right at the elecroweak scale, which has important implications for the LHC.[27]

3. X-ray Detection of Relic Sterile Neutrinos

The relic sterile neutrinos can decay into the lighter neutrinos and an the X-ray photons,[28] which can be detected by the X-ray telescopes.[21] The X-ray flux depends on the sterile neutrino abundance. If all the dark matter is made up of sterile neutrinos ($\Omega_s \approx 0.2$), then the limit on the mass and the mixing angle is given by the dashed line in Fig. 1. However, the interactions in the lagrangian (1) cannot produce such an $\Omega_s = 0.2$ population of the sterile neutrinos for the masses and mixing angles along this dashed line, unless the universe has a relatively large lepton asymmetry.[13] If the lepton asymmetry is small, the interactions in eq. (1) can produce the relic sterile neutrinos via the neutrino oscillations off-resonance at some sub-GeV temperature.[11] This mechanism provides the lowest possible abundance (except for the low-temperature cosmologies, in which the universe is never reheated above a few MeV after inflation[29]). The model-independent bound[20,23] based on this scenario is shown as a solid (purple) region in Fig. 1. It is based on the flux limit from X-ray observations[21] and the state-of-the-art calculation of the sterile neutrino production by oscillations.[30]

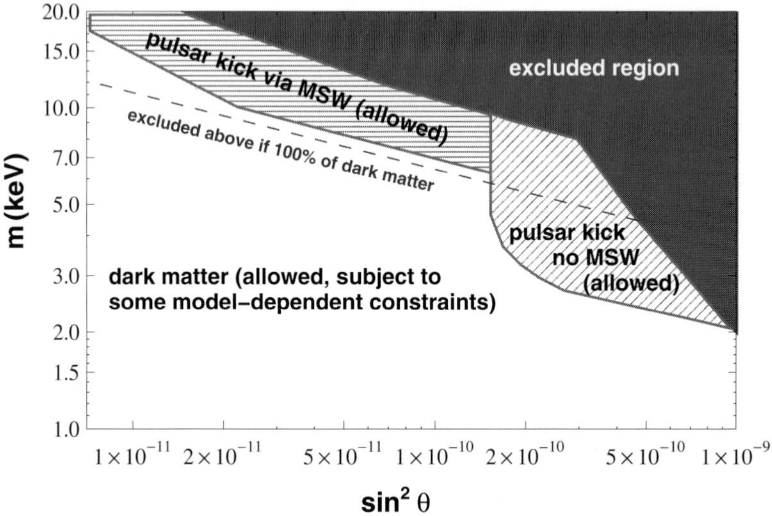

Fig. 1. The solid excluded region is based on a combination of the X-ray and the small-scale structure bounds;[23] it applies even if sterile neutrinos constitute only a fraction of dark matter. The dashed line shows the X-ray bound under the assumption that sterile neutrinos make up all the dark matter. Additional bounds from structure formation may apply, depending on the free-streaming length, whose relation with the particle mass depends on the production scenario. The region for the pulsar kicks shown here is based on the re-analyzes of the earlier results,[7] which will be reported in an upcoming paper.

4. X-rays and the Formation of the First Stars

The X-ray photons from sterile neutrino decays in the early universe could have affected the star formation. Although these X-rays alone are not sufficient to reionize the universe, they can catalyze the production of molecular hydrogen and speed up the star formation,[18] which, in turn, would cause the reionization. Molecular hydrogen is a very important cooling agent, necessary for the collapse of primordial gas clouds that gave birth to the first stars. The fraction of molecular hydrogen must exceed a certain minimal value for the star formation to begin.[31] The reaction $\mathrm{H}+\mathrm{H}\rightarrow\mathrm{H}_2+\gamma$ is very slow in comparison with the combination of reactions

$$\mathrm{H}^+ + \mathrm{H} \rightarrow \mathrm{H}_2^+ + \gamma, \tag{5}$$

$$\mathrm{H}_2^+ + \mathrm{H} \rightarrow \mathrm{H}_2 + \mathrm{H}^+, \tag{6}$$

which become possible if the hydrogen is ionized. Therefore, the ionization fraction determines the rate of molecular hydrogen production. If dark matter is made up of sterile neutrinos, their decays produce a sufficient flux of photons to increase the ionization fraction by as much as two orders of magnitude.[18] This has a dramatic effect on the production of molecular hydrogen and the subsequent star formation.

Decays of the relic sterile neutrinos during the dark ages could produce an observable signature in the 21-cm background.[32] It can be detected and studied by such instruments as the Low Frequency Array (LOFAR), the 21 Centimeter Array (21CMA), the Mileura Wide-field Array (MWA) and the Square Kilometer Array (SKA).

4.1. *Sterile Neutrinos and the Supernova*

Sterile neutrinos with masses below several MeV can be produced in the supernova explosion; they can play an important role in the nucleosynthesis,[6] as well as in generating the supernova asymmetries and the pulsar kicks.[7] Since the sterile neutrinos interact with nuclear matter very weakly, they can be very efficient at transporting the heat in the cooling protoneutron star, altering the dynamics of the supernova.[9] This could lead to an enhancement of the supernova explosion. An additional enhancement can come from the increase in convection in front of the neutron star propelled by the asymmetric emission of sterile neutrinos.[10]

4.1.1. *The Pulsar Kicks*

The observations of neutrinos from SN1987A constrain the amount of energy that the sterile neutrinos can take out of the supernova, but they are still consistent with the sterile neutrinos that carry away as much as a half of the total energy of the supernova. A more detailed analysis shows that the emission of sterile neutrinos from a cooling newly born neutron star is anisotropic due to the star's magnetic field.[7] The anisotropy of this emission can result in a recoil velocity of the neutron star as high as $\sim 10^3$km/s. While both the active and the sterile neutrinos are produced with some anisotropy, the asymmetry in the amplitudes of active neutrinos is quickly washed out in multiple scatterings as these neutrinos diffuse out of the star in the approximate thermal equilibrium.[8] In contrast, the sterile neutrinos are emitted from the supernova with the asymmetry equal to their production asymmetry. Hence, they give the recoiling neutron star a momentum, large enough to explain the pulsar kicks for the neutrino emission anisotropy as small as a few per cent.[7] This mechanism can be the explanation of the observed pulsar velocities. The range of masses and mixing angles required to explain the pulsar kicks is shown in Fig. 1.

The pulsar kick mechanism based on the sterile neutrino emission has several additional predictions:[7]

- the kick velocities were predicted to correlate with the axis of rotation;[7] recently, this spin-kick correlation was confirmed by the observations[34]
- the kick should last 10 to 15 seconds, while the protoneutron star is cooling by the emission of neutrinos, but the onset of the kick can be delayed by a few seconds, depending on the mass and mixing angles;[7] this delayed kick can be tested using the observational data[35]
- neutrino-driven kicks can deposit additional energy behind the supernova shock,[9,10] and they are expected to produce asymmetric jets with the stronger jet pointing *in the same direction* as the neutron star velocity.[10]

5. Conclusions

The fundamental physics responsible for the neutrino masses is likely to involve some additional SU(2)-singlet fermions, or sterile neutrinos. The Majorana masses of these states can range from a few eV to some values well above the electroweak scale. A sterile neutrino with a keV mass is a viable dark matter candidate. One can discover the relic sterile neutrinos using the X-ray observations. The same neutrinos can be produced in the supernova explosions, and the anisotropy in their emission can explain the observed pulsar velocities. The X-rays from the sterile neutrino decays can play an important role in the production of molecular hydrogen, which is necessary for the formation of the first stars.

This work was supported in part by the DOE grant DE-FG03-91ER40662 and by the NASA ATP grants NAG 5-10842 and NAG 5-13399. The author appreciates the hospitality of Aspen Center for Physics, where part of this work was done.

References

1. For a review, see, *e. g.*, A. Strumia and F. Vissani, arXiv:hep-ph/0606054.
2. P. Minkowski, Phys. lett. **B67** , 421 (1977); M. Gell-Mann, P. Ramond, and R. Slansky, *Supergravity* (P. van Nieuwenhuizen et al. eds.), North Holland, Amsterdam, 1980, p. 315; T. Yanagida, in *Proceedings of the Workshop on the Unified Theory and the Baryon Number in the Universe* (O. Sawada

and A. Sugamoto, eds.), KEK, Tsukuba, Japan, 1979, p. 95; S. L. Glashow, *The future of elementary particle physics*, in *Proceedings of the 1979 Cargèse Summer Institute on Quarks and Leptons* (M. Lévy et al. eds.), Plenum Press, New York, 1980, pp. 687; R. N. Mohapatra and G. Senjanović, Phys. Rev. Lett. **44**, 912 (1980).

3. B. Pontecorvo, JETP, 53, 1717 (1967).
4. A. Kusenko, S. Pascoli and D. Semikoz, JHEP **0511**, 028 (2005). S. Choubey, N. P. Harries and G. G. Ross, Phys. Rev. D **74**, 053010 (2006); Phys. Rev. D **76**, 073013 (2007).
5. P. H. Frampton, S. L. Glashow, T. Yanagida, Phys. Lett. B **548**, 119 (2002).
6. G. C. McLaughlin et al., Phys. Rev. C **59**, 2873 (1999); D. O. Caldwell, G. M. Fuller and Y. Z. Qian, Phys. Rev. D **61**, 123005 (2000); J. Fetter et al., Astropart. Phys. **18**, 433 (2003).
7. A. Kusenko and G. Segrè, Phys. Lett. B **396**, 197 (1997); A. Kusenko and G. Segrè, Phys. Rev. D **59**, 061302 (1999). G. M. Fuller, A. Kusenko, I. Mocioiu, and S. Pascoli, Phys. Rev. D **68**, 103002 (2003); M. Barkovich, J. C. D'Olivo and R. Montemayor, Phys. Rev. D **70**, 043005 (2004); A. Kusenko, Int. J. Mod. Phys. D **13**, 2065 (2004). L. C. Loveridge, Phys. Rev. D **69**, 024008 (2004).
8. A. Kusenko, G. Segrè and A. Vilenkin, Phys. Lett. B **437**, 359 (1998).
9. J. Hidaka and G. M. Fuller, Phys. Rev. D **74**, 125015 (2006).
10. C. L. Fryer and A. Kusenko, Astrophys. J. Suppl. **163**, 335 (2006).
11. S. Dodelson and L. M. Widrow, Phys. Rev. Lett. **72**, 17 (1994).
12. K. Abazajian, G. M. Fuller and M. Patel, Phys. Rev. D **64**, 023501 (2001); A. D. Dolgov and S. H. Hansen, Astropart. Phys. **16**, 339 (2002).
13. X. d. Shi and G. M. Fuller, Phys. Rev. Lett. **82**, 2832 (1999). C. T. Kishimoto, G. M. Fuller and C. J. Smith, arXiv:astro-ph/0607403.
14. A. Y. Smirnov and R. Zukanovich Funchal, Phys. Rev. D **74**, 013001 (2006).
15. T. Asaka, S. Blanchet and M. Shaposhnikov, Phys. Lett. B **631**, 151 (2005); T. Asaka, M. Shaposhnikov and A. Kusenko, Phys. Lett. B **638**, 401 (2006); W. l. Guo, arXiv:0709.1632 [hep-ph].
16. A. de Gouvea, Phys. Rev. D **72**, 033005 (2005). A. de Gouvea, J. Jenkins and N. Vasudevan, arXiv:hep-ph/0608147.
17. Y. Chikashige, G. Gelmini, R. D. Peccei and M. Roncadelli, Phys. Lett. B **94**, 499 (1980). Y. Chikashige, R. N. Mohapatra and R. D. Peccei, Phys. Lett. B **98**, 265 (1981).
18. P. L. Biermann and A. Kusenko, Phys. Rev. Lett. **96**, 091301 (2006); M. Mapelli, A. Ferrara and E. Pierpaoli, Mon. Not. Roy. Astron. Soc. **369**, 1719 (2006); J. Stasielak, P. L. Biermann and A. Kusenko, Astrophys. J. **654**, 290 (2007); E. Ripamonti, M. Mapelli and A. Ferrara, Mon. Not. Roy. Astron. Soc. **375**, 1399 (2007); J. Stasielak, P. L. Biermann, A. Kusenko, arXiv:astro-ph/0701585; arXiv:0710.5431 [astro-ph].
19. F. Munyaneza, P.L. Biermann, P. L., Astron and Astrophys., 436, 805 (2005)
20. A. Kusenko, Phys. Rev. Lett. **97**, 241301 (2006).
21. K. Abazajian, G. M. Fuller and W. H. Tucker, Astrophys. J. **562**, 593 (2001); A. Boyarsky et al., arXiv:astro-ph/0512509; A. Boyarsky et al., JETP

Lett. **83**, 133 (2006); A. Boyarsky et al., arXiv:astro-ph/0603368; A. Boyarsky et al., arXiv:astro-ph/0603660; S. Riemer-Sorensen, S. H. Hansen and K. Pedersen, arXiv:astro-ph/0603661; K. Abazajian and S. M. Koushiappas, arXiv:astro-ph/0605271; C. R. Watson, J. F. Beacom, H. Yuksel and T. P. Walker, arXiv:astro-ph/0605424; K. N. Abazajian et al., Phys. Rev. D **75**, 063511 (2007); A. Boyarsky, J. Nevalainen and O. Ruchayskiy, Astron. Astrophys. **471**, 51 (2007); A. Boyarsky, O. Ruchayskiy and M. Markevitch, arXiv:astro-ph/0611168; S. Riemer-Sorensen et al., Phys. Rev. D **76**, 043524 (2007); A. Boyarsky et al., Astropart. Phys. **28**, 303 (2007); H. Yuksel, J. F. Beacom and C. R. Watson, arXiv:0706.4084 [astro-ph]; A. Boyarsky et al., arXiv:0709.2301 [astro-ph].

22. M. Viel, et al., Phys. Rev. D **71**, 063534 (2005); U. Seljak, A. Makarov, P. McDonald and H. Trac, arXiv:astro-ph/0602430; M. Viel et al., Phys. Rev. Lett. **97**, 071301 (2006); M. Viel et al., W. L. W. Sargent, arXiv:0709.0131 [astro-ph].
23. A. Palazzo, D. Cumberbatch, A. Slosar et al., arXiv:0707.1495 [astro-ph].
24. X. Hernandez, G. Gilmore, *MNRAS* 297, 517 (1998); J. Sommer-Larsen, A. D. Dolgov, *Astrophys. J.* **551**, 608 (2001); F. Governato *et al.*, *Astrophys. J.* **607**, 688 (2004); M. Fellhauer *et al.*, *Astrophys. J.* **651**, 167 (2006); B. Allgood *et al.*, *MNRAS* **367**, 1781 (2006); T. Goerdt et al., *ibid.*, **368**, 1073 (2006); G. Gilmore et al., *Astrophys. J.*, **663**, 948 (2007); L. E. Strigari et al., P. Madau, arXiv:0704.1817 [astro-ph]; R. F. G. Wyse, G. Gilmore, arXiv:0708.1492 [astro-ph]; D. Boyanovsky, H. J. de Vega and N. Sanchez, arXiv:0710.5180 [astro-ph]; D. Boyanovsky, arXiv:0711.0470 [astro-ph].
25. M. Shaposhnikov and I. Tkachev, Phys. Lett. B **639**, 414 (2006).
26. K. Kadota, arXiv:0711.1570 [hep-ph].
27. T. Binoth and J. J. van der Bij, Z. Phys. C **75**, 17 (1997); A. Datta et al., Z. Phys. C **72**, 449 (1996). H. Davoudiasl, T. Han and H. E. Logan, Phys. Rev. D **71**, 115007 (2005). M. L. Graesser, arXiv:0705.2190 [hep-ph]; S. Profumo, M. J. Ramsey-Musolf and G. Shaughnessy, JHEP **0708**, 010 (2007); V. Barger et al., arXiv:0706.4311 [hep-ph].
28. P. B. Pal and L. Wolfenstein, Phys. Rev. D **25**, 766 (1982).
29. G. Gelmini et al., Phys. Rev. Lett. **93**, 081302 (2004).
30. K. Abazajian, Phys. Rev. D **73**, 063506 (2006); T. Asaka, M. Laine and M. Shaposhnikov, JHEP **0606**, 053 (2006); JHEP **0701**, 091 (2007); D. Boyanovsky and C. M. Ho, JHEP **0707**, 030 (2007); Phys. Rev. D **76**, 085011 (2007); D. Boyanovsky, Phys. Rev. D **76**, 103514 (2007).
31. M. Tegmark et al., Astrophys. J. **474**, 1 (1997).
32. M. Valdes et al., Mon. Not. Roy. Astron. Soc. **377**, 245 (2007).
33. C. Y. Ng and R. W. Romani, Astrophys. J., **660**, 1357 (2007).
34. D. J. Helfand, E. V. Gotthelf and J. P. Halpern, Astrophys. J., **556**, 380 (2001); S. Johnston et al., Mon. Not. Roy. Astron. Soc. **364**, 1397 (2005); S. Johnston et al., arXiv:0708.4251 [astro-ph]. O. Kargaltsev, G. G. Pavlov and G. P. Garmire, Astrophys. J., **660**, 1413 (2007).
35. H. Spruit and E. S. Phinney, Nature **393**, 139 (1998); C. Y. Ng and R. W. Romani, Astrophys. J., **660**, 1357 (2007).

SN 1987 A — 20th ANNIVERSARY OF REGISTRATION OF THE NEUTRINO SIGNAL WITH BAKSAN, KAMIOKANDE II and IMB DETECTORS

IRINA V. KRIVOSHEINA

Radiophysical-Research Institute, Nishnij-Novgorod, Russia,
E-mail: irinaKV@web.de, http : //klapdor − k.de

The detection of neutrinos from SN 1987A opened a new era in neutrino astrophysics in the last century. We present a historical view about registration of the neutrino signal from supernova SN 1987A in the Large Magellanic Cloud by the BAKSAN liquid scintillator detector and by the two water Cherenkov detectors - Kamiokande-II and IMB. All three detectors together observed a total neutrino signal of 24 events at 7:35 UT February 23, 1987. I will concentrate mostly on the BAKSAN supernova group analysis of the neutrino signal, which was done already in the years 1987-1988. The results of this analysis (determination of average properties of the neutrino signal: the total energy of neutrino emission, the effective neutrino temperature, the total luminosity of the neutrino signal, duration of the neutrino burst) are presented. The common analysis of all three detectors shows, that these 'parameters' have good agreement with the general theoretical description of explosions of supernovae. The analysis shows that including the Baksan data is very important for the understanding of the SN87A event. The latest results of 20 years of observation of our Galaxy by the BAKSAN scintillation telescope show, that the upper limit of the mean frequency of gravitational collapses is $<0.13\,\mathrm{yr}^{-1}$ at a 90% confidence level.

1. Introduction

Exactly 10 years passed after explosion of supernova SN 1987A in the Large Magellanic Cloud (LMC), and more than 70 years ago the neutrino, as particle, was introduced in our life. Now we well know, that neutrinos play a very important role in supernova explosions, in $0\nu\beta\beta$ decay and other physical processes.

How neutrino astronomy was prepared for such a millennium event?

From the theoretical side, one of the pioneer works was done by some Russian colleagues.[26–28] From the article[28] we have learnt, that the duration of the neutrino signal should be around 20 sec and have learnt about the form of the spectra of ν_e, ν_μ, ν_τ which has to be expected after explosion of a supernova II. Another fundamental, heavy theoretical work with a calculation of the neutrino transport in different star models was performed in the eighties and nineties, (see[30]), and later the leadership was taken by a group from München.[31] In the beginning of the eighties the 'r-processes' in stars were better understood on the basis of theoretical work for matrix elements of beta decay of nuclei far from stability.[13–16] One can conclude, that the theory was more or less ready.

What about experiments? At that time (time of explosion of SN 1987A) the neutrino community operated with five detectors. Let us call them "second generation of collapse-neutrino detectors". The detectors were located in different countries (different continents and all of them in the northern hemispere of our Earth). They were operated with different targets and used different fiducial masses (see Table 1).

Table 1. Important technical parameters of the main neutrino telescopes (BAKSAN, KII, IMB, LSD), running at the time of explosion of the SN87A.

Detector	Fiducial mass (t) of target	Energy Thresh. (MeV)	Backgr. rate (sec^{-1})	Number of events	Duration of signal (sec.)	Time of the signal (UT)
			(in the time of explosion of SN 1987A)			
BAKSAN	200 of C_9H_{20}	10	0.034	5	9.1	7:35
KII	2140 of H_2O	8.5	0.022	11	12.5	7:35
IMB	5000 of H_2O	35	0.077	8	6.0	7:35
LSD	90 of C_10H_{22}	5.5	0.012	5	7.0	2:52

(1) BAKSAN Neutrino Scintillation Telescope (BAKSAN)[1–4,8] (Baksan valley, Andyrchi Mountain (near Mount Elbrus), Northern Caucasus, Russia), target - 330 tons of liquid scintillator (Fig. 1).
(2) Kamiokande II (KII)[23] (Kamioka valley, Central Japan, 300 km west of Tokyo), a water Cherenkov detector using 3000 m^3 of clean water (Fig.1).
(3) Irvine-Michigan-Brookhaven (IMB)[24] (Morton-Thiokal mine, near Cleveland, USA), a water Cherenkov detector using 7000 tons (full volume $17\times18\times22.5\,m^3$) of water (Fig. 1).
(4) LSD (Large Scintillation Detector)[25] (Mont Blanc, Italy-France), scintillator detector with active mass of 90 tons.
(5) Artemovsk scintillator detector (the total mass was 105 tons), in Artemovsk mine (Ukraine) (did not operate at time of explosion of SN 1987A).

So-called "first generation of neutrino detectors" were constructed in the USA in 1973, but unfortunately all three Cherenkov detectors ran only some years. No burst candidates (with one not confirmed exclusion) were

Fig. 1. View of The BAKSAN Underground Scintillation Neutrino Telescope (right - upper horizontal plane, left - vertical plane), location - valley Andyrchi, Caucasus, Russia. Bottom: Left: KAMIOKANDE II underground neutrino telescope, location - Kamioka mine, Japan. Right: IMB neutrino detector, Morton-Thiokal mine, near Cleveland, USA.

found. But anyway it was a *first experience* to use such type of detectors in the future.

One could see, that experimentalists had a good chance to be able to detect a neutrino-antineutrino signal from explosion of a supernova II in our Galaxy, or near to it.

It should also be noted that in the last 20 years, after the SN 1987A, collaborations like Kamiokande, SNO, MACRO constructed a new scale (Megatons of target) of experiments, with the plan to also look for the neutrino signal from a supernova in our Galaxy or outside of the Galaxy. Not all of them survived until today, like MACRO - it was a nice 'tracking' detector (Italy, Gran Sasso), or (partly) SuperKamiokande (Japan), which reduced the sensitivity (the number of the photomultiplier tubes), after a heavy accident in the laboratory. At present the neutrino community *has only three running experiments*: 1 - SuperKamiokande (Japan), 2 - the LVD detector in Italy with smaller size than originally proposed by the collaboration, and 3 - the BAKSAN scintillation telescope (Russia).

What kind of criteria should be fulfilled by an experiment searching for supernova explosion?

(1) A "Supernova" detector should be able to detect the $\nu, \bar{\nu}$, as well as all types of neutrinos ν_e, ν_μ, ν_τ. **(2)** The time distribution of the events (neutrinos-antineutrinos) and perfect *absolute* time. **(3)** The energy of each event. **(4)** All detectors *must* have a 'low' energy threshold (not bigger than 10 MeV). **(5)** The detector must be located in a deep underground laboratory, to reduce the background from the muons. **(6)** The detector must have a big fiducial mass, to be able to see a neutrino signal from our Galaxy or from the Universe with sufficiently good statistics. **(7)** The detector must run permanently and stable before and after a SN signal.

2. Detectors

2.1. *BAKSAN Neutrino Scintillation Telescope*

The BAKSAN neutrino scintillation telescope was proposed in 1964-1965, and as site was chosen the Andyrchi valley, Caucasus, Russia near the highest mountain of Europe - peak Elbrus, in a depth of 850 m.w.e. (meter water equivalent). From the year 1979 the BAKSAN Neutrino Telescope is running with various programs: search for the magnetic monopole, for the fluxes of muons, for high energy neutrino flux from extragalactic sources, and also our program was included in the scientific plans (in autumn 1979) under the title: "Search for medium energies of the neutrinos after explosion of supernova in our Galaxy".

The BAKSAN neutrino scintillation telescope is based on an organic (oil-based) liquid scintillator C_nH_{2n+2}, which consists of hydrogen (H) and carbon (C). The high concentration of hydrogen (n∼9) allowed us to be mostly efficient to detect the electron antineutrino, via the charged current reaction on protons $\tilde{\nu}_e + p \longrightarrow n + e^+$ (1).

The cross section is $\sigma \sim E_{\tilde{\nu}}^2$. The positron gives additional visible energy in the scintillator (∼1 MeV) because of the detection of 2 gammas from e^+e^- annihilation. The neutron which is produced in the reaction (1) could also be detected by radiative captures: $n + p \longrightarrow D^* \longrightarrow D + \gamma, E_\gamma \sim 2.2\,MeV$, the average time of the interaction being τ_c=170 μs. Electron-neutrinos with $E_{\nu_e} >\sim 30$ MeV can be detected through their capture by carbon $\nu_e +^{12} C \longrightarrow^{12} N + e^-$.

2.1.1. Some technical features

Figure 2 shows *a schematical view* of the BAKSAN underground scintillation Neutrino Telescope. The total mass of liquid scintillator is 330 tonnes, which is constructed in 3150 standard counters (detectors) (see Fig. 2). The 3150 counters are constructed in eight planes - four horizontal and four ver-

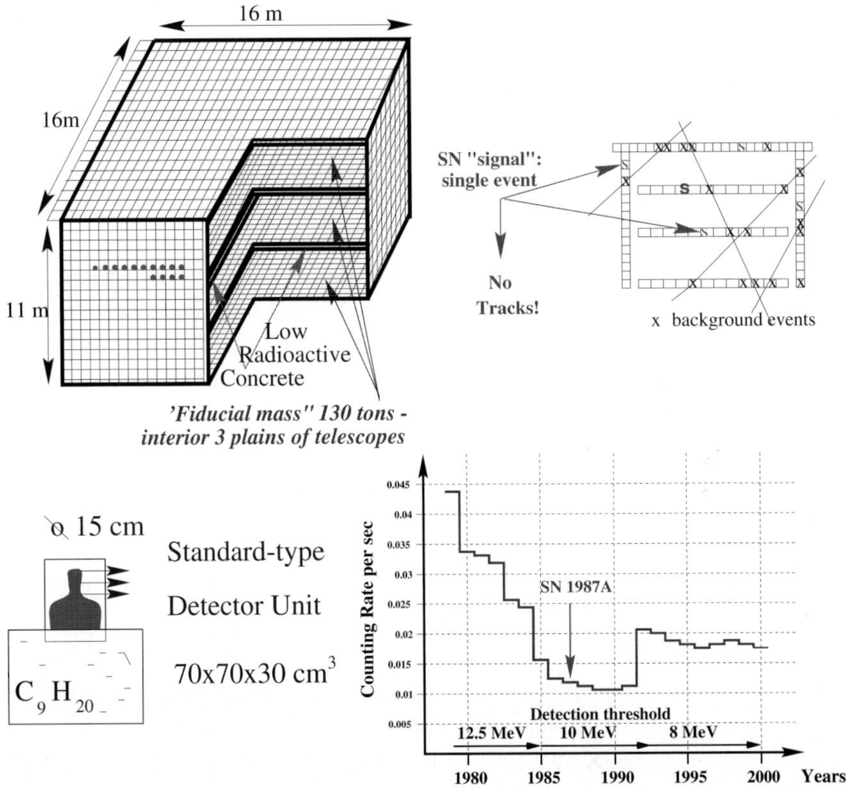

Fig. 2. Upper part: Left: A schematical view of the BAKSAN underground scintillation Neutrino Telescope, location - Baksan valley, Caucasus, Russia. 3150 detectors have been installed, the telescope has an extra external 'Low Background' shielding using 'special' low background concrete, as outer (in the rock) and inner shield. Right: A schematic view of Supernova signals ('s' - single tracks) and background ('x' - normal tracks - muons, which pass a telescope) for the BAKSAN Neutrino Telescope. Bottom part: Left: A schematic view of standard detector units used in the BAKSAN Neutrino Telescope. The size of an individual detector is $(70 \times 70 \times 30)\,\mathrm{cm}^3$. The diameter of the photomultiplier (on the upper part of each detector) is 15 cm. The threshold of each detector was 10 MeV in the time of the SN87A explosion. Right: The counting rate of single events of the 1200 detectors in the three internal planes of the telescope (the target mass is 130 t) during the period 1980 - 2000. The improvement of the threshold with time is shown. The arrow indicates the time of explosion of SN 1987A on February 23, 1987 (from[32]).

tical. The telescope has three internal planes, with the size $(14\times14)\ \mathrm{m}^2$, and each consists of 400 standard modules (see Fig. 2). The upper plane (see also Fig. 1) has a size of $(17\times17)\,\mathrm{m}^2$, and consists of 576 modules. The four vertical planes consist of 330×2 counters (two planes) and another two planes - 360×2 counters.

In Fig. 2 (bottom left) is shown *a standard detector* from the BAKSAN Neutrino Telescope which has a size of $(70 \times 70 \times 30)\,\mathrm{cm}^3$, and is filled with the liquid scintillator C_nH_{2n+2} ($n \sim 9$). The detector is viewed by one standard photomultiplier (F-49B), which has a photocathode with diameter of 15 cm. *The energy thresholds of the individual detectors* in the time of explosion of supernova SN87A were 8 MeV and 10 MeV for the horizontal and for the vertical planes, respectively. *The energy calibration* was performed using the amplitude of the most probable energy (50 MeV) release when cosmic-ray muons pass through the telescope. *The detector efficiencies* were ~8% for 8 MeV electrons, and ~ 50% for 10 MeV electrons and ~ 80% for electron energies >12 MeV. *The counting rate* of the *background* of the telescope was $\sim 0.034\,\mathrm{sec}^{-1}$, for a fiducial mass of 200 tons and for events with energies < 50 MeV. *Background*: The main source of the background are cosmic ray muons crossing the telescope (see Fig. 2- upper right), and discharges in the photomultiplier tubes.

It should be noted, that cosmic ray muons crossing the telescope produce usually tracks passing several planes. In case of the 'collapse' program we used a 'special' events selection, which we called "single detector mode". In the Fig. 2- (upper right) they are indicated by "s". By the "single detector mode" we mean, that all events, which could be reconstructed into tracks were taken out from the analysis. Only events, which are correlated *only with a single detector* (all other 3149 standard counters of the telescope 'slept' in that time, used as a anticoincidence shield), *are accepted for the analysis.* The reason is that the neutrino energies from a SN are rather low and tracks of produced positrons will fully be inside the volume of one individual detector. The dead time was ~4 ms, and the usual accuracy of our clock was ± 1 sec. The information from all detectors (the amplitude, the coordinates and the absolute time of the event) were collected in the memory of the on-line computer, and each 900 sec automatically written on magnetic tape.

Another possibility to reduce the background was, to reduce the 'fiducial volume' of the telescope. From the work to understanding the background of the BAKSAN neutrino scintillation telescope, which was performed earlier in the eighties, was determined, that in the final analysis should be taken data only from the three internal horizontal planes (see Fig. 2). The internal horizontal planes are mostly shielded from the muons. Then the fiducial mass is ~ 130 tonnes, and the background was excellent, $\sim 0.013\,\mathrm{sec}^{-1}$!

In addition it was found that the lowest parts of the vertical planes of the telescope also have a relatively small background (see[33]). This allowed us to increase the fiducial mass up to 200 tonnes.

The traditional routine work with the individual detectors (regular 'war' with discharges in the photomultiplier tubes) allowed as to reduce the background of the individual detectors and the total background for the collapse ("single detector mode") program from 0.043 count s^{-1} (in beginning of the eighties) to 0.0127 counts s^{-1} (in the time of the SN 1987A) (see Fig. 2 - bottom right).

2.1.2. Rules of Selection of the Data

By using a "single detector mode", events were selected into 'groups' of events over a time interval of 20 seconds.[28] If the number of events in such time cluster was equal or more than four, the computer informed about a 'candidate' for collapse detection.

The rate expectations for the rate of a such 'group' with different numbers of background events could be found by using a sliding time interval method:

$$N_{exp} = n_o \cdot \exp^{-m\tau} \cdot (m \cdot \tau)^{k-1} \cdot [(k-1)!]^{-1}, \tag{1}$$

where $n_o = m \cdot T$ is the total number of background events during the observation time T; m is the average background count rate per second; τ is the duration of sliding time window; k is the number of events spreaded over τ.

3. Water Cherenkov Detectors

First generation of water Cherenkov detectors. Another important type of underground detectors are water Cherenkov detectors (target - H_2O). Already from beginning of the seventies in the 20th century, they were used for possible registration of the neutrino-antineutrino signal from supernovae, in the Homestake mine (South Dakota) at a depth of 4400 m.w.e., in Barbetone (Ohio) at a depth of 1800 m.w.e., and in Mont-Blanc (Italy) at a depth of 4270 m.w.e. Unfortunately all these small detectors (with a total mass of not more than 15 tons) ran only short time. But any way, by the first one was detected one cluster, during 3 ms, consisting of 24 events in the energy range of ≤ 100 MeV. But at that time (1974) only one detector was operating for search of a supernova neutrino signal, and this very narrow 'pocket' impulse was unfortunately not confirmed.

Second, present generation of water Cherenkov detectors. The second generation includes such big projects as IBM (Morton-Thiokol mine, Ohio) and Kamiokande (Kamiokande II) (Japan). Later since April 1st, 1996 started to operate SuperKamiokande (Japan), 50 000 tons imaging water Cherenkov detector and the SNO observatory (Creighton mine near

Sudbury, Ontario, Canada), a 1000 ton heavy-water detector, 2 km underground, operating since November 1999.

Both detectors (IMB and Kamiokande II) were designed to search for proton decay at the depth of 1570 m.w.e. and 2400 m.w.e.,[37] respectively. Both detectors in 1987 *just were on the way* to be able to search for a neutrino flux from a supernova. We remind, that, in contrast, the BAKSAN neutrino scintillation telescope has been observing the Galaxy *permanently* since June 30, 1980.

Some important technical parameters (fiducial mass of the target, energy threshold, background counting rate) of Kamiokande II and IMB detectors are summarised in the Table 1.

Cherenkov detectors are very efficient to detect the electron antineutrino (see equation (1)) by reaction with the free protons in the water. The energy threshold is typically 5-10 MeV, but at the time of SN 1987A, Kamiokande II had a threshold of $\sim$8MeV and IMB of $\sim$35 MeV. The efficiency at threshold typically is $\sim$ 50% (for more details about the light-water Cherenkov detectors we refer to[23,24,38]).

Let us discuss now the neutrino signal which was observed by the three underground detectors BAKSAN (Russia), Kamiokande II (Japan) and IMB (USA) on 24 February 1987.

4. Supernova SN1987A

Both - scintillation and Cherenkov detectors - were and are the flagships of this research, other approaches for supernova neutrino detection are in the stadium of development. A visible explosion of the supernova in our Galaxy was observed 400 years ago, but at that time scientists had only optical telescopes, but no neutrino detectors .....

On the basis of the optical observation of SN 1987A (name of the exploded star was Sanduleak -69° 2002) in the nearby Large Magellanic Cloud (LMC) on 24 February 1987, was performed the search in the data of all (at that time - four) running neutrino underground detectors.

But only three of them - BAKSAN neutrino scintillation telescope, Kamiokande II, and IMB were able to detect a signal on 23 February 1987 at 07:35 UT. The LSD detector under the Mont Blanc at that time was not able to see a neutrino signal.

4.1. *Some Important Parameters of the Neutrino Signal*

The total neutrino signal from all three underground detectors consists of 24 events (see Table 2 and Fig. 3): 5 events were registrated by the BAKSAN neutrino scintillation telescope, 11 events came from Kamiokande II and 8 events were claimed by the IMB detector.

The energy window of the detected signal: events seen by the BAKSAN neutrino scintillation telescope collaboration are located in the energy interval (12. ÷ 23.3) MeV, by Kamiokande II collaboration - (7.5 ÷ 36.) MeV, and the IMB collaboration reported (20. ÷ 40.) MeV (see Tables 2 and 3).

The time interval of each signal is, respectively, 9.1 sec from the BAKSAN neutrino scintillation telescope data, 13 sec from the Kamiokande II and 6 sec from the data set of the IMB detector (see Table 2).

The time sequences of events detected by BAKSAN, Kamiokande II, IMB and by LSD at 7:35 UT on February 23, 1987 are shown in Fig. 3 and listed in Table 2.

4.2. *Problems*

Unfortunately the BAKSAN neutrino scintillation telescope, Kamiokande II, and the IMB had some problems during the time of the measurements, which finally disturbed such *a landmark event in astrophysics!*

The BAKSAN neutrino scintillation telescope hade some not pleasant situation with the determination of the absolute time of the signal, but the situation was taken under control. When carefully looking into our data taken in the weeks *before* the SN87A event we found an *absolute shift* of 54 sec in one jump, since February 17, 1987.[3,4] We conservatively stated in spite of this, that the accuracy of our telescope clock was (-54 sec, +2 sec). *All other important parameters* of the BAKSAN neutrino scintillation telescope, like stability of detectors (the discharges in the photomultiplier tubes), on-line system for identification of the parameters of each signal - were not corrupted during the long term of the measurements!

Kamiokande II had also a problem. The accuracy of the Japanese clock was ±1 min.[23,37] As reason that a better time calibration was not made after the SN event, an abrupt power outage in the Kamioka mine on 25 February 1987 was reported. But in general the trigger rate of the Kamiokande II detector at that time was not so stable, only at the end of May the counting rate of the instrument was taken under control (by improvement of the isolation from the mine air both in the 3000 ton tank and in the water-purification system[38]), see Fig. 4 from,[38] and later in,[39] and also as fig. 3.20 of the Nobel Laureate Talk by M. Koshiba with the caption "The detector performance at the beginning of 1987".[39]

The Fig. 4 can be compared to Fig. 2, which shows the background development of the BAKSAN neutrino scintillation telescope. Both figures

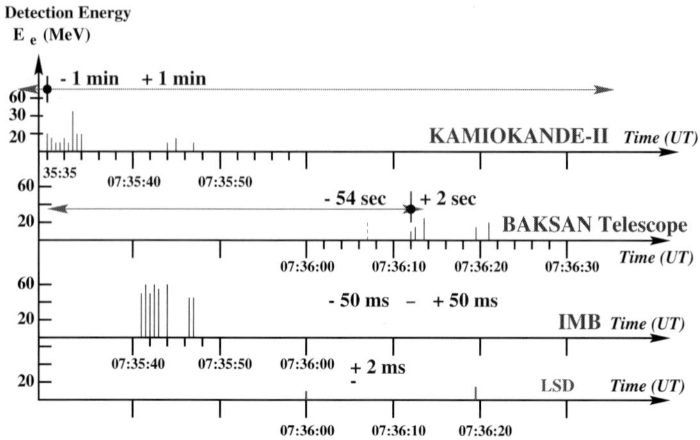

Fig. 3. The time sequences of events detected by the BAKSAN telescope (the second axis), Kamiokande-II (the first axis), IMB (the third axis) and LSD (the last axis) detectors at 07:35 UT on February 24, 1987.[1,3,4,23–25] Vertical axis - the detection energy of each event (MeV). Horizontal axis - the absolute time (UT), from 07:35:40 UT in steps of 2 sec, and in steps of 1 sec between 07:35:35 - 07:35:40 UT (only for Kamiokande II events). The uncertainties of the time determination in the different experiments is indicated by long right-left arrows (for KII and BAKSAN telescopes).

Table 2. Table of the neutrino events registered by the BAKSAN neutrino scintillation telescope,[1,3] Kamiokande II,[23] IMB,[24] and LSD[25] in the time of explosion of the SN87A.

Detector	Time (UT) on February 23 1987	Electron en. (MeV)	Detector	Time (UT) on February 23 1987	Electron en. (MeV)
KII[23]	7:35:35.00±1 min	20.0 ± 2.9	IMB[24]	7:35:41.37±50 ms	38 ± 9.5
	7:35:35.11 ± 1 min	13.5 ± 3.2		7:35:41.79 ± 50 ms	37 ± 9.3
	7:35:35.30 ± 1 min	7.5 ± 2.0		7:35:42.02 ± 50 ms	40 ± 10
	7:35:35.32 ± 1 min	9.2 ± 2.7		7:35:42.52 ± 50 ms	35 ± 8.8
	7:35:35.51 ± 1 min	12.8 ± 2.9		7:35:42.92 ± 50 ms	29 ± 7.3
	7:35:36.54 ± 1 min	35.4 ± 8.0		7:35:44.06 ± 50 ms	37 ± 9.3
	7:35:36.73 ± 1 min	21.0 ± 4.2		7:35:46.38 ± 50 ms	20 ± 5.0
	7:35:36.92 ± 1 min	19.8 ± 3.2		7:35:46.38 ± 50 ms	20 ± 5.0
	7:35:44.22 ± 1 min	8.6 ± 2.7		7:35:46.96 ± 50 ms	24 ± 6.0
	7:35:44.43 ± 1 min	13.0 ± 2.6			
	7:35:47.44 ± 1 min	8.9 ± 1.9			
BAKSAN[1,3]	7:36:11.818 ($^{+2}_{-54}$) sec	12. ± 2.4	LSD[25]	2:52:36.79 ± 2 ms	7 ± 1.4
	7:36:12.253 ($^{+2}_{-54}$) sec	18. ± 3.6		2:52:40.65 ± 2 ms	8 ± 1.6
	7:36:13.528 ($^{+2}_{-54}$) sec	23.3. ± 4.7		2:52:41.01 ± 2 ms	11 ± 2.2
	7:36:19.505 ($^{+2}_{-54}$) sec	17. ± 3.4		2:52:42.70 ± 2 ms	7 ± 1.4
	7:36:20.917 ($^{+2}_{-54}$) sec	20.1 ± 4.0		2:52:43.80 ± 2 ms	9 ± 1.8

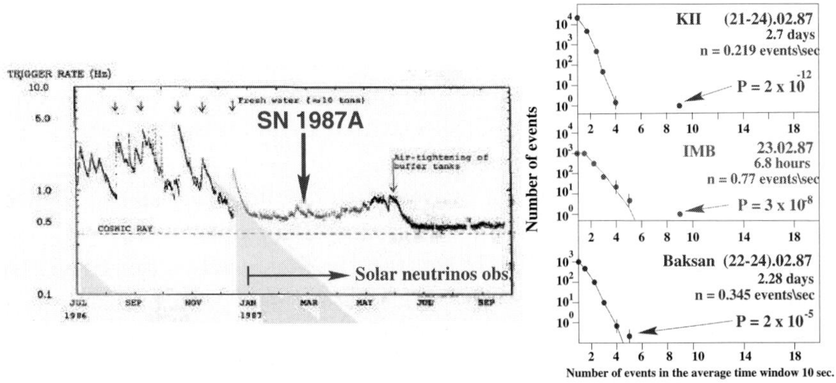

Fig. 4. Left: The background development of the Kamiokande II detector.[38,39] It was improved only *after* the SN87A signal, by installing isolation structures from the mine air both in the 3000 ton tank and in the Water-purification system,[38] and kept under control since end of May 1987. Right: Poisson distribution for events within 10 sec intervals detected in the period surrounding the time of Supernovae 1987A. n is the average of the background counting rate per second. The analysis was performed from the data sample for Kamiokande II - 2.7 days, for IMB - during 6.8 hours, and for the BAKSAN neutrino scintillation telescope - 2.28 days.

cannot be compared to the situation at IMB, since their background situation before and after SN87A has never been published. Also we know of *no detailed* paper from IMB.

It is important to remember that the IMB collaboration at that time had an excellent clock with the accuracy of $\pm$ 50 ms in universal time (UT)[24]! But unfortunately one quarter of the PMT's were off,[24] and the presented background was relatively larger than for the Kamiokande II and BAKSAN neutrino scintillation telescopes (see Table 1). The trigger efficiency to detect the events with energy 20 MeV was very small, only 0.14, instead of 80% for the BAKSAN neutrino scintillation telescope (see Fig. 7 left).

Fig. 7 (left) shows a comparison of the trigger efficiencies of the three detectors.[9,10,22] It is visible from the figure, that the BAKSAN neutrino scintillation telescope and the Kamiokande II detector had a similar efficiency to detect the neutrino flux with energy more than $\sim$ 8 MeV. The IMB detector mostly was sensitive for 'high' energy events with energies more than 30 MeV.

All three groups (we will discuss only detectors which detected a neutrino signal) reported results of an analysis of some time period before and after the signal. The BAKSAN group analysed data in the period: 1-23 February 1987,[1,3,4] Kamiokande II the time interval from the 21 - 24

February 1987,[23] IMB performed an analysis of only ~ 6.4 h around 7:35 UT at UT 23 February.[24]

All collaborations claimed in their publications,[1,3,23,24] that they could not explain the observed significance at that period in such a narrow time window of 9 to 13 sec, by background.

All other observed 'clusters' of events in 10 sec intervals can be explained naturally as fluctuations of the background. They follow nicely a Poisson distribution (see Fig. 4-right). In contrast the bursts with 5 (for BAKSAN) and 9 events (for KII and IMB) should occur only with probabilities of $2{\cdot}10^{-5}$ (BAKSAN), $2{\cdot}10^{-12}$ (Kamiokande II), and $3{\cdot}10^{-8}$ (IMB). We thus can conclude from these numbers that these 24 events are *not accidental*!

4.3. *Unexpected Calibration of the BAKSAN Clock*

On the basis of our work which was done in the one-two years after explosion of the supernova, we are happy to say, that we performed an *independent* calibration of our clock. The calibration was done during the work to search correlations between the LSD detector data, the BAKSAN neutrino scintillation telescope data and the data from the Rome and the Maryland gravitational wave antennas during the time period of February 23, 1987. As results of that search we observed an unexpected phenomenon — a temporal correlation between events recorded by all underground detectors (the data were kindly given to us by all collaborations) and two gravitational wave antennas[12] during a one-hour-period of 01:45 - 2:45 UT on February 23, 1987. The data recorded by the LSD detector and by the BAKSAN neutrino scintillation telescope show a correlation between low-energy events, the main source of which is natural radioactivity, and cosmic ray high energy muons, respectively.

This effect was observed after shifting the absolute time of the BAKSAN neutrino scintillation telescope clock by 30 sec. Now if we assume, that the effect (correlations) observed at 01:45 - 2:45 UT provided us with an independent calibration of the BAKSAN neutrino scintillation telescope clock, with this calibration the beginning of the neutrino signal detected by the BAKSAN neutrino scintillation telescope at 07:35 UT will *coincide* with the beginning of the IMB neutrino signal[3,10] (see Fig. 3!). For more details and very unexpected conclusions we refer to our published articles.[11,12]

4.4. *Time Profile of the Neutrino Signal*

The observation in the BAKSAN neutrino scintillation telescope, Kamiokande II and IMB detectors of the neutrino signal from the explosion of the

SN 1987A, *allowed us (on the basis of very low statistics) to make a first step* to understand the time profile of the neutrino signal.

This was done by our BAKSAN "collapse" group and presented and published in 1988 in.[3,10] Let us remind about this very interesting and original work.

Obviously, because the absolute time for two of the three detectors was *not very accurate*, we could not put directly together the signals from all three detectors, in the proper time sequence. But, registration of the neutrino signal from the explosion of a supernova is such a *fantastic!* and *so rare* event (only one per 400 years?), that it gave us a strong push to perform such an analysis with some assumptions.

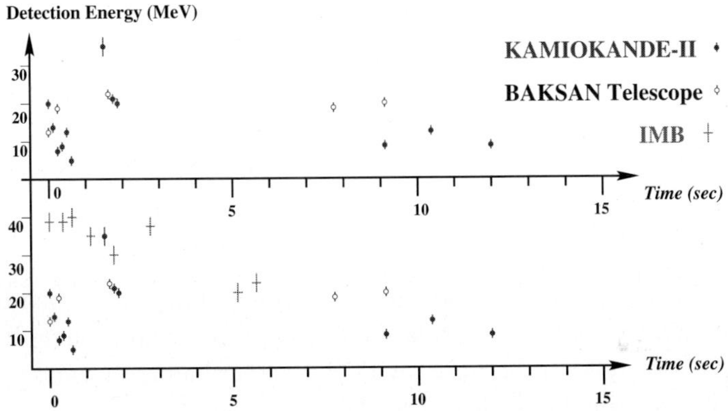

Fig. 5. Time profiles of the events recorded by the BAKSAN telescope (white circles), Kamikande-II (black circles), setting zero time to be the time of the first events for Baksan and KII detectors at 07:35 UT on 24.02.1987 (upper part), and assuming coincidence of the first IMB (crosses) signal with the first signal for Baksan and KII detectors (lower part).[1,3]

Let us *first assume* that the time profiles of the events recorded by the BAKSAN neutrino scintillation telescope and by Kamiokande II (see Fig. 5 (upper part)), and by all three detectors (see Fig. 5 (lower part)) will be put together in the correct time sequence, when setting the time equal for the first event of each detector, at 07:35 UT on February 23, 1987. In the figure 5 (upper part) are presented the signals only from the BAKSAN neutrino scintillation telescope and Kamiokande II. The vertical axis corresponds to the energy of the individual events, and the horizontal axis is a relative time scale, from 0 bis 15 sec. Point '0' is related to the first event of the BAKSAN neutrino scintillation telescope and of the Kamiokande II detectors. Fig. 5 (lower part) shows in addition to the BAKSAN neutrino scintillation telescope and to the Kamiokande II signal, the neutrino signal claimed by the IMB collaboration, also attributing time '0' to the first event of IMB.

From the Fig. 5 (upper part) one can see from the presented (11+5) events in both, BAKSAN neutrino scintillation telescope and Kamiokande II, *a bunch structure in time. The first bunch* - between 0 and 1 sec, includes 2 events from the BAKSAN telescope data and 6 events from the data of Kamiokande II telescope. *The second bunch* - duration less than 1 second - consists from one event from the BAKSAN neutrino scintillation telescope data and 3 events from Kamiokande II. And the *third bunch* includes 3 events from the BAKSAN neutrino scintillation telescope and 4 events from Kamiokande II. One could see even, that *the energies of the events for both telescopes are similar in the same bunches.*

Between each bunch we observed *gaps. The first gap* (between first and second bunch) has a duration of ~1 sec and the second gap of (6-7) seconds between the second and the third bunches. The probability to construct such gap structure from small statistics for both detectors is rather small (level of 10^{-3} - 10^{-2}) or less.

Let us, now, include in our 'analytical business' the 8 events, which were reported by the IMB collaboration (see Fig. 5 (lower part)).

It is obvious that the IMB signal has another time structure. The events are not coincident (by energies) with the common first and second bunches of the BAKSAN neutrino scintillation telescope and Kamiokande II! Why? Not correct energies, or because the IMB detector had not a good efficiency (only 14%) for small energies (~20 MeV), but only for energies more than 30 MeV (where the efficiency was ~50%) (see Fig. 7) and.[24] Then one could conclude, that the IMB detector sees another effect, which was 'provocated' by SN 1987A, and the Kamiokande II detector sees only one of such 'high energy' events (number 6, with the energy 35.4 MeV).

Let us set now the time of the first event from the IMB signal equal with that of the first 'high energy' event of the second bunch of the Kamiokande II signal. Then we could see, that the second gap after including the IMB events, is somehow 'missing'. The third bunch got a more broad structure (in addition two more events) together with 5 events from the BAKSAN neutrino scintillation telescope and the Kamiokande II telescope (see Fig. 6 (upper part)).

Or another variant of the setting of the time: let us set the first 'high energy' event of the second bunch of the Kamiokande II signal equal to that of the first event from the second bunch of the IMB signal. Then we could say, that IMB sees the beginning of the 'high energy' part of the common neutrino signal, and the second IMB bunch is perfectly correlated with the common second bunch of the BAKSAN neutrino scintillation telescope and the Kamiokande II experiments (see Fig. 6 (bottom part)). Then the total duration of the neutrino burst will be more than 15 sec!

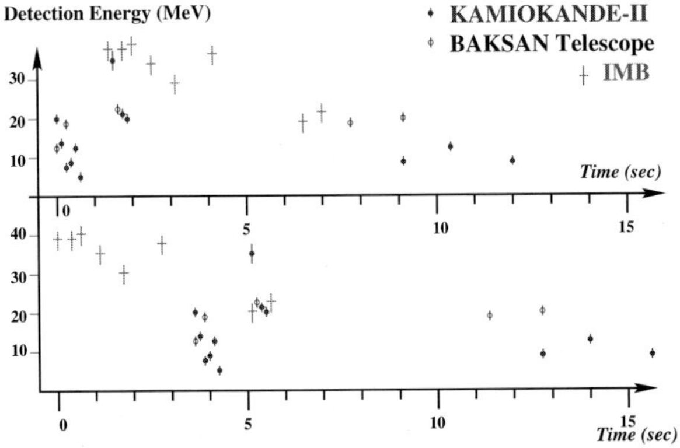

Fig. 6. Time profiles of the events recorded by the BAKSAN telescope (white circles), Kamiokande-II (black circles), and IMB detector (crosses), setting zero time for BAKSAN neutrino scintillation telescope and Kamiokande II to be the time of the first events for Baksan and KII detectors at 07:35 UT on 24.02.1987, and assuming coincidence of the first IMB signal with the first 'high' energy event of KII (upper part). Assuming coincidence of the first event from the second gap of the IMB signal, with the first high energy event of the KII (lower part). The total shift to the left side is ~ 3 sec.

After all these discussions, we have to say, that the time profile of the neutrino signal still today is an open question. We have to wait till a new explosion of a supernova.

The most recent analysis of the temporal structure of the neutrino signal from the SN 1987A was done in 2001,[22] where the authors used a Bayesian analysis (which is very adequate to work with *very* low statistics - also in our recent work about the evidence of a signal from $0\nu\beta\beta$ beta decay we used this method, see[18,20,21]). This analysis shows, *'a strong evidence for two components in the neutrino signal ....'* on the basis of an analysis of the energies and arrival times of the neutrinos from supernova SN 1987A detected by Kamiokande II, IMB and the BAKSAN neutrino scintillation telescopes.

Personally for me it was very pleasant to see, that for the *first time* after 14 years of explosion of the SN 1987A, was so directly and strongly expressed **"... that the BAKSAN neutrino scintillation telescope data are fully consistent with the Kamiokande II and IMB data. ...."**[22]

5. Main Parameters of the Neutrino Signal from SN 1987A

Let us on the basis of our understanding of the Supernova calculate the main parameters of the neutrino signal.

We shall take into account, that the distance to the SN 1987A $\sim$ is 50 Kpc, and the neutrino spectrum has a Fermi-Dirac shape (2)

$$F(E_{\tilde{\nu}_e,\alpha}) = E^2{}_{\tilde{\nu}_e} \cdot \exp^{[-\alpha(E_{\tilde{\nu}_e}/kT)^2]} / (1 + \exp(E_{\tilde{\nu}_e}/kT)), \qquad (2)$$

where kT is the effective neutrino temperature (MeV),
$E_{\tilde{\nu}_e}$ - neutrino energy (MeV),
α - absorption coefficient of the thermal neutrino spectrum.

Now, knowing the target (number of free protons in the target) we can perform our estimations.[3] One example, for the absorption coefficient $\alpha = 0$, is presented in table 3.

Table 3. The final values of the neutrino signal,[10] deduced from the experimental data of the neutrino telescopes BAKSAN, KII, IMB.

Detector	Average detected energy(MeV)	Neutrino temperature (MeV)	$E_\nu \times 10^{52}$ (ergs)
BAKSAN	19.4 ± 1.7	3.3 ± 0.4	18.6 ± 8.5
KII	16.7 ± 1.1	2.8 ± 0.3	5.8 ± 1.8
IMB (5000)	33.8 ± 2.9	4.5 ± 0.7	2.9 ± 1.1

What we could learn from that analysis?

First: the average detected neutrino energy for the BAKSAN neutrino scintillation telescope is (19.4 ± 1.7) MeV and for Kamiokande II (16.7 ± 1.1) MeV. Conclusion on that point: the energies for both detectors are consistent within 1.5σ. But a *big difference* exists between the IMB data and the other two detectors.

Second: the neutrino temperature. From the BAKSAN neutrino scintillation telescope data the neutrino temperature is located in the interval of $(2.9 \div 3.7)$ MeV, for the Kamiokande II data it is $(2.5 \div 3.1)$ MeV. Again, the neutrino temperatures *overlap for both - the BAKSAN neutrino scintillation telescope and the Kamiokande II detector* - on the level of one sigma! For the IMB data the neutrino temperature was found to be $(3.8 \div 5.2)$ MeV, consistent within 2σ with the BAKSAN neutrino scintillation telescope and the Kamiokande II data.

Third: the total antineutrino energy. For the BAKSAN neutrino scintillation telescope data (unexpected!) this value is in $\sim$ 3 times larger than from the Kamiokande II data and $\sim$ 6 times larger than from IMB.

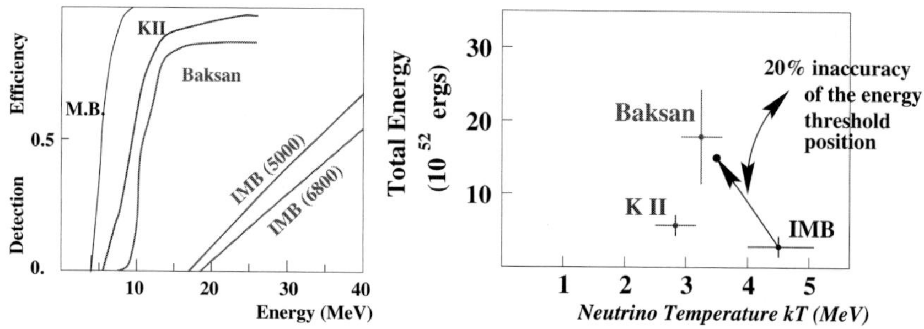

Fig. 7. Left: Comparison of trigger efficiencies of all SN87A detectors.[9,10,22] Right: Effective parameters calculated from the experimental data for the three detectors, which see the neutrino signal from the SN87A: BAKSAN Neutrino Telescope, Kamiokande II and IMB (see[10]).

Such a big difference for the total antineutrino energy between the BAKSAN neutrino scintillation telescope data and the KII and IMB detectors *can be corrected only after full understanding of all parameters* of the two latter neutrino detectors: the energy threshold and the fiducial mass of the detectors.

Let us come back to our 'old' idea about correction of the threshold in the IMB experiment. If we assume 20% of inaccuracy of the energy threshold position for the IMB detector, *than we could get a total antineutrino energy and neutrino temperature for the IMB detector which is very close to the BAKSAN neutrino scintillation telescope data* (see the blue solid line from IMB to Baksan in Fig. 7). It is well known (see[22]), that the calibration spectra of the IMB detector for the time of explosion of the SN 1987A were never published.

The fiducial mass of the Kamiokande II detector (in our opinion) also should be revised (on the basis of very low statistics). Instead of 2140 tons, the mass *could be corrected* to the mass which the collaboration later used for the solar neutrino program[36] (the fiducial volume for the solar neutrinos was 680 tons, i.e. for similar energies, of more than 10 MeV). If this correction is made, than this reduction of the fiducial volume exactly gives a difference by a factor of 2.5 (!!!) for the total antineutrino energy, which we can calculate from the Kamiokande II data (see Fig. 7). So, after this corrections of the KII and the IMB data full consistensy is obtained between the observation of the three detectors.

Anyway, one can, even, on the level of the 2-3 sigma existing differences between the data from the BAKSAN neutrino scintillation telescope, the Kamiokande II and IMB telescopes, conclude, that the observation of the neutrino flux from Supernova SN1987A *is the first direct evidence of a neutrino burst from a Supernova and opens for us the era of neutrino astronomy.* The BAKSAN/KII/IMB neutrino burst fits well the standard assumptions of star collapse, so that the collapse which created that burst on February 23rd 1987 near 07:35 UT *really happened!*

Acknowledgement

The author would like to thank first of all to his colleagues Prof. E.N. Alexeyev and Dr. L.N. Alexeyeva for fruitful collaboration during my work together with them under this program at the Baksan Neutrino Observatory. Thanks to Soviet (at that time) Government, that they built such an excellent underground laboratory (for more than 10 years the BAKSAN Neutrino Observatory was the biggest underground laboratory in the world) and allowed us to make our research.

I would like to thank to Prof. Dr. habil H.V. Klapdor-Kleingrothaus who from the time of the Supernova 87A till today encouraged me to remember, that *I am a member of one of the groups detecting such a 'Millennium event'.*

References

1. E.N. Alekseev, L.N. Alekseeva, V.I. Volchenko, I.V. Krivosheina, *JETP Lett.* **45** (1987) 589-592.
2. E.N. Alekseev, L.N. Alekseeva, V.I. Volchenko, I.V. Krivosheina, *Pisma Zh. Eksp. Teor. Fiz.* **45** (1987) 461-464.
3. E.N. Alekseev, L.N. Alekseeva, I.V. Krivosheina, V.I. Volchenko, *Phys. Lett.* **B205** (1988) 209-214.
4. E.N. Alekseev, L.N. Alekseeva, I.V. Krivosheina, V.I. Volchenko, Heidelberg, Germany, 1987, 'Neutrino Physics' 288-298, in Proc. of *WEIN 1986*, ed. H.V. Klapdor, and in Proc. of *ICRC*, Moscow, USSR, 1987, **9** (1988) 59-67.
5. E.N. Alekseev et al., *JETP Lett.* **49** (1989) 548-552.
6. E.N. Alekseev et al., *Pisma Zh. Eksp. Teor. Fiz.* **49** (1989) 480-483.
7. E.N. Alekseev et al., *Sov. Phys. JETP* **72** (1991) 585-589, *Zh. Eksp. Teor. Fiz.* **99** (1991) 1057-1065.
8. I.V. Krivosheina, *Int. J. Mod. Phys.* **D13** (2004) pp. 2085-2105.
9. I.V. Krivosheina et al., *talk at Europ. ICRC*, Balaton, Hungary, 1988.
10. L.N. Alekseeva et al., *XXIAU General Assembly*, Aug. 1988, Baltimore, USA.
11. E.N. Alekseev, L.N. Alekseeva, V.N. Zakidyshev, N.F. Klimenko, I.M. Kogai, I.V. Krivosheina and V.A. Poddubnyi, *JETP Lett.* **49** (1989) 548-552, and *Pisma Zh. Eksp. Teor. Fiz.* **49** (1989) 480-483.
12. E.N. Alekseev, L.N. Alekseeva, N.F. Klimenko, I.M. Kogai, I.V. Krivosheina et al., Proc. of *TAUP'89* (1999) 323; Proc. of *Int. Symposium on Weak and Electromagnetic Interact. in Nuclei (WEIN-89)*, Montreal, Quebec, Canada, 15-19 May 1989, 'Weak and Electromagn. Inter. in Nuclei' (1990) 757.

13. H.V. Klapdor, *Prog. Part. Nucl. Phys.* **10** (1983) 131-225.
14. H.V. Klapdor, *Fortsch. Phys.* **33** (1985) 1-55
15. H.V. Klapdor, *Prog. Part. Nucl. Phys.* **17** (1986) 419-455.
16. K. Grotz and H.V. Klapdor, "The Weak Interaction in Nuclear, Particle and Astrophysics", *Stuttgart, F.R. Germany: TEUBNER* (1989) 1-458, *Adam Hilger, IOP Publ., Bristol, New York* (1990) 1-488, *Mir, Moskau* (1992) 1-446; *Shandong Science and Technology Press, Jinan, China* (1998) 1-577.
17. H.V. Klapdor-Kleingrothaus, "60 Years of Double Beta Decay - From Nuclear Physics to Beyond the Standard Model", *World Scientific, Singapore* (2001) pp. 1-1281, and second edition "70 Years of Double Beta Decay", *World Scientific, Singapore* (2008).
18. H.V. Klapdor-Kleingrothaus, I.V. Krivosheina *Mod. Phys. Lett.* **A21** (2006) 1547-1566, H.V. Klapdor-Kleingrothaus, I.V. Krivosheina, A Dietz et al., *Phys. Lett.* **B586** (2004) 198-212, and *NIM* **A522** (2004) 371-406.
19. H.V. Klapdor-Kleingrothaus et al. *Mod. Phys. Lett.* **A16** (2001) 2409-2420.
20. H.V. Klapdor-Kleingrothaus, A. Dietz, I.V. Krivosheina, *Part. and Nucl.* **110** (2002) 57-79.
21. H.V. Klapdor-Kleingrothaus, A. Dietz, I.V. Krivosheina, *Foundations of Physics* **31** (2002) 1181-1223 and Corrigenda, 2003.
22. T.J. Loredo and D.Q. Lamb, *Phys. Rev.* **D65** (2002) 063002.
23. K.S. Hirata et al., *Phys. Rev. Lett.* **58** (1987) 1490.
24. R.M. Bionta et al., *Phys. Rev. Lett.* **58** (1987) 1494.
25. M. Aglietta et al, *IAU Circular* **No. 4323** (1987) V.L. Dadukin et al., *Pisma Zh. Eksp. Teor. Fiz.* **49** (1989) 464-466.
26. Ya. B. Zel'dovich and O.Kh. Gusseinov, *Dokl. Akad. Nauk SSSR* **162** (1965) 791, *(Sov. Phys. Dokl.* **10** (1965) 524).
27. G.V. Domogatsky, G.T. Zatsepin, Proc., *9th ICRC, London* **V.2** (1965) 1030.
28. V.S. Imshennik, D.K Nadyozhin, *Itogi. Nauk. Tekhn. ser. Astr.* **21**(1982)63.
29. A. Burrows, R. Gandhi and M.S. Turner, *Phys. Rev. Lett.* **68** (1992) 3834.
30. A. Burrows, *astro-ph/* **9706137**.
31. H.-T. Janka, *Astron. Astrophys.* **368** (2001) 527, H.-T. Janka et al., *Lect. Notes Phys.* **578** (2001) 333; K. Langanke et al., *Phys. Rev. Lett.* **90** (2003) 241102 K. Kifonidis et al., *astro-ph/* **0302239** and there ref.
32. E.N. Alekseev, L.N. Alekseeva, *J. Exp. Theor. Phys.* **95** (2002) 5-10, *Zh. Eksp. Teor. Fiz.* **95** (2202) 10-16, and astro-ph/0212499
33. L.N. Alekseeva, *PHD* (1995), Moscow INR
34. E.N.Alekseev, L.N.Alekseeva etal., *17thICRC, Paris*, 1981, **7** (1982)110.
35. Y. Fukuda et al. (Super-Kamiokande Collab.), *Phys. Lett.* **B467** (1999) 185.
36. Super-Kamiokande Collab. (K.S. Hirata et al.), *Phys. Rev. Lett.* **63**(1989)16.
37. K.S. Hirata, T. Kajita et al., *Phys. Rev.* **D38** (1988) 448-458.
38. Suzuki A., in "Physics and Astrophysics of Neutrinos", eds. M. Fukugita and A. Suzuki, *Springer, Heidelberg* (1994) 388-447.
39. M. Koshiba, *Phys. Rep.* **220** (1998) 229-402; *Talk of Nobel Price for Astrophysics Search* (2002).
40. GNO Coll., *Nucl. Phys. Proc. Suppl.* **110** (2002) 311 and **77** (1999) 20.

COSMOLOGICAL CONSTRAINT ON "NEUTRINO MASS" FROM CMB ANISOTROPIES WITH PRIMORDIAL MAGNETIC FIELD

K. KOJIMA[a,*], D. G. YAMAZAKI[b] and T. KAJINO[a,b]

[a] *Department of Astronomy, University of Tokyo,*
7-3-1 Hongo, Bunkyo-ku, Tokyo 113-0033, Japan
[b] *National Astronomical Observatory, Division of Theoretical Astrophysics,*
2-21-1 Osawa, Mitaka, Tokyo 181-8588, Japan
[*] *E-mail: kojima@th.nao.ac.jp*

It is one of the biggest challenge in particle physics and cosmology to determine the mass of neutrinos. Current neutrinoless double beta decay experiment show that the total neutrino mass is smaller than 1–6 eV.

In recent years, precise CMB data make it possible to constrain neutrino masses cosmologically. It says that the total neutrino mass is smaller than 2 eV in standard ΛCDM model.[1] But if there are other physical processes which affects the CMB anisotropies, such constraint may be changed. A primordial magnetic field (PMF) is one of such physical processes. In this work, we constrain the total neutrino mass from CMB anisotropies with the PMF effect.

Keywords: Massive neutrino; Neutrino mass; Primordial magnetic field.

1. Neutrino Effect on the CMB

Neutrinos are one of the dark matter candidates, and massive neutrinos have virtually a large effect on the formation of the large scale structure (LSS) as a hot dark matter because of their free streaming. Neutrinos also have large three effects on the CMB anisoropies at mutipole $\ell > 200$. First, neutrinos enforces the Integrated Sachs-Wolfe effect. This effect is caused by the change of the gravitational potential and the curvature. The change of anisotropic stress also increases the ISW effect. Massive neutrinos are not perfect non-relativistic particles, and they become non-relativistic ones gradually. Such gradual process reduces the anisotropic stress, and increases the ISW effect. The second effect is the free-streaming effect on the primordial fluid of the universe. The typical scale affected by this effect is scale which is smaller than the conformal time when massive neutrinos become non-relativistic. The multipoles affected by free-streaming is about $\ell > 300$ when $\sum m_\nu \sim 2\text{eV}$. This effect decays the gravitational potential and the curvature rapidly, and enforces the acoustic oscillation. This process upshifts the zero-point of the acoustic oscillation. The third effect is the horizontal peak shift of CMB power spectrum. It was caused by the increase of the sound horizon, as we can know from the approxmate solution of the acoustic oscillation. The above neutrino effects on the power spectrum make it possible to constrain neutrino masses from the observed anisotropies of the CMB. This method of constraint show that $\sum m_\nu < 2\text{eV}$ under standard ΛCDM model[1] with WMAP-3yr data.[2]

2. Primordial Magnetic Field

Magnetic fields in clusters of galaxies have been observed with a strength of 0.1–1.0 μG. The existense of a primordial magnetic field (PMF) of order 1 nG whose field lines collapse as structure forms is one possible explanation for such magnetic fields in galactic clusters. Also the PMF could influence a variety of phenomena in the early universe such as the CMB, and the matter density field. Therefore, in order to constrain the neutrino mass with more precise physical process, we must take into consideration the PMF effect.

We parametarize the strength of the magnetic field with the B_λ.[3] B_λ is the smoothed magnetic field over scale λ. In this work, we set $\lambda = 1\text{Mpc}$. To calculate CMB anisotropies with the PMF, we make following assumptions. *First*, electrons and protons distribute uniformly in the early universe. It means that the conductance of the universe is infinite, and the electric field is approximately 0. *Second*, the magnetic field has the power law spectrum with a spectrum index n_B. In this work, we want to fix n_B to calculate faster. From Ref. 4,5, we can know that $n_B = -2.9$ is valid whenever the PMF was generated. It is not known when the PMF was generated, so we use this value.

The effect of the PMF can be separated into three modes. The scalar mode[6] increases anisotropies dramatically at $\ell < 10$. This is because of the photon fluctuation and Sachs-Wolffe effect caused by the scalar PMF. The vector mode enforces power spectrum at $\ell > 1000$, which can be neglected in the WMAP region. The tensor mode is much smaller than other two modes, so we neglect it.

3. Result and Discussion

We constrain the neutrino mass from CMB anisotropies with the PMF effect. We used the Markov chain Mote Carlo method (COSMOMC[7]) and calculated likelihood with 600,000 samples with the WMAP-3yr data. Free parameters are the density parameters for cold dark matter, baryonic matter, and massive neutrinos, the hubble parameter, the optical depth, the scalar spectral index of primordial density perturbation, their amplitude, and the magnetic field B_λ. The fixed parameter is the spectram index of the PMF ($n_B = -2.9$). Fig. 1 is the result. This figure says that the neutrino mass and the strength of the PMF have no correlation. The constraint obtained in this calculation is $\sum m_\nu < 1.35$eV, and $B_\lambda < 2.67$nG (2σ C.L.).

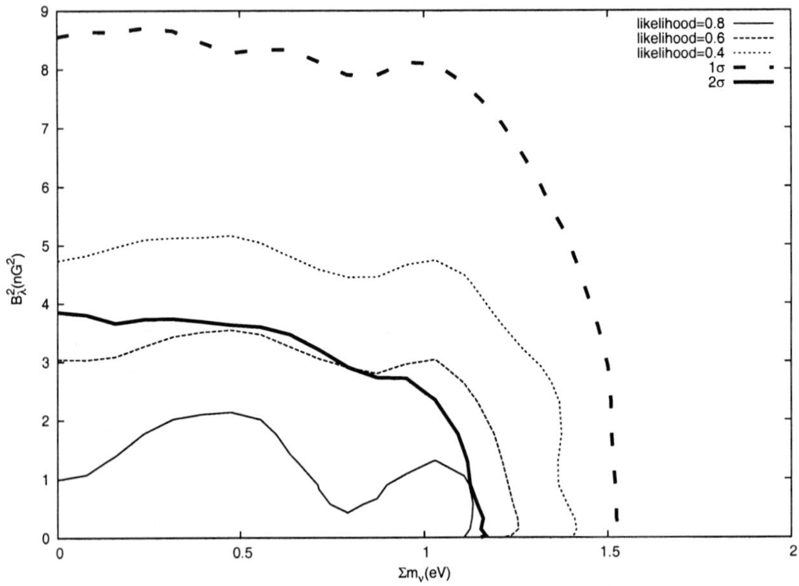

Fig. 1. The likelihood contour of $B_\lambda^2 - \sum m_\nu$ plane. The narrow solid, long-dashed and short-dashed curves represent likelihood contour 0.8, 0.6 and 0.4, respectively. The thick solid and dotted curves are 1σ and 2σ, respectively.

We obtained smaller m_ν constraint from CMB power spectrum with PMF. $\sum m_\nu$ and B_λ have no correlation because massive neutrino and the PMF have effect at separate region of mutipole ℓ. But in higher multipoles, $\ell > 1000$, massive neutrinos and the vector PMF both have the large effect on the CMB. So we should calculate including vector PMF, using observational data such as the CBI and the ACBAR, which have high mutipole data up to $\ell \sim 2500$. We also obtained the B_λ constraint. But n_B is constant in this calculation. We should vary n_B to obtain more general result.

References

1. M. Fukugita et al., *Phys. Rev. D* **74**, 027302 (2006).
2. WMAP Explanatory Supplement: http://lambda.gsfc.nasa.gov/.
3. A. Mack et al., *Phys. Rev. D* **65**, 123004 (2002).
4. D. G. Yamazaki et al., *ApJ.* **646** 719 (2006).
5. C. Caprini and R. Durrer, *Phys. Rev. D* **65** 023517 (2002).
6. M. Giovannini, *astro-ph*, 0604134v1.
7. A. Lewis and S. Bridle, *Phys. Rev. D* **66**, 103511 (2002).

PART VIII

Axion and Axino, Exotics

AXION AND AXINO CONTRIBUTION TO DARK MATTER

JIHN E. KIM

Department of Physics and Astronomy and Center for Theoretical Physics, Seoul National University, Seoul 151-747, Korea
jekim@phyp.snu.ac.kr

I discuss the essential features of the QCD axion: the strong CP solution and hence its theoretical necessity. I also review the effects of the QCD axion on astrophysics and cosmology, in particular with emphasis on its role in the dark matter component together with its supersymmetric partner axino.

Keywords: Dark matter, Axion, Strong CP problem, Superstring axions, Axion mixing; PACS Nos.: 12.20.Fv,14.80.Mz,95.35.+d,96.60.Vg.

1. Introduction

Cosmology with cold dark matter (CDM) was the leading candidate of the early universe in 1980s. But this view has changed completely just before beginning the new mileniam. The current view of the dominant components of the universe is $\Omega_{\rm CDM} \simeq 0.23$ and $\Omega_\Lambda \simeq 0.73$ [1]. The plausible DM candidates at present are the lightest SUSY particle (LSP), axion, axino, and gravitino, where we will review mostly on axion and its CDM-related possibility [2].

There are numerous cases supporting the nonluminous dark matter in the universe: flat rotation curves, Chandra satellite photo, and gravitational lensing effects. The DM density around us here is usually taken as $\rho_{\rm DM} \simeq 0.3\ {\rm GeV/cm^3}$.

The proposed CDM candidates belong to either incoherent particles or coherent oscillations of spin-0 fields. The incoherent CDM particles are usually called weakly interacting massive particles(WIMP) or decay products of WIMPs. A coherent oscillation arises from the potential describing a bosonic particle. WIMPs are the massive particles with weak interaction cross sections [3]. For a WIMP to be a successful CDM candidate, usually the WIMP interaction cross section at the time of decoupling is needed around [4] $\langle\sigma_{\rm int}v\rangle|_{\rm at\ decoupling} \approx 0.2\times 10^{-26}\ {\rm cm^3 s^{-1}}$ where $\Omega_m h^2 \simeq 0.113 \pm 0.009$. This is roughly explained by the LSP with the low energy SUSY. Currently, there are experimental efforts to discover the LSP as predicted by SUSY models. Also, direct cosmological searches are going on [5].

It is known that the density perturbation grew much earlier than the time of recombination. If it grew after the recombination time, the density perturbation grown afterward was not enough to make galaxies. For galaxy formation, therefore, DM is needed since proton density perturbation could not grow before the recombination time, but DM could. With DM, the equality point of radiation and matter energy densities can occur much earlier than the recombination time since DM is not prohibited in collapsing by Silk damping [6]. If the WIMP mass and interaction cross section falls in the region mentioned above, it can be CDM. In this talk, we consider CDM as the LSP and in addition axion also. But if the LSP is the only the CDM component, then the LSP mass would give one number for the DM density, which may not be the case. Thus, even if the LSP is contributing to the CDM density, we may need axion to account for the right amount of CDM around us.

All the discussion leading to axion started with the discovery of instanton solutions in nonabelian gauge theories, which let to an intrinsic additional parameter θ, the so-called vacuum angle, in nonabelian gauge theories. In the θ vacuum, we must consider the P and T (or CP) violating interaction parametrized by $\bar\theta$,

$$\mathcal{L} = \bar\theta\{F\tilde F\} \equiv \frac{\bar\theta}{64\pi^2}\epsilon^{\mu\nu\rho\sigma}F^a_{\mu\nu}F^a_{\rho\sigma} \tag{1}$$

where $\bar\theta = \theta_0 + \theta_{weak}$ is the final value including the effects of the electroweak CP violation. It is known that $\bar\theta$ is a physical parameter contributing to the dipole moment of neutron, d_n. From the upper bound of $|d_n| < 3\times 10^{-26} e{\rm cm}$ [7], we obtain a bound on $\bar\theta$, $|\bar\theta| < 10^{-9}$. The strong CP problem is the problem on ". Why is this $\bar\theta$ so small?" There are three explanations for the smallness of $\bar\theta$:

1. Calculable $\bar\theta$, 2. Massless up quark, 3. Axion.

The Nelson-Barr type CP violation is mostly discussed for the calculable $\bar\theta$ models since it is designed to allow the Kobayashi-Maskawa type weak CP violation at the electroweak scale. It introduces extra fields and corresponding interactions beyond the standard model. For example, vectorlike heavy quarks are introduced at high energy scales. The scheme is designed such that at low energy the Yukawa couplings are real, which is needed anyway from the beginning to set $\theta_0 = 0$. This solution is possible with the specific forms for the couplings and their phases and in addition the assumption on VEVs of Higgs doublets [8].

The next possibility had been the massless up quark case. Suppose that we chiral-transform a quark as $q \to e^{i\gamma_5 \alpha} q$. Then, the QCD Lagrangian changes as

$$\int d^4x[-m_q\bar{q}q - \bar{\theta}\{F\tilde{F}\}] \to \int d^4x[-m_q\bar{q}e^{2i\gamma_5\alpha}q - (\bar{\theta} - 2\alpha)\{F\tilde{F}\}] \quad (2)$$

If $m_q = 0$, it is equivalent to changing $\bar{\theta} \to \bar{\theta} - 2\alpha$. Thus, there exists a shift symmetry $\bar{\theta} \to \bar{\theta} - 2\alpha$. It is known that the tunneling amplitude due to instanton solutions with a zero mass quark vanishes [9], which implies that the shift symmetry is an exact symmetry. In this case, $\bar{\theta}$ is not physical, and hence there is no strong CP problem if the lightest quark (i. e. the up quark) is massless. The recent compilation of the light quark masses [10], $m_u = 3 \mp 1$ MeV and $m_d = 6 \pm 1.5$ MeV, is convincing enough to rule out the massless up quark possibility.

2. Axions

Peccei and Quinn (PQ) tried to mimic the above symmetry $\bar{\theta} \to \bar{\theta} - 2\alpha$ of the massless quark case, by considering the full electroweak theory Lagrangian [11]. They found such a symmetry if H_u is coupled only to up-type quarks and H_d couples only to down-type quarks. In this way, Peccei and Quinn found the $\bar{\theta}$ shift symmetry as in the massless quark case, which is called the PQ global symmetry $U(1)_{PQ}$. But unlike the massless quark case, here $\bar{\theta}$ is physical. Even though the coefficient of $\{F\tilde{F}\}$ changes in the same way, these two cases differ in that the tunneling amplitude vanishes with a massless quark but does not vanish without a massless quark. Since the tunneling amplitude is not vanishing, physics depends on the value of $\bar{\theta}$. The phase β of the Higgs fields disappears in the classical Lagrangian. But its coupling to $\{F\tilde{F}\}$ is generated at one loop level, which is the $U(1)_{PQ}$-QCD-QCD anomaly. With this one loop term, the Lagrangian is not invariant under the phase shift symmetry β. It is explicitly broken and the phase field β does not have a flat potential. Weinberg and Wilczek interpreted this phenomenon using the spontaneous symmetry breaking of the global symmetry $U(1)_{PQ}$. The phase β turns out to be the Goldstone boson (*axion*) of that spontaneous symmetry breaking of $U(1)_{PQ}$ [12]. This is said that $\bar{\theta}$ is made dynamical, but in the theory the component was there from the beginning. The free energy depending on $-\cos\bar{\theta}$ is the potential for the axion where $a = \bar{\theta}F_a$. Since it is proportional to $-\cos\bar{\theta}$, the minimum of the potential is at $\bar{\theta} = 0$ [11,13], and the vacuum chooses $\bar{\theta} = 0$. Thus, the axion solution of the strong CP problem is a kind of cosmological solution.

Nowadays, cosmologically considered axions are very light, which arises from the phase of SU(2)×U(1) singlet scalar field σ. The simplest case is the Kim-Shifman-Vainstein-Zakharov (KSVZ) axion model [14] which incorporates a heavy quark Q with the following coupling and the resulting chiral symmetry

$$\begin{aligned}\mathcal{L} &= -\bar{Q}_L Q_R \sigma + (\text{h.c.}) - V(|\sigma|^2) - \bar{\theta}\{F\tilde{F}\},\\ \mathcal{L} &\to -\bar{Q}_L e^{i\gamma_5\alpha} Q_R e^{i\beta}\sigma + (\text{h.c.}) - V(|\sigma|^2) - (\bar{\theta} - 2\alpha)\{F\tilde{F}\}.\end{aligned} \tag{3}$$

Here, Higgs doublets are neutral under $U(1)_{PQ}$. By coupling σ to H_u and H_d, one can introduce a PQ symmetry also not introducing heavy quarks necessarily, and the resulting axion is called the Dine-Fischler-Srednicki-Zhitnitskii (DFSZ) axion [15]. In string models, most probably both heavy quarks and Higgs doublets contribute to the σ field couplings. The VEV of σ is much above the electroweak scale and the axion is a *very light axion* [16]. The σ field may contain very tiny SU(2) doublet components, and hence practically we can consider the axion as the phase of σ, $\sigma = [(V+\rho)/\sqrt{2}]e^{ia/F_a}$ with the identification $a \equiv a + 2\pi N_{DW}F_a$. Since the domain wall (DW) number appears in the phase of σ, the magnitude F_a can be somewhat smaller than the singlet VEV v, $F_a = v/N_{DW}$.

The couplings of σ determine the axion interactions. Because the axion a is a Goldstone boson, the axion couplings are of the derivative form except for the anomaly term. In this talk, we will concentrate on this anomaly coupling. Axion is directly related to $\bar{\theta}$. Its birth was from the PQ symmetry whose spontaneous breaking introduced a. Generally, however, we can define a as a pseudoscalar field without potential terms except for the one arising from the gluon anomaly,

$$\frac{a}{F_a}\left\{\frac{g^2}{32\pi^2}F^a_{\mu\nu}\tilde{F}^{a\mu\nu}\right\}. \tag{4}$$

Then, we note that this kind of nonrenormalizable term can arise in several ways: string theory, M-theory [17], large extra (n) dimensions [18], composite models [19], and renormalizable theories. The axion decay constant is given with the scale defining the model, except in renormalizable models where it is given by the PQ symmetry breaking scale. In any case, the essence of the axion solution (wherever it originates) is that $\langle a\rangle$ seeks $\bar{\theta} = 0$

whatever happened before. In this sense it is a cosmological solution. The potential arising from the anomaly term after integrating out the gluon field is the axion potential. The height of the potential is $\sim O(\Lambda_{\rm QCD}^4)$. Two important properties of axions are: (i) periodic potential with the period $2\pi F_a$, and (ii) the minima are at $a = 0, 2\pi F_a, 4\pi F_a, \cdots$. This cosine form of the potential is determined by considering the axion mixing with pions as

$$V(a) = \frac{Z}{(1+Z)^2} f_\pi^2 m_\pi^2 \left[1 - \cos\frac{a}{F_a}\right] \tag{5}$$

where $Z = m_u/m_d$. Its expansion gives the axion mass $m_a \simeq 0.6 \left(\frac{10^7 \text{ GeV}}{F_a}\right)$ eV. There can be the axion-photon-photon anomalous coupling of the form $a\mathbf{E}\cdot\mathbf{B}$ which can be checked in laboratory, astrophysical and cosmological tests. The old laboratory bound of $F_a > 10^4$ GeV has been obtained from meson decays ($J/\Psi \to a\gamma, \Upsilon \to a\gamma, K^+ \to a\pi^+$), beam dump experiments ($p(e)N \to aX \to \gamma\gamma X, e^+e^-X$), and nuclear deexcitation ($N^* \to Na \to N\gamma\gamma, Ne^+e^-$) [20].

3. Axions from Superstring

If axion is present and string theory is setting up the framework for the standard model, it must appear from string theory. So, it is important to realize axions in string theory. There is a nagging problem here: in general it is difficult to introduce a detectable QCD axion with an exact PQ symmetry. So, an approximate PQ symmetry is a possibility for the detectable QCD axion.

Superstring tells us that there is no global symmetry, incliuding the bosonic degrees from the antisymmetric tensor field B_{MN}. If an axion is present, it is better to be one component in B_{MN}. Massless pseudoscalar degrees from B_{MN} are classified as the model independent (MI) axion [21] and the model dependent (MD) axion [22]. It is known that MD axions are generally heavy [23], but it may be a model dependent statement.

The superstring axion decay constants are expected near the string scale which is too large compared to 10^{12} GeV. For the MI axion it has been calculated in [24], $F_a \sim 10^{16}$ GeV. So, a key question in superstring axion models toward the discovery *a la* the Sikivie type detectors is, "How can one obtain a low value of F_a?"

An idea is the following. In some compactification schemes, an anomalous U(1) gauge symmetry results [25], where the U(1) gauge boson eats the MI axion so that the U(1) gauge boson becomes heavy. Then, a global symmetry survives down the anomalous U(1) gauge boson scale, and F_a the breaking scale of this global symmetry, may be put in the axion window, which was stressed early in [26]. However, this idea on the decay constant does not work necessarily.

MD-axion decay constants were tried to be lowered by localizing them at fixed points [27,28]. It uses the flux compactification idea and is possible to have a small F_a compared to the string scale by localizing the axion at a fixed point. Here, one needs a so-called Giddings-Kachru-Polchinski throat [29]. However, even if one lowered some F_a, we must consider the hidden sector also in estimating the axion masses and decay constants. With the hidden confining force, we need two θs which have to be settled to zero and hence we need at least two axions. In this case, axion mixing must be considered. Here, there is an important (almost) theorem: the *cross theorem* on decay constants and condensation scales. Suppose that there are two axions a_1 with F_1 and a_2 with F_2 ($F_1 \ll F_2$) which couple to two nonabelian groups with scales Λ_1 and Λ_2 ($\Lambda_1 \ll \Lambda_2$). The theorem states that [28,30]: according to the diagonalization process in most cases with general couplings, the larger condensation scale Λ_2 chooses the smaller decay constant F_1, and the smaller condensation scale Λ_1 chooses the larger decay constant F_2. So, just obtaining a small decay constant is not enough. The hidden sector may steal the smaller decay constant. It is likely that the QCD axion is left with the larger decay constant.

So, we must look for another possibility for a detectable QCD axion. This can be possible with an approximate PQ symmetry in string models. After all, the topologically attractive B_{MN} may not be the QCD axion we want. There exists an earlier field theoretic work regarding an approximate PQ symmetry, starting with a discrete symmetry [31] where Z_9 was used. In string models, such approximate PQ symmetry was not calculated before. Since now we have an explicit model for the MSSM [32,33], it is possible to consider an approximate symmetry. We can check whether this idea of approximate global symmetry is realized. For this idea to work, it is better that the PQ symmetry is preserved up to sufficiently higher orders. In this sense, Z_{12} is helpful. In this vein, we calculated the $a\gamma\gamma$ coupling from the Z_{12-I} superstring model [32] for the first time [34]. This string calculation is shown as a line in Figs. 1 and 3. Here, there are so many Yukawa couplings to be considered in a string derived model. For example, we encountered $O(10^4)$ terms for the $d = 7$ superpotential and it is not a trivial task to find an approximate PQ symmetry direction. Noting that the MI axion with anomalous U(1) always has a large decay constant since all fields are

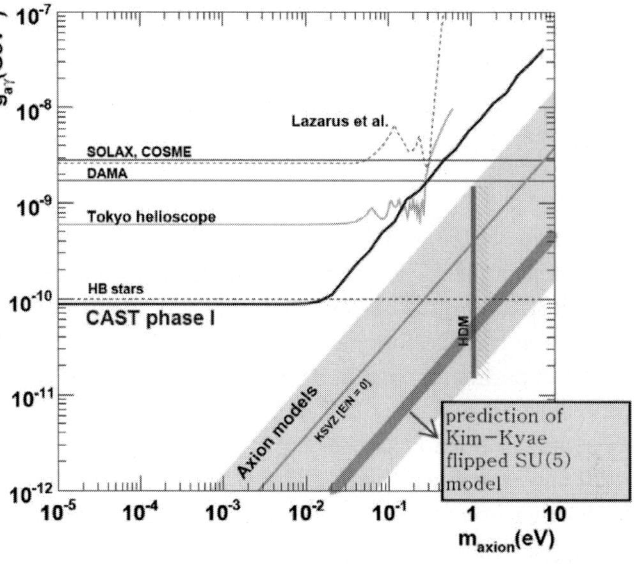

Fig. 1. The CAST bound with other constraints. Here, we show some field theoretic models and the prediction from the string Z_{12-I} model [32].

charged under this anomalous U(1). So, a phenomenologically observable QCD axion must need an approximate PQ symmetry.

4. Axions from Stars

We use the axion couplings to e, p, n, and photon to study the core evolution of a star. The important process is the Primakoff process for which the following coupling $c_{a\gamma\gamma}$ is assumed,

$$\mathcal{L} = -c_{a\gamma\gamma}\frac{a}{F_a}\{F_{\rm em}\tilde{F}_{\rm em}\}, \quad c_{a\gamma\gamma} = \bar{c}_{a\gamma\gamma} - 1.93$$
$$\bar{c}_{a\gamma\gamma} = {\rm Tr}Q^2_{\rm em}|_{E\gg M_Z}. \tag{6}$$

Axion helioscopes of Tokyo [35] and CAST (CERN axion solar telescope) [36] and also the geomagnetic conversion use this photon coupling.

In the hot plasma in stars, axions once produced most probably escape the core of the star and take out energy. This contributes to the energy loss mechanism of star and should not dominate the luminocity. From the Primakoff process which is present in any model, we obtain $F_a > 10^7$ GeV except the $m_a > 200$ keV region for which a is too heavy to be produced in the core of a star. From the nucleon-nucleon-a coupling, SN1987A gave

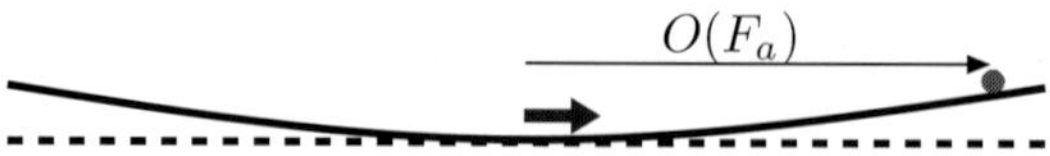

Fig. 2. The almost flat axion potential. The misalignment angle is expected to be of order 1 but can also be very small as shown by the thick blue arrow.

a strong bound $F_a > 0.6 \times 10^9$ GeV [37] with the correct nucleon state taken into account in [38]. The recent refined analysis gives $F_a > 10^{10}$ GeV [39]. For the DFSZ axion, the aee coupling is present and the Compton-like scattering ($\gamma e \to ae$) in the core gave the bound $g_{aee} < 1.3 \times 10^{-13}$ [39].

Laboratory experiments can perform more than just the energy loss mechanism in the core of a star. The early Tokyo experiment [35] could not give a more stringent bound than the supernova limit, but the CAST could compete with the supernova bound. The CAST result is shown in Fig. 1 together with other bounds [36]. In this figure, hypothetical field theoretic models are schematically shown also.

5. Axions in the Universe

The potential of the very light axion is of almost flat. Therefore, a chosen vacuum point (the red bullet in Fig. 2) stays there for a long time, and starts to oscillate when the Hubble time H^{-1} is comparable to the oscillation period (the inverse axion mass), $H < m_a$. This occurs when the temperature of the universe is about 1 GeV [40]. The DW problem in the standard Big Bang cosmology [41] disappears when the reheating temperature after inflation is below the PQ symmetry breaking scale. Since the reheating temperature is $T_{\rm RH} < 10^9$ GeV or 10^7 GeV in some models [42], here we will not worry about the DW problem any more.

The axion is created at $T = F_a$, but the universe $\langle a \rangle$ does not begin to roll until $H = m_a$, i.e. at $T = 1$ GeV. From then, the classical field $\langle a \rangle$ starts to oscillate. Then, the axion CDM energy fraction is given by $\Omega_a = \frac{1}{2}\left(0.6 \times 10^{-5}{\rm eV}/m_a\right)^{7/6}$. If F_a is large($> 10^{12}$ GeV), then the axion energy density dominates the universe. Summarizing the astro and cosmological constraints, we take the axion F_a window as,

$$10^{10}\ {\rm GeV} \le F_a \le 10^{12}\ {\rm GeV}. \tag{7}$$

But there is an anthropic argument even beyond $F_a > 10^{12}$ GeV.

The first anthropic argument on axion was given by Pi in the new inflationary model [44] and the recent more refined version is given by Tegmark et. al. [45]. The homogeneous axion field value (with $a \to -a$ symmetry) right after inflation can take any value between 0 and πF_a or $\bar{\theta}_{\rm mis} = [0, \pi]$. So Ω_a may be at the required value by an appropriate initial misalignment angle for $F_a > 10^{12}$ GeV by taking the needed $\bar{\theta}_{\rm mis}$ in the new, chaotic or hybrid inflationary scenarios. Tegmark et. al. argue that their prior probability function is calculable for axion models, which is rather obvious. For the axion with the almost flat potential, it is almost equally probable for $\langle a \rangle$ to sit anywhere in the region $\langle a \rangle = [0, F_a]$ since its location does not affect the other anthropic arguments conspicuously. For axion, one relevant figure for our purpose is scalar fluctuation $Q \simeq \delta\rho/\rho$ vs ξ (matter density per CMB photon), which is shown in [45]. If a WIMP is the sole candidate for CDM, one obtains just one number for $\delta\rho/\rho$. So, we may need the CDM favored WIMP and in addition the axion with $F_a > 10^{12}$ GeV. This fits very well to the anthropic principle since any extra $\delta\rho/\rho$ can be provided by axion with an appropriate $\bar{\theta}_{\rm mis}$ with $F_a > 10^{12}$ GeV. Namely, WIMPs may be dominantly the CDM, and the rest amount of DM is provided by axions using some of the anthropic arguments. The Compton wave length of micro eV axion is roughly 20 cm. So, the optimal size for detecting a micro eV axion is of dimension (20 cm)^3. The optimal dimension of the detector depends on the mass of axion, which is the key difficulty in the cosmic axion search.

If axion is the CDM component of the universe, then they can be detected even though it may be very difficult. Sikivie suggested the cavity axion detector [46]. To detect axion masses in the region 10^{-6} eV, one needs low temperature cavity with dimension $O(> 10^4\ {\rm cm}^3)$ and the magnetic field strength of $O(10$ Tesla). The current status of cosmic axion search is shown in Fig. 3.

6. Axinos as CDM

Not to have too many thermally produced gravitinos after inflation, the reheating temperature must be bounded [42]: $T_{\rm RH} < 10^9$ GeV (old bound) or $T_{\rm RH} < 10^7$ GeV (new bound if $M_{\rm gluino} < m_{3/2}$). Thus, in SUSY theories we must consider a relatively small reheating temperature. The axion solution of the strong CP problem with SUSY implies the superpartner of axion, *axino*, with a low reheating temperature. The neutralino LSP seems the most attractive candidate for DM simply because the TeV order SUSY breaking scale introduces the LSP as a WIMP. This scenario needs an exact or an effective R-parity for proton to be sufficiently long lived.

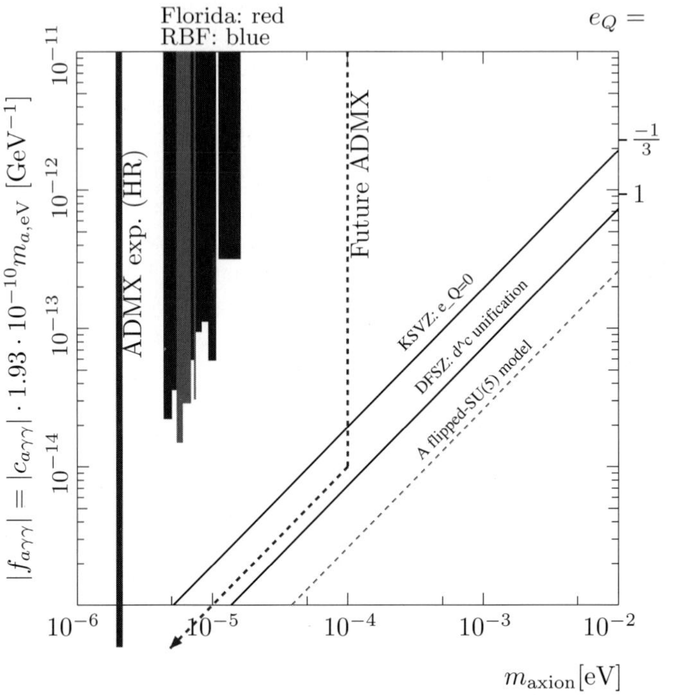

Fig. 3. The bounds on cosmic axion searches with some theoretical expectations.

Axion's scalar partner, saxion, can also affect the cosmic evolution, but its effect is not so dramatic as the effect of axino [47].

For axino to be the LSP, it must be lighter than the lightest neutralino and gravitino. Thus, the axino mass estimate is of prime importance. The conclusion is that there is no theoretical upper bound on the axino mass [48], and we take it any value between 100 GeV and keV. Thus, there are two axino DM possibilities: warm DM with keV axinos [49] and CDM with GeV axinos [50]. The gravitino problem is absent if gravitino is the next LSP (NLSP), $m_{\tilde a} < m_{3/2} < m_\chi$, since the thermally produced (TP) gravitinos would decay to axino and axion which would not affect the BBN produced light elements [51]. On the other hand, if the lightest neutralino is the NLSP, the TP mechanism restricts the reheating temperature after inflation. At the high reheating temperature, the TP is dominant in the axino production. If the reheating temperature is below the critical energy density line, there exists another axino CDM possibility by the nonthermally produced (NTP) axinos by the neutralino decay [52]. This situation

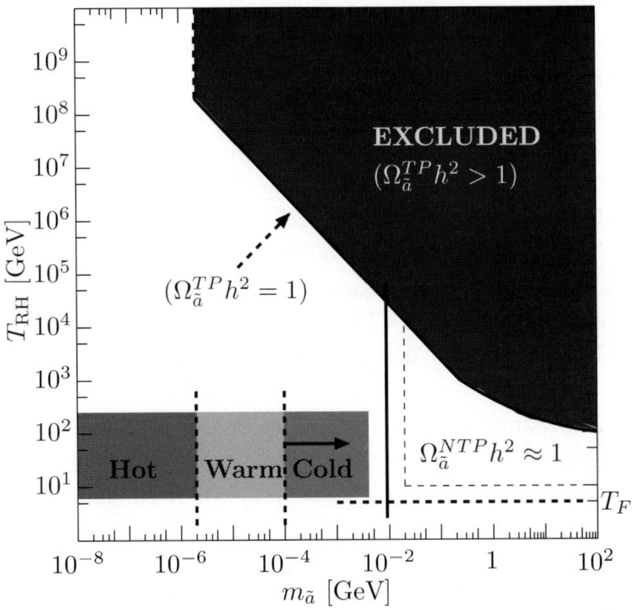

Fig. 4. The solid line is the upper bound from TP. The yellow region is the region where NTP can give cosmologically interesting results ($\Omega_{\tilde{a}}^{\rm NTP}h^2 \simeq 1$). The freezeout temperature is $T_F \approx \frac{m_\chi}{20}$.

is shown in Fig. 4. Since the final axino energy fraction is reduced by the mass ratio, $\Omega_{\tilde{a}}h^2 = (m_{\tilde{a}}/m_\chi)\Omega_\chi h^2$ for $m_{\tilde{a}} < m_\chi < m_{3/2}$, the stringent cosmologically constrained MSSM parameter space for m_χ can be expanded. As shown in this figure, the NTP axinos can be CDM for relatively low reheating temperature ($<$ 10 TeV) for 10 MeV $< m_{\tilde{a}} < m_\chi$. The yellow region corresponds to the MSSM models with $\Omega_\chi h^2 < 10^4$, and a small axino mass renders the possibility of axino forming 23% of the closure density. If all SUSY mass parameters are below 1 TeV, then probably $\Omega_\chi h^2 < 100$ (the grey region) but a sufficient axino energy density requires $m_{\tilde{a}} > 1$ GeV. Thus, if the LHC does not detect the neutralino needed for its closing of the universe, the axino closing is a possibility.

In the gauge mediated SUSY breaking (GMSB) scenario [53], the gravitino mass is generally smaller than the neutralino mass and possibly smaller than the axino mass, for which case cosmology has been studied [54]. Indeed, recently it has been shown that the GMSB is possible in the compactification of the heterotic string [33].

7. Conclusion

The popular CDM candidates are WIMPs and very light axions. We know that they must fill the universe based on the observational grounds and by the fact that we exist here in a planet. Direct searches for WIMPs in the universe use the WIMP cross section in our environment. The LHC machine will tell whether the LSP mass falls in the CDM needed range or not. The other candidate a very light axion, whether or not it is the dominant CDM component, is believed to exist from the need for a solution of the strong CP problem. So, I reviewed the axion and the related issues: solutions of the strong CP problem, the axion CDM possibility, the axino CDM possibility with order GeV axino mass, and an observable QCD axion possibility from superstring.

Maybe solar axions are easier to be detected than cosmic axions. Then, axion is not the dominant component of CDM. Most exciting however would be that axion is discovered and its discovery confirms instanton physics of QCD (by experiments).

Acknowledgments

This work is supported in part by the KRF Grants, No. R14-2003-012-01001-0 and No. KRF-2005-084-C00001.

References

1. D. N. Spergel *et. al.* (WMAP Collaboration), Astrophys. J. Suppl. Ser. **170**, 377 (2007).
2. For more references, see, J. E. Kim, arXiv.0711.1708.
3. B. W. Lee and S. Weinberg, Phys. Rev. Lett. **39**, 165 (1977).
4. D. N. Spergel *et. al*, Astrophys. J. Suppl. Ser. **148**, 175 (2003).
5. R. Bernabei *et. al.* (DAMA collaboration), Riv. Nuovo Cim. **26**, 1 (2003); H. S. Lee *et. al.* (KIMS Collaboration), Phys. Rev. Lett. **99**, 091301 (2007).
6. J. Silk, Astrophys. J. **151**, 459 (1968).
7. C. A. Baker *et. al.*, Phys. Rev. Lett. **97**, 131801 (2006).
8. A. E. Nelson, Phys. Lett. **B136**, 387 (1984); S. M. Barr, Phys. Rev. Lett. **53**, 329 (1984).
9. G. 't Hooft, Phys. Rev. **D14**, 3472 (1976).
10. A. V. Manohar and C. T. Sachrajda, in Particle Data Book [W.-M. Yao *et. al.*, J. of Physics **G33**, 1 (2006)].
11. R. D. Peccei and H. R. Quinn, Phys. Rev. Lett. **38**, 1440 (1977).
12. S. Weinberg, Phys. Rev. Lett. **40**, 223 (1978); F. Wilczek, Phys. Rev. Lett. **40**, 279 (1978).
13. C. Vafa and E. Witten, Phys. Rev. Lett. **53**, 535 (1984).

14. J. E. Kim, Phys. Rev. Lett. **43**, 103 (1979); M. A. Shifman, V. I. Vainstein, V. I. Zakharov, Nucl. Phys. **B166**, 4933 (1980).
15. M. Dine, W. Fischler and M. Srednicki, Phys. Lett. **B104**, 199 (1981); A. P. Zhitnitskii, Sov. J. Nucl. Phys. **31**, 260 (1980).
16. Once it was called an invisible axion but is better to be called a very light axion due to the possibility of its detection.
17. K. Choi, Phys. Rev. **D56**, 6588 (1997).
18. L. Di Lella, A. Pilaftsis, G. Raffelt, and K. Zioutas, Phys. Rev. **D62**, 125011 (2000); M. Chaichian and A. B. Kobakhidze, Phys. Rev. Lett. **87**, 171601 (2001); K. Choi, Phys. Rev. Lett. **92**, 101602 (2004).
19. J. E. Kim, Phys. Rev. **D31**, 1733 (1985); K. Choi and J. E. Kim, Phys. Rev. **D32**, 1828 (1985); K. S. Babu, K. Choi, J. C. Pati, and X. Zhang, Phys. Lett. **B333**, 364 (1994).
20. For a review, see, J. E. Kim, Phys. Rep. **150**, 1 (1987).
21. E. Witten, Phys. Lett. **B149**, 351 (1984).
22. E. Witten, Phys. Lett. **B155**, 151 (1985).
23. X. G. Wen and E. Witten, Phys. Lett. **B166**, 397 (1986).
24. K. Choi and J. E. Kim, Phys. Lett. **B154**, 393 (1985).
25. M. Dine, N. Seiberg, and E. Witten, Nucl. Phys. **B289**, 317 (1987); J. J. Atick, L. J. Dixon, and A. Sen, Nucl. Phys. **B292**, 109 (1987); M. Dine, I. Ichinose, and N. Seiberg, Nucl. Phys. **B293**, 253 (1987).
26. J. E. Kim, Phys. Lett. **B207**, 434 (1988).
27. J. P. Conlon, JHEP **0605**, 078 (2006).
28. I.-W. Kim and J. E. Kim, Phys. Lett. **B639**, 342 (2006).
29. S. B. Giddings, S. Kachru and J. Polchinski, Phys. Rev. **D66**, 106006 (2002).
30. J. E. Kim, JHEP **9905**, 022 (1999).
31. G. Lazarides and Q. Shafi, Phys. Rev. Lett. **56**, 432 (1986).
32. J. E. Kim and B. Kyae, Nucl. Phys. **B770**, 47 (2007).
33. J. E. Kim, Phys. Lett. **B651**, 407 (2007); Phys. Lett. **B656**, 207 (2007).
34. K.-S. Choi, I.-W. Kim and J. E. Kim, JHEP **0703**, 116 (2007).
35. W. Ootani *et. al.*, Phys. Lett. **B461**, 371 (1999); Y. Inoue *et. al.*, Phys. Lett. **B536**, 18 (2002).
36. S. Andriamonje *et. al.* (CAST Collaboration), JCAP **0704**, 010 (2007).
37. G. Raffelt and D. Seckel, Phys. Rev. Lett. **60**, 1793 (1988); M. S. Turner, Phys. Rev. Lett. **60**, 1797 (1988); R. Mayle *et. al.*, Phys. Lett. **B203**, 188 (1988); T. Hatsuda and M. Yoshimura, Phys. Lett. **B203**, 469 (1988).
38. K. Choi, K. Kang and J. E. Kim, Phys. Rev. Lett. **62**, 849 (1989).
39. G. Raffelt, hep-ph/0611350 (2006).
40. J. Preskill, M. B. Wise and F. Wilczek, Phys. Lett. **B120**, 127 (1983); L. F. Abbot and P. Sikivie, Phys. Lett. **B120**, 133 (1983); M. Dine and W. Fischler, Phys. Lett. **B120**, 137 (1983).
41. P. Sikivie, Phys. Rev. Lett. **48**, 1156 (1982).
42. J. Ellis, J. E. Kim and D. V. Nanopoulos, Phys. Lett. **B145**, 181 (1984); M. Kawasaki, K. Kohri and T. Moroi, Phys. Rev. **D71**, 083502 (2005).
43. S. J. Asztalos, L. J. Rosenberg, K. van Bibber, P. Sikivie, and K. Zioutas, Ann. Rev. Nuc. Part. Sci. **56**, 293 (2006).

44. S.-Y. Pi, Phys. Rev. Lett. **52**, 1725 (1984).
45. M. Tegmark, A. Aguirre, M. Rees, and Wilczek, Phys. Rev. **D73**, 023505 (2006).
46. P. Sikivie, Phys. Rev. Lett. **51**, 1415 (1983).
47. J. E. Kim, Phys. Rev. Lett. **67**, 3465 (1991); S. Chang and H. B. Kim, Phys. Rev. Lett. **77**, 591 (1996).
48. E. J. Chun, J. E. Kim and H. P. Nilles, Phys. Lett. **B287**, 123 (1992); E. J. Chun and A. Lukas, Phys. Lett. **B357**, 43 (1995). See, also, J. E. Kim, A. Masiero and D. V. Nanopoulos, Phys. Lett. **B139**, 346 (1984).
49. K. Rajagopal, M. S. Turner and F. Wilczek, Nucl. Phys. **B358**, 447 (1991).
50. L. Covi, J. E. Kim and L. Roszkowski, Phys. Rev. Lett. **82**, 4180 (1999).
51. T. Asaka and T. Yanagida, Phys. Lett. **B494**, 297 (2000).
52. L. Covi, H. B. Kim, J. E. Kim and L. Roszkowski, JHEP **0105**, 033 (2001).
53. M. Dine and A. E. Nelson, Phys. Rev. **D48**, 1277 (1993); M. Dine, A. E. Nelson and Y. Shirman, Phys. Rev. **D51**, 1362 (1995).
54. E. J. Chun, H. B. Kim and J. E. Kim, Phys. Rev. Lett. **72**, 1956 (1994); H. B. Kim and J. E. Kim, Nucl. Phys. **B433**, 421 (1995).

NEUTRINO DARK ENERGY, Λ DOMINANCE AND RESOLUTION OF THE FIFTH FORCE PROBLEM

E. I. GUENDELMAN* and A. B. KAGANOVICH**

Physics Department, Ben Gurion University,
Beer Sheva 84105, Israel
**E-mail: guendel@bgu.ac.il*
***E-mail: alexk@bgu.ac.il*

A field theory is presented where the behavior of the matter may be completely different depending on its density. In normal conditions, i.e. when the local energy density of the matter is many orders of magnitude larger than the dark energy density, the General Relativity is reproduced including the absence of the long range scalar force. In the opposite limit, where the order of magnitude of the (local) matter energy density becomes comparable with the dark energy density, the behavior of the matter drastically changes. In such Cosmo-Low-Energy-Physics (CLEP) states, the dust possesses a pressure and for example, gas of nonrelativistic neutrinos behaves as the dark energy. The latter consists of the cosmological constant (CC) and the quintessence-like scalar field. A transition to a dominance of the CC is demonstrated.

Keywords: Dark energy; Cosmological Constant; Neutrino; Fifth force.

The proposed field theory model is an alternative theory which is a generally coordinate invariant with the effective action of the form[1] $S = \int L_1 \Phi d^4x + \int L_2 \sqrt{-g} d^4x$ including two Lagrangians L_1 and L_2 and two volume measures $\sqrt{-g}$ and Φ. One is the usual measure of integration $\sqrt{-g}$ in the 4-dimensional space-time manifold equipped with the metric $g_{\mu\nu}$. Another is the new measure of integration $\Phi = \varepsilon^{\mu\nu\alpha\beta}\varepsilon_{abcd}\partial_\mu\varphi_a\partial_\nu\varphi_b\partial_\alpha\varphi_c\partial_\beta\varphi_d$, $a = 1,2,3,4$ in the same 4-dimensional space-time manifold defined by means of four scalar fields φ_a. It is assumed that in such a Two Measures Field Theory (TMT) the Lagrangian densities L_1 and L_2 are functions of all matter fields, the metric, the connection but not of the "measure fields" φ_a. In such a case the theory possesses an infinite dimensional symmetry.[1] One can hope that this symmetry should prevent emergence of a measure fields dependence in L_1 and L_2 after quantum effects are taken into account. TMT is formulated in the first order formalism, that is all fields, including also metric, connection and the measure fields

φ_a are taken as independent dynamical variables. All the relations between them are results of equations of motion. In particular, the independence of the metric and the connection means that the relation between them is not necessarily according to Riemannian geometry. Except for the listed assumptions we do not make any changes as compared with principles of the standard field theory in curved space-time. In other words, all the freedom in constructing different models in the framework of TMT consists of the choice of the concrete matter content and the Lagrangians L_1 and L_2 that is quite similar to the standard field theory.[3-5]

The matter content of our simplified model includes the dilaton scalar field ϕ and the so-called primordial neutrino field Ψ. We allow in both L_1 and L_2 all the usual terms considered in field theory models in curved space-time. Keeping the general structure of the TMT action it is convenient to represent it in the following form:[3]

$$S = \int d^4x e^{\alpha\phi/M_p}(\Phi + b\sqrt{-g})\left[-\frac{1}{\kappa}R(\omega, e) + \frac{1}{2}g^{\mu\nu}\phi_{,\mu}\phi_{,\nu}\right]$$
$$- \int d^4x e^{2\alpha\phi/M_p}[\Phi V_1 + \sqrt{-g}V_2] - \int d^4x e^{\frac{3}{2}\alpha\phi/M_p}(\Phi + h\sqrt{-g})\mu\overline{\Psi}\Psi$$
$$+ \int d^4x e^{\alpha\phi/M_p}(\Phi + k\sqrt{-g})\frac{i}{2}\overline{\Psi}\left(\gamma^a e_a^\mu \overrightarrow{\nabla}_\mu^{(i)} - \overleftarrow{\nabla}_\mu^{(i)}\gamma^a e_a^\mu\right)\Psi \qquad (1)$$

where $\overrightarrow{\nabla}_\mu^{(N)} = \vec{\partial} + \frac{1}{2}\omega_\mu^{cd}\sigma_{cd}$; $R(\omega, V) = e^{a\mu}e^{b\nu}R_{\mu\nu ab}(\omega)$ is the scalar curvature; e_a^μ and ω_μ^{ab} are the vierbein and spin-connection; $g^{\mu\nu} = e_a^\mu e_b^\nu \eta^{ab}$ and $R_{\mu\nu ab}(\omega) = \partial_{mu}\omega_{\nu ab} + \omega_{\mu a}^c\omega_{\nu cb} - (\mu \leftrightarrow \nu)$. V_1 and V_2 are constants with the dimensionality $(mass)^4$. Constants b, k, h are non specified dimensionless real parameters and we will only assume that they have close orders of magnitude $b \sim k \sim h$; α is a real parameter which we take to be positive.

The action (1) is invariant under the global scale transformations:

$$e_\mu^a \to e^{\theta/2}e_\mu^a, \quad \varphi_a \to \lambda_{ab}\varphi_b, \quad \phi \to \phi - \frac{M_p}{\alpha}\theta, \quad \Psi \to e^{-\theta/4}\Psi, \qquad (2)$$

where $\theta = const$, $\lambda_{ab} = const$, $\det(\lambda_{ab}) = e^{2\theta}$ and $\omega_{ab}^\mu \to \omega_{ab}^\mu$.

Variation of the measure fields φ_a yields the equation $L_1 = sM^4$ where L_1 is now defined as the part of the integrand of the action (1) coupled to the measure Φ. The appearance of the integration constant sM^4 spontaneously breaks the global scale invariance (2). In what follows we choose $s = +1$. All other equations of motion resulting from (1) in the first order formalism depend on the measure fields φ_a only via the scalar field $\zeta \equiv \Phi/\sqrt{-g}$ and contain terms proportional to $\partial_\mu\zeta$ that makes the space-time non-Riemannian and equations of motion - non canonical. However, with the new set of variables (ϕ remains unchanged) $\tilde{e}_{a\mu} = e^{\frac{1}{2}\alpha\phi/M_p}(\zeta + b)^{1/2}e_{a\mu}$, $\tilde{g}_{\mu\nu} = e^{\alpha\phi/M_p}(\zeta + b)g_{\mu\nu}$, $\Psi' = e^{-\frac{1}{4}\alpha\phi/M_p}(\zeta + k)^{1/2}(\zeta + b)^{-3/4}\Psi$, which we call the *Einstein frame*, the spin-connections become those of the Einstein-Cartan space-time. One can show that equations for the primordial fermions in the Einstein frame take the standard form of fermionic equations in the Einstein-Cartan space-time, and fermion 4-currents are not only covariantly conserved but also have the canonical form without any ϕ and ζ dependence. Since $\tilde{e}_{a\mu}$, $\tilde{g}_{\mu\nu}$ and Ψ' are invariant under the scale transformations (2), spontaneous breaking of the scale symmetry (by means of the equation $L_1 = sM^4$) is reduced in the new variables to the *spontaneous breaking of the shift symmetry* $\phi \to \phi + const$. Notice that the Goldstone theorem generically is not applicable in this theory (see the second reference in Ref.[2])).

After the change of variables to the Einstein frame and some simple algebra, the gravitational equations take the standard GR form $G_{\alpha\beta}(\tilde{g}_{\mu\nu}) = \frac{\kappa}{2}T^{eff}_{\alpha\beta}$ where $G_{\alpha\beta}(\tilde{g}_{\mu\nu})$ is the Einstein tensor in the Riemannian space-time with the metric $\tilde{g}_{\mu\nu}$ and $T^{eff}_{\alpha\beta} = \phi_{,\alpha}\phi_{,\beta} - \frac{1}{2}\tilde{g}_{\alpha\beta}\tilde{g}^{\mu\nu}\phi_{,\mu}\phi_{,\nu} + \tilde{g}_{\alpha\beta}V_{eff}(\phi;\zeta) + T^{(\nu,can)}_{\alpha\beta} + T^{(\nu,noncan)}_{\mu\nu}$; $V_{eff}(\phi;\zeta) = (\zeta + b)^{-2}[b\left(M^4e^{-2\alpha\phi/M_p} + V_1\right) - V_2]$; $T^{(\nu,can)}_{\alpha\beta}$ and $T^{(\nu,noncan)}_{\alpha\beta}$ are the canonical and *noncanonical* contributions to the energy momentum tensor of the neutrino $T^{(\nu,noncan)}_{\alpha\beta} = -\tilde{g}_{\alpha\beta}Z(\zeta)m(\zeta)\overline{\Psi'}\Psi' \equiv -\tilde{g}_{\alpha\beta}\Lambda^{(\nu)}_{dyn}$ where $Z(\zeta)$ and $m(\zeta)$ are respectively

$$Z(\zeta) \equiv \frac{(\zeta - \zeta_1)(\zeta - \zeta_2)}{2(\zeta + k)(\zeta + h)}, \qquad m(\zeta) = \frac{\mu(\zeta + h)}{(\zeta + k)(\zeta + b)^{1/2}} \tag{3}$$

where $\zeta_{1,2} = \frac{1}{2}\left[k - 3h \pm \sqrt{(k - 3h)^2 + 8b(k - h) - 4kh}\right]$.

The noncanonical contribution $T^{(\nu,noncan)}_{\mu\nu}$ of the fermions into the energy-momentum tensor has the transformation properties of a cosmological constant (CC) term but it is proportional to $\overline{\Psi}'\Psi'$. This is why we refer to it as "dynamical fermionic Λ term". This fact is displayed explicitly by defining $\Lambda^{(\nu)}_{dyn}$. Notice the very nontrivial and important result[3] that due to $T^{(\nu,noncan)}_{\alpha\beta}$, the nonrelativistic fermions may possess a pressure. Similar result is also obtained for a dust matter described phenomenologically.[4]

The dilaton ϕ field equation in the Einstein frame reads

$$\Box\phi - \frac{\alpha}{M_p(\zeta + b)}\left[M^4e^{-2\alpha\phi/M_p} - \frac{(\zeta - b)V_1(v) + 2V_2(v)}{\zeta + b}\right] = -\frac{\alpha}{M_p}\Lambda^{(ferm)}_{dyn}, \tag{4}$$

where $\Box\phi = (-\tilde{g})^{-1/2}\partial_\mu(\sqrt{-\tilde{g}}\tilde{g}^{\mu\nu}\partial_\nu\phi)$.

The scalar field ζ is defined by the constraint which is the consistency condition of equations of motion. In the Einstein frame the constraint takes the form

$$-\frac{1}{(\zeta+b)^2}\left\{(\zeta-b)\left[sM^4e^{-2\alpha\phi/M_p}+V_1\right]+2V_2\right\}=Z(\zeta)m(\zeta)\overline{\Psi'}\Psi'. \quad (5)$$

Generically the constraint (5) determines ζ as a very complicated function of ϕ and $\overline{\Psi'}\Psi'$. However, in a few most important situations the theory allows exact solutions of great interest.[3–5]

In the case of the complete absence of massive fermions, the constraint determines ζ as the function of ϕ alone. Then the potential of the ϕ-field describing the dark energy in the vacuum reads

$$V^{(0)}_{eff}(\phi)\equiv V_{eff}(\phi;\zeta)|_{\overline{\Psi'}\Psi'=0}=\frac{(V_1+sM^4e^{-2\alpha\phi/M_p})^2}{4[b(V_1+sM^4e^{-2\alpha\phi/M_p})-V_2]}. \quad (6)$$

One should now pay attention to the interesting and very important result that the explicit $\overline{\Psi'}\Psi'$ dependence involving **the same form of ζ dependence** $Z(\zeta)m(\zeta)\overline{\Psi'}\Psi'\equiv\Lambda^{(\nu)}_{dyn}$ appears simultaneously[a] in the neutrino contribution to the pressure (through $T^{(\nu,noncan)}_{\alpha\beta}$), in the effective dilaton to dust coupling (in the r.h.s. of Eq. (4)) and in the r.h.s. of the constraint (5). With the made above assumption $b\sim k\sim h$, the detailed analysis of the constraint (5) shows that it describes a balance between the dark energy density (which is of the order close to the l.h.s. of (5) and $m(\zeta)\overline{\Psi'}\Psi'$ which is the energy density of the nonrelativistic fermion matter. When the latter is many orders of magnitude larger than the dark energy density (that is the *normal condition* for particle theory and GR) the balance is provided only if $Z(\zeta)$ is extremely close to zero. This explains why in the normal conditions simultaneously the fermion masses become constant, the pressure of both nonrelativistic fermions[3] and dust[4] becomes zero, the fifth force Yukawa coupling turns out to be zero automatically, and the GR is restored.[3–4]

Let us study a toy model[3] where in addition to the homogeneous scalar field ϕ, the spatially flat universe is filled also with the cold gas of uniformly distributed nonrelativistic neutrinos. It turns out that in such a model, a regime opposite to the normal conditions becomes possible and further it is energetically more preferable than the one in the absence of fermions case. Spreading of the neutrino wave packets during their free motion lasting

[a]Note that analogous result has been observed in the model[4] where the macroscopic (dust) matter has been studied instead of fermionic matter in the present model.

a long time yields extremely small values of $\overline{\Psi}'\Psi' = u^\dagger u$ (u is the large component of the Dirac spinor Ψ'). There is a solution where the decaying fermion contribution $u^\dagger u \sim \frac{const}{a^3}$ to the constraint (5) is compensated by approaching $\zeta \to -k$, see Eq.(3). The change of ζ from its ϕ-dependence in the fermion vacuum to the regime $\zeta \to -k$ leads to a reorganization of the ϕ-field dynamics. Using the constraint, the neutrino contribution to the total averaged energy-momentum tensor $< T_{\alpha\beta}^{(tot)} >_{cosm.av.} = < T_{\alpha\beta}^{(\nu)} >_{cosm.av.}$ $+T_{\alpha\beta}^{(\phi)}$ can be expressed in terms of the ϕ-field alone. Then the total dark energy energy density and pressure take the form $\rho_{d.e.} = \frac{1}{2}\dot{\phi}^2 + U_{eff}^{(tot)}(\phi)$ and $p_{d.e.} = \frac{1}{2}\dot{\phi}^2 - U_{eff}^{(tot)}(\phi)$ respectively where $U_{eff}^{(tot)}(\phi) \equiv \Lambda + V_{q-like}(\phi)$, $\Lambda = (V_2 - kV_1)(b-k)^{-2}$ is the CC and $V_{q-like}(\phi) = -k(b-k)^{-2}M^4 e^{-2\alpha\phi/M_p} + \mathcal{O}(\zeta + k)$ is the quintessence-like potential.

For a particular value of the parameter $\alpha = \sqrt{3/8}$, the cosmological equations allow the following analytic solution: $\phi(t) = \frac{M_p}{2\alpha}\varphi_0 + \frac{M_p}{2\alpha}\ln(M_p t)$, $a(t) \propto t^{1/3}e^{\lambda t}$, where $\lambda = M_p^{-1}\sqrt{\Lambda/3}$, $e^{-\varphi_0} = \frac{2(b-k)^2 M_p^2}{\sqrt{3}|k|M^4}\sqrt{\Lambda}$. The neutrino mass in such Cosmo-Low Energy Physics (CLEP) state increases exponentially in time and its ϕ dependence is double-exponential: $m_N|_{clep} \sim (\zeta + k)^{-1} \sim a^{3/2}(t) \sim t^{1/2}e^{\frac{3}{2}\lambda t} \sim \exp\left[\frac{3\lambda e^{-\varphi_0}}{2M_p}\exp\left(\frac{2\alpha}{M_p}\phi\right)\right]$. The moment $t = t_c$ when decaying $V_{q-like}(\phi)$ becomes equal to Λ has an acceptable order $t_c = M_P/[\sqrt{3\Lambda}(\sqrt{2}-1)] \approx 4.8 \cdot 10^9 yrs$ where we have assumed that the present value of the dark energy density is dominated by Λ.

Acknowledgments

We would like to thank D. Eichler, R. Shaisultanov and R. Viollier for very interesting conversations.

References

1. See: E.I. Guendelman and A.B. Kaganovich, *Phys. Rev.* **D60**, 065004 (1999) and references therein.
2. E.I. Guendelman, *Mod. Phys. Lett.* **A14**, 1043 (1999); *Class. Quant. Grav.* **17**, 361 (2000); gr-qc/0004011; *Mod. Phys. Lett.* **A14**, 1397 (1999); gr-qc/9901067; hep-th/0106085; *Found. Phys.* **31**, 1019 (2001).
3. E.I. Guendelman and A.B. Kaganovich, *Int. J. Mod. Phys.* A**17**, 417 (2002); *Mod. Phys. Lett.* A**17**, 1227 (2002); Int.J.Mod.Phys. A**21**, 4373 (2006).
4. E.I. Guendelman and A.B. Kaganovich, arXiv: 0704.1998 [gr-qc], to appear in Ann. Phys.
5. E.I. Guendelman and A.B. Kaganovich, *Phys. Rev.* D**75**, 083505 (2007).

The FIRST RESULTS OF THE GammeV AXION-LIKE PARTICLES SEARCH EXPERIMENT

JONGHEE YOO
(for GammeV collaboration)
Fermi National Accelerator Laboratory, PO Box 500
Batavia, IL 60510, USA
E-mail: yoo@fnal.gov

We report the first results of the GammeV experiment, a search for meV mass particles converted from photons. The experiment uses a strong external magnetic field to induce oscillations between incident photons from a pulsed laser and axion-like particles. A mirror is placed at a position within the magnetic field such that photons can not pass through the second part of the magnetic field. On the other hand, the weakly interacting axion-like particles can pass through beyond the mirror and may convert back into the photons. The regenerated photons then detected with a photomultiplier tube. The oscillation baseline of the apparatus is designed to vary, allowing probes of different values of particle mass. We find no excess of events above background. The two-photon couplings of possible new scalar (pseudoscalar) particles to be less than 3.2×10^{-7} GeV^{-1} (3.2×10^{-7} GeV^{-1}) in the limit of massless particles.

Keywords: Axion, photon, dark matter.

1. Introduction

Modern cosmology indicates that 23% of the Universe consists of dark matter. There are no lack of dark matter candidate particles such as the LSP, LKP, axion, axino, and gravitino. Any stable neutral non-baryonic massive particles can be a dark matter candidate. The axion is a dark matter candidate within the Standard Model regime. The axion was originally suggested to solve the strong CP problem in QCD by breaking $U(1)_{PQ}$ symmetry.[1,2]

The axion-like interaction Lagrangian for a pseudoscalar particle is given by:

$$\mathcal{L}_{\rm int} = -\frac{g}{4}\phi F_{\mu\nu}\widetilde{F}^{\mu\nu} = g\phi(\vec{E}\cdot\vec{B}) \tag{1}$$

while that for a scalar particle is:

$$\mathcal{L}_{\text{int}} = -\frac{g}{4}\phi F_{\mu\nu}F^{\mu\nu} = g\phi(\vec{E}\cdot\vec{E} - \vec{B}\cdot\vec{B}). \tag{2}$$

Micro cavity experiment is currently the most sensitive experiment to detect the cosmogenic cold-dark matter axions. ADMX experiment may be able to detect the photons from the resonant conversion of the axion in the magnetic field permeated microwave cavity using an ultra-low-noise receiver. The earlier results using an 1 liter cavity and HFET amplifier consistent with absence of the axion-dark matter in the mass range of $m_a \simeq 10^{-4} \sim 10^{-6}$ eV and axion-photon coupling scale of $g_{\gamma aa} \simeq 10^{-13} \sim 10^{-15}$ GeV^{-1}.[3,4]

Couplings of light particles to photons are constrained by considering stellar cooling. Cosmological, astrophysical and nuclear physical constraints restrict the allowed axion mass ranges 10^{-6} eV $\leq m_a \leq 10^{-2}$ eV and 3 eV $\leq m_a \leq$ 20 eV.[5,6]

Recent reports from the PVLAS photon oscillation experiment showed a positive signal which may be interpreted as an evidence of a coupling of axion-like particle to photons.[7] Any weakly interacting two photon vertex particles produced by an oscillation would have escaped through the cavity mirrors if they are sufficiently weakly coupled to matter. PVLAS was able to determine a relative attenuation of laser polarization components transverse to and parallel to the external magnetic field. Newer PVLAS results[8] included the measurements of non-zero polarization rotation with 532 nm photons, and non-zero ellipticity.

The PVLAS measurements gave a consistent picture of a scalar mass $m_\phi \sim 1.2$ meV and a two-photon coupling scale $g \sim 2.5 \times 10^{-6}$ GeV^{-1}. The region is not excluded by the earlier BFRT experiment.[9,10] However, the PVLAS signals could not be reproduced after the experimental apparatus was rebuilt in an effort to improve the detection.[11] The reasons for the negative results are not clearly explained. The BMV experiment recently searched for regenerated photons and excluded at 99.9% C.L. the pseudoscalar interpretation of PVLAS signal.[12]

The GammeV experiment is designed to probe the region in parameter space suggested by the PVLAS results. The experiment is similar to that originally suggested by van Bibber et.al.[13] A positive signal would provide an unambiguous evidence of oscillations of photons into new weakly interacting particles.[14]

2. Experiment Setup

Figure 1 shows schematics of the GammeV experiment setup. There are three major component of the experiment. The photon source (laser), external magnetic field (Tevatron magnet) and photon read out (PMT). A Continuum Surelite I-20 Nd:YAG laser is used as the light source. The laser emits vertically polarized 160 mJ, 532 nm pulses of 5 ns width with a rate of 20 Hz. The vertical mode is used to probe pseudo-scaler particle search. For a scaler particle search, a half wave plate is used to obtain horizontal polarization mode. The laser lights are sent through a vacuum tight warm bore inserted into a 6 m Tevatron superconducting dipole magnet. The magnet produces a 5 T vertical field uniform across the aperture of the $1\frac{7}{8}$". A mirror located in the warm bore reflect back the beam to a power meter in the laser box in order not to heat up the magnet core. The mirror, on the end of a long hollow stainless steel "plunger", is inserted into the warm bore. The mirror may be placed at various positions within the magnet. In this experiment we have used two positions of plunger (center 2.9 m and edge 1 m) to cover entire PVLAS parameter space.

The beam passing through the end of the plunger is a pure (pseudo-)scalar beam. These particles can then oscillate back into photons through the magnetic field region after the mirror. The regenerated photons then propagates $\sim$ 5 m into a dark PMT box. A Hamamatsu H7422P PMT module is used to read out the photons.

The axion-photon oscillation probability can be written in GammeV engineering unit as :

$$P_{\gamma\to\phi} = \frac{4B^2\omega^2}{M^2(\Delta m^2)^2}\sin^2\left(\frac{\Delta m^2 L}{4\omega}\right) \tag{3}$$

$$\approx \frac{4B^2\omega^2}{M^2 m_\phi^4}\sin^2\left(\frac{m_\phi^2 L}{4\omega}\right) \tag{4}$$

$$= 1.5\times 10^{-11}\frac{(B/\text{Tesla})^2(\omega/\text{eV})^2}{(M/10^5\ \text{GeV})^2(m_\phi/10^{-3}\ \text{eV})^4} \times \sin^2\left(1.267\frac{(m_\phi/10^{-3}\ \text{eV})^2(L/\text{m})}{(\omega/\text{eV})}\right) \tag{5}$$

where B is the strength of the external magnetic field, ω is the inital photon energy, L is the magnetic oscillation baseline. $\Delta m^2 = m_\phi^2 - m_\gamma^2$, the mass-squared difference between the scalar mass and the effective photon mass. The phase advance of photons in the vacuum region may be modelled with an effective mass $m_\gamma^2 = -2\omega^2(n-1)$, where n is the index of

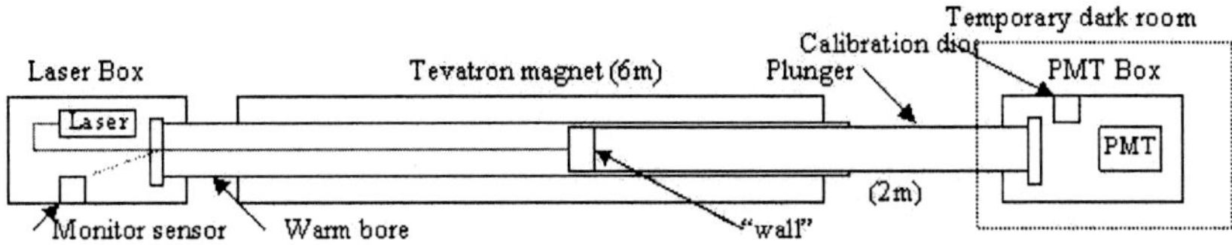

Fig. 1. Experiment setup of the GammeV.

refraction.[15] The warm bore and the interior of the plunger are pumped to vacuum pressures of 10^{-3} torr, and upper limit of the effective photon mass is estimated $\sqrt{-m_\gamma^2} < 4 \times 10^{-4}$ eV. The contribution from the effective photon mass is negligible the PVLAS allowed scaler mass region. The regeneration probability contains two factors of equation 4 and varies as $\sin^2\left(\frac{m_\phi^2 L_1}{4\omega}\right)\sin^2\left(\frac{m_\phi^2 L_2}{4\omega}\right)$ where $L_1 + L_2 = 6$ m.

The plunger guided into the PMT box and 2" lens to focus the beam onto the 5 mm diameter GaAsP photocathode of the PMT. The laser alignment is performed using a low power helium-neon laser. The alignment is verified before and after data-taking by replacing the sealed plunger with an open-ended plunger, and firing the Nd:YAG laser onto a flash paper target. An optical transport efficiency of 92% is measured using the ratio of laser power transmitted through the open-ended plunger through the various optics and vacuum windows to the initial laser power, using the same power meter in both cases to remove systematic effects. The quantum efficiency of the photocathode is specified to be 40% while the collection efficiency of the metal package PMT is believed to be near 100%. The PMT pulses are amplified by 46 dB and then sent into a NIM discriminator. Using a highly attenuated LED flasher as a single photon source, the discriminator threshold is optimized to give 99.4% efficiency for triggering on single photo-electron (spe) pulses while also efficiently rejecting the sub-spe noise sources. With this threshold, and using the built-in cooler to cool the tube to 0° C, a typical measured dark count rate of ~ 130 Hz is observed. The total photon counting efficiency is estimated to be $33 \pm 3.3\%$.

For timing coincidence measurements, we use modified Quarknet boards[16,17] for both laser box and PMT box signal read out. The Quarknet boards determines the time of rising and falling edges of time-over-threshold triggers from the PMT and from a monitoring photodiode that is located inside the laser box. The boards provide 1.25 ns timing precision. The clocks on the laser board and on the PMT board are synchronized using an external trigger from a signal generator. The absolute timing between the laser

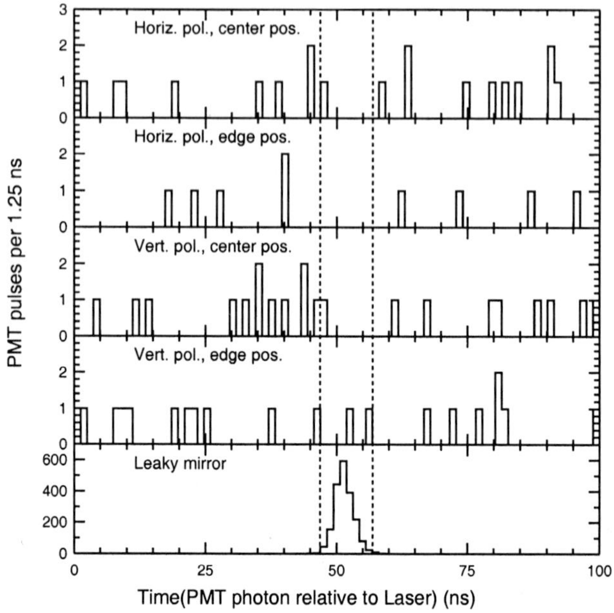

Fig. 2. PMT trigger times for the four run configurations, shown relative to the expected time distribution of photons as calibrated from the leaky mirror data.

pulses and the PMT traces is established by removing the plunger with the beam dump mirror, and allowing the laser to shine on the PMT through several attenuation stages consisting of two partially reflective mirrors, a pinhole, and multiple absorptive filters mounted directly on the aperture of the PMT module. The 10^{19} photons per second emitted by the laser are attenuated to a corresponding PMT trigger rate of less than 0.1 Hz for this timing calibration and to provide an in situ test of the data acquisition system.

For coincidence counting between the laser pulses and the PMT, a 10 ns gate is chosen to include 95% of the measured photon time distribution shown in Fig. 2. The coincident dark count rate is then estimated to be $R_{noise} = 20\ \text{Hz}\times 120\ \text{Hz}\times 10\ \text{ns} = 2.4\times 10^{-5}$ Hz. This noise rate is negligible to the expected signal rate of $\sim 2 \times 10^{-3}$ Hz estimated from the central values of the PVLAS parameters.

3. Data Collection

In order to probe the entire PVLAS range of m_ϕ and g, we took four configurations of plunger position and polarization. The plunger is placed either in the center of the magnet with $L_1 = 3.1$ m and $L_2 = 2.9$ m or near the far edge of the magnet with $L_1 = 5.0$ m and $L_2 = 1.0$ m. For each plunger position, vertically and horizontally polarized laser light is sent to into the magnet in order to test pseudoscalar and scalar couplings. The central plunger position covers most of the PVLAS signal region. Moving the plunger to the far position simultaneously changes the baseline for both the initial oscillation and the regeneration, and shifts the two regions of diminished sensitivity away. The two plunger positions thus cover the entire PVLAS signal region.

The operating conditions are monitored during each run. The reflected beam from the plunger mirror is directed into a calorimetric power meter in the laser box. The number of incident photons is determined with 10% accuracy. The beam spot is monitored using a camera which takes 30 Hz of images of the reflected laser spot on the pick-off mirror. The total pathlength to the pick-off mirror is comparable to the pathlength to the PMT; so transverse deviations seen in the images are closely matched to the deviations at the PMT. ~mm scale transverse deviations of beam offsets are seen during the course of run due to small changes in the orientation of the plunger mirror as the plunger slowly cools through heat leaks in the warm bore insulation. In addition, the focusing lens at the PMT makes the final light collection system insensitive to these potential sub-mrad angular deviations. Nevertheless, the alignment is double-checked using the open-ended plunger after collecting data in each configuration, and no misalignment has ever exceeded our tolerances. An LED flasher fires every 5 minutes to confirm the integrity of the light collection system. The operation of the PMT is monitored using its dark rate.

4. Results

The 10 ns coincidence time window is defined using the leaky mirror data. The background in the signal region is estimated by measuring the dark count rate outside of the coincidence window. We observe no statistically significant excess above the background in any of the four configurations. Table 1 shows summary of the results. The PMT timing data, along with the leaky-mirror calibration data, are shown in an expanded time scale in Fig. 2.

The Rolke-Lopez method[18] is used to obtain limits on the regeneration probabilities, and use Eqn. 4 to obtain the corresponding 3σ upper bounds on the coupling g as a function of m_ϕ. The laser power measurement and

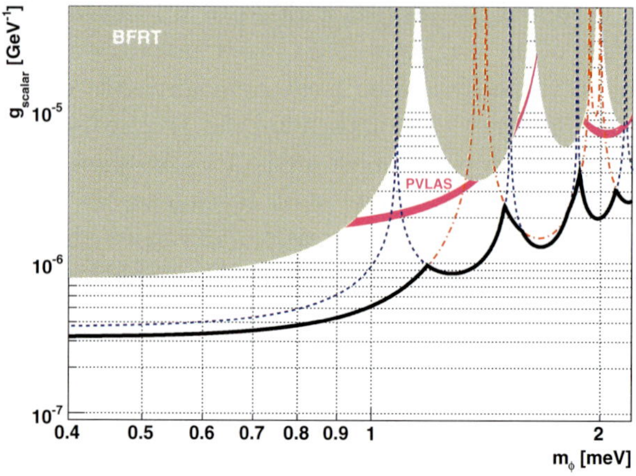

Fig. 3. 3σ limit contours for scalar particles. The solid black line is the union of the exclusion limits from the central (red dot-dashed) and the edge (blue dashed) plunger positions. The 3σ PVLAS rotation signal (pink/dark grey) and the 3σ BFRT regeneration limit (tan/light grey) are also shown.

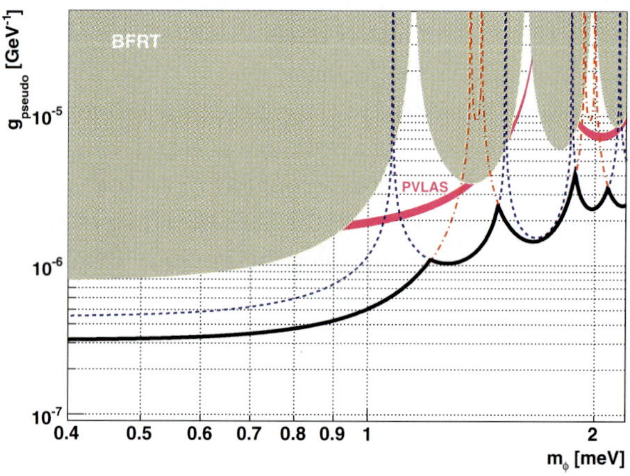

Fig. 4. 3σ limit contours for pseudoscalar particles.

the photon transport/detection efficiency each contribute 10% systematic uncertainty which is incorporated in the limits. These limit plots are shown in Figs. 3 and 4 along with the PVLAS 3σ signal region and the BFRT 3σ regeneration limits. As expected, the regions of insensitivity for one plunger position are well-covered by using the other plunger position. The weakly-interacting axion-like particle interpretation of the PVLAS data is excluded at more than 5σ by GammeV data for both scalar and pseudoscalar par-

Table 1. Summary of data in each of the 4 configurations.

Configuration	# photons	Est.Bkgd	Candidates	g[GeV^{-1}]
Horiz.,center	6.3×10^{23}	1.6	1	3.2×10^{-7}
Horiz.,edge	6.4×10^{23}	1.7	0	3.7×10^{-7}
Vert.,center	6.6×10^{23}	1.6	1	3.2×10^{-7}
Vert.,edge	7.1×10^{23}	1.5	2	4.5×10^{-7}

ticles. The asymptotic 3σ upper bounds on g for small m_ϕ are listed in Tab. 1. The GammeV exclusion region extends slightly beyond the previous best limit set by BFRT and sets limits in regions where BFRT had reduced sensitivity.

Acknowledgements

We would like to thank the staff of the Fermilab magnet test facility for their tireless efforts, and the technical staff of the Fermilab Particle Physics Division design group who aided in the design and construction of the apparatus. This work is supported by the U.S. Department of Energy under contract No. DE-AC02-07CH11359. AC is also supported by NSF-PHY-0401232.

References

1. R. D. Peccei and H. R. Quinn, Phys. Rev. Lett. **38**, 1440 (1977).
2. J. E. Kim, Phys. Rep. **150**, 1 (1987).
3. S. J. Asztalos, Astrophys. J. Lett. **571**, L27 (2002).
4. S. J. Asztalos, Phys. Rev. **D69**, 011101(R) (2004).
5. A. Burrows, M. T. Russel and M. S. Turner, Phys. Rev. **D42** 3297 (1990).
6. J. Engel, D. Seckel and A. C. Hayes, Phys. Rev. Lett. **65** 960 (1990).
7. E. Zavattini *et al.* [PVLAS Collaboration], Phys. Rev. Lett. **96**, 110406 (2006).
8. U. Gastaldi *et al.* [PVLAS Collaboration], Proc. XXXIII Int. Conf. on High Energy Physics, Moscow, Russia (2006).
9. R. Cameron *et al.*, Phys. Rev. D **47**, 3707 (1993).
10. G. Ruoso *et al.*, Z. Phys. C **56**, 505 (1992).
11. E. Zavattini *et al.* [PVLAS Collaboration], arXiv:0706.3419 [hep-ex].
12. C. Robilliard, R. Battesti, M. Fouche, J. Mauchain, A. M. Sautivet, F. Amiranoff and C. Rizzo, arXiv:0707.1296 [hep-ex].
13. K. Van Bibber, N. R. Dagdeviren, S. E. Koonin, A. Kerman and H. N. Nelson, Phys. Rev. Lett. **59**, 759 (1987).
14. G. Raffelt and L. Stodolsky, Phys. Rev. D **37**, 1237 (1988).
15. K. van Bibber, P. M. McIntyre, D. E. Morris and G. G. Raffelt, Phys. Rev. D **39**, 2089 (1989).
16. S. Hansen et.al, IEEE Trans.Nucl.Sci., vol. 51, no. 3, pp. 926-930 (2004).
17. H. G. Berns, et.al, IEEE Trans.Nucl.Sci., vol. 2, pp. 789-792 (2003).
18. W. A. Rolke and A. M. Lopez, Nucl. Instrum. Meth. A **458**, 745 (2001).

NEGATIVELY-CHARGED MASSIVE PARTICLES AS PROGENITORS FOR CDM, AND SOLUTION TO THE ^{6}Li AND ^{7}Li PROBLEM IN BIG BANG NUCLEOSYNTHESIS

M. KUSAKABE[1,†*], T. KAJINO[1,2], R. N. BOYD[3], T. YOSHIDA

Division of Theoretical Astronomy, National Astronomical Observatory of Japan, Mitaka, Tokyo 181-8588, Japan

[1]*Department of Astronomy, Graduate School of Science, University of Tokyo, Hongo, Bunkyo-ku, Tokyo 113-0033, Japan*

[2]*Department of Astronomical Science, The Graduate University for Advanced Studies, Mitaka, Tokyo 181-8588, Japan*

[3]*Present Address: Lawrence Livermore National Laboratory, University of California, Livermore, CA 94550, USA*

[†]*E-mail: kusakabe@th.nao.ac.jp*

G. J. MATHEWS

Department of Physics and Center for Astrophysics, University of Notre Dame, Notre Dame, IN 46556, USA

E-mail: gmathews@nd.edu

The ^{6}Li abundance observed in metal-poor halo stars exhibits a possible plateau similar to that for ^{7}Li. The observed abundance of ^{6}Li is a factor of 10^3 larger and that of ^{7}Li is a factor of 3 lower than the abundances predicted in the standard big bang when the baryon-to-photon ratio is fixed by WMAP, so that some mechanism would have produced ^{6}Li and deplete ^{7}Li. We show that both of these abundance anomalies can be explained by the existence of a long-lived massive, negatively-charged leptonic particle during nucleosynthesis. Such particles would capture onto the synthesized nuclei thereby reducing the reaction Coulomb barriers and opening new transfer reaction possibilities, and catalyzing a second round of big bang nucleosynthesis. We perform a fully dynamical calculation of big bang nucleosynthesis taking account of the recombination and many possible nuclear reactions and confirmed that this solution to both of the Li problems can be achieved with or without the additional effects of stellar destruction.

Keywords: Big Bang nucleosynthesis; Dark matter; Early universe; Stars.

*Research Fellow of the Japan Society for the Promotion of Science.

1. Introduction

The golden age in observational cosmology has started with a recent successful mission of the Wilkinson Microwave Anisotropy Probe (WMAP) of the cosmic microwave background (CMB). The CMB-WMAP data, with information from various other astronomical observations, strongly suggest that our universe is likely flat and cosmic expansion seems accelerating, which indicates the finite values of cosmological energy density of dark energy, dark matter as well as baryon [1]. It is the biggest challenge to resolve what is the nature of the dark energy and the dark matter in modern cosmology and particle and nuclear astrophysics.

Unknown dark matters thus exist in the Universe as a natural interpretation of observations. It is suggested that stable particles in supersymmetry or universal extra-dimension models possibly have Ω_m-value consistent with observation, therefore they are candidates of dark matters in the universe [2] Unstable particles, on the other hand, would have decayed a long time ago. But they might have existed in the early universe, and they might include negatively-charged particles like supersymmetric partner of leptons, i.e. slepton. A gravitino-slepton-lepton interaction is given by a coupling of an inverse of Planck mass $M_{\rm Pl}$, and decay lifetime of slepton is given by $\sim M_{\rm pl}^2/M^3$, where M is the related mass scale. When one suppose that dark matters of the measured energy density are stable particles in those models, the week scale mass is thought to be appropriate for the mass of particles. With that motive, supposing that the mass is about 100 GeV to 1 TeV, the decay life of slepton comes to $10^5 - 10^8$ s. Since this value is much larger than the time scale of big bang nucleosynthesis (BBN) of about 200 seconds, the relic exotic particles would survive the BBN epoch to have effect on BBN.

Recent spectroscopic observations of metal-poor halo stars (MPHSs) have indicated a primordial plateau of the ^{6}Li abundance as a function of the metallicity which is similar to that for ^{7}Li. The standard BBN model has a serious defect that the predicted primordial abundance of ^{6}Li is about three orders of magnitude smaller than the observed abundance level, and that of ^{7}Li is factor of $\sim$ three overproduced when one adopts a value of the universal baryonic density parameter, $\Omega_B = 0.04$, inferred from the recent WMAP-3 data.

Many mechanisms have been proposed for ^{6}Li production existing in MPHSs [3–6]. We consider a cosmological process of non-standard BBN by exotic particles. The BBN model with hypothetical negatively-charged massive particles is studied much recently, and several important reactions were suggested [7–10]. This model is considered to be a candidate of solution to the Li problems. The charged particle, denoted by X^-, recombines with a nucleus, and resulting exotic states might experience exotic nuclear reactions. We therefore try to calculate BBN fully dynamically, taking account of recombination of X^- as well as many possible nuclear reactions of X-bound nuclei, and check if lithium problems are resolved in a realistic fully dynamical calculation. Details of this study will be published soon.

2. Model

We estimate binding energies of nuclides with X^-. We assume that the X^- particle has spin 0, charge $-e$, and mass m_X much larger than nuclear mass of the order of 1 GeV. The charge distribution of nuclides is assumed to be the Gaussian type. We solve two-body Shrödinger equations involving the coulomb potential by variational calculation, and obtain binding energies. In estimating binding energies, we used root mean square charge radii determined by experiments, when experimental data exist. The binding energies are of the order of 1 MeV.

The recombination of a nucleus with a X^- particle is the same as that of an electron with a proton. A difference is that the mass of X is much larger than that of electron, and therefore binding energies are of the order of MeV. We included new BBN reactions of X-bound nuclides, and solved abundance evolutions in the expanding universe.

3. Result

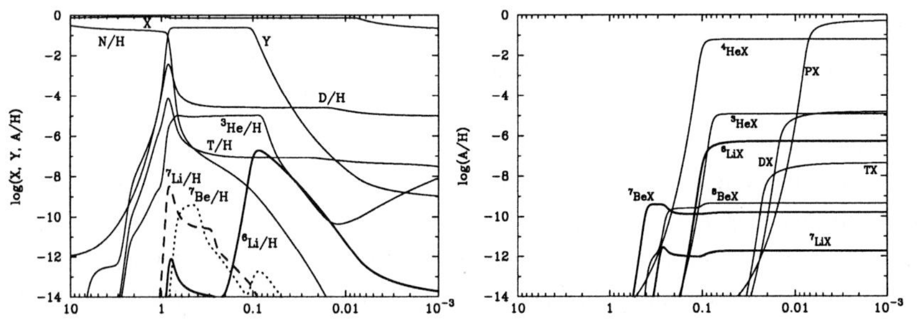

Fig. 1. Nuclear abundance flows of normal nuclei (left) and X-nuclei (right) as a function of T_9 in the case of parameters $m_X = 50$ GeV, $Y_X = 0.8$ and $\tau_X = \infty$.

The panels in Fig. 1 show light element abundances as a function of

$T_9 \equiv T/(10^9$ K) with the cosmic photon temperature T, in a calculation with certain input parameter, where $m_X = 50$ GeV, X abundance is $Y_X = n_X/n_b = 0.8$ with number densities of X^- and baryon, n_X and n_b, respectively, and decay lifetime of X is $\tau_X = \infty$. The left part of the panels (high temperatures) is the same as in standard BBN. However, when the temperature of the universe decreases to $T_9 \sim 0.5$, the recombination of ^{7}Be proceeds, and other nuclides also start recombining with X^- particles. One can find the destruction of $^7\mathrm{Be}_X$ at $T_9 \sim 0.3 - 0.2$. The destruction mechanism is $^7\mathrm{Be}_X(p,\gamma)^8\mathrm{B}_X$ through an atomic resonance [8]. In our previous calculation with rough inputs on the charge radii of nuclides, the same $^7\mathrm{Be}_X(p,\gamma)^8\mathrm{B}_X$ reaction but through the 1^+ nuclear excited state of ^{8}B is found to have a possibility of destruction of $^7\mathrm{Be}_X$ (Kusakabe et al. submitted to Phys. Rev. D). But we confirmed that this resonant channel is not important in this calculation using more realistic charge radii, or binding energies. We confirmed that ^{6}Li is produced abundantly by the $^4\mathrm{He}_X(d,X^-)^6$Li reaction [7].

We calculate nucleosynthesis with parameters of the X^- particle; abundance Y_X and decay lifetime τ_X. We fixed baryon-to-photon ratio to be $\eta = 6.1 \times 10^{-10}$. For a set of parameters, light element abundances are obtained. This X-catalyzed BBN has differences in abundances of ^{6}Li and ^{7}Li, and other nuclear abundances were not influenced significantly. We obtain the calculated Li abundance relative to the observed abundances, which we designate as d-factor, since this factor should be explained by a stellar depletion in atmospheres of MPHSs. d=1 satisfies the requirement that resulting lithium isotopic abundances are consistent with observations. We found a consistent ^{6}Li abundance region and consistent ^{7}Li abundance region. Furthermore, there is a parameter region leading to ^{7}Li destruction and ^{6}Li production simultaneously, that is $Y_X \gtrsim 1$ and $\tau_X \approx (1-2) \times 10^3$ s.

4. Summary

We calculate light element nucleosynthesis during BBN with negatively-charged X^- particles in a fully dynamical manner. X-catalyzed BBN model provides a solution to the ^{6}Li problem. As suggested in previous studies [7,10], X^- particles greatly enhance the production of ^{6}Li through the recombination of ^{4}He followed by the X^- transfer reaction, i.e. $^4\mathrm{He}(X^-,\gamma)^4\mathrm{He}_X(d,X^-)^6$Li. The ^{7}Li overproduction problem is also resolved simultaneously in the parameter region of $Y_X \gtrsim 1$ and $\tau_X \approx (1-2) \times 10^3$ s. The dominant destruction reaction is the resonant reaction $^7\mathrm{Be}_X + p \rightarrow {}^8\mathrm{B}_X^{*a} \rightarrow {}^8\mathrm{B}_X + \gamma$ through the atomic excited state of $^8\mathrm{B}_X$ [8].

This parameter region could give a constraint to the mass of cold dark matter particles if the ^{6}Li and ^{7}Li problems are caused by negatively charged X^- particles as analyzed in this study. If X^- finally decays to a dark matter Y^0 and any residues, the WMAP-CMB constraint of $\Omega_{\rm CDM} = 0.2$ leads to $m_Y \lesssim 5$ GeV for $Y_X \gtrsim 1$. If m_X is close to m_Y, m_X must be less than the order of 10 GeV.

Acknowledgments

This work has been supported in part by the Mitsubishi Foundation, the Grant-in-Aid for Scientific Research (17540275) of the Ministry of Education, Science, Sports and Culture of Japan, and JSPS Core-to-Core Program of International Research Network for Exotic Femto Systems (EFES). MK acknowledges the support by the Japan Society for the Promotion of Science. Work at the University of Notre Dame was supported by the U.S. Department of Energy under Nuclear Theory Grant DE-FG02-95-ER40934. RNB gratefully acknowledges the support of the National Astronomical Observatory of Japan during his stay there.

References

1. D. N. Spergel *et al.* [WMAP Collaboration], Astrophys. J. Suppl. **170**, 377 (2007)
2. J. L. Feng, A. Rajaraman and F. Takayama, Phys. Rev. Lett. **91**, 011302 (2003)
3. M. Kusakabe, T. Kajino and G. J. Mathews, Phys. Rev. D **74**, 023526 (2006)
4. K. Jedamzik, K. Y. Choi, L. Roszkowski and R. Ruiz de Austri, JCAP **0607**, 007 (2006)
5. T. K. Suzuki and S. Inoue, Astrophys. **573**, 168 (2002)
6. E. Rollinde, E. Vangioni-Flam and K. A. Olive, Astrophys. J. **627**, 666 (2005)
7. M. Pospelov, Phys. Rev. Lett. **98**, 231301 (2007)
8. C. Bird, K. Koopmans and M. Pospelov, arXiv:hep-ph/0703096.
9. K. Kohri and F. Takayama, Phys. Rev. D **76**, 063507 (2007)
10. K. Hamaguchi, T. Hatsuda, M. Kamimura, Y. Kino and T. T. Yanagida, Phys. Lett. B **650**, 268 (2007)

DARK ELECTRIC MATTER OBJECTS: HISTORY OF DISCOVERY, MODES OF INTERACTION WITH MATTER, SOME INFERENCES AND PROSPECTS

E. M. DROBYSHEVSKI and M. E. DROBYSHEVSKI

Ioffe Physico-Technical Institute, Russian Academy of Sciences
St.Petersburg, 194021 Russia; E-mail: emdrob@mail.ioffe.ru

Experiments with thin ZnS(Ag) scintillators provide evidence with C.L. > 99.99% for the existence of DArk Electric Matter Objects - daemons (presumably negatively charged Planckian particles with $M \sim 10^{-5}$ g) captured from the Galactic disk into near-Earth, almost circular heliocentric orbits (NEACHOs). Their flux at $V \approx$ 10-15 km/s was found to be as high as $f_\oplus > 10^{-7}$ cm^{-2}s^{-1} and vary with $P = 0.5$ y, with maxima in March and September. A daemon flux $f_\oplus \sim 10^{-7}$-10^{-6} cm^{-2}s^{-1} is capable of accounting for the Troitsk anomaly in the ^{3}T β-spectrum and suggests its more pronounced manifestation in future KATRIN experiment. In view of the channeling effect on iodine recoil nuclei in the NaI(Tl) crystal, the DAMA/NaI experiment is also apparently detecting a flux of daemons, $f_\oplus \sim 6\times10^{-7}$ cm^{-2}s^{-1}, but in this case of those falling with V = 30-50 km/s from strongly elongated, Earth-crossing heliocentric orbits (SEECHOs) oriented in the antapex direction, as a result of which the number of events detected in the 2-6-keV interval varies with $P = 1$ y.

1. Planckian DM objects and possibility of their detetion

Our Universe started from Planckian scales, and it appears only reasonable to assume that the larger part of its mass, Dark Matter (DM), remains confined in Planckian objects, elementary black holes whose gravitational radius cannot be smaller than their Compton wavelength. Such objects with $M \sim 10^{-5}$ g and $r_g \sim 10^{-33}$ cm can carry an electric charge of up to $Ze \approx G^{1/2}M \approx 10$e. While the properties of similar DArk Electric Matter Objects, *daemons*, were considered by a number of authors [1-8], but the possibility of their detection was perceived skeptically (e.g., [1]).

We started from a notion that negative daemons are nuclear-active particles. Their capture of atomic nuclei should give rise to a release of energy $W \approx 1.8ZZ_nA^{-1/3}$ MeV ($\sim 10^2$ MeV), with the attendant emission of scintillation-active electrons and nucleons [9]. Becoming confined in the remainder of the nucleus, the daemon should bring about successive decay of nucleons [10], thus lowering the charge of the remainder to $|Z_n| - |Z| < 0$, which makes it possible for the daemon to capture another nucleus with emission of new particles, etc. We noted also that a negative daemon is capable of catalyzing fusion of light nuclei, up to $Z_n \sim$ 6-9 (an analog of the muon catalysis of deuterons [11]). It should be pointed out that our understanding of the properties and behavior of daemons developed parallel to and in intimate connection with the experiments we were conducting.

Adopting $\sim$0.3 GeV/cm^3 for the DM density in the galactic halo, we arrive at $\sim 5\times10^{-12}$ cm^{-2}s^{-1} for the halo daemon flux intercepted by the Earth. The same applies to the galactic disk DM population with velocity dispersion of 4-30 km/s. The Sun moves relative to the disk objects with $V \approx 20$ km/s (in the direction of the apex, which does not lie in the plane of ecliptic [12]). For such a velocity, the effective cross section of the Sun, amplified by gravitational focusing, exceeds its geometric cross section $\sim 10^3$ times.

In crossing the Sun, daemons suffer slowing down, so that a sizable part of them falls back to move along strongly elongated and rapidly shrinking rosette-shaped trajectories, whose perihelia lie inside the Sun. These daemons settle eventually toward its center to form there a compact daemon kernel, which is capable of accounting for many specific features in solar physics, including generation of non-electron neutrinos [13]. If a daemon propagating along a rosette trajectory crosses the Earth's gravitational sphere of action, the perihelion of this trajectory has a high probability to leave the body of the Sun. This is how daemons enter and build up in stable SEECHOs (until their next encounter with the Earth). Straightforward estimates made in 1996 [14] in the gas-kinetic approximation of the mean free path suggest that the SEECHO daemon flux may reach as high as $f_\oplus \sim 3\times10^{-7}$ cm^{-2}s^{-1} for a velocity of 30-50 km/s.

Recent celestial mechanics calculations [15] showed that SEECHOs crowd primarily in the "shadow" zone, i.e., the zone on the antapex side relative the Sun, because the petals of the rosette trajectories emerging into the apex hemisphere can no longer reach the Earth's orbit as a result of the slowing down the daemons undergo after several crossings of the Sun. Whence it follows that the flux of SEECHO daemons should exhibit a distinct one-year modulation with a maximum some time in June (it is appropriate to point out here that the direction to the apex is slightly different for different galactic disk populations, namely, stars of various classes, interstellar gas etc.; therefore, pinpointing this direction relative to the disk daemons is a task for the future).

Subsequent crossings by SEECHO objects of the Earth's gravitational sphere result in their gradual transfer to NEACHOs and accumulation there. It is from NEACHOs that daemons fall on the Earth with $V \approx$ 10(11.2)-15 km/s.

2. The history of the daemon discovery

The above considerations, which were continually refined with due account of the building up experience, served as a basis for development of several generations of detectors.

In the very beginning (October 1996 - October 1998), we made an attempt at visualizing daemon-assisted catalysis of the fusion of light nuclei by detecting the sound wave generated by a large energy release along the trajectory (up to $\sim 10^3$ erg/cm [14]). The experiments performed with lithium detectors demonstrated, however, extremely strong damping (along a distance of 2-3 cm) of sound with a characteristic frequency of ~25 MHz. The experiments in which 45-mm-thick Be plates were used also did not produce promising results, the more so that the unavoidable increase here of the characteristic frequency up to ~600 MHz would require development of unique methods of detection of such ultrasound. Therefore, both sides of the Be plates were coated with a ~10-μm-thick layer of ZnS(Ag) scintillator, which could register jets of electrons and up to $\sim 10^4$ ^{18}O nuclei (produced in fusion of Be nuclei) at the points of daemon entry into and exit out of the plates. Each side of the plate was viewed by a FEU-167 PM tube. Two such plates, each of 600 cm^2 area, were assembled into a telescope oriented continuously in the direction of the expected arrival of SEECHO daemons ($\sim 35^o$ leading the Sun). A 700-h exposure accumulated in April-May 1999 did not yield the expected results; indeed, no scintillations with the shape characteristic of the passage of heavy particles (HPSs) and shifted by Δt = 0.5-1.5 μs with respect to one another were observed [16]. An analysis of reasons for this failure led us to the understanding that the catalytic action of daemons should unavoidably be poisoned by their capture of heavy nuclei which are present in our beryllium produced by powder metallurgy (~0.1 at.% impurities, including Si, Fe etc.). This suggested immediately the possibility of daemon-stimulated nucleon decay [10].

It is this new ideology that underpinned the development in November 1999 of a four-module detector [17]. Each module consisted of a cubic iron-sheet container (0.3-mm-thick Fe sheet coated on both sides with 2 μm Sn layer), 51 cm on a side, covered on top with black paper. In the middle of

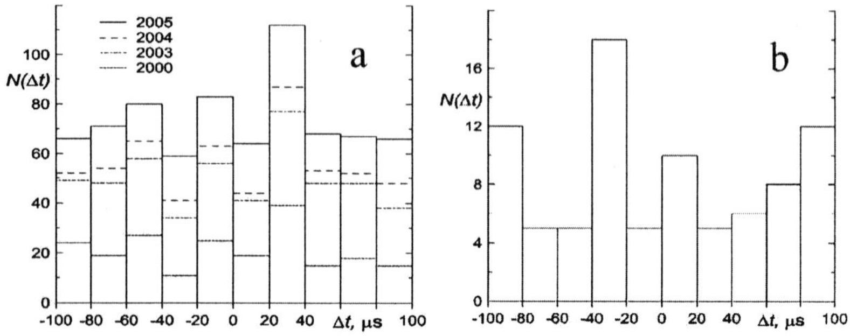

Fig. 1. Distributions $N(\Delta t)$ of pair events on their time shift relative to the upper screen HPSs. (a) The sum of Marches 2000, 2003, 2004, and 2005. The $+30\mu s$ maximum exceeds the mean level by 33 events; its significance is 3.63σ (C.L. $>$ 99.97 %) [18]. (b) The underground Baksan experiment [21]. The photo-cathode of the bottom PMT, which is sensitive to daemon passages due to its thick $\sim$1 μm inner Al coating, is screened with Al foil. Observations during September 3-11, 2005; 86 events altogether. Significance of the -30 μs maximum is 2.2σ.

the module, spaced by 7 cm, were fixed two transparent polystyrene plates, 4 mm thick, separated by black paper. Their bottom faces were coated by a $\sim$3.6 mg/cm^2 ZnS(Ag) powder layer. Each plate was viewed on its side by a FEU-167 PM tube. We purposefully made the system sensitivity asymmetric for enhancing the differences in the up/down distributions in the number of time-shifted events. Initially (November-December 1999), all modules were also oriented in the direction of the expected arrival of SEECHO daemons from the Sun (including those crossing the Earth in the evening time).

2.1. *Observation of NEACHO daemons in St. Petersburg*

In January 2000, we discontinued orienting the system; from that time on, the scintillators were always kept horizontally. In February, the $N(\Delta t)$ distribution revealed some features with a hint at statistical significance, so that while previously the detector was switched on for 10-12 h in the daytime, starting with March 1, 2000, it was operated on the round-the-clock basis.

Figure 1a displays $N(\Delta t)$ distributions with HPSs on the top PM tube obtained in the March experiments in 2000, 2003, 2004, and 2005 [18]. The peak at $20 < \Delta t < 40$ μs stands out with a 99.97% confidence. Accepting 29 cm as a base distance from the top ZnS(Ag) layer to the bottom lid of the container (see also Sec. 2.2 below), we arrive at a velocity of 10-15 km/s,

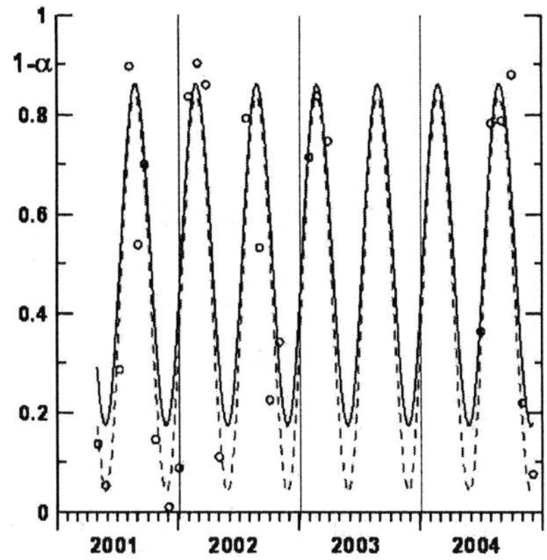

Fig. 2. Seasonal variation of 1 - α, the extent to which the distribution $N(\Delta t)$ (at $-100 < \Delta t < +100$ μs) deviates from the constant level produced by background events. (- - - -) Weights of all the points are equal, the correlation coefficient of the sine curve ($P =$ 0.5 yr) with the points is $r = 0.87$, its C.L. > 99.9%. (——) Weights of the points are proportional to 1 - α; $r = 0.73$, its C.L. = 99.3%.

a figure characteristic of objects falling from NEACHOs [17-19]. Assuming $\Delta t \approx 30$ μs to be the time taken up by "digestion" of a Zn nucleus captured in the ZnS(Ag) layer, we come to $\Delta\tau_{ex} \sim 10^{-6}$ s for the time of daemon-stimulated decay of a nucleon (proton or neutron) [19]. For such a value of $\Delta\tau_{ex}$, the dimensions of our detector make it insensitive to the passage of daemons with $V > 30$ km/s, even if they capture S nuclei in passing through the ZnS(Ag) layer.

After a full year of observations and testing of the equipment, we came to the conclusion that the NEACHO flux varies with $P = 0.5$ y, with maxima in March and September (Fig. 2) [20]. This correlates well with our recent calculations of the capture of galactic disk daemons crossing the Sun. As the Sun moves in the apex direction relative to the daemon population of the galactic disk, daemons with an impact parameter of $\approx 9.2 R_\odot$ in the plane that contains the direction to the apex and is tilted at the minimum angle to the plane of ecliptic, cross the Sun with the corresponding slowing down and become captured into distinctive cross-shaped rosette trajectories (Fig. 3). Their petals, which are normal to the direction to the apex, cross *repeatedly* the Earth's orbit close (*sic*!) to the equinox zones, so that the daemons moving along them can eventually transfer to SEECHOs and, subsequently, to NEACHOs with a probability which is the highest compared with the other possible scenarios. It is these NEACHO daemons that are detected

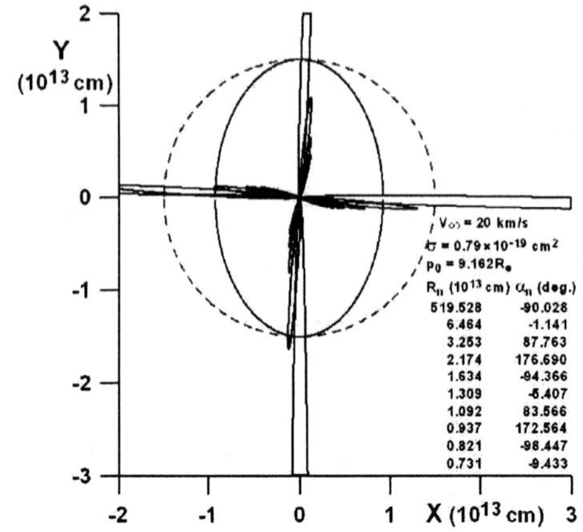

Fig. 3. An example of multi-loop (cross-like) trajectory of an object being braked by the Solar matter at repeated passages through the Sun (its center is at X = 0, Y = 0) [15]. Object of 3×10^{-5} g mass and cross-section $\sigma = 0.79\times10^{-19}$ cm^{2}falls from infinity (X = -∞; $V_\infty = 20$ km/s) with an impact parameter $p_0 = \mathrm{Y}(-\infty) = 9.162\mathrm{R}_\odot$. The figure plane contains the apex direction and a normal to it lying in the ecliptic plane. An ellipse with the semi-major axis of 1 AU is the Earth's orbit projection; the dotted circle of 1 AU radius is given as a scale for the reader's orientation.

in March and September. Defining the force that slows down the daemon in the Sun by $\sigma\rho V^2$ (ρ is the density of solar material), observation of the maximum NEACHO daemon flux in these months (together with the maximum registration of SEECHOs by DAMA/NaI in June, see Sec. 4) yields for the daemon slowing-down cross section $\sigma \approx 1\times10^{-19}$ cm^2 [15].

2.2. *Underground experiments in Baksan with a daemon-sensitive PM tube*

Experiments performed with FEU-167 PM tubes (photocathode ∅100 mm, the bulb photocathode section ∅125 mm) of different manufacturers (Svetlana in St. Petersburg and Ekran in Novosibirsk) showed that tubes with a thicker Al coating (~1 μm rather than ~0.1 μm as specified) of the inner surface of the bulb photocathode section turn out to be sensitive to the passage of daemons with $V \sim 10$ km/s [18]. This is easy to understand because (*i*) the path length of a complex with Z = -1 to capture of an Al nucleus is ~1 μm, and capture of a nucleus in the metal culminates in emission of many refilling electrons, and (*ii*) passage of the complex, which consists of the daemon plus the remainder of the nucleus captured previously and now being digested, through 4-5 cm of vacuum inside the PM tube, is highly probable to give rise to the appearance and increase of the negative charge of the complex, because in vacuum this charge is not balanced by the positive charge of the nuclei captured for Z = -1 in a path of 1-5 mm through air.

The module, in which only the top scintillator was left and the role of the bottom sensor was played by a FEU-167 with a thick inner Al coating whose photocathode was screened by aluminized Lavsan film, was first tried in the Baksan underground laboratory in September 2005 [21]. To reduce the Rn background, the detector was flushed by liquid N_2 vapor.

The experiment revealed a ~hundredfold increase in detector sensitivity per unit area; indeed, the measured daemon flux with V = 10-15 km/s was as high as $\sim 10^{-7}$ $cm^{-2}s^{-1}$, and it propagated from the interior of the Earth upward, with the maximum in the $N(\Delta t)$ distribution observed for -40 < Δt < -20 μs (Fig. 1b). The underground measurements conducted in March 2006 (but without nitrogen flushing) revealed a standard peak at 20 < Δt < 40 μs. This substantiated our choice of 29 cm between the top scintillator and the bottom lid of the module as the base distance (Sec. 2.1).

These results find a ready explanation if we assume that daemons build up in NEACHOs external to the Earth's orbit and, touching it on the outside, catch up with the Earth close to the equinoxes (actually, one to two weeks before that). This buildup in the external NEACHOs appears only natural, because in order for an object to be ejected out of the Earth's orbit its velocity has to be increased by $\Delta V \geq 12.3$ km/s, whereas a single interaction with the Earth produces $\Delta V \leq 11.2$ km/s.

It thus follows that it is desirable to perform synchronous measurements in the Northern and Southern hemispheres; in the Southern hemisphere, in March the primary daemon flux will pass from bottom upward, and in September, from top downward.

3. NEACHO & GESCO daemons and the Troitsk anomaly in ^{3}T β-spectrum

Attempts at deriving the electron antineutrino mass from the end-point of the ^{3}T β-spectrum (E_0 = 18574-18590 eV) revealed the presence near the end-point (at $E_0 - E \approx 5$ eV) of a step with an amplitude ~1-3 mHz, which appears and varies slightly in position with P = 0.5 y. This step was observed in the Troitsk experiment making use of an extended gaseous-tritium source of total horizontal area 0.35 m^2 (dia. 5 cm, 3 m long plus

$\sim$4 m-long channel leading to the spectrometer) [22]. No step is observed in Mainz experiments with a well devised solid tritium source.

A significant feature of the gas source, as we see it, is the use of superconducting windings of Nb, $NbTi_2$ and Nb_3Sn alloys needed to create the magnetic field channeling the β electrons. Daemons crossing the windings capture Nb-containing atomic clusters and carry them across the gas source channel. Excitation of these clusters by nuclear-active daemons may conceivably bring about ejection of the inner K electron of niobium (its binding energy $E_k = 18990$ eV), with a resultant emission of niobium Auger electrons, in particular, of a very compact group of five lines (18566, 18568, 18569, 18570 and 18572 eV), which are close to E_0. The half-year periodicity and the phase of appearance of the step are clearly related with the variation of the low-energy NEACHO daemon flux (and the subsequent flux of objects captured from NEACHOs to GESCOs, geocentric Earth-surface crossing orbits). Adding the velocity of 10-15 km/s to that of Auger electrons accounts for the dispersion in their energy $\Delta E \approx 5$ eV [23].

Note that fine details in the emission of Nb K electrons from a cluster containing Nb atoms remain unclear and require further consideration. The mechanism involved would be easier to understand if the daemon carrying 9, rather than 10, electronic charges would capture Sn nuclei in the windings ($Z_{Sn} = 50$). In these conditions, the positive charge of the daemon/Sn complex would be exactly equal to that of the Nb nucleus ($Z_{Nb} = 41$). Shakeup of the electronic shells of such a complex, be it as a result of transitions of the Sn nucleus to lower-lying Rydberg levels in the daemon field (they lie within the electronic shells) or as a result of daemon-assisted catalysis of predominantly neutrons' decays in the Sn nucleus (see Sec. 5), would obviously give rise to emission of Auger electrons typical of Nb.

Even if only a third of the daemon-stimulated nucleon decays separated in time by $\sim 10^{-6}$ s in the cluster nuclei are accompanied by emission of Auger electrons in the above-mentioned lines, the observed step amplitude should be reached for $f_\oplus \sim 10^{-7}$-10^{-6} cm^{-2}s^{-1}. It may be predicted that the KATRIN version with a gaseous T_2 source (and Nb-Sn containing windings) will reveal a still stronger "Troitsk anomaly".

4. SEECHO daemons and the DAMA/NaI experiment

There are grounds to believe that the one-year modulation of the number of scintillation signals at the 0.04 cpd/kg/keV level observed at a C.L. $= 6.3\sigma$ in the DAMA/NaI experiment in Gran Sasso [24] can likewise be accounted for by daemons rather than by WIMPs as surmised by the authors.

This conclusion is argued persuasively for by the fact that the results obtained in DAMA/NaI are not reproduced in other experiments designed for detection of WIMPs with detectors of other types, although the sensitivity of the latter sometimes even exceeds noticeably that of the DAMA/NaI. To account for this, one is forced to devise rather exotic WIMPs [25].

Interestingly, the one-year modulation is observed only within a narrow range of scintillation amplitudes in NaI(Tl), which correspond to electron energies of 2-6 keV. But these are exactly the energies of the iodine recoils knocked out elastically by SEECHO daemons which fall on the Earth with V = 30-50 km/s. The SEECHO flux reaches a maximum sometime in early June (see Sec. 1), which likewise conforms to the DAMA/NaI data.

At first glance, it might seem that the coincidence of the iodine recoil energies E_r with the observed energy of scintillations produced by 2-6-keV electrons is at odds with neutron beam calibrations which yield for $E_r >$ 10 keV of iodine a quenching factor $q < 10\%$ (q is the ratio of scintillation signal intensities due to ions and electrons of equal energies). Taking into account ion channeling along some crystallographic directions changes the situation, however, because such ions impart nearly all of their energy to valence electrons of the crystal, so that for them $q \to 1$. The channeling probability grows with decreasing E_r, so that for the iodine ion with $E_r =$ 4 keV emitted in an arbitrary direction it is $\eta \approx 20\%$. Thus, the efficiency of registration of 4-keV iodine ions by a NaI(Tl) crystal for $q = 1$ should be ≈20% [26].

Whence it follows that for producing the observed number of 2-6 keV events at the 0.04 cpd/kg/keV level in the 96-kg NaI(Tl) detector with an effective area of ≈1500 cm^2 one would have to assume a SEECHO daemon flux of about $0.04\times96\times4/86400\times1500\times\eta = 6\times10^{-7}$ cm^{-2}s^{-1} [26]. Obviously enough, a number of attendant points still remain unclear, for instance, those associated with the mechanisms involved in the elastic knocking out of ions by daemons carrying remainders of previously captured nuclei, with a single-hit criterion application, the case where one takes into account events that happened in one out of the nine crystals in the DAMA/NaI experiment only, and so on. Let us hope that LIBRA experiment will provide information capable of shedding light on the situation.

5. Attempts to reveal the daemon-stimulated proton decay

Besides the existence itself of daemons, another basic assumption underpinning our experiment was gradual decomposition of a nucleus captured by the negative daemon residing inside it. We believed that the decomposition involved successive daemon-stimulated proton decays. The time taken up by a daemon-containing Zn nucleus to cross our detector ($\approx$30 μs) limits the mean proton decay time to $\Delta\tau_{ex} \sim 1$ μs (Sec. 2.1).

To detect such a series of successive events, we assembled a detector of 4.3-cm-thick CsI(Tl) crystals of 560 cm^2 total area. We hoped that for $f_\oplus \sim 10^{-7}$ $cm^{-2}s^{-1}$ we would observe once every 5 h a trail of scintillations spaced, on the average, by about $\sim$1 μs. Adopting a velocity V = 10-15 km/s, we would observe 3 to 4 such scintillations.

Decay of a proton should release an energy of 938 MeV. If the proton decomposes into π^+ and π^0 mesons, they will leave only 100 MeV in a 4.3-cm-thick CsI crystal.

However, at the threshold level of about 50 MeV, we did not observe triple events during weeks!

We thus come to a tentative conclusion that the daemon, a negatively charged Planckian black hole, decomposes the captured nucleus in a somewhat different way. It would be difficult to add something definite on this point at this time. One cannot exclude the possibility, for instance, that a nucleus decomposes by consecutive capture by the daemon of electrically neutral neutrons (by "winding" them on its gravitational horizon, as it were) accompanied by emission of non-detectable gravitons etc. In these conditions, the nucleus would preserve for a certain time a constant (the original) number of protons and its positive charge. The neutron-deficient nuclear remainder would then get rid of the large proton excess by rare (once every few tens of μs) fission events and/or practically simultaneous ejection of products of a new proton fusion (α particles and so on), a process similar to the one occurring in daemon-assisted catalysis of many ($\sim$10) protons in the Sun considered by us elsewhere [27]. The scenario of consecutive daemon-assisted decay of neutrons in a nucleus is argued for possibly by the multiple emission of Auger electrons by the daemon/captured Sn nucleus complex at $Z = 41 = constant$ (see Sec. 3). The daemon itself also may free itself, but much more rarely, of the neutron matter wound on it by flash-evaporating it in the form of hard radiation, to regain its original Planckian mass.

6. Concluding remarks and future prospects

The daemon approach to the nature of DM which we are developing has turned out to be valid, self-consistent and fruitful. Fairly simple experiments not only substantiate the existence of daemons (by now, with a C.L. > 99.99%) but suggest possible directions in which the daemon paradigm can be extended (the need of taking into account the Sun's motion relative to the galactic disk population, daemon-stimulated nucleon decay etc.).

It is essential that the existence of daemons and their buildup in the Solar system in Earth-crossing orbits of different types are argued for not only by our experiments but by the DAMA/NaI project as well. By the way, it becomes now clear why its results are not reproduced in other experiments aimed at detection of WIMPs, as well as that the unavoidable effect of channeling of low-energy iodine ions (<10 keV) should play a substantial part in interpretation of the DAMA/NaI results. All the features of the Troitsk anomaly in the tail of the ^{3}T β-spectrum, whose existence was already being questioned by the very people who had discovered it because of its appearing to be impossible to explain, also find ready interpretation in terms of the daemon paradigm (we have in mind the amplitude and width of the step, the half-year periodicity and the phase of its appearance).

The agreement between the ground-level daemon fluxes estimated theoretically in 1996 ($\sim 3\times 10^{-7}$ cm^{-2}s^{-1} [14]), measured by us experimentally in St. Petersburg and Baksan ($>10^{-7}$ cm^{-2}s^{-1}, 2005 [18,21]), derived from the Troitsk anomaly (10^{-7}-10^{-6} cm^{-2}s^{-1}, 2000 [23]), and following from DAMA/NaI studies ($\sim 6\times 10^{-7}$ cm^{-2}s^{-1}, 2003 [26]) is truly remarkable. Celestial mechanics calculations based on our and DAMA/NaI figures have yielded the cross section of daemon braking by solar matter of 10^{-19} cm^2.

Measurements of the direction of propagation of the primary daemon flux suggest daemon buildup in NEACHOs external to the Earth's orbit. It would thus be reasonable to carry out synchronous studies of the March and September fluxes in the Northern and Southern hemispheres. In the Northern hemisphere, in March, the primary daemon flux goes from above downward, while in September, as follows from the Baksan measurements, it passes upward, whereas in the Southern hemisphere the daemons should move in opposite directions relative to the observer.

The conclusions that can be drawn from the daemon paradigm concerning the existence of daemon kernels in the Earth and in the Sun permit us to understand or explain in a new way many geo- and heliophysical observations. The crowding of daemons toward the Galactic center clarifies the reason for the observed appearance of positrons there as a product of daemon-assisted catalysis of proton decay and/or fusion [27-28]. The same may apply to generation of the excess GeV radiation (see Sec. 5).

We have suddenly and completely unexpectedly found ourselves at the threshold of Planckian physics and painfully far away from an adequate understanding of the properties of daemons and of their interaction with matter. It appears therefore only natural that many of the above considerations may turn out in need of further refinement and critical analysis.

Acknowledgements

EMD is greatly indebted to the Dark-2007 Organizers and to the Russian Foundation for Basic Research for financial support of his attending the Conference.

References

(1) M.A.Markov, *Sov. Phys. JETP*, **24**, 584-592 (1967).
(2) K.P.Stanyukovich, *Sov. Phys. Doklady*, **168**, 781-784 (1966).
(3) S.Hawking, *Mon. Not. Roy. Astron. Soc.*, **152**, 75-78 (1971).
(4) M.Turner, in "*Dark Matter in the Universe*" (IAU Symp. No. **117**), J.Kormendy & G.R.Knapp (eds.), Dordrecht: Reidel, pp.445-488 (1987).
(5) J.D.Barrow, E.J.Copeland, A.R.Liddle, *Phys. Rev.*, **D46**, 645-657 (1992).
(6) P.Ivanov, P.Naselsky, I.Novikov, *Phys. Rev.*, **D50**, 7173-7178 (1994).
(7) S.Alexeyev *et al.*, *Class. Quantum Grav.*, **19**, 4431-4443 (2002).
(8) E.M.Prodanov, R.I.Ivanov, V.G.Gueorguiev, *Astropart. Phys.*, **27**, 150-154 (2007); *hep-ph*/0703005.
(9) E.M.Drobyshevski, *Preprint PhTI*-**1663**, St-Petersburg (1996).
(10) E.M.Drobyshevski, *Mon. Not. Roy. Astron. Soc.*, **311**, L1-L3 (2000).
(11) S.S.Gershtein, L.I.Ponomarev, in "*Muon Physics*", V.W.Hughes & C.S.Wu (eds.), Acad. Press, v.**III**, pp.142-233 (1977).
(12) C.W.Allen, *Astrophysical Quantities*, 3rd ed., Univ. London, The Athlone Press (1973).
(13) E.M.Drobyshevski, *Astron. Astrophys. Trans.*, **23**, 173-183 (2004); *astro-ph*/0205353.
(14) E.M.Drobyshevski, in "*Dark Matter in Astro- and Particle Physics (DARK'96)*", H.V.Klapdor-Kleingrothaus & Y.Ramachers (eds.), World. Sci., pp.417-424 (1997).
(15) E.M.Drobyshevski, M.E.Drobyshevski, *Astron. Astrophys. Trans.*, **26** (in press) (2007); *arXiv*:0704.0982.
(16) E.M.Drobyshevski, *Physics of Atomic Nuclei*, **63**, 1112-1117 (2000).
(17) E.M.Drobyshevski, *Astron. Astrophys. Trans.*, **21**, 65-73 (2002); see also *astro-ph*/0007370.
(18) E.M.Drobyshevski, *Astron. Astrophys. Trans.*, **25**, 43-55 (2006); *astro-ph*/0605314.

(19) E.M.Drobyshevski *et al.*, *Astron. Astrophys. Trans.*, **22**, 19-32 (2003); *astro-ph*/0108231.
(20) E.M.Drobyshevski *et al.*, *Astron. Astrophys. Trans.*, **22**, 263-271 (2003); *astro-ph*/0305597.
(21) E.M.Drobyshevski, M.E.Drobyshevski, *Astron. Astrophys. Trans.*, **25**, 57-73 (2006); *astro-ph*/0607046.
(22) V.M.Lobashev, *Physics of Atomic Nuclei*, **63**, 1037-1043 (2000).
(23) E.M.Drobyshevski, *hep-ph*/0502056.
(24) R.Bernabei *et al.*, *Riv. Nuovo Cim.*, **20**, 1-73 (2003); *astro-ph*/0307403.
(25) R.Bernabei *et al.*, *Int. J. Mod. Phys.*, **A21**, 1445-1469 (2006).
(26) E.M.Drobyshevski, *arXiv*:0706.3095.
(27) E.M.Drobyshevski, *Mon. Not. Roy. Astron. Soc.*, **282**, 211-218 (1996).
(28) E.M.Drobyshevski, *astro-ph*/0402367.

PART IX

Cosmological Variations and Fundamental Constants

DARK ENERGY WITHOUT DARK ENERGY*

DAVID L. WILTSHIRE

Department of Physics and Astronomy, University of Canterbury,
Private Bag 4800, Christchurch 8140, New Zealand
E-mail: David.Wiltshire@canterbury.ac.nz
http://www2.phys.canterbury.ac.nz/~dlw24/

An overview is presented of a recently proposed "radically conservative" solution to the problem of dark energy in cosmology. The proposal yields a model universe which appears to be quantitatively viable, in terms of its fit to supernovae luminosity distances, the angular scale of the sound horizon in the cosmic microwave background (CMB) anisotropy spectrum, and the baryon acoustic oscillation scale. It may simultaneously resolve key anomalies relating to primordial lithium abundances, CMB ellipticity, the expansion age of the universe and the Hubble bubble feature. The model uses only general relativity, and matter obeying the strong energy condition, but revisits operational issues in interpreting average measurements in our presently inhomogeneous universe, from first principles. The present overview examines both the foundational issues concerning the definition of gravitational energy in a dynamically expanding space, the quantitative predictions of the new model and its best–fit cosmological parameters, and the prospects for an era of new observational tests in cosmology.

Keywords: Dark energy, theoretical cosmology, observational cosmology.

1. Introduction

Dark energy is widely described as the biggest problem in cosmology today; one which may demand new physics and a theoretical paradigm shift.[1] In this paper I suggest that the solution to the problem of dark energy is intimately related to the correct understanding of observational anomalies – in particular, the observed abundance and emptiness of voids, which has also elicited separate calls for a paradigm shift.[2] I propose that the paradigm shift that is required to understand both these issues actually entails no "new" physics, but a revisiting of fundamental issues relating to the sub-

*Based on talks presented at the NZIP2007, GRG18 and Dark2007 conferences.

tlety of the definition of gravitational energy within general relativity, and its relation to the careful modelling of the distribution of matter that we actually observe.

The punchline is that cosmic acceleration can be understood as an apparent effect, and dark energy as a misidentification of those aspects of cosmological gravitational energy which by virtue of the strong equivalence principle cannot be localized,[3] namely gradients in the quasilocal gravitational energy associated with spatial curvature gradients, and the kinetic energy of expansion, between bound systems and the volume–average position in freely expanding space. With this interpretation, a two–scale model can be constructed,[3] and a simple exact solution[4] of the relevant equations of cosmic evolution, the Buchert equations,[5] can be found. This solution provides a fit to type Ia supernovae (SneIa) luminosity distances which is statistically indistinguishable from that of the standard Lambda Cold Dark Matter (ΛCDM) model, while simultaneously satisfying key independent cosmological tests and offering the potential to resolve significant observational anomalies.[6]

Over the past decade we have come to think of "dark energy" as a homogeneous isotropic fluid–like quantity, with a local pressure related to its energy density by $P = w\rho$, where $w < -1/3$, so that the strong energy condition is violated. In fact, observations appear to be most consistent with a pure vacuum energy, $w = -1$, and determination of the value of the parameter, $w(z)$, as a function of redshift, z, is seen as the goal of "precision cosmology". What is proposed here, if correct, will turn this situation on its head. We can look forward to an era of precision cosmology, but one in which the focus will be on the complex hierarchical structure of the universe, rather than any one simple fluid equation of state. Nonetheless, while the proposed solution is intimately connected to the growth of inhomogeneities, and their backreaction on the geometry of the universe,[7] at its heart it addresses the question of the normalization of gravitational energy relative to observers within the observed structure. Thus I claim that the solution to the central foundational question does concern energy, and since "nothing" is "dark" the terminology "dark energy" is actually quite apt for the new proposal, if the community will allow the liberty of a change to the assumed definition of those words.

In this paper I will give an overview of the proposed solution to the problem of dark energy, the extent of its present quantitative successes, and more importantly the directions for future work. I will present fewer technical details than may be found in papers already published[3,4,6] or in preparation.[8,9] I aim to give a general overview to researchers in astrophysics, particle physics or general relativity, who have no specific prior experience with the averaging of inhomogeneous cosmological models.

2. The Universe We Observe

Our most widely tested "concordance model" of the universe relies on the assumption of an isotropic homogeneous geometry at all epochs of cosmic evolution. By the evidence of the cosmic microwave background (CMB) radiation, the universe was very smooth at the time of last scattering, and these assumptions were completely well–justified then. The departure from homogeneity was of order $\delta\rho/\rho \sim 10^{-5}$ in photons and the baryons that couple to them, and perhaps of order $\delta\rho/\rho \sim 10^{-3}$ in non–baryonic dark matter, which gives rise to the potential wells responsible for the dominant Sachs–Wolfe effect. By the Copernican principle, the assumption of global isotropy and homogeneity is completely justified at the epoch of last scattering, and it is safe to assume that the evolution of the universe was therefore extremely closely modelled by the Friedmann–Lemaître–Robertson–Walker (FLRW) solutions at that epoch. Furthermore, if we consider the spectrum of CMB anisotropies[10,11] then overall it appears that globally the universe is very close to spatially flat. Its initial evolution at the time of last scattering would therefore have been very close to that of an Einstein–de Sitter model at that epoch, even in the case of the ΛCDM paradigm since "dark energy" only becomes appreciable at late epochs.

At the present epoch, however, the distribution of matter is far from homogeneous on scales less than 150–300 Mpc. The actual universe has a sponge–like structure, dominated by huge voids. These voids are surrounded by bubble walls, and threaded by filaments, within which clusters of galaxies are located. Recent surveys suggest[12] that some 40–50% of the present volume of the universe is in voids of a characteristic scale $30h^{-1}$ Mpc, where h is the dimensionless Hubble parameter, $H_0 = 100h\,\mathrm{km\ sec^{-1}\,Mpc^{-1}}$. If smaller minivoids[13] and larger supervoids[14] are included, then our observed universe is presently void–dominated by volume.

Quite apart from the fact that this observed structure appears emptier than the vistas that Newtonian N–body simulations typically produce, the mere fact that the universe is presently inhomogeneous means that the assumptions implicit in the FLRW approximation can no longer be justified at the present epoch in the almost exact sense that they were justified at the epoch of last scattering. Homogeneity only applies at the present day in an *average sense*. The manner in which we take the average, and the

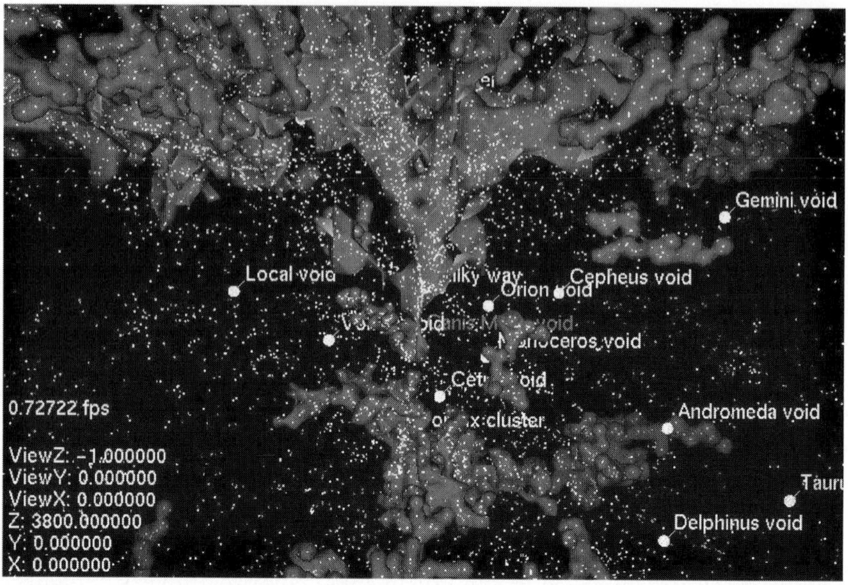

Fig. 1. Local voids and bubbles from the 6df survey. Courtesy of A. Fairall.

operational issues associated with this are not trivial, since the problem of fitting the local geometry of bound systems to the full dynamical geometry of the evolving universe[15,16] is a complicated one.

Since a broadly isotropic Hubble flow is observed, an almost FLRW geometry would appear to be a good approximation at some level of averaging, if our position is a typical one. I will demonstrate, however, that the assumption that the observed universe evolves exactly as a smooth featureless dust fluid means that we factor out central physically important questions which need to be understood to correctly relate our own observations to the average geometry.

2.1. *The Sandage–de Vaucouleurs Paradox*

There is a central foundational paradox concerning the expansion of the universe, which others have called the "Hubble–de Vaucouleurs paradox"[17,18] or the "Hubble–Sandage paradox",[19] but which I will call the *Sandage–de Vaucouleurs paradox* since it was originally raised by Sandage and collaborators[20] in objection to de Vaucouleurs' hierarchical cosmology.[21]

The problem is that in the conventional way of thinking about cosmological averages, below the scale of homogeneity we should expect large statistical scatter in the peculiar velocities of galaxies. In fact, if we were to average on scales of order 20Mpc, which is about 10% of the scale of homogeneity, then the statistical scatter should be so large that no linear Hubble law can be extracted. Yet 20Mpc is the scale on which Hubble originally extracted his famous linear law. By conventional understanding the statistical quietness of the local Hubble flow does not make sense.

One might attempt to explain the Sandage–de Vaucouleurs paradox as a consequence of dark energy, since it is well–known that in any FLRW model which expands forever – with or without dark energy – peculiar velocities eventually decay. However, if one tries to numerically model 12.5 Gyr of motion of very local galaxies, then it turns out that the initial conditions of each galaxy appear to have the most bearing on the problem.[19] In particular, one can fit a realistic motion with a cosmological constant, Λ, or alternatively in an open universe without Λ. Similar conclusions are also reached in constrained dark matter simulations of the Local Group.[22] Furthermore, the ΛCDM parameters required for the velocity dispersion predicted by structure formation to match the observed velocity dispersion, do not coincide with the concordance parameters.[23] Evidently a cosmological constant alone cannot explain the quietness and linearity of the local Hubble flow.

A related, but not exactly equivalent, issue is that using a conventional kinematic approach, the peculiar velocities of local galaxies appear to be at least a few factors too small to have arisen from a smooth distribution of matter at the time of recombination.[24]

3. Averaging, Backreaction and Dark Energy: The Debate

The present distribution of matter is clearly very complex, and since we cannot solve the Einstein equations for this distribution of matter analytically, there is an important question as to how we operationally match the average geometry of this distribution to the simple FLRW models that we know how to solve. Given that the nonlinear growth of structure appears to be roughly correlated to the epoch when cosmic acceleration is inferred to begin, a number of cosmologists have questioned whether the introduction of a smooth dark energy is a mistaken interpretation of the observations.[7] Attention has focused on the possibility that effects attributed to cosmic acceleration may actually be due to the backreaction from the growth of inhomogeneities in determining the geometry of the observed universe, without exotic dark energy. Different interpretations of a host of complex technical issues have led to a vigorous debate.

There are two large streams of research, which I will not focus on, for physical reasons. The first is backreaction in perturbation theory about an FLRW background, which became the focus of debate in 2005 following public attention generated by papers of Kolb and collaborators.[25,26] Within perturbation theory one may demonstrate that there is a potentially significant effect from backreaction.[26] However, this argument cannot be conclusive – if the second–order terms in the perturbation expansion exhibit an effect which might be interpreted as cosmic acceleration, such an effect may go away when the third–order terms are considered, and so on. Perturbation theory is very relevant near the epoch of last scattering, when the assumption of homogeneity was extremely good; but by the present epoch the nonlinear structures are so numerous and complex that we are beyond the regime of its applicability. Thus perturbation theory cannot provide a complete solution, and will not be discussed here.

The second stream of research I will not consider are those that involve exact inhomogeneous solutions of the Einstein equations, at the expense of introducing matter distributions which are unlikely. The spherically symmetric Lemaître–Tolman–Bondi (LTB) solutions are perhaps the most well–studied class of such models. While they may be very realistic descriptions of single voids, to apply them to the universe as a whole violates the Copernican principle, which I shall retain. One may obtain LTB solutions within which one can obtain reasonable fits to supernovae luminosity distances.[27] However, in my view, given their high degree of symmetry, these are at best toy models, which one cannot hope to reproduce in structure formation scenarios based on our knowledge of the power spectrum of density perturbations at the time of last–scattering.

To confront the actual inhomogeneous universe, which has no particular spatial symmetries below the scale of homogeneity, we must deal with schemes that average the full non–linear Einstein equations. There are many schemes for constructing averages, including those of Zalaletdinov[28] and Buchert.[5] There are various grounds for favouring the approach of Zalaletdinov,[28] which is fully covariant and averages all of the Einstein equations. However, Zalaletdinov's scheme is a general one, and for the cosmological problem additional assumptions are required. In Buchert's scheme one just average scalar quantities, and an additional integrability condition is required for the equations to close. However, with reasonable cosmological assumptions, the correlation tensor in Zalaletdinov's scheme takes the form of a spatial curvature,[29] and Buchert's scheme can be realized as a consistent limit.[30] Furthermore Buchert's scheme yields equations which are close to the Friedmann equations. Given that the Friedmann equations have worked so well to date, Buchert's average would appear to give a natural framework within which corrections to the Friedmann evolution can be quantitatively examined for the universe we observe. I shall adopt Buchert's scheme, with caveats to be discussed.

Buchert's scheme strictly deals with irrotational dust cosmologies, characterized by an energy density, $\rho(t,\mathbf{x})$, expansion, $\vartheta(t,\mathbf{x})$, and shear, $\sigma(t,\mathbf{x})$, on a compact domain, $\mathcal{D}$, of a suitably defined spatial hypersurface of constant average time, t, and spatial 3–metric, ${}^3g_{ij}$. Angle brackets are taken to denote the spatial volume averages, e.g., for the scalar curvature

$$\langle\mathcal{R}\rangle \equiv \left(\int_{\mathcal{D}} \mathrm{d}^3x\sqrt{\det{}^3g}\,\mathcal{R}(t,\mathbf{x})\right)/\mathcal{V}(t)$$

with $\mathcal{V}(t) \equiv \int_{\mathcal{D}} \mathrm{d}^3x\sqrt{\det{}^3g}$. The important lesson of Buchert averaging is that time evolution and averaging to do not commute.[5] Generally for any scalar Ψ,

$$\frac{\mathrm{d}}{\mathrm{d}t}\langle\Psi\rangle - \langle\frac{\mathrm{d}\Psi}{\mathrm{d}t}\rangle = \langle\Psi\vartheta\rangle - \langle\vartheta\rangle\langle\Psi\rangle \tag{1}$$

The fact that the r.h.s. of (1) does not vanish, as is the case for the FLRW cosmologies, is a manifestation of *backreaction.*

Applied to the equations of cosmic evolution one obtains the exact *Buchert equations*

$$3\frac{\dot{\bar{a}}^2}{\bar{a}^2} = 8\pi G\langle\rho\rangle - \tfrac{1}{2}\langle\mathcal{R}\rangle - \tfrac{1}{2}\mathcal{Q}, \tag{2}$$

$$3\frac{\ddot{\bar{a}}}{\bar{a}} = -4\pi G\langle\rho\rangle + \mathcal{Q}, \tag{3}$$

$$\partial_t\langle\rho\rangle + 3\frac{\dot{\bar{a}}}{\bar{a}}\langle\rho\rangle = 0, \tag{4}$$

where $\bar{a}(t) \equiv \left[\mathcal{V}(t)/\mathcal{V}(t_0)\right]^{1/3}$,

$$\mathcal{Q} \equiv \frac{2}{3}\left(\langle\vartheta^2\rangle - \langle\vartheta\rangle^2\right) - 2\langle\sigma\rangle^2 \tag{5}$$

and following integrability condition follows from (2)–(5):

$$\partial_t\left(\bar{a}^6\mathcal{Q}\right) + \bar{a}^4\partial_t\left(\bar{a}^2\langle\mathcal{R}\rangle\right) = 0. \tag{6}$$

It is observed that the kinematic backreaction term, $\mathcal{Q}$, enters (3) with the same sign as a cosmological constant in the equivalent Raychaudhuri equation in the FLRW paradigm, but enters (2) with the *opposite* sign to a cosmological constant in the Friedmann equation. Given this fact, together with the domain-dependence of the average, the question of whether Buchert's scheme can generate sufficient backreaction to give apparent acceleration for realistic initial conditions has been the subject of much debate. On reasonable grounds one might conclude[31] that while the effects are real, they are too small in magnitude to give departures sufficiently large from the FLRW expectation to register as cosmic acceleration.

It is at this point in the argument caution must be exercised. As every student of general relativity knows, one cannot simply write down a time parameter and assume that it is the parameter one measures on one's clock, without specifying how it is related to local invariants. We actually measure luminosity *distances*, and the deduction of cosmic acceleration involves two time derivatives. We therefore have to be extremely careful to operationally specify how t is to be related to our own clocks. It must be observed that the scale–factor $\bar{a}(t)$ is not related to an exact local metric, that has been substituted in Einstein's equations. Rather we must solve the Buchert equations, and determine how the best–fit almost–FLRW geometry that is obtained relates operationally to our own measurements.

4. Finite Infinity and Gravitational Energy

In a completely arbitrary inhomogeneous universe the calibration of rods and clocks at one point relative to another can vary arbitrarily. However, our clocks and rods and those of stars in distant galaxies we observe, appear to a very good approximation to be determined by geodesics of ideal solutions – the Kerr and Schwarzschild geometries – in which space is asymptotically flat, with an exact time symmetry at spatial infinity. This time symmetry, mathematically described by a timelike Killing vector, must be an idealization which breaks down at some level, since the universe is expanding.

In his pioneering work on the fitting problem, Ellis[15] suggested its solution should involve the notion of *finite infinity*[a], "*fi*", namely a timelike surface within which the dynamics of an isolated system such as the solar system can be treated without reference to the rest of the universe. After all, the matter in our galaxy and other typical galaxies broke away from the Hubble flow to be come a bound system well over 10 billion years ago. With sufficient computing resources, we can integrate the motions of stars

---

[a]As finite infinity adds a new notion of infinity to the concepts timelike, spatial and null infinity, a new mathematical symbol, *fi*, is appropriate - in LaTeX: \def\finfty{\mathop{\hbox{\it fi}}}.

within our galaxy for billions of years without considering the dynamics of the universe outside the galaxy. Thus within finite infinity spatial geometry might be considered to be effectively asymptotically flat, and governed by "almost" Killing vectors.

The concept of finite infinity does not appear to have been further developed in the intervening two decades. However, given that the normalization of our clocks in the idealized Schwarzschild and Kerr geometries is related to the timelike Killing vector at spatial infinity, finite infinity would appear to be the appropriate reference surface for the definition of gravitational energy, which in stationary spacetimes is tied to the asymptotically flat region.[32] In Newtonian terms, it is the scale at which we set the zero of the gravitational potential.

The definition of gravitational energy in general is an extremely difficult problem, on account of the fact that space itself is dynamical and can carry energy and momentum. By the strong equivalence principle, since the laws of physics must coincide with those of special relativity at a point, it is only internal energy that can be localized in an energy–momentum tensor on the r.h.s. of the Einstein equations. Any uniquely relativistic aspects of gravitational energy associated with spatial curvature and geometrodynamics cannot be included in the energy momentum tensor, but are at best described by a quasilocal formulation.[33]

Einstein himself worried about the problem of quasilocal gravitational energy, in terms of energy–momentum complexes, and many mathematical relativists since Einstein have also considered the problem. It is quite possible that the general problem of a definition of quasilocal energy does not have a solution, since it depends on the split of space and time. Here I will not be interested in the general problem for an arbitrary solution of the Einstein equations, but in the specific problem for a universe which was effectively homogeneous and isotropic at last scattering, with nearly scale–invariant perturbations.

The fact that quasilocal energy is not part and parcel of theoretical framework of the FLRW cosmology, is easily illustrated by the Newtonian perspective.[34] The l.h.s. of the standard Friedmann equation with $\Lambda = 0$ can be regarded as the difference of a kinetic energy density per unit rest mass, $E_{\rm kin} = \dot{\bar{a}}^2/(2\bar{a}^2)$ and a total energy density per unit rest mass $E_{\rm tot} = -k/(2\bar{a}^2)$ of the opposite sign to the Gaussian curvature, k, while the energy–momentum tensor on the r.h.s. is the Newtonian potential energy per unit rest mass. The terms in the Einstein tensor represent forms of gravitational energy, but since they are identical for all observers in an isotropic homogeneous geometry, one can always synchronize clocks and calibrate rods of ideal isotropic observers unambiguously.

As soon as inhomogeneity enters the game, however, one will have gradients in both the kinetic energy of expansion, and in spatial curvature. Given the present value of the Hubble constant, it is likely that the spatial curvature gradient is the most significant effect. Quasilocal energy gradients between a bound system and finite infinity could conceivably be relevant for asymptotic galactic dynamics, or even the solution of the Pioneer anomaly. This speculation is left for future work. The proposal of Refs. [3,4] is concerned principally with dynamically evolving spatial curvature gradients. In this context, one should observe that the idea that negative spatial curvature is associated with positive gravitational energy, evident already in the Newtonian framework, remains true in the LTB models, where the Gaussian curvature is replaced by an inhomogeneous energy function $E(r)$.

Cosmological gravitational energy is largely uncharted territory in the more rigorous quasilocal framework[33] within which the problem should ultimately be framed. In any quasilocal formulation results depend crucially on the reference spacetime and surface of integration. Recently Chen, Liu and Nester[35] have obtained a result over which they expressed surprise, but which is consistent with the present proposal. They find that for an isotropic observer in synchronous gauge in a $k = -1$ Friedmann universe the quasilocal energy in their particular Hamiltonian formalism is negative. A similar result is obtained using a different approach by Garecki.[36] These results are expected in the current approach, since one is effectively subtracting a fiducial flat spacetime in each case, and the *relative* sign of energy depends on the observer. An isotropic $k = 0$ Friedmann observer has zero quasilocal energy in the approach of Chen, Liu and Nester; thus relative to the $k = -1$ geometry the $k = 0$ geometry has negative quasilocal energy, but conversely relative to the $k = 0$ geometry the $k = -1$ has positive quasilocal energy. Our viewpoint here will be that the fiducial reference point is the $k = 0$ geometry of the finite infinity region. This agrees with the Newtonian version of energy in the Friedmann equation, the LTB energy function, and with the idea that binding energy is negative.[32]

5. Finite Infinity and the True Critical Density

In the standard FLRW cosmology the critical density is proportional to the square of the Hubble parameter. However, in an inhomogeneous cosmology this is no longer true,

$$\rho_{\rm cr} \neq \frac{3H_{\rm av}^2}{8\pi G} \tag{7}$$

as the average Hubble parameter, $H_{\rm av}$, is related to the spatial scale of averaging and the clock chosen. In the spirit of the spherical collapse model, we are familiar with the idea that a perturbation near the time of last scattering can evolve as an FLRW model of different density, until it turns around and collapses, by which point other regions closer to the true mean density are within the past light cone.

In a universe which grew from a nearly scale–invariant spectrum of perturbations, there will always be spatial correlations of some density perturbation which is relatively far from the mean, once one samples on scales close to the spatial scale of the particle horizon. This is a simple consequence of cosmic variance, and is illustrated by the following analogy. Take a bucket filled with grains of sand of identical diameter, whose density is Gaussian distributed about a mean. Provided the container is much larger than the individual grains then the mean density of the bucket can be expected to be extremely close to the mean density of all the sand from the beach on which it was collected. If one repeats the exercise for stones of larger and larger diameter, the mean density of the bucket will on average differ more and more from the mean density of the beach as the diameter of the stones becomes comparable to the size of the bucket, by a $\sqrt{N}$ statistic.

Cosmic variance[b] in the actual universe is complicated by three things. Firstly, since general relativity is causal, the bucket in the analogy is the past light cone which grows with time. Secondly, the perturbations are not uniform density but contain other perturbations embedded within them, like Russian dolls. Thirdly, below a scale of apparent homogeneity the initial perturbations can no longer be thought of as perturbations. They have undergone a non-linear transition to form the structure that we see: stars, globular clusters, and clusters of galaxies strung in filaments and bubbles around voids.

Every inflationary cosmologist who has thought about the foundations of the subject realises that the present density of the observable universe need not be the same as the whole universe, observable and unobservable, and in general these should differ on account of cosmic variance in a scale–

[b]I use the term *cosmic variance* in a general sense, rather than restricting it to one statistical aspect of the CMB temperature fluctuations which arise as an observable consequence of the variance in the density perturbation spectrum.

free density perturbation spectrum. Nonetheless, in the FLRW paradigm we unwittingly end up violating this. As long as we assume that the average density evolves by the Friedmann equation, we implicitly assume that the mean density in our past light cone today is the same as the whole ensemble of observable universes. As long as we evolve the average geometry purely by the Friedmann equation, with equality in (7), we will miss an important aspect of actual cosmic evolution. Furthermore, as long we study inhomogeneity in the context of a Swiss cheese model[37] by cutting and pasting holes into a background which still evolves by the Friedmann equation alone, we will still fail to uncover essential features of cosmic evolution.

There is perhaps no obvious way to define finite infinity in an arbitrary inhomogeneous background. To proceed I make the crucial observation that since our universe was effectively homogeneous and isotropic at last scattering, a notion of a universal critical density scale did exist then. It was the density required for gravity to overcome the initial uniform expansion velocity of the dust fluid. I will assume, as is consistent with primordial inflation, that the present horizon volume of the universe was very close to the critical density at last scattering, with scale–invariant perturbations.

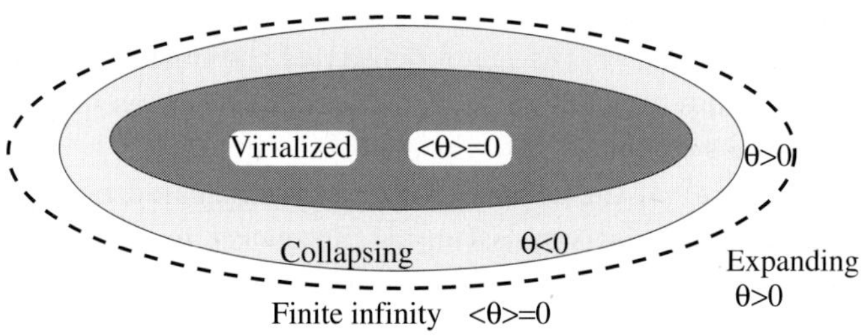

Fig. 2. Finite infinity, *fi*.

In view of the existence of backreaction, what is required is a notion of true critical density without assuming evolution via the Friedmann equation. This should be a dynamically evolving demarcation scale between systems which will become bound and those that are unbound, given available data within the past light cone at any epoch. As outlined in Ref. [3], and depicted in Fig. 2, finite infinity is defined in terms of a scale over which the average expansion, including virialized, collapsing and marginally expanding regions is zero, $\langle\vartheta\rangle_{fi} = 0$, while being positive outside. Then $\rho_{\rm cr} \equiv \langle\rho\rangle_{fi}$. Finite infinity is a non–static boundary analogous to the spheres cut out in the Swiss cheese model,[37] but it involves average geometry rather than matching exact solutions, and no assumptions about homogeneity are made outside finite infinity. Finite infinity represents a physical scale expected to lie outside virialized galaxy clusters, but within the filamentary walls surrounding voids. Since it is a scale related to the true critical density, space at finite infinity boundaries can be described by the spatially flat metric

$$\mathrm{d}s^2_{\mathit{fi}} = -\mathrm{d}\tau_{\mathrm{w}}^2 + a_{\mathrm{w}}^2(\tau_{\mathrm{w}}) \left[\mathrm{d}\eta_{\mathrm{w}}^2 + \eta_{\mathrm{w}}^2 \mathrm{d}\Omega^2\right] , \tag{8}$$

where $\mathrm{d}\Omega^2 \equiv \mathrm{d}\theta^2 + \sin^2\theta\, \mathrm{d}\phi^2$. While I use the term "bubble wall" it should be recognized that an ideal bubble wall would consist of finite infinity regions touching exactly at their boundaries. Due to the existence of minivoids one should take care in identifying the scale of local walls empirically.

6. Average Homogeneity in a Lumpy Universe

An important issue that has to be faced is the split of space and time. Density is not a covariant quantity, but will depend on the foliation of hypersurfaces chosen to average on, and the specification of the observers within the hypersurfaces. The hypersurfaces chosen by Buchert are the standard ones based on surfaces of constant, t, where this is the affine parameter on geodesics of ideal comoving observers. This choice, which is analogous to the choice of synchronous gauge in FLRW models, is always possible for an energy–momentum tensor with dust in the absence of vorticity.

While it is entirely reasonable that Buchert's choice of gauge is consistent at the volume–average position, provided one averages over suitably large spatial volumes, there is a problem of physical interpretation since observers in bound systems are located in places where the physical assumptions in Buchert's gauge choice no longer apply. As soon as regions start to collapse, geodesics will cross, vorticity comes into play and comoving coordinates cannot be chosen globally. Actual observers are in locations where the geometry is better modelled by the vacuum Schwarzschild and Kerr solutions, rather than by a rotationless dust fluid.

Since a nearly isotropic Hubble flow is observed, it is clear that at some level a uniform Hubble flow should be obtained, despite the large–scale inhomogeneity. I will make a choice of surfaces of homogeneity which *implicitly solves the Sandage–de Vaucouleurs paradox* by the assumption that on suitably small scales of averaging *the bare Hubble flow is uniform below the scale of apparent homogeneity*,

$$\frac{1}{\ell_r(t)}\frac{\mathrm{d}\ell_r(t)}{\mathrm{d}t} = \frac{1}{3}\langle\theta\rangle_{\mathcal{D}_1} = \frac{1}{3}\langle\theta\rangle_{\mathcal{D}_2} = \cdots = \bar{H}(t), \tag{9}$$

t being the local proper time in each averaging region $\mathcal{D}_I$, and $\ell_r \equiv \mathcal{V}^{1/3} = \left[\int_{\mathcal{D}_I} \mathrm{d}^3x\sqrt{^3g}\right]^{1/3}$ a relevant proper distance. Here no $\mathcal{D}_I$ can be contained within a finite infinity region. The important point is that although it will appear to any one observer that voids expand faster than bubble walls, the clocks of observers within voids will also tick faster than those of observers at finite infinity on account of the relative gravitational energy gradient. Thus, provided the gravitational energy and spatial curvature gradients are correlated, apparent variance in the Hubble flow for any one observer is nonetheless perfectly consistent with uniformity of the bare, or "quasilocally measurable" Hubble flow.

The uniform expansion gauge, which is equivalent to one of the standard gauges in cosmological perturbation theory,[38] is given as an ansatz for defining the surfaces of homogeneity. However, it should be noted that this ansatz is *less restrictive* than the standard ansatz implicit in the FLRW models, where expansion is uniform, and all ideal isotropic observers also have synchronous clocks and measure the same spatial curvature on the relevant surfaces of homogeneity. Here uniformity in the bare Hubble expansion is retained, but spatial curvature and gravitational energy of isotropic observers vary in a correlated fashion.

It should be emphasized that the Copernican principle is retained here. If we compare ideal isotropic observers, some in galaxies and some in voids, they will also measure an isotropic CMB. However, they will potentially measure different mean CMB temperatures, and different angular scales in the CMB anisotropies. These two observations respectively relate to relative gravitational energy and relative spatial curvature. While the clocks of all ideal observers are synchronized initially, when the FLRW approximation was a good one, ultimately they will diverge; by 38% on average at the present epoch. Since we exchange photons with other bound systems, which are within finite infinity regions, such a large clock–rate variance is not directly detectable in experiments conceived to date.

The uniform bare Hubble constant gauge is more than simply an ansatz which implicitly solves the Sandage–de Vaucoulers paradox while dealing with the question of cosmological gravitational energy. As is argued in a separate paper,[8] it may be realized as a consequence of the strong equivalence principle applied to expanding space. Among the class of all possible motions of a timelike geodesic congruence there is a class of conformal motions which *do not isolate any direction in space as preferred*, in contrast to individual boosts. The statement then is that for such motions we cannot distinguish the circumstance in which the particles are *moving* from a common origin in a static Minkowski space, or alternatively are *at rest* in the underlying expanding universe. The two situations are equivalent; giving a *Cosmological Principle of Equivalence.*[8] In terms of the conceptual basis of relativistic cosmology, this amounts to a refinement of the Principle of Inertia. Not only can we not distinguish which one of Galileo's ships[c], or of Einstein's trains is the one that is moving; for conformal motions we can also not distinguish whether it is the collection of ships or trains that are moving, or the 'sea' or 'railyard' expanding in between.

7. The Fractal Bubble Model

In Ref. [3] I have written down a Buchert average of the Einstein equations based on two scales, finite infinity regions within which the geometry is given by (8), and voids within which the geometry is negatively curved with local scale factor, $a_{\rm v}$. The geometry near the centres of voids is given by

$$\mathrm{d}s^2_{\mathcal{D}_C} = -\mathrm{d}\tau_{\rm v}^2 + a_{\rm v}^2(\tau_{\rm v}) \left[\mathrm{d}\eta_{\rm v}^2 + \sinh^2(\eta_{\rm v})\mathrm{d}\Omega^2\right] . \tag{10}$$

The Buchert equations are not written in terms of the local geometry (8) or (10) at either finite infinity or the void centres, but in terms of an intermediate volume–average location with a volume–average time parameter, t, and an average scale factor, $\bar{a}$, defined by

$$\bar{a}^3 = f_{\rm vi}a_{\rm v}^3 + f_{\rm wi}a_{\rm w}^3, \tag{11}$$

$f_{\rm vi}$ and $f_{\rm wi} = 1 - f_{\rm vi}$ being the respective initial void and wall volume fractions at last scattering. Initially $f_{\rm vi} \ll 1$, and $f_{\rm wi} \simeq 1$, since the universe is homogeneous and isotropic at last–scattering, evolving like an Einstein–de Sitter one. The Buchert average is constructed over the entire present particle horizon volume.

Although the actual surfaces of homogeneity will not coincide with Buchert's ones on small scales, we can still make use of Buchert's scheme, provided that we take care in defining the relationship between the volume–average quantities of Buchert's scheme, and those in a finite infinity region. Ultimately, in order to deal with actual gradients in spatial curvature between finite infinity and void centres, it may be necessary to use a scheme such as that of Zalaletdinov,[28] or to look at the problem in terms of Ricci

[c]Actually Galileo just talked about a single ship, in a context more like Einstein's closed elevator, but the principle is the same.

flow.[39] This may be necessary, for example, in determining the most appropriate gauge for the background to structure formation simulations. The approach adopted in Refs. [3,4,6] is that the Buchert equations are solved in volume average time, but the uniform expansion ansatz is used to calibrate observable quantities relative to observers at finite infinity. This involves a dressing of cosmological parameters over and above the volume factors considered by Buchert and Carfora.[39,40]

The independent Buchert equations, including the integrability condition that ensures their closure, may be written

$$\bar{\Omega}_M + \bar{\Omega}_k + \bar{\Omega}_Q = 1, \tag{12}$$

$$\bar{a}^{-6}\partial_t\left(\bar{\Omega}_Q \bar{H}^2 \bar{a}^6\right) + \bar{a}^{-2}\partial_t\left(\bar{\Omega}_k \bar{H}^2 \bar{a}^2\right) = 0\,, \tag{13}$$

where the volume–average matter, curvature and kinematic backreaction parameters are respectively

$$\bar{\Omega}_M = \frac{8\pi G\bar{\rho}_{M0}\bar{a}_0^3}{3\bar{H}^2\bar{a}^3}\,;\ \bar{\Omega}_k = \frac{-k_{\rm v} f_{\rm vi}{}^{2/3} f_{\rm v}{}^{1/3}}{\bar{a}^2\bar{H}^2}\,;\ \bar{\Omega}_Q = \frac{-\dot{f}_{\rm v}{}^2}{9f_{\rm v}(1-f_{\rm v})\bar{H}^2}\,. \tag{14}$$

The average curvature is due to the voids only, which are assumed to have $k_{\rm v} < 0$, an overdot denotes a derivative w.r.t. volume–average time, t, and $\bar{H} \equiv \dot{\bar{a}}/\bar{a}$ is the volume–average or bare Hubble parameter.

Although the quantity, $\bar{H}$, has no particular significance for a general Buchert average, for our particular assumptions it represents the underlying uniform quasilocally measured Hubble flow. The quantities $\bar{\Omega}_M$, $\bar{\Omega}_k$ and $\bar{\Omega}_Q$ are then interpreted as the bare cosmological parameters, relevant to a comoving isotropic observer at an average position in freely expanding space, which will be within a void, but not at its centre. These quantities are the closest analogues of the standard FLRW density parameters.

It must be recalled that the scale factor $\bar{a}$ does not correspond to an exact geometry substituted into the Einstein equations, and then solved. Rather we integrate the Buchert equations and then reconstruct the average spherically symmetric geometry

$$ds^2 = -dt^2 + \bar{a}^2(t)\,d\bar{\eta}^2 + A(\bar{\eta},t)\,d\Omega^2, \tag{15}$$

the area function A being defined by a horizon-volume average.[3] The fact that (15) is spherically symmetric reflects the fact that the geometry is reconstructed by taking an average on our radial null geodesics. It is therefore *not* an LTB model, since exact spherical symmetry has not been assumed in the Einstein equations.

Since (15) does not correspond to the local geometry of a finite infinity observer, if we assume that our rods and clocks differ little in calibration from those at finite infinity, we still have to be careful in relating our local geometry (8) to (15). We do this by conformal matching of the radial null geodesics of (8) and (15), which requires that $d\eta_w = {f_{wi}}^{1/3} d\bar{\eta}/[\bar{\gamma}(1-f_v)^{1/3}]$. It then turns out that the finite infinity geometry may be rewritten

$$ds^2_{fi} = -d\tau_w^2 + \frac{\bar{a}^2}{\bar{\gamma}^2}\left[d\bar{\eta}^2 + r_w^2(\bar{\eta},\tau_w)\, d\Omega^2\right] \qquad (16)$$

where $r_w \equiv \bar{\gamma}(1-f_v)^{1/3} {f_{wi}}^{-1/3} \eta_w(\bar{\eta},\tau_w)$. The geometry (16), which now again has a more general spherically symmetric form, is effectively the closest match to the FLRW geometry we usually attempt to fit to the whole universe with the assumption that spatial curvature everywhere matches our own, and clocks everywhere are synchronized to our own wall time, τ_w.

In place of the bare cosmological parameters (14), one can define conventional dressed parameters with respect to the geometry (16), relevant to "wall observers" such as ourselves. In particular, the dressed density parameter, Ω_M, defined according to $\Omega_M = \bar{\gamma}^3\bar{\Omega}_M$ is effectively the conventional parameter, whose numerical value will be similar to that which we infer in FLRW models. Since (16) is not an FLRW metric it does not make particular sense to supplement Ω_M by additional dressed parameters, as they will not sum to unity.

What is more significant cosmologically is that the dressed geometry (16) will yield a dressed Hubble parameter,

$$H \equiv \frac{1}{a}\frac{da}{d\tau_w} = \bar{\gamma}\bar{H} - \tfrac{d}{dt}\bar{\gamma} = \bar{\gamma}\bar{H} - \bar{\gamma}^{-1}\tfrac{d}{d\tau_w}\bar{\gamma}\,. \qquad (17)$$

where $a \equiv \bar{a}/\bar{\gamma}$, which differs from the bare Hubble parameter, $\bar{H}$. Similarly, a wall observer will determine a dressed deceleration parameter $q = -\ddot{a}/(H^2 a)$, which differs from the bare deceleration parameter $\bar{q} = -\ddot{\bar{a}}/(\bar{H}^2\bar{a})$.

The general solution of the two scale Buchert equations was found in Ref. [4]. The general solution is specified by four independent parameters, $\bar{H}_0$, $\bar{\Omega}_{M0}$, f_{v0} and ϵ_i, which are respectively the bare Hubble constant, the bare matter density, the present epoch void volume fraction, and a small parameter $\epsilon_i \ll 1$ related to the initial void fraction. However, two of the four parameters are greatly restricted by taking priors[6] at the surface of last scattering consistent with the CMB. Furthermore, there is a tracker solution to which all solutions with these priors approach to within 1% by a redshift $z \sim 37$. This effectively reduces the number of free parameters to two; we take these to be $\bar{H}_0$ and f_{v0}. The tracker solution is given by

$$\bar{a} = \frac{\bar{a}_0 (3\bar{H}_0 t)^{2/3}}{2 + f_{v0}} \left[3f_{v0}\bar{H}_0 t + (1 - f_{v0})(2 + f_{v0}) \right]^{1/3} \tag{18}$$

$$f_v = \frac{3f_{v0}\bar{H}_0 t}{3f_{v0}\bar{H}_0 t + (1 - f_{v0})(2 + f_{v0})}, \tag{19}$$

in terms of the volume–average cosmic time, t. Since the lapse function and bare Hubble parameter are related to the void fraction by $\bar{\gamma} = 1 + \frac{1}{2} f_v = \frac{3}{2}\bar{H}t$, the relation $d\tau_w = \bar{\gamma}^{-1} dt$ may be readily integrated to obtain the wall time, τ_w, relevant to observers in galaxies,

$$\tau_w = \tfrac{2}{3}t + \frac{2(1 - f_{v0})(2 + f_{v0})}{27 f_{v0}\bar{H}_0} \ln\left(1 + \frac{9 f_{v0}\bar{H}_0 t}{2(1 - f_{v0})(2 + f_{v0})}\right). \tag{20}$$

As is observed in Ref. [4], expressions for many relevant observable quantities, including effective dressed luminosity and angular diameter distances, d_L and d_A, are readily obtained. It is particularly interesting to compare the bare volume–average tracker solution deceleration parameter,

$$\bar{q} = \frac{2\,(1 - f_v)^2}{(2 + f_v)^2}, \tag{21}$$

with the corresponding dressed deceleration parameter

$$q = \frac{-\,(1 - f_v)\,(8 f_v^{\,3} + 39 f_v^{\,2} - 12 f_v - 8)}{\left(4 + f_v + 4 f_v^{\,2}\right)^2}, \tag{22}$$

At early times, when $f_v \ll 1$, both deceleration parameters begin at the Einstein–de Sitter value, $\bar{q} \simeq q \simeq \frac{1}{2}$. The bare deceleration parameter always remains positive, meaning that a volume–average observer in freely expanding space detects no cosmic acceleration, in accord with the intuition of the critics.[31] However, the dressed deceleration parameter changes sign at epoch when $f_v \simeq 0.5867$, at a zero of the cubic in (22), so that a wall observer *does* detect apparent acceleration.

The fractal bubble (FB) model provides a solution to the observational coincidence that the onset of cosmic acceleration appears to roughly coincide with the growth of large structures. Cosmic acceleration is an apparent effect that arises from gravitational energy gradients related to spatial curvature gradients. It begins when the void fraction reaches 59% at a redshift $z \simeq 0.9$. This statement itself should be able to be tested in future. Traditionally observational cosmologists define voids in terms of a density contrast threshold, when quoting void volume fractions.[12] For example, one might argue as to whether one should take $\delta\rho/\rho < -0.9$ or some other bound. In the present model, the void volume fraction is related technically to those regions within the past light cone which are not within finite infinity. With some further refinement, it should be possible to relate this to an observational definition of void fractions. Then as surveys improve, the claims about the properties of $f_v(z)$ can be compared with observation.

This illustrates the power of the FB model as compared to the spatially flat ΛCDM model. Both models depend on the Hubble parameter and one other free parameter. However, Ω_Λ is not directly observable, whereas the void–volume fraction, f_v, is empirically observable in principle.

8. Observational Tests

In Ref. [6] we have examined the fit of the FB model to the Riess07 gold data set[41] of SneIa. We find that with 182 data points and two degrees of freedom the best–fit $\chi^2 = 162.7$, i.e., a χ^2 of approximately 0.9 per degree of freedom, which is a good fit. While a marginally lower $\chi^2 = 158.7$ may be obtained for the best–fit flat ΛCDM model (with $H_0 = 62.6\,\text{km sec}^{-1}\,\text{Mpc}^{-1}$, $\Omega_{M0} = 0.34$, $\Omega_{\Lambda 0} = 0.66$), the difference is not significant. In fact, a Bayesian model comparison of the FB model against a flat ΛCDM model with priors $55 \le H_0 \le 75\,\text{km sec}^{-1}\,\text{Mpc}^{-1}$, $0.01 \le \Omega_{M0} \le 0.5$. gives a Bayes factor of $\ln B = 0.27$ in favour of the FB model. Statistically, such a small margin is "not worth more than a bare mention"[42] or "inconclusive";[43] that is to say, the two models are statistically indistinguishable.

The residual difference $\Delta\mu = \mu_{\text{FB}} - \mu_{\text{empty}}$, in the standard distance modulus, $\mu = 5\log_{10}(d_L) + 25$, of the best–fit FB model from that of a coasting Milne universe of the same Hubble constant, $H_0 = 61.7\,\text{km sec}^{-1}\,\text{Mpc}^{-1}$, is plotted in Fig. 3 and compared with binned data from the Riess07 gold data set. Apparent acceleration occurs for positive residuals in the range, $z \lesssim 0.9$. (Note, however, that the exact range of redshifts corresponding to apparent acceleration also depends on the value of the Hubble constant of the Milne universe used to compute the residual.) The equivalent theoretical residual for the spatially flat ΛCDM model with the same values of H_0 and Ω_{M0} is also shown. The difference in the gradients of the two curves should lead to observable differences if observing programs which are trying to measure the "equation of state, $P = w\rho$, $w(z)$, of dark energy" can achieve sufficient accuracy.

In Fig. 4 we address three major independent cosmological tests. In addition to statistical confidence limits in the (H_0, Ω_{M0}) parameter space we also consider the angular scale of the sound horizon in the CMB anisotropy

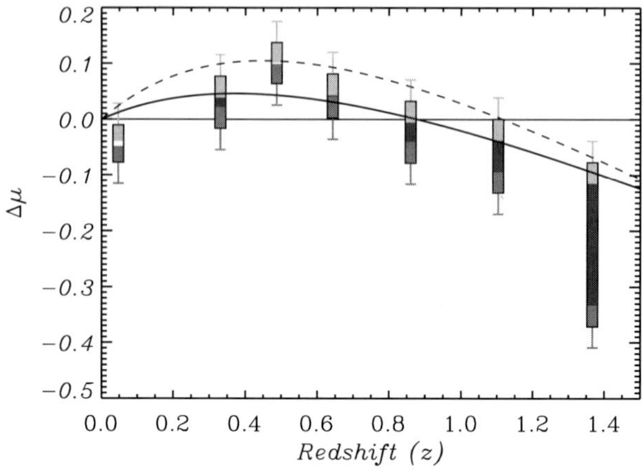

Fig. 3. The difference in the distance modulus, $\mu = 5\log_{10}(d_L) + 25$, with d_L in units Mpc, of the FB model with $H_0 = 61.7\,\text{km sec}^{-1}\,\text{Mpc}^{-1}$, $\Omega_{M0} = 0.326$ from that of an empty coasting Milne universe, with the same value of H_0. The solid curve shows the FB model expectation, and the dashed curve the expectation for a spatially flat ΛCDM model with the same values of H_0 and Ω_{M0}. Whiskers indicate how the statistical uncertainties, shown as boxes, move when the background value of H_0 for the Milne universe which is subtracted is varied within the 2σ limits. For further details see Ref. [6].

spectrum,[10,11] and the comoving scale of the baryon acoustic oscillation (BAO) seen in galaxy clustering[44] using the relevant dressed geometry (16).

Ideally we should recompute the spectrum of Doppler peaks for the FB model. However, this is a mammoth task, as the standard numerical codes have been written solely for FLRW models, and every step has to be carefully reconsidered. For this reason we first ask whether parameters exist for which the effective angular diameter scale of the sound horizon matches the angular scale of the sound horizon, $\delta = 0.01$ rad, of the ΛCDM model, as determined by WMAP.[10] Since there is no change to the physics of recombination, but just an overall change to the *calibration* of cosmological parameters, this is entirely reasonable. In Fig. 4 we plot parameter ranges which match the $\delta = 0.01$ rad sound horizon scale to within 2%, 4% and 6%, using the calculation of the sound horizon given in Ref. [3], Sec. 7.2. The 2% contour would roughly correspond to the 2σ limit if the WMAP uncertainties for the ΛCDM model are maintained. As this can only be confirmed by detailed computation of the Doppler peaks, the additional levels have been chosen cautiously.

In the case of the BAO scale, as we do not yet have the resources to

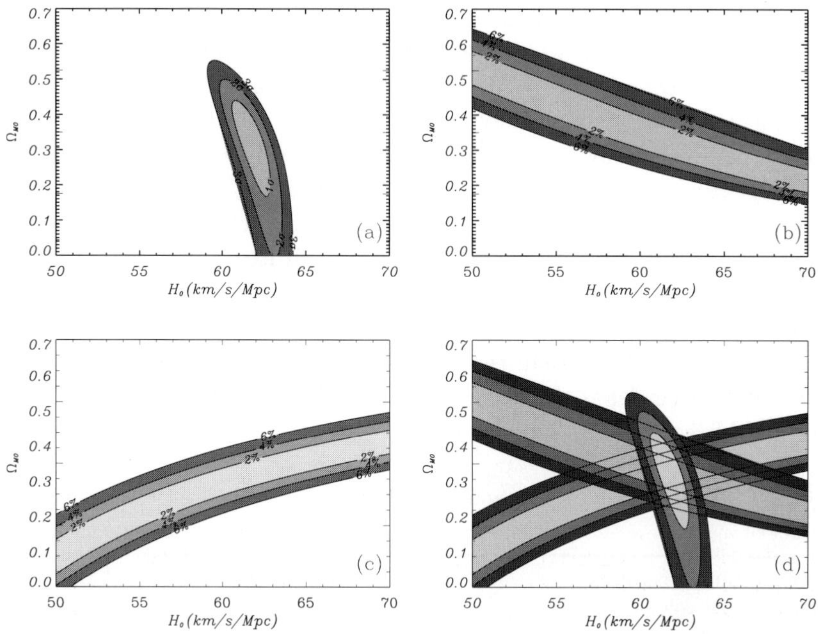

Fig. 4. Various cosmological tests in the parameter space of the dressed Hubble constant, H_0, and the dressed matter density parameter, Ω_{M0}. **(a)** Statistical confidence limits on SneIa in the Riess07 gold data set;[41] **(b)** Parameters which match $\delta = 0.01$ angular scale of sound horizon[10,11] of ΛCDM model to within 2%, 4%, 6%; **(c)** Parameters which match effective comoving $104h^{-1}$ Mpc scale of baryon acoustic oscillation[44] to within 2%, 4%, 6%; **(d)** Overlay of panels (a),(b),(c).

analyse the galaxy clustering data directly, we also begin with a simple but effective check. Since the dressed geometry (16) does provide an effective almost–FLRW metric adapted to our clocks and rods in spatially flat regions, the effective comoving scale in this geometry should match the corresponding observed BAO scale of $104h^{-1}$Mpc. In Fig. 4 we therefore also plot parameter values which match this scale to within 2%, 4% or 6%.

It should be noted that even if SneIa are disregarded, the parameters which fit the two independent tests relating to the sound horizon and the BAO scale agree with each other, to the accuracy shown, for values of the Hubble constant which include the value of Sandage *et al.*;[48] but *not* for the values of H_0 greater than $70\,\text{km sec}^{-1}\,\text{Mpc}^{-1}$ which best–fit the WMAP data[10,11] with the FLRW model. The value $H_0 = 62.3 \pm 1.3\,(\text{stat}) \pm 5.0\,(\text{syst})\,\text{km sec}^{-1}\,\text{Mpc}^{-1}$ determined by the Hubble Key Team of Sandage *et al.*[48] has been controversial, given the 14% difference from values which best-fit the WMAP data with the ΛCDM model.[10,11] However, the WMAP analysis only constitutes a direct measurement of CMB temperature anisotropies; the determination of cosmological parameters involves model assumptions. In the FB model, the model assumptions are different. In particular, on account of the differences in gravitational energy and local spatial curvature between observers in bound systems, and those at the volume average in freely expanding space, the calibration of quantities related to the CMB has to be carefully reconsidered. Observers at the volume average detect a cooler mean CMB temperature[3] $\bar{T}_0 = \bar{\gamma}_0^{-1} T_0 = 1.98$ K than the $T_0 = 2.73$ K we measure. Accounting for these differences leads to concordance for parameter values which include the Hubble constant of Sandage *et al.*[48]

Table 1. Best-fit cosmological parameters derived from the independent parameters, H_0, f_{v0}.

dressed Hubble constant	$H_0 = 61.7^{+1.2}_{-1.1}$ km sec^{-1} Mpc^{-1}
present void volume fraction	$f_{v0} = 0.76^{+0.12}_{-0.09}$
mean lapse function	$\bar{\gamma}_0 = 1.381^{+0.061}_{-0.046}$
bare density parameter	$\bar{\Omega}_{M0} = 0.125^{+0.060}_{-0.069}$
conventional dressed density parameter	$\Omega_{M0} = 0.33^{+0.11}_{-0.16}$
mass ratio of	
non–baryonic dark matter to baryonic matter	$(\bar{\Omega}_{M0} - \bar{\Omega}_{B0})/\bar{\Omega}_{B0} = 3.1^{+2.5}_{-2.4}$
bare Hubble constant	$\bar{H}_0 = 48.2^{+2.0}_{-2.4}$ km sec^{-1} Mpc^{-1}
effective dressed deceleration parameter	$q_0 = -0.0428^{+0.0120}_{-0.0002}$
age of universe measured in a galaxy	$\tau_{w0} = 14.7^{+0.7}_{-0.5}$ Gyr.

Note: [a] 1 σ statistical uncertainties from SneIa alone are shown.

Best–fit cosmological parameters are shown in Table 1. As statistical bounds are not available for the sound horizon and BAO tests, we show 1σ uncertainties from SneIa only. However, given that there are many independent estimates of the dressed matter density, Ω_{M0}, we expect that the uncertainties quoted can be significantly reduced if such priors are imposed.

8.1. *Resolving the Lithium Abundance Anomaly*

The angular scale of the sound horizon and the BAO tests have been applied assuming a *volume–average* baryon–to–photon ratio in the range $\eta_{B\gamma} = 4.6$–5.6×10^{-10} adopted by Tytler *et al.*[45] prior to the release of WMAP1. With this range it is possible to achieve concordance with lithium abundances, while also better fitting helium abundances. This may resolve the primordial lithium abundance anomaly.[49,50]

With the 2003 WMAP1 release,[10] the baryon–to–photon ratio was increased to the very upper range of values that had previously been considered, due to its effect on the ratio of the heights of the first two Doppler peaks. This ratio of peak heights is sensitive to the mass ratio of baryons to non–baryonic dark matter – rather than directly to the baryon–to–photon ratio – as it depends physically on baryon drag in the primordial plasma. The fit to the Doppler peaks required more baryons than the range of Tytler *et al.*[45] admitted, when calibrated with the FLRW model. In the FB calibration, on account of the difference between the bare and dressed density parameters, a bare value of $\bar{\Omega}_{B0} \simeq 0.03$ nonetheless corresponds to a conventional dressed value $\Omega_{B0} \simeq 0.08$, and an overall mass ratio of baryonic matter to non–baryonic dark matter typically of about 1:3, which is larger than for ΛCDM. This would certainly indicate sufficient baryon drag to accommodate the ratio of the first two peak heights.

8.2. *Spatial Curvature and the Ellipticity Anomaly*

Since the release of the Boomerang experiment[46] in 2000, it has been generally assumed that the angular position of the first Doppler peak is a measure of the spatial curvature of the universe. However, this inference assumes that the spatial curvature is the same everywhere, as is appropriate for the FLRW models. In the present model there are spatial curvature gradients, and we must revisit the calculation from first principles, as outlined in Ref. [3]. Insofar as the angular scale of the sound horizon is reproduced for the FB calibration, the overall angular scale cannot be regarded as a measure of average spatial curvature.

If there is a sizable negative average spatial curvature at the present epoch, then there must be ways of detecting it other than via the measurement of the angular scale of the Doppler peaks. Such effects do indeed arise when one considers more subtle measurements associated with the average geodesic deviation of null geodesics. Indeed one prediction is that there should be nontrivial ellipticity in the CMB anisotropies on account of greater geodesic mixing. This effect is in fact observed,[47] and is an important anomaly for the standard ΛCDM paradigm, overlooked by the majority of the community to date.

A detailed quantification of the degree of ellipticity expected, and its comparison with the observed signal, will be an important test of the FB model in future.

8.3. *The WMAP3 Normalization*

It has been recently noted that the values of the normalization of the primordial spectrum $\sigma_8 \sim 0.76$ and matter content $\Omega_{M0} \sim 0.24$ implied by WMAP3 are barely compatible with the abundances of massive clusters determined from X–ray measurements.[51] In fact, SneIa also best fit the ΛCDM model with higher values of Ω_{M0} comparable to the the best–fit value for the FB model, namely $\Omega_{M0} \sim 0.33$. In the FB model the best–fit void fraction, $f_{v0} \sim 0.76$, appears to be the quantity that is mimicked by dark energy fraction, $\Omega_{\Lambda 0}$, as far as the WMAP normalization to the ΛCDM model is concerned. The flat ΛCDM model constrains $\Omega_{M0} = 1 - \Omega_{\Lambda 0}$. For the FB model, the corresponding constraint, $\Omega_{M0} \simeq \frac{1}{2}(1 - f_{v0})(2 + f_{v0})$, gives quite different larger values, consistent with many other observational determinations of the conventional matter density parameter.

8.4. *The Expansion Age*

Structure formation scenarios in standard ΛCDM model have some difficulty in explaining the observed apparent very early formation of galaxies.[52] Of course, structure formation contains many model–dependent assumptions, so direct measurements of the ages of metal poor stars in old globular clusters are ultimately a better indicator. The observational bounds are not very tight, because of the large uncertainties involved in particular nuclear processes. Individual ages are generally consistent with the accepted 13.7Gyr ΛCDM age of the universe, but are sometimes in tension with it,[53] given our uncertainties in knowing how early on the first stars could form.

It is interesting to note that the FB model not only adds a billion years to the age of the universe as measured in a galaxy, but that the percentage difference is larger at earlier times. For concordance ΛCDM $\tau = 0.85$ Gyr at $z = 6.4$ when distant quasars are seen, and $\tau = 0.365$ Gyr at reionization at $z = 11$. By contrast for the best–fit FB model $\tau = 1.14$ Gyr at $z = 6.4$ and $\tau = 0.563$ Gyr at $z = 11$, making the universe 34% and 54% older than concordance ΛCDM at the respective epochs.

The fact that the expansion age of the universe determined in a galaxy is greater than that of concordance ΛCDM may alleviate but not completely solve various aspects of the age problem. However, the fact that the age of the universe is 18.6Gyr at the volume average is a basic indicator that the whole issue of structure formation, including calibrations, needs to be re-examined from first principles.

8.5. *The Hubble Bubble*

Recent analysis of SneIa data by Jha, Riess and Kirshner[54] confirms an effect which has been known about for some time, and has been interpreted as our living in a local void,[55] the "Hubble bubble". If one excludes SneIa within the Hubble bubble at redshifts $z \lesssim 0.025$, then the value of the Hubble constant obtained is lower by $6.5 \pm 1.8\%$.[54] It is for this reason that SneIa at $z \leq 0.023$ have been excluded from the Riess07 gold data set.[41] The Hubble bubble is very problematic for the standard dark energy paradigm, as it can contribute a systematic error to the estimate of the equation of state parameter, w, which is much larger than the statistical uncertainties that precision cosmologists hope to attain.[54,56]

In the FB model the Hubble bubble is expected as a feature that results from the implicit resolution of the Sandage–de Vaucouleurs paradox. Since the bare Hubble parameter characterizes the uniform "quasilocally measured" Hubble flow, eq. (17) also quantifies the apparent variance in the Hubble flow below the scale of homogeneity. The present bare Hubble constant, $\bar{H}_0$, is lower than the global average, H_0. It represents the value we as observers in galaxies would obtain for measurements averaged solely within the plane of an ideal local bubble wall, on scales dominated by finite infinity regions. For us in particular, measurements of the Hubble constant towards the Virgo cluster would represent a scale over which we might hope to detect the bare Hubble constant, $\bar{H} \sim 48\,\mathrm{km\ sec^{-1}\,Mpc^{-1}}$, as this appears to be the direction that most closely represents a local bubble wall. Ultralocal minivoids[13] complicate any empirical measurement.

'Local' measurements across single voids of the dominant $30h^{-1}$ Mpc diameter,[12] probe the scale over which photons on null geodesics encounter the fewest finite infinity domains. Such measurements should give a Hubble 'constant' which exceeds the global average H_0 by an amount commensurate to $H_0 - \bar{H}_0$. As voids are dominant by volume, an isotropic average will generally produce a Hubble 'constant' *greater than* H_0 for local averages until we sample on large enough scales that the volume average of walls and voids is the same as the global one. Thus the average will steadily decrease from its maximum at $\sim 30h^{-1}$ Mpc until the scale of homogeneity ($\sim 100h^{-1}$ Mpc) is reached: the Hubble bubble feature. This expectation is dramatically confirmed by the data points in Fig. 5, reproduced from a recent paper of Li and Schwarz.[57] With $h = 0.617$, the maximum Hubble 'constant', up to about 20% higher than average, should be attained at a scale of about 48Mpc, thereafter steadily decreasing until leveling out at the scale of homogeneity, which must be reached before the BAO scale of about 168 Mpc.

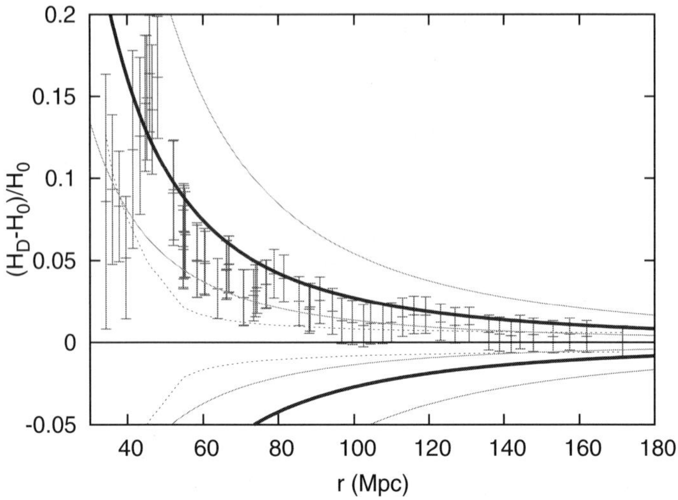

Fig. 5. Scale dependence of the normalized difference of the Hubble rate, averaged within a domain, H_D, and the global average, H_0, with data of Freedman *et al.*[58] Courtesy of Li and Schwarz,[57] who provide further details.

Fig. 5 also illustrates how the resolution of the Sandage-de Vaucouleurs paradox gives unique predictions. The two thick lines in Fig. 5 show Li and Schwarz's estimates of the bounds between which the data should lie using Buchert averaging, but not accounting for the clock effect. Given that the data indicates a consistently higher Hubble constant below the scale of homogeneity, the only explanation for the lack of great statistical scatter between the two lines – if all clocks are effectively synchronous – is that by a statistical fluke we happen to be in a large local void which is expanding faster into the surrounding medium.[27,55] However, if the conventional FLRW assumptions apply beyond this Hubble bubble, then by the standard structure formation scenarios such a circumstance seems impossible.

In the FB model the Hubble bubble is not a statistical fluke, but like cosmic acceleration is an apparent effect that arises from clock rate variance. Distant observers in galaxies beyond our own Hubble bubble will also detect a Hubble bubble centred on their location. The theoretical derivation of the curve that should pass though the data points in Fig. 5 will provide an important test of the FB model. To determine this curve, and its variance, requires some knowledge of void statistics, and probably some Monte–Carlo simulations of the manner in which the scale of homogeneity is filled on average. Particular anisotropies which look like "large" voids[14] might be expected from the variance in alignments of dominant $30h^{-1}$ diameter voids below the scale of homogeneity. A systematic error of 1–2% in the CMB dipole subtraction[61] may well be the consequence of a Rees–Sciama dipole resulting from such foreground inhomogeneities.[62]

While the decades long debate among astronomers about the value of the Hubble constant has been mainly dominated by arguments about systematics,[48,59] the question of the scale of averaging has no doubt also played a role. Lower values of H_0 will be expected if one specifically selects directions and scales which approximate our own bubble wall.[60] However, distance determinations on scales $\lesssim 50$ Mpc will generally give higher values of H_0. Since many of the first steps on the cosmic distance ladder are calibrated on such scales, many related issues require careful reconsideration.

8.6. *Prospects for Future Cosmological Tests*

The FB model provides a strong new competitor to the standard ΛCDM cosmology, as Table 2 shows. Indeed, since it steps out of a paradigm in which everything rests on a single equation of state, it will provide new and unique predictions. Current observational programs of course focus on the measurement of $w(z)$. The difference in expectations of the FB model for those programs will be highlighted in a forthcoming paper.[9]

Table 2. Model comparison.

Observation	ΛCDM	Fractal bubble model[a]
SneIa luminosity distances	Yes	Yes
BAO scale (clustering)	Yes	Yes
Sound horizon scale (CMB)	Yes	Yes
Doppler peak fine structure	Yes	Still to calculate
Integrated Sachs–Wolfe effect	Yes	Still to calculate
Primordial ^{7}Li abundances	No	Yes
CMB ellipticity	No	[Yes]
CMB low multipole anomalies	No	[Foreground void(s): Rees–Sciama dipole]
CMB Sunyaev–Zel'dovich signal[63,64]	No	Still to calculate
Hubble bubble	No	Yes
Nucleochronology of globular clusters	Tension	Yes
X-ray cluster abundances	Marginal	Yes
Emptiness of voids	No	[Yes]
Sandage-de Vaucouleurs paradox	No	Yes
Coincidence problem	No	Yes

Note: [a] Square brackets indicate cases where the present indication is suggestive, but much detailed calculation remains to be done.

The quantities that remain to be calculated in Table 2 relate largely to the CMB. Detailed construction of the Doppler peaks, to enable comparison to WMAP, and determination of parameters such as σ_8, is a matter of urgency. It is a rather non–trivial exercise in recalibration of standard quantities, in which all steps need to be carefully reconsidered. Detailed quantitative calculations of CMB ellipticity and the integrated Sachs–Wolfe effect can only be performed in conjunction with such an analysis.

While the calculation of the integrated Sachs–Wolfe (ISW) effect will differ in the FB model, one must be careful not to base intuition on that of perturbations on FLRW backgrounds. In particular, the observed ISW signal is believed to confirm dark energy since its consequence is a suppression of the gravitational collapse of matter at relatively recent times.[65] If one replaces the words "dark energy" by "voids" in the standard qualitative explanation, then a probable description of the ISW effect in the FB model emerges. The same correlation of the ISW signal with clumped structures[65] is expected; what is important is the magnitude of the effect. Mattsson[66] has recently proposed an extension of the Dyer–Roeder formalism, which may have some relevance for such calculations.

In offering a new paradigm for cosmology, every observational test naturally has to be revisited from first principles. Thus there are potential tests which relate to both strong and weak gravitational lensing, for example, which have not been listed in Table 2. Qualitatively one expects voids to be emptier in the FB model than in LCDM structure formation simulations. However, the specification of the background average surfaces of homogeneity, in terms of a uniform bare Hubble expansion gauge, and a correct post–Newtonian approximation on such a background, need to be examined before structure formation simulations are attempted.

Ultimately, apparent variance in the Hubble flow below the scale of homogeneity will give predictions which cannot be reproduced in the FLRW scenario. Not only should we determine the average curve that passes through the data points in Fig. 5, we should ultimately collect thousands of SneIa measurements or other distance measurements on scales up to 200Mpc, and test the correlation of the apparent Hubble flow with actual structure: a sort of Hubble flow tomography of the nearby universe.

9. Conclusion

A true "concordance cosmology" should agree with all reliable observations, and not just a carefully selected subset. A glance at Table 2 reveals that there are many anomalies in the standard ΛCDM model. Much attention has been focused on the CMB anisotropies, on account of the spectacular success of the WMAP mission.[10,11] However, it must be recalled that many ingredients go into the analysis of the CMB temperature fluctuations. In looking at these ingredients we must ask *"what is the weakest link"*?

In my view, the weakest link is not primordial nucleosynthesis,[49] which is based on nuclear physics that we understand very well, but the cosmological model. The weakest link comes from abandoning the theoretical principles that we are careful to apply in other circumstances. In particular, in general relativity one should model the universe with the matter distribution one observes, rather than trying to impose onto the universe a simple mathematical solution based on some simplifying symmetry.

It is unfortunate that general relativists have been obsessed by exact solutions of Einstein's equations, whether they involve likely or unlikely approximations for the matter distribution. We should face up to the fact that the solution for the actual matter distribution is analytically intractable, and therefore the question of cosmological averaging[5,7,15,16,28–30] is paramount. Furthermore, once we do take an average we must address the fundamental problem that the relationship of rods and clocks at one point to those at a distant point, a conceptual centrepiece of general relativity, is highly non–trivial once gradients in spatial curvature and gravitational energy are considered. In an expanding universe these involve subtle dynamical aspects of general relativity, which cannot be localized at a point on account of the equivalence principle.

Some colleagues when presented with Table 2, and the fact that the right hand column is based entirely on general relativity with no new or exotic physics (beyond a need for non–baryonic dark matter), suggest I should invoke Ockham's Razor at this point and declare the cosmological constant dead. Other colleagues, who sometimes confess to Newtonian intuition, are of the view that a clock–rate difference of 38% accumulated between bound systems and the volume average over the lifetime of the universe is nonetheless so great that I must be mistaken somewhere, in spite of Table 2.

In my view caution should always be exercised, but this includes caution with the conceptual basis of our theory and the operational interpretation of measurements. To those who are uncomfortable with my proposal about cosmological quasilocal gravitational energy let me ask the following: Without reference to an asymptotically flat static reference scale, which does not exist given the universe is expanding, and without reference to a background which evolves by the Friedmann equation at some level,[37] an assumption which is manifestly violated by the observed inhomogeneities, *What keeps clocks synchronized in cosmic evolution*? Please explain.

In the FB proposal the fact that our FLRW approximation has served so well is understood as a consequence of the fact that the bare Hubble flow is uniform, despite large–scale inhomogeneities. Uniformity of the quasilocally measured Hubble flow does not imply uniformity of spatial curvature, nor of gravitational energy, however. If we examine the Hubble flow over scales on which the average gradients are not large, we will not see large statistical scatter in the Hubble flow. But averaged over larger scales, below the scale of homogeneity, we will see a variance in the apparent Hubble flow, that agrees with observations of the statistical properties of voids,[12] and which seems to accord with Fig. 5. The fact that there is a statistically dominant void scale[12] may well be associated with the evolution of the scale corresponding to the second (rarefaction) Doppler peak in the CMB anisotropy spectrum, just as evolution of the first (compression) peak appears to statistically define the cutoff for the scale of homogeneity.

Any proposal which seeks to genuinely shift the paradigm against accepted intuition, even a conservative proposal based on a principled approach[3,8] to our best theory of gravitation, naturally faces much initial scepticism. Nonetheless, I suspect that in future cosmologists will wonder how we ever could have expected that two decades of precision measurements would simply confirm the naïve cosmological models constructed in the 1920s, long before we knew what the universe actually looked like. I suspect that the coming decades will see us explore and finally understand the arcane territory of general relativity relating to quasilocal gravitational energy,[32,33,35,36] informed directly by new cosmological observations.

Acknowledgements

I am grateful to many people for discussions and correspondence including, in particular, Thomas Buchert. I also thank Anthony Fairall and Dominik Schwarz for granting me permission to reproduce their Figures 1 and 5 respectively.

References

1. R. Durrer and R. Maartens, arXiv:0711.0077.
2. P.J.E. Peebles, Astrophys. J. **557**, 495 (2001).
3. D.L. Wiltshire, New J. Phys. **9**, 377 (2007).
4. D.L. Wiltshire, Phys. Rev. Lett. **99**, 251101 (2007).

5. T. Buchert, Gen. Relativ. Grav. **32**, 105 (2000); Gen. Relativ. Grav. **33**, 1381 (2001).
6. B.M. Leith, S.C.C. Ng and D.L. Wiltshire, Astrophys. J. **672**, L91 (2008).
7. S. Räsänen, JCAP **11**, 003 (2006); M.N. Célérier, New Adv. Phys. **1** (2007) 29 [astro-ph/0702416]; N. Li and D.J. Schwarz, Phys. Rev. **D 76**, 083011 (2007); T. Buchert, arXiv:0707.2153; and references therein.
8. D.L. Wiltshire, *"Cosmological principle of equivalence"*, in preparation.
9. D.L. Wiltshire, *"Observational quantities in the fractal bubble universe"*, in preparation.
10. C.L. Bennett *et al.*, Astrophys. J. Suppl. **148**, 1 (2003).
11. D.N. Spergel *et al.*, Astrophys. J. Suppl. **170**, 377 (2007).
12. F. Hoyle and M.S. Vogeley, Astrophys. J. **566**, 641 (2002); Astrophys. J. **607**, 751 (2004).
13. A.V. Tikhonov and I.D. Karachentsev, Astrophys. J. **653**, 969 (2006).
14. L. Rudnick, S. Brown and L.R. Williams, arXiv:0704.0908.
15. G.F.R. Ellis, in B. Bertotti, F. de Felice and A. Pascolini (eds), *General Relativity and Gravitation*, (Reidel, Dordrecht, 1984) pp. 215–288.
16. G.F.R. Ellis and W. Stoeger, Class. Quantum Grav. **4**, 1697 (1987).
17. F. Sylos Labini, M. Montuori and L. Pietronero, Phys. Rep. **293**, 61 (1998).
18. Y. Baryshev, AIP Conf. Proc. **822**, 23 (2006).
19. A.D. Chernin, I.D. Karachentsev, M.J. Valtonen, V.P. Dolgachev, L.M. Domozhilova and D.I. Makarov, Astron. Astrophys. **415**, 19 (2004).
20. A. Sandage, G.A. Tammann and E. Hardy, Astrophys. J. **172**, 253 (1972).
21. G. de Vaucouleurs, Science **167**, 1203 (1970).
22. Y. Hoffman, L.A. Martinez-Vaquero, G. Yepes and S. Gottlöber, arxiv:0711.4989.
23. M. Axenides and L. Perivolaropoulos, Phys. Rev. **D 65**, 127301 (2002).
24. A.B. Whiting, Astron. J. **131**, 1996 (2006).
25. E.W. Kolb, S. Matarrese, A. Notari and A. Riotto, hep-th/0503117.
26. E.W. Kolb, S. Matarrese and A. Riotto, New J. Phys. **8**, 322 (2006).
27. M.N. Célérier, Astron. Astrophys. **353**, 63 (2000); J.W. Moffat, JCAP **05**, 001 (2006); H. Alnes, M. Amarzguioui and O. Gron, Phys. Rev. **D 73**, 083519 (2006); T. Biswas, R. Mansouri and A. Notari, astro-ph/0606703.
28. R.M. Zalaletdinov, Gen. Relativ. Grav. **24**, 1015 (1992); **25**, 673 (1993).
29. A.A. Coley, N. Pelavas and R.M. Zalaletdinov, Phys. Rev. Lett. **95**, 151102 (2005); A.A. Coley and N. Pelavas, Phys. Rev. **D 74**, 087301 (2006); **D 75**, 043506 (2007).
30. A. Paranjape and T.P. Singh, Phys. Rev. **D 76**, 044006 (2007).
31. A. Ishibashi and R.M. Wald, Class. Quantum Grav. **23**, 235 (2006).
32. J. Katz, Class. Quantum Grav. **22**, 5169 (2005); J. Katz, D. Lynden-Bell and J. Bičák, Class. Quantum Grav. **23**, 7111 (2006); D. Lynden-Bell, J. Katz and J. Bičák, Phys. Rev. **D 75**, 024040 (2007); Err. ibid. **D 75**, 044901 (2007).
33. L.B. Szabados, Living Rev. Rel. **7**, 4 (2004).
34. H. Bondi, *"Cosmology"*, (Cambridge Univ. Press, 1960)
35. C.M. Chen, J.L. Liu and J.M. Nester, Mod. Phys. Lett. **A 22**, 2039 (2007).
36. J. Garecki, arXiv:0708.2783.

37. A. Einstein and E.G. Straus, Rev. Mod. Phys. **17**, 120 (1945); Err. ibid. **18**, 148 (1946).
38. J.M. Bardeen, Phys. Rev. **D 22**, 1882 (1980).
39. T. Buchert and M. Carfora, Class. Quantum Grav. **19**, 6109 (2002).
40. T. Buchert and M. Carfora, Phys. Rev. Lett. **90**, 031101 (2003).
41. A.G. Riess *et al.*, Astrophys. J. **659**, 98.
42. R.E. Kass and A.E. Raftery, J. Am. Statist. Assoc. **90**, 773 (1995).
43. R. Trotta, Mon. Not. R. Astr. Soc. **378**, 72 (2007).
44. D.J. Eisenstein *et al.*, Astrophys. J. **633** (2005) 560; S. Cole *et al.*, Mon. Not. R. Astr. Soc. **362** (2005) 505.
45. D. Tytler, J.M. O'Meara, N. Suzuki, and D. Lubin, Phys. Scripta **T85**, 12 (2000).
46. P. de Bernardis *et al.* [Boomerang Collaboration], Nature **404** (2000) 955.
47. V.G. Gurzadyan, C.L. Bianco, A.L. Kashin, H. Kuloghlian and G. Yegorian, Phys. Lett. **A 363** (2007) 121; V.G. Gurzadyan, P. de Bernardis, G. De Troia, C.L. Bianco, A.L. Kashin, H. Kuloghlian, S. Masi, F. Piacentini, G. Polenta and G. Yegorian, Mod. Phys. Lett. **A 20** (2005) 813.
48. A. Sandage, G.A. Tammann, A. Saha, B. Reindl, F.D. Macchetto and N. Panagia, Astrophys. J. **653** (2006) 843.
49. G. Steigman, Int. J. Mod. Phys. **E 15** (2006) 1.
50. M. Asplund, D.L. Lambert, P.E. Nissen, F. Primas and V.V. Smith, Astrophys. J. **644** (2006) 229.
51. G. Yepes, R. Sevilla, S. Gottloeber, and J. Silk, Astrophys. J. **666**, L61 (2007).
52. K. Glazebrook *et al.*, Nature **430**, 181 (2004); A. Cimatti *et al.*, Nature **430**, 184 (2004).
53. L.M. Krauss and B. Chaboyer, Science **299**, 65 (2003).
54. S. Jha, A.G. Riess and R.P. Kirshner, Astrophys. J. **659**, 122 (2007).
55. K. Tomita, Mon. Not. R. Astr. Soc. **326**, 287 (2001); Prog. Theor. Phys. **106**, 929 (2001).
56. W.M. Wood-Vasey *et al.*, Astrophys. J. **666**, 694 (2007).
57. N. Li and D.J. Schwarz, arXiv:0710.5073.
58. W.L. Freedman *et al.*, Astrophys. J. **553**, 47 (2001).
59. F. van Leeuwen, M. W. Feast, P. A. Whitelock and C. D. Laney, Mon. Not. R. Astr. Soc. **379**, 723 (2007).
60. M. Federspiel, G.A. Tammann and A. Sandage, Astrophys. J. **495**, 115 (1998).
61. P.E. Freeman, C.R. Genovese, C.J. Miller, R.C. Nichol and L. Wasserman, Astrophys. J. **638** (2006) 1.
62. A. Rakic, S. Räsänen and D.J. Schwarz, Mon. Not. R. Astr. Soc. **369** (2006) L27.
63. R. Lieu, J.P.D. Mittaz and S.-N. Zhang, Astrophys. J. **648**, 176 (2006).
64. R.M. Bielby and T. Shanks, Mon. Not. R. Astr. Soc. **382**, 1196 (2007).
65. S. Boughn and R. Crittenden, Nature **427**, 45 (2004).
66. T. Mattsson, arXiv:0711.4264.

PART X

Underground Laboratories

ILIAS AND THE EUROPEAN DEEP LABORATORIES

N.J.C. SPOONER (On behalf of ILIAS)
Department of Physics and Astronomy, University of Sheffield
Hicks Building, Hounsfield Road, Sheffield, S3 7RH
England, UK

A brief summary is given of the status and prospects for the underground laboratories in Europe as covered by the ILIAS project. Particular emphasis is given to the possibility of expansion in relation to the likely demand for space, and specifically on the prospects for very large, megaton scale, excavations.

Keywords: ILIAS, Underground Science.

1. Introduction

ILIAS (Integrated Large Infrastructure for Astroparticle Science), comprises a set of networks and joint R&D projects funded by the EU, the prime focus being to improve cooperation between the deep underground laboratories in Europe: Gran Sasso, Frejus, Canfranc and Boulby. Table 1 gives summary details of the main characteristics of these laboratories. Included here is the CUPP laboratory in Finland. Though not a contract member of ILIAS this site has developed close links with the main grouping.

A prime motivation of the ILIAS cooperation is to improve the service to users of underground laboratories and assist in determining future directions for these facilities at a European level, but in an international context. As part of this work ILIAS has set out to assess the demand for underground space and provide a means for communicating ideas for meeting this demand across the laboratories, including the prospects for constructing new very large caverns. The following sections provide a brief overview of this activity.

2. Future Science Demand, Space and Availability

Within the framework of ILIAS progress has been made to review the demand and availability of deep underground space in Europe and worldwide.

Table 1. Comparison of major features of the deep underground laboratories covered by ILIAS.

Infrastructure	LNGS Gran Sasso	LSM Frejus	LSC Canfranc	IUS Boulby	CUPP Pyhasalmi
Country	Italy	France	Spain	UK	Finland
Year of completion	1987	1982	1986,1995	1989,2003	1993,2001
Area (m^2)	6000	500	150+600	500+1000	500-1000
Volume (m^3)	180000	3500	8000	3000	100-10000
Access	Horizontal	Horizontal	Horizontal	Vertical	Slated road
Depth (m.w.e.)	3700	4800	2450	2800	1050, 1444, 4060
Surface profile	Mountain	Mountain	Mountain	Flat	Flat
Muon flux ($m^{-2}day^{-1}$)	24	4	406	34	8.6@4060m
${}^1_0n$ flux (>1 MeV) (106 $cm^{-2}s^{-1}$)	0.8	1.6	3.8	1.3	unknown
Number of visiting scientists	700	100	50	30	15

Information on this is also contained in the ApPEC roadmap review [1] and ILIAS reports .[2] Information also comes from the DUSEL process in the US .[3] The general conclusion is that demand for space internationally is very high and exceeds availability by at least a factor two. A particular aspect is the clear and growing demand for space for applications outside physics. Major applications are listed below with estimated proportions of demand expressed as percentage of potential projects in each area of the total demand:

- Geology, geosciences, hydrology, hydromechanics, rock mechanics, engineering (30%)
- Physics and astrophysics (25%)
- Microbiology biogeochemistry, geomicrobiology, biology and bio-prospecting (20%)
- Earth and environmental Sciences (10%)
- Education and Outreach (5%)
- Security, society, health and defense (5%)

Though this reflects broad interest across many fields the percentages do not reflect the volume of space required, physics projects in general requiring the largest demand for space and resources. The question of depth requirement is also not reflected here. In particular a relatively small number of experiments require depths greater than that of Frejus. A draft report on the space demand can be found within the ILIAS annual reports.[2]

The US DUSEL process has been highly valuable as a source of information for demand and assimilation into this work. The Document 'Deep Science' contains background information for this (obtainable from the DUSEL web site and NSF).[3] The SNOLAB process for selection of possible experiments at that site has also yielded useful input information.[4] Some preliminary conclusions note the following points:

(1) there are at least 100 potential experiments seeking or expecting to seek underground space throughout the world in the next 5-10 years including upgrades or continuation of existing experiments.
(2) A majority of these have specialist needs and most are not physics experiments but involve geological or biological applications
(3) The European laboratories do not currently have sufficient capacity but could also, if expanded, accommodate further overseas experiments if new space is not found outside Europe
(4) Mine sites have certain advantage in the non-physics areas due to the need from some of those disciplines to have contact with newly excavated rock

A first European meeting on Underground Science as a field covering many disciplines (not just physics) was held in on Oct 21-22nd in York UK, jointly with ILIAS.[5] Whilst mainly aimed at possible users of the Boulby site the meeting was successful in attracting interest from across Europe and beyond and has provided a useful source of information on the demand for underground space and interest in underground science in general. Over 60 people attended the workshop. Delegates included researchers from a wide range of science disciplines (astroparticle physics, nuclear astrophysics, low background studies, geo-microbiology, geology/geophysics). Also present were affiliates of NSF and DUSEL (the push for a national underground laboratory in the USA), representatives from the other European Labs from Modane and Gran Sasso.

3. Demand for Upgrade of Laboratory Facilities

There is a clear need also for upgrades of existing infrastructures underground in Europe. There is a particular demand for upgrades and space for dark matter experiments and for new neutrino physics ideas. Particular issues are:

(1) the need to continue upgrading general safety features of the all the laboratories
(2) the need for more space in the current facilities to accommodate upgrades of existing science programmes
(3) the longer term need for access to more space at deeper levels, in particular for potential large scale dark matter experiments
(4) the need to provide new facilities for more non-physics applications including geophysics and biology
(5) the need to consider design options for very large underground caverns for possible 100 Kton scale multi-purpose physics experiments

General expansion and service improvements are being studies or undertaken by all the European laboratories and are summarised here:

- Boulby: laboratory expansion is being considered to deeper levels at Boulby below the current salt layer where current laboratories are sited. A joint feasibility study is underway with the mine company CPL.
- Canfranc: with the new laboratories at Canfranc just completed there are no firm plans for further expansion at present. However, the rock is becoming well understood and there are no intrinsic issues that would prevent investigation of expansion. Any deeper levels would require down-excavation and thought placed on the issues of horizontal vs. vertical access.
- Frejus: there are clear needs, plans and prospects for expansion with potential for a variety of size of cavern in association with excavation of a new safety fire evacuation tunnel parallel to the existing road tunnel at Frejus. There is demand form local researchers here, for instance through the NEMO double beta decay experiment and for dark matter. There is justified need for an integrated multi-shielded ELB facility. Such a laboratory could be made deep underground with *submarine* technology, so that the whole laboratory is surrounded by a water shielding. There is a time window to construct this when the new safety tunnel is ready, presumably by 2012. The new laboratory would be built on the other side of the safety tunnel (not between the tunnels as previously planned). A by-pass from the safety tunnel is required. The plans include the low background room of 12,000 m^3 and a larger hall of 30,000 m^3, or alternatively a cylindrical cavity of 41,000 m^3. The volume of the tunnels, including the by-pass may be 14,000 m^3.

- Gran Sasso: considerable upgrades of the features at LNGS have recently been completed. This includes new water systems, containment, ventilation and safety upgrades. Although there are plans for expansion within the site and also for possible new caverns in the nearby area, these are at an early stage and are subject to discussion regarding the environmental impacts in particular.
- Phyasalmi: this site, though not a full partner in ILIAS, has been participating in discussions by invitation. The site is exploring the prospect for new caverns at various depths in hard rock down to 1.4 km depth.

4. Very Large Scale Caverns in Europe

The new LAGUNA project, covering proton decay, neutrino astrophysics and other science needing a very large (100-1000 Kton) detector, includes a dedicated workpackage on deep site large excavations, and is becoming a source of information and demand relevant to very large excavations. Particular issues for large scale detectors at candidate sites in Europe are:

Boulby: The Boulby laboratories have good potential to expand into harder rock with prospects for cavities as large as those seen in existing hard rock locations (20-30 m span). So far no detailed technical studies have been made. CUPP/Pyhasalmi: at this site in Finland several deep underground constructions have been completed, some of which are of size and shape interesting for large laboratories, demonstrating the possibility for excavating cavities of width at least 20 m. There is a positive view on the feasibility of large caverns based on preliminary engineering studies. Frejus: the rock structure, geology and history of underground development at Frejus is well understood. The existing constructed cavities of 11,000 m^3 behave well. According to engineering studies the rock in the middle of the road, right below the peak of Frejus mountain, has the best rock quality, with the smallest convergence. The preliminary study demonstrated that for a megaton scale detector such as MEMPHYS the best option is a cylindrical shaft, which behaves much better than a horizontal cavity. Spans of up to 65 m are possible, hence a megaton-sized detector must be divided into five cavities. For a 100 kton liquid argon detector (GLACIER) the only possibility is to divide it into two parts. There is a time window to start construction of this after 2012. Proposal for a new laboratory in the Gran Sasso area: the idea for a new laboratory in Gran Sasso has recently emerged, basically a green field laboratory at shallow depth, suitable for example for liquid argon experiments searching for neutrino oscillation in the CNGS neutrino beam. It would be several kilometers away from the present laboratory, on either side of the mountain. The rock is composed either of sandstone or limestone, depending on the location. Sunlab: this new proposed site, the Sieroszowice Underground Laboratory in South-Western Poland, is being planned in a huge copper mine, at a 70 m thick inclined salt layer extending from 600 to below 1000 metres. There is already a cavern of size at least 100 x 15 x 20 m^3, at the depth of 900-950 m. According to pre-feasibility studies it may be possible to construct a very large cavity with 70 m span in 650-700 m levels. The radioactivity of the salt was found to be extremely low. The humidity is also very low, but the temperature is high, 35 degrees in 950 m level.

Fig. 1. Visit to the Pyhäsalmi underground lab of members of the joint underground safety group of ILIAS.

5. Conclusion and Inter-Laboratory Cooperation and Management

There is a clear demand for underground space in Europe and an exciting array of experiments requiring this space from non-physics areas as well as the traditional areas of underground physics. In this light, within ILIAS a significant start has been made to develop deeper cooperation between the laboratories. Of particular note is cooperation in outreach and safety. The former no includes common outreach events and open days, the latter joint safety training days and specialist joint safety visits across the laboratories (see Fig. 1). A further deepening of relations is currently underway through the CoMag team of directors that meets regularly to develop an approach to common management issues. Building in these developments there is clearly a rich future for the field of underground science in Europe.

Acknowledgements

We would like to thank the EU framework 6 organisation for support of this work through the ILIAS project contract no. RII3-CT-2004-506222.

References

1. http://www.ifh.de/ csspier/Roadmap-oct19.doc
2. http://ilias.in2p3.fr
3. http://www.deepscience.org
4. http://www.snolab.ca
5. http://www.shef.ac.uk/physics/bus2006

SNOLAB: A NEW INTERNATIONAL FACILITY FOR UNDERGROUND ASTROPARTICLE PHYSICS

A. NOBLE

Department of Physics, Queen's University, Kingston, Ontario, K7L-3N6,
E-mail: potato@snolab.ca

In the following we present the status of SNOLAB, a new international facility for Underground Physics currently nearing completion in Sudbury, Canada. We describe the infrastructure available, as well as the scientific programme envisioned.

Keywords: Astroparticle Physics, Underground Laboratory, Dark Matter, Double Beta Decay, Neutrinos, SuperNovae.

1. Introduction

The past few years have been witness to great advances in the field of Astroparticle Physics. We now understand that neutrinos have a small, but finite mass, and flavour oscillations have been observed. A complete understanding of the neutrino sector, and its integration into the standard model, is now a major industry world wide. Experiments currently under design or operation aim to measure the mixing angles to high precision, determine the absolute masses or search for a CP violating phase that may give insight into the matter-antimatter asymmetry observed in the Universe today. Neutrinoless double beta decay experiments are searching for very rare nuclear decay processes which if observed may determine the absolute masses of neutrinos and will demonstrate that neutrinos are Majorana in nature.

Precision observational astronomy has also led to a new understanding of the Universe and its constituents. The Universe is now understood to be comprised of $\approx$25% Dark Matter and $\approx$70% Dark Energy, the nature of which remain complete mysteries today. The Dark Matter is widely believed to be a new form of massive particle which will interact with normal material through weak interactions. These weakly interacting massive particles (WIMPs) may be observed through the detection of low energy nuclear recoils produced via the WIMP nucleon interactions. Numerous efforts are currently underway which attempt to observe Dark Matter particles through this direct detection. These are very challenging experiments, as experiments are looking for events of very low energy, in a potentially large background, at a rate of order events per year per tonne of material. The next generation detectors currently under design aim to be able to probe these very low cross-sections using a variety of targets, and should challenge the theoretical predictions for the expected rates.

There is an interesting synergy between direct Dark Matter searches and searches for new forms of matter at the LHC. In many supersymmetric extensions to the standard model, there is a lightest supersymmetric particle which is forbidden to decay by R-parity. This may be identified with the relic Dark Matter particles left over from the big bang.

The experiments that will search for Dark Matter, Neutrinoless Double Beta Decay, as well as further solar neutrino experiments have in common the need for very large detector volumes with very low energy thresholds. The control of backgrounds in such experiments is paramount, and hence need to be performed in a deep underground laboratory in clean room conditions.

SNOLAB was conceived in consideration of the need for such space in a deep clean underground laboratory. As an outgrowth of the very successful SNO programme, there was already one very large cavern, an established relationship with the mine owners (Inco), a highly skilled technical workforce and experienced physicists from SNO wanting to continue with an expanded astrophysics programme.

2. Description

SNOLAB was designed to be able to accommodate three to four large scale (order 1000 tonne) experiments, as well as a number of small to mid sized experiments. The large spaces are comprised of the existing SNO cavern, a new "Rectangular Hall" and a new "Cryopit". The latter has been designed with high pressure bulkheads and a dedicated ventilation system that will allow the space to be operated safely with large volumes of cryogenic fluids or noxious gasses. Mid-sized experiments will be supported in the Ladder labs, and small experiments can be installed in a number of utility areas. Due to timing of capital funding, the Ladder labs and Rectangular Hall were excavated first in Phase I of SNOLAB, and the excavation of the Cryopit followed a few years later in Phase II of the project. The layout of the new facility is depicted in figure 1.

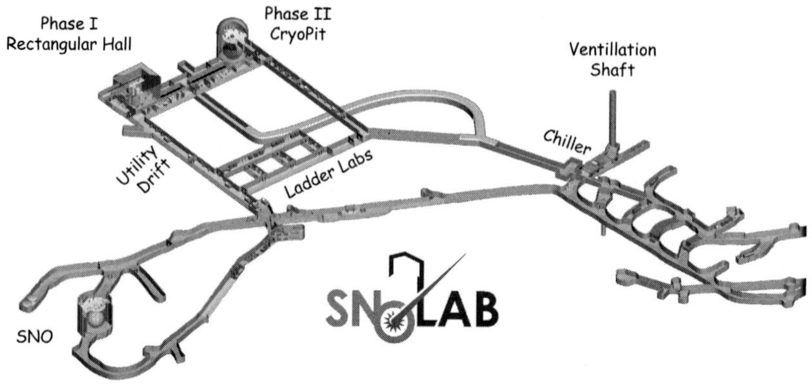

Fig. 1. Schematic view of the SNOLAB facility. Only the cavern and utility room marked as SNO existed prior to the SNOLAB expansion.

Table 1 gives the total floor space and volume available in the existing SNO facility, as well as when combined with phases I and II.

Table 1. Underground Spaces in the new SNOLAB.

Phase	Clean Room Space		Experimental Area	
	Surface Area (m^2)	Volume (m^3)	Surface Area (m^2)	Volume (m^3)
SNO Original	1,133	13,321	752	11,679
SNO + Phase I	3,899	29,719	2,427	23,721
SNOLAB Total	4,942	37,241	3,055	29,555

The main experimental halls have the dimensions as given in table 2. In addition to this there are numerous utility areas. For full details, see the user manual located at `http://www.snolab.ca/files/docs/snolab_users_handbook_rev02.pdf`

SNOLAB also has a new building on surface, which has been occupied since 2005. This above ground facility supports experiments by providing research space including; four large clean laboratories, a clean assembly room, a low background counting facility, and a chemistry lab on the main floor. On the second floor, the facility has offices and meeting rooms, and on the third floor there are more offices, meeting rooms and experimental control rooms. In addition, there is a large auditorium suitable for collaboration meetings, a changing/showering area for personnel working underground, and an attached warehouse and machine shop.

Table 2. Sizes of main experimental halls and caverns in SNOLAB.

Phase	Length (m)	Width (m)	Height to Shoulder (m)	Height to Back (m)
SNO Cavern	$22^{a}\phi$		24.2	30.0
Rectangular Hall	18.3	15.1	15.1	19.7
Cryopit	$15.1^{b}\phi$		15.1	19.7
Ladder Area 1	32.0	5.9	3.6	5.6
Ladder Area 2	22.9	7.5	5.1	7.6

Note: [a]Dimension at deck height. Waist is 22 m.
Note: [b]Dimension at deck height.

A photo of the new surface building is shown in figure 2.

Fig. 2. Photograph of the new Surface building at SNOLAB.

To complement the physical lab spaces underground, SNOLAB also provides a significant amount of basic infrastructure. In addition to basic power, lighting, ventilation, water and fire protection, SNOLAB can provide a few specialty services, as outlined below:

- *Ultra Pure Water.* The light water purification plant used in for the SNO project will be reconfigured to supply ultra pure water to the experiments for use in shielding. The water is purified by two stages of reverse osmosis, ion-exchange, biological sterilization with UV light, and radon removal with a large process degasser.
- *Background Counting Facilities.* Various counting facilities, including a High Purity Germanium Counter, Electrostatic Chambers, and radon board extraction systems with associated Lucas Cell counters.

- *Chilled Water and Cooling.* A large chiller system provides air-conditioning for the lab. At SNOLAB depths, the normal rock temperature is over 40 °C. The air-conditioning will maintain the lab at a comfortable 20 °C. The chilled water will be at about 6 °C and can be used by the experiments for local cooling.
- *Monitoring and Slow Controls for Lab Systems.* Measure, record and provide alarms for enviromental parameters and equipment status.
- *Liquid Nitrogen.* Brought underground several times per week in 250 ℓ dewars.
- *Some Cranes and Lifting Devices.* Designed to fit the needs of the experiments identified for those areas.
- *Chemistry Lab.* There is a well equiped chemistry lab on surface, and plans for a similar lab underground.
- *Transportation of Equipment and Supplies.* Transportation to the underground lab will usually be via so called "blue boxes" which will be packed on surface in clean room conditions. These will then travel underground maintaining the cleanliness as they accommodate the increased air pressure. At the lab, they will interface to the lab entrance in a way that will allow the contents to be removed directly into the clean SNOLAB environment.
- *Basic Machine Shop Facilities.* Both underground and on surface. The underground facility will be operated as a clean room machine shop.

Finally, the facility will be operated by a staff of about 40 personnel, including research scientists, engineers and technical staff. These personnel are highly skilled in underground operations and the requirements for operating a facility like SNOLAB, having gained considerable experience working on the SNO project.

3. Status of the Facility

The underground excavation of the two phases is well underway. The first phase comprises the large Rectangular Hall, the Ladder labs, and the utility drifts. The second phase was to add the Cryopit. The excavation of phase I is 100% complete. Excavation of the Cryopit and it's utility drifts began in August 2007 and is now well advanced, with excavation to be complete by May 2008.

The outfitting of Phase I (Installation of steel work, walls, painting, power, ventilation, lighting and plumbing) is progressing very well, and should be complete by February, 2008. Once this is complete, the contractor will begin the outfitting of the Cryopit.

Fig. 3. Photograph of the new Rectangular Hall after completion of excavation.

When the Cryopit excavation is complete the dirtiest work will be finished. At that point the Phase I outfitting will also be complete, and this defines the time at which the first beneficial occupancy in the lab can take place. We anticipate the first experiments will begin installation in May/June of 2008. The following table summarizes the construction status and schedule.

Table 3. Construction Status and Expected Completion Date.

	Construction Status	
	Excavation	Outfitting
Phase I	100% Complete	75%. Feb 2008
Phase II	40% May 2008	0% August 2008
First Occupancy Underground	May 2008	
Completion of Laboratory	October 2008	
Surface Facility	Occupied Since July 2005	

The two photographs show the Rectangular Hall (see figure 3) shortly after completion of the excavation and installation of the excavation and installation of the stairs, and a smaller room in the Ladder Lab area (see figure 4). The latter was taken at the completion of excavation and painting of the walls.

Fig. 4. Photograph of one room in the Ladder Lab section of SNOLAB.

4. Status of the Experimental Program

SNOLAB has held a series of workshops to help define the experimental program, and, conversely, to get input from prospective experiments to ensure that the lab design can cater to the needs of the experiments. Approximately 25 letters of intent were received. These have been examined by an international experimental advisory committee, and an initial suite of experiments has been selected. The experiments presently being considered for installation or prototyping within the next two years are summarized in table 4. Roughly 50% of the space in the lab has been assigned for use by these experiments.

5. Conclusions

SNOLAB, a new International Facility for Undergound Astroparticle Physics, is nearing completion. The new space will be able to house several large next generation Dark Matter, Neutrinoless Double Beta Decay,

Table 4. Proposed Initial Experimental Program.

Experiment	Physics	Status
PICASSO Ib	**Dark Matter**. Superheated Droplet Detector with Spin-Dependent sensitivity	Prototype in Operation. Planning to scale up.
DEAP I	**Dark Matter** with Liquid Argon. 7Kg Mass Prototype. Scintillation and PSD	Installed in SNOLAB. Commissioning in progress.
MiniClean	**Dark Matter** with 360 Kg of Liquid Argon.	Installation foreseen for 2008.
DEAP/CLEAN	Large scale. **Dark Matter** with 3600 Kg Liquid Argon.	Installation of infrastructure in 2008.
SuperCDMS	**Dark Matter** with Germanium ZIP detectors.	World leading experiment currently operational in Sudan Mine. Plan to upgrade and relocate to SNOLAB to reach next level of sensitivity.
SNO+	**Neutrinoless Double Beta Decay** with Neodymium Loaded Liquid Scintillator. Also **Solar Neutrinos, Geoneutrinos** and **Reactor Neutrinos**	Will use existing SNO detector. Currently working on design issues. Installation to begin in 2008
EXO	**Neutrinoless Double Beta Decay** with Xenon Gas. Method hopes to exploit Ba tagging to reduce backgrounds and to look at tracks.	R&D in progress. Prototype envisioned for 2009.
HALO	**Supernovae** watch experiment. Will use existing lead, neutron detectors from SNO, DAQ and electronics from NCD phase of SNO	Hardware largely in place. Seeking modest funding for installation and operation.
PUPS	**Seismicity.** An experiment to monitor seismicity deep in the earth away from surface where signal is confused by reflections and surface effects.	Installed and Operational.

Solar Neutrino, or Supernova Search experiments. SNOLAB also has space for several small or medium sized experiments. Finally, SNOLAB welcomes interests from other disciplines.

SNOLAB has approved a number of experiments for installation in 2008, and several experiments are already running prototypes in the original SNO lab. The physics program looks rich and exciting. SNOLAB is well poised to act as host to a number of potentially world leading astroparticle physics experiments in the deepest and cleanest facility in the world. Discoveries made at SNOLAB have the potential to be of the greatest fundamental importance. The future for astroparticle physics at SNOLAB is very bright.

LIST OF PARTICIPANTS

Ahluwalia, Dharam Vir
Department of Physics and Astronomy
Rutherford Building
Univ. of Canterbury, Private Bag 4800
Christchurch 8020
NEW ZEALAND
Tel: +64 3 364 2563 (Internal 6563)
dharamvir.ahluwalia-khalilova@canterbury.ac.nz
http://www2.phys.canterbury.ac.nz/editorial/

Allen, Roland E.
Department of Physics
Texas A&M Unive., College Station
Texas 77843-4242, USA
Tel: 1-979-845-4341
Fax: 1-979-845-2590
Email: allen@tamu.edu
http://faculty.physics.tamu.edu/allen

Araujo, Henrique
Astroph. Group, Blackett Lab.
Imperial College London
Prince Consort Road
London SW7 2BW
ENGLAND
Tel: +44 (0)20 759 47549
Fax: +44 (0)20 759 47541
Email: H.Araujo@imperial.ac.uk

Balazs, Csaba
School of Physics, Monash Univ.
Melbourne, VIC 3800
AUSTRALIA
Email: Csaba.Balazs@sci.monas.edu.au

Bell, Nicole
School of Physics, Univ. of Melbourne
Victoria 3010, AUSTRALIA
Tel: + 61 3 8344 3112
Fax: + 61 3 9347 4783
Email: n.bell@unimelb.edu.au
http://www.ph.unimelb.edu.au/ nfb

Berian, James
Inst. for Astronomy, Blackford Hill
Edinburgh EH9 3HJ
UNITED KINGDOM
Tel: +44 (0)131 6688100
Email: jbj@roe.ac.uk
http://www.roe.ac.uk/ jbj

Blake, Chris
Mail H39
Swinburne University of Technology
P.O. Box 218, Hawthorn, Victoria 3122
AUSTRALIA
Email: cblake@astro.swin.edu.au

Bodaghee, Arash
INTEGRAL Science Data Centre
University of Geneva Observatory
Versoix, 1290
SWITZERLAND
Tel: +4122 369 8117
Email: arash.bodaghee@obs.unige.ch
http://obswww.unige.ch

Bradac, Marusa
Univ. of California, Dep. of Physics/MC9530
Broida Hall 2015H
Santa Barbara, CA 93106-9530
USA
Tel: 1-805-893-2246
Fax: 1-805-893-3307
Email: marusa@physics.ucsb.edu
http://www.physics.ucsb.edu/ marusa

Brewer, Brendon
Inst. of Astronomy, School of Physics A28
University of Sydney, NSW 2006
AUSTRALIA
Tel: +61 2 9351 2546
Fax: +61 2 9351 7726
Email: brewer@physics.usyd.edu.au
http://www.physics.usyd.edu.au/ brewer/

Cerulli, Ricardo
INFN - LNGS
S.S. 17/bis, km 18+910
I-67010 Assergi (AQ), ITALY
Tel: +39-0862 437 296
Fax: +39-0862 437 570
Email: riccardo.cerulli@lngs.infn.it
http://www.people.roma2.infn.it
/dama/web/home.html

Colless, Matthew
Anglo-Australian Observatory
P.O. Box 296
Epping, NSW 1710
AUSTRALIA
Tel: +61-2-9372-4812
Fax: +61-2-9372-4880
Email: colless@aao.gov.au

de Boer, Wim
Universität Karlsruhe (TH)
Physikhochhaus/IEKP
Wolfgang-Gaede-Str. 1
D-76131 Karlsruhe
GERMANY
Tel: + 49 (721) 608 3593
Fax: + 49 (721) 608 7930
Fax: CERN: +41 22 767 3529
Email: wim.de.boer@cern.ch
deboer@ekp.physik.uni-karlsruhe.de

Drobyshevski, Edward
Ioffe Physico-Technical Institute
Russian Academy of Sciences
194021 St.Petersburg, RUSSIA
Email: emdrob@mail.ioffe.ru

Eckart, Andreas
I. Physikalisches Institut
Zülpicher Str. 77, Köln, 50937
GERMANY
Tel: +49 221/470-3546
Fax: +49 221/470-5162
Email: eckart@ph1.uni-koeln.de

Famaey, Benoit
Institut d'Astronomie et d'Astrophysique
Universite Libre de Bruxelles
C.P. 226, Boulevard du Triomphe
B-1050 Bruxelles, BELGIUM
Tel: +32 2 6502833
Fax: +32 2 6504226
Email: bfamaey@astro.ulb.ac.be
http://www.astro.ulb.ac.be/ bfamaey/

Francis J., Matthew
Inst. for Astronomy, School of Physics A28
University of Sydney
Sydney, NSW 2006
AUSTRALIA
Tel: +61 2 9351 2546
Fax: +61 2 9351 7726
Email: mfrancis@physics.usyd.edu.au
http://www.physics.usyd.edu.au/ francis/

Fuchs, Burkhard
Astronomisches Rechen-Istitut am
Zentrum für Astronomie
der Universität Heidelberg
Mönchhofstraße 12-14
69120 Heidelberg
GERMANY

Tel: +49 6221 54 1826
Fax: +49 6221 54 1888
Email: fuchs@ari.uni-heidelberg.de
http://www.ari.uni-heidelberg.de

Georgevits, George
Centre for Astroph. & Supercomp.
Swinburne Univ. of Technol.
Mail #39, P.O. Box 218
Hawthorn, Victoria, 3122
AUATRALIA
Email: georgg@bigpond.net.au
http://www.phys.unsw.edu.au/astro/

Guendelman, Eduardo
Physics Department
Ben Gurion University of the Negev
Beer Sheva 84105
ISRAEL
Email: guendel@bgu.ac.il

Hazama, Ryuta
Graduate School & Faculty of Eng.
Hiroshima University
1-4-1 Kagamiyama, Higashi-Hiroshima
739-8527, JAPAN
Tel: +81-82-424-7521
Fax: +81-82-422-7193
Email: hazama@hiroshima-u.ac.jp

Khosroshahi, Habib G.
Astrophysics Research Institute
Liverpool John Moores University
Birkenhead CH41 1LD
UNITED KINGDOM

Killedar, Madhura
School of Physics, Univ. of Sydney
NSW 2006, AUSTRALIA
Tel: +61 2 9351 5184 (7726)
Fax: +61 2 9351 7726
Email: madhura@physics.usyd.edu.au

Kim, Jihn E.
Department of Physics & Astronomy
Center for Theoretical Physics
Seoul National University
Seoul 151-747
KOREA
Tel: +82 2 880 6605
Fax: +82 2 884 7167
Email: jekim@phyp.snu.ac.kr
http://phya.snu.ac.kr/~jekim/

Kinion, Darin
Fermilab, P.O. Box 500
Batavia, IL 60510-5011
USA
Email: kinion1@llnl.gov

Klapdor-Kleingrothaus, Hans Volker
Stahlbergweg 12, Lobbach
D-74931 GERMANY
Tel: +49-6226-41088
prof.klapdor-kleingrothaus@hotmail.de
Home-page: http://www.klapdor-k.de

Knollmann, Steffen R.
Astrophysikalisches Institut Potsdam
An der Sternwarte 16, D-14482 Potsdam
GERMANY
Tel: +49 331 7499 517
Email: sknollmann@aip.de

Kojima, Kazuhiko
National Astronomical Observatory
Division of Theoretical Astrophysics
2-21-1 Osawa, Mitaka, Tokyo 181-8588
JAPAN
Tel: +81 422 34 3735
Fax: +81 422 34 3746
Email: kojima@th.nao.ac.jp

Krause, Axel
Arnold Sommerfeld Center for Theor. Phys.
Department für Physik
Ludwig-Maximilians-Universität München,
Theresienstr. 37, 80333 München, GERMANY
Tel: +49 89 2180 4374
Fax: +49 89 2180 4186
axel.krause@physik.uni-muenchen.de

Krivosheina, Irina Vladimirovna
Radiophysical Research Institute
ul. Bolshaja-Pecherskaja 25
603005, Nishnij-Novgorod, RUSSIA
Tel: +49-6226-41088
Email: irinaKV@web.de
Home-page: http://www.klapdor-k.de

Kusakabe, Motohiko
NINS, National Astronom. Observ. of Japan
2-21-1 Osawa, Mitaka
Tokyo, 181-8588, JAPAN
Tel: +81 422 34 3745
Fax: +81 422 34 3746
Email: kusakabe@th.nao.ac.jp

Kusenko, Alexander
Department of Physics & Astron.
UCLA-154705, Los Angeles
CA 90095-1547, USA
Tel: 310-825-4814
Fax: 310-794-7528
Email: kusenko@ucla.edu
http://www.physics.ucla.edu/ kusenko/

Lewis, Geraint
School of Physics, Univ. of Sydney
NSW 2006, AUSTRALIA
Tel: +61 2 9351 5184 (7726)
Fax: +61 2 9351 7726
Email: gfl@physics.usyd.edu.au
http://www.physics.usyd.edu.au/ gfl

Leubner, Manfred P.
Institute for Astro- & Particle Phys.
Univ. of Innsbruck, Technikerstr. 25
A-6020, Innsbruck, AUSTRIA
Tel: +43-(0)512-507-6054
Fax: +43-(0)512-507-2923
Email: manfred.leubner@uibk.ac.at
http://homepage.uibk.ac.at/ c706102/

Longhitano, Marco
Institute of Astronomy
Department of Physics and Astronomy
University of Basel
Venusstrasse 7
CH-4102, Binningen, SWITZERLAND
Email: marco.longhitano@uinibas.ch

Mankoč Borštnik, Norma Susana
University of Ljubljana, FMF
Jadranska 19, 1000 Ljubljana
SLOVENIA
Email: norma.mankoc@fmf.uni-lj.si

Montgomery, Colin R.L.
46 Daley Av., Daleys Point
NSW 2257, AUSTRALIA
Tel: + (02) 4344 7600
Email: ColinRL@bigpond.net.au
Colin.Montgomery@jcu.edu.au

Murphy, Michael
Centre for Astroph. and Supercomputing
Swinburne University of Technology
Cnr Serpells Ln & Burwood Rd
Hawthorn, Victoria 3122, AUSTRALIA
Tel: +61 (0)3 9214 5818
Fax: +61 (0)3 9214 8797
Email: mim@ast.cam.ac.uk
http://astronomy.swin.edu.au/ mmurphy

Nakahata, N.
Kamioka Observatory, ICRR, Univ. of Tokyo
Higashi-Mozumi, Kamioka-cho
Hida-shi, Gifu-ken, JAPAN, 506-1205
Tel: +81-578-85-9603
Fax: +81-578-85-2121
Email: nakahata@suketto.icrr.u-tokyo.ac.jp
http://www-sk.icrr.u-tokyo.ac.jp/ nakahata/

Navas, Sergio
Departamento Física Teórica
y del Cosmos & C.A.F.P.E.
Univ. of Granada, Granada E-18071
SPAIN
Tel: 34-958-244152, *Fax:* 34-958-248529
Email: navas@ugr.es
http://www.ugr.es/~navas

Neupane, Ishwaree
Dep. of Phys. & Astronomy, Univ. of Canterbury
Private Bag 4800, Christchurch 8020
NEW ZEALAND
Email: ishwaree.neupane@canterbury.ac.nz

Noble, Tony
Dep. of Physics, Queen's Univ.
Kingston, Ontario, K7L-3N6
CANADA
Tel: 613-533-2679 (Queen's)
Tel: 705-692-7000 x 2236 (SNOLAB)
Email: potato@snolab.ca

Ohlsson, Tommy
Department of Theoretical Physics
Royal Institute of Technology (KTH)
AlbaNova University Center
Roslagstullsbacken 21
SE-106 91 Stockholm, SWEDEN
Tel: +46-8-55378161
Fax: +46-8-55378216

Email: tommy@theophys.kth.se
www.theophys.kth.se/ tommy/

Otten, Ernst
Institute of Physics
Staudinger Weg 7, Univ. of Mainz
D-55099 Mainz, GERMANY
Tel: +49 (0)6131 - 39 - 22518
Email: ernst.otten@uni-mainz.de
http://www.physik.uni-mainz.de/exakt/
neutrino/en_gruppe.html

Power, Chris
Centre for Astroph. & Supercomp.
Swinburne Univ. of Techn.
Mail #39, P.O. Box 218
Hawthorn, Victoria 3122, AUSTRALIA
Tel: +61 (3) 9214 8302
Fax: +61 (3) 9214 8797
Email: cpower@astro.swin.edu.au

Prodanov, Emil M.
School of Mathematical Sciences
Dublin Inst. of Technology
Kevin Street, Dublin 8
IRELAND
Tel: +353 87 9895345
Fax: +353 1 402 4994
Email: Emil.Prodanov@dit.ie

Rebelo, Margarida Nesbitt
Dep. de Física & Centro de Física
Teórica de Partículas (CFTP)
Instituto Superior Técnico (IST)
Av. Rovisco Pais, P-1049-001 Lisboa
PORTUGAL or
CERN, Department of Physics
Theory Division, CH-1211 Genève 23
SWITZERLAND
Tel: + 351 - 21 8419010

Fax: + 351 - 21 841 9143
Email: rebelo@alfa.ist.utl.pt
Margarida.Rebelo@cern.ch

Rizzo, Alfio
Vrije Universiteit Brussel
Faculteit Wetenschappen
Dienst ELEM, Pleinlaan, 2
B-1050 Brussels
BELGIUM
Tel: + 32(0)26293220
Fax: + 32(0)26293816
Email: alfio.rizzo@vub.ac.be
http://w3.iihe.ac.be/ rizzo

Schrabback, Tim
Leiden Observatory
PO Box 9513, NL-2300 RA
Leiden
NETHERLANDS
Tel: +31 71 527 5877
Fax: +31 71 527 5743
Email: tim@astro.uni-bonn.de
http://www.strw.leidenuniv.nl/ schrabback/

Schritt, Dimitri
Department of Physics & Astronomy
Rutherford Building
Univ. of Canterbury, Private Bag 4800
Christchurch 8020, NEW ZEALAND
Tel: +64 3 364 2563
Email: dsc35@student.canterbury.ac.nz

Scott, Patrick
Dep. of Physics, Stockholm University
AlbaNova High Energy Astrophysics
and Cosmology Centre (HEAC)
SE-106 91, Stockholm, SWEDEN
Tel: +46 8 5537 8734
Fax: +46 8 5537 8601
Email: pat@physto.se

Spooner, Neil
Dep. of Physics & Astronomy, Univ. of Sheffield
Hicks Building Hounsfield Road
Sheffield S3 7RH, ENGLAND
Tel: +114 222 4422
Fax: +114 272 8079
Email: n.spooner@sheffield.ac.uk
ww.shef.ac.uk/ phys/research/pa/

Tisserand, Patrick
Research School of Astron. & Astroph.
Australian National University
Cotter Road, Weston Creek, ACT 2611
AUSTRALIA
Tel: +61 2 6125 0274
Fax: +61 2 6125 0260
Email: tisseran@mso.anu.edu.au
http://www.mso.anu.edu.au

Viollier, Raoul D.
Centre for Theoretical Physics
and Astrophysics
Dep. of Physics, University of Cape Town
Private Bag, Rondebosch 7701
Cape Town, SOUTH AFRICA
Tel: 00 27 21 685 5510
Fax: 00 27 21 650 3352
Email: raoul.viollier@uct.ac.za

Wiltshire, David L.
Dep. of Phys. & Astron., Rutherford Building
Univ. of Canterbury, Private Bag 4800
Christchurch 8020, NEW ZEALAND
Tel: +64 3 364 2473, *Fax:* +64 3 364 2469
Email: David.Wiltshire@canterbury.ac.nz
http://www2.phys.canterbury.ac.nz/~dlw24/

Winer, Brian L.
Dep. of Phys., Ohio State University
191 West Woodruff Ave

Columbus, Ohio 43210, USA
Tel: (614) 292-8996 (Office)
Tel: (614) 292-7370 (Lab)
Email: winer@mps.ohio-state.edu
www.physics.ohio-state.edu

Yamazaki, Dai G.
Division of Theoretical Astronomy
National Astronomical Observatory
2-21-1 Osawa, Mitaka, Tokyo 181-8588
JAPAN
Tel: +81-422-34-3727
Fax: +81-422-34-3746
Email: yamazaki@th.nao.ac.jp
http://th.nao.ac.jp/ yamazaki/

Yoo, Jonghee
MS209 POBox 500, Kirk and Pine street
Batavia, IL, 60510, USA
Tel: 1-630-840-3665
Fax: 1-630-840-8231
Email: yoo@fnal.gov

Zakharov, Alexander
Inst. of Theor. & Exper. Physics
B. Cheremushkiskaya, 25
Moscow, 117259, RUSSIA
Tel: +7-495-1299759
Fax: +7-495-1270833
Email: zakharov@itep.ru

INDEX